Mass Production of Beneficial Organisms

Invertebrates and Entomopathogens

Mass Production of Beneficial Organisms

Invertebrates and Entomopathogens

Edited by

Juan A. Morales-Ramos
USDA-ARS
National Biological Control Laboratory
Stoneville, MS, USA

M. Guadalupe Rojas
USDA-ARS
National Biological Control Laboratory
Stoneville, MS, USA

David I. Shapiro-Ilan
USDA-ARS, SAA
SE Fruit and Tree Nut Research Unit
Byron, GA, USA

AMSTERDAM • BOSTON • HEIDELBERG • LONDON
NEW YORK • OXFORD • PARIS • SAN DIEGO
SAN FRANCISCO • SINGAPORE • SYDNEY • TOKYO

Academic Press is an imprint of Elsevier

Academic Press is an imprint of Elsevier
32 Jamestown Road, London NW1 7BY, UK
225 Wyman Street, Waltham, MA 02451, USA
525 B Street, Suite 1800, San Diego, CA 92101-4495, USA

British Library Cataloguing-in-Publication Data
A catalogue record for this book is available from the British Library

Library of Congress Cataloging-in-Publication Data
A catalog record for this book is available from the Library of Congress

ISBN: 978-0-12-391453-8

For information on all Academic Press publications visit our website at elsevierdirect.com

Typeset by TNQ Books and Journals Pvt Ltd.
www.tnq.co.in

Dedication

We dedicate this book to Dr Edgar G. King Jr. His vision for the development and application of augmentative biological control was one of the most relevant driving forces to make the field what it is today. In the early 1980s, he focused his efforts to promote the development of mass production technology for *Trichogramma* spp., and through these efforts, it became clear to him that a more important aspect of the augmentative biological control field needed to be addressed. Specifically, it became apparent that if augmentation of natural enemies was going to be successful, it would have to become an economically viable commercial enterprise. To achieve this goal, the field required efficient mass production techniques. The key component to the production process, in Dr King's view, was to develop artificial diets for key entomophagous arthropods. Working simultaneously as the Laboratory Director of the then Subtropical Agricultural Research Laboratory and as a Research Leader of the Biological Control Research Unit, he planned an ambitious project targeting a primary pest of an important annual crop using augmentative biological control. The target was the boll weevil and the biocontrol agent was *Catolaccus grandis*. He put together a team to develop mass rearing techniques, including mechanization, artificial diets, and release strategies. The project was successful, and by 1996, it was the first example of a primary pest being successfully controlled by an in vitro reared parasitoid. Such achievement has not been replicated to the date of publication of this book. Later in his career, he conceived the idea of a research facility dedicated entirely to the study and development of technology for mass production and application of biological control agents. With the help of other visionaries like him, he realized this dream and the National Biological Control Laboratory became a reality, opening its door in December 2004. The idea for this book was born, in part, from Edgar Kings's vision for successful implementation and commercialization of biological control agents.

Edgar G. King

Biographic information:

Born October 3, 1943, at Corpus Christi, TX

Obtained Bachelor of Science in Biology from McNeese State College at Lake Charles, LA in 1966

Obtained Master of Science in Entomology from Louisiana State University at Baton Rouge, LA in 1968

Obtained Doctor of Philosophy in Entomology from Louisiana State University at Baton Rouge, LA in 1971

Became Director of the Southern Insect Management Research Laboratory, Stoneville, MS in 1978

Became Director of the then Subtropical Agricultural Research Laboratory and later Kika de la Garza Subtropical Agricultural Research Center, Weslaco, TX in 1988

Became the Associate Area Director of USDA-ARS Mid South Area in 1998

Became Director of USDA–ARS Mid South Area in 2001

Contents

Section III

16. Insect Protein as a Partial Replacement for Fishmeal in the Diets of Juvenile Fish and Crustaceans

Eric W. Riddick

17. Insects as Food for Insectivores

Mark D. Finke and Dennis Oonincx

This page intentionally left blank

Contributors

Derek R. Artz US Department of Agriculture, Agricultural Research Service (USDA-ARS), Pollinating Insects Research Unit, Logan, Utah, USA

Robert Behle USDA-ARS-NCAUR, Crop Bioprotection Research Unit, Peoria, Illinois, USA

Tim Birthisel The Andersons Inc., Turf and Specialties Group, Asheville, North Carolina, USA (Retired)

Carlos A. Blanco USDA Animal and Plant Health Inspection Service, Riverdale, Maryland, USA

Kevin R. Butt School of Built and Natural Environment, University of Central Lancashire, Preston, UK

Leslie Chan Australian Institute of Bioengineering and Nanotechnology, The University of Queensland, St Lucia, Queensland, Australia

Hongyin Chen Sino-American Biological Control Laboratory, OIRP, ARS, USDA, Beijing, China

Matthew A. Ciomperlik US Department of Agriculture, Animal and Plant Health Inspection Service, Plant Protection and Quarantine, Center for Plant Health Science and Technology (USDA-APHIS-PPQ-CPHST), Edinburg, Texas, USA

Allen C. Cohen Insect Rearing Education and Research Program, Department of Entomology, North Carolina State University, Raleigh, North Carolina, USA

Terry L. Couch Becker Microbial Products Inc., Parkland, Florida, USA

Thomas A. Coudron Biological Control of Insects Research Laboratory, US Department of Agriculture, Agricultural Research Service (USDA-ARS), Columbia, Missouri, USA

Patrick De Clercq Department of Crop Protection, Ghent University, Ghent, Belgium

Maria Luisa Dindo Dipartimento di Scienze Agrarie (DipSA), Alma Mater Studorium Università di Bologna, Bologna, Italy

Aaron T. Dossey All Things Bugs, Gainesville, Florida, USA

Mark D. Finke Mark Finke LLC, Rio Verde, Arizona, USA

John A. Goolsby USDA-ARS, Knipling-Bushland Livestock Insects Research Laboratory, Cattle Fever Tick Research Laboratory, Edinburg, Texas, USA

Juli A. Gould USDA-APHIS-PPQ-CPHST, Buzzards Bay, Massachusetts, USA

Simon Grenier UMR BF2I, INRA/INSA de Lyon, current address: 6 rue des Mésanges, 69680 Chassieu, France (Retired)

David Grzywacz Natural Resources Institute, University of Greenwich, Central Avenue, Chatham, Kent, UK

Richou Han Guangdong Entomological Institute, Guangzhou, China

Kim A. Hoelmer US Department of Agriculture, Agricultural Research Service, European Biological Control Laboratory (USDA-ARS-EBCL), Montpellier, France

Stefan T. Jaronski US Department of Agriculture, Agricultural Research Service, Sidney, Montana, USA

Juan Luis Jurat-Fuentes Department of Entomology and Plant Pathology, University of Tennessee, Knoxville, Tennessee, USA

Alan A. Kirk USDA-ARS, European Biological Control Laboratory, Montpellier, France (Retired)

Norman C. Leppla Entomology and Nematology Department, University of Florida, Institute of Food and Agricultural Sciences, Gainesville, Florida, USA

Christopher N. Lowe School of Built and Natural Environment, University of Central Lancashire, Preston, UK

David Moore CABI, Bakeham Lane, Egham, Surrey, UK

Juan A. Morales-Ramos National Biological Control Laboratory, US Department of Agriculture, Agricultural Research Service (USDA-ARS), Stoneville, Mississippi, USA

Patrick J. Moran US Department of Agriculture, Agricultural Research Service (USDA-ARS), Exotic and Invasive Weeds Research Unit, Albany, California, USA

Dennis Oonincx Laboratory of Entomology, Department of Plant Sciences, Wageningen University, Wageningen, The Netherlands

Stephen S. Peterson AgPollen LLC, Visalia, California, USA

Charles J. Pickett California Department of Food and Agriculture, Sacramento, California, USA

Maribel Portilla USDA-Agricultural Research Service, Stoneville, Mississippi, USA

Xuehong Qiu Guangdong Entomological Institute, Guangzhou, China

R.J. Rabindra College of Post Graduate Studies, Central Agricultural University, Umiam, Shillong, Meghalaya, India

Alexis E. Racelis Department of Biology, University of Texas–Pan American, Edinburg, Texas, USA

Steve Reid School of Chemistry and Molecular Biosciences, The University of Queensland, Brisbane, Queensland, Australia

Eric W. Riddick National Biological Control Laboratory, US Department of Agriculture, Agricultural Research Service (USDA-ARS), Stoneville, Mississippi, USA

M. Guadalupe Rojas National Biological Control Laboratory, US Department of Agriculture, Agricultural Research Service (USDA-ARS), Stoneville, Mississippi, USA

Don P.A. Sands Commonwealth Science, Industry and Research Organization (CSIRO), Brisbane, Queensland, Australia (Retired)

David I. Shapiro-Ilan USDA-ARS, SAA, SE Fruit and Tree Nut Research Unit, Byron, Georgia, USA

Rhonda L. Sherman Biological and Agricultural Engineering, North Carolina State University, Raleigh, North Carolina, USA

Marianne Shockley Department of Entomology, University of Georgia, Athens, Georgia, USA

Gregory S. Simmons USDA-APHIS-PPQ-CPHST, Salinas, California, USA

K. Rod Summy Department of Biology, University of Texas–Pan American, Edinburg, Texas, USA

Monique M. van Oers Laboratory of Virology, Wageningen University, Wageningen, The Netherlands

This page intentionally left blank

Section I

This page intentionally left blank

Chapter 1

Introduction

Norman C. Leppla,[1] Juan A. Morales-Ramos,[2] David I. Shapiro-Ilan[3] and M. Guadalupe Rojas[2]

[1]Entomology and Nematology Department, University of Florida, Institute of Food and Agricultural Sciences, Gainesville, FL, USA, [2]National Biological Control Laboratory, USDA-Agricultural Research Service, Stoneville, MS, USA, [3]USDA-ARS, SAA, SE Fruit and Tree Nut Research Unit, Byron, GA, USA

1.1. CHALLENGES OF MASS PRODUCING BENEFICIAL ORGANISMS

Mass Production of Beneficial Organisms contains chapters on producing selected organisms useful to humankind, including arthropods, microorganisms, bees, and earthworms. It is comprised of a series of comprehensive descriptions of the industrial-level production of insects, mites, and pathogens for biological control, and beneficial invertebrate organisms for food, pollination, and other purposes. Additionally, there are reports on artificial diet development and quality assurance for arthropods, as well as entomopathogen production and formulation. The final section covers insects as food for animals, insectivores, and humans, along with solitary bees for pollination and earthworm mass culture. This is a unique assemblage of topics organized around the goal of producing large amounts of organisms for a variety of useful purposes.

Mass production of these organisms is somewhat arbitrary to define in terms of the number produced per time interval. Rather, it is characterized by the magnitude and degree of separation of the rearing processes, usually involving a single species. It takes place in large, multiroom mass rearing facilities or "biofactories" specially designed for this purpose. There is a trained labor force with at least one employee assigned to each independent rearing process, such as diet preparation or another single production activity. Depending on the species being produced, large amounts of host material, artificial diet ingredients, or growth media are used and there usually is some mechanization to make the rearing more efficient. Thus, mass production of beneficial organisms can be considered an industrial process with all of the associated logistical requirements, including substantial quantities of production materials, continuous maintenance of facilities and equipment, and distribution of high-quality products (see Chapter 9).

Mass Production of Beneficial Organisms. http://dx.doi.org/10.1016/B978-0-12-391453-8.00001-7

Principles and procedures for mass producing beneficial arthropods and microbes have developed independently, although there are commonalities. Both kinds of organisms are produced in biological systems that depend on genetically suitable founding populations, uncontaminated diets or media, mechanized equipment, controlled environments, quality assurance, packaging, and delivery to customers as effective products. The subjects encompassed in principles and procedures for developing and operating production systems for these organisms can be divided into the following: facility design and management, including health and safety; environmental biology; management of microbial contamination; nutrition and diet; population genetics; and quality control (Schneider, 2009). Unlike general principles, however, procedures are typically species-specific in terms of diet or substrate and associated culturing methods. A suitable host organism must be used in the absence of an artificial diet to rear an arthropod, and similarly beneficial microorganisms often are cultured on a defined artificial medium or, when in vitro culture is not feasible, on susceptible hosts. Regardless of species, procedures for mass rearing any beneficial organism are divided into a series of steps based on its life cycle.

Insect mass production progressed naturally from relatively small-scale rearing of insects for human and animal food, such as honey or mealworms, *Tenebrio molitor* L., or for their products that historically have included silk, cochineal dye, lac, and beeswax. Reliable supplies of insects that behaved normally also were needed for research and teaching (Needham et al., 1937). Blowflies, horseflies, several filth flies, mosquitoes, and the common bed bug, *Cimex lectularius* L., have been essential for the advancement of medical and veterinary research. These insects and vectors of human and animal pathogens, such as mosquitoes and the tsetse flies, were used to screen chemical compounds for repellency and toxicity, as well as to develop useful formulations. *Drosophila melanogaster* Meigen became the standard insect model for genetic research. For crop protection, large numbers of insect species were needed for studies on host plant resistance to insects, including certain pest Heteroptera, Diptera, Coleoptera, and Lepidoptera, such as the European corn borer, *Ostrinia nubilalis* (Hubner). Commodity treatments were developed for the Khapra beetle, *Trogoderma granarium* Everts, additional grain-infesting beetles and moths, several kinds of tephritid fruit flies, and many other insect species. Large quantities of the boll weevil, *Anthonomus grandis* Boheman, noctuid moths, the pink bollworm, *Pectinophora gossypiella* (Saunders), and other Coleoptera and Lepidoptera were used to develop attractants and traps. Some of these insects also provided hosts for rearing imported natural enemies in quarantine and prior to release in the field. Moreover, most predators and parasitoids used in augmentation biological control require massive amounts of natural and factitious hosts. These hosts typically are more difficult to rear consistently than the natural enemy itself. Due to this host rearing limitation, large populations of insects for release in autocidal control must be produced on artificial diets.

More than 50 species of arthropod natural enemies are produced in large enough numbers to be marketed widely in the United States (Leppla and Johnson, 2011). Many of these and additional species are sold in Europe (van Lenteren, 2003). Popular predators include several phytoseiid mites, coccinellids, cecidomyiids, and chrysopids, and the most commonly used parasitic wasps are in the taxonomic families Aphelinidae, Braconidae, Pteromalidae, and Trichogrammatidae. Predaceous mites are used extensively for biological control of phytophagous mites, fungus gnats, and thrips on potted and bedding plants in protected culture and interiorscapes. They are particularly useful for two-spotted spider mite control on ornamental, fruit, and vegetable crops. Depending on the species, lady beetles are released to control a variety of scales, mealybugs, aphids, thrips, and whiteflies. The cecidomyiid, *Aphidoletes aphidimyza* (Rondani), is often used for aphid biological control, as are *Chrysoperla* spp. Probably the most popular aphelinid is *Encarsia formosa* Gahan, released extensively in greenhouses to control whiteflies; *Eretmocerus* spp. also are available for this purpose. Applications of braconids, *Aphidius* spp., are made to manage a wide range of aphids. A specialized purpose for mass-reared natural enemies is the use of pteromalids for biological control of filth flies in manure and compost.

Numerous species of the lepidopteran egg parasitoid, *Trichogramma* spp., have been mass produced on factitious hosts in semi-mechanized rearing facilities for decades, becoming the most prevalent augmentative parasitoid in both number of production facilities and quantities produced. They typically are reared on eggs of the angoumois grain moth, *Sitotroga cerealella* (Oliver), or Mediterranean flour moth, *Ephestia kuehniella* Zeller, that infest stored grain on which they are reared, such as wheat and barley. Several biofactories in the former Soviet Union each consistently produced millions of *Trichogramma* spp. per day for years to control pest Lepidoptera in field crops. A highly successful European corn borer biological control project has been conducted in Germany, Switzerland, and France since about 1992 (Kabiri and Bigler, 1996). Producers of commercial natural enemies and collaborative government/grower groups throughout the world have developed a variety of simple, highly productive rearing systems for *Trichogramma* spp. In every situation, large containers of grain are infested with host eggs, yielding larvae that feed and eventually molt into adults that deposit eggs. The eggs are harvested, exposed to *Trichogramma* spp. adults, and used to maintain the colony or attached to a substrate and distributed in a crop. Periodically, eggs of the target pest are used instead of the factitious host to maintain high levels of pest parasitism.

Arthropod mass rearing reached an industrial level with the development of autocidal control and eradication of the New World screwworm fly, *Cochliomyia hominivorax* (Coquerel). The first large-scale production facility was established near Sebring, Florida in a converted surplus U.S. Air Force hangar. During the early period of screwworm eradication in the southeastern United States, 50 million flies were produced weekly from larvae reared on meat of cattle, horses, pigs, whales, and nutria, *Myocaster coypus* (Molina). Production was moved to Mission, Texas and increased to 75–200 million per week to support eradication of the fly from the southwestern United States. As eradication progressed further south, the biofactory in Texas was closed and a new one established at Tuxtla Gutierrez, Mexico to produce 250–300 million flies per week (Meyer, 1987). Mechanization was increased and meat was replaced with a liquid larval diet composed of various formulations of lean ground beef, citrated beef blood, milk, water, and formalin dispensed onto cotton linters at first and subsequently cellulose acetate blankets in shallow trays. The final and most complex liquid larval diet contained dried whole chicken egg, dried whole bovine blood, calf suckle (powdered milk substitute), sucrose, dried cottage cheese, and formalin (Taylor, 1992). Mature larvae left the trays, fell into a water stream, and were collected for pupation, sterilization, and release from airplanes. Larval rearing became more efficient by incorporating the liquid diet ingredients into a gelling agent and eliminating the acetate mats. However, the flies continued to be fed ground beef mixed with honey. Overlapping shingle-like egg masses were oviposited on wooden frames treated with spent larval medium. Screwworm eggs were scraped from the frames, incubated, and placed on small pieces of lean meat before being transferred to the gelled diet as first instar larvae. Because mass production of the screwworm enabled this pest to be eradicated from virtually all of North America and Mexico, except some Caribbean islands, the rearing facility was relocated to Pacora, Panama. There, state-of-the-art insect mass rearing facilities, equipment, materials and methods have been established and continuously improved.

Another pioneering insect mass rearing program was developed for the Mediterranean fruit fly, *Ceratitis capitata* (Wiedemann), a global pest of tropical fruit and citrus. The purpose of the program in North America and Mexico was to apply the sterile insect technique by adapting concepts and methods proven successful for eradicating the screwworm. A biofactory was built at Metapa, Mexico based on a mass rearing system developed primarily at Seibersdorf, Austria and allied research in Costa Rica, Hawaii, and a few other locations. The flies were held in large cages and fed granulated sugar and protein hydrolysate formed into dry cakes. Water was provided in tubes fitted with absorbent wicks. The flies oviposited on nylon cloth sheets from which the eggs dropped into a water bath for collection and incubation. The initial larval diet was a suspension of soy flour, wheat bran, granulated sugar, torula yeast, methyl para-hydroxybenzoate, and water mixed into sugarbeet bagasse. The bagasse often was of poor quality and sources became unreliable, so it was replaced later with a variety of starch materials, such as corncob grits. As with the screwworm, mature medfly larvae leave the diet but, for mass rearing, the medium containing medfly larvae was transferred to large, cylindrical rotating larval separation machines. The larvae then were accumulated, placed into pupation trays, and held for adult emergence. The goal was achieved of producing 500 million pupae per week for distribution by air as flies. The medfly was eradicated from the United States and Mexico, except for periodic outbreaks caused by breeches of quarantine. Eventually, at least 14 medfly mass rearing facilities were built throughout the world, the largest at El Pino, Guatemala with a maximum production of 5×10^9 per week.

Mass rearing of insects for human and animal food is in its infancy but will become increasingly important as shortages of animal and plant protein become more frequent. It eventually will not be prudent to use massive amounts of grain suitable for human consumption to produce animals for food. Because natural populations of insects generally are not adequate, mass production of suitable species will be necessary. Candidates will have some of the following characteristics: reasonably large, easy to culture, ability to thrive on organic waste, high rate of reproduction, not significantly cannibalistic, and not prone to become diseased. The black soldier fly, *Hermetia illucens* L., is being studied in several countries for industrial-level bioconversion of food waste. Dried house fly, *Musca domestica* L., larvae or pupae contain 50–60% protein and the species has a very high reproductive rate, a minimum life cycle of 7–10 days, and the pupae can weigh 20 mg or more, depending on the type of waste consumed by the larvae. Perhaps efficient and reliable mass rearing systems can be developed for normally abundant types of insects, such as termites and ants. Another approach would be to adapt systems developed for the pink bollworm, codling moth, *Cydia pomonella* (L.), the silkworm, *Bombyx mori* (L.), and several noctuid moths by developing artificial diets containing minimal amounts of protein. Larvae of some Lepidoptera are relatively large and rich in protein and lipids.

1.2. CHALLENGES OF ARTHROPOD MASS PRODUCTION FOR BIOLOGICAL CONTROL

Mass production and release of arthropod natural enemies is the foundation of augmentative biological control (King, 1993; Elzen and King, 1999; Morales-Ramos and Rojas, 2003). Mass production of beneficial arthropods is a complex process that often involves a multidisciplinary effort. This approach requires a substantial economic investment to develop the technology, construct adequate facilities, and hire and train personnel. The investment needed to establish mass production systems for new arthropod species can only be met by large organizations such as government or large industry. Effective use of biological control agents to control major pests in commercial crops involves the release of tens of thousands to millions of individuals (King et al., 1985). Knowledge has accumulated over decades on small-scale rearing of arthropod natural enemies, but new technologies are needed to push production capabilities from thousands to millions of organisms per week.

Government-supported arthropod mass production has been successful for autocidal control programs, such as those for the screwworm, pink bollworm, and Mediterranean fruit fly. Autocidal control programs usually aim to eradicate the target pest and therefore tend to be temporary. Nevertheless, these programs have contributed essential methods and expertise for advancing insect mass production.

The resulting mass production technology has been used in government-supported augmentative biological control of some key pests. Parasitoid and predator production technologies developed in these programs have helped the biological control industry to slowly emerge during the 1990s. This industry aims to commercialize arthropod natural enemies for augmentative biological control of pests.

The commercialization of arthropods as biological control agents has dramatically changed the direction in which mass production technology evolved. In a free market economy, mass-produced natural enemies must compete with other pest control technologies to become commercially viable and sustainable. Biological control agents must effectively control the target pests and their cost must be competitive (King et al., 1985). The first section of this book contains chapters that describe commercial successes and failures of mass-produced arthropods intended for biological control of pests. The chapters deal with different arthropod groups and technologies for their mass production and explain the difficulties in bringing them to commercial application.

Only a limited number of arthropod natural enemies can be economically mass produced with the present state of mass rearing technology. One of the major obstacles to produce natural enemies is the intrinsic need to mass produce their host or prey. This situation doubles the costs by producing two species while generating revenues from only one (Van Driesche and Bellows, 1996). Further complications arise from the need to also grow the host plant for production of the herbivores. As a result, with few exceptions, herbivore species that have been reared on artificial diets have been preferred as sources of food for mass-produced natural enemies. Consequently, the range of species of mass-produced natural enemies has been influenced by their capacity to develop and reproduce on a diet composed of those herbivore species that have been reared on artificial diets. Predatory arthropods capable of developing and reproducing on easy-to-rear factitious prey also have been more frequently mass produced (Van Driesche and Bellows, 1996). Many commercially produced natural enemies are omnivore predators capable of feeding on plant materials. Some examples include the phitoseiid mites *Neoseiulus californicus* (McGregor), *N. fallacies* (Garman) (Croft et al., 1998), and *Amblyseius swirkii* Athias-Henriot (Messelink et al., 2008); the insidious flower bug *Orius insidiosus* (Say); *Harmonia axyridis* (Pallas); *Coleomegilla maculata* (deGeer); and *Hippodamia convergens* Guérin-Méneville (Lundgren, 2009).

Developing artificial diets for parasitoids and predators can be a way to simplify their mass production, making it more cost-effective. However, artificial diets are often inferior to natural prey or hosts as sources of nutrition for entomophagous species (Grenier, 2009). As a result, parasitoids and predators grown on artificial diets often possess diminishing biological characteristics, which reduce their quality as biological control agents (Grenier and De Clercq, 2003; Riddick, 2009). Directions for future development of artificial diets are presented in Chapter 7 and methods to evaluate their quality are described in Chapters 8 and 9.

1.3. CHALLENGES OF MASS PRODUCING PATHOGENS FOR BIOLOGICAL CONTROL

Microbial control can be defined as the use of pathogens for control of pests. Thus, microbial control is a branch of the broader discipline of biological control and may be thought of as applied epizootiology (Shapiro-Ilan et al., 2012). Most researchers also include microbial byproducts in their definition of microbial control, such as toxins or metabolites. Furthermore, some consider natural suppression of pests without any human intervention to be included in microbial control, but in this volume we limit microbial control to intentional manipulation of the targeted system. Microbial control agents (e.g. pathogenic viruses, bacteria, fungi, and protists) can be applied for suppression of weeds, plant diseases, or insects (Tebeest, 1996; Montesinos, 2003; Janisiewicz and Korsten, 2002; Vega and Kaya, 2012). This volume includes chapters on production of microbial control agents for suppression of insect pests in the second section. Specifically, it covers mass production of four major groups of entomopathogens: nematodes (Chapter 10), fungi (Chapter 11), bacteria (Chapter 12), and viruses (Chapter 13). Another group of entomopathogens, the protists, is not included in this book because currently there is no commercial production of these agents.

Chemical pesticides can be harmful to humans and the environment and may cause secondary pest outbreaks and resistance (Debach, 1974). In contrast, microbial control agents (similar to arthropod biocontrol agents) are safe to humans and the environment and generally have little or no effect on other nontarget organisms; microbial control agents also generally have a substantially reduced risk of inducing resistance (Lacey and Shapiro-Ilan, 2008). On the other hand, relative to chemical insecticides, microbial control agents have certain disadvantages, such as susceptibility to environmental degradation, such as ultraviolet light. Additionally, the narrow host range of certain microbials may be perceived as a drawback, such as if a grower is trying to target a variety of pests at one time (Fuxa, 1987; Shapiro-Ilan et al., 2012).

In many systems, another disadvantage to implementing microbial agents in biological pest suppression is that they cost more to use than chemical pesticides. However, unlike arthropod parasitoids, some microbial agents can be produced in vitro, which can substantially increase cost. For example, in vitro production systems have been developed for various species of entomopathogenic bacteria, fungi, and nematodes. To date, commercial in vitro production of entomopathogenic viruses has not been accomplished, but research is underway to achieve that goal. In the meantime, production of entomopathogenic virus relies exclusively on in vivo technology. Although commercial entomopathogenic nematodes usually are produced in vitro, some companies still produce nematodes in vivo, which results in higher production costs for labor and insect hosts.

Production efficiency and cost are critical factors impacting the success or failure of commercial ventures involving entomopathogens

(Lacey et al., 2001; Shapiro-Ilan et al., 2012). The series of chapters on entomopathogens review and analyze various factors that affect production of each group. A number of factors that impact efficiency are shared across the entomopathogen groups, including choice of species or strain, strain stability and improvement, environmental factors (e.g. temperature, humidity, and aeration), inoculation rates, and production densities. Also, regardless of pathogen group, media composition is a critical factor for in vitro production, and host species and quality is crucial for in vivo production. On the other hand, certain factors pertain only to some groups but not others and may be highly specific. For example, bioreactor design and fermentation parameters only pertain to entomopathogens produced under liquid in vitro conditions. For production of heterorhabditid nematodes in liquid culture, recovery (the initiation of molting from the dauer stage) is an important issue specific to this group. The cost of pesticide registration can also be a major consideration prior to commencing production of most entomopathogens, but this issue is generally not relevant for entomopathogenic nematodes because in most countries they are exempt from registration requirements that apply to other pathogen groups (Ehlers, 2005).

In addition to discussing factors that affect production efficiency and cost, recent advances in production technology are reviewed for each pathogen group. For example, recent advances in production of entomopathogenic nematodes include production and application of infected host cadavers, automated technology for in vivo production, and the use of inbred lines to stabilize beneficial traits in production strains (Shapiro-Ilan et al., 2003, 2010; Bai et al., 2005; Morales-Ramos et al., 2011; see Chapter 10). A recent innovation in fungus production was based on the discovery that *Metarhizium* spp. can produce microsclerotia (compact melanized bodies that conidiate upon rehydration); thus, production technology for these novel propagules has ensued (Jackson and Jaronski, 2012; see Chapter 11). As another example, in Chapter 13, improvement of baculovirus production through the study of genomics/transcriptomics of insect cell lines is discussed.

The section on entomopathogens offers an analysis of state-of-the-art production in various systems. Considerations for production may vary in different markets, countries, and economies. For example, Chapter 15 presents a perspective on production technology in less industrialized countries. Production technologies used in these situations may not be viable elsewhere—for example, because of reduced labor costs and less mechanized production systems. However, production ventures in less industrialized countries may face different hurdles, such as reduced levels of capital, infrastructure, or technology (see Chapter 15). Regardless of the production system, one critical factor to the success of all entomopathogen products is formulation. Chapter 14 is devoted entirely to issues related to formulation of entomopathogens. Formulations are not only required as simple carriers for most microbial control agents, but they may also provide other benefits such as improved shelf life, protection from environmental degradation, ease of handling, and enhanced efficacy (see Chapter 14). Clearly, the combination of production and formulation technology is paramount to the successful implementation of entomopathogens in microbial control. Thus, the collection of chapters on microbial control agents brings together the challenges facing the industry and offers perspectives and potential solutions on how to enhance commercialization in the future.

1.4. CHALLENGES OF MASS REARING INVERTEBRATES FOR THEIR PRODUCTS AND ECOLOGICAL SERVICES

Commercial products from invertebrates have consisted mostly of silk, honey, wax, and dye, and ecological services have include pollination, food for animals, waste decomposition, and biological control. Silk is mostly produced by culturing the mulberry silk moth, *Bombix mori* L., but other species are also commercially grown to produce silk, including the Chinese Tussah moth, *Antheraea pernyi* (Guénerin-Méneville); the Assam silk moth, *A. assamensis* (Helfer); the tensan silk moth, *A. yamamai* (Guénerin-Méneville); and the eri silk moth *Samia cynthia* (Drury) (Hill, 2009). In addition to the production of silk, the commonly discarded pupae of the silk moth can be used as a source of food for stock animals and people because the pupae have a high nutritional value (Lin et al., 1983; Mishra et al., 2003; Khatun et al., 2005; Longvah et al., 2011). Silk moth production has provided resources for the development of biological control in China. For example, production of *Trichogramma* spp. parasitoids to control lepidopteran pests is commonly accomplished by using silk moth eggs. In vitro production of *Trichogramma* spp. in China also is based on the use of pupal silkworm hemolymph (Grenier, 1994). Thus, the biological control industry benefits from another insect production capability. By developing new industries that produce materials from invertebrates, the biocontrol industry also can be advanced.

In relatively recent years, a new set of invertebrate species has been produced commercially for food, pollination, and soil restoration. Commercially produced insects sold for feed include the house cricket, *Acheta domesticus* L.; the rusty red roach, *Blatta lateralis* Walker; the greater wax moth, *Galleria mellonella* L.; the butter worm, *Chilecomadia moorei* Silva; the mealworm, *T. molitor*; the super worm, *Zophobas morio* F.; the soldier fly, *H. illucens*; and the house fly, *M. domestica*. (Finke, 2002, 2013). Many of these species have been studied for use as poultry feed (Calvert et al., 1969; Klasing et al., 2000; Ramos-Elorduy et al., 2002; Zuidhof et al., 2003; Anand et al., 2008). Cultured insects have also been investigated for use as feed in aquaculture (Bondari and Sheppard, 1987; Ng et al., 2001; Fasakin et al., 2003). Moreover, some of the commercially produced insect species have even been proposed as potential food for humans (Gordon, 1998; Ramos-Elorduy, 1998; DeFoliart, 1999). Applications for commercially produced insects for feed are discussed in Chapters 16, 17, and 18.

Cultured earthworm species commercialized for fish bait and fish feed include *Lumbricus terrestris* L., *Lumbricus rubellus* Hoffmeister, *Eudrilus eugeniae* (Kinberg), and *Eisenia foetida* (Savigny) (Harper and Greaser, 1994;

Mason et al., 2006). In addition, earthworms have been produced commercially for soil restoration, including *Aprrectodea longa* (Ude), *A. caliginosa* (Savigny), *Allobophora chlorotica* (Savigny), and *L. terrestris* (Lowe and Butt, 2005). Culture techniques and applications for earthworms are discussed in Chapter 20.

Traditionally, pollination of high value crops has been accomplished by managing the honey bee, *Apis melifera* L. (O'Toole, 2008). The culture and use of solitary bees has increased recently in North America, however. One example is the leaf cutting bee, *Megachile rotunda* F., which is used in alfalfa pollination (Stephen, 2003). Other species cultured for alfalfa pollination include the alkali bee, *Nomia melanderi* Cockerell (Cane, 2008). The blue orchard bee, *Osmia lingaria* Say, is used as a pollinator in many high-value crops, including almonds, apples, pears, and cherries (Bosch and Kemp, 2002; Torchio, 2003; Cane, 2005; Bosch et al., 2006). The culture and use of solitary bees for pollination is reviewed in Chapter 19.

In conclusion, the challenges of mass producing beneficial organisms, particularly arthropods and pathogens for biological control, are being addressed by this book's authors and others. Their production technologies often are based on systems originally developed for area-wide pest management, but they have improved with advances in science and technology. The systems created by these innovators have many comparable production processes, although each is unique to the biology of the species being mass produced. The new methods and materials they incorporate into a production system for one species often can be adapted for use with another, thereby advancing the entire field of producing beneficial organisms. Thus, the primary purpose of this book is to assemble examples of production systems for arthropods, pathogens, and other organisms that have components to be compared and adapted for use in efficiently mass producing a variety of beneficial organisms.

REFERENCES

Anand, H., Ganguly, A., Haldar, P., 2008. Potential value of acridids as high protein supplement for poultry feed. Int. J. Poult. Sci. 7, 722–725.

Bai, C., Shapiro-Ilan, D.I., Gaugler, R., Hopper, K.R., 2005. Stabilization of beneficial traits in *Heterorhabditis bacteriophora* through creation of inbred lines. Biol. Control 32, 220–227.

Bondari, K., Sheppard, D.C., 1987. Soldier fly, *Hermetia illucens* L., larvae as feed for channel catfish, *Ictalurus punctatus* (Rafinesque), and blue tilapia *Oreochromis aureus* (Steindachner). Aquacult. Fisheries Manage. 18, 209–220.

Bosch, J., Kemp, W.P., 2002. Developing and establishing bee species as crop pollinators: the example of *Osmia* spp. and fruit trees. Bull. Entomol. Res. 92, 3–16.

Bosch, J., Kemp, W.P., Trostle, G.E., 2006. Bee population returns and cherry yields in an orchard pollinated with *Osmia lignaria* (Hymenoptera: Megachilidae). J. Econ. Entomol. 99, 408–413.

Calvert, C.C., Martin, R.D., Morgan, N.O., 1969. House fly pupae as food for poultry. J. Econ. Entomol. 62, 938–939.

Cane, J.H., 2005. Pollination potential of the bee *Osmia aglaia* for cultivated raspberries and blackberries (*Rubus*: Rosaceae). HortScience 40, 1705–1708.

Cane, J.H., 2008. A native ground-nesting bee (*Nomia melanderi*) sustainably managed to pollinate alfalfa across an intensively agricultural landscape. Apidologie 39, 315–323.

Croft, B.A., Monetti, L.N., Pratt, P.D., 1998. Comparative life histories and predation types: are *Neoseiulus californicus* and *N. fallacies* (Acari: Phytoseiidae) similar type II selective predators of spider mites? Environ. Entomol. 27, 531–538.

Debach, P., 1974. Biological Control by Natural Enemies. Cambridge University Press, London.

DeFoliart, G.R., 1999. Insects as food: why the western attitude is important. Annu. Rev. Entomol. 44, 21–50.

Ehlers, R.U., 2005. Forum on safety and regulation. In: Grewal, P.S., Ehlers, R.U., Shapiro-Ilan, D.I. (Eds.), Nematodes as Biological Control Agents. CABI, Wallingford, Oxon, UK, pp. 107–114.

Elzen, G.W., King, E.G., 1999. Periodic releases and manipulation of natural enemies. In: Bellows, T.S., Fisher, T.W. (Eds.), Handbook of Biological Control, Academic Press, San Diego, CA, pp. 253–270.

Fasakin, E.A., Balogun, A.M., Ajayi, O.O., 2003. Evaluation of full fat and defatted maggot meals in the feeding of clariid fish *Clarias gariepinus* fingerlings. Acuacult. Res. 34, 733–738.

Finke, M.D., 2002. Complete nutrient composition of commercially raised invertebrates used as food for insectivores. Zoo Biol. 21, 286–293.

Finke, M.D., 2013. Complete nutrient content of four species of feeder insects. Zoo Biol. 32, 27–36.

Fuxa, J.R., 1987. Ecological considerations for the use of entomopathogens in IPM. Annu. Rev. Entomol. 32, 225–251.

Gordon, D.G., 1998. The Eat-a-Bug Cookbook. Ten Speed Press, Berkeley, CA, 101 pp.

Grenier, S., 1994. Rearing of *Trichogramma* and other egg parasitoids on artificial diets. In: Wajnbergiand, E., Hassan, S.A. (Eds.), Biological Control with Egg Parasitoids, CAB International, Willingford Oxon, UK, pp. 73–92.

Grenier, S., 2009. *In vitro* rearing of entomophagous insects—past and future trends: a minireview. Bull. Insectology 62, 1–6.

Grenier, S., De Clercq, P., 2003. Comparison of artificially vs. naturally reared natural enemies and their potential for use in biological control. In: van Lenteren, J.C. (Ed.), Quality Control and Production of Biological Control Agents Theory and Testing Procedures, CABI Publishing, Willingford Oxon, UK, pp. 115–131.

Harper, J.K., Greaser, G.L., 1994. Agricultural Alternatives: Earthworm Production. Publication No. R3M994ps1918 Pennsylvania State University, University Park, PA.

Hill, J.E., 2009. Through the Jade Gate to Rome: A Study of the Silk Routes during the Later Han Dynasty 1st and 2nd Centuries CE: An Annotated Translation Chronicle on the 'Western Regions' in the Hou Hanshu. Book Surge Publishing, Charleston, SC 689 pp.

Jackson, M.A., Jaronski, S.T., 2012. Development of pilot-scale fermentation and stabilisation processes for the production of microsclerotia of the entomopathogenic fungus *Metarhizium brunneum* strain F52. Biocontrol. Sci. Technol. 22, 915–930.

Janisiewicz, W.J., Korsten, L., 2002. Biological control of post-harvest diseases of fruits. Annu. Rev. Phytopathol. 40, 411–441.

Kabiri, F., Bigler, F., 1996. Development and use of *Trichogramma* in the field for insect control in Western Europe. In: Silvy, C. (Ed.), Technology Transfer in Biological Control: From Research to Practice, IOBC/OILB, Montpellier, France, pp. 145–146.

Khatun, R., Azamal, S.A., Sarker, M.S.K., Rashid, M.A., Hussain, M.A., Miah, M.Y., 2005. Effect of silkworm pupae on the growth and egg production performance of Rhode Island red (RIR) pure line. Int. J. Poult. Sci. 4, 718–720.

King, E.G., 1993. Augmentation of parasites and predators for suppression of arthropod pests. In: Lumsden, R.D., Vaughn, J.L. (Eds.), Pest Management: Biologically Based Technologies, Conference Proceedings Series, American Chemical Society, Washington, DC, pp. 90–100.

King, E.G., Hopper, K.R., Powell, J.E., 1985. Analysis of systems for biological control of crop arthropod pests in the U.S. by augmentation of predators and parasites. In: Hoy, M.A., Herzog, D.C. (Eds.), Biological Control in Agricultural IPM Systems, Academic Press, Orlando, FL, pp. 201–227.

Klasing, K.C., Thacker, P., Lopez, M.A., Calvert, C.C., 2000. Increasing the calcium content of mealworms (*Tenebrio molitor*) to improve their nutritional value for bone mineralization of growing chicks. J. Zoo. Wildl. Med. 31, 512–517.

Lacey, L.A., Shapiro-Ilan, D.I., 2008. Microbial control of insect pests in temperate orchard systems: potential for incorporation into IPM. Annu. Rev. Entomol. 53, 121–144.

Lacey, L.A., Frutos, R., Kaya, H.K., Vail, P., 2001. Insect pathogens as biological control agents: do they have a future? Biol. Control. 21, 230–248.

Leppla, N.C., Johnson II, K.L., 2011. Guidelines for Purchasing and Using Commercial Natural Enemies and Biopesticides in Florida and Other States. University of Florida/IFAS, EDIS IPM-146 (IN849) http://edis.ifas.ufl.edu/in849.

Lin, S.W., Njaa, L.R., Eggum, B.O., Shen, H.Y., 1983. Chemical and biological evaluation of silk worm chrysalid protein. J. Sci. Food Agric. 34, 896–900.

Longvah, T., Mangthya, K., Ramulu, P., 2011. Nutrient composition and protein quality evaluation of eri silkworm (*Samia ricinii*) prepupae and pupae. Food Chem. 128, 400–403.

Lowe, C.N., Butt, K.R., 2005. Culture techniques for soil dwelling earthworms: a review. Pedobiologia 49, 401–413.

Lundgren, J.G., 2009. Relationship of Natural Enemies and Non-prey Foods. Progress in Biological Control Series. vol. 7. Springer, 453 pp www.springer.com.

Mason, W.T., Rottmann, R.W., Dequine, J.F., September 2006. Culture of Earthworms for Bait or Fish Food. University of Florida IFAS Extension Publication #CIR1053/FAO16.

Messelink, G.J., Ramakers, P.M.J., Cortez, J.A., Janssen, A., 2008. How to enhance pest control by generalist predatory mites in greenhouse crops. In: Mason, P.G., Gillespie, D.R., Vincent, C. (Eds.), Proceeding of the Third International Symposium of Biological Control of Arthropods, Christchurch, New Zealand, pp. 309–318.

Meyer, N.L., 1987. History of the Mexico–United States Screwworm Eradication Program. Vantage Press, NY.

Mishra, N., Hazarika, N.C., Narain, K., Mahanta, J., 2003. Nutritive value of non-mulberry and mulberry silkworm pupae and consumption pattern in Assam, India. Nutr. Res. 23, 1303–1311.

Montesinos, E., 2003. Development, registration and commercialization of microbial pesticides for plant protection. Int. Microbiol. 6, 245–252.

Morales-Ramos, J.A., Rojas, M.G., 2003. Natural enemies and pest control: an integrated pest management concept. In: Koul, O., Dhaliwal, G.S. (Eds.), Predators and Parasitoids, Taylor & Francis, London, U.K. and New York, NY, pp. 17–39.

Morales-Ramos, J.A., Rojas, M.G., Shapiro-Ilan, D.I., Tedders, W.L., 2011. Automated Insect Separation System. US Patent No. 8025027.

Needham, J.G., Galtsoff, P.S., Lutz, F.E., Welch, P.S., 1937. Culture Methods for Invertebrate Animals. Comstock Publishing, Ithaca, NY.

Ng, W.K., Liew, F.L., Ang, L.P., Wong, K.W., 2001. Potential of mealworm (*Tenebrio molitor*) as an alternative protein source in practical diets for African catfish, *Clarias gariepinus*. Aquacul. Res. 32, 273–280.

O'Toole, C., 2008. Forward. In: James, R.R., Pitts-Singer, T.L. (Eds.), Bee Pollination in Agricultural Ecosystems, Oxford University Press, NY, pp. v–vii.

Ramos-Elorduy, J., 1998. Creepy Crawly Cuisine, the Gourmet Guide to Edible Insects. Park Street Press, Rochester, VT 150 pp.

Ramos-Elorduy, J., Avila González, E., Rocha Hernandez, A., Pino, J.M., 2002. Use of *Tenebrio molitor* (Coleoptera: Tenebrionidae) to recycle organic wastes and as feed for broiler chickens. J. Econ. Entomol. 95, 214–220.

Riddick, E.W., 2009. Benefit and limitations of factitious prey and artificial diets on life parameters of predatory beetles, bugs, and lacewings: a mini-review. BioControl 54, 325–339.

Schneider, J.C., 2009. Principles and Procedures for Rearing High Quality Insects. Mississippi State University, Mississippi State, MS.

Shapiro-Ilan, D.I., Lewis, E.E., Tedders, W.L., Son, Y., 2003. Superior efficacy observed in entomopathogenic nematodes applied in infected-host cadavers compared with application in aqueous suspension. J. Invertebr. Pathol. 83, 270–272.

Shapiro-Ilan, D.I., Morales, Ramos, J.A., Rojas, M.G., Tedders, W.L., 2010. Effects of a novel entomopathogenic nematode-infected host formulation on cadaver integrity, nematode yield, and suppression of *Diaprepes abbreviatus* and *Aethina tumida* under controlled conditions. J. Invertebr. Pathol. 103, 103–108.

Shapiro-Ilan, D.I., Bruck, D.J., Lacey, L.A., 2012. Principles of epizootiology and microbial control. In: Vega, F.E., Kaya, H.K. (Eds.), Insect Pathology, second ed. Elsevier, Amsterdam, pp. 29–72.

Stephen, W.P., 2003. Solitary bees in North American agriculture: a perspective. In: Strickler, K., Cane, J.H. (Eds.), For Nonnative Crops, Whence Pollinators of the Future?, Entomological Society of America, Lanham, MD, pp. 41–66.

Taylor, D.B., 1992. Rearing systems for screwworm mass production. In: Anderson, T.E., Leppla, N.C. (Eds.), Advances in Insect Rearing for Research and Pest Management, Westview Press, Boulder, CO, pp. 393–403.

Tebeest, D.O., 1996. Biological control of weeds with plant pathogens and microbial pesticides. Adv. Agron. 56, 115–137.

Torchio, P.F., 2003. Development of *Osmia lignaria* (Hymenoptera: Megachilidae) as a managed pollinator of apple and almond crops: a case history. In: Strickler, K., Cane, J.H. (Eds.), For Nonnative Crops, Whence Pollinators of the Future?, Entomological Society of America, Lanham, MD, pp. 67–84.

Van Driesche, R.G., Bellows Jr., T.S., 1996. Biological Control. Chapman & Hall, New York, NY pp 539.

Van Lenteren, J.C., 2003. Commercial availability of biological control agents. In: van Lenteren, J.C. (Ed.), Quality Control and Production of Biological Control Agents, Theory and Testing Procedures, CABI Publishing, Cambridge, MA, pp. 167–179.

Vega, F.E., Kaya, H.K. (Eds.), 2012. Insect Pathology, second ed. Elsevier, Amsterdam.

Zuidhof, M.J., Molnar, C.L., Morley, F.M., Wray, T.L., Robinson, F.E., Hhan, B.A., Al-Ani, L., Goonewardene, L.A., 2003. Nutritive value of house fly (*Musca domestica*) larvae as a feed supplement for turkey poults. Anim. Feed Sci. Technol. 105, 225–230.

FURTHER READING

Costello, S.L., Pratt, P.D., Rayachhetry, M.B., Center, T.D., 2002. Morphology and life history characteristics of *Podisus mucronatus* (Heteroptera: Pentatomidae). Fla. Entomol. 85, 344–350.

De Bortoli, S.A., Otuka, A.K., Vacari, A.M., Martins, M.I.E.G., Volpe, H.X.L., 2011. Comparative biology and production costs of *Podisus nigrispinus* (Heteroptera: Pentatomidae) when fed different types of prey. Biol. Control 58, 127–132.

De Clercq, P., Merlevede, F., Tirry, L., 1998. Unnatural prey and artificial diets for rearing *Podisus maculiventris* (Heteroptera: Pentatomidae). Biol. Control 12, 137–142.

Grundy, P.R., Maelzer, D.A., Bruce, A., Hassan, E., 2000. A mass-rearing method for the assassin bug *Pristhesancus plagipennis* (Hemiptera: Reduviidae). Biol. Control 18, 243–250.

Lemos, W.P., Ramalho, F.S., Serrão, J.E., Zanuncio, J.C., 2003. Effects of diet on development of *Podisus nigrispinus* (Dallas) (Het., Pentatomidae), a predator of the cotton leafworm. J. Appl. Entomol. 127, 389–395.

Rhodes, E.M., Liburd, O.E., 2009. Featured Creatures. Entomology and nematology, FDACS/DPI, EDIS, University of Florida. http://entnemdept.ufl.edu/creatures/beneficial/Neoseiulus_californicus.htm.

Saint-Cyr, J.F., Cloutier, C., 1996. Prey preference by the stinkbug *Perillus bioculatus*, a predator of the Colorado potato beetle. Biol. Control 7, 251–258.

Shapiro-Ilan, D.I., Lewis, E.E., Behle, R.W., McGulre, M.R., 2001. Formulation of entomopathogenic nematode-infected cadavers. J. Invertebr. Pathol. 78, 17–23.

Shapiro-Ilan, D., Rojas, M.G., Morales-Ramos, J.A., Lewis, E.E., Tedders, W.L., 2008. Effects of host nutrition on virulence and fitness of entomopathogenic nematodes: lipid- and protein-based supplements in *Tenebrio molitor* diets. J. Nematol. 40, 13–19.

Chapter 2

Production of Coleopteran Predators

Eric W. Riddick[1] and Hongyin Chen[2]

[1]*National Biological Control Laboratory, USDA-ARS, Stoneville, MS, USA,* [2]*Sino-American Biological Control Laboratory, OIRP, ARS, USDA, Beijing, China*

2.1. INTRODUCTION

2.1.1. Aims of this Chapter

This chapter reviews the scientific literature to highlight research on the production of coleopteran predators, with a primary focus on lady beetles (family Coccinellidae). Mass production is necessary for the permanent establishment of predators, periodic colonization of predators in an area to augment existing populations, and inundative releases for short-term control of a pest (Etzel and Legner, 1999). Mass production will also benefit researchers interested in testing the side effects of various toxins on beneficial, nontarget arthropods (Martos et al., 1992; Li et al., 2011). Finally, mass production could also benefit scientists interested in restoring populations of rare, threatened, or endangered coleopteran predators (Harmon et al., 2007; Gwiazdowski et al., 2011).

The main emphasis of this chapter is on reviewing the published literature over the last few decades to develop new and creative insights on predator production for applied biological control. Transfer of these insights to the biological control industry, especially the natural enemy industry, should be a major outcome of this work. Why dedicate an entire chapter solely to predatory beetles? One of the most successful biological control programs involved the use of a lady beetle, namely the vedalia beetle *Rodolia cardinalis* (Mulsant), to control cottony cushion scale *Icerya purchasi* Maskell in California citrus groves in 1989 (Caltagirone and Doutt, 1989). The outstanding performance of this lady beetle led to a worldwide fascination with coccinellids and other natural enemies that could provide control of crop pests on a spectacular scale (Sawyer, 1996). Today, one challenge to greater use of coccinellids and other less heralded coleopteran predators in applied biological control is to create cost-effective techniques to rear and stockpile (store) species. Many of the current rearing methods continue to depend on a tritrophic system of rearing: the host plant, natural prey

Mass Production of Beneficial Organisms. http://dx.doi.org/10.1016/B978-0-12-391453-8.00002-9
2014 Published by Elsevier Inc.

(herbivorous pest), and predator. This system is labor intensive and not cost-effective. Therefore, the aim of this work is to highlight what we know about rearing predatory coleopterans and to provide ideas and suggestions that might stimulate more advances in this important field of study.

2.1.2. Predatory Beetles in Culture

Predatory beetles (order Coleoptera) have a long history of importance as natural enemies of plant-feeding insects and mites in natural and managed ecosystems. Many coleopteran predators, especially those in the family Coccinellidae (representing the lady beetles, ladybird beetles, or ladybugs), have been used with moderate success in managing pest populations throughout the world (Hodek and Honěk, 1996; Obrycki and Kring, 1998; Hagen et al., 1999; Hodek and Evans, 2012). Some of the coleopteran predators covered in this chapter are or have been commercially available for biological control of crop pests. In van Lenteren (2012), a list of species available at present or in the past is provided, along with information on the region of the world the insects were/are released (for augmentative biological control), the estimated year of first use, and the market value (small, medium, large) of each species. Many of these taxa have been subjects of research on nutritional ecology and rearing as it relates to biological control (Hagen, 1987; Hodek, 1996; Thompson and Hagen, 1999; van Lenteren, 2003). In this chapter, we provide a list of exemplary coleopteran predators that have been subjects of rearing-related research (Table 2.1).

2.1.3. Overview of the Content

The coverage of this chapter is in favor of coccinellids because most rearing-related work has historically focused on this group of predatory beetles. The most detailed section of this book chapter addresses the food of predatory beetles. Knowledge of the nutritional requirements of a predator is central to devising methods to produce the insect in sufficient quantities for biological control. We cover the feeding preferences of predators when presented with natural prey. Then we provide evidence for feeding on factitious food/prey (i.e. alternatives to natural prey) and artificial diets.

The next section of this chapter reviews the importance of rearing density (population density) on the production of predatory beetles. Two critical topics of concern here are the effects of crowding and cannibalism on production. Next, we examine the design of oviposition substrates and enclosures or rearing cages that support most efficient population growth. Issues of rearing scale and the possibility of large-scale production of predators also are reviewed. Controlling and maintaining temperatures inside enclosures that maximize the growth of predatory beetles is very important. We discuss some methods that scientists use to estimate the optimum temperature for population growth.

TABLE 2.1 List of Coleopteran Predators as Subjects of Research Related to Mass Production

Family	Species*	Natural/ Target Prey	References
Carabidae	*Abax parallelepipedus* (Piller and Mitterpacher)	Snails, slugs	Symondson (1994)
Carabidae	*Calosoma sycophanta* (L.)	Gypsy moth larvae, pupae	Weseloh (1998)
Carabidae	*Lebia grandis* (Hentz)	Colorado potato beetle immatures	Weber et al. (2006), Weber and Riddick (2011)
Carabidae	*Poecilus chalcites* (Say)	Dipterans, lepidopterans	Lundgren et al. (2005)
Cleridae	*Thanasimus dubius* (F.)	Bark beetles	Reeve et al. (2003), Costa and Reeve (2012)
Cleridae	*Thanasimus formicarius* (L.)	Bark beetles	Faulds (1988), Lawson and Morgan (1992)
Coccinellidae	*Adalia bipunctata* (L.)*	Aphids	Kariluoto et al. (1976), Bonte et al. (2010)
Coccinellidae	*Brumoides suturalis* (F.)	Scales, mealybugs	Bista et al. (2012)
Coccinellidae	*Chilocorus bipustulatus* (L.)*	Scales	Henderson et al. (1992)
Coccinellidae	*Chilocorus nigritus* (F.)*	Scales	Hattingh and Samways (1993)
Coccinellidae	*Cleobora mellyi* (Mulsant)	Psyllids, chrysomelids	Bain et al. (1984)
Coccinellidae	*Clitostethus arcuatus* (Rossi)*	Whiteflies	Mota et al. (2008), Yazdani and Zarabi (2011)
Coccinellidae	*Coccinella septempunctata* (L.)*	Aphids	Sarwar and Saqib (2010)

Continued

TABLE 2.1 List of Coleopteran Predators as Subjects of Research Related to Mass Production—cont'd

Family	Species*	Natural/ Target Prey	References
Coccinellidae	*Coccinella undecimpunctata* (L.)	Aphids	Farag et al. (2011)
Coccinellidae	*Coleomegilla maculata* (DeGeer)	Aphids, lepidopterans, coleopterans	Atallah and Newsom (1966), Silva et al. (2010)
Coccinellidae	*Cryptolaemus montrouzieri* (Mulsant)*	Scales, mealybugs	Heidari and Copland (1993), Attia et al. (2011)
Coccinellidae	*Delphastus catalinae* (Horn)*	Whiteflies	Pickett et al. (1999), Kutuk and Yigit (2007)
Coccinellidae	*Diomus terminatus* (Say)	Aphids	Tifft et al. (2006)
Coccinellidae	*Eriopis connexa* (Germar)	Aphids	Martos et al. (1992), Silva et al. (2009)
Coccinellidae	*Harmonia axyridis* (Pallas)*	Aphids	Dong et al. (2001), Sighinolfi et al. (2008)
Coccinellidae	*Hippodamia convergens* (Guerin)*	Aphids	Hussein and Hagen (1991), Kato et al. (1999)
Coccinellidae	*Hippodamia variegata* (Goeze)*	Aphids	van Lenteren (2012)
Coccinellidae	*Menochilus sexmaculatus* (F.)	Aphids, scales, psyllids	Khan and Khan (2002)
Coccinellidae	*Nephus includens* (Kirsch)*	Mealybugs	Canhilal et al. (2001)
Coccinellidae	*Propylea dissecta* (Mulsant)	Aphids	Omkar and Pathak (2009)

TABLE 2.1 List of Coleopteran Predators as Subjects of Research Related to Mass Production—cont'd

Family	Species*	Natural/ Target Prey	References
Coccinellidae	*Propylea japonica* (Thunberg)	Aphids	Hamasaki and Matsui (2006), Wang et al. (2008)
Coccinellidae	*Rhyzobius lophanthae* (Blaisdell)*	Scales	Stathas (2000)
Coccinellidae	*Rodolia cardinalis* (Mulsant)*	Scales	Matsuka et al. (1982), Grafton-Cardwell et al. (2005)
Coccinellidae	*Sasajiscymnus tsugae* (Sasaji and McClure)	Adelgids	Flowers et al. (2007), Conway et al. (2010)
Coccinellidae	*Stethorus punctillum* (Weise)*	Mites	Riddick et al. (2011a, b)
Colydiidae	*Dastarcus helophoroides* (Fairmaire)	Cerambycid larvae	Ogura et al. (1999)
Histeridae	*Carcinops pumilio* (Erichson)*	Dipterans	Achiano and Giliomee (2006)
Rhizophagidae	*Rhizophagus grandis* (Gyllenhal)	Bark beetles	Couillien and Grégoire (1994)
Staphylinidae	*Holobus flavicornis* (Lacordaire)*	Mites	van Lenteren (2012)
Staphylinidae	*Aleochara bilineata* (Gyllenhal)*	Dipterans	Whistlecraft et al. (1985)
Staphylinidae	*Dalotia coriaria* (Kraatz)*	Dipterans, thrips, coleopterans	Birken and Cloyd (2007)
Trogossitidae	*Temnochila virescens* (F.)	Bark beetles	Lawson and Morgan (1992)
Trogossitidae	*Trogossita japonica* (Reitter)	Cerambycid larvae	Ogura and Hosoda (1995)

**Species sold commercially in the past or present for biological control of pests (see van Lenteren, 2012).*

The next section of this chapter concerns checks and balances to ensure that unwanted pathogens and parasites do not infiltrate cultures. We address other concerns, such as preventing the deterioration of cultures through inbreeding and other genetic means. We touch upon the importance of assessing the health (quality) of cultures at the preshipment, in-shipment, and prerelease stages. The final section serves to conclude this chapter, synthesize data, and provide recommendations for future research.

2.2. FOODS AND PRODUCTION OF PREDATORS

2.2.1. Feeding Preferences and Natural Prey

The feeding preferences of coleopteran predators vary greatly. Feeding preferences in the context of this review refer to the innate restrictions, or lack thereof, that predators display toward potential prey. Scientists often group species into the categories of polyphagous, oligophagous, and monophagous, indicating broad to narrow feeding preferences. Truly monophagous species, which specialize on one prey species or genus, are rare in nature. Most coleopteran predators likely fall along a continuum between oligophagous and polyphagous. Some oligophagous predators include a carabid *Lebia grandis* Hentz (Weber et al., 2006; Weber and Riddick, 2011), coccinellids *Sasajiscymnus tsugae* Sasaji and McClure (Flowers et al., 2007; Jetton et al., 2011), *Stethorus punctillum* Weise (Riddick et al., 2011a, b), *R. cardinalis* (Matsuka et al., 1982; Grafton-Cardwell et al., 2005) and *Clitostethus arcuatus* Rossi (Mota et al., 2008; Yazdani and Zarabi, 2011). Most coleopteran predators used previously in biological control and reared in the laboratory are moderately polyphagous (Table 2.1).

Researchers often use the terms *essential prey* and *alternative* (nonessential) *prey* when they refer to the food preferences of coccinellids. Essential prey are required for multigenerational development and reproduction, whereas alternative prey typically sustain the lifespan of a predator during shortages of essential prey (Hodek and Honěk, 1996). Intuitively, the structure of the mouthparts, digestive tract, and the spectrum of enzymes in the digestive tract of a predator should correlate with feeding preferences, in a general sense. Some correlations do exist, although generally at the family or genus level rather than the species level in coccinellids (Minelli and Pasqual, 1977; Samways et al., 1997). Giorgi et al. (2009) provide a review of the purported evolution of food preferences in coccinellids.

Researchers have experimented with rearing coccinellids and other predatory beetles for biological control over the years. Several coccinellids, staphylinids, and a histerid have been commercially available (Table 2.1). The technology of producing these predators remains in an infant stage; rearing on natural prey is commonplace. Nevertheless, researchers have tried to rear several coccinellid and noncoccinellid coleopteran predators on artificial diets with or without arthropod components.

Most rearing-related research on coleopteran predators to date has involved coccinellids. Detailed reviews of the feeding preferences of coccinellids are available for species that prefer aphids (Obrycki et al., 2009), scales, mealybugs, whiteflies, and psyllids (Hodek and Honěk, 2009), mites (Biddinger et al., 2009), and nonhemipteran insects, such as other coleopterans and lepidopterans (Evans, 2009). By far, most of this work has concentrated on coccinellids that prefer aphids (aphidophagous) as prey, despite the belief that species that use other insects as prey (e.g. scales) are more effective biocontrol agents (Magro et al., 2002). Some aphidophagous coccinellids are polyphagous and are capable of reproducing on other insects, besides aphids. Coccinellids that prefer scales, mealybugs, whiteflies, psyllids, or mites (rather than aphids) apparently are more restricted in their prey preferences (Hodek and Evans, 2012). Researchers are currently testing various techniques to increase mass production of the coccinellid *S. tsugae* on woolly adelgids in several laboratories in the eastern North America for augmentative releases into forests to suppress populations of the hemlock woolly adelgid (Cohen et al., 2008; Conway et al., 2010). Development of methods of rearing *S. tsugae* using natural prey or target prey is not cost-effective over the long term. Alternative (i.e. factitious) prey would expedite and streamline the rearing of these important coccinellids. Factitious prey should be easier or less expensive to rear than natural or target prey of coleopteran predators.

2.2.2. Feeding on Factitious Foods and Plant Products

Factitious prey/foods are typically insects, mites, or crustaceans that support the development and reproduction of predators (although often at suboptimal levels) in lieu of natural or target prey. Researchers have presented factitious prey/food to predators in various forms: live, irradiated, frozen, or lyophilized. Lepidopteran eggs represent a nutrient-rich food for some generalist coccinellids (Herrera, 1960; Riddick, 2009) but not other species (Michaud and Qureshi, 2005; Hodek and Evans, 2012). Abdel-Salam and Abdel-Baky (2001) fed the multicolored Asian lady beetle *Harmonia axyridis* (Pallas) larvae and adults fresh versus frozen eggs of the grain moth, *Sitotroga cerealella* Olivier; fresh *S. cerealella* eggs were more nutritious or beneficial for predator development and reproduction. Specty et al. (2003) found that *Ephestia* eggs contained a greater percentage of amino acids and lipids compared to the pea aphid *Acyrthosiphum pisum* (Harris), a natural prey of *H. axyridis*; pea aphids contained a greater percentage of glycogen. The food intake of *H. axyridis* was threefold greater when feeding on pea aphids than *Ephestia* eggs. According to Specty et al. (2003), *Ephestia* eggs had a positive effect on *H. axyridis* growth and reproduction (see Table 2.2). Ettifouri and Ferran (1993) and Ferran et al. (1997) found that rearing *H. axyridis* on *Ephestia* eggs over many generations could reduce the capacity of larvae to intensely search for and recognize chemical cues (tracks) of its prey, the pea aphid, *A. pisum*. Caution must be used when rearing predators

TABLE 2.2 Coccinellid Predators* and the Effects of Factitious Food on Life History

Lady Beetle	Factitious Food	Results	References
Cryptolaemus montrouzieri (larvae, adults)	*Ephestia kuehniella* eggs	Development (↓), fecundity (↓), longevity (✓)	Attia et al. (2011)
Harmonia axyridis (larvae, adults)	*E. kuehniella* eggs	Survival (↑), size (↑), fecundity (↑)	Specty et al. (2003)
Hippodamia convergens (larvae, adults)	*E. kuehniella* eggs	Development (↑), oviposition period (✓), fecundity (✓), longevity (✓)	Kato et al. (1999)
Propylea japonica (larvae, adults)	*E. kuehniella* eggs	Survival (✓), fecundity (↓)	Hamasaki and Matsui (2006)
Adalia bipunctata (larvae, adults)	*E. kuehniella* eggs plus pollen	Development (✓), survival (✓)	De Clercq et al. (2005a)
	E. kuehniella eggs plus pollen	Development (↓), oviposition (↓), fecundity (↓)	Jalali et al. (2009a)
A. bipunctata (larvae)	*E. kuehniella* eggs plus pollen	Development (↑), survival (✓), size (✓), consumption (↑)	Jalali et al. (2009b)
A. bipunctata (larvae, adults)	*E. kuehniella* eggs plus pollen	Development (↓), survival (↓), size (↓), fecundity (↑), egg hatch (↓), longevity (✓)	Bonte et al. (2010)
	Brine shrimp cysts (eggs) plus pollen	Development (↓), survival (↓), size (↓), fecundity (✓), egg hatch (↓), longevity (↑)	Bonte et al. (2010)

Data on noncoccinellid beetles included in text only.
Symbols indicate that factitious foods have no significant effect (✓), increase (↑), or decrease (↓) on a life parameter when compared against a control (natural prey).

solely on factitious prey without a period of exposure to natural prey prior to releasing them for biological control.

Attia et al. (2011) proved that it was possible to rear larvae and adults of the mealybug destroyer *Cryptolaemus montrouzieri* Mulsant on *Anagasta* (*Ephestia*) *kuehniella* (Zeller) eggs. *Cryptolaemus montrouzieri* larvae developed faster on *Ephestia kuehniella* eggs than on mealybug *Planococcus citri* (Risso) eggs (Table 2.2). Although fecundity was 33% less (on average), adult longevity was approximately the same for females fed *Ephestia* eggs in place of *P. citri* eggs. Pilipjuk et al. (1982) succeeded in rearing *C. montrouzieri* on *S. cerealella* eggs. However, mealybug ovisacs or an extract of mealybugs or their ovisacs were required to stimulate oviposition.

When offered *Ephestia* eggs, the convergent lady beetle *Hippodamia convergens* Guerin-Meneville took a few days longer to complete development than when fed aphids, *Schizaphis graminum* (Rondani), and *Brachycaudus schwartzi* Börner (Kato et al., 1999). Other biological parameters relating to adult lifespan and reproduction were comparable for predators fed aphids (*S. graminum*) versus *Ephestia* eggs (Table 2.2). Hamasaki and Matsui (2006) found that larvae and adults of a coccinellid *Propylea japonica* (Thunberg) developed and reproduced on *Ephestia* eggs in place of pea aphids; however, fecundity was lower for adults reared on the moth eggs.

Researchers are revealing the benefits of combining insect and plant products for improving the fitness of natural enemies (Smith, 1960, 1965; Lundgren, 2009, 2010; Farag et al., 2011). Some predators are capable of compensating for a less-than-optimal diet of factitious prey by feeding on plant protein, carbohydrates, and sterols (pollen) when it is available. Consumption of plant material during periods of prey shortage is a survival strategy of some predators (Berkvens et al., 2008, 2010). De Clercq et al. (2005a) discovered that supplementing flour moth eggs with frozen moist bee pollen, rather than natural prey (live pea aphids), allowed for adequate egg hatch, development, and survival of *Adalia bipunctata* L. larvae. *Ephestia* eggs plus fresh bee pollen combination is as nutritionally suitable as aphids for *A. bipunctata* larval development but not for reproduction (Jalali et al., 2009a, b). Bonte et al. (2010) found that *A. bipunctata* development as well as reproduction (fecundity) improved when fed a mixture of moist bee pollen plus *E. kuehniella* eggs rather than pea aphids (see Table 2.2).

Brine shrimp (*Artemia* spp.; Branchiopoda), a product of the aquaculture industry, is relatively inexpensive to produce in comparison to *Ephestia* moth eggs (De Clercq et al., 2005b). Hongo and Obayashi (1997) were among the first to test *Artemia* eggs as factitious food for *H. axyridis*. They discovered that *Artemia salina* (L.) eggs alone supported the development of larvae but not adult emergence, which was just 18%. When combining *A. salina* eggs with sucrose and Y2A (brewer's yeast; Asahi Breweries, Japan), adult emergence improved significantly (61%). The weight of adult females was about the same regardless of whether they fed on *A. salina* eggs alone or eggs with the additives (Hongo and Obayashi, 1997). A combination of brine shrimp (*Artemia franciscana* Kellogg) powder and synthetic bee pollen supported the development and reproduction of *Coleomegilla maculata* (DeGeer) in laboratory experiments and toxicity bioassays (Lundgren and Weber, 2010; Pilorget et al., 2010; Li et al., 2011; Lundgren et al., 2011; Weber and Lundgren, 2011). However, these studies did not compare the benefits of using brine shrimp versus natural prey or *Ephestia* eggs as food for *C. maculata*. Bonte et al. (2010) compared the benefits of brine shrimp eggs (cysts) plus moist bee pollen versus natural prey (pea aphids) on the life history of *A. bipunctata* (see Table 2.2). They found that development time increased and size decreased but longevity increased when predators consumed brine shrimp eggs rather than pea aphids.

Heidari and Copland (1993) found that honeydew was a food source or at least an arrestant for *C. montrouzieri* larvae and adults. Adults spent more time foraging on portions of leaves coated with honeydew but spent less time foraging on clean surfaces. However, egg production was very low on a diet of honeydew alone.

Birken and Cloyd (2007) observed the feeding preferences of the staphylinid *Dalotia coriaria* (Kraatz), a predator of fungus gnats and thrips in greenhouses. They discovered that larvae and adults consume oatmeal in the presence or absence of fungus gnat larvae in laboratory arenas. Oatmeal can serve as an inexpensive supplemental food in rearing of *D. coriaria*, which is commercially available for control of fungus gnats in greenhouses.

2.2.3. Feeding on Artificial Diets

The use of an artificial diet may represent a step toward more cost-effective rearing of coccinellids and other coleopteran predators. Researchers have classified artificial diets by the identification of chemicals within them. These include holidic diets, in which all constituents are known in chemical (molecular) structure; meridic diets, in which most of the constituents are known chemically; and oligidic diets, in which a few of the constituents are known chemically (Vanderzant, 1974). An alternative system of classification distinguishes artificial diets based on the presence or absence of insect components (i.e. tissues, hemolymph, cells, protein, amino acids, etc.) within them (Grenier and De Clercq, 2003; Grenier, 2009). We prefer using the latter system in this chapter. However, we expand "insect components" to "arthropod components" to reflect the usage of protein from crustaceans in artificial diets.

Artificial diets containing arthropod components. Artificial diets that contain insect matter (tissues, protein, cells, etc.) are useful when predators need chemical cues and other feeding stimulants found in live prey (De Clercq, 2004). There has been only limited success in rearing predatory beetles with artificial diets containing arthropod components. To date, most research has centered on aphidophagous coccinellids. Smirnoff (1958) was one of the first to formulate artificial diets (with arthropod components) for coccinellids. He reported that an artificial medium for rearing adults was essentially composed of cane sugar, agar, royal jelly, honey, and pulverized dry insects (natural prey of the predator species in rearing). To rear larval stages, he supplemented this medium with royal jelly and beef jelly. Although 19 species (in the genera *Coccinella*, *Harmonia*, *Thea*, *Rhizobius*, *Rodolia*, *Exochomus*, *Scymnus*, *Stethorus*, *Chilocorus*, *Clitostethus*, *Pharoscymnus*, and *Mycetaea*) were reared on these two media, Smirnoff (1958) did not present any data on the effects of these media on predator development and reproduction.

He does mention that gravid females would not produce eggs when fed the media inside the confines of a Petri dish; placement of females in larger cages containing twigs of a plant that the predator forages on under natural conditions induced oviposition. This study shows that several coccinellids (*Mycetaea tafilaletica* Smirnoff, *Scymnus kiesenwetteri* Mulsant, *Scymnus pallidivestis* Mulsant, and *Thea vigintiduopunctata* L.) lived from three to five times longer when fed the media rather than natural prey (Smirnoff, 1958).

In a more recent study, Sarwar and Saqib (2010) attempted to rear the aphidophagous seven-spot ladybird *Coccinella septempunctata* L. using an artificial diet somewhat similar to that used by Smirnoff (1958). Their diet consisted of cane sugar, honey, protein hydrolysate, royal jelly, yeast, and pulverized dried mustard aphid (*Lipaphis erysimi* Kaltenbach). In comparison to natural prey (mustard aphid), the artificial diet was substandard; a low percentage of larvae completed development and females produced much fewer eggs when fed the artificial diet (Table 2.3).

The aphidophagous coccinellid *P. japonica* completed development when fed crude artificial diets consisting of powdered formulations of defatted yellow mealworm (*Tenebrio molitor* L.) pupae with yeast and sugar or housefly (*Musca domestica* L.) larvae with yeast and sugar versus soybean aphid *Aphis glycines* Matsumura (Wang et al., 2008). However, when compared to a diet of soybean aphids, the two powdered diets had negative effects on larval development and body size (Table 2.3).

Silva et al. (2009) found that an artificial diet containing a commercial cat food did not support the development of the aphidophagous coccinellid *Eriopis connexa* (Germar) to the adult stage. Note that simply adding eggs of the moth *E. kuehniella*, frozen for one day, to this artificial diet resulted in a development rate of 72% from the larval to adult stage. In a similar study, Silva et al. (2010) reared the pink spotted lady beetle *C. maculata* on a combination of a yeast-based artificial diet plus moth eggs, *E. kuehniella* in comparison to *Ephestia* eggs alone, *Spodoptera frugiperda* (J. E. Smith) eggs alone, or aphids *S. graminum* alone (Table 2.3). They found that predators took less time to complete larval development when fed aphids rather than the other food treatments, but total development from larva to adult stage did not differ between food treatments. The percentage of larvae surviving to the adult stage was high—greater than 85% (mean value) in all food treatments—with no significant differences between any of the treatments. *Coleomegilla maculata* males and females fed the artificial diet with *Ephestia* eggs weighed more than individuals fed *Ephestia* eggs alone. Adult males and females fed the artificial diet with *Ephestia* eggs were not heavier than individuals fed aphids alone (Silva et al., 2010).

A study on the convergent lady beetle, *H. convergens*, examined the effect of color morphs of the green peach aphid, *Myzus persicae* (Sulzer) on predator egg production (Wipperfürth et al., 1987). The authors gave female predators an artificial diet, ad libitum, and less-than-optimal quantities of wingless virginoparous versus winged gynoparous forms of *M. persicae*. Females failed to

TABLE 2.3 Coccinellid Predators* and the Effects of Artificial Diets on Life History

Lady Beetle	Artificial Diet—Protein Base	Results	References
Adalia bipunctata (larvae, adults)	Ground beef, beef liver, chicken egg yolk plus pollen	Development (↑), size (↓), survival (↓), fecundity (↓), longevity (✓)	Bonte et al. (2010)
	Ground beef, beef liver, chicken egg yolk plus pollen, brine shrimp (*Artemia*) eggs	Development (↑), size (↓), survival (✓), fecundity (↓), longevity (✓)	Bonte et al. (2010)
Chilocorus nigritus (larvae, adults)	Honeybee brood (powdered)	Development (↑), adult emergence (↓), size (↓), fecundity (↓)	Hattingh and Samways (1993)
Clitostethus arcuatus (larvae, adults)	Yeast, honey, pollen	Fecundity (↓), fertility (↓), sex ratio (✓), longevity (↑)	Yazdani and Zarabi (2011)
Coccinella septempunctata (larvae, adults)	Yeast, protein hydrolysate, aphids (powdered)	Development (↑), survival (↓), oviposition (↓)	Sarwar and Saqib (2010)
Coleomegilla maculata (larvae)	Yeast plus *Ephestia* eggs	Development (✓), survival (✓), size (✓)	Silva et al. (2010)
Eriopis connexa (larvae)	Cat food plus *Ephestia* eggs	Development (↑), adult emergence (↓), size (↓)	Silva et al. (2009)
Harmonia axyridis (larvae, adults)	Chicken egg yolk, chicken liver, casein hydrolysate	Development (↑), survival (✓), size (↓), oviposition (↓)	Dong et al. (2001)
	Chicken whole egg, chicken liver, casein hydrolysate	Development (↑), survival (✓), size (↓), oviposition (↓)	Dong et al. (2001)
	Pork liver, amino acid solution	Development (↑), adult emergence (↓), size (↓), pre-oviposition period (↑), fecundity (↓)	Sighinolfi et al. (2008)

TABLE 2.3 Coccinellid Predators* and the Effects of Artificial Diets on Life History—cont'd

Lady Beetle	Artificial Diet—Protein Base	Results	References
Menochilus sexmaculatus (larvae, adults)	Chicken liver, powdered	Development (↑), survival (↓), fecundity (↓), longevity (↓)	Khan and Khan (2002)
Propylea japonica (larvae, adults)	Housefly (powdered)	Development (↑), adult emergence (✓), size (↓)	Wang et al. (2008)
	Mealworm (powdered)	Development (↑), adult emergence (✓), size (↓)	Wang et al. (2008)

Symbols indicate that diets have no significant effect (✓), increase (↑), or decrease (↓) on a life parameter when compared against a control (usually natural prey).
**Data on noncoccinellid beetles included in text only.*

produce eggs when fed the artificial diet, which consisted of an 85:5:10 mixture of protein (lyophilized potato tuberworm, *Phthorimaea operculella* Zeller), yeast autolysate, and honey, respectively. Females produced more eggs when fed wingless virginoparous *M. persicae*.

Matsuka and Okada (1975) evaluated a series of artificial food sources to support the development of *H. axyridis* larvae. These included yeast, chicken liver, banana, royal jelly, pollen, and dried milk. The addition of pulverized drone honeybee powder to these diets slightly enhanced diet efficacy. In a follow-up study, a diet consisting of drone honeybee brood was suitable for *H. axyridis* larval development. After fractionating the diet, a cationic fraction, soluble in water, was determined to be essential for larval development. Consequently, they discovered that potassium was important for successful larval development (Matsuka and Takahashi, 1977). Niijima et al. (1997) experimented with rearing the multicolored Asian lady beetle *H. axyridis* on a diet of powdered honeybee brood. Approximately 90% and 80% of *H. axyridis* larvae fed powdered honeybee versus powdered aphid *Lachnus tropicalis* (van der Goot), respectively, reached the adult stage. Development time averaged 19 and 20 days for individuals fed powdered honeybee versus powdered aphids. However, adults reared on honeybee (rather than aphids) weighed less. Unfortunately, Niijima et al. (1997) did not report effects on fecundity or on longevity of adults reared from birth on this honeybee diet.

Bonte et al. (2010) fed the two-spot ladybird, *A. bipunctata,* another aphidophagous species, an artificial diet based on a beef and chicken protein combination with pollen and brine shrimp cysts (Table 2.3). They found that development time increased, body size decreased, and fecundity decreased in comparison to natural prey (pea aphids). Larval survival rate and adult longevity were unaffected by the artificial diet in relation to natural prey.

A partially coccidophagous coccinellid, *Cleobora mellyi* Mulsant, was introduced into New Zealand from Australia for biological control of the eucalyptus tortoise beetle *Paropsis charybdis* Stål, a major defoliator of *Eucalyptus* spp. in forests (Bain et al., 1984). An experiment compared freeze-dried pork liver, potato tuberworm larvae, and cerambycid larvae (Huhu grubs) as food for rearing *C. mellyi* larvae. Predators completed development on all three foods, but development was fastest with the tuberworm diet. Note that adults produced from all three foods still required feeding on natural prey (i.e. scales, psyllids, aphids) for approximately a week before any oviposition occurred.

Chilocorus beetles are important predators of scale insects. Two artificial diets—one for adults and the other for larvae—were the best of several ones for maintaining *Chilocorus nigritus* (F.) in the laboratory during periods of shortages of natural prey (Hattingh and Samways, 1993). Note that the diet deemed best for maintenance of *C. nigritus* adults contained powdered honeybee (*Apis mellifera* L.) brood (83.5%), royal jelly (1.7%), and glucose (4.5%). The diet considered best for *C. nigritus* larvae contained powdered honeybee brood (55%), royal jelly (3%), honey (4.5%), pulverized Oleander scale *Aspidiotus nerii* Bouché (3%), sucrose (4%), glucose (9%), and fructose (11%). Both diets contained the same percentages of wheat germ, brewer's yeast and ascorbic acid. However, both were not as suitable as natural (or target) prey for culturing this predator over consecutive generations. Predators fed artificial diets rather than scales took longer to develop, were smaller in size, and were less fecund (Table 2.3).

Henderson et al. (1992) tested the potential of using freeze-dried then powdered artificial diets to rear three *Chilocorus* species. The main protein constituents in these diets were pupae of the light brown apple moth *Epiphyas postvittana* Walker, pupae of the German yellowjacket wasp *Vespula germanica* (F.), and larvae/pupae of the honeybee *A. mellifera* L. The researchers discovered that the wasp diet was marginally suitable for larval development of *Chilocorus infernalis* Mulsant, *Chilocorus bipustulatus* (L.), and *Chilocorus cacti* (L.). Body weight of individuals reaching the adult stage (e.g. *C. cacti*) was significantly less when the wasp diet rather than natural prey (scales) was the food source since birth. Adults fed artificial diets alone did not oviposit (Henderson et al., 1992).

Research also considered the potential of artificial diets to rear the vedalia beetle *R. cardinalis*, an important predator of scale insects (Matsuka et al., 1982). Adult females lived longer but produced significantly fewer eggs when fed an artificial diet of powdered honeybee brood with sucrose rather than live prey, *I. purchasi*. Interestingly, feeding adults sucrose for 2 days and *I. purchasi* for 1 day resulted in oviposition rates that exceeded that of the control group (i.e. adults fed *I. purchasi* alone). The authors stated that this feeding routine would permit continuous rearing of *R. cardinalis*.

Rearing of predatory beetles other than coccinellids has met limited success thus far. *Dastarcus helophoroides* (Fairmaire) is a colydiid beetle that attacks wood-boring insects, including cerambycid beetles and xylocopid bees. Ogura et al. (1999) evaluated artificial diets containing some insect components as food for this predator. Rearing newly hatched larvae on diets composed of silkworm pupa powder, dry yeasts, yeast extract, sucrose, peptone, squid liver oil, preservatives, and distilled water resulted in very low adult emergence rates (i.e. less than 10%). However, if larvae were fed with paralyzed cerambycid larvae until achieving a body length of approximately 8 mm then reared on an artificial diet, adult emergence rates were much improved (i.e. greater than 50%). *Dastarcus helophoroides* larvae did not commonly cannibalize siblings even when reared in groups. Ogura and Hosoda (1995) tried to rear *Trogossita* (=*Temnochila*) *japonica* Reitter, a trogossitid predator of bark and wood-boring beetles, on artificial diets. A diet consisting of silkworm pupa powder, dry yeasts, sucrose, peptone, squid liver oil, agar, and distilled water supported the development of *Trogossita japonica* larvae. Although diet-reared larvae developed into smaller adults, they mated and oviposited at normal levels when provided with natural prey (Ogura and Hosoda, 1995).

Artificial diets without arthropod components. Research has shown very limited success in rearing predators on artificial diets devoid of any arthropod components. Atallah and Newsom (1966) found that incorporating a filtered extract of cotton leaves (containing carotenoids and sterols) into an artificial diet based on protein from casein, soybean hydrolysate, and a liver fraction was satisfactory for rearing the pink-spotted lady beetle *C. maculata* for eight consecutive generations with low mortality after the second generation. Unfortunately, the authors did not provide a control (e.g. natural or factitious prey) to compare with the results of the artificial diet. Addition of vitamin E to the adult diet increased copulation of *C. maculata* adults. The observation that a diet that lacked this filtered extract (of cotton leaves) was not satisfactory for development of this predator (Atallah and Newsom, 1966) suggests the presence of a stimulatory substance in cotton leaves that could possibly be isolated and incorporated into new and improved diets.

Experiments considered the potential of three formulations of a chicken liver-based artificial diet on development, weight gain, and survival of *H. convergens* larvae (Hussein and Hagen, 1991). Results indicated that *H. convergens* larvae survived on a formulation containing powdered chicken liver, yeast hydrolysate, and sucrose at a ratio of 1:1:2. However, a natural diet of pea aphids, *A. pisum*, was best for optimum weight gain and development of this predator. The authors suggest that the growth and development of *H. convergens* larvae could improve by adding egg yolk to the diet formulation. Note that *H. convergens* adults are predators of nonaphid prey such as adults of the silverleaf whitefly *Bemisia argentifolii* Bellows and Perring and eggs of the pink bollworm *Pectinophora gossypiella* (Saunders) (Hagler, 2009).

Researchers evaluated the potential of three commercial liver extracts (i.e. labeled as S, L, and 2) in combination with an experimental (oligidic) diet, on the growth and development of *H. convergens* larvae (Racioppi et al., 1981). They compared the performance of their diet with a previously established Vanderzant diet. Time required to complete development was shorter and body weight of adults was greater for individuals reared on the experimental diet (with or without liver extracts) in comparison to the Vanderzant diet. However, when fed live aphids (*S. graminum*), development time was shorter and the body weight (i.e. body mass) of pupae and adults were greater than cohorts fed the experimental diet containing the "S" liver extract. Removal of tryptophan and cystine from the diet resulted in deformed adults, as evidenced by underdeveloped tibia and tarsi.

A series of artificial diet formulations were developed and compared with natural prey as suitable food for rearing *A. bipunctata* (Kariluoto et al., 1976). The main ingredients in the diet formulations included variable amounts of wheat germ, brewer's yeast, sucrose, honey, chicken eggs, casein, liver, a salt mixture, vitamins, antibiotics, agar, and water. Even when using the most effective formulation, development time was 20–30% longer than that required by individuals fed natural prey (aphids, *M. persicae*). Adult emergence was 60–80% versus 65–95% when reared on the best formulations versus natural prey, respectively. Although females did not oviposit when reared on even the best diet formulations, some adults were able to live for 6 months.

Bonte et al. (2010) compared an artificial diet based on protein from beef and chicken supplemented with pollen with natural prey (pea aphids) on the life parameters of *A. bipunctata*. Their diet had negative effects on the life cycle of this predator (Table 2.3), but adult longevity was unaffected. Experiments were conducted with two antibiotics, namely sorbic acid (SA) and methyl-4-hydroxybenzoate (MpHB), in an effort to retard the growth of mold (e.g. *Aspergillus niger*, *Penicillium* spp.) on an artificial diet used for rearing *A. bipunctata* (Kariluoto, 1978). Five concentrations (ranging from 0 to 2000 ppm) of both compounds were tested in 25 combinations. SA was more effective than MpHB and a 1500-ppm concentration of SA controlled mold on the artificial diet for 3 weeks at 25 °C. Unfortunately, an increase in concentration of both compounds tended to cause an increased retardation of larval development. Note that adult emergence, however, was the highest (i.e. 76%) when a concentration of 1000 ppm of SA was used in the larval diet. No concentration of either antibiotic negatively affected adult weight.

Khan and Khan (2002) experimented with an artificial diet based on powdered chicken liver and sucrose to rear the coccinellid *Menochilus sexmaculatus* (F.)—a predator of aphids, psyllids, and scale insects in south East Asia. Unfortunately, the chicken liver diet was inferior to natural prey such as the green peach aphid *M. persicae* (Table 2.3).

Sighinolfi et al. (2008) compared the effects of a pork liver-based artificial diet to factitious prey (*E. kuehniella* eggs) on *H. axyridis* larvae and adults.

Although complete development from first instar larva to mature adults was possible when using the artificial diet, differences of both food sources on predator life history parameters evidenced some nutritional deficiencies in the artificial diet (Table 2.3). Dong et al. (2001) tested four artificial diets for rearing *H. axyridis* in the laboratory; these included egg yolk diet, whole egg diet, gelatin diet, and starch diet. All four diets contained the following base ingredients (in equivalent quantities): chicken liver, cane sugar, honey, Brewer's yeast, casein enzymatic hydrolysate, soy oil, salt, and vitamins. The whole egg diet produced a survival rate of 82.5% in the first generation and three generations were tested. The starch diet produced a first generation survival rate of 97.5% and two generations were tested. The gelatin diet produced the lowest survival rate. In comparison to factitious prey (i.e. eggs of the grain moth or the pink bollworm), the starch diet and the whole egg diet met similar expectations in terms of survival rate, development rate, adult body weight, and fecundity (Table 2.3).

Yazdani and Zarabi (2011) had limited success rearing *C. arcuatus*, a coccinellid that attacks whiteflies, on an artificial diet consisting of yeast, honey, and pollen (Table 2.3). When compared against live prey such as the greenhouse whitefly *Trialeurodes vaporariorum* (Westwood), this artificial diet resulted in predators that were less fecund and fertile. However, adult males and females lived longer when fed the artificial diet rather than whiteflies (*T. vaporariorum*).

Research to develop artificial diets for predators of the hemlock woolly adelgid *Adelges tsugae* (Annand) is ongoing. Preliminary work by Cohen et al. (2008) demonstrated the possibility of developing an artificial diet devoid of arthropod components for the coccinellid *S. tsugae*, which is one of the most important predators of *A. tsugae*. *Sasajiscymnus tsugae* will feed on a chicken egg-based artificial diet when hemlock woolly adelgids are in low supply or of inferior quality (Cohen et al., 2008). An artificial diet for the coccinellid would greatly expand mass-rearing operations of this predator for releases into forests in the northeastern United States.

Several researchers have attempted to rear noncoccinellids on artificial diets devoid of arthropod components. Weseloh (1998) tested an artificial diet for the carabid *Calosoma sycophanta* (L.), which is a predator of larvae and pupae of gypsy moth *Lymantria dispar* (L.) and other lepidopterans in forest ecosystems. Chicken meat and beef liver constituted the base protein in his diet. He found that larval development and survival were similar between the artificial diet and natural prey, *L. dispar* pupae. However, adults were smaller when reared as larvae on the artificial diet rather than natural prey.

Lundgren et al. (2005) developed a rearing protocol and observed aspects of the life history of a carabid *Poecilus chalcites* (Say) in the laboratory. *Poecilus chalcites* is a common, generalist predator of ground-dwelling arthropods in several agroecosystems. Larvae completed development (with 80% pupation rates) when using a meridic diet based on a commercially produced cat food. Note that the addition of other components to the diet (such as chicken egg or chicken liver) increased the body weight of *P. chalcites* larvae. Unfortunately, diet-reared adults failed to produce eggs.

An artificial diet based on veal, veal gravy, hen's eggs, potted meat, infant formula, and casein hydrolysate in Parafilm® capsules was developed for the clerid, *Thanasimus dubius* (F.), an important predator of the scolytids *Dendroctonus frontalis* Zimmermann and *Ips grandicollis* (Eichhoff) (Reeve et al., 2003). The authors developed this diet to match the chemical composition of the tissues of *I. grandicollis*. Diet-reared individuals were similar in quality to those taken directly from the wild in regards to survival rate, longevity, and predation capacity. Diet-reared individuals maintained their capacity to kill natural or factitious prey even after five generations on the artificial diet. Costa and Reeve (2012) found that sorbic acid, a preservative to increase shelf life of the diet, reduced the fecundity of *T. dubius* females by 20–40% but had no measurable effect on adult longevity or body size. The authors concluded that sorbic acid was not an essential component in their diet because removal of this preservative did not affect rearing efficiency.

Carcinops pumilio (Erichson) is a histerid predator of dipterans such as the housefly *M. domestica*. Research involved the testing of an artificial diet containing a protein-rich food additive versus natural prey (Achiano and Giliomee, 2006). Although predators could develop from egg to adult on the artificial diet, it took longer. Additionally, predators weighed less, mortality rate was higher, and oviposition rate was lower for individuals fed the artificial diet rather than natural prey (Achiano and Giliomee, 2006).

2.3. REARING DENSITY AND PRODUCTION

2.3.1. Crowding

Crowding is a behavioral response to changes in the density of populations. It can have positive or negative consequences on rearing insects in confined spaces. Although crowding would appear to be advantageous for mating, the designation of a density that is not too restricting is situational and may differ depending on the lifestyles of the species under consideration. Peters and Barbosa (1972) reviewed the literature on density effects on life history of insects in culture. They admit that overcrowding is often detrimental to development, body size, and fecundity of reared insects. Omkar and Pathak (2009) found that crowding affected the coccinellid *Propylea dissecta* (Mulsant) immatures when reared in plastic beakers (11 cm high by 9 cm wide). Rearing densities of 8, 16, 25, or 35 (rather than 4) larvae per beaker had negative effects on development time, larval survival, and adult emergence. Bista et al. (2012) considered the influence of crowding and diet (aphid species) on the development and survival of the coccinellid *Brumoides suturalis* (F.) in Petri dishes (14 cm wide by 1.5 cm high) in the laboratory. Rearing larvae of *B. suturalis* singly versus in groups (1 larva versus 10 larvae per dish) did not affect rearing success. However, prey species had a more significant impact; development was faster on *A. pisum* and larval survival greater on *Aphis gossypii* Glover (Bista et al., 2012).

2.3.2. Cannibalism

Cannibalism is a normal behavioral event in the population dynamics of numerous animals living under natural conditions (Fox, 1975). Investigators have more often observed cannibalistic behavior in the laboratory in arthropod colonies confined in enclosures (Peters and Barbosa, 1972). In respect to predatory beetles, Banks (1956) was one of the first scientists to show that cannibalism is a significant source of mortality of the egg stage in coccinellids; emerging larvae consume the unhatched eggs of their siblings, even at low population density. Common costs associated with larval cannibalism of eggs rather than feeding on moth (*Ephestia*) eggs is longer development time and reduced adult size in at least two coccinellids, *H. axyridis* and *Olla v-nigrum* (Mulsant) (Michaud, 2003). Adult coccinellids cannibalize their own eggs and unrelated eggs, especially when natural prey is in low supply (Cottrell, 2005). Cannibalism of eggs has beneficial effects on the nutrition of coccinellids. Dimetry (1974) found that *A. bipunctata* larvae could develop to the adult stage solely on a diet of conspecific eggs. *Adalia bipunctata* may even live longer when feeding on conspecific eggs rather than their prey (aphids), especially during periods of prey shortage (Agarwala, 1991). Egg cannibalism allowed larval development of the coccinellid *H. axyridis* on low-quality prey, such as nutrient-deficient aphids (Snyder et al., 2000). The recognition of siblings (kin) may have some bearing on the occurrence of cannibalism. Agarwala and Dixon (1993) found that *A. bipunctata* adult females were reluctant to eat their own eggs but adult males did not appear to avoid the eggs they sired, which suggests that adult females recognized their own offspring.

Despite the nutritional benefits of cannibalism and the capacity of some adults to recognize their own siblings, the propensity of coccinellids to cannibalize their siblings or offspring could hamper attempts at developing space-saving techniques for mass rearing coccinellids in support of augmentative biological control (Allen and Riddick, 2012). Assuredly, cannibalism is a key obstacle to efficient mass production of coccinellids (Hodek, 1996). Researchers have attempted to curb cannibalism in laboratory arenas and cages over the years. For example, Shands et al. (1970) used excelsior (slender shavings of wood) to increase travelling space in the base of oviposition cages for *C. septempunctata* adults so that they would be less inclined to come in contact with and cannibalize eggs that had been oviposited on corrugated pasteboard.

Cannibalism is a common behavior of other predators, such as carabids, under confined conditions in the laboratory. Weseloh (1996) used moist peat moss (3 cm deep, at the base of plastic rearing containers) to reduce cannibalism amongst larvae of the carabid *C. sycophanta* by 50%. Symondson (1994) did not observe cannibalism of eggs or larvae when rearing larvae of the carabid *Abax parallelepipedus* Piller and Mitterpacher in batches, in containers with earthworms as prey, and with peat and a small amount of clay as the substrate. This predator was a candidate biocontrol agent of slugs such as *Deroceras reticulatum* (Muller) in protected horticulture in the United Kingdom (Symondson, 1994). Manipulating the rearing substrate, providing refugia to reduce contact between siblings, and ensuring that food is of suitable quality and quantity can reduce cannibalistic behavior of some predators.

2.3.3. Design of Oviposition Substrates and Rearing Enclosures

Natural host plants are typical oviposition substrates for coleopteran predators in the field (Vatansever et al., 2003). Natural and target prey are often reared on host plants, which provide oviposition sites in close proximity to prey, in cages in the field, or in the laboratory. For example, researchers produced the lady beetle *S. tsugae* on target prey (the hemlock woolly adelgid) on hemlock twigs in cages in rearing facilities in the eastern United States (Conway et al., 2010); adults oviposited on gauze as often as hemlock twigs, at least when adult females are confined inside 3.8 l glass jars. This study demonstrates the feasibility of using nonplant material as oviposition substrates in lieu of host plants. Eliminating live plants from cages would reduce costs associated with rearing three trophic levels (host plants, prey, and predator). Eliminating plants could reduce the harmful effects of plant defenses on developing coccinellids. Research has shown that plant trichomes and plant toxins (allelochemicals) can reduce the growth and development of coccinellids (Riddick and Wu, 2010b; Riddick et al., 2011a).

Some alternative substrates suitable for oviposition include synthetic products. For example, the pink-spotted lady beetle *C. maculata* preferred ovipositing on textured surfaces (Kimwipes®) rather than smooth surfaces (Parafilm paper backing) in laboratory cages (Allen and Riddick, 2012). The two-spot ladybird *A. bipunctata* laid more eggs per cluster on filter paper than on a natural substrate (spruce needles) in the laboratory (Timms and Leather, 2007). A coccinellid *Diomus terminatus* (Say), a predator of the corn leaf aphid, *Rhopalosiphum maidis* (Fitch), oviposited on wax paper strips and on sorghum leaves in laboratory containers (Tifft et al., 2006).

For some coccinellids, application of a chemical cue from natural prey or host plants stimulates oviposition onto alternative substrates. Merlin et al. (1996a) found that chemical cues in the wax filaments of the mealybug *P. citri* stimulated oviposition in *C. montrouzieri*. Smirnoff (1958) stated that several mass-produced coccinellids do not oviposit in large cages unless a twig from the natural host plant is available. Extracts from *Juniperus virginiana* (L.) wood stimulated oviposition in *C. maculata* in the laboratory (Boldyrev et al., 1969). Ethanol extracts from cinnamon, clove, and teak also stimulated oviposition in *C. maculata* (Smith and Williams, 1976). Treatment of surfaces with extracts from fennel and cypress, in comparison to untreated surfaces, stimulated oviposition in *A. bipunctata* (Iperti and Prudent, 1986). In a preliminary study, Allen and Riddick (2012) found that methyl salicylate did not improve oviposition of *C. maculata* inside cages. Methyl salicylate is an herbivore-induced plant volatile known to attract beneficial arthropods including the coccinellid *Stethorus punctum picipes* (Casey) in the field (James, 2005).

The physical presence of other predators (conspecific or heterospecific) or the odors that they leave behind (in feces or tracks) can deter oviposition (see Seagraves, 2009). Chemical cues produced by conspecific larvae deter oviposition in *C. montrouzieri* (Merlin et al., 1996b). Little is known about the compounds that deter oviposition of coccinellids and other coleopteran predators in mass production systems. More research is needed to determine how to enhance oviposition while discouraging oviposition deterrence in a range of coccinellids within enclosures and to determine the effect of cage size on the efficacy of oviposition stimulants for coccinellids.

There is no standard enclosure (cage) design for mass rearing of coleopteran predators. The design of cages could have important implications for large-scale production and automation of the rearing process (Smith and Nordlund, 1999). Space is usually limited in many rearing facilities, so using small rather than large cages would seem logical. However, the biology of the species under rearing is critical for selecting an optimal cage size and shape. Cannibalism could become a problem as population density increases with decreasing or no change in cage size. Investigators have constructed rearing cages from various materials such as plastic, polypropylene, poly(methyl methacrylate), glass, or wood. The only common factor amongst the different styles and materials used in designing these cages is the use of screens in one or more sides or the top (lid) of the cages. The screens are metal or aluminum wire, nylon, or muslin fabric of variable mesh sizes to prevent escape of developing predators and prey while providing air circulation. The orientation of cages can also affect production of coleopteran predators. A simple change from vertical to horizontal orientation of hemlock twigs and artificial oviposition substrates in 3.8 l glass jars increased egg production of the coccinellid *S. tsugae* (Conway et al., 2010).

2.3.4. Rearing Scale

The ability to envision large-scale production of any natural enemy should be an integral part of the initial stages of planning a rearing program (Nordlund, 1998). To our knowledge, published research that attempts to translate rearing results for coccinellids or other coleopteran predators into large-scale application is limited. Whistlecraft et al. (1985) developed a technique to rear the staphylinid *Aleochara bilineata* (Gyllenhal), a predator-parasitoid of the root maggots *Delia radicum* (L.) and *Delia antiqua* (Meigen). Adult beetles accepted second and third instar *D. antiqua* larvae as prey in plastic containers lined with a layer of moist sand under complete darkness at 14 °C. Adults also oviposited in the sand at a rate of 6 eggs per female per day. First instar *A. bilineata* larvae searched for and parasitized maggot pupae at a relatively high rate (71%) when held in containers also lined with moist sand at 22 °C and long photoperiod (16 hours light and 8 hours dark) to prevent *A. bilineata* larvae from entering diapause. Whistlecraft et al. (1985) found that approximately 10,000 *A. bilineata* adults resulted from just 5 h of human labor per week.

Lawson and Morgan (1992) described a method of rearing large numbers of a clerid beetle *T. dubius* and a trogossitid beetle *Temnochila virescens* F. using pine billets (0.8 m long, 22.5 cm wide) infested with the bark beetle *I. grandicollis*. At a harvest rate of 12 predator larvae per billet with 3 billets per week, approximately 1850 mature larvae were produced per year. The authors achieved this level of production by maintaining the colonies at 25 °C in an insectary rather than a laboratory.

Greene (1996) reared the predatory staphylinid *Creophilus maxillosus* L. on blowfly *Calliphora* spp. larvae in complete darkness in plastic boxes (11 cm^2, 3.5 cm deep) lined with damp sand. Females laid an average of 500 eggs per day at a temperature of 28 °C in this system with a 30% survival from egg to adult producing 150 adults. In theory, nearly 422,000 females (third generation) would result within approximately 90 days (Greene, 1996).

Although production of coccinellids with natural or target prey can be expensive, Pickett et al. (1999) developed a method of rearing *Delphastus catalinae* (Horn) in the laboratory using whiteflies *B. argentifolii* as prey. Using poinsettia as host plants, more than 174,000 *D. catalinae* adults were harvested in a 20-week time period. They discovered that harvesting 50% of the adults per week stabilized the colony and maximized production. The cost of producing this beetle was $0.22 USD, which included mostly labor costs (Pickett et al., 1999).

Salom et al. (2012) described the sequence of developments that led to their current system for rearing the derodontid beetle *Laricobius nigrinus* Fender, another important predator of the adelgid *A. tsugae*. With careful consideration to environmental conditions inside the rearing cages, the authors eventually produced over 19,000 beetles per year. The availability of live prey was critical to the mass production of this predator. The authors thought it was necessary to introduce wild (field-collected) beetles into the colony each year.

2.4. TEMPERATURE AND PRODUCTION

2.4.1. Optimizing Temperature for Rearing

Temperature is an important abiotic factor influencing population growth and fitness of predatory arthropods in nature and in confinement (Hallman and Denlinger, 1998; Stathas, 2000). Numerous studies attest to the importance of manipulating temperature (along with other factors such as photoperiod and humidity) for rearing predators under conditions that maximize growth and fitness (Schüder et al., 2004; Schneider, 2009). Exemplary studies demonstrate the usefulness of life table analyses in determining the optimal temperature (or optimal temperatures within a narrow range) to rear coccinellids that prey upon aphids, scales, and whiteflies (Canhilal et al., 2001; Mota et al., 2008; Jalali et al., 2009a). Atlihan and Chi (2008) found that *Scymnus subvillosus* (Goeze) had its highest net reproductive rate at 25 °C rather than 20, 30, or 35 °C under laboratory conditions. *Scymnus subvillosus* is an aphidophagous predator in agroecosystems, especially orchards, in Turkey. Another aphidophagous coccinellid, *H. axyridis*, had its highest net reproductive rate and population doubling time at 15 °C rather than 20 °C or 25 °C (Castro et al., 2011); nevertheless, the authors claimed that a temperature of 25 °C was most suitable for rearing *H. axyridis*. The aphidophagous coccinellid *P. dissecta* had its highest net reproductive rate at 27 °C rather than 25, 30, or 35 °C under laboratory conditions in North India and the authors considered this temperature the optimal for mass rearing (Pervez and Omkar, 2004). The whitefly predator *Nephaspis oculatus* (Blatchley) displayed best population growth when reared inside incubators at 26 °C, rather than 20, 23, 29, 31, or 33°C (Ren et al., 2002). Kutuk and Yigit (2007) determined that the whitefly predator *D. catalinae* reproduced at its highest rate at a temperature of 25 rather than 30°C or alternating between 25 and 35 °C (on a 12 h/12 h light/dark cycle); the authors concluded that a range between 25 and 30 °C was optimum for mass producing *D. catalinae*. In a study using the same predator (*D. catalinae*), Legaspi et al. (2008) found that population growth was best at 22 and 26 rather than 30°C.

2.4.2. Reducing Temperature for Cold Storage

Cold storage of natural enemies is one of the most important strategies used to facilitate the mass production and utilization of natural enemies for biological control (Glenister and Hoffman, 1998; Leopold, 1998). Short-term storage could help the producer balance the differences in costs between product supply and demand; long-term storage could reduce production costs if rearing occurred only during the growing seasons. The perceived advantages of cold storage are (1) prolongation of lifespan by reducing metabolic rate, (2) shipment over long distances without high mortality from starvation, and (3) synchronization of natural enemy abundance with prey/host abundance (Riddick, 2001). In the context of this chapter, cold storage involves techniques that reduce the metabolic rate of coleopteran predators to prolong their lifespan. Some knowledge of the physiology of the predator during the winter season of the year (in temperate zones of the world) can provide clues to more efficient cold storage. Dormancy (diapause, quiescence) is a major strategy used by arthropods, including predatory and parasitic species, to survive during harsh winter conditions (Leather et al., 1993). Denlinger (2008) stated that diapause merits more attention from the scientific community.

Researchers interested in long-term cold storage have attempted to simulate the winter season in the laboratory for several coleopteran predators (Table 2.4).

TABLE 2.4 Coccinellid Predators* and the Effect of Cold Storage on Survival

Predatory Beetle	Storage Conditions	Results	References
Adalia bipunctata (adults)	6 °C, 70–90% relative humidity (RH)	70% alive, 6 months	Hamalainen (1977)
A. bipunctata (eggs)	10 °C, 80–90% RH	55–65% hatched, 2 weeks	Hamalainen and Markkula (1977)
Coccinella septempunctata (eggs)	10 °C, 80–90% RH	65% hatched, 1 week	Hamalainen and Markkula (1977)
Coccinella undecimpunctata (adults)	6 °C, 60% RH	50% alive, 1.5 months, with feeding	Abdel-Salam and Abdel-Baky (2000)
C. undecim-punctata (eggs)	6 °C, 60% RH	65% hatched, 1 week	Abdel-Salam and Abdel-Baky (2000)
C. undecim-punctata (larvae)	6 °C, 60% RH	15% alive, 2 weeks	Abdel-Salam and Abdel-Baky (2000)
C. undecimpunctata (pupae)	6 °C, 60% RH	65% alive, 2 weeks	Abdel-Salam and Abdel-Baky (2000)
Coleomegilla maculata (adults)	4.4 °C, 60–70% RH	69% alive, 6 months	Neel and Solomon (1985)
C. maculata (larvae)	8 °C, 60–70% RH	90–100% alive, 2 weeks; predation (✓)	Gagné and Coderre (2001)
Eriopis connexa (eggs)	4 °C, RH not reported	86% alive, 1 day	Miller (1995)
E. connexa (pupae)	4 °C, RH not reported	100% alive, 3 weeks	Miller (1995)
Harmonia axyridis (adults)	4 and 8 °C, RH not reported	84% alive, 4 months	Seo and Youn (2002)
H. axyridis (adults, parasite-free)	8 °C, 60% RH	100% alive, 4.5 months	Riddick (2010)
H. axyridis (adults)	3 and 6 °C; 60–70% RH	80% alive, 5 months; fecundity (✓)	Ruan et al. (2012)

Symbol (✓) indicates no significant effect on life parameter.
**Data on noncoccinellid beetles included in text only.*

Adults of the two-spot ladybird, *A. bipunctata*, were field-collected in August–September and placed inside artificial hibernacula for cold storage during the winter months in the laboratory (Hamalainen, 1977). Survival of *A. bipunctata* was good, with 70% or more of the individuals surviving until March and 50% surviving until May. Neel and Solomon (1985) removed *C. maculata* adults from winter aggregations between December and March in two consecutive seasons, placed them inside cloth or paper bags, then cold-stored them in incubators in the laboratory. The authors periodically (every 2–3 weeks) removed the bags to mist the inhabitants to prevent dehydration. Up to 69% of adults survived in cold storage, depending on the month taken from aggregations. Survival was greatest for those individuals removed from aggregations after January each year. Faulds (1988) described methods of storing a clerid beetle, *Thanasimus formicarius* (L.), released for control of bark beetles *Hylastes ater* (F.) and *Hylurgus ligniperda* (F.). Storage of *T. formicarius* adults was feasible when adults were fed prior to storage and removed periodically (at 3-month intervals) for feeding. Within-storage mortality was only 4% and storage had only negligible adverse effects on fecundity. Abdel-Salam and Abdel-Baky (2000) had some success at storing *Coccinella undecimpunctata* L. in the laboratory when adults consumed aphids prior to entering storage. Seo and Youn (2002) stored prewintering adults of the coccinellid *H. axyridis* without food for more than 120 days with a survival rate of 84%. Riddick (2010) found that 100% of parasite-free *H. axyridis* adults (males and females) survived under simulated winter conditions in a laboratory refrigerator for 4.5 months. Ruan et al. (2012) subjected field-collected prewintering *H. axyridis* adults to long-term cold storage with good survival rates and reproduction after removal from storage. Finally, Whistlecraft et al. (1985) found that the staphylinid *A. bilineata* could be stored as diapausing larvae for up to 6 months with no appreciable reduction in survival rate.

In attempts to facilitate short-term cold storage of coleopteran predators, researchers have subjected life stages that do not typically undergo winter conditions at low temperatures (see Table 2.4). Hamalainen and Markkula (1977) found that eggs of *A. bipunctata* stored for 2 weeks had a survival rate of 55–65% when returned to room temperature. The hatch rate of nonstored *A. bipunctata* eggs was approximately 68%. *Coccinella septempunctata* eggs could be stored for just 1 week at a hatch rate of 65% when returned to room temperature. The hatch rate of nonstored *C. septempunctata* eggs was 75–80% (Hamalainen and Markkula, 1977). Abdel-Salam and Abdel-Baky (2000) determined the potential of storing egg, larval, or pupal stages of the coccinellid *C. undecimpunctata* in the laboratory. They discovered that eggs could be effectively stored at this temperature for up to 1 week, with a hatch rate of 65%; no eggs hatched after 2 weeks of storage. Larvae and pupae could safely be stored for 2 weeks and 1 week, respectively, without significant declines in larval survival and adult emergence from pupae. The potential of storing larval stages of *Coleomegilla maculata lengi* Timberlake was assessed in the laboratory by

Gagné and Coderre (2001) using insects cultured for more than 10 consecutive generations on a diet of beef liver and moth (*E. kuehniella*) eggs. They found that second and third instars stored successfully for 2 weeks, without considerable adverse effects on development and predation potential (i.e. voracity) of larvae when returned to rearing conditions. Miller (1995) showed that eggs of the coccinellid *E. connexa* could be stored for 1 day without negative effects on hatch rate; pupae, on the other hand, were stored at the same temperature for 3 weeks with no appreciable mortality.

Much more research is necessary on the response of coleopteran predators to low temperatures in relation to production. Other strategies that could enhance the potential to cold store predators might involve the use of cryoprotectant and carbohydrate molecules (see Riddick and Wu, 2010a) and other molecules that protect the cells and tissues of predators during harsh conditions in the field. Storage of some insect eggs (such as heteropteran and lepidopteran) destined to serve as artificial hosts for parasitoids is possible in liquid nitrogen (Leopold, 1998). To our knowledge, no coleopteran predator or any other natural enemy has been stored successfully in liquid nitrogen to date. Diapause induction is another technique that holds promise in enhancing storage potential of predators (Chang et al., 1996). Researchers could also explore the usefulness of inducing coleopteran predators to transition into a state of diapause in subsequent studies.

2.5. QUALITY CONTROL AND PRODUCTION

2.5.1. Safeguards against Unwanted Pathogens and Parasites

Coccinellids and other coleopteran predators have the potential to harbor a diversity of parasites and pathogens under natural conditions in the field and artificial conditions in the laboratory (Ceryngier and Hodek, 1996; Bjørnson and Schütte, 2003; Riddick et al., 2009; Riddick, 2010; Ceryngier et al., 2012). These pathogens can hinder the development of immature stages and lead to the demise of some individuals. Parasites and pathogens could inhibit the reproductive performance of adults. They could undermine biocontrol efforts as well as lead to inadvertent introduction of parasites and pathogens into new locations. Bjørnson (2008a) found that *H. convergens* adults, shipped from three companies and destined for field releases, harbored several pathogens including microsporidia, eugregarines, and a braconid endoparasitoid. To prevent the inadvertent spread of pathogens from one locality to the next, workers should screen the coccinellids prior to field releases.

2.5.2. Preventing Colony Deterioration

Continuous mating between closely related individuals (i.e. inbreeding) in a population of insects over many generations can have detrimental effects on the health of a colony and on the performance of individuals released in the field (Mackauer, 1976; Roush and Hopper, 1995). Inbreeding may reduce the fitness of coccinellids such as *A. bipunctata*, *C. maculata*, *H. axyridis*, and *Propylea quatuordecimpunctata* (L.) (Kidd, 1993; Hurst et al., 1996; Morjan et al., 1999; Facon et al., 2011; Seko et al., 2012). Methods to reduce or prevent the deterioration of colonies may include manipulative breeding between hybrids to restore normal growth, survival, and reproductive capacities of populations in culture. Mating between siblings should be avoided by mixing adults with different parenthood origins. The practice of periodically adding wild (field) adults into well-established laboratory cultures to mate with domesticated adults could prevent inbreeding depression and subsequent reduction in fitness. Continuous monitoring of colonies for any undesirable changes in the behavior of adults, in comparison to adults found in the field, is necessary (Boller, 1972; Huettel, 1976).

2.5.3. In-Shipment, Post Shipment, and Prerelease Assessments

The in-shipment effect of shipping containers on beneficial arthropods has received limited attention to date. A study by Bjørnson and Raworth (2005) showed that Styrofoam boxes do not adequately maintain internal temperatures. Consequently, beneficials are often subject to external temperatures during transport from commercial companies and release site. The authors suggest refrigerating the Styrofoam boxes (with beneficials inside) during shipment.

The sex ratio of individual species purchased from commercial companies can represent a measure of quality in a mass production system. Heimpel and Lundgren (2000) purchased predators (coccinellids) and parasitoids from insectaries and used log-likelihood goodness-of-fit tests (Sokal and Rohlf, 1981) to determine if the sex ratio of adults of each species differed from a proportion of 0.5 males. Of the six species of coccinellids (*C. montrouzieri*, *D. pusillus*, *H. axyridis*, *H. convergens*, *Rhyzobius* (*Lindorus*) *lophanthae* (Blaisdell), *S. punctillum*), none deviated significantly from the expected sex ratio. O'Neil et al. (1998) assessed the post shipment quality of the convergent lady beetle, *H. convergens*, from several commercial companies. The percentage of adults alive upon arrival at their laboratory ranged from 50% to 81%, with no significant differences between companies. The percentage of unparasitized adults was 78–91% and females represented from 58% to 68% of the shipments, without any significant differences between companies. Finally, 67–80% of females produced eggs after arrival in the laboratory, irrespective of company. Bjørnson (2008b) found that *H. convergens* adults, shipped from three commercial companies, did not commence oviposition until approximately 5 days later, when fed aphids in experimental arenas in the laboratory.

Researchers have shown that the ability of mass-produced insects to take flight (i.e. flight capacity) is a potentially useful measure of quality of a beneficial insect prior to its release (Couillien and Grégoire, 1994; van Lenteren, 2003). The longer that the rhizophagid beetle *Rhizophagus grandis* Gyllenhal remained in cold storage (at 3–7 °C), the less likely adults were capable of taking flight in a wind tunnel (Couillien and Grégoire, 1994). Thus, prolonged cold storage beyond several months may have negative consequences on the ability of *R. grandis* to disperse from release sites in search of prey (bark beetles). The capacity to fly is not always a useful measure of quality. Individuals of a nonflying strain rather than the normal (wild-type) strain of the coccinellid *H. axyridis* remained and oviposited on plants in greenhouses for a longer time (Ferran et al., 1997); fitness does not seem to differ between wild-type and nonflying strains (Tourniaire et al., 2000).

Quality control guidelines are established or currently being developed for a range of natural enemies (van Lenteren et al., 2003). These guidelines often include prerelease assessments of the performance of a natural enemy in a laboratory setting. Thus far, provisional quality control guidelines are available for just one predatory beetle, namely *C. montrouzieri* (van Lenteren et al., 2003). A more detailed discussion on quality control of mass produced arthropods is presented in Chapter 9 of this book.

2.6. CONCLUSIONS AND RECOMMENDATIONS

2.6.1. Synthesis

In this chapter, we have highlighted research related to producing coleopteran predators primarily for augmentative biological control. Rearing coleopteran predators on natural prey reared on host plants is the mainstay in production research to date. Research shows that some alternative prey/foods, such as lepidopteran moth eggs and brine shrimp eggs, are stand-alone factitious foods for a few coccinellids and supplements in artificial diets for several others. Plant-based products such as honey, pollen, and royal jelly are supplements to or additives in some artificial diets and represent beneficial sources of nutrients. Novel methods that incorporate feeding stimulants into diets to incite feeding on nonprey or artificial diets by specialists need further exploration. Most of the artificial diets based on vertebrate protein, with or without arthropod components, are not as effective as natural prey or factitious food for production on coleopteran predators.

The capacity to rear coleopteran predators within the confines of a laboratory cage is amazing in itself given that many species travel considerable distances in search of food, shelter, and mates in nature. We are gathering knowledge on how to design cages and establish relevant rearing densities within the confines of these cages to encourage and stimulate feeding, mating, and oviposition while discouraging cannibalism. The ability to manipulate temperature to maximize growth and development or to decrease metabolic rates for stockpiling or overwintering is possible for several species. Several studies point out the necessity to monitor colonies on a regular schedule for unwanted pathogens and parasites as well as to ensure that we are producing coleopteran predators of high quality for the market. Preshipment and postshipment assessments of the health of products are important to the sustainability of any production system.

2.6.2. Future Research

- The current criteria to measure the success or failure of factitious prey or an artificial diet for coleopteran predators may need modification. These criteria typically include the following life parameters: development time, larval survival, adult emergence, body size, oviposition (fecundity), and adult longevity. Previous research has shown that most artificial diets created thus far have less than ideal affects on these life parameters when compared against what is possible using natural or target prey. By far, one of the most important life parameters that is often overlooked is predation potential (Grenier and De Clercq, 2003). After rearing on factitious prey or an artificial diet for multiple generations, the capacity of a coleopteran predator to locate, capture, kill, and then consume target prey upon initial contact, after minimal conditioning before release, is a rational estimate of success.
- There is evidence that generalist (polyphagous) coleopteran predators can undergo adaptation to specific prey after several generations of exposure, at least in the laboratory (Rana et al., 2002). Adaptation to new prey also occurs in specialist (oligophagous) coccinellids (Ragab, 1995; Causton et al., 2004) and carabids (Weber et al., 2006; Weber and Riddick, 2011). More research is necessary to predict which species are most likely to adapt to an arthropod-free artificial diet.
- Detailed research is necessary to determine how to manipulate rearing (population) densities relative to food quality/quantity, cage size, oviposition, and mating to reduce the negative effects of crowding and cannibalism in colonies. Although cannibalism can have benefits on the health of an individual, excessive levels of cannibalism are often detrimental to overall colony health.
- More research is needed to establish rigorous measures of quality control (i.e. regular monitoring of products, before or after shipment to customers, for acceptable fitness and unwanted symbionts) for more coleopteran predators. Then, we can develop specific standards that allow producers to assess the fitness of their products before sale.
- Development of automated systems for mass rearing is critical to the long-term utilization of natural enemies in augmentative biological control. Utilization of machinery to reduce manual labor in the rearing process might increase cost effectiveness. Smith and Nordlund (1999) claim that automation (i.e. replacement of human labor with mechanical and/or electrical components) would lower costs, increase production, and improve the quality of reared insects. Scaling up the rearing operation to produce millions of predators is necessary (Nordlund, 1998; Smith and Nordlund, 2000).

ACKNOWLEDGMENTS

We thank Juan Morales-Ramos for inviting us to write this review chapter. Keiji Takasu and Zhixin Wu translated key passages in literature published in Japanese and Chinese, respectively. The U.S. Government has the right to retain a nonexclusive, royalty-free license in and to any copyright of this article. This article reports the results of research only. Mention of a commercial or proprietary product does not constitute an endorsement of the product by the U.S. Department of Agriculture.

REFERENCES

Abdel-Salam, A.H., Abdel-Baky, N.F., 2000. Possible storage of *Coccinella undecimpunctata* (Col., Coccinellidae) under low temperature and its effect on some biological characteristics. J. Appl. Entomol. 124, 169–176.

Abdel-Salam, A.H., Abdel-Baky, N.F., 2001. Life table and biological studies of *Harmonia axyridis* Pallas (Col.: Coccinellidae) reared on the grain moth eggs of *Sitotroga cerealella* Olivier (Lep.: Gelechiidae). J. Appl. Entomol. 125, 455–462.

Achiano, K.A., Giliomee, J.H., 2006. Rearing the house fly predator *Carcinops pumilio* (Erichson) (Coleoptera: Histeridae) on an artificial diet. Afr. J. Biotechnol. 5, 1161–1166.

Agarwala, B.K., 1991. Why do ladybirds (Coleoptera: Coccinellidae) cannibalize? J. Biosci. 16, 103–109.

Agarwala, B.K., Dixon, A.F.G., 1993. Kin recognition: egg and larval cannibalism in *Adalia bipunctata* (Coleoptera: Coccinellidae). Eur. J. Entomol. 90, 45–50.

Allen, M.A., Riddick, E.W., 2012. A system for harvesting eggs from the pink-spotted lady beetle. Psyche 2012, 6, Article ID 923653.

Atallah, Y.H., Newsom, L.D., 1966. Ecological and nutritional studies on *Coleomegilla maculata* De Geer (Coleoptera: Coccinellidae). I. The development of an artificial diet and a laboratory rearing technique. J. Econ. Entomol. 59, 1173–1179.

Atlihan, R., Chi, H., 2008. Temperature-dependent development and demography of *Scymnus subvillosus* (Coleoptera: Coccinellidae) reared on *Hyalopterus pruni* (Homoptera: Aphididae). J. Econ. Entomol. 101, 325–333.

Attia, A.R., El Arnaouty, S.A., Afifi, A.I., Abd Alla, A.E., 2011. Development and fecundity of the coccinellid predator, *Cryptolaemus montrouzieri* Mulsant on different types of prey. Egypt. J. Biol. Pest Control 21, 283–289.

Bain, J., Singh, P., Ashby, M.D., Van Boven, R.J., 1984. Laboratory rearing of the predatory coccinellid *Cleobora mellyi* (Col.: Coccinellidae) for biological control of *Paropsis charybdis* (Col.: Chrysomelidae) in New Zealand. Entomophaga 29, 237–244.

Banks, C.J., 1956. Observations on the behaviour and mortality in Coccinellidae before dispersal from the egg shells. Proc. R. Entomol. Soc. London 31, 56–60.

Berkvens, N., Bonte, J., Berkvens, D., Deforce, K., Tirry, L., De Clercq, P., 2008. Pollen as an alternative food for *Harmonia axyridis*. BioControl 53, 201–210.

Berkvens, N., Landuyt, C., Deforce, K., Berkvens, D., Tirry, L., De Clercq, P., 2010. Alternative foods for the multicoloured Asian lady beetle *Harmonia axyridis* (Coleoptera: Coccinellidae). Eur. J. Entomol. 107, 189–195.

Biddinger, D.J., Weber, D.C., Hull, L.A., 2009. Coccinellidae as predators of mites: Stethorini in biological control. Biol. Control 51, 268–283.

Birken, E.M., Cloyd, R.A., 2007. Food preference of the rove beetle, *Atheta coriaria* Kraatz (Coleoptera: Staphylinidae) under laboratory conditions. Insect Sci. 14, 53–56.

Bista, M., Mishra, G., Omkar, 2012. Influence of crowding and diet on the development and survival of the ladybird *Brumoides suturalis* (Coleoptera: Coccinellidae) reared on two aphid species. Int. J. Trop. Insect Sci. 32, 64–68.

Bjørnson, S., 2008a. Natural enemies of the convergent lady beetle, *Hippodamia convergens* Guérin-Méneville: their inadvertent importation and potential significance for augmentative biological control. Biol. Control 44, 305–311.

Bjørnson, S., 2008b. Fecundity of commercially available convergent lady beetles, *Hippodamia convergens*, following shipment. Biocontrol Sci. Technol. 18, 633–637.

Bjørnson, S., Raworth, D.A., 2005. Styrofoam boxes used for shipping mass-reared beneficial arthropods: internal temperature responds rapidly to ambient temperature. Biocontrol Sci. Technol. 15, 755–760.

Bjørnson, S., Schütte, C., 2003. Pathogens of mass-produced natural enemies and pollinators. In: van Lenteren, J.C. (Ed.), Quality Control and Production of Biological Control Agents: Theory and Testing Procedures, CABI Publ., Oxon, UK, pp. 133–165.

Boldyrev, M.I., Wilde, W.H.A., Smith, B.C., 1969. Predaceous coccinellid oviposition responses to *Juniperus* wood. Can. Entomol. 101, 1199–1206.

Boller, E., 1972. Behavioral aspects of mass-rearing of insects. Entomophaga 17, 9–25.

Bonte, M.M., Samih, A., De Clercq, P., 2010. Development and reproduction of *Adalia bipunctata* on factitious and artificial foods. BioControl 55, 485–491.

Canhilal, R., Uygun, N., Carner, G.R., 2001. Effects of temperature on development and reproduction of a predatory beetle, *Nephus includens* Kirsch (Coleoptera: Coccinellidae). J. Agric. Urban Entomol. 18, 117–125.

Caltagirone, L.E., Doutt, R.L., 1989. The history of the vedalia beetle importation to California and its impact on the development of biological control. Annu. Rev. Entomol. 34, 1–16.

Castro, C.F., Almeida, L.M., Penteado, S.R.C., 2011. The impact of temperature on biological aspects and life table of *Harmonia axyridis* (Pallas) (Coleoptera: Coccinellidae). Fla. Entomol. 94, 923–932.

Causton, C.E., Lincango, M.P., Poulson, T.G.A., 2004. Feeding range studies of *Rodolia cardinalis* (Mulsant), a candidate biological control agent of *Icerya purchasi* Maskell in the Galápagos Islands. Biol. Control 29, 315–325.

Ceryngier, P., Hodek, I., 1996. Enemies of Coccinellidae. In: Hodek, I., Honěk, A. (Eds.), Ecology of Coccinellidae, Kluwer Acad. Publ., Dordrecht, Netherlands, pp. 319–350.

Ceryngier, P., Roy, H.E., Poland, R.L., 2012. Natural enemies of ladybird beetles. In: Hodek, I., van Emden, H.F., Honěk, A. (Eds.), Ecology and Behaviour of the Ladybird Beetles, Blackwell Publ. Ltd, Oxford, UK, pp. 375–443.

Chang, Y.F., Tauber, M.J., Tauber, C.A., 1996. Reproduction and quality of F_1 offspring in *Chrysoperla carnea*: differential influence of quiescence, artificially induced diapause, and natural diapause. J. Insect Physiol. 42, 521–528.

Cohen, A.C., Cheah, C.A.S.J., Strider, J., Hain, F., 2008. Diet development for hemlock woolly adelgids and their predators. In: Onken, B., Reardon, R. (Eds.), Fourth Symposium on Hemlock Woolly Adelgid in the Eastern United States, USDA-Forest Service, Forest Health Technology Enterprise Team (FHTET), Morgantown, WV, pp. 150–156.

Conway, H.E., Culin, J.D., Burgess, L.W., Allard, C., 2010. Maximizing oviposition efficiency when mass rearing the coccinellid, *Sasajiscymnus tsugae*, a predator of the hemlock woolly adelgid, *Adelges tsugae*. J. Insect Sci. 10, 152, Available online: insectscience.org/10.152.

Costa, A., Reeve, J.D., 2012. The effect of larval predators *Thanasimus dubius* (Coleoptera: Cleridae), produced by an improved system of rearing, against the southern pine beetle *Dendroctonus frontalis* (Coleoptera: Curculionidae). Biol. Control 60, 1–6.

Cottrell, T.E., 2005. Predation and cannibalism of lady beetle eggs by adult lady beetles. Biol. Control 34, 159–164.

Couillien, D., Gregoire, J.C., 1994. Take-off capacity as a criterion for quality control in mass-produced predators, *Rhizophagus grandis* (Coleoptera: Rhizophagidae) for the biocontrol of bark beetles, *Dendroctonus micans* (Coleoptera: Scolytidae). Entomophaga 39, 385–395.

De Clercq, P., 2004. Culture of natural enemies on factitious foods and artificial diets. In: Capinera, J.L. (Ed.), Encyclopedia of Entomology, vol. 1. Kluwer Acad. Publ., Dordrecht, The Netherlands, pp. 650–652.

De Clercq, P., Bonte, M., Van Speybroeck, K., Bolckmans, K., 2005a. Development and reproduction of *Adalia bipunctata* (Coleoptera: Coccinellidae) on eggs of *Ephestia kuehniella* (Lepidoptera: Phycitidae) and pollen. Pest Manage. Sci. 61, 1129–1132.

De Clercq, P., Arijs, Y., Van Meir, T., Van Stappen, G., Sorgeloos, P., Dewettinck, K., Rey, M., Grenier, S., Febvay, G., 2005b. Nutritional value of brine shrimp cysts as a factitious food for *Orius laevigatus* (Heteroptera: Anthocoridae). Biocontrol Sci. Technol. 15, 467–479.

Denlinger, D.L., 2008. Why study diapause? Entomol. Res. 38, 1–9.

Dimetry, N.Z., 1974. The consequences of egg cannibalism in *Adalia bipunctata* (Coleoptera: Coccinellidae). Entomophaga 19, 445–451.

Dong, H., Ellington, J.J., Remmenga, M.D., 2001. An artificial diet for the lady beetle *Harmonia axyridis* Pallas (Coleoptera: Coccinellidae). Southwest Entomol. 26, 205–213.

Ettifouri, M., Ferran, A., 1993. Influence of larval rearing diet on the intensive searching behaviour of *Harmonia axyridis* (Col.: Coccinellidae) larvae. Entomophaga 38, 51–59.

Etzel, L.K., Legner, E.F., 1999. Culture and colonization. In: Bellows, T.S., Fisher, T.W. (Eds.), Handbook of Biological Control. Academic Press, San Diego, pp. 125–197, Chapter 7.

Evans, E.W., 2009. Lady beetles as predators of insects other than Hemiptera. Biol. Control 51, 255–267.

Facon, B., Hufbauer, R.A., Tayeh, A., Loiseau, A., Lombaert, E., Vitalis, R., Guillemaud, T., Lundgren, J.G., Estoup, A., 2011. Inbreeding depression is purged in the invasive insect *Harmonia axyridis*. Curr. Biol. 21, 424–427.

Farag, N.A., Abd El-Wahab, T.E., Abdel-Moniem, A.S.H., 2011. The influence of some honeybee products as a diet substitute on the different stages of *Coccinella undecimpunctata* L. in Egypt. Arch. Phytopathol. Plant Prot. 44, 253–259.

Faulds, W., 1988. Improved techniques for the laboratory rearing of *Thanasimus formicarius*. N. Z. J. For. Sci. 18, 187–190.

Ferran, A., Gambler, J., Parent, S., Legendre, K., Tourniere, R., Giuge, L., 1997. The effect of rearing the ladybird *Harmonia axyridis* on *Ephestia kuehniella* eggs on the response of its larvae to aphid tracks. J. Ins. Behav. 10, 129–144.

Flowers, R.W., Salom, S.M., Kok, L.T., Mullins, D.E., 2007. Behaviour and daily activity patterns of specialist and generalist predators of the hemlock woolly adelgid, *Adelges tsugae*. J. Insect Sci. 7, Article 44, available online: insectscience.org/7.44.

Fox, L.R., 1975. Cannibalism in natural populations. Annu. Rev. Ecol. Syst. 6, 87–106.

Gagné, I., Coderre, D., 2001. Cold storage of *Coleomegilla maculata* larvae. Biocontrol Sci. Technol. 11, 361–369.

Giorgi, J.A., Vanderberg, N.J., McHugh, J.V., Forrester, J.A., Ślipiński, S.A., Miller, K.B., Shapiro, L.R., Whiting, M.F., 2009. The evolution of food preferences in Coccinellidae. Biol. Control 51, 215–231.

Glenister, C.S., Hoffman, M.P., 1998. Mass-reared natural enemies: scientific, technological, and informational needs and considerations. In: Ridgway, R.L., Hoffmann, M.P., Inscoe, M.N., Glenister, C.S. (Eds.), Mass-Reared Natural Enemies: Application, Regulation, and Needs, Thomas Say Publ. Entomol., Entomol. Soc. Am., Lanham, MD, pp. 242–267.

Grafton-Cardwell, E.E., Gu, P., Montez, G.H., 2005. Effects of temperature on development of vedalia beetle, *Rodolia cardinalis* (Mulsant). Biol. Control 32, 473–478.

Greene, G.L., 1996. Rearing techniques for *Creophilus maxillosus* (Coleoptera: Staphylinidae), a predator of fly larvae in cattle feedlots. J. Econ. Entomol. 89, 848–851.

Grenier, S., 2009. In vitro rearing of entomophagous insects—past and future trends: a minireview. Bull. Insectol. 62, 1–6.

Grenier, S., De Clercq, P., 2003. Comparison of artificially vs naturally reared natural enemies and their potential for use in biological control. In: van Lenteren, J.C. (Ed.), Quality Control and Production of Biological Control Agents: Theory and Testing Procedures, CABI Publ., Oxon, UK, pp. 115–131.

Gwiazdowski, R.A., Gillespie, S., Weddle, R., Elkinton, J.S., 2011. Laboratory rearing of common and endangered species of North American tiger beetles (Coleoptera: Carabidae: Cicindelinae). Ann. Entomol. Soc. Am. 104, 534–542.

Hagen, K.S., 1987. Nutritional ecology of terrestrial insect predators. In: Slansky Jr., F., Rodriguez, J.G. (Eds.), Nutritional Ecology of Insects, Mites, Spiders, and Related Invertebrates, Wiley and Sons, New York, pp. 533–577.

Hagen, K.S., Mills, N.G., Gordh, G., McMurtry, J.A., 1999. Terrestrial arthropod predators of insect and mite pests. In: Bellows, T.S., Fisher, T.W. (Eds.), Handbook of Biological Control: Principles and Applications of Biological Control, Academic Press, San Diego, pp. 383–503.

Hagler, J., 2009. Comparative studies of predation among feral, commercially-purchased, and laboratory-reared predators. BioControl 54, 351–361.

Hallman, G.J., Denlinger, D.L., 1998. Temperature Sensitivity in Insects and Application in Integrated Pest Management. Westview Press, Boulder, Colorado.

Hamalainen, M., 1977. Storing dormant *Coccinella septempunctata* and *Adalia bipunctata* (Col., Coccinellidae) adults in the laboratory. Ann. Agric. Fenn. 16, 184–187.

Hamalainen, M., Markkula, M., 1977. Cool storage of *Coccinella septempunctata* and *Adalia bipunctata* (Col., Coccinellidae) eggs for use in the biological control in greenhouses. Ann. Agric. Fenn. 16, 132–136.

Hamasaki, K., Matsui, M., 2006. Development and reproduction of an aphidophagous coccinellid, *Propylea japonica* (Thunberg) (Coleoptera: Coccinellidae), reared on an alternative diet, *Ephestia kuehniella* Zeller (Lepidoptera: Pyralidae) eggs. Appl. Entomol. Zool. 41, 233–237.

Harmon, J.P., Stephens, E., Losey, J., 2007. The decline of native coccinellids (Coleoptera: Coccinellidae) in the United States and Canada. J. Ins. Conserv. 11, 85–94.

Hattingh, V., Samways, M.J., 1993. Evaluation of artificial diets and two species of natural prey as laboratory food for *Chilocorus* spp. Entomol. Exper. Appl. 69, 13–20.

Heidari, M., Copland, M.J.W., 1993. Honeydew: a food resource or arrestant for the mealybug predator *Cryptolaemus montrouzieri*? Entomophaga 38, 63–68.

Heimpel, G.E., Lundgren, J.G., 2000. Sex ratios of commercially reared biological control agents. Biol. Control 19, 77–93.

Henderson, R.C., Hill, M.G., Wigley, P.J., 1992. Freeze-dried artificial diets for three species of *Chilocorus* ladybirds. N. Z. Entomol. 15, 83–87.

Herrera, A.J.M., 1960. Investigaciones sobre la cria artificial del coccinélido *Coleomegilla maculata* (De Geer). Rev. Per. Entomol. Agric. 3, 1–6.

Hodek, I., 1996. Food relationships. In: Hodek, I., Honěk, A. (Eds.), Ecology of Coccinellidae, Kluwer Acad. Publ., Dordrecht, The Netherlands, pp. 143–238, Chapter 6.

Hodek, I., Evans, E.W., 2012. Food relationships. In: Hodek, I., van Emden, H.F., Honěk, A. (Eds.), Ecology and Behaviour of the Ladybird Beetles (Coccinellidae), Blackwell Publ. Ltd, UK, pp. 141–274, Ch. 8.

Hodek, I., Honěk, A., 1996. Ecology of Coccinellidae. Kluwer Acad. Publ., Dordrecht, Netherlands.

Hodek, I., Honěk, A., 2009. Scale insects, mealybugs, whiteflies and psyllids (Hemiptera, Sternorrhyncha) as prey of ladybirds. Biol. Control 51, 232–243.

Hongo, T., Obayashi, N., 1997. Use of diapause eggs of brine shrimp, *Artemia salina* (Linné) for artificial diet of coccinellid beetle, *Harmonia axyridis* (Pallas). Jpn. J. Appl. Entomol. Zool. 41, 101–105 (in Japanese).

Huettel, M.D., 1976. Monitoring the quality of laboratory-reared insects: a biological and behavioral perspective. Environ. Entomol. 5, 807–814.

Hurst, G.D.D., Sloggett, J.J., Majerus, M.E.N., 1996. Estimation of the rate of inbreeding in a natural population of *Adalia bipunctata* (Coleoptera: Coccinellidae) using a phenotypic indicator. Eur. J. Entomol. 93, 145–150.

Hussein, M.Y., Hagen, K.S., 1991. Rearing of *Hippodamia convergens* on artificial diet of chicken liver, yeast and sucrose. Entomol. Exp. Appl. 59, 197–199.

Iperti, G., Prudent, P., 1986. Effect of the substrate properties on the choice of oviposition sites by *Adalia bipunctata*. In: Hodek, I. (Ed.), Ecology of Aphidophaga, Junk Publ., Dordrecht, The Netherlands, pp. 143–149.

Jalali, M.A., Tirry, L., De Clercq, P.D., 2009a. Effects of food and temperature on development, fecundity and life-table parameters of *Adalia bipunctata* (Coleoptera: Coccinellidae). J. Appl. Entomol. 133, 615–625.

Jalali, M.A., Tirry, L., De Clercq, P., 2009b. Food consumption and immature growth of *Adalia bipunctata* (Coleoptera: Coccinellidae) on a natural prey and factitious food. Eur. J. Entomol. 106, 193–198.

James, D.G., 2005. Further field evaluation of synthetic herbivore-induced plant volatiles as attractants for beneficial insects. J. Chem. Ecol. 31, 481–495.

Jetton, R.M., Monahan, J.F., Hain, F.P., 2011. Laboratory studies of feeding and oviposition preference, developmental performance, and survival of the predatory beetle, *Sasajiscymnus tsugae* on diets of the woolly adelgids, *Adelges tsugae* and *Adelges piceae*. J. Insect Sci. 11, Article 68, available online: insectscience.org/11.68.

Kariluoto, K.T., 1978. Optimum levels of sorbic acid and methyl-*p*-hydroxy-benzoate in an artificial diet for *Adalia bipunctata* (Coleoptera, Coccinellidae) larvae. Ann. Entomol. Fenn. 44, 94–97.

Kariluoto, K.T., Junnikkala, E., Markkula, M., Atallah, Y.H., Newsom, L.D., 1976. Attempts at rearing *Adalia bipunctata* L. (Col., Coccinellidae) on different artificial diets. Ann. Entomol. Fenn. 42, 91–97.

Kato, C.M., Vanda, V.H.P., Moraes, J.C., Auad, A.M., 1999. Rearing of *Hippodamia convergens* Guérin-Meneville (Coleoptera: Coccinellidae) on eggs of *Anagasta kuehniella* (Zeller) (Lepidoptera: Pyralidae). An. Soc. Entomol. Bras. 28, 455–459.

Khan, M.R., Khan, M.R., 2002. Mass rearing of *Menochilus sexmaculatus* Fabricius (Coleoptera: Coccinellidae) on natural and artificial diets. Int. J. Agric. Biol. 4, 107–109.

Kidd, G.S., 1993. Effect of inbreeding in *Coleomegilla maculata* (Coleoptera: Coccinellidae) and *Galerucella calmariensis* (Coleoptera: Chrysomelidae) (MS Thesis). Cornell U., Ithaca, New York.

Kutuk, H., Yigit, A., 2007. Life table of *Delphastus catalinae* (Horn) (Coleoptera: Coccinellidae) on cotton whitefly, *Bemisia tabaci* (Genn.) (Homoptera: Aleyrodidae) as prey. J. Plant Dis. Protect. 114, 20–25.

Lawson, S.A., Morgan, F.D., 1992. Rearing of two predators, *Thanasimus dubius* and *Temnochila virescens*, for the biological control of *Ips grandicollis* in Australia. Entomol. Exp. Appl. 65, 225–233.

Leather, S.R., Walters, K.F.A., Bale, J.S., 1993. The Ecology of Insect Overwintering. Cambridge Univ. Press., UK.

Legaspi, J.C., Legaspi Jr., B.C., Simmons, A.M., Soumare, M., 2008. Life table analysis for immatures and female adults of the predatory beetle, *Delphastus catalinae*, feeding on whiteflies under three constant temperatures. J. Insect Sci. 8 (07), Available online: insectscience.org/8.07.

Leopold, R.A., 1998. Cold storage of insects for integrated pest management. In: Hallman, G.J., Denlinger, D.L. (Eds.), Temperature Sensitivity in Insects and Application in Integrated Pest Management, Westview Press, Boulder, Colorado, pp. 235–267, Ch. 9.

Li, Y., Ostrem, J., Romeis, J., Chen, M., Liu, X., Hellmich, R.L., Shelton, A.M., Peng, Y., 2011. Development of a tier-1 assay for assessing the toxicity of insecticidal substances against *Coleomegilla maculata*. Environ. Entomol. 40, 496–502.

Lundgren, J.G., 2009. Relationships of Natural Enemies and Non-Prey Foods. Progress in Biological Control. vol. 7. Springerhttp://www.springer.com.

Lundgren, J.G., 2010. Nutritional aspects of non-prey foods in the life histories of predaceous Coccinellidae. Biol. Control 51, 294–305.

Lundgren, J.G., Weber, D.C., 2010. Changes in digestive rate of a predatory beetle over its larval stage: implications for dietary breadth. J. Ins. Physiol. 56, 431–437.

Lundgren, J.G., Duan, J.J., Paradise, M.S., Wiedenmann, R.N., 2005. Rearing protocol and life history traits for *Poecilus chalcites* (Coleoptera: Carabidae) in the laboratory. J. Entomol. Sci. 40, 126–135.

Lundgren, J.G., Moser, S.E., Hellmich, R.L., Seagraves, M.P., 2011. The effects of diet on herbivory by a predaceous lady beetle. Biocontrol Sci. Technol. 21, 71–74.

Mackauer, M., 1976. Genetic problems in the production of biological control agents. Annu. Rev. Entomol. 21, 369–385.

Magro, A., Hemptinne, J.L., Codreanu, P., Grosjean, S., Dixon, A.F.G., 2002. Does the satiation hypothesis account for the difference in efficacy of coccidophagous and aphidophagous ladybird beetles in biological control? A test with *Adalia bipunctata* and *Cryptolaemus montrouzieri*. BioControl 47, 537–543.

Martos, A., Givovich, A., Niemeyer, H.M., 1992. Effect of DIMBOA, an aphid resistance factor in wheat, on the aphid predator *Eriopis connexa* Germar (Coleoptera: Coccinellidae). J. Chem. Ecol. 18, 469–479.

Matsuka, M., Okada, I., 1975. Nutritional studies on an aphidophagous coccinellid, *Harmonia axyridis*. I. Examination of artificial diets for the larval growth with special reference to drone honeybee powder. Bull. Fac. Agric., Tamagawa Univ. 15, 1–9.

Matsuka, M., Takahashi, S., 1977. Nutritional studies of an aphidophagous coccinellid *Harmonia axyridis*. II. Significance of minerals for larval growth. Appl. Entomol. Zool. 12, 325–329.

Matsuka, M., Watanabe, M., Niijima, K., 1982. Longevity and oviposition of vedalia beetles on artificial diets. Environ. Entomol. 11, 816–819.

Merlin, J., Lemaitre, O., Grégoire, J.C., 1996a. Oviposition in *Cryptolaemus montrouzieri* stimulated by wax filaments of its prey. Entomol. Exp. Appl. 79, 141–146.

Merlin, J., Lemaitre, O., Grégoire, J.C., 1996b. Chemical cues produced by conspecific larvae deter oviposition by the coccidophagous ladybird beetle, *Cryptolaemus montrouzieri*. Entomol. Exp. Appl. 79, 147–151.

Michaud, J.P., 2003. A comparative study of larval cannibalism in three species of ladybird. Ecol. Entomol. 28, 92–101.

Michaud, J.P., Qureshi, J.A., 2005. Induction of reproductive diapause in *Hippodamia convergens* (Coleoptera: Coccinellidae) hinges on prey quality and availability. Eur. J. Entomol. 102, 483–487.

Miller, J.C., 1995. A comparison of techniques for laboratory propagation of a South American ladybeetle, *Eriopis connexa* (Coleoptera: Coccinellidae). Biol. Control 5, 462–465.

Minelli, A., Pasqual, C., 1977. The mouthparts of ladybirds: structure and function. Boll. Zool. 44, 183–187.

Morjan, W.E., Obrycki, J.J., Krafsur, E.S., 1999. Inbreeding effects of *Propylea quatuordecimpunctata* (Coleoptera: Coccinellidae). Ann. Entomol. Soc. Am. 92, 260–268.

Mota, J.A., Soares, A.O., Garcia, P.V., 2008. Temperature dependence for development of the whitefly predator *Clitostethus arcuatus* (Rossi). BioControl 53, 603–613.

Neel, W.W., Solomon, J.D., 1985. Collection, storage, and release of predaceous coccinellids in young cottonwood plantations. J. Agric. Entomol. 2, 212–214.

Niijima, K., Wataru, A., Matsuka, M., 1997. Development of low-cost and labor saving artificial diet for mass production of an aphidophagous coccinellid, *Harmonia axyridis* (Pallas). Bull. Fac. Agric., Tamagawa Univ. 37, 63–74.

Nordlund, D.A., 1998. Capacity and quality: keys to success in the mass rearing of biological control agents. Nat. Enemies Ins. 20, 169–179.

Obrycki, J.J., Kring, T.J., 1998. Predaceous Coccinellidae in biological control. Annu. Rev. Entomol. 43, 295–321.

Obrycki, J.J., Harwood, J.D., Kring, T.J., O'Neil, R.J., 2009. Aphidophagy by Coccinellidae: application of biological control in agroecosystems. Biol. Control 51, 244–254.

Ogura, N., Hosoda, R., 1995. Rearing of a coleopterous predator, *Trogossita japonica* (Col.: Trogossitidae), on artificial diets. Entomophaga 40, 371–378.

Ogura, N., Tabata, K., Wang, W., 1999. Rearing of the colydiid beetle predator, *Dastarcus helophoroides*, on artificial diet. BioControl 44, 291–299.

Omkar, Pathak, S., 2009. Crowding affects the life attributes of an aphidophagous ladybird beetle, *Propylea dissecta*. Bull. Insectol. 62, 35–40.

O'Neil, R.J., Giles, K.L., Obrycki, J.J., Mahr, D.L., Legaspi, J.C., Katovich, K., 1998. Evaluation of the quality of four commercially available natural enemies. Biol. Control 11, 1–8.

Pervez, A., Omkar, 2004. Temperature-dependent life attributes of an aphidophagous ladybird, *Propylea dissecta*. Biocontrol Sci. Technol. 14, 587–594.

Peters, T.M., Barbosa, B., 1972. Influence of population density on size, fecundity, and developmental rate of insects in culture. Annu. Rev. Entomol. 22, 431–450.

Pilipjuk, V.I., Bugaeva, L.N., Baklanova, E.V., 1982. On the possibility of breeding of the predatory beetle *Cryptolaemus montrouzieri* Muls. (Coleoptera, Coccinellidae) on the eggs of *Sitotroga cerealella* Ol. Entomol. Obozr. 1, 50–52 (in Russian).

Pickett, C.H., Casanave, K.A., Schoenig, S.E., Heinz, K.M., 1999. Rearing *Delphastus catalinae* (Coleoptera: Coccinellidae): practical experience and a modeling analysis. Can. Entomol. 131, 115–129.

Pilorget, L., Buckner, J., Lundgren, J.G., 2010. Sterol limitation in a pollen-fed omnivorous lady beetle (Coleoptera: Coccinellidae). J. Ins. Physiol. 56, 81–87.

Racioppi, J.V., Burton, R.L., Eikenbary, R., 1981. The effects of various oligidic synthetic diets on the growth of *Hippodamia convergens*. Entomol. Exper. Appl. 30, 68–72.

Ragab, M.E., 1995. Adaptation of *Rodolia cardinalis* (Mulsant) (Col., Coccinellidae) to *Icerya aegyptiaca* (Douglas) (Hom., Margarodidae) as compared with *Icerya purchasi* Mask. J. Appl. Entomol. 119, 621–623.

Rana, J.S., Dixon, A.F.G., Jarosik, V., 2002. Costs and benefits of prey specialization in a generalist insect predator. J. Anim. Ecol. 71, 15–22.

Reeve, J.D., Rojas, M.G., Morales-Ramos, J.A., 2003. Artificial diet and rearing methods for *Thanasimus dubius* (Coleoptera: Cleridae), a predator of bark beetles (Coleoptera: Scolytidae). Biol. Control 27, 315–322.

Ren, S.X., Stansly, P.A., Liu, T.X., 2002. Life history of the whitefly predator *Nephaspis oculatus* (Coleoptera: Coccinellidae) at six constant temperatures. Biol. Control 23, 262–268.

Riddick, E.W., 2001. Effect of cold storage on emergence, longevity, fertility, and survival of *Cotesia marginiventris* (Hymenoptera: Braconidae). J. Entomol. Sci. 36, 366–379.

Riddick, E.W., 2009. Benefits and limitations of factitious prey and artificial diets on life parameters of predatory beetles, bugs, and lacewings: a mini-review. BioControl 54, 325–339.

Riddick, E.W., 2010. Ectoparasitic mite and fungus on an invasive lady beetle: parasite coexistence and influence on host survival. Bull. Insectol. 63, 13–20.

Riddick, E.W., Wu, Z., 2010a. Potential long-term storage of the predatory mite *Phytoseiulus persimilis*. BioControl 55, 639–644.

Riddick, E.W., Wu, Z., 2010b. Lima bean–lady beetle interactions: hooked trichomes affect survival of *Stethorus punctillum* larvae. BioControl 56, 55–63.

Riddick, E.W., Cottrell, T.E., Kidd, K.A., 2009. Natural enemies of the Coccinellidae: parasites, pathogens, and parasitoids. Biol. Control 51, 306–312.

Riddick, E.W., Rojas, M.G., Wu, Z., 2011a. Spider mite mediates sublethal effects of its host plant on growth and development of its predator. Arthropod-Plant Interact. 5, 287–296.

Riddick, E.W., Rojas, M.G., Morales-Ramos, J., Allen, M., Spencer, B., 2011b. *Stethorus punctillum* (Coleoptera: Coccinellidae). In: Shelton, A. (Ed.), Biological Control: A Guide to Natural Enemies in North America, Cornell Univ., College of Agriculture and Life Sciences, Ithaca, New York, Online http://www.biocontrol.entomology.cornell.edu/predators/spunctillum.html.

Roush, R.T., Hopper, K.R., 1995. Use of single family lines to preserve genetic variation in laboratory colonies. Ann. Entomol. Soc. Am. 88, 713–717.

Ruan, C.C., Du, W.M., Wang, X.M., Zhang, J.J., Zhang, L.S., 2012. Effect of long-term cold storage on the fitness of pre-wintering *Harmonia axyridis* (Pallas). BioControl 57, 95–102.

Salom, S.M., Kok, L.T., Lamb, A.B., Jubb, C., 2012. Laboratory rearing of *Laricobius nigrinus* (Coleoptera: Derodontidae): a predator of the hemlock woolly adelgid (Hemiptera: Adelgidae). Psyche 2012, 9, Article ID 936519.

Samways, M.J., Osborn, R., Saunders, T.L., 1997. Mandible form relative to the main food type in ladybirds (Coleoptera: Coccinellidae). Biocontrol Sci. Technol. 7, 275–286.

Sarwar, M., Saqib, S.M., 2010. Rearing of predatory seven spotted ladybird beetle *Coccinella septempunctata* L. (Coleoptera: Coccinellidae) on natural and artificial diets under laboratory conditions. Pakistan J. Zool. 42, 47–51.

Sawyer, R.C., 1996. To Make a Spotless Orange: Biological Control in California. Iowa State Univ. Press, Ames, Iowa.

Schneider, J.C., 2009. Environmental biology of insect rearing. In: Schneider, J.C. (Ed.), Principles and Procedures for Rearing High Quality Insects, Mississippi State University, Mississippi State, MS, pp. 97–120, Chapter 6.

Schüder, I., Hommes, M., Larink, O., 2004. The influence of temperature and food supply on the development of *Adalia bipunctata* (Coleoptera: Coccinellidae). Eur. J. Entomol. 101, 379–384.

Seagraves, M.P., 2009. Lady beetle oviposition behavior in response to the trophic environment. Biol. Control 51, 313–322.

Seko, T., Miyatake, T., Miura, K., 2012. Assessment of hybrid vigor between flightless lines to restore survival and reproductive characteristics in the ladybird beetle *Harmonia axyridis*. BioControl 57, 85–93.

Seo, M.J., Youn, Y.M., 2002. Effective preservation methods of the Asian ladybird, *Harmonia axyridis* (Coleoptera: Coccinellidae), as an application strategy for the biological control of aphids. J. Asia-Pac. Entomol. 5, 209–214.

Shands, W.A., Holmes, R.L., Simpson, G.W., 1970. Improved laboratory production of eggs of *Coccinella septempunctata*. J. Econ. Entomol. 63, 315–317.

Sighinolfi, L., Febvay, G., Dindo, M.L., Rey, M., Pageaux, J.-F., Baronio, P., Grenier, S., 2008. Biological and biochemical characteristics for quality control of *Harmonia axyridis* (Pallas) (Coleoptera, Coccinellidae) reared on a liver-based diet. Arch. Ins. Biochem. Physiol. 68, 26–39.

Silva, R.B., Zanuncio, J.C., Serrão, J.E., Lima, E.R., Figueiredo, M.L.C., Cruz, I., 2009. Suitability of different artificial diets for development and survival of stages of the predaceous ladybird beetle *Eriopis connexa*. Phytoparasitica 37, 115–123.

Silva, R.B., Cruz, I., Figueiredo, M.L.C., Tavares, W.S., 2010. Development of *Coleomegilla maculata* De Geer (Coleoptera: Coccinellidae) with prey and artificial diet. Rev. Bras. Milho Sorgo 9, 13–26.

Smirnoff, W.A., 1958. An artificial diet for rearing coccinellid beetles. Can. Entomol. 90, 563–565.

Smith, B.C., 1960. A technique for rearing coccinellid beetles on dry foods, and influence of various pollens on the development of *Coleomegilla maculata* lengi Timb. (Coleoptera: Coccinellidae). Can. J. Zool. 38, 1047–1049.

Smith, B.C., 1965. Growth and development of coccinellid larvae on dry foods (Coleoptera: Coccinellidae). Can. Entomol. 97, 760–768.

Smith, R.A., Nordlund, D.A., 1999. Automation of insect rearing—a key to the development of competitive augmentative biological control. Nat. Enemies Ins. 21, 70–81.

Smith, R.A., Nordlund, D.A., 2000. Mass rearing technology for biological control agents of *Lygus* spp. Southwest. Entomologist (Suppl. 23), 121–127.

Smith, B.C., Williams, R.R., 1976. Temperature relations of adult *Coleomegilla maculata lengi* and *C. m. medialis* (Coleoptera: Coccinellidae) and responses to ovipositional stimulants. Can. Entomol. 108, 925–930.

Snyder, W.E., Joseph, S.B., Preziosi, R.F., Moore, A.J., 2000. Nutritional benefits of cannibalism for the lady beetle *Harmonia axyridis* (Coleoptera: Coccinellidae) when prey quality is poor. Environ. Entomol. 29, 1173–1179.

Sokal, R.R., Rohlf, F.J., 1981. Biometry: The Principles and Practice of Statistics in Biological Research, second ed. Freeman, New York.

Specty, O., Febvay, G., Grenier, S., Delobel, B., Piotte, C., Pageaux, J.-F., Ferran, A., Guillaud, J., 2003. Nutritional plasticity of the predatory ladybeetle *Harmonia axyridis* (Coleoptera: Coccinellidae): comparison between natural and substitution prey. Arch. Ins. Biochem. Physiol. 52, 81–91.

Stathas, G.J., 2000. The effect of temperature on the development of the predator *Rhyzobius lophanthae* and its phenology in Greece. BioControl 45, 439–451.

Symondson, W.O.C., 1994. The potential of *Abax parallelepipedus* (Col.: Carabidae) for mass breeding as a biological control agent against slugs. Entomophaga 39, 323–333.

Thompson, S.N., Hagen, K.S., 1999. Nutrition of entomophagous insects and other arthropods. In: Bellows, T.S., Fisher, T.W. (Eds.), Handbook of Biological Control: Principles and Applications of Biological Control. Academic Press, San Diego, CA, pp. 594–652.

Tifft, K.H., Leppla, N.C., Osborne, L.S., Cuda, J.P., 2006. Rearing *Diomus terminatus* (Coleoptera: Coccinellidae) on the corn leaf aphid, *Rhopalosiphum maidis* (Homoptera: Aphididae). Fla. Entomol. 89, 263–265.

Timms, J.E.L., Leather, S.R., 2007. Ladybird egg cluster size: relationships between species, oviposition substrate and cannibalism. Bull. Entomol. Res. 97, 613–618.

Tourniaire, R., Ferran, A., Giuge, L., Piotte, C., Gambier, J., 2000. A natural flightless mutation in the ladybird, *Harmonia axyridis*. Entomol. Exper. Appl. 96, 33–38.

Vanderzant, E.S., 1974. Development, significance, and application of artificial diets for insects. Annu. Rev. Entomol. 19, 139–160.

van Lenteren, J.C., 2003. Commercial availability of biological control agents. In: van Lenteren, J.C. (Ed.), Quality Control and Production of Biological Control Agents: Theory and Testing Procedures, CABI Publishing, Oxon, UK, pp. 167–178.

van Lenteren, J.C., 2012. The state of commercial augmentative biological control: plenty of natural enemies, but a frustrating lack of uptake. BioControl 57, 1–20.

van Lenteren, J.C., Hale, A., Klapwik, J.N., van Schelt, J., Steinberg, S., 2003. Guidelines for quality control of commercially produced natural enemies. In: van Lenteren, J.C. (Ed.), Quality Control and Production of Biological Control Agents: Theory and Testing Procedures, CABI Publishing, Oxon, UK, pp. 265–303.

Vatansever, G., Ulusoy, M.R., Erkilic, L.B., 2003. Improving the mass rearing possibilities of *Serangium montazerii* Fürsch (Coleoptera: Coccinellidae) on different host plants of *Bemisia tabaci* (Genn.) (Homoptera: Aleyrodidae). Turk. J. Agric. For. 27, 175–181.

Wang, L., Chen, H., Zhang, L., Wang, S., Yang, H., 2008. Improved formulations of an artificial diet for larvae of *Propylea japonica*. Chin. J. Biol. Control 24, 306–311 (in Chinese).

Weber, D.C., Lundgren, J.G., 2011. Effect of prior diet on consumption and digestion of prey and non-prey food by adults of the generalist predator *Coleomegilla maculata*. Entomol. Exp. Appl. 140, 146–152.

Weber, D.C., Riddick, E.W., 2011. *Lebia grandis* (Coleoptera: Carabidae). In: Shelton, A. (Ed.), Biological Control: A Guide to Natural Enemies in North America, Cornell Univ., College of Agriculture and Life Sciences, Ithaca, New York, Online http://www.biocontrol.entomology.cornell.edu/predators/Lebia.html.

Weber, D.C., Rowley, D.L., Greenstone, M.H., Athanas, M.M., 2006. Prey preference and host suitability of the predatory and parasitoid carabid beetle, *Lebia grandis*, for several species of *Leptinotarsa* beetles. J. Insect Sci. 6 (9), 14, Available online: insectscience.org/6.09.

Weseloh, R.M., 1996. Rearing the cannibalistic larvae of *Calosoma sycophanta* (Coleoptera: Carabidae) in groups. J. Entomol. Sci. 31, 33–38.

Weseloh, R.M., 1998. An artificial diet for larvae of *Calosoma sycophanta* (Coleoptera: Carabidae), a gypsy moth (Lepidoptera: Lymantriidae) predator. J. Entomol. Sci. 33, 233–240.

Whistlecraft, J.W., Harris, C.R., Tolman, J.H., Tomlin, A.D., 1985. Mass-rearing technique for *Aleochara bilineata* (Coleoptera: Staphylinidae). J. Econ. Entomol. 78, 995–997.

Wipperfürth, T., Hagen, K.S., Mittler, T.E., 1987. Egg production by the coccinellid *Hippodamia convergens* fed on two morphs of the green peach aphid, *Myzus persicae*. Entomol. Exp. Appl. 44, 195–198.

Yazdani, M., Zarabi, M., 2011. The effect of diet on longevity, fecundity, and sex ratio of *Clitostethus arcuatus* (Rossi) (Coleoptera: Coccinellidae). J. Asia-Pac. Entomol. 14, 349–352.

FURTHER READING

Dixon, A.F.G., 2001. Insect Predator-Prey Dynamics: Ladybird Beetles and Biological Control. Cambridge Univ. Press., UK.

Dixon, A.F.G., Hemptinne, J.L., Kindlmann, P., 1997. Effectiveness of ladybirds as biological control agents: patterns and processes. Entomophaga 42, 71–83.

This page intentionally left blank

Chapter 3

Production of Heteropteran Predators

Patrick De Clercq,[1] Thomas A. Coudron[2] and Eric W. Riddick[3]
[1]Department of Crop Protection, Ghent University, Ghent, Belgium, [2]Biological Control of Insects Research Laboratory, US Department of Agriculture, Agricultural Research Service (USDA-ARS), Columbia, MO, USA, [3]National Biological Control Laboratory, US Department of Agriculture, Agricultural Research Service (USDA-ARS), Stoneville, MS, USA

3.1. INTRODUCTION

True bugs (order Hemiptera, suborder Heteroptera) are characterized by piercing and sucking mouthparts with which they pierce plants or animals to suck up their fluids or liquefied tissues. Feeding mechanisms of phytophagous and carnivorous bugs have been described in more detail by Hori (2000) and Cohen (2000b), respectively. In short, predatory bugs use a feeding method termed "solid-to-liquid" feeding: they inject digestive enzymes into the prey's body and then suck up the digested and liquefied tissues (Cohen, 1990, 2000b). This extra-oral digestion allows them to utilize the high-nutrient prey tissues besides hemolymph and predisposes them to attack relatively large prey (Cohen, 2000b). In addition, many predatory heteropterans inject venom through their mouthparts that quickly immobilizes their prey.

It has been estimated that 65% of all families within the Heteroptera are partially or entirely composed of carnivorous species (Henry and Froeschner, 1988; Cohen, 2000b), and thus many of these may have potential for use in the biological control of arthropod pests. In fact, several species of predatory bugs are economically important biological control agents of key agricultural pests. Species from the families Anthocoridae, Miridae, Lygaeidae, Nabidae, Reduviidae, and Pentatomidae (subfamily Asopinae) have been or are being used in augmentative biological control programs in various agroecosystems (Table 3.1).

Most of the carnivorous heteropterans used for biological control are polyphagous predators feeding on a wide array of arthropod prey. Flower bugs (Anthocoridae) of the genera *Orius* and *Anthocoris* have primarily been produced for the control of thrips and psyllids, respectively, but they also feed on a range of other small arthropods, including aphids. *Macrolophus*, *Dicyphus*, and *Nesidiocoris* spp. are plant bugs (Miridae) feeding on whiteflies, but they

Mass Production of Beneficial Organisms. http://dx.doi.org/10.1016/B978-0-12-391453-8.00003-0

TABLE 3.1 Main Species of Predatory Heteroptera Used in Augmentative Biological Control Programs, with Indication of the Primary Target Pests and the Area Where They Have Been or Are Currently Used

Species	Family	Main Targets*	Area**
Anthocoris nemoralis	Anthocoridae	Psyllids	Europe and North America
Anthocoris nemorum	Anthocoridae	Psyllids and thrips	Europe
Orius albidipennis	Anthocoridae	Thrips	Europe and Africa
Orius insidiosus	Anthocoridae	Thrips	North America, Latin America and Europe
Orius laevigatus	Anthocoridae	Thrips	Europe, North Africa, and Asia
Orius majusculus	Anthocoridae	Thrips	Europe
Orius sauteri	Anthocoridae	Thrips	Asia
Orius strigicollis	Anthocoridae	Thrips	Asia
Orius tristicolor	Anthocoridae	Thrips	North America and Europe
Xylocoris flavipes	Anthocoridae	Stored product pests	North America, Africa, and Asia
Nabis americoferus and *Nabis roseipennis*	Nabidae	Lepidopterans	North America
Nabis pseudoferus	Nabidae	Lepidopterans	Europe
Dicyphus hesperus	Miridae	Whiteflies	North America and Europe
Dicyphus tamaninii	Miridae	Whiteflies	Europe
Macrolophus melanotoma (=M. caliginosus)	Miridae	Whiteflies	Europe
Macrolophus pygmaeus	Miridae	Whiteflies	Europe and Africa
Nesidiocoris tenuis	Miridae	Whiteflies and lepidopterans	Europe, North Africa, and Asia
Brontocoris tabidus	Pentatomidae	Lepidopteran defoliators	Latin America

TABLE 3.1 Main Species of Predatory Heteroptera Used in Augmentative Biological Control Programs, with Indication of the Primary Target Pests and the Area Where They Have Been or Are Currently Used—cont'd

Species	Family	Main Targets*	Area**
Picromerus bidens	Pentatomidae	Lepidopteran and coleopteran defoliators	Europe
Perillus bioculatus	Pentatomidae	Colorado potato beetle	North America and Europe
Podisus maculiventris	Pentatomidae	Lepidopteran and coleopteran defoliators	North America, Europe, and Asia
Podisus nigrispinus	Pentatomidae	Lepidopteran defoliators	Latin America
Geocoris punctipes	Lygaeidae	Whiteflies, thrips, and lepidopterans	North America
Rhynocoris marginatus	Reduviidae	Lepidopterans and heteropterans	Asia
Pristhesancus plagipennis	Reduviidae	Lepidopterans and heteropterans	Australia

**Primary target pests are listed, although additional targets may occur.*
***Both commercial and experimental releases are included.*
Source: Based in part on van Lenteren (2011).

also attack thrips, mites, aphids, and other arthropod pests. Damsel bugs (Nabidae) of the genus *Nabis* have been evaluated against lepidopteran pests in various field and greenhouse crops, but their predation on aphids, psyllids, leaf hoppers, and other insect pests has also received attention. The big-eyed bugs (Lygaeidae) of the genus *Geocoris* are generalist predators feeding on whiteflies, thrips, mites, aphids, and eggs and small larvae of lepidopterans. Members of the stinkbug subfamily Asopinae are all predators feeding primarily on larvae of leaf-feeding lepidopterans, coleopterans, and hymenopterans; economically important species are found in the genera *Podisus*, *Picromerus*, *Arma*, and *Perillus*. Finally, several assassin bugs (Reduviidae), including *Rhynocoris*, *Pristhesancus*, and *Zelus* spp., have been identified to be predators of key pests in diverse crop systems, but very few have been actively used in augmentation programs. More details on the biology and biological control potential of the main families and species of predatory Heteroptera can be found in Schaefer and Panizzi (2000).

According to van Lenteren (2011), the Palearctic species *Macrolophus pygmaeus* (Rambur) and *Orius laevigatus* (Fieber) are among the 12 economically most important invertebrate biological control agents, based on the number of countries in which the species are marketed. The commercial use of both species did not start before the early 1990s. It is worth noting that the former species was distributed commercially for some time in Europe under the erroneous identity of *Macrolophus caliginosus* Wagner, which is a different species (in fact, the presently accepted name of this species is *Macrolophus melanotoma* (Costa)). This was most probably due to a misidentification based on unreliable morphological characteristics (Martinez-Cascales et al., 2006a, b; Machtelinckx et al., 2009). The species can be reliably discriminated only by using molecular tools. As a result, several studies using insects purchased from commercial sources under the name *M. caliginosus* may in fact have studied *M. pygmaeus*.

Several heteropteran families display trophic omnivory (also called zoophytophagy), allowing them to exploit both plant and prey resources (Coll, 1998; Coll and Guershon, 2002; Cohen, 1990, 2004; Torres and Boyd, 2009). Plants may be used to varying degrees to extract water or supplementary nutrients from them. Besides being a vital nutrient, the acquired water may be essential for the process of prey feeding (Gillespie and McGregor, 2000). Their mixed feeding strategy may further allow them to perform better in terms of development and reproduction or survive periods of prey scarcity. Although in many cases zoophytophagous bugs do not produce damage when piercing plant tissues, several predatory mirids such as *Nesidiocoris tenuis* (Reuter), *Dicyphus tamaninii* Wagner, and *M. caliginosus* do damage plants (Sampson and Jacobson, 1999; Wheeler, 2001; Coll and Guershon, 2002; Calvo et al., 2009; Sanchez, 2009; Arnó et al., 2010) or transfer plant pathogens (Burgess et al., 1983). Furthermore, the adult females of predatory bugs may require plants to deposit eggs on (e.g. Pentatomidae, Geocoridae, and Reduviidae) or in (e.g. Anthocoridae, Miridae, and Nabidae). Their omnivory and need for plants as oviposition substrates may not only have implications for the practical use of predatory bugs in the field but also have bearing on their mass production. In this chapter, we investigate how the zoophytophagous feeding habit of heteropteran predators contributes to the complexity of their mass production.

We will start by providing an overview of natural, factitious, and artificial foods that have been used for the production of heteropteran predators. In the following section, the use of plants and plant materials as sources of water and supplementary nutrients, and as living and oviposition substrates, will be discussed, as well as the potential of alternative substrates. The impacts of crowding, cannibalism, and the presence of microorganisms on the performance of rearing systems are also addressed. Finally, rearing systems for a selection of economically important species will be described, with special attention to housing and rearing densities, and aspects of production scale. Future challenges and recommendations for research will be presented in a concluding section.

3.2. FOODS

3.2.1. Natural Prey

Natural rearing systems for predators use the natural or target prey. This is often an herbivorous species, which has to be maintained on plants or plant materials or in some cases can be reared on artificial diets (e.g. lepidopteran larvae). Natural rearing systems are essentially tritrophic, that is, they comprise three trophic levels: the predator, the herbivorous prey, and the prey's food. Although such systems have proven to be effective for the mass production of certain arthropod natural enemies, they may face problems of discontinuity and usually involve high costs related to labor and facilities for plant production (e.g. greenhouses) (Etzel and Legner, 1999; De Clercq, 2008; Riddick, 2009). Several studies have reported the use of natural prey for sustaining small-scale cultures of heteropteran predators in research laboratories, but for the reasons mentioned here this type of prey is rarely used in mass production systems. Natural prey is often used as a benchmark to evaluate the nutritional value of unnatural or artificial foods. However, as many predatory heteropterans are generalist feeders, there may be large differences in the nutritional quality of prey that is attacked by a given predator in its natural habitat. Therefore, the use of natural prey may not always result in the production of superior predators. In this context, the key to success in the rearing of predatory insects is to provide variation in the food offered, even when rearing on natural foods (Richman and Whitcomb, 1978; Evans et al., 1999; Lundgren, 2011).

Predatory stinkbugs of the genus *Podisus* are highly polyphagous and are relatively easily reared on larval forms of various insect prey. Although in some cases larvae of hymenopterans or coleopterans have been used for laboratory culturing, *Podisus* bugs have been primarily produced on lepidopteran larvae (De Clercq, 2000). Nymphs and adults of *Podisus maculiventris* (Say) and *Podisus nigrispinus* (Dallas) have been cultured on live caterpillars of several noctuids, including *Spodoptera* spp. (Yu, 1987; De Clercq et al., 1988; De Clercq and Degheele, 1993b), *Pseudoplusia includens* (Walker) (Orr et al., 1986), *Trichoplusia ni* (Hübner) (Biever and Chauvin, 1992), *Heliothis virescens* (F.) (Pfannenstiel et al., 1995), *Helicoverpa zea* (Boddie) (Warren and Wallis, 1971), *Alabama argillacea* (Hübner), and *Anticarsia gemmatalis* (Hübner) (Lemos et al., 2003; Torres et al., 2006). Other predatory pentatomids were reared on noctuid larvae, including *Eocanthecona furcellata* (Wolff) (Yasuda and Wakamura, 1992) and *Picromerus bidens* L. (Mayné and Breny, 1948; Mahdian et al., 2006). Most of these noctuid caterpillars can be reared on artificial diets devoid of living plant material, precluding the necessity for space for the production of host plants. Occasionally, caterpillars from other families have been used as food for rearing predatory pentatomids, like *Hyphantria cunea* (Drury) and *Malacosoma americanum* (F.) (Warren and Wallis, 1971). Frozen caterpillars proved to be a suitable diet for several predatory stinkbugs (Warren and Wallis, 1971; De Clercq et al., 1988; Sipayung et al., 1992; Yasuda and Wakamura, 1992);

deep freezing surpluses from caterpillar cultures may thus allow for some flexibility in production systems using natural prey.

The two-spotted stinkbug, *Perillus bioculatus* (F.), is more restricted in its diet: although it reportedly attacks insects from various orders, including Lepidoptera, it is usually associated with chrysomelid prey in the field (see De Clercq, 2000, and references therein). Several workers reared *P. bioculatus* on eggs and (young) larvae of its target prey, the Colorado potato beetle, *Leptinotarsa decemlineata* (Say) (e.g. Franz and Szmidt, 1960; Tremblay, 1967; Tamaki and Butt, 1978; Heimpel and Hough-Goldstein, 1994; Adams, 2000a), but this requires the continuous availability of potato foliage as no adequate artificial diet is available for the prey. Biever and Chauvin (1992), Coudron and Kim (2004), and De Clercq (unpublished data) found that consecutive generations of *P. bioculatus* can be successfully reared on coddled larvae of *T. ni* or *Spodoptera littoralis* (Boisduval). Adams (2000a, b) used frozen larvae of *H. virescens* to rear the two-spotted stinkbug. The use of these noctuid prey species renders production more cost-effective given the relatively easy rearing of these lepidopterans on artificial diets. Yocum and Evenson (2002) pointed out that for optimal results, a diet of caterpillars had to be supplemented with eggs of the Colorado potato beetle, particularly for the early instars of the predator.

Noctuid prey was also found suitable for rearing predatory bugs from other families. The big-eyed bug, *Geocoris punctipes* (Say), was fed on coddled larvae of *Spodoptera exigua* (Hübner) by Champlain and Sholdt (1966, 1967) and on eggs of *H. zea* or *Lygus hesperus* (Knight) by Cohen and Debolt (1983). Cohen and Urias (1986) reported superior results on a combination of heat-killed *S. exigua* larvae and *H. virescens* eggs. The damsel bugs *Nabis americoferus* (Carayon), *Nabis roseipennis* Reuter, and *Nabis rufusculus* Reuter were maintained in the laboratory on fresh *H. virescens* eggs (Braman and Yeargan, 1988). Kiman and Yeargan (1985) used frozen eggs of the latter noctuid to maintain their colony of the anthocorid *Orius insidiosus* (Say). Nadgauda and Pitre (1987), on the other hand, reported that feeding *N. roseipennis* on larger *H. virescens* larvae effectively sustained the predator's development. Ables (1978) reared the assassin bug *Zelus renardii* (Kolenati) on *H. virescens* larvae, whereas Grundy et al. (2000) provided *Pristhesancus plagipennis* Walker with larvae of *Helicoverpa armigera* (Hübner). *Rhynocoris marginatus* (F.) and other reduviids were successfully reared on larvae of the target prey *Spodoptera litura* (F.) in India (Venkatesan et al., 1997; Sahayaraj and Paulraj, 2001).

Occasionally, other natural prey types have been used to feed stock colonies of predatory bugs. For instance, aphids, including the pea aphid, *Acyrthosiphon pisum* (Harris), and green peach aphid, *Myzus persicae* (Sulzer), have been used for culturing various predatory bugs like nabids (Guppy, 1986) and anthocorids (Ruth and Dwumfour, 1989). Yokoyama (1980) presented a rearing method for *Geocoris pallens* Stål using eggs and nymphs of the milkweed bug, *Oncopeltus fasciatus* (Dallas), as food.

The warehouse pirate bugs *Xylocoris* spp. can be reared on various insects found in stored-product habitats, where these generalist predatory bugs are naturally found. Chu (1969) was able to mass rear *Xylocoris galactinus* (Fieber) on larvae of *Tribolium castaneum* (Herbst) that were infesting cracked corn. Dunkel and Jaronski (2003) used similar methods to maintain a stock culture of *Xylocoris flavipes* (Reuter) on larvae of *T. castaneum* fed on wheat flour and brewer's yeast. Arbogast et al. (1971) showed that *X. flavipes* could be reared on eggs and first instars of *Plodia interpunctella* (Hübner). Press et al. (1973) later demonstrated that *X. flavipes* can be reared on *P. interpunctella* eggs that are killed by gamma irradiation or freezing, inspiring later workers to use eggs of lepidopteran storage pests as factitious prey for various predatory bugs.

3.2.2. Factitious Prey

Costs of production may be reduced when predatory insects can be produced on unnatural or factitious prey that is easier and less expensive to rear than the natural prey. Factitious prey are comprised of organisms that are not normally attacked by the predator, mostly because they do not occur in its natural habitat, but do sustain its development in a laboratory environment (De Clercq, 2008). Factitious prey may be offered fresh, but in many cases, they are frozen, irradiated, or lyophilized for improved storage or use in predator cultures (Riddick, 2009). As many heteropteran predators used in augmentative biological control programs are highly polyphagous, they are usually amenable to rearing on factitious prey (Riddick, 2009).

Lepidopteran eggs, particularly those of the Mediterranean flour moth *Ephestia* (*Anagasta*) *kuehniella* Zeller, the Indian meal moth *P. interpunctella*, the rice moth *Corcyra cephalonica* (Stainton), and the Angoumois grain moth *Sitotroga cerealella* (Olivier), have been extensively used as factitious food for various insect predators, including predatory bugs. Lopez et al. (1987) reported that frozen eggs of *S. cerealella* were suitable to temporarily support the rearing of the lygaeid *G. punctipes*, although they were considered inferior to fresh *Heliothis* eggs. Alauzet et al. (1992) and Blümel (1996) showed that the anthocorid *Orius majusculus* (Reuter) could be reared successfully on eggs of *Ephestia kuehniella*. Fauvel et al. (1987) and Cocuzza et al. (1997) found that *M. caliginosus* and *O. laevigatus*, respectively, performed even better on *E. kuehniella* eggs than on some of their natural prey. Zhou and Wang (1989) reported similar findings for *Orius sauteri* (Poppius) fed on rice moth eggs. Since the 1990s, lepidopteran eggs, and in particular eggs of *E. kuehniella*, have become the standard food for commercially available heteropteran predators such as *Orius* spp. (Fig. 3.1) (Richards and Schmidt, 1995; van den Meiracker, 1999; Tommasini et al., 2004; Kakimoto et al., 2005; Bueno et al., 2006; Bonte and De Clercq, 2008; Venkatesan et al., 2008), *Macrolophus* spp. (Fauvel et al., 1987; Castañé et al., 2006), and *N. tenuis* (Calvo et al., 2009; Sanchez et al., 2009). Eggs of *E. kuehniella* were also suitable to sustain development of the

FIGURE 3.1 Adult of *Orius naivashae* feeding on *Ephestia kuehniella* eggs. *Photo: J. Bonte.* (For color version of this figure, the reader is referred to the online version of this book.)

predatory pentatomids *P. maculiventris* and *P. bidens*, but the latter species could not produce eggs on this food (Mahdian et al., 2006). To avoid hatching of the eggs in the predator cultures, they are usually killed by deep freezing, or by gamma or UV irradiation (Etzel and Legner, 1999). In addition, eggs of *E. kuehniella* can be used to support populations of *Macrolophus* and *Nesidiocoris* bugs in the crop (Lenfant et al., 2000; Urbaneja-Bernat et al., 2013).

The continuous use of lepidopteran eggs as a factitious food in mass-rearing systems does have some drawbacks, the most important of which is their high cost. Although the moths are easily produced on inexpensive foods (flour or grains of cereals), there are substantial monetary investments for the mechanization of rearing procedures (handling of the rearing medium, harvesting the moths, and harvesting and cleaning the eggs), for climate management and the associated energy budget, and for the health of workers (repeated inhalation exposure to scales is known to cause allergies). This has led to high market prices, especially for *E. kuehniella* eggs, which are currently about US$500 per kilogram (De Clercq, 2008; K. Bolckmans and P. Couwels, personal communication). Urbaneja-Bernat et al. (2013) reported that supplementing a diet of *E. kuehniella* with a sucrose solution at 0.5 M increased fecundity of *N. tenuis* and lowered the number of *E. kuehniella* eggs consumed by the predator, which ultimately may reduce production cost. In contrast, the anthocorid *O. laevigatus* did not benefit from a supplementary 5% sucrose solution (Bonte and De Clercq, 2010a).

Noninsect materials may hold potential as factitious foods for rearing insect predators. Brine shrimp of the genus *Artemia* (Branchiopoda) are routinely used as a feed in aquaculture (Lavens and Sorgeloos, 2000). Arijs and De Clercq (2001) were the first to test cysts (diapausing eggs) of the brine shrimp (*Artemia franciscana* Kellogg) as a food for a heteropteran predator. They compared the development and reproduction of *O. laevigatus* on

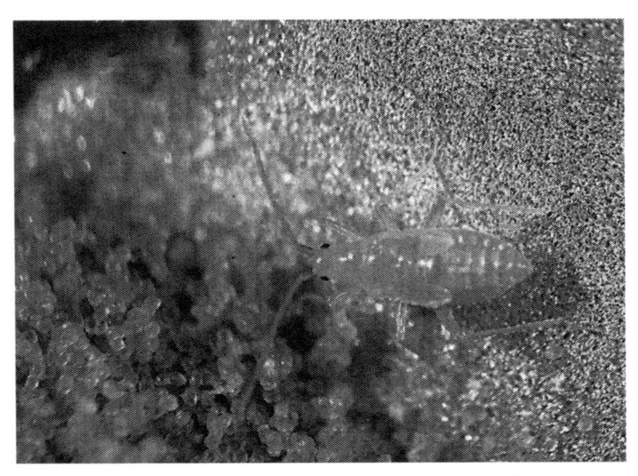

FIGURE 3.2 Nymph of *Macrolophus pygmaeus* with decapsulated *Artemia franciscana* cysts. *Photo: L. Parmentier.* (For color version of this figure, the reader is referred to the online version of this book.)

A. franciscana cysts versus *E. kuehniella* eggs. Feeding this predator during a single generation on hydrated, decapsulated cysts (which had been stored in dry form) resulted in development, fecundity, and oviposition rates comparable to those of predators fed frozen *Ephestia* eggs. Later studies indicated that brine shrimp cysts are suitable food for several *Orius* spp. (Riudavets et al., 2006; Bonte and De Clercq, 2008) and *Macrolophus* spp. (Fig. 3.2) (Callebaut et al. 2004; Castañé et al., 2006; Riudavets et al., 2006; Vandekerkhove et al., 2009), but less so for the larger pentatomid predators, *P. maculiventris* and *P. bidens* (Mahdian et al., 2006). Results obtained with different predatory bugs have been known to vary as a function of decapsulation, hydration, and origin of the cysts (De Clercq et al., 2005; Vandekerkhove et al., 2009). In addition, some batch-to-batch variability in performance has been observed. Whereas the predators usually can handle the cysts better when they are hydrated to some degree, the use of fully hydrated cysts leads to problems with molds in the rearing containers (Vandekerkhove et al., 2009). Moreover, prolonged rearing on cysts as a sole food has been associated with fitness losses in *Orius* bugs (De Clercq et al., 2005). Although the practical value of brine shrimp cysts as a feed does not equal that of *E. kuehniella* eggs, they did find their way to practice as a supplementary food in the mass rearing of different predatory heteropterans. An important consideration here is that *Artemia* cysts are at least an order of magnitude cheaper than *E. kuehniella* eggs (Arijs and De Clercq, 2001). Currently, dry *Artemia* cysts are routinely mixed with *E. kuehniella* eggs to reduce inputs of expensive lepidopteran eggs in the production process of different predatory heteropterans. Lu et al. (2011) found that feeding *Orius strigicollis* (Poppius) *Artemia* cysts during the nymphal stages and *E. kuehniella* eggs during the adult stage yielded good results for the mass rearing of this predator.

Another potential source of factitious foods for predatory heteropterans produced by the billions in mass-rearing facilities for sterile insect techniques is fruit flies (Tephritidae). Takara and Nishida (1981) reared *O. insidiosus* on a diet of eggs of the oriental fruit fly, *Dacus (Bactrocera) dorsalis* (Hendel), with similar developmental and reproductive performance as compared to a natural diet. Eggs of the medfly, *Ceratitis capitata* (Wiedemann), have been used to produce the mirid predators *Cyrtorhinus lividipennis* Reuter (Liquido and Nishida, 1985) and *M. pygmaeus* (Nannini and Souriau, 2009). Hough-Goldstein and McPherson (1996) reported that colonies of the two-spotted stinkbug *P. bioculatus* could be maintained on a combination of heat-killed larvae of the Mexican fruit fly, *Anastrepha ludens* (Loew), and the European corn borer, *Ostrinia nubilalis* (Hübner).

Predatory bugs of the families Pentatomidae and Reduviidae have been reared using larvae and pupae of several insect orders as factitious prey. These include the larvae of the lepidopterans *Galleria mellonella* L., *C. cephalonica*, *Diatraea saccharalis* (F.), and *Bombyx mori* L.; larvae and pupae of the coleopterans *Tenebrio molitor* L. and *Zophobas confusa* Gebien; and larvae of the dipterans *Musca domestica* L. and *Calliphora erythrocephala* Meigen (Li et al., 1997; De Clercq, 2000, and references therein; Grundy et al., 2000; Sahayaraj, 2002; Zanuncio et al., 2001, 2005; Torres et al., 2006). De Clercq and Degheele (1993b) reported better success when rearing *Podisus* bugs with larvae of the greater wax moth, *G. mellonella*, as compared with those of the beet armyworm, *S. exigua*, and attributed this to the higher protein and fat content and lower water content of the former prey type. De Clercq et al. (1998b) found that nymphal development of *P. maculiventris* was faster on pupae of the yellow mealworm *T. molitor* than on larvae of *T. molitor* or on larvae of *G. mellonella*. Male or female body weight was not affected by prey treatment, but total egg production was greatest for females reared on *G. mellonella* larvae. However, the authors pointed out that the expense for producing yellow mealworms is lower than that for wax moth larvae. Moreover, live wax moth larvae also proved more difficult to handle by the predators (mainly due to their web-spinning behavior), such that they need to be inactivated or killed before being introduced in the predator cultures. In Brazil, larvae and pupae of mealworms and larvae of houseflies have been used successfully for the rearing of different pentatomid predators, including *P. nigrispinus*, *Supputius cinctipes* (Stål), and *Brontocoris tabidus* (Signoret) (Zanuncio et al., 1996, 2001; Jusselino-Filho et al., 2003; Zanuncio et al., 2005). De Bortoli et al. (2011) stated that the larval stage of the housefly *M. domestica* was the most cost-effective factitious prey for mass rearing of *P. nigrispinus*. Torres et al. (2006) pointed out that the cost of producing yellow mealworms and houseflies is low, given the relatively low cost of the diet components, labor for handling the insects, rearing containers, and space. Zanuncio et al. (2001) studied the development and reproduction of the Neotropical pentatomid predator *P. nigrispinus*, fed with combinations of yellow mealworm pupae and housefly larvae, either presented together or presented on alternate days. They noted that heavier females were produced from a diet of *T. molitor* pupae alone, or from a diet of both prey on alternate days, rather than from both prey simultaneously or *M. domestica* alone. The authors further found that female body weight did not correlate with fecundity. Nonetheless, fecundity of *P. nigrispinus* was superior when the predator was maintained on mixed prey (given simultaneously or on alternate days) than when reared on *T. molitor* alone. The authors concluded that a mixed diet of *T. molitor* pupae and *M. domestica* larvae was the most suitable for the production of *P. nigrispinus*. Grundy et al. (2000) came to similar conclusions when rearing the assassin bug *P. plagipennis* on coddled larvae of *T. molitor* and *H. armigera*. When reared on *T. molitor* alone, *P. plagipennis* had the highest body weight, but its fecundity was substantially inferior to that on a mix of both prey species. These findings again indicate the benefits of providing predatory heteropterans with mixed diets.

3.2.3. Artificial Diets

3.2.3.1. Definitions

The availability of an artificial diet that optimizes the growth and reproduction of a natural enemy could advance the automation of its mass production (De Clercq, 2008; see also Chapter 7). The literature on artificial diets for predatory bugs prior to 1998 was covered by the reviews of Thompson (1999) and Thompson and Hagen (1999). The literature published between 1998 and 2007 was reviewed by Riddick (2009). Artificial diets for insects have traditionally been classified as holidic (chemically defined), meridic (most components are known chemically), or oligidic (mainly composed of crude organic materials) (Dougherty, 1959). As the distinction between these three categories is not always clear, Grenier and De Clercq (2003) proposed a classification system that separates artificial diets for insect natural enemies (predators and parasitoids) based on whether they contain insect components (e.g. tissues, hemolymph, cells, protein, and amino acids) or not. Artificial diets containing insect components may be useful when predators require certain nutrients, feeding stimulants, and other chemical cues found in arthropod prey (De Clercq, 2008; Riddick, 2009).

3.2.3.2. Effects on Development and Reproduction

Table 3.2 provides an overview of exemplary studies that compared the life parameters of heteropteran predators fed on artificial diets versus factitious or natural prey. Cohen (1985) proposed a meat-based diet for the rearing of the big-eyed bug, *G. punctipes*, which served as a basis for several subsequent studies using meat diets for a variety of predatory bugs. The diet was a paste composed of fatty ground beef and beef liver supplemented with a 5% sucrose solution and wrapped in stretched Parafilm. The gross nutritional composition of this diet reflected that of the predator's natural prey, *H. virescens* eggs. The diet was supplemented with free water only, and no plant materials were offered.

The weights of adults and eggs of diet-fed *G. punctipes* were comparable to those of their counterparts fed insect eggs, but fecundity was lower, which was mainly attributed to egg cannibalism. The diet was cheap (c. US$3 per kg) and allowed continuous production of the predator for over 150 generations over a time span of nearly 15 years (Cohen et al., 1999; Cohen, 2000a).

De Clercq and Degheele (1992, 1993a) reported that consecutive generations of the pentatomid predators *P. maculiventris* and *P. nigrispinus* (misidentified as *Podisus sagitta*) could be reared on a meat-based diet modified from Cohen's (1985) diet. The main modification consisted of the addition of fresh chicken's egg yolk. In comparison to rearing on wax moth larvae, the development of the predators was prolonged by 15–40%, and adult weights were 18–28% lower. Fecundity on the artificial diet was only about 30–50% of that on *G. mellonella* larvae, but egg weights and egg hatch were similar. In a follow-up study on *P. maculiventris*, De Clercq et al. (1998a) confirmed that adults reared on an artificial diet composed of meat and egg yolk were smaller than those produced on the larvae of two factitious prey, *G. mellonella* and *T. molitor*. However, whereas fecundity of predators on the artificial diet remained inferior to that on wax moth larvae, it was similar to that on yellow mealworms. Mahdian et al. (2006) used a diet that was slightly modified from that proposed by De Clercq et al. (1998a) (ascorbic acid was omitted) for *P. maculiventris*. Their results indicated that nymphal survival and longevity on the diet were acceptable, but female body weight and oviposition rate were reduced when using the artificial diet rather than live prey such as *S. littoralis* or *G. mellonella* larvae.

Greany and Carpenter (1998) reported that artificial diets containing ground liver and fresh egg yolk encapsulated in Parafilm allow egg-to-adult development of several predatory heteropterans, including *P. maculiventris*, *P. bioculatus*, *G. punctipes*, and *X. flavipes*. Wittmeyer and Coudron (2001) reported that an artificial diet for *P. maculiventris* based on beef liver and whole egg encapsulated in Mylar–Parafilm domes was inferior to coddled cabbage looper (*T. ni*) larvae. The actual cost required to double the size of the *P. maculiventris* population, in culture, was two times greater when predators were fed exclusively on the artificial diet in the nymphal and adult stages, as compared to using *T. ni* larvae. Supplying *T. ni* larvae to adults and artificial diet to nymphs tended to improve oviposition, reproductive rate, and intrinsic rate of increase, but not enough to offset the loss in egg production. Wittmeyer et al. (2001) provided further evidence that the same artificial diet was ineffective, in comparison to *T. ni* larvae, as a sole food source for maximizing egg production in *P. maculiventris*. Their study highlighted the impact of nymphal and adult food sources on oogenesis (egg maturation) and vitellogenesis (yolk deposition) in this pentatomid predator.

Coudron et al. (2002) evaluated the effects of a blended, buffered mixture of a plant-based diet for *Lygus* bugs with beef liver and whole egg, encapsulated in Mylar–Parafilm domes, on *P. maculiventris*. Nymphal survival was similar, but other life parameters (see Table 3.2) were negatively impacted when predators

TABLE 3.2 **Exemplary Studies on the Influence of Consuming Artificial Diets versus Factitious or Natural Prey on Life History Parameters of Predatory Heteropterans**

Predator	Artificial Diet – Animal Protein Base	Results and Outcome	Reference
Orius laevigatus nymphs and adults (Anthocoridae)	Beef liver and ground beef – in Parafilm packets	Development time (↑), survival (↓), size (√), and oviposition (√)	Arijs and De Clercq (2004)
O. laevigatus nymphs and adults	Hen's egg yolk – in Parafilm domes	Development time (↑), survival (↓), size (↓), oviposition (√), oogenesis (↓), and longevity (√)	Bonte and De Clercq (2008)
Geocoris punctipes adults (Lygaeidae)	Beef liver and ground beef – in Parafilm®	Size (√) and oviposition (↓)	Cohen (1985)
Dicyphus tamaninii nymphs and adults (Miridae)	Beef liver, fatty ground beef, and hen's egg yolk – in Parafilm	Development time (↑), survival (√) by fourth or fifth generation, size (↓), and oviposition (√)	Iriarte and Castañé (2001)
D. tamaninii nymphs and adults	Beef liver, fatty ground beef, hen's egg yolk, and casein – reformulated diet	Development time (↑), survival (↑), size (√), and oviposition (√)	Zapata et al. (2005)
Hyaliodes vitripennis nymphs and adults (Miridae)	Pork liver and whey powder – in Parafilm	Survival (↑), longevity (↑), and oviposition (↑)	Firlej et al. (2006)
H. vitripennis nymphs and adults	Beef liver, ground beef, and whole hen's egg – in Parafilm	Survival (↑) and longevity (↑)	Firlej et al. (2006)
Macrolophus caliginosus nymphs and adults (Miridae)	Beef liver, fatty ground beef, hen's egg yolk, and casein – in Parafilm	Development time (↑), survival (↓), and size (↓)	Castañé and Zapata (2005)
Macrolophus pygmaeus (caliginosus) nymphs and adults (Miridae)	Hen's egg yolk – in Parafilm domes	Development time (↑), survival (√), size (↓), oviposition (↓), and oogenesis (↓)	Vandekerkhove et al. (2006, 2011)

Continued

TABLE 3.2 Exemplary Studies on the Influence of Consuming Artificial Diets versus Factitious or Natural Prey on Life History Parameters of Predatory Heteropterans—cont'd

Predator	Artificial Diet – Animal Protein Base	Results and Outcome	Reference
Perillus bioculatus adults (Pentatomidae)	Pork liver and fatty ground beef – in Parafilm domes	Oogenesis (↓)	Adams (2000a)
P. bioculatus adults	Pork liver – in Parafilm domes	Oviposition (↓) and longevity (√)	Adams (2000b)
P. bioculatus nymphs and adults	Beef and whole hen's egg – in Mylar–Parafilm	Development time (↑), survival (↓), total oviposition (↓), and longevity (↓)	Coudron and Kim (2004)
P. bioculatus nymphs and adults	Chicken liver and tuna fish – in Parafilm capsules	Development time (↑), survival (↓), and oviposition (↓)	Rojas et al. (2000)
Picromerus bidens nymphs and adults (Pentatomidae)	Beef liver, fatty ground beef, and hen's egg yolk – in Parafilm	Development time (↑), survival (↓), and size (↓)	Mahdian et al. (2006)
Podisus maculiventris nymphs and adults (Pentatomidae)	Beef liver, fatty ground beef, and hen's egg yolk – in Parafilm	Development time (↑), survival (√), size (↓), oviposition (↓), and longevity (√)	De Clercq et al. (1998a) and Mahdian et al. (2006)
P. maculiventris nymphs and adults	Beef liver and whole hen's egg – in Mylar–Parafilm	Development time (↑), size (↓), and oviposition (↓)	Wittmeyer and Coudron (2001)
P. maculiventris nymphs and adults	Beef liver and whole hen's egg – in Mylar–Parafilm	Body size (↓), oviposition (↓), oogenesis (↓), and vitellogenesis (↓)	Wittmeyer et al. (2001)
P. maculiventris nymphs and adults	Beef liver and whole hen's egg – in Mylar–Parafilm	Development time (↑), size (↓), survival (√), and oviposition (↓)	Coudron et al. (2002)
Podisus nigrispinus nymphs and adults (Pentatomidae)	Beef liver and ground beef	Development time (↑), size (↓), and ovarian weight (↓)	Lemos et al. (2003)
P. nigrispinus adults	Beef liver and ground beef	Ovarian development (↓) and oogenesis (↓)	Lemos et al. (2005)

Symbols indicate decrease (↓), increase (↑), or no significant effect (√) on a given life parameter, in comparison to a control of natural or factitious prey, for the designated predator.
Source: Modified from Riddick (2009).

were fed the zoophytogenous artificial diet for one or 11 generations compared to cohorts fed natural prey (*T. ni* larvae) for one generation. However, a gradual improvement in life parameters occurred from the first to the 11th generation on the artificial diet, possibly as a consequence of adaptation. Based on a comparison of the cost to double the population size, the authors estimated that after 11 generations, the cost of rearing *P. maculiventris* on the artificial diet was 1.2 times higher than on natural prey.

The value of artificial diets for the production of another pentatomid predator, the two-spotted stinkbug, *P. bioculatus*, was investigated by Adams (2000a, b), Rojas et al. (2000), and Coudron and Kim (2004). Adams (2000a) reported that females fed an artificial diet rather than a control diet of freeze-killed *H. virescens* larvae had fewer developing follicles. Ten-day-old females fed the artificial diet contained chorionated (mature) follicles in just 40% of their ovarioles, whereas 9-day-old females fed the control diet contained chorionated follicles in 100% of their ovarioles. An average of 42 and 138 eggs were oviposited by mated females fed the artificial diet and frozen *H. virescens* larvae, respectively (Adams, 2000a). Apparently, the artificial diet did not provide an adequate supply of nutrients essential for egg development in *P. bioculatus*.

In another study on *P. bioculatus*, Rojas et al. (2000) formulated two meridic diets to reflect the nutritional composition of eggs of the Colorado potato beetle, *L. decemlineata*. They modified the first diet (diet 1) with a substitution of chemically defined components to produce the second diet (diet 2). In comparison to a control of *L. decemlineata* eggs, both artificial diets were not satisfactory for timely development and survival of *P. bioculatus* nymphs or oviposition and longevity of adults. After 11 generations, females fed diet 2 weighed more than cohorts fed the natural prey. The authors observed that larger predators resulted after feeding on artificial diet for many generations, suggesting that some adaptation had occurred resulting in greater ingestion of the diet.

Coudron and Kim (2004) cultured *P. bioculatus* on an artificial diet based on animal protein and undisclosed plant protein for 11 generations, but the diet was substandard to natural prey (*T. ni* larvae). Nevertheless, the cost associated with culturing *P. bioculatus* on an artificial diet or natural prey was about the same. A meat-based artificial diet was inferior for rearing another pentatomid, *P. bidens*, as compared to natural and factitious prey (Mahdian et al., 2006). Adult females were heavier on lepidopteran larvae (the cotton leafworm, *S. littoralis*, and the greater wax moth, *G. mellonella*) than on the artificial diet. Total fecundity of *P. bidens* on *S. littoralis* larvae was superior to that on *G. mellonella*; *P. bidens* did not produce eggs when fed the artificial diet.

The Neotropical pentatomid *P. nigrispinus* benefited less when fed an artificial diet rather than when given natural or factitious prey (Lemos et al., 2003). Nymphs of the predator reared on an artificial diet weighed significantly less than their counterparts fed cotton leafworm (*A. argillacea*) larvae. Furthermore, the fresh weight of ovaries from newly emerged *P. nigrispinus* females was lower for individuals reared on an artificial diet than for those produced on yellow mealworm (*T. molitor*), housefly (*M. domestica*), or cotton leafworm larvae. Lemos et al. (2005) confirmed that food source has a profound impact on ovarian development; this study showed that newly emerged *P. nigrispinus* females had ovaries containing oocytes in an advanced, intermediate, or early stage of development when fed *A. argillacea* larvae, *T. molitor* or *M. domestica* larvae, or an artificial diet, respectively.

Given their economic importance as biocontrol agents, there has been extensive work on artificial diets for *Orius* spp. Zhou and Wang (1989) designed a diet to sustain development of *O. sauteri* composed of brewer's yeast, egg yolk, a soy protein acid hydrolysate, sucrose, honey, palmitic acid, and water. The diet yielded good nymphal survival, but development was prolonged as compared to natural and factitious prey. Ferkovich and Shapiro (2004a) reared *O. insidiosus* on a diet based on that of Zhou and Wang (1989) supplemented with cells from an embryonic cell line, IPLB–PiE, of the Indian meal moth *P. interpunctella*. The diet was offered to the bugs in Parafilm domes produced with an encapsulation device (Greany and Carpenter, 1998). It was found that supplementation using the PiE cell line increased egg production. Oviposition rate increased as the concentration of cells added to the diet increased from 0.25 to 0.75 ml of PiE cells/ml of artificial diet. A follow-up study investigated the effects of insect prey-derived and non-insect-derived supplements, added to the Zhou and Wang (1989) diet, on the egg production and oviposition rates of *O. insidiosus* (Ferkovich and Shapiro, 2004b). Proteins derived from *P. interpunctella* eggs were found to be superior for *O. insidiosus* in terms of egg production and oviposition as compared to proteins from vertebrates (bovine serum albumin, chicken liver, beef liver, and chicken egg albumin). The concentration of *P. interpunctella* egg proteins necessary for satisfactory oviposition rates was 80- to 800-fold less than that of vertebrate-derived proteins needed for the same purpose (Ferkovich and Shapiro, 2004b).

Ferkovich and Shapiro (2005) fractionated *E. kuehniella* eggs to identify the most active fraction via feeding bioassays with *O. insidiosus*. They noted that egg production of the predator increased only when a fraction with a pH of 5 was included. According to the authors, this finding points to the presence of a specific nutritional factor in certain lepidopteran eggs that enhances the fecundity of predators. Ferkovich and Lynn (2005) supplemented the artificial diet of Zhou and Wang (1989) with cells from an embryonic cell line, Ek-x4V, from *E. kuehniella* eggs. Adults of *O. insidiosus* fed the supplemented diet had higher oviposition rates than those fed the unmodified artificial diet, but their oviposition rates were inferior to those of adults fed *E. kuehniella* eggs. Likewise, oviposition of *O. insidiosus* was enhanced when an artificial diet was supplemented with protein from *E. kuehniella* eggs in comparison to being unmodified (Ferkovich et al., 2007). Yet the modified artificial diet was substandard to intact *E. kuehniella* eggs.

Lee and Lee (2004) reared the anthocorid *O. strigicollis* on an artificial diet composed of powdered pupae of the Chinese oak silkworm, *Antheraea pernyi*

(Guérin-Méneville); beef liver; beef powder; and chicken egg yolk. When fed this diet, development time and survival of the predator were 14 days and 68.5%, respectively. Total fecundity averaged 82.5 eggs during an oviposition period of 18 days. Unfortunately, the authors did not compare the performance of the predator on the artificial diet versus diets of natural or factitious prey.

Arijs and De Clercq (2004) used a deletion and addition approach to modify the meat-based diet developed by De Clercq et al. (1998a) for *P. maculiventris* and used it for rearing *O. laevigatus* (Fig. 3.3). They found that beef liver was the most important component to support development and reproduction of the anthocorid, whereas egg yolk, ground beef, ascorbic acid, and sucrose were of minor importance. Predators supplied with the different artificial diets tested in this study developed slower than those given *E. kuehniella* eggs, but their adult weights were similar. Fecundity of females provided with diets containing beef liver was similar to or somewhat lower than that of females fed lepidopteran eggs. Oviposition rate and egg hatch were not affected by diet. Bonte and De Clercq (2008) compared the values of different artificial diets to those of two factitious foods, *E. kuehniella* eggs and *A. franciscana* cysts, for the rearing of *O. laevigatus*. Overall, the factitious foods proved superior to the artificial diets, and a meat and liver diet yielded better results than three diets based on egg yolk. Within the egg yolk diets, the developmental fitness of the anthocorid varied proportionally with the amount of egg yolk present in the diet. Bonte and De Clercq (2011) reported that access to a bean pod had a positive influence on the development of *O. laevigatus* nymphs reared on an egg yolk-based artificial diet.

Artificial diets have been tested for mirid predators, with variable results. Grenier et al. (1989) modified an artificial diet developed for *Trichogramma*

FIGURE 3.3 Adult of *Orius laevigatus* feeding on a meat-based artificial diet wrapped in Parafilm. *Photo: Y. Arijs.* (For color version of this figure, the reader is referred to the online version of this book.)

egg parasitoids for the rearing of *M. caliginosus* nymphs. They did so by supplementing a basic diet containing organic acids, amino acids, sugars, minerals, vitamins, albumin, and egg yolk with different vertebrate sources of proteins and lipids. The artificial diets were offered in floral foam enveloped in Parafilm. Best results were obtained when the basic diet was supplemented with chicken embryo extract, yielding a nymphal survival of 21% (versus 33% in the control on *E. kuehniella* eggs). However, when the latter artificial diet was supplemented with leaves of *Pelargonium peltatum* L., 62% of the nymphs reached the adult stage. Similarly, development and fecundity of the related species *M. pygmaeus* were better on an egg yolk-based diet supplemented with honeybee pollen than on the artificial diet alone, indicating the importance of plant feeding for these mirids.

Castañé and Zapata (2005) tested the value of a meat-based artificial diet for culturing *M. caliginosus*. Predators reared on the meat diet had smaller hind tibiae, weighed less, and took longer to develop than their counterparts reared on *Ephestia* eggs. Nevertheless, the artificial diet allowed the production of continuous generations of the mirid in the absence of plant material (as a moisture source or oviposition substrate). Vandekerkhove et al. (2006) investigated whether artificial diets based on hen's egg yolk could provide adequate nutrition for the development and reproduction of *M. pygmaeus* (named *M. caliginosus*). They reported that the artificial diets, encapsulated in Parafilm domes (Fig. 3.4), resulted in prolonged development and lower adult weights as compared with *E. kuehniella* eggs, but survival was similar. Further, dissection of females at day 7 indicated that females fed *Ephestia* eggs had more developing eggs (oocytes) in their ovaries than those fed an artificial diet. Callebaut et al. (2004) also found that *Ephestia* eggs were more suitable than artificial diets for *M. caliginosus* reproduction via an examination of egg load.

FIGURE 3.4 Nymph of *Macrolophus pygmaeus* feeding on a Parafilm dome containing an egg yolk diet. *Photo: T. Machtelinckx.* (For color version of this figure, the reader is referred to the online version of this book.)

In the studies by Callebaut et al. (2004) and Vandekerkhove et al. (2006) on *Macrolophus* bugs and by Bonte and De Clercq (2008) on *O. laevigatus*, egg load correlated significantly with oviposition rate. This result indicated that counting oocytes in females in the early adult stage constitutes a reliable and fast tool for assessing the reproductive potential of these and perhaps other predatory bugs.

Iriarte and Castañé (2001), Castañé et al. (2002), and Zapata et al. (2005) conducted diet-related research on *D. tamaninii*, another mirid predator of whiteflies and other small arthropods found in the Mediterranean area. Iriarte and Castañé (2001) found that *D. tamaninii* is amenable to rearing, during its nymphal and adult stages, on a meat-based artificial diet devoid of plant material. Despite an initial reduction of survival of individuals in the first generation, an increase in survival occurred with succeeding generations reared on the artificial diet. In an attempt to increase the suitability of a meat-based artificial diet for the mirid, Zapata et al. (2005) conducted a biochemical analysis of the carcasses of diet-reared adults. They detected nutritional deficiencies in the original diet formulation by comparison of the total, free amino acid, and lipid profiles of females reared on the meat diet versus control females fed *E. kuehniella* eggs. To remediate the nutritional deficiencies, Zapata et al. (2005) added new sources of protein (aspartate and casein) and fatty acids (soybean oil) to the original diet, and more water to lower the concentration of all components and improve diet acceptability. This reformulated artificial diet was more beneficial to the predators, as evidenced by an increase in survival of nymphs and body weight of adults. Adults fed the reformulated diet had hind tibia of similar lengths as cohorts reared on *Ephestia* eggs. Oviposition rates did not differ among control and artificial diets.

Firlej et al. (2006) observed promising effects of artificial diets on another mirid, *Hyaliodes vitripennis* (Say), which is an indigenous generalist predator of several arthropod pests, including mites in apple orchards in North America. Two meat-based diets (originally designed for the coccinellid *Coleomegilla maculata* (DeGeer) and the chrysopid *Chrysoperla rufilabris* (Burmeister)) increased nymphal survival and adult longevity of the predator in comparison to individuals reared on (one of its) natural prey, the two-spotted spider mite *Tetranychus urticae* Koch. Adults reared on the coccinellid diet had higher oviposition rates than those reared on the chrysopid diet or on *T. urticae*. It is not clear whether better performance of the mirid on the artificial diet as compared with *T. urticae* means that the diet is optimal for reproduction or that *T. urticae* is a poor natural prey. The addition of a possible phagostimulant, the plant sterol β-sistosterol, to the coccinellid diet decreased nymphal mortality and slightly increased weight gain in comparison to cohorts reared on the coccinellid diet alone. The addition of sucrose to the coccinellid diet did not have beneficial effects. The authors concluded that further diet improvements are necessary to increase the egg hatch and fecundity of females reared on either one of the meat diets from the first instar.

Finally, reduviid predators have been cultured on artificial diets. Sahayaraj et al. (2006) reared *Rhynocoris* spp. on a basic artificial diet containing beef extract, milk, powder and egg yolk supplemented with different proportions of insect materials, pig liver, or pig blood. Behavioral experiments showed that diets containing pig liver were preferred by the predators.

In summary, artificial diets tested for culturing predatory bugs were variable in composition, but it usually consisted of vertebrate protein from beef, chicken, and, infrequently, pig. One study tested a combination of chicken liver and tuna fish base components (Rojas et al., 2000). Beef liver was a component in the artificial diet of eight of nine species, and hen's egg (whole or yolk only) was a component in the diet of six of nine species. In most cases, artificial diets were inferior as a stand-alone food source of similar effectiveness when compared to factitious or natural prey for predatory bugs. Rarely were artificial diets as suitable as (e.g. Zapata et al., 2005), let alone more suitable than (Firlej et al., 2006), factitious or natural prey to support the development and reproduction of predatory bugs.

3.2.3.3. *Effect on Predation Potential*

Although developmental and reproductive parameters are useful to assess the quality of a natural enemy produced on an artificial food, the ultimate quality parameter is its effectiveness as a biological control agent. Demonstrating that a predator retains the capacity to capture, kill, and consume live (target) prey, despite continuous culturing on an (inanimate) artificial diet, is indeed one of the most relevant measures of predator quality (Grenier and De Clercq, 2003).

De Clercq and Degheele (1993a) cultured the pentatomids *P. maculiventris* and *P. nigrispinus* (*P. sagitta*) for over 15 consecutive generations exclusively on artificial larvae consisting of a meat-based artificial diet encapsulated in stretched Parafilm. When presented with live prey, the nymphs and adults of both species had similar predation rates on caterpillars of *S. exigua* as their counterparts maintained on *G. mellonella* larvae. Similarly, Chocorosqui and De Clercq (1999) reared *P. maculiventris* for three consecutive generations on a similar meat-based artificial diet. Despite the smaller size of nymphs and adults reared on an artificial diet rather than factitious prey (*G. mellonella* larvae), the predation rate was unaffected. Diet-reared predators killed as many or even slightly more beet armyworm (*S. exigua*) larvae compared to cohorts on factitious prey. Higher predation rates of artificial diet-reared predators versus those reared on factitious prey were also reported for *P. nigrispinus* (Saavedra et al., 1997).

Predation potential of the big-eyed bug, *G. punctipes*, was compared between "domesticated" (i.e. laboratory-reared on an artificial diet for 60 continuous generations) versus F_1 progeny of feral (wild) females (Cohen, 2000a). Progeny of wild females were fed *H. virescens* eggs and heat-killed *S. exigua* larvae in the laboratory. In the predation experiments, females in both treatments were supplied with *H. virescens* larvae or pea aphids (*A. pisum*). Although domesticated females were smaller than wild females, consumption rates were approximately the same between the two treatment groups, regardless of the prey species (Cohen, 2000a).

Castañé et al. (2002) proved that *D. tamaninii* cohorts on a meat diet were just as effective as those reared on *E. kuehniella* eggs at killing two of its main target prey, the greenhouse whitefly (*Trialeurodes vaporariorum* (Westwood)) and the cotton aphid (*Aphis gossypii* Glover). Castañé and Zapata (2005) reported that *M. caliginosus* adults of the seventh generation of rearing on a meat-based diet and on artificial moisture and oviposition sources were as adept at killing the greenhouse whitefly (*T. vaporariorum*), sweet potato whitefly (*Bemisia tabaci* (Gennadius)), and two-spotted spider mite (*T. urticae*) as those reared on *E. kuehniella* eggs. Vandekerkhove et al. (2011) found that fifth instars of another mirid, *M. pygmaeus*, had lower body weights when fed on egg yolk-based artificial diets, but killed similar numbers of prey (*M. persicae nicotianae* Blackman) as their counterparts fed *E. kuehniella* eggs. The predators consumed most of the contents of their prey. Likewise, Bonte and De Clercq (2010b) reported that despite a lower body weight, *O. laevigatus* fifth instars and adults reared on an egg yolk-based artificial diet killed as many second instars of the target prey, *Frankliniella occidentalis*, as their peers cultured on *E. kuehniella* eggs.

Consumption of a meat-based artificial diet did not reduce the capacity of the reduviid *R. marginatus* to kill its natural prey *S. litura* in the laboratory (Sahayaraj and Balasubramanian, 2009). Although the authors did not describe the artificial diet, pig liver, beef extract, and egg yolk were the likely sources of protein in it, since one of the authors at the same research center used a meat-based diet of these components to rear *R. marginatus* (see Sahayaraj et al., 2006).

These studies suggest that even after long-term rearing on inanimate artificial diets, heteropteran predators have little difficulty switching to natural or target prey and do not have impaired predation abilities as compared with counterparts maintained on nutritionally optimal factitious prey. Several of these studies also indicate that body size (weight) is not a reliable predictor of a predator's killing capacity. However, it is noteworthy that in most of the studies discussed here, the factitious food also consisted of immobile insect materials. Henaut et al. (2000) reported that adults of the anthocorid *O. majusculus*, which were reared in their nymphal stages on *E. kuehniella* eggs, were less effective as predators of pea aphids than their aphid-reared peers, indicating that experience with an inanimate alternative food may generate behavioral changes that affect the predator's efficacy as a biological control agent.

3.2.3.4. Challenges for the Practical Use of Artificial Diets

In 1999, Cohen et al. pointed out that despite several decades of research on artificial diets for entomophagous insects and mites, no commercial mass-rearing programs were in fact using them. In spite of additional research and a few patent applications for artificial diets that could also be used for the production of predatory bugs (e.g. Cohen, 1999, 2003; White et al., 2001), the situation has remained largely unchanged. This is in sharp contrast to the extensive rearing of phytophagous insects on artificial diets. Besides the lower quality of predators reared on artificial diets, as outlined in this chapter, there are many practical issues that complicate the adoption of artificial diets in commercial production systems for entomophagous insects (aptly reviewed by Cohen et al., 1999; see also Chapter 7).

A major practical downside of artificial diets proposed for culturing predatory heteropterans is their high propensity for chemical and microbial spoilage. Cohen et al. (1999) noted that the extra-oral digestion of predatory bugs tends to accelerate spoilage because of microbial and enzymatic contamination. Removal of spoiled feeds from the rearing containers is time-consuming and increases production cost. The problem can be alleviated by sterilizing the diet or by incorporating antimicrobials and antioxidants (see Cohen, 2004, for detailed methods). For instance, packages of a meat and liver diet using Parafilm as wrapping material require replacement on a daily basis to prevent decline in nymphal survival in *O. laevigatus*. When gentamycin sulfate was added to the diet at 0.05%, diet packages could be kept in the rearing containers for 3 days without detrimental effects on the development of the predator (Arijs et al., 2002). Inglis and Cohen (2004) reported that microbial spoilage of meat diets was primarily caused by lactic acid bacteria; these authors also found that adding the antibacterial agents streptomycin sulfate and chlortetracycline and the antifungal agents propionic acid and potassium sorbate to a meat-based diet was highly effective in reducing growth of the microbial populations associated with the different organic components of the diet. Also, fresh egg yolk-based diets encapsulated in Parafilm (e.g. Vandekerkhove et al., 2006) easily spoil and need to be replaced at least every 2 days for optimal results with *Macrolophus* and particularly *Orius* bugs. In order to extend its shelf life, Goussaert (2003) lyophilized an egg yolk-based diet and rehydrated it before feeding to *O. laevigatus*. Developmental performance of the anthocorid was similar on fresh and rehydrated diets.

More work is needed on presentation of diets for predatory heteropterans. Most, if not all, of the diets tested for predatory bugs were encased in (stretched) Parafilm. This waxy film is readily accepted by various hemipterans (carnivorous and herbivorous species alike) as a feeding membrane. The Parafilm coating provides a firm surface that facilitates the penetration and stabilization of the piercing–sucking mouthparts of predatory bugs (Cohen, 1990; De Clercq et al., 1998a). Furthermore, it appears to slow down spoilage and dehydration of the encapsulated diet. However, packaging of diets with Parafilm has proven labor intensive even when using semiautomated systems, like the encapsulation device described by Greany and Carpenter (1998). More effective forms of encapsulation could stimulate the use of artificial diets, but few successes have been reported for heteropteran predators. De Clercq et al. (1998a) developed a gelled form of a meat-based diet for *P. maculiventris*, but results were inferior to packaging the diet in Parafilm sheets. Tan et al. (2010) microencapsulated an artificial diet for *O. sauteri* using sodium alginate and chitosan. They reported the best microencapsulation results with 1% sodium alginate, 0.8% chitosan, and a 1:1 proportion of core material to wall-forming material. The predator readily accepted the microcapsules, but nymphs developed slower and females were shorter lived and less fecund than those fed on a control diet of spider mites. Hydrocapsules (Agricultural Research Service, Gainesville, FL, USA) have been proposed for encapsulating artificial media, but besides a patent (Toreki et al., 2004), there is no published record of their efficacy for rearing predatory bugs. Finally, Ferkovich et al. (2007) pointed out that besides the diet containment method, also the size and shape of individual diet units may have a direct effect on the performance of a predator in an artificial rearing system.

Several of the studies discussed in this chapter indicate that phagostimulant components derived from insect prey or plants may enhance the acceptability of artificial diets to heteropteran predators. Indeed, a lack of feeding stimulants may be at least as crucial as a lack or poor balance of nutrients in a predator's diet (De Clercq et al., 1998a; Cohen, 2004; Ferkovich et al., 2007). Interestingly, Torres et al. (1997) and Saavedra et al. (2001) showed that brushing over the Parafilm casing of artificial larvae with a crude extract of the predator's (male) dorso-abdominal glands (DAG) or with synthetic DAG–pheromone reduced the time for location and feeding on the artificial larvae by *P. nigrispinus*. This discovery illustrates the potential of semiochemicals to improve the attractiveness and acceptability of artificial diets for heteropteran predators and enhance their practical value in production systems.

3.3. PLANT MATERIALS AND ALTERNATIVES

3.3.1. Plant Substrates

The zoophytophagous predators of the Heteroptera order display varying degrees of plant feeding. They probe plants in the first place to fulfill their need for water: in many cases, water extracted from prey is insufficient to meet the predators' metabolic needs, and water may be required to enable extra-oral digestion. In addition, some predatory bugs acquire supplementary nutrients from plants (Naranjo and Gibson, 1996; Coll, 1998; Gillespie and McGregor, 2000; Coll and Guershon, 2002; Lundgren, 2009, 2011). Furthermore, the female adults of predatory heteropterans may use plants as oviposition substrates (Coll, 1998; Lundgren, 2011).

The influence of supplementary plant materials on the developmental and reproductive performance of predatory bugs varies among studies (Naranjo and Gibson, 1996; Coll, 1998; Lundgren, 2009, 2011). The benefit of plant feeding appears to be dependent on the quality of the insect prey and plant material used, and it can be species or stage specific (Naranjo and Gibson, 1996). The suboptimal nutritional value of certain artificial diets may be compensated in part by feeding on plant material, like *Pelargonium* leaves or potato sprouts with *M. caliginosus* (Grenier et al., 1989; Castañé and Zapata, 2005), pollen with *M. pygmaeus* (Vandekerkhove and De Clercq, 2010), and bean pods with *O. laevigatus* (Bonte and De Clercq, 2011).

Overall, supplementing a prey-only (insect) diet with plant material accelerates nymphal development, increases nymphal survival and adult longevity, and enhances fecundity (Coll, 1998; Malaquias et al., 2010; Lundgren, 2011; Zanuncio et al., 2011). However, it is less clear whether plant material is a source of nutrients versus a source of free water. For instance, no differences in the developmental and reproductive parameters of *G. punctipes* were detected between individuals fed bean pods as a supplement to insect eggs and those provided with a supplementary source of free water (Cohen and Debolt, 1983). Gillespie and McGregor (2000), however, found that nymphs of the mirid *Dicyphus hesperus* Knight provided with *E. kuehniella* eggs and tomato leaves had shorter developmental times than those supplied with flour moth eggs and wet cotton. Pods of green bean (*Phaseolus vulgaris* L.) are routinely used as a moisture source and oviposition substrate, in addition to insect prey, in *Orius* spp. cultures. Note that green bean pods (as compared to free water) can have positive, neutral, or negative effects on the development and/or reproduction of *Orius* spp. (Naranjo and Gibson, 1996; Bonte and De Clercq, 2010a, 2011; Bonte et al., 2012); negative effects may, in part, result from chemical treatment of the beans.

Some species of anthocorids and mirids benefit from feeding on pollen (Salas-Aguilar and Ehler, 1977; Kiman and Yeargan, 1985; Richards and Schmidt, 1996a; Lundgren, 2009; Vandekerkhove and De Clercq, 2010; Bonte et al., 2012). Pollen (usually bee-collected pollen) is included in the diet of some cultured species (e.g. Ferkovich et al., 2007; Venkatesan et al., 2008; Bonte et al., 2012). However, research shows that there is considerable variation in the performance of anthocorid predators (mainly *Orius* spp.) on pollen. This could be due to differences in the nutritional quality and defensive properties of the different pollen species involved in the studies (Richards and Schmidt, 1996a; Lundgren, 2009). In addition, the use of pollen in insect cultures has some practical drawbacks as its quality tends to deteriorate quickly, and, particularly when offered fresh, it is prone to fungal contamination.

For heteropteran predators that insert their eggs into the plant tissue (anthocorids, mirids, and nabids), plants or plant parts are an essential part of the production system. Several plant materials are oviposition substrates for anthocorid, mirid, and nabid predators. These include pods and stems of green bean, plant seedlings (e.g. cotton, soybean, and sharp pepper), sprouts (e.g. potato, soybean, and alfalfa), and inflorescences (e.g. farmer's friend, *Bidens pilosa* L.) for anthocorids and nabids (Richards and Schmidt, 1996b; Coll, 1998; Bueno et al., 2006; Ito, 2007; Bonte and De Clercq, 2010a). Tobacco plants or leaves have been used for predatory mirids (e.g. Fauvel et al., 1987; Constant et al., 1996; Sanchez et al., 2009; Vandekerkhove et al., 2011). A number of studies have directly compared the suitability of different plant materials as oviposition substrates for predatory bugs. Braman and Yeargan (1988) noted that *Nabis* bugs tended to deposit more eggs in soybean seedlings than in green beans. Zhou et al. (1991) noted a higher oviposition rate of *O. sauteri* in soybean sprouts than in bean pods or shoots of *Forsythia suspensa* (Thunb.) Vahl. Alauzet et al. (1992) reported that *O. majusculus* deposited more eggs in leaves of geranium (*P. peltatum*) than in those of ivy (*Hedera helix* L.). Constant et al. (1996) reported similar oviposition rates by *M. caliginosus* in different parts of tobacco and geranium. Richards and Schmidt (1996b) found that although bean stems, bean pods, and potato sprouts are all suitable oviposition substrates for *O. insidiosus*, bean stems were preferred by the predator in choice tests. Moreover, the bean stems were less susceptible to mold than bean pods. The mold negatively affected the survival of hatchlings. Nonetheless, pods of green bean remain the most widely used material for mass rearing of *Orius* spp.

Besides their beneficial role as moisture sources or oviposition substrates, plant materials in predator cultures may reduce cannibalism by providing hiding places (Coll, 1998; Bonte and De Clercq, 2010a; and discussed further in this chapter). Plant parts may also be useful for storage, shipping, or application of predatory bugs (Coll, 1998). For instance, buckwheat hull is used in cultures of *Orius* bugs (e.g. Thomas et al., 2012) and as a carrier (in a mixture with vermiculite) in commercial packagings of these predators. Ito (2007) furnished rearing cages for *Orius* spp. with wheat grains to prevent cannibalism and excessive moisture.

3.3.2. Artificial Substrates

In order to meet the need for plant materials, large surfaces of greenhouses or open fields must be available for growing plants, or plant materials have to be purchased on the market. Moreover, plant materials must be of good quality and free of pesticides. Thus, the requirement for plant materials reduces the cost effectiveness, reliability, and simplicity of a production system (Castañé and Zalom, 1994; Bolckmans, 2007). The elimination of plants from the rearing system of heteropteran predators requires the availability of alternative moisture sources and artificial living and oviposition substrates (Vandekerkhove et al., 2011). Most heteropterans that deposit eggs on plant surfaces in nature (pentatomids, reduviids, and geocorids) readily accept artificial substrates for oviposition, like absorbent paper toweling, mesh fabric, cotton balls, or cotton wadding, precluding the need for host plants in the rearing system.

However, the situation is more complex for heteropterans that naturally insert their eggs into plant tissues. A number of workers have proposed artificial oviposition substrates for anthocorids, mirids, and nabids. Shimizu and Hagen (1967) observed that two anthocorids (*Anthocoris antevolens* White and *Orius tristicolor* (White)), a nabid (*N. americoferus*), and a mirid (*L. hesperus*) oviposited into Parafilm-wrapped blocks of water-soaked cellulose sponge. The Parafilm had to be stretched for the anthocorids, but not for the nabid. No data were given regarding hatching rates. Parajulee and Phillips (1992) tested water-soaked cotton dental rolls as oviposition substrates for the anthocorid *Lyctocoris campestris* (F.). The predator oviposited into the rolls, but hatching of the eggs was poor. Better results were obtained with stacks of water-saturated Whatman filter papers. Castañé and Zalom (1994) developed an artificial oviposition substrate for *O. insidiosus* by covering a medium of carrageenan salt of potassium chloride (Gelcarin) with paraffin wax or Parafilm. The Parafilm coating proved inadequate, but positive results were obtained when the Gelcarin medium was covered with paraffin wax. Females successfully oviposited in the latter substrate, and rates of oviposition, egg hatch, and nymphal development were similar to those on green beans. However, the authors noted that the thickness of the paraffin layer was critical, and no eggs were found when the layer was thicker than 0.045 mm. Richards and Schmidt (1996b) reported similar problems when their agar-based oviposition packets for *O. insidiosus* were covered with either stretched or unstretched Parafilm. When the Parafilm was not stretched, the bugs were unable to penetrate the surface with their ovipositor, leaving any eggs laid on the surface to dessicate. On the other hand, when the Parafilm was stretched, the eggs did not attach to the Parafilm layer and sank into the medium, causing the nymphs to drown upon hatching. Shapiro and Ferkovich (2006) reported that water-filled domes from Parafilm prepared with an encapsulation apparatus sufficed to harvest small quantities of viable eggs of *O. insidiosus*; however, the eggs had to be extracted from the domes to allow hatching. Constant et al. (1996), Iriarte and Castañé (2001), and Castañé and Zapata (2005) used moistened dental cotton rolls wrapped in stretched Parafilm as oviposition substrates for mirid predators. Constant et al. (1996) reported that egg laying by *M. caliginosus* was substantially increased when spraying the substrates with an ethanol extract of *Inula viscosa* (L.) Ait., a preferred host plant of this mirid. Iriarte and Castañe (2001) produced more than five continuous generations of *D. tamaninii* using artificial larvae containing a meat diet as food and Parafilm-wrapped dental cotton rolls dipped in green bean extract as oviposition substrates. Castañé and Zapata (2005) succeeded in rearing seven generations of *M. caliginosus* using the same artificial oviposition substrate and diet, without any detrimental effects on the bug's predation abilities. Vandekerkhove et al. (2011) cultured *M. pygmaeus* on *E. kuehniella* eggs but without plants, using an oviposition substrate modified from that of Constant et al. (1996) for over 30 consecutive generations. Parameters of development, reproduction, and predation capacity of predators from the plantless rearing system were similar to those of predators from a control group maintained on tobacco leaves. Using similar methods, De Puysseleyr et al. (2013) succeeded in producing continuous generations of another mirid, *N. tenuis*, without plants (Fig. 3.5). Plantless rearing of this mirid for five generations using *E. kuehniella* eggs as food led to a prolonged preoviposition period and lower egg-hatching rates and adult weights, but other developmental and reproductive parameters were not adversely affected.

FIGURE 3.5 *Nesidiocoris tenuis* female ovipositing in an artificial substrate. *Photo: V. De Puysseleyr and T. Machtelinckx.* (For color version of this figure, the reader is referred to the online version of this book.)

Free water is often supplied via moistened cotton wool or paper toweling. Parafilm domes produced with an encapsulation apparatus (Greany and Carpenter, 1998) have also been used as a source of free water for *Orius* and *Macrolophus* bugs (e.g. Ferkovich and Shapiro, 2004a,b; Bonte and De Clercq, 2010a; Vandekerkhove et al., 2011; Bonte et al., 2012). An elegant way of supplying water is via hydrocapsules, which are capsules with a polymeric outer coating produced in different sizes (Toreki et al., 2004). Hydrocapsules with a diameter of 1–2 mm have been used to supply water or sucrose solutions to anthocorid and mirid bugs (Fig. 3.6) (e.g. Shapiro and Ferkovich, 2006; Shapiro et al., 2009; Urbaneja-Bernat et al., 2013).

Under natural conditions, most predatory heteropterans used in biological control live (i.e. forage, feed, mate, and rest) on plant surfaces, whereas these may (in part) be lacking in cultures. Absorbent paper toweling has been widely used as a living substrate in culture containers of predatory bugs. When wadded or shredded, the paper toweling increases the surface area and spatial complexity of the rearing environment, providing hiding places for the predators. Furthermore, the paper toweling may allow the absorption of fluids secreted by the predators or their prey (Cohen, 1985; De Clercq et al., 1988). Other paper-based materials, like honeycomb paper and corrugated cardboard strips, have been used to create a living environment for predatory bugs in culture (e.g. Parajulee and Phillips, 1992; Blümel, 1996; Bueno et al., 2006). Absorbent paper materials may not always be an ideal living substrate for predatory bugs. Bonte and De Clercq (2010a) demonstrated that survival of *O. laevigatus* on lipophilic surfaces such as wax paper was higher than on absorbent household paper. Waxy paper materials were also used in the rearing of other *Orius* spp. (e.g. Ferkovich

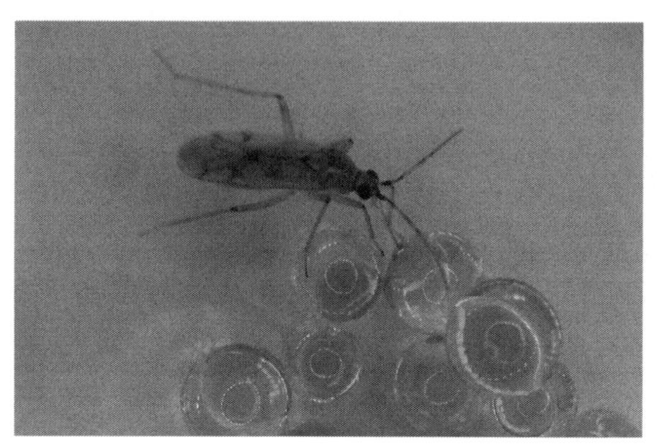

FIGURE 3.6 *Nesidiocoris tenuis* female feeding on a water-containing hydrocapsule. *Photo: P. Urbaneja-Bernat.* (For color version of this figure, the reader is referred to the online version of this book.)

and Shapiro, 2004a; Bonte et al., 2012) and of *M. pygmaeus* (Vandekerkhove et al., 2011). Finally, Blümel (1996) also reported good results when using vermiculite as a substrate in both culture and storage containers of *Orius* spp.

3.4. CROWDING AND CANNIBALISM

Another important consideration for the production of predatory bugs is rearing density. Crowding leads to a higher degree of competition for food and space and increased stress, which in turn may lead to lower developmental or reproductive success and eventually result in mortality, which may or may not be due to cannibalism. Particularly in populations with overlapping life stages, cannibalism is not uncommon, and older or larger predators may look upon younger and smaller conspecifics (including eggs) as potential prey (Fig. 3.7) (Arbogast, 1979; Tommasini et al., 2002; Rudolf, 2007; Bonte and De Clercq, 2011; Bueno and van Lenteren, 2012). Age synchronization is therefore an important first measure to reduce cannibalism in mass cultures of predatory bugs (De Clercq and Degheele, 1993b; van den Meiracker, 1999; Bueno and van Lenteren, 2012). Cannibalism can also be avoided to some extent by lowering rearing density in each subsequent developmental stage (i.e. less individuals per container) or by providing more refugia in the rearing arena, using plant materials or artificial substrates (as discussed here) (i.e. less individuals per unit of rearing space). Whereas some studies have reported a reduction of cannibalism in predatory bugs with increasing levels of food supply (Arbogast, 1979; Grundy et al., 2000; Tommasini et al., 2002), cannibalism has been observed even when nutritionally adequate food is abundant. Bonte and De Clercq (2011) reported that providing ad libitum *E. kuehniella* eggs did not prevent cannibalistic behavior at high nymphal densities in *O. laevigatus*. The higher attractiveness of the mobile conspecific food versus that of the immobile heterospecific (factitious) food was believed to be the

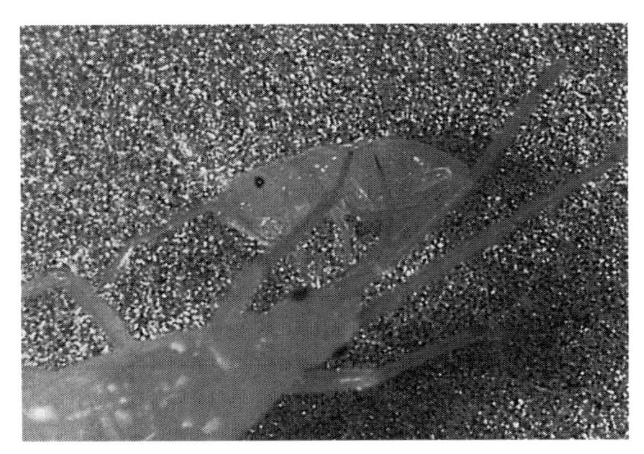

FIGURE 3.7 Cannibalism in *Macrolophus pygmaeus. Photo: V. De Puysseleyr and B. Vandekerkhove.* (For color version of this figure, the reader is referred to the online version of this book.)

reason for this behavior in this and other species of predatory bugs (Grundy et al., 2000; Bonte and De Clercq, 2011). Whereas some studies have noted higher levels of cannibalism with nutritionally suboptimal food sources like artificial diets (e.g. Cohen, 1985; De Clercq and Degheele, 1992, 1993b), other studies have not observed a clear relationship between diet quality and cannibalism in predatory bugs (Leon-Beck and Coll, 2007; Bonte and De Clercq, 2011). Bueno and van Lenteren (2012) concluded that cannibalism is usually not a serious obstacle to mass culturing predatory heteropterans when food and moisture are available.

3.5. MICROORGANISMS

There is very little published information on the impact of microorganisms on the performance of rearing systems for heteropteran predators. Entomopathogenic fungi of the genus *Entomophthora* may cause problems in the production and application of *Macrolophus* bugs (e.g. Nedstam, 2002). However, effective humidity control is sufficient to prevent most fungal problems in the production line (J. Klapwijk, Koppert BV, personal communication).

There is an increase in research on the role of bacterial endosymbionts in beneficial insects. An extensive survey of the presence of *Wolbachia* in 134 species of Japanese terrestrial heteropteran bugs revealed a high infection frequency of approximately 35%, including nabid and lygaeid predators (Kikuchi and Fukatsu, 2003). Machtelinckx et al. (2012) reported that *M. pygmaeus* and *M. caliginosus* were infected with *Wolbachia* and *Rickettsia. Wolbachia* was responsible for strong cytoplasmic incompatibility (CI) effects in *M. pygmaeus* (Machtelinckx et al., 2009). Removal of the endosymbionts in *M. pygmaeus* was, however, not associated with significant fitness effects (Machtelinckx et al., 2012). Watanabe et al. (2011) reported that the anthocorid *O. strigicollis* had two strains of CI-inducing *Wolbachia.*

The presence of CI-inducing endosymbionts may have important practical implications for the production of these predatory bugs: incompatibility may lead to a strong suppression of population growth when populations with a different infection status interact. This may happen at the production level, when a mass culture is infused with insect material of a different infection status, or in the field, when growers release predators from different (commercial) sources in their crop. Machtelinckx et al. (2009) concluded that a survey for the presence of *Wolbachia* in commercial and wild populations of economically important predatory heteropterans is therefore highly warranted.

In conclusion, when establishing a colony of predatory heteropterans or infusing it with field-collected insects to increase genetic heterogeneity, it is imperative to check the health status of the feral material. Colonies of predatory bugs face relatively few problems with diseases, but sanitation procedures should be part of the rearing routine. Besides paying attention to entomopathogens, it may be wise to check for the presence of CI-inducing endosymbionts. Finally, a more profound knowledge of the gut microbiota may assist in optimizing foods for production of predatory heteropterans (see Chapter 7).

3.6. BREEDING AND COLONY MAINTENANCE

To date, production of heteropteran predators has focused on producing mass numbers wherein the quality, or fitness is as close as possible to that found in nature. Selection for improved performance and reduced production costs would be most desirable, and improving host ranges may be attainable. However, production traits are likely to be complex, multigene traits. Information on genes underpinning these desirable traits is lacking. This will require knowledge of the genetic diversity, gene sequence, and developing phenotypic and genetic markers that identify traits associated with specialized lines.

The study by De Clercq et al. (1998b) indicated heritable genetic variation in developmental rate and body size of *P. maculiventris*. Selection for faster development was useful for reducing costs of production, but selection for larger body size was less rewarding because body size was unrelated to fecundity or predation capacity. Mohaghegh-Neyshabouri et al. (1996) observed a lack of correlation between body weight of females and their fecundity in *P. maculiventris* and *P. nigrispinus*.

Amplified fragment-length polymorphism (AFLP), used to assess genetic variability within or between populations, showed substantial variability within a field population of *P. maculiventris*, which suggests opportunity for trait selection (Kneeland et al., 2012). A significant amount of the genetic variability found in the field population was retained after it had been maintained in the laboratory for over 100 generations, resulting in little genetic difference between the field and laboratory populations. Retention of genetic variability after extended laboratory rearing provides gene-based evidence for retention of fitness attributes (such as fecundity, longevity, and predation capacity) in mass-produced predators. The greatest variability was between populations from different geographical locations (Kneeland et al., 2012). These results indicate that sufficient genetic variation does exist within a field or laboratory population to support breeding efforts for this heteropteran predator, and that crosses among populations may provide additional opportunities to enhance breeding efforts.

Knowledge of the changes in gene expression associated with specific traits can assist in breeding programs. For example, a *Drosophila* cDNA microarray comparison of gene expression between adult females of the heteropteran predator *P. bioculatus* identified six genes associated with decreased fecundity and one gene associated with increased fecundity (Coudron et al., 2006). The most pronounced differentially expressed gene was actin, which was upregulated in low-fecundity females. Other upregulated genes in low-fecundity females included a protein kinase, a transcription factor, and several genes involved in cellular development. The one upregulated gene in high-fecundity females was an RNA-processing gene, which is consistent with the finding that high-fecundity females had nearly twice the amount of RNA. Perhaps molecular markers such as these will serve as early indicators of positive or negative responses that correlate with physiological traits such as fecundity, and be used to direct and accelerate breeding programs.

Mackauer (1976) recommends that one should initiate a colony with a sufficient number of founders and maintain a minimum population level to limit the effects of inbreeding or genetic drift. De Clercq et al. (1998b) studied the effect of inbreeding in the pentatomid predator *P. maculiventris* throughout 30 generations following introduction in the laboratory. There were consistent differences in body size among two inbred lines and a reference line in most generations, but fecundity did not differ among populations. Developmental, reproductive, and behavioral parameters indicated no deleterious effects of inbreeding. In generation 30, the fecundity of all lines had dropped to about half of that in generation 15, which was attributed to nongenetic factors.

Mohaghegh et al. (1998a, b) studied the effects of maternal age and egg weight on offspring performance of *P. maculiventris* and *P. nigrispinus*. They observed that in *P. maculiventris*, offspring from large eggs developed faster, weighed more, and in turn yielded larger eggs than offspring from small eggs. Furthermore, they also found that in both species, the offspring of young parents took less time to complete development and achieved heavier body weights compared to offspring of old parents. Workers recommended the use of young parents (2–4 weeks old) for egg production.

Bonte and De Clercq (2010c) found that the age of males influenced the reproductive output of *O. laevigatus* females: on nutritionally suboptimal diets, mating with newly emerged males resulted in slower oocyte development. On such suboptimal foods, they advised placing female adults together with sufficient numbers of males older than 5 days to ensure successful mating. Leon-Beck and Coll (2009), on the other hand, suggest removing all males from the rearing within 2–3 days after male emergence, as females maintained with males have lower rates of fecundity than those without males for the remainder of the oviposition period. Based on similar findings, Torres et al. (2006) recommend maintaining female-biased sex ratios in colonies of *P. nigrispinus* to reduce the costs of male maintenance.

3.7. MASS-REARING SYSTEMS

For reasons of confidentiality, there is little published information on commercial mass-rearing systems for predatory heteropterans. However, several papers have described systems for medium- to large-scale production of economically important predatory heteropterans. The procedures described in a selection of exemplary studies are summarized in this section. Major factors that were indicated to determine the success of a rearing system were climate, food quality and quantity, living and oviposition substrate, type of container, and rearing density.

Schmidt et al. (1995) produced nymphs and adults of *O. insidiosus* at 24 °C in 4.5 l zip-lock plastic bags, containing 200–600 individuals of similar age. The insects remained in the same bag from egg to reproducing adults. Eggs of both *S. cerealella* and *E. kuehniella* were offered to the bugs coated on moistened cardboard strips. Green bean pods and stems were placed in the bags to provide water and oviposition sites. Excess condensation was removed with an absorbent material. Blümel (1996) reared *O. laevigatus* and *O. majusculus* at 25 °C on frozen *E. kuehniella* eggs in polypropylene boxes with corrugated board scraps, and used cyclamen leaves as an oviposition substrate. Tommasini et al. (2004) describe a rearing procedure for *O. laevigatus* at 26 °C. The procedure started by placing about 1500 eggs deposited in green beans in 3.6 l plastic boxes furnished with buckwheat hulls. Nymphs and adults were fed twice per week with (frozen) *E. kuehniella* eggs glued on paper strips. The whole life cycle was completed in the same container. Adults were kept for oviposition for about 4 weeks, and bean pods were changed twice per week. Ito (2007) developed a simple method to mass rear *O. strigicollis* and *O. laevigatus* at 22 °C in petri dishes (9 cm diameter, 2 cm high) containing 30 g of heat-sterilized wheat grains, using (frozen) *E. kuehniella* eggs as food and a fresh soybean seedling as a water supply and oviposition substrate. This method was used to produce both species at initial nymphal densities of 400 individuals per petri dish. The insects were easily separated from the wheat grains using a sieve. Bueno et al. (2006) and Bueno and van Lenteren (2012) provide descriptions of rearing systems for *O. insidiosus*. Bueno et al. (2006) tested plastic bags (4 l), petri dishes (0.8 l), and glass jars (1.7 l) as containers at different rearing densities and a mean temperature of 26 °C. They found that both the immature and adult stages could be reared in glass jars using *E. kuehniella* eggs as food and farmer's friend inflorescences as moisture source and oviposition substrate. The amount of lepidopteran eggs to be supplied to nymphal and adult containers was based on a study by Yano et al. (2002), who calculated the minimum number of eggs needed per individual for optimal development and reproduction of the related (but smaller) anthocorid *O. sauteri*. Further improvements of this rearing method allowed the production of approximately 8000 adults per glass jar from an initial 250 eggs (Bueno and van Lenteren, 2012). Thomas et al. (2012) described the mass rearing of *Orius pumilio* (Champion) in 6.41 containers with buckwheat hulls; they were fed *E. kuehniella* eggs and hydrocapsules containing water at 25 °C, and green beans were used as a substrate for oviposition. New colonies were started weekly with approximately 18,000 eggs. Based on oviposition data and the analysis of yolk protein accumulation by enzyme-linked immunosorbent assay (ELISA), they recommended that the maximum period to maintain an adult colony for egg production should be about 1 month.

Grundy et al. (2000) present a mass-rearing method for the assassin bug *P. plagipennis*. When reared at 25 °C on a diet of heat-killed yellow mealworm (*T. molitor*) larvae, the optimum rearing density was 20–27 nymphs per 51 container, based on an assessment of nymphal mortality, developmental time, and adult body weight. The optimum density for oviposition was 16 adults per 51 container. Fecundity was better on a mixed diet of *T. molitor* and *H. armigera* than on *T. molitor* alone.

Rearing methods for *Podisus* bugs were described by De Clercq et al. (1988), De Clercq and Degheele (1993b), and Torres et al. (2006). Torres et al. (2006) provided a description of mass production methods for *P. nigrispinus* in Brazil. Cage designs for nymphs and adults were described in detail and illustrated. Up to 600 nymphs could be raised in 111 plastic trays. Adults were kept in 81 Plexiglas cages at a density of 18 individuals per cage. In order to reduce costs for the maintenance of males, a 2:1 female:male sex ratio was maintained in the adult cages. Several natural and factitious prey were used, but best results, also in terms of cost efficiency, were obtained with mealworms and housefly larvae. Rearing methods for these factitious prey were also presented.

3.8. CONCLUSIONS

Heteropteran predators probe and feed on a wide variety of insects, plants, and organic matter in nature; many are omnivorous, some polyphagous, and some specialists (Torres and Boyd, 2009). Our ability to rear these insects in small or large numbers varies from species to species for reasons that remain largely unknown. This chapter highlights the progression of efforts to rear heteropteran predators and demonstrates that researchers have amassed a large amount of information, ranging from feeding behavior to the chemical makeup of natural and artificial foods. Most information is for those insects that are effective predators or successfully reared.

Performance on alternative food sources has been compared with the insect's natural prey in regard to growth and other commonly accepted fitness parameters. Life history parameters are the most common types of information, and, in contrast, genetic information is relatively rare. Not all information is proving to be as useful as first thought. For example, several inconsistencies were found when associating increased weight and size with improved fecundity or fitness. By comparison, some information with seemingly high correlative value is rare, such as ovarian development as an indicator of dietary effect and the chemical composition of the predator to aid in diet formulation. Consequently, these parameters may have been underutilized. In all cases, we should measure nutritional parameters over several generations to detect adaptation or selection trends as well as chronic effects only detectable after prolonged feeding.

Most heteropteran predators are reared on natural or factitious prey. Yet the lure of an artificial diet remains because of desired advantages and because of numerous successes of pest insects reared on artificial diets, including some heteropteran species (e.g. *Lygus* spp.). Although several heteropteran predators have been reared for consecutive generations on artificial diets devoid of living material, no predatory heteroperans are currently mass reared commercially on an artificial diet.

Arguably, we are missing important knowledge regarding genetics, behavior, and developmental and reproductive biology in the case of heteropteran predators that would assist our production efforts on factitious prey as well as artificial food. A case could also be made for shortfalls in nutritional information (e.g. information about trace nutrients). Optimizing artificial oviposition substrates for predatory bugs will require a more thorough understanding of their reproductive ecology, with specific details of their oviposition behavior (Lundgren, 2011). Likewise, the microbiota in heteropterans is important, but analysis and preservation of existing microbiota have eluded researchers. Each of these areas represents future opportunities for improving and expanding our ability to produce heteropteran predators.

ACKNOWLEDGMENTS

We thank Juan Morales-Ramos, M. Guadalupe Rojas, and David Shapiro-Ilan for inviting us to write this chapter. Thanks also go to Yves Arijs, Jochem Bonte, Veronic De Puysseleyr, Thijs Machtelinckx, Laurian Parmentier, Pablo Urbaneja-Bernat, and Bjorn Vandekerkhove for kindly providing photographic material. The US government has the right to retain a non-exclusive, royalty-free license in and to any copyright of this chapter. This chapter reports the results of research only. Mention of a commercial or proprietary product does not constitute an endorsement of the product by the US Department of Agriculture or Ghent University.

REFERENCES

Ables, J.R., 1978. Feeding behavior of an assassin bug, *Zelus renardii*. Ann. Entomol. Soc. Am. 71, 476–478.

Adams, T.S., 2000a. Effect of diet and mating status on ovarian development in a predaceous stink bug *Perillus bioculatus* (Hemiptera: Pentatomidae). Ann. Entomol. Soc. Am. 93, 529–535.

Adams, T.S., 2000b. Effect of diet and mating on oviposition in the two spotted stink bug *Perillus bioculatus* (F.) (Heteroptera: Pentatomidae). Ann. Entomol. Soc. Am. 93, 1288–1293.

Alauzet, C., Dargagnon, D., Hatte, M., 1992. Production d'un hétéroptère prédateur: *Orius majusculus* (Het., Anthocoridae). Entomophaga 37, 249–252.

Arbogast, R.T., 1979. Cannibalism in *Xylocoris flavipes* (Hemiptera: Anthocoridae), a predator of stored-product insects. Entomol. Exp. Appl. 25, 128–135.

Arbogast, R.T., Carthon, M., Roberts Jr., J.R., 1971. Developmental stages of *Xylocoris flavipes* (Hemiptera: Anthocoridae), a predator of stored-product insects. Ann. Entomol. Soc. Am. 64, 1131–1134.

Arijs, Y., De Clercq, P., 2001. Rearing *Orius laevigatus* on cysts of the brine shrimp *Artemia franciscana*. Biol. Control 21, 79–83.

Arijs, Y., De Clercq, P., 2004. Liver-based artificial diets for the production of *Orius laevigatus*. BioControl 49, 505–516.

Arijs, Y., Vanthournout, L., De Clercq, P., 2002. Preservation of an artificial diet for rearing the predator *Orius laevigatus*. Med. Fac. Landbouwk. Toeg. Biol. Wet. Univ. Gent. 67 (3), 467–471.

Arnó, J., Castañé, C., Riudavets, J., Gabarra, R., 2010. Risk of damage to tomato crops by the generalist zoophytophagous predator *Nesidiocoris tenuis* (Reuter) (Hemiptera: Miridae). Bull. Entomol. Res. 100, 105–115.

Biever, K.D., Chauvin, R.L., 1992. Suppression of the Colorado potato beetle (Coleoptera: Chrysomelidae) with augmentative releases of predaceous stinkbugs (Hemiptera: Pentatomidae). J. Econ. Entomol. 85, 720–726.

Blümel, S., 1996. Effect of selected mass-rearing parameters on *O. majusculus* (Reuter) and *O. laevigatus* (Fieber). IOBC/WPRS Bull. 19 (1), 15–18.

Bolckmans, K., 2007. Reliability, quality and cost: the basic challenges of commercial natural enemy production. In: van Lenteren, J.C., De Clercq, P., Johnson, M. (Eds.), Proceedings of the 11th Workshop of the IOBC Global Working Group on Arthropod Mass Rearing and Quality Control. Bull. IOBC Global (3), 8–11.

Bonte, J., Vangansbeke, D., Maes, S., Bonte, M., Conlong, D., De Clercq, P., 2012. Moisture source and diet affect development and reproduction of *Orius thripoborus* and *Orius naivashae*, two predatory anthocorids from southern Afr. J. Insect Sci. 12, 1.

Bonte, M., De Clercq, P., 2008. Developmental and reproductive fitness of *Orius laevigatus* reared on factitious and artificial diets. J. Econ. Entomol. 101, 1127–1133.

Bonte, M., De Clercq, P., 2010a. Impact of artificial rearing systems on the developmental and reproductive fitness of the predatory bug, *Orius laevigatus*. J. Insect Sci. 10, 104.

Bonte, M., De Clercq, P., 2010b. Influence of diet on the predation rate of *Orius laevigatus* on *Frankliniella occidentalis*. BioControl 55, 625–629.

Bonte, M., De Clercq, P., 2010c. Influence of male age and diet on reproductive potential of *Orius laevigatus* (Hemiptera: Anthocoridae). Ann. Entomol. Soc. Am. 103, 597–602.

Bonte, M., De Clercq, P., 2011. Influence of predator density, diet and living substrate on developmental fitness of *Orius laevigatus*. J. Appl. Entomol. 135, 343–350.

Braman, S.K., Yeargan, K.V., 1988. Comparison of developmental and reproductive rates of *Nabis americoferus*, *N. roseipennis*, and *N. rufusculus* (Hemiptera: Nabidae). Ann. Entomol. Soc. Am. 81, 923–930.

Bueno, V.H.P., Mendes, S.M., Carvalho, L.M., 2006. Evaluation of a rearing method for the predator *Orius insidiosus*. Bull. Insectol. 59, 1–6.

Bueno, V.H.P., van Lenteren, J.C., 2012. Predatory bugs (Heteroptera). In: Panizzi, A.R., Parra, J.R.P. (Eds.), Insect Bioecology and Nutrition for Integrated Pest Management, CRC Press, Boca Raton, FL, pp. 539–569.

Burgess, L., Dueck, J., McKenzie, D.L., 1983. Insect vectors of the yeast *Nematospora coryli* in mustard, *Brassica juncea*, crops in southern Saskatchewan. Can. Entomol. 115, 25–30.

Callebaut, B., Van Baal, E., Vandekerkhove, B., Bolckmans, K., De Clercq, P., 2004. A fecundity test for assessing the quality of *Macrolophus caliginosus* reared on artificial diets. Parasitica 60, 9–14.

Calvo, J., Bolckmans, K., Stansly, P.A., Urbaneja, A., 2009. Predation by *Nesidiocoris tenuis* on *Bemisia tabaci* and injury to tomato. BioControl 54, 237–246.

Castañé, C., Iriarte, J., Lucas, E., 2002. Comparison of prey consumption by *Dicyphus tamaninii* reared conventionally, and on a meat-based diet. BioControl 47, 657–666.

Castañé, C., Quero, R., Riudavets, J., 2006. The brine shrimp *Artemia* sp. as alternative prey for rearing the predatory bug *Macrolophus caliginosus*. Biol. Control 38, 405–412.

Castañé, C., Zalom, F.G., 1994. Artificial oviposition substrate for rearing *Orius insidiosus*. Biol. Control 4, 88–91.

Castañé, C., Zapata, R., 2005. Rearing the predatory bug *Macrolophus caliginosus* on a meat-based diet. Biol. Control 34, 66–72.

Champlain, R.A., Sholdt, L.L., 1966. Rearing *Geocoris punctipes*, a lygus bug predator, in the laboratory. J. Econ. Entomol. 59, 1301.

Champlain, R.A., Sholdt, L.L., 1967. Life history of *Geocoris punctipes* (Hemiptera: Lygaeidae) in the laboratory. Ann. Entomol. Soc. Am. 60, 883–885.

Chocorosqui, V.R., De Clercq, P., 1999. Developmental and predatory performance of *Podisus maculiventris* (Say) (Heteroptera: Pentatomidae) reared on a meat-based artificial diet. Med. Fac. Landbouww. Univ. Gent. 64, 229–234.

Chu, Y.I., 1969. On the bionomics of *Lyctocoris beneficus* (Hiura) and *Xylocoris galactinus* (Fieber) (Anthocoridae, Heteroptera). J. Fac. Agric. Kyushu Univ. 15, 1–136.

Cocuzza, G.E., De Clercq, P., Van de Veire, M., De Cock, A., Degheele, D., Vacante, V., 1997. Reproduction of *Orius laevigatus* and *Orius albidipennis* on pollen and *Ephestia kuehniella* eggs. Entomol. Exp. Appl. 82, 101–104.

Cohen, A.C., 1985. Simple method for rearing the insect predator *Geocoris punctipes* on a meat diet. J. Econ. Entomol. 78, 1173–1175.

Cohen, A.C., 1990. Feeding adaptations of some predaceous *Hemiptera*. Ann. Entomol. Soc. Am. 83, 1215–1223.

Cohen, A.C., 1999. Artificial media for rearing entomophages comprising sticky, cooked whole egg. US Patent 5945271.

Cohen, A.C., 2000a. Feeding fitness and quality of domesticated and feral predators: effects of long-term rearing on artificial diet. Biol. Control 17, 50–54.

Cohen, A.C., 2000b. How carnivorous bugs feed. In: Schaefer, C.W., Panizzi, A.R. (Eds.), Heteroptera of Economic Importance, CRC Press, Boca Raton, FL, pp. 563–570.

Cohen, A.C., 2003. Artificial diets for arthropods. US Patent 6506597.

Cohen, A.C., 2004. Insect Diets: Science and Technology. CRC Press, Boca Raton, FL.

Cohen, A.C., Debolt, J.W., 1983. Rearing *Geocoris punctipes* on insect eggs. Southwest. Entomol. 8, 61–64.

Cohen, A.C., Nordlund, D.A., Smith, R.A., 1999. Mass rearing of entomophagous insects and predaceous mites: are the bottlenecks biological, engineering, economic, or cultural? Biocontrol News Inform. 20, 85N–90N.

Cohen, A.C., Urias, N.M., 1986. Meat-based artificial diets for *Geocoris punctipes* (Say). Southwest. Entomol. 11, 171–176.

Coll, M., 1998. Living and feeding on plants in predatory Heteroptera. In: Coll, M., Ruberson, J.R. (Eds.), Predatory Heteroptera: Their Ecology and Use in Biological Control, Entomological Society of America, Lanham, MD, pp. 89–130.

Coll, M., Guershon, M., 2002. Omnivory in terrestrial arthropods: mixing plant and prey diets. Ann. Rev. Entomol. 47, 267–297.

Constant, B., Grenier, S., Bonnot, G., 1996. Artificial substrate for egg laying and embryonic development by the predatory bug *Macrolophus caliginosus* (Heteroptera: Miridae). Biol. Control 7, 140–147.

Coudron, T.A., Kim, Y., 2004. Life history and cost analysis for continuous rearing of *Perillus bioculatus* (Heteroptera: Pentatomidae) on a zoophytogenous artificial diet. J. Econ. Entomol. 97, 807–812.

Coudron, T.A., Wittmeyer, J., Kim, Y., 2002. Life history and cost analysis for continuous rearing of *Podisus maculiventris* (Say) (Heteroptera: Pentatomidae) on a zoophytophagous artificial diet. J. Econ. Entomol. 95, 1159–1168.

Coudron, T.A., Yocum, G.D., Brandt, S.L., 2006. Nutrigenomics: a case study in the measurement of insect response to nutritional quality. Entomol. Exp. Appl. 121, 1–14.

De Bortoli, S.A., Otuka, A.K., Vacari, A.M., Martins, M.I.E.G., Volpe, H.X.L., 2011. Comparative biology and production costs of *Podisus nigrispinus* (Hemiptera: Pentatomidae) when fed different types of prey. Biol. Control 58, 127–132.

De Clercq, P., 2000. Predaceous stinkbugs (Pentatomidae: Asopinae). In: Schaefer, C.W., Panizzi, A.R. (Eds.), Heteroptera of Economic Importance, CRC Press, Boca Raton, FL, pp. 737–789.

De Clercq, P., 2008. Culture of natural enemies on factitious foods and artificial diets, second ed. In: Capinera, J.L. (Ed.), Encyclopedia of Entomology, vol. I. Springer, Dordrecht, pp. 1133–1136.

De Clercq, P., Arijs, Y., Van Meir, T., Van Stappen, G., Sorgeloos, P., Dewettinck, K., Rey, M., Grenier, S., Febvay, G., 2005. Nutritional value of brine shrimp cysts as a factitious food for *Orius laevigatus* (Heteroptera: Anthocoridae). Biocontrol Sci. Technol. 15, 467–479.

De Clercq, P., Degheele, D., 1992. A meat-based diet for rearing the predatory stinkbugs *Podisus maculiventris* and *Podisus sagitta* (Het.: Pentatomidae). Entomophaga 37, 149–157.

De Clercq, P., Degheele, D., 1993a. Quality assessment of the predatory bugs *Podisus maculiventris* (Say) and *Podisus sagitta* (Fab.) (Heteroptera: Pentatomidae) after prolonged rearing on a meat-based artificial diet. Biocontrol Sci. Technol. 3, 133–139.

De Clercq, P., Degheele, D., 1993b. Quality of predatory bugs of the genus *Podisus* (Heteroptera: Pentatomidae) reared on natural and artificial diets. In: Nicoli, G., Benuzzi, M., Leppla, N.C. (Eds.), Proceedings of the 7th Workshop of the IOBC Global Working Group on Quality Control of Mass Reared Arthropods, pp. 129–142.

De Clercq, P., Keppens, G., Anthonis, G., Degheele, D., 1988. Laboratory rearing of the predatory stinkbug *Podisus sagitta* (Fab.) (Heteroptera: Pentatomidae). Med. Fac. Landbouww. Rijksuniv. Gent. 53, 1213–1217.

De Clercq, P., Merlevede, F., Tirry, L., 1998a. Unnatural prey and artificial diets for rearing *Podisus maculiventris* (Heteroptera: Pentatomidae). Biol. Control 12, 137–142.

De Clercq, P., Vandewalle, M., Tirry, L., 1998b. Impact of inbreeding on performance of the predator *Podisus maculiventris*. BioControl 43, 299–310.

De Puysseleyr, V., De Man, S., Höfte, M., De Clercq, P., 2013. Plantless rearing of the zoophytophagous bug *Nesidiocoris tenuis*. BioControl 58 (2), 205–213. http://dx.doi.org/10.1007/s10526-012-9486-7.

Dougherty, E.C., 1959. Introduction to axenic culture of invertebrate Metazoa: a goal. Ann. N. Y. Acad. Sci. 77, 27–54.

Dunkel, F.V., Jaronski, S.T., 2003. Development of a bioassay system for the predator, *Xylocoris flavipes* (Heteroptera: Anthocoridae), and its use in subchronic toxicity/pathogenicity studies of *Beauveria bassiana* strain GHA. J. Econ. Entomol. 96, 1045–1053.

Etzel, L.K., Legner, E.F., 1999. Culture and colonization. In: Bellows, T.S., Fisher, T.W. (Eds.), Handbook of Biological Control. Principles and Applications of Biological Control, Academic Press, San Diego, CA, pp. 125–197.

Evans, E.W., Stevenson, A.T., Richards, D.R., 1999. Essential versus alternative foods of insect predators: benefits of a mixed diet. Oecologia 121, 107–112.

Fauvel, G., Malausa, J.C., Kaspar, B., 1987. Etude en laboratoire des principales caractéristiques biologiques de *Macrolophus caliginosus* (Heteroptera: Miridae). Entomophaga 32, 529–543.

Ferkovich, S.M., Lynn, D.E., 2005. Enhanced egg laying in adult predators fed artificial diet supplemented with an embryonic cell line derived from eggs of *Ephestia kuehniella* Zeller (Lepidoptera: Pyralidae). Fla. Entomol. 88, 329–331.

Ferkovich, S.M., Shapiro, J.P., 2004a. Increased egg-laying in *Orius insidiosus* (Hemiptera: Anthocoridae) fed artificial diet supplemented with an embryonic cell line. Biol. Control 31, 11–15.

Ferkovich, S.M., Shapiro, J.P., 2004b. Comparison of prey-derived and non-insect supplements on egg-laying of *Orius insidiosus* maintained on artificial diet as adults. Biol. Control 31, 57–64.

Ferkovich, S.M., Shapiro, J.P., 2005. Enhanced oviposition in the insidious flower bug, *Orius insidiosus* (Hemiptera: Anthocoridae), with a partially purified nutritional factor from prey eggs. Fla. Entomol. 88, 253–257.

Ferkovich, S.M., Venkatesan, T., Shapiro, J.P., Carpenter, J.E., 2007. Presentation of artificial diet: effects of composition and size of prey and diet domes on egg production by *Orius insidiosus* (Heteroptera: Anthocoridae). Fla. Entomol. 90, 502–508.

Firlej, A., Chouinard, G., Coderre, D., 2006. A meridic diet for the rearing of *Hyaliodes vitripennis* (Hemiptera: Miridae), a predator of mites in apple orchards. Biocontrol Sci. Technol. 16, 743–751.

Franz, J., Szmidt, A., 1960. Beobachtungen beim Züchten von *Perillus bioculatus* (Fabr.) (Heteropt., Pentatomidae), einem aus Nordamerika importierten Räuber des Kartoffelkäfers. Entomophaga 5, 87–110.

Gillespie, D.R., McGregor, R.R., 2000. The functions of plant feeding in the omnivorous predator *Dicyphus hesperus*: water places limits on predation. Ecol. Entomol. 25, 380–386.

Goussaert, K., 2003. Kunstmatige voedingsbodems voor de massakweek van de roofwants *Orius laevigatus*. MSc thesis, Ghent University.

Greany, P.D., Carpenter, J.E., 1998. Culture medium for parasitic and predaceous insects. US Patent 5799607.

Grenier, S., De Clercq, P., 2003. Comparison of artificially vs naturally reared natural enemies and their potential for use in biological control. In: van Lenteren, J.C. (Ed.), Quality Control and Production of Biological Control Agents: Theory and Testing Procedures, CABI Publishing, Wallingford, UK, pp. 115–131.

Grenier, S., Guillaud, J., Delobel, B., Bonnot, G., 1989. Nutrition et élevage du prédateur polyphage *Macrolophus caliginosus* (Heteroptera, Miridae) sur milieux artificiels. Entomophaga 34, 77–86.

Grundy, P.R., Maelzer, D.A., Bruce, A., Hassan, E., 2000. A mass-rearing method for the assassin bug *Pristhesancus plagipennis* (Hemiptera: Reduviidae). Biol. Control 18, 243–250.

Guppy, J.C., 1986. Bionomics of the damsel bug *Nabis americoferus* Carayon (Hemiptera: Nabidae), a predator of the alfalfa blotch leafminer (Diptera: Agromyzidae). Can. Entomol. 118, 745–751.

Heimpel, G.E., Hough-Goldstein, J.A., 1994. Components of the functional response of *Perillus bioculatus* (Hemiptera: Pentatomidae). Environ. Entomol. 23, 855–859.

Henaut, Y., Alauzet, C., Ferran, A., Williams, T., 2000. Effect of nymphal diet on adult predation behavior in *Orius majusculus* (Heteroptera: Anthocoridae). J. Econ. Entomol. 93, 252–255.

Henry, T.J., Froeschner, R.C. (Eds.), 1988. Catalog of the Heteroptera, or True Bugs, of Canada and the Continental United States, E.J. Brill, New York.

Hori, K., 2000. Possible causes of disease symptoms resulting from the feeding. In: Schaefer, C.W., Panizzi, A.R. (Eds.), Heteroptera of Economic Importance, CRC Press, Boca Raton, FL, pp. 11–35.

Hough-Goldstein, J., McPherson, D., 1996. Comparison of *Perillus bioculatus* and *Podisus maculiventris* (Hemiptera: Pentatomidae) as potential control agents of the Colorado potato beetle (Coleoptera: Chrysomelidae). J. Econ. Entomol. 89, 1116–1123.

Inglis, G.D., Cohen, A.C., 2004. Influence of antimicrobial agents on the spoilage of a meat-based entomophage diet. J. Econ. Entomol. 97, 235–250.

Iriarte, J., Castañé, C., 2001. Artificial rearing of *Dicyphus tamaninii* (Heteroptera: Miridae) on a meat-based diet. Biol. Control 22, 98–102.

Ito, K., 2007. A simple mass-rearing method for predaceous *Orius* bugs in the laboratory. Appl. Entomol. Zool. 42, 573–577.

Jusselino-Filho, P., Zanuncio, J.C., Fragoso, D.B., Serrão, J.E., Lacerda, M.C., 2003. Biology of *Brontocoris tabidus* (Heteroptera: Pentatomidae) fed with *Musca domestica* (Diptera: Muscidae) larvae. Braz. J. Biol. 63, 463–468.

Kakimoto, K., Urano, S., Noda, T., Matuo, K., Sakamaki, Y., Tsuda, K., Kusigemati, K., 2005. Comparison of the reproductive potential of three *Orius* species, *O. strigicollis*, *O. sauteri*, and *O. minutes* (Heteroptera: Anthocoridae), using eggs of the Mediterranean flour moth as a food source. Appl. Entomol. Zool. 40, 247–255.

Kikuchi, Y., Fukatsu, T., 2003. Diversity of *Wolbachia* endosymbionts in heteropteran bugs. Appl. Environ. Microbiol. 69, 6082–6090.

Kiman, Z.B., Yeargan, K.V., 1985. Development and reproduction of the predator *Orius insidiosus* (Hemiptera: Anthocoridae) reared on diets of selected plant material and arthropod prey. Ann. Entomol. Soc. Am. 78, 464–467.

Kneeland, K., Coudron, T.A., Lindroth, E., Stanley, D., Foster, J.E., 2012. Genetic variation in field and laboratory populations of the spined soldier bug, *Podisus maculiventris*. Entomol. Exp. Appl. 143, 120–126.

Lavens, P., Sorgeloos, P., 2000. The history, present status and prospects of the availability of *Artemia* cysts for aquaculture. Aquaculture 181, 397–403.

Lee, K.S., Lee, J.H., 2004. Rearing of *Orius strigicollis* (Heteroptera: Anthocoridae) on artificial diet. Entomol. Res. 34, 299–303.

Lemos, W.P., Ramalho, F.S., Serrão, J.E., Zanuncio, J.C., 2003. Effects of diet on development of *Podisus nigrispinus* (Dallas) (Het., Pentatomidae), a predator of the cotton leafworm. J. Appl. Entomol. 127, 389–395.

Lemos, W.P., Ramalho, F.S., Serrão, J.E., Zanuncio, J.C., 2005. Morphology of female reproductive tract of the predator *Podisus nigrispinus* (Dallas) (Heteroptera: Pentatomidae) fed on different diets. Braz. Arch. Biol. Technol. 48, 129–138.

Lenfant, C., Ridray, G., Schoen, L., 2000. Biopropagation of *Macrolophus caliginosus* Wagner for a quicker establishment in southern tomato greenhouses. IOBC/WPRS Bull. 23, 247–251.

Leon-Beck, M., Coll, M., 2007. Plant and prey consumption cause a similar reductions in cannibalism by an omnivorous bug. J. Insect Behav. 20, 67–76.

Leon-Beck, M., Coll, M., 2009. The mating system of the flower bug *Orius laevigatus*. Biol. Control 50, 199–203.

Li, L.Y., Zhu, D.F., Zhang, M.L., Guo, M.F., 1997. Biology and rearing methods of *Eocanthecona furcellata* (Wolff) (Hemiptera: Pentatomidae). In: Li, L.Y. (Ed.), Parasitoids and Predators (Insecta) of Agricultural and Forestry Arthropod Pests, Guangdong High Education Press, Guangzhou, pp. 234–240.

Liquido, N.J., Nishida, T., 1985. Population parameters of *Cyrthorhinus lividipennis* Reuter (Heteroptera: Miridae) reared on eggs of natural and factitious prey. Proc. Hawaii. Entomol. Soc. 25, 87–93.

Lopez Jr., J.D., House, V.S., Morrison, R.K., 1987. Suitability of frozen *Sitotroga cerealella* (Olivier) eggs for temporary rearing of *Geocoris punctipes*. Southwest. Entomol. 12, 223–228.

Lu, C.T., Chiu, Y.C., Hsu, M.Y., Wang, C.L., Lin, F.C., 2011. Using cysts of brine shrimp, *Artemia franciscana*, as an alternative source of food for *Orius strigicollis* (Poppius) (Hemiptera: Anthocoridae). J. Taiwan Agric. Res. 60, 300–308.

Lundgren, J.G., 2009. Relationships of Natural Enemies and Non-Prey Foods. Springer Science + Business Media, New York.

Lundgren, J.G., 2011. Reproductive ecology of *predaceous Heteroptera*. Biol. Control 59, 37–52.

Machtelinckx, T., Van Leeuwen, T., Vanholme, B., Gehesquière, B., Dermauw, W., Vandekerkhove, B., Gheysen, G., De Clercq, P., 2009. *Wolbachia* induces strong cytoplasmic incompatibility in the predatory bug *Macrolophus pygmaeus*. Insect Mol. Biol. 18, 373–381.

Machtelinckx, T., Van Leeuwen, T., Van De Wiele, T., Boon, N., De Vos, W.H., Sanchez, J.A., Nannini, M., Gheysen, G., De Clercq, P., 2012. Microbial community of predatory bugs of the genus *Macrolophus* (Hemiptera: Miridae). BMC Microbiol. 12 (Suppl. 1), S9.

Mackauer, M., 1976. Genetic problems in the production of biological control agents. Annu. Rev. Entomol. 21, 369–385.

Mahdian, K., Kerckhove, J., Tirry, L., De Clercq, P., 2006. Effects of diet on development and reproduction of the predatory pentatomids *Picromerus bidens* and *Podisus maculiventris*. BioControl 51, 725–739.

Malaquias, J.B., Ramalho, F.S., Souza, J.V.S., Rodrigues, K.C.V., Wanderley, P.A., 2010. The influence of fennel feeding on development, survival, and reproduction in *Podisus nigrispinus* (Dallas) (Heteroptera: Pentatomidae). Turk. J. Agric. For. 34, 235–244.

Martinez-Cascales, J.I., Cenis, J.L., Cassis, G., Sanchez, J.A., 2006a. Species identity of *Macrolophus melanotoma* (Costa 1853) and *Macrolophus pygmaeus* (Rambur 1839) (Insecta: Heteroptera: Miridae) based on morphological and molecular data and bionomic implications. Insect Syst. Evol. 37, 385–404.

Martinez-Cascales, J.I., Cenis, J.L., Sanchez, J.A., 2006b. Differentiation of *Macrolophus pygmaeus* (Rambur 1839) and *Macrolophus melanotoma* (Costa 1853) (Heteroptera: Miridae) based on molecular data. IOBC/WPRS Bull. 29 (4), 223–228.

Mayné, R., Breny, R., 1948. *Picromerus bidens* L.: Morphologie. Biologie. Détermination de sa valeur d'utilisation dans la lutte biologique contre le doryphore de la pomme de terre—La valeur économique antidoryphorique des Asopines indigènes belges. Parasitica 4, 189–224.

Mohaghegh, J., De Clercq, P., Tirry, L., 1998a. Effects of maternal age and egg weight on developmental time and body weight of offspring in *Podisus maculiventris* (Heteroptera: Pentatomidae). Ann. Entomol. Soc. Am. 91, 315–322.

Mohaghegh, J., De Clercq, P., Tirry, L., 1998b. Maternal age and egg weight affect offspring performance in the predatory stink bug *Podisus nigrispinus*. BioControl 43, 163–174.

Mohaghegh-Neyshabouri, J., De Clercq, P., Degheele, D., 1996. Influence of female body weight on reproduction in laboratory-reared *Podisus nigrispinus* and *Podisus maculiventris* (Heteroptera: Pentatomidae). Med. Fac. Landbouwk. Toeg. Biol. Wet. Univ. Gent. 61, 693–696.

Nadgauda, D., Pitre, H.N., 1987. Feeding and development of *Nabis roseipennis* (Hemiptera: Nabidae) on *Heliothis virescens* (Lepidoptera: Noctuidae) larvae. J. Entomol. Sci. 21, 45–50.

Nannini, M., Souriau, R., 2009. Suitability of *Ceratitis capitata* (Diptera, Tephritidae) eggs as food source for *Macrolophus pygmaeus* (Heteroptera, Miridae). IOBC/WPRS Bull. 49, 323–328.

Naranjo, S.E., Gibson, R.L., 1996. Phytophagy in predaceous Heteroptera: effects on life history and population dynamics. In: Alomar, O., Wiedenmann, R.N. (Eds.), Zoophytophagous Heteroptera: Implications for Life History and Integrated Pest Management, Entomological Society of America, Lanham, MD, pp. 57–93.

Nedstam, B., 2002. *Macrolophus caliginosus* affected by a fungal pathogen. IOBC/WPRS Bull. 25 (1), 205–208.

Orr, D.B., Russin, J.S., Boethel, D.J., 1986. Reproductive biology and behavior of *Telenomus calvus* (Hymenoptera: Scelionidae), a phoretic egg parasitoid of *Podisus maculiventris* (Hemiptera: Pentatomidae). Can. Entomol. 118, 1063–1072.

Parajulee, M.N., Phillips, T.W., 1992. Laboratory rearing and field observations of *Lyctocoris campestris* (Heteroptera: Anthocoridae), a predator of stored-product insects. Ann. Entomol. Soc. Am. 85, 736–743.

Pfannenstiel, R.S., Hunt, R.E., Yeargan, K.V., 1995. Orientation of a hemipteran predator to vibrations produced by feeding caterpillars. J. Insect Behav. 8, 1–9.

Press, J.W., Flaherty, B.R., Davis, R., Arbogast, R.T., 1973. Development of *Xylocoris flavipes* (Hemiptera: Anthocoridae) on eggs of *Plodia interpunctella* (Lepidoptera: Phycitidae), killed by gamma radiation or by freezing. Environ. Entomol. 2, 335–336.

Richards, P.C., Schmidt, J.M., 1995. A rearing method for the production of large numbers of the insidious flower bug, *Orius insidiosus* (Say) (Hemiptera: Anthocoridae). Can. Entomol. 127, 445–447.

Richards, P.C., Schmidt, J.M., 1996a. The effects of selected dietary supplements on survival and reproduction of *Orius insidiosus* (Say) (Hemiptera: Anthocoridae). Can. Entomol. 128, 171–176.

Richards, P.C., Schmidt, J.M., 1996b. The suitability of some natural and artificial substrates as oviposition sites for the insidious flower bug, *Orius insidiosus*. Entomol. Exp. Appl. 80, 325–333.

Richman, D.B., Whitcomb, W.H., 1978. Comparative life cycles of four species of predatory stinkbugs (Hemiptera: Pentatomidae). Fla. Entomol. 61, 113–119.

Riddick, E.W., 2009. Benefits and limitations of factitious prey and artificial diets on life parameters of predatory beetles, bugs, and lacewings: a mini-review. BioControl 54, 325–339.

Riudavets, J., Arno, J., Castañe, C., 2006. Rearing predatory bugs with the brine shrimp *Artemia* sp. as alternative prey food. IOBC/WPRS Bull. 29 (4), 235–240.

Rojas, M.G., Morales-Ramos, J.A., King, E.G., 2000. Two meridic diets for *Perillus bioculatus* (Heteroptera: Pentatomidae), a predator of *Leptinotarsa decemlineata* (Coleoptera: Chrysomelidae). Biol. Control 17, 92–99.

Rudolf, V.H.W., 2007. Consequences of stage-structured predators: cannibalism, behavioral effects, and trophic cascades. Ecology 88, 2991–3003.

Ruth, J., Dwumfour, E.F., 1989. Laboruntersuchungen zur Eignung einiger Blattlausarten als Beute der räuberischen Blumenwanze *Anthocoris gallarum-ulmi* (DeG.) (Het., Anthocoridae). J. Appl. Entomol. 108, 321–327.

Saavedra, J.L.D., Zanuncio, J.C., Torres, J.B., Vinha, E., 2001. Resposta de ninfas de *Podisus nigrispinus* (Dallas, 1851) (Heteroptera: Pentatomidae) a presa e dieta artificial com diferentes componentes. Rev. Cientif. 29, 19–31.

Saavedra, J.L.D., Zanuncio, J.C., Zanuncio, T.V., Guedes, R.N., 1997. Prey capture ability of *Podisus nigrispinus* (Dallas) (Het., Pentatomidae) reared for successive generations on a meridic diet. J. Appl. Entomol. 121, 327–330.

Sahayaraj, K., 2002. Small scale laboratory rearing of a reduviid predator, *Rhynocoris marginatus* Fab. (Hemiptera: Reduviidae), on *Corcyra cephalonica* Stainton larvae by larval card method. J. Cent. Eur. Agric. 3, 137–147.

Sahayaraj, K., Balasubramanian, R., 2009. Biological control potential of artificial diet and insect hosts reared *Rhynocoris marginatus* (Fab.) on three pests. Arch. Phytopathol. Plant Prot. 42, 238–247.

Sahayaraj, K., Paulraj, M.G., 2001. Rearing and life table of reduviid predator *Rhynocoris marginatus* Fab. (Het., Reduviidae) on *Spodoptera litura* Fab. (Lep., Noctuidae) larvae. J. Appl. Entomol. 125, 321–325.

Sahayaraj, K., Martin, P., Selvaraj, P., Ragu, G., 2006. Feeding behaviour of reduviid predators on meat and insect-based artificial diets. Belg. J. Entomol. 8, 55–65.

Salas-Aguilar, J., Ehler, L.E., 1977. Feeding habits of *Orius tristicolor*. Ann. Entomol. Soc. Am. 70, 60–62.

Sampson, C., Jacobson, R.J., 1999. *Macrolophus caliginosus* Wagner (Heteroptera: Miridae): a predator causing damage to UK tomatoes. IOBC/WPRS Bull. 22, 213–216.

Sanchez, J.A., 2009. Density thresholds for *Nesidiocoris tenuis* (Heteroptera: Miridae) in tomato crops. Biol. Control 51, 493–498.

Sanchez, J.A., Lacasa, A., Arnó, J., Castañe, C., Alomar, O., 2009. Life history parameters for *Nesidiocoris tenuis* (Reuter) (Het., Miridae) under different temperature regimes. J. Appl. Entomol. 133, 125–132.

Schaefer, C.W., Panizzi, A.R. (Eds.), 2000. Heteroptera of Economic Importance. CRC Press, Boca Raton, FL.

Schmidt, J.M., Richards, P.C., Nadel, H., Ferguson, G., 1995. A rearing method for the production of large numbers of the insidious flower bug, *Orius insidiosus* (Say) (Hemiptera: Anthocoridae). Can. Entomol. 127, 445–447.

Shapiro, J.P., Ferkovich, S.M., 2006. Oviposition and isolation of viable eggs from *Orius insidiosus* in a parafilm and water substrate: comparison with green beans and use in enzyme-linked immunosorbent assay. Ann. Entomol. Soc. Am. 99, 586–591.

Shapiro, J.P., Reitz, S.R., Shirk, P.D., 2009. Nutritional manipulation of adult female *Orius pumilio* (Hemiptera: Anthocoridae) enhances initial predatory performance. J. Econ. Entomol. 102, 500–506.

Shimizu, J.T., Hagen, K.S., 1967. An artificial ovipositional site for some Heteroptera that insert their eggs into plant tissue. Ann. Entomol. Soc. Am. 60, 1115–1116.

Sipayung, A., Desmier de Chenon, R., Sudharto, P., 1992. Study of the *Eocanthecona-Cantheconidea* (Hemiptera: Pentatomidae, Asopinae) predator complex in Indonesia. J. Plant Prot. Tropics 9, 85–103.

Takara, J., Nishida, T., 1981. Eggs of the oriental fruit fly for rearing the predacious anthocorid, *Orius insidiosus* (Say). Proc. Hawaii Entomol. Soc. 23, 441–445.

Tamaki, G., Butt, B.A., 1978. Impact of *Perillus bioculatus* on the Colorado potato beetle and plant damage. USDA Techn. Bull. No. 1581.

Tan, X.L., Wang, S., Li, X.L., Zhang, F., 2010. Optimization and application of microencapsulated artificial diet for *Orius sauteri* (Hemiptera: Anthocoridae). Acta Entomol. Sin. 53, 891–900.

Thomas, J.M.G., Shirk, P., Shapiro, J.P., 2012. Mass rearing of a tropical minute pirate bug, *Orius pumilio* (Hemiptera: Anthocoridae). Fla. Entomol. 95, 202–204.

Thompson, S.N., 1999. Nutrition and culture of entomophagous insects. Annu. Rev. Entomol. 44, 561–592.

Thompson, S.N., Hagen, K.S., 1999. Nutrition of entomophagous insects and other arthropods. In: Bellows, T.S., Fisher, T.W. (Eds.), Handbook of Biological Control: Principles and Applications of Biological Control. Academic Press, San Diego, CA, pp. 594–652.

Tommasini, M.G., Burgio, G., Mazzoni, F., Maini, S., 2002. On intraguild predation and cannibalism in *Orius insidiosus* and *Orius laevigatus* (Rhynchota: Anthocoridae): laboratory experiments. Bull. Insectol. 55, 49–54.

Tommasini, M.G., van Lenteren, J.C., Burgio, G., 2004. Biological traits and predation capacity of four *Orius* species on two prey species. Bull. Insectol. 57, 79–93.

Toreki, W., Manukian, A., Strohschein, R., 2004. Hydrocapsules and method of preparation thereof. US Patent 6780507.

Torres, J.B., Zanuncio, J.C., Moura, M.A., 2006. The predatory stinkbug *Podisus nigrispinus*: biology, ecology and augmentative releases for lepidopteran larval control in I forests in Brazil. CAB Reviews, Perspect. Agric. Vet. Sci. Nutr. Nat. Resour. 1 (15), 1–18.

Torres, J.B., Boyd, D.W., 2009. Zoophytophagy in predatory Hemiptera. Braz. Arch. Biol. Technol. 52, 1199–1208.

Torres, J.B., Zanuncio, J.C., Saavedra, J.L.D., Aldrich, J.R., 1997. Extrato da glândula de feromônio na atração e estimulação alimentar de ninfas de *Podisus nigrispinus* (Dallas) e *Supputius cinctipes* (Stal). An. Soc. Entomol. Bras. 26, 463–469.

Tremblay, E., 1967. Primi tentativi di introduzione in Italia (Campania) del *Perillus bioculatus* (Fabr.), nemico naturale della dorifora della patata (*Leptinotarsa decemlineata* Say). Ann. Fac. Sci. Agr. Univ. Napoli 1, 1–13.

Urbaneja-Bernat, P., Alonso, M., Tena, A., Bolckmans, K., Urbaneja, A., 2013. Sugar as nutritional supplement for the zoophytophagous predator *Nesidiocoris tenuis*. BioControl 58, 57–64.

Vandekerkhove, B., De Clercq, P., 2010. Pollen as an alternative or supplementary food for the mirid predator *Macrolophus pygmaeus*. Biol. Control 53, 238–242.

Vandekerkhove, B., De Puysseleyr, V., Bonte, M., De Clercq, P., 2011. Fitness and predation potential of *Macrolophus pygmaeus* reared under artificial conditions. Insect Sci. 18, 682–688.

Vandekerkhove, B., Parmentier, L., Van Stappen, G., Grenier, S., Febvay, G., Rey, M., De Clercq, P., 2009. *Artemia* cysts as an alternative food for the predatory bug *Macrolophus pygmaeus*. J. Appl. Entomol. 133, 133–142.

Vandekerkhove, B., van Baal, E., Bolckmans, K., De Clercq, P., 2006. Effect of diet and mating status on ovarian development and oviposition in the polyphagous predator *Macrolophus caliginosus* (Heteroptera: Miridae). Biol. Control 39, 532–538.

van den Meiracker, R.A.F., 1999. Biocontrol of Western Flower Thrips by Heteropteran Bugs. PhD thesis, Amsterdam University, The Netherlands.

van Lenteren, J.C., 2011. The state of commercial augmentative biological control: plenty of natural enemies, but a frustrating lack of uptake. BioControl 57, 1–20.

Venkatesan, S., Seenivasagam, R., Karuppasamy, G., 1997. Influence of prey species on feeding response, development and reproduction of the reduviid, *Cydnocoris gilvus* Burm. (Reduviidae: Heteroptera). Entomon 22, 21–27.

Venkatesan, T., Jalali, S.K., Srinivasamurthy, K., Bhaskaran, T.V., 2008. Development, survival and reproduction of an anthocorid predator (*Orius tantillus*) on artificial and natural diets. Ind. J. Agric. Sci. 78, 102–105.

Warren, L.O., Wallis, G., 1971. Biology of the spined soldier bug, *Podisus maculiventris* (Hemiptera: Pentatomidae). J. Ga. Entomol. Soc. 6, 109–116.

Watanabe, M., Miura, K., Hunter, M.S., Wajnberg, E., 2011. Superinfection of cytoplasmic incompatibility-inducing *Wolbachia* is not additive in *Orius strigicollis* (Hemiptera: Anthocoridae). Heredity 106, 642–648.

Wheeler, A., 2001. Biology of the Plant Bugs (Hemiptera: Miridae). Cornell University Press, London.

White, J.H., Stauffer, L.A., Gallagher, K.A., 2001. Materials for rearing insects, mites, and other beneficial organisms. US Patent 6291007.

Wittmeyer, J.L., Coudron, T.A., 2001. Life table parameters, reproductive rate, intrinsic rate of increase and realized cost of rearing *Podisus maculiventris* (Say) (Heteroptera: Pentatomidae) on an artificial diet. J. Econ. Entomol. 94, 1344–1352.

Wittmeyer, J.L., Coudron, T.A., Adams, T.S., 2001. Ovarian development, fertility and fecundity in *Podisus maculiventris* Say (Heteroptera: Pentatomidae): an analysis of the impact of nymphal, adult, male and female nutritional source on reproduction. Invertebr. Reprod. Dev. 39, 9–20.

Yano, E., Watanabe, K., Yara, K., 2002. Life history parameters of *Orius sauteri* (Poppius) (Het., Anthocoridae) reared on *Ephestia kuehniella* eggs and the minimum amount of the diet for rearing individuals. J. Appl. Entomol. 126, 389–394.

Yasuda, T., Wakamura, S., 1992. Rearing of the predatory stink bug, *Eocanthecona furcellata* (Wolff) (Heteroptera: Pentatomidae), on frozen larvae of *Spodoptera litura* (Fabricius) (Lepidoptera: Noctuidae). Appl. Entomol. Zool. 27, 303–305.

Yocum, G.D., Evenson, P.L., 2002. A short term auxiliary diet for the predaceous stinkbug *Perillus bioculatus* (Hemiptera: Pentatomidae). Fla. Entomol. 85, 567–571.

Yokoyama, V.Y., 1980. Method for rearing *Geocoris pallens* (Hemiptera: Lygaeidae), a predator in California cotton. Can. Entomol. 112, 1–3.

Yu, S.J., 1987. Biochemical defense capacity in the spined soldier bug (*Podisus maculiventris*) and its lepidopterous prey. Pestic. Biochem. Physiol. 28, 216–223.

Zanuncio, J.C., Beserra, E.B., Molina-Rugama, A.J., Zanuncio, T.V., Pinon, T.B.M., Maffia, V.P., 2005. Reproduction and longevity of *Supputius cinctipes* (Het.: Pentatomidae) fed with larvae of *Zophobas confusa*, *Tenebrio molitor* (Col.: Tenebrionidae) or *Musca domestica* (Dip.: Muscidae). Braz. Arch. Biol. Technol. 48, 771–777.

Zanuncio, J.C., Ferreira, A.M.R.M., Tavares, W.S., Torres, J.B., Serrão, J.E., Zanuncio, T.V., 2011. Rearing the predator *Brontocoris tabidus* (Heteroptera: Pentatomidae) with *Tenebrio molitor* (Coleoptera: Tenebrionidae) pupa on *Eucalyptus grandis* in the field. Am. J. Plant Sci. 2, 449–456.

Zanuncio, J.C., Molina-Rugama, A.J., Serrão, J.E., Pratissoli, D., 2001. Nymphal development and reproduction of *Podisus nigrispinus* (Heteroptera: Pentatomidae) fed with combinations of *Tenebrio molitor* (Coleoptera: Tenebrionidae) pupae and *Musca domestica* (Diptera: Muscidae) larvae. Biocontrol Sci. Technol. 11, 331–337.

Zanuncio, J.C., Saavedra, J.L.D., Oliveira, H.N., Degheele, D., De Clercq, P., 1996. Development of the predatory stinkbug *Brontocoris tabidus* (Signoret) (Heteroptera: Pentatomidae) on different proportions of an artificial diet and pupae of *Tenebrio molitor* L. (Coleoptera: Tenebrionidae). Biocontrol Sci. Technol. 6, 619–625.

Zapata, R., Specty, O., Grenier, S., Febvay, G., Pageaux, J.F., Delobel, B., Castañé, C., 2005. Carcass analysis to improve a meat-based diet for the artificial rearing of the predatory mirid bug *Dicyphus tamaninii*. Arch. Insect Biochem. Physiol. 60, 84–92.

Zhou, W., Wang, R., 1989. Rearing of *Orius sauteri* (Hemiptera: Anthocoridae) with natural and artificial diets. Chin. J. Biol. Control 5, 9–12.

Zhou, W., Wang, R., Qiu, S., 1991. Use of soybean sprouts as the oviposition material in mass rearing of *Orius sauteri* (Het.: Anthocoridae) with natural and artificial diets. Chin. J. Biol. Control 7, 7–9.

FURTHER READING

Vacante, V., Cocuzza, G.E., De Clercq, P., Van de Veire, M., Tirry, L., 1997. Development and survival of *Orius albidipennis* (Reuter) and *O. laevigatus* (Fieber) (Het.: Anthocoridae) on various diets. Entomophaga 42, 493–498.

Zanuncio, J.C., Zanuncio, T.V., Guedes, R.N.C., Ramalho, F.S., 2000. Effect of feeding on three *Eucalyptus* species on the development of *Brontocoris tabidus* (Het.: Pentatomidae) fed with *Tenebrio molitor* (Col.: Tenebrionidae). Biocontrol Sci. Technol. 10, 443–450.

Chapter 4

Production of Dipteran Parasitoids

Maria Luisa Dindo[1] and Simon Grenier[2,*]
[1]*Dipartimento di Scienze Agrarie (DipSA), Alma Mater Studiorum Università di Bologna, Bologna, Italy,* [2]*UMR BF2I, INRA/INSA de Lyon, current address: 6 rue des Mésanges, 69680 Chassieu, France*

4.1. INTRODUCTION

Dipteran parasitoids are underestimated and often forgotten with regard to their role in biological control. This is partially due to their relatively low number, as they represent about 20% of all insect parasitoids, the majority of which are hymenopterans (Feener and Brown, 1997). Nevertheless, they constitute the second most important group of parasitoid insects due to their biological, physiological, and behavioral characteristics, which include their various parasitization modes that allow them to take possession of the hosts in different environments and thus reduce insect pest populations. Many representatives of this group (especially Tachinidae and Bombyliidae) are unique because of their ability to attack hosts hidden in vegetables or in the soil (Mellini, 1997).

Dipteran parasitoids are far less studied than parasitic wasps from all points of view (e.g. biology, the host–parasitoid relationship, and also applied biocontrol). Yet, the research performed has shown that they play a major role in regulating phytophagous insect populations, and their importance is increasingly recognized, especially in natural biological control (Sampaio et al., 2009). The general image of "flies" in biocontrol, however, remains unpopular.

Compared to hymenopteran parasitoids, far less research has been performed on dipterans, including the development of rearing technology, and, to date, "mass" production has been performed for only a few species. Moreover, their large size relative to wasp parasitoids is unfavorable for mass production, which increases the needs for food and space to obtain a production level comparable to what has been achieved with hymenopterans. Yet, several studies concerning the rearing of these beneficial insects have been carried out. The purpose of this

*Retired.

Mass Production of Beneficial Organisms. http://dx.doi.org/10.1016/B978-0-12-391453-8.00004-2

chapter is to review the work done with some new perspectives and bring this group of parasitoids to light.

Eggleton and Belshaw (1992) acknowledged parasitoids in many families of Diptera, but only a few (i.e. Tachinidae, Sarcophagidae, Phoridae, Cryptochaetidae, and Bombyliidae) include species of interest for applied biological control that would also be of interest for production. This chapter will be mainly (but not exclusively) devoted to the Tachinidae, which comprise about 8,500 species described worldwide (O'Hara, 2013), and represent the largest and most important family of nonhymenopteran parasitoids. Some examples of tachinid and other dipteran parasitoids implicated in biological control will be presented in the first section. Subsequently, the most important aspects of their biology relevant for rearing will be described; then, production techniques, in vivo as well as in vitro, from different points of view will be considered. Finally, some perspectives with the aim of stimulating new ideas and new research efforts on these topics will be presented.

4.2. DIPTERAN PARASITOIDS AS BIOCONTROL AGENTS

We will briefly summarize some examples of the use of dipteran parasitoids in classical and augmentative biocontrol programs (sensu van Lenteren, 2012), as the success of these types of biological control (especially augmentative) is widely based on the development of efficient techniques for culturing the entomophagous species involved (Greathead, 1986). Several species have shown the potential to control target insect pests, but fewer have actually been utilized in biocontrol programs. According to recent available information, some of these dipteran parasitoids have been or are mass cultured in state-funded facilities (as discussed in this chapter), but production in commercial biofactories has only very rarely occurred. On the contrary, dipteran predators, such as *Aphidoletes aphidimyza* Rondani and *Episyrphus balteatus* (De Geer), have been marketed and used as biocontrol agents (van Lenteren, 2012). Fly parasitoids could potentially be better exploited in biological control if technology for their commercial production is developed. Examples of dipteran parasitoids utilized in applied biological control programs are presented in Table 4.1.

4.2.1. Tachinidae

There are many examples of attempts (either successful or not) to utilize Tachinidae in classical and augmentative biological control programs, especially in the Nearctic and Neotropical regions and, to a lesser extent, in the Australian and Oceanian regions. It is interesting to note that examples in the Palearctic region are almost completely absent. The results achieved by these programs until the late 1980s were reviewed by Grenier (1988a), who has also analyzed the reasons for successes and failures. Failures were often related to lack of knowledge regarding the biology, behavior, and ecology of the species used as well as improper rearing techniques and shipment procedures.

Grenier (1988a) reported different examples of classical and augmentative biological control programs performed against forest lepidopterous defoliators and sawflies in North America using tachinid flies. One of the best successes was achieved with *Cyzenis albicans* Fallen, which was imported from Europe and released in Canada (15,000 flies in 1979–1980) to control the winter moth *Operophtera brumata* L. The fly established and helped to control the target insect pest over several years (Hulme and Green, 1984; Horgan et al., 1999). In the early 1900s, 16 tachinid species were introduced from Europe to the northern United States (New England) to control the gypsy moth *Lymantria dispar* (L.) and the brown tail moth *Nygmia phaerrhoea* Don. Two species—namely, *Blepharipa pratensis* Meigen and *Compsilura concinnata* Meigen—became established in many US states and were later used for augmentative releases of 83,658 flies against the gypsy moth (Blumenthal et al., 1979). Two other species, *Exorista larvarum* (L.) and *Parasetigena sylvestris* (Robineau-Desvoidy),well known as antagonists of *L. dispar* in Europe, were used in inoculative releases in the northern United States and became established (Sabrosky and Reardon, 1976; Kenis and Lopez Vaamonde, 1998). Another tachinid of European origin, *Aphantor haphopsis* (=*Ceranthia*) *samarensis* (Villeneuve), is considered a good candidate for biological control of the gypsy moth due to its limited host range, which may result in a lower risk to nontarget lepidopterous species (Fuester et al., 2001). This tachinid was repeatedly introduced in the northern United States and Canada, but establishment has not been confirmed (McManus and Csóka, 2007). Similarly, *Lydella jalisco* Woodley, a parasitoid of Mexican origin, was repeatedly introduced from its native Mexico into Texas for the classical control of the Mexican rice borer, *Eoreuma loftini* (Dyar) (a key pest of sugarcane in Texas), with limited results (Lauziére et al., 2001). Conversely, *Lixophaga diatraeae* Townsend, originating from Cuba, is one of the species more effectively utilized, in both classical and augmentative biological control, against lepidopterous sugarcane borers of the genera *Diatraea* and *Chilo*, especially *Diatraea saccharalis* (F.). As reviewed by Grenier (1988a), between 1915 and 1971 *L. diatraeae* was extensively introduced into different sugarcane-producing countries (e.g. in the southern United States, West Indies islands, South America, and Southeast Asia) and established in some of these regions (including Louisiana, Guadalupe, St. Kitts, Colombia, and Brazil). Successful establishment was not dependent on the number of flies released but rather on the climatic and ecological conditions in the countries of introduction. Usually, the best successes were obtained in islands. Grenier (1988a) also reported various examples of effective control of *D. saccharalis* populations following augmentative releases of *L. diatraeae* in different areas of introduction, including Louisiana and Florida. These results were achieved despite the high mobility of the flies and the continental position of these states (King et al., 1981). Augmentative releases of *L. diatraeae* to control *D. saccharalis* and

xamples of Dipteran Parasitoids Utilized in Classical (C) or Augmentative (A) Biological Control Progra
the Text*

Parasitoid Species	Target Insect	Type of Biological Control	Country of A
Cyzenis albicans	*Operophtera brumata*	C	Canada
Aphantorhaphopsis samarensis	*Lymantria dispar*	C	Canada and N
Exorista larvarum			Northern USA
Parasetigena sylvestris			
Blepharipa pratensis	*Lymantria dispar* and *Nygmia phaerrhoea*	C and A	
Compsilura concinnata			
Lydella jalisco	*Eoreuma loftini*	C	USA (Texas)
Archytas marmoratus	*Helicoverpa zea* *Spodoptera frugiperda*	A**	Southern USA
Lixadmontia franki	*Metamasius callizona*	C	Southern USA
Ormia depleta	*Scapteriscus* spp.	C	
Lydella minense *Paratheresia claripalpis*	*Diatraea* spp.	C and A	Brazil, Colom

ʟixophaga diatraeae	*Diatraea saccharalis* and other lepidopterous sugarcane borers	C and A	Southern USA, West islands, South Ameri and Southeast Asia
		A	Cuba
	Chilo infuscatellus and other lepidopterous sugarcane borers	A**	China (Guangxi)
Trichopoda pennipes	*Nezara viridula*	C	Different countries
Trichopoda giacomelli		C	Australia
ʒessa remota	*Levuana iridescens*	C	Fiji islands
ʌgria housei	*Choristoneura fumiferana*	C	Canada
ᑕryptochaetum iceryae	*Icerya purchasi*	C	USA (California) Mauritius Israel
ᑕryptochaetum jorgepastori	*Paleococcus fuscipennis*	C	Israel
ᑭseudacteon spp.	*Solenopsis* spp.	C	Southern USA

.l examples in different regions of the world.

other lepidopterous sugarcane borers have been (and are currently) widely and successfully carried out in Cuba, which is the country of origin of the tachinid (Nicholls et al., 2002; Montes, 2008). The control of *D. saccharalis* is one of the main objectives of Cuba's National Biological Control Program, which was created by the Sugar Ministry (MINAZ) in 1980 and is still in progress. Medina (2002) reported that, inside this program, more than 100 million *L. diatraeae* flies were produced per year in a network of 54 reproduction centers (CREEs) of entomophagous insects and entomopathogens. They were established in all 15 provinces of the island (Massó Villalón, 2007). Fly releases over 1.6 million ha have led to successful biocontrol of *D. saccharalis*. Moreover, *L. diatraeae* was also imported into China (Guangxi) to control *Chilo infuscatellus* Snellen and other sugarcane lepidopterous pests; high levels of efficacy in controlling these target insects were observed in laboratory and field trials (Deng et al., 2008, 2010).

In Latin America, other tachinids, including *Lydella minense* Townsend from Brazil and *Paratheresia claripalpis* van de Wulp from Mexico, have been utilized against sugarcane borers of the genus *Diatraea*, in either classical or augmentative biological control programs (Grenier, 1988a; van Lenteren and Bueno, 2003; Weir et al., 2007; Bustillo Pardey, 2009). In different South American countries (e.g. Brazil, Colombia, and Peru), there are many units funded by the government for the mass production of insect natural enemies, including these two species (van Lenteren and Bueno, 2003). Arrigoni (1992) reports that in the 1980s, 5.7 million flies of both species were produced in 17 facilities of Copersucar, which is the biggest Brazilian sugar and ethanol company. *Paratheresia claripalpis* and *L. minense* have also been produced by commercial biofactories in Peru (Bioinsa, Trujillo) and Columbia (Biodefensas, Santiago de Cali).

During the 1980s and 1990s in the southern United States, *Archytas marmoratus* (Townsend), a native larval-pupal parasitoid of noctuids, was successfully mass produced and used in augmentative field release experiments against the corn earworm *Helicoverpa zea* (Boddie) and the fall armyworm *Spodoptera frugiperda* (J. E. Smith) in whorl-stage corn (Gross, 1990a; Proshold et al., 1998). The parasitoid releases were also combined with other techniques such as inherited sterility (Mannion et al., 1995). Two other tachinids, *Eucelatoria bryani* Sabrosky and *Palexorista laxa* Curran, were used in field cage release experiments against *Heliothis virescens* (F.) and showed potential for controlling this target pest (King and Coleman, 1989). However, despite these promising results, further investigation or implementation of biocontrol using *A. marmoratus*, *E. bryani*, and *P. laxa* was apparently not pursued.

Trichopoda pennipes (F.), which is native to the southern United States and South America, is a parasitoid of the green vegetable bug *Nezara viridula* (L.) and has been introduced into different countries throughout the world to control the target insect, with mixed success (Grenier, 1988a; O'Hara, 2008). In Italy its introduction, followed by establishment, was fortuitous (Colazza et al., 1996).

A few years later, attempts to introduce *Trichopoda giacomelli* (Blanchard) in Australia (Queensland) were successful as the parasitoid has established and proved capable of decreasing populations of *N. viridula* (Coombs and Sands, 2000).

In recent years, a wide classical biological control program was carried out in Florida with the tachinid *Lixadmontia franki* Wood and Cave being used against the Mexican bromeliad weevil *Metamasius callizona* (Chevrolat), which is a native insect of Mexico and Guatemala. *Lixadmontia franki*, originally found in Honduras on a congeneric species of bromeliad-eating weevil, was imported, mass reared, and repeatedly released (2279 flies in total) throughout 2007–2009 in four sites. However, postrelease monitoring has recovered only two F_2 flies at one site (Cooper et al., 2011). Among the suggested reasons for the minimal success achieved were the differences in climate and elevation between the areas of the parasitoid's origin and its introduction, and also a lack of knowledge on parasitoid biology and difficulty in rearing, which limited the number of flies available for releases. Research aimed at improving the performance of this tachinid as a biocontrol agent of *M. callizona* deserves to be continued. In Florida, a classical biological control program was also conducted, with more success, against mole crickets of the genus *Scapteriscus* (an exotic pest from South America). A natural enemy, the tachinid *Ormia depleta* (Wiedemann), was imported from Brazil and released at different sites in Florida. The parasitoid became established, spread throughout southern Florida, and effectively contributed to controlling the target insects (Frank et al., 1996; Frank and Parkman, 1999).

At the laboratory of entomology of the University of Bologna, *E. larvarum*, a polyphagous parasitoid of Lepidoptera, has been the subject of extensive research concerning mass production, especially on artificial media, as will be reported in this chapter. The results obtained in laboratory studies show promise that augmentative releases of *E. larvarum* may contribute to lowering populations of noctuid species of agricultural interest (Simões et al., 2004; Depalo et al., 2010, 2012).

4.2.2. Other Dipteran Parasitoids

Besides tachinids, other dipteran parasitoids have shown potential as biocontrol agents. In 1971, 2800 *Agria housei* Sewell (Sarcophagidae), which were obtained from a laboratory culture, were released (unsuccessfully) in Canada against the spruce budworm *Choristoneura fumiferana* Freeman (Hulme and Green, 1984). Sarcophagid flies of the genus *Blaesoxipha* are known as natural enemies of grasshoppers, but their potential as biocontrol agents is not well known (Mason and Erlandson, 1994; Danyk et al., 2005). Conversely, encouraging results have been achieved with Cryptochaetidae, a small family parasitizing Margarodidae. Greathead (1986) reported the successful introduction of two Cryptochaetidae of Australian origin into California and Mauritius to control *Icerya purchasi* Maskell and *Icerya seychellarum* (Westwood), respectively; however, these introductions were later neglected due to the superior performance and ease of handling of *Rodolia cardinalis* Mulsant. In the 1990s, *Cryptochaetum iceryae* (Williston) was introduced from California to Israel for biological control of *I. purchasi*, suppressing the target insect in 2 years at some release sites (Mendel and Blumberg, 1991). Another species, *Cryptochaetum jorgepastori* Cadalha, was introduced to Israel from Spain and contributed to control of the homopteran pest, *Palaeococcus fuscipennis* (Burmeister) (Mendel et al., 1998).

Extensive research conducted in the southern United States has focused on the potential of introducing phorid species of the genus *Pseudacteon* (from South America) as biocontrol agents of the imported *Solenopsis* fire ant (Orr et al., 1995). Some introductions (e.g. *Pseudacteon novenas* Borgmeier at a site in Texas) were successful despite difficulties in rearing the parasitoid (Plowes et al., 2012). The effectiveness of introduced *Pseudacteon* spp. in regulating population densities is, however, doubtful (Morrison, 2012).

Although they are difficult to handle in captivity, some bombyliids of the genus *Exhyalanthrax* could be promising biological control agents against *Glossina* spp. and merit further consideration (Greathead, 1980). In special habitats, *Villa cana* (Meigen) appeared to regulate some populations of *Thaumetopoea pityocampa* (Denis and Schiffermuller) (Biliotti et al., 1965).

4.2.3. Side Effects

As it is well known, classical biological control has raised some concern about potential negative effects by exotic natural enemies, mainly polyphagous predators and parasitoids, on indigenous nontarget species (Howarth, 1991). One of the best known examples is the tachinid *Bessa remota* (Aldrich), which was successfully introduced from Malaysia into the Fiji Islands in the 1920s, where it provided effective control of the coconut moth *Levuana iridescens* Bethune-Baker but was subsequently believed to have caused the extinction of both the target and some nontarget species. Extinction is, however, a controversial issue, since the shortage of reports of *L. iridescens* was mainly due to the declining value of copra, which resulted in less research on coconuts, and lack of a comprehensive campaign for explorations of coconut palm fronds (Hoddle, 2006). However, these concerns about classical biological control do not suggest a need to prohibit the importation of natural enemies. Classical biocontrol remains the best way to control nonintentional introductions of insect pests. Nevertheless, more careful studies on the impact of introductions on indigenous species are needed (van Lenteren et al., 2006). Regulations to control the introductions of macroorganisms in many countries are now commonplace (Mes Démarches, 2005). The possibility of rearing an exotic parasitoid may enhance one's ability to obtain more knowledge on its biology, host range, and host–parasitoid interaction, thus providing useful information regarding the risks and benefits of natural enemy introductions: also in this view, research concerning the production of parasitoids, including tachinids and other dipterans, is very important and needs to be encouraged.

4.3. ASPECTS OF DIPTERAN PARASITOID BIOLOGY OF SPECIAL INTEREST FOR PRODUCTION

A deep knowledge of parasitoid biology (especially host range, oviposition strategies, and host–parasitoid interaction) is a key tool to increase chances of rearing success. More exhaustive information on the biology of fly parasitoids may be found in specialized reviews, such as Clausen (1962) and Feener and Brown (1997) for dipteran parasitoids in general; Mellini (1991), Belshaw (1994), Stireman et al. (2006), and Dindo (2011) for tachinids; Pape (1987) for sarcophagids; Queezada and DeBach (1973) for cryptochaetids (e.g. *C. iceryae*); Disney (1994) for phorids; and Yeates and Greathead (1997) for bombyliids.

4.3.1. Host Range

Tachinids parasitize an extensive range of plant-feeding insects. Hosts are primarily larval Lepidoptera, but a variety of insects of other taxa are also hosts, including Heteroptera, Dermaptera, Phasmatodea, Orthoptera (nymphs and adults), Coleoptera (larvae and adults), and larval Hymenoptera Symphyta (Cerretti and Tschorsnig, 2010). About 70% of the known host species are Lepidoptera, and among them more than 50% belong to four families of economical importance (Noctuidae, Geometridae, Tortricidae, and Pyralidae) (Grenier and Liljestrhöm, 1991). The hosts of most tachinid species are, however, still unidentified. In 1991, over one-half of the European species of tachinids had no known hosts (Mellini, 1991). No tachinid species are known to parasitize eggs or pupae, although many are larval-pupal parasitoids. Many tachinids are polyphagous (i.e. *E. larvarum* and *C. concinnata*), whereas others have been found to exhibit relatively narrow host ranges (i.e. *T. giacomelli* and *Phryxe caudata* Rondani). In some cases, oviposition occurs preferably on hosts of one sex, males (e.g. *T. pennipes*) or females (e.g. *Istocheta aldrichi* (Mesnil)) (Grenier and Liljestrhöm, 1991).

Of the families considered in this chapter, sarcophagids are known largely as saprophagous, but some species are parasitoids. Some genera (e.g. *Agria*, *Sarcophaga*, and *Blaesoxipha*) contain parasitoid species of forest or agricultural insect pests (Lepidoptera for *Agria* and *Sarcophaga*, Hymenoptera Symphyta for *Agria*, and Orthoptera and Coleoptera for *Blaesoxipha*) (Pape, 1994). Millipedes as well as nonarthropod invertebrates (e.g. helicid snails) are also known as hosts of sarcophagid parasitoids (Pape, 1990; Shoaib and Cagan, 2004). Cryptochaetids are parasitoids of Homoptera Margarodidae, whereas many phorids are parasitoids of ants. For example, in South America, about 20 species of the genus *Pseudacteon* are natural enemies of fire ants (Porter and

Pesquero, 2001). The recorded host range of bombyliids includes seven insect orders and also some Araneae. Almost one-half of the records are from Hymenoptera (bees and wasps), but also tsetse flies and some lepidopterous larvae are attacked, whereas a few species are hyperparasitoids of tachinids. A record also exists of a bombyliid attacking the fruit fly *Rhagoletis pomonella* Walsh (Muniz-Reyes et al., 2011).

4.3.2. Oviposition Strategies

Knowledge of the oviposition strategy adopted, and of the mechanisms involved in host selection, is critical for the parasitization process and for successful continuous in vivo culture of parasitoids (dipterans in our case). This aspect is also crucial for in vitro production because it influences the way that eggs (or larvae) must be obtained to be placed on the artificial medium. In particular, achieving the goal of continuous in vitro culture without the host requires accurate information on the parasitoid oviposition strategy and factors triggering oviposition to induce direct oviposition on the artificial substrate.

Regarding the dipteran parasitoids considered in this chapter, tachinids may adopt direct or indirect oviposition strategies, depending on the species. In direct strategies, oviposition may occur in diverse modes, for example eggs may be deposited or projected on the host integument (e.g., *Exorista japonica* Townsend, *E. larvarum*, or *Carcelia gnava* (Meigen)). More rarely, eggs are inserted into the host via the genital–anal or buccal cavity (e.g. *Rondania cucullata* Robineau-Desvoidy) or even injected into the host body (e.g., Blondeliini and *E. bryani*). In the latter mode, some sternites of the female are modified as a piercing organ utilized to perforate the host integument and guide the ovipositor, which inserts an egg (or a larva) into the cut. In indirect strategies, females are usually ovoviviparous and the eggs may be placed close to a host. First instars are generally of the planidium type and have to wait for a host to pass by (e.g. *A. marmoratus*) or have to search for a host on their own (e.g. *L. diatraeae*). This latter strategy allows parasitoids to reach hosts that live in concealed places that are inaccessible to adult flies. Most Goniini deposit minute microtype eggs on the host food-plant, and these eggs hatch only after being accidentally eaten by a host. About 40% of Palearctic species show indirect modes, but ovipositing on the host body is the most common strategy used by tachinids. Mixed strategies are also possible; for instance, the females of ormiine tachinids may larviposit on or near their hosts (Fowler, 1987).

Sarcophagid flies of the genus *Blaesoxipha* are ovoviviparous or viviparous and also display either direct or indirect oviposition (or larviposition) strategies, similar to those described for tachinids (Allen and Pape, 1996). In Cryptochaetidae, eggs are oviposited directly in the host haemocoel through a "false" piercing ovipositor formed by the terminal segments of the female abdomen. *Pseudacteon* phorids that parasitize fire ants exhibit a direct oviposition strategy, since females hover near hosts and then drop down to insert an egg into the host thorax with a specialized ovipositor (Porter, 1998). The oviposition in bombyliids could be peculiar, as observed in *V. cana* by Biliotti et al. (1965). The female *V. cana* collects sand or dust from the soil in a special structure (perivaginal "pouch") located at the tip of its abdomen. The female broadcasts its eggs while flying over an area where the host has been detected. The eggs pass through the pouch where they are coated with sand or dust and thus resemble the environment.

The mechanisms of host selection in dipterans are related to their oviposition strategy and rely on chemical and physical stimuli. Many aspects, however, remain unknown as dipteran parasitoids have been far less studied than hymenopterans. In species with direct strategies, females use chemical cues to locate the habitat (plant or other host-food sources) and host or cues derived from interaction between the host and the plant, such as "frass" (Godfray, 1994; Aldrich et al., 2006). For instance, a fire ant alarm pheromone has been shown to attract the phorid fly parasitoid *Pseudacteon tricuspis* Borgmeier (Sharma et al., 2011), whereas host-induced plant volatiles were found to be crucial host localization cues for the tachinids *E. japonica* and *E. larvarum* (Ichiki et al., 2011; Depalo et al., 2012). Physical stimuli, including visual cues, also play a role in host habitat and host location in dipteran parasitoids, especially tachinids (Stireman, 2002). In particular, host size, color, texture, and movements can affect the oviposition behavior of a number of dipteran parasitoids, including *E. japonica* (Ichiki et al., 2011). Host location by phonotaxis has been demonstrated for *O. depleta* and other ormiine tachinids, in which the female finds hosts by being attracted by their mating calls (Fowler, 1987). In the same way, sarcophagids parasitizing cicadas are attracted by singing males (Schniederkötter and Lakes-Harlan, 2004).

In the case of an indirect oviposition or larviposition strategy, adult females use cues (physical and chemical) for locating only the host habitat or microhabitat. For instance, a substance isolated from *H. virescens* larvae was found to elicit larviposition in the host environment by *A. marmoratus* (Nettles and Burks, 1975). Physical factors associated with the host food plant (such as shape, size, and leaf color) were found to be essential for successful host microhabitat location and oviposition by the tachinid *Pseudogonia rufifrons* (Wiedemann), which oviposits microtype eggs on the leaves of certain plants (Mellini et al., 1980). In the tribe Goniini, no stimuli from parasitoid eggs attract the host larvae. The contacts between the host and parasitoid eggs laid on host food occur passively and accidentally, and the eggs hatch only when ingested by the host. Higher mortality of eggs or first instars occurs in indirect oviposition strategies than direct ones, and thus the former is associated with higher parasitoid fecundity (several thousand eggs). When host location and acceptance are directed to first instars, the specific cues involved in detection of a suitable host are largely unknown.

Associative learning, defined as the establishment through experience of an association between two cues or between a cue and a behavioral response

(Vet et al., 1990), has been well documented for hymenopteran parasitoids. For example, *Microplitis croceipes* (Cresson) females, experienced by contact with host feces, learned to recognize and fly to different volatile odors, even novel ones like vanilla, that were associated with the feces (Lewis and Tumlison, 1988). This learning also occurs in dipterans, and so far it has been observed in two tachinid species, *Drino bohemica* Mesnil (Monteith, 1963) and *Ormia ochracea* (Bigot) (Paur and Gray, 2011). Similarly, the capacity to avoid superparasitism by recognizing previously parasitized hosts has been documented in two solitary tachinids, *Myiopharus doryphorae* (Riley) and *Myiopharus aberrans* (Townsend), but the mechanisms involved are presently unclear (Lopez et al., 1995). Many tachinids are optionally gregarious in part depending on the host size. However, even in gregarious species, excessive superparasitism should be avoided in favor of optimizing production, as it may result in lower size and even the death of all individuals (Mellini and Campadelli, 1997; Baronio et al., 2002). All of the biological aspects described here deserve more consideration and more research as they may be important to improve the rearing procedures of dipteran parasitoids as well as the quality and performances of the flies obtained in captivity.

4.3.3. Host–Parasitoid Interactions

Unlike hymenopterans, dipteran parasitoids cannot suppress the host immune system or affect host physiology through secretions injected by ovipositing females (dipteran parasitoids lack a "true" piercing ovipositor) or derived from teratocytes. Yet, dipteran parasitoids display good strategies, which are especially known for tachinids, to avoid host encapsulation. Many tachinid larvae escape the host immune response and turn the host response to their advantage by forming respiratory funnels, which are sclerotized folders around the hind part of their body. Funnels may be primary, built in the host integument by first instars as soon as they enter, or secondary, formed by late-first or early-second instars in the host integument or tracheae. The tachinids that form primary funnels (*E. larvarum*) breathe atmospheric air from the beginning of their development, which permits them to grow rapidly, while those forming secondary funnels (*E. bryani* and *P. rufifrons*) breathe through their integument, and thus grow slowly, until funnel formation. The early larval stages of the tachinids that do not form primary funnels may escape host immune response by moving into a specific host tissue, such as muscle (*P. rufifrons*), or between the peritrophic membrane and gut wall (*C. concinnata*) (Baronio and Campadelli, 1979; Ichiki and Shima, 2003).

Another interesting aspect is host development following parasitoid attack. Parasitoids may be classified as "koinobionts" or "idiobionts"; koinobionts allow their host to continue to feed and grow beyond parasitization, whereas idiobionts permanently paralyze or kill the host before the parasitoid egg hatches (Haeselbarth, 1979; Askew and Shaw, 1986). However, dipteran parasitoids do not fit well into this classification. Many species show characteristics of both strategies, for example tachinids (Dindo, 2011) and bombyliids (Yeates and Greathead, 1997). In Tachinidae, many species (*P. rufifrons*, *Pseudoperichaeta nigrolineata*, and *A. marmoratus*) exhibit a high degree of physiological integration with their host. The development of these species is dependent on host hormones because their first-instar larvae have to wait until the host larva has reached maturity or is in the pupal stage before molting to second instar (Baronio and Sehnal, 1980; Grenier, 1988b). As a consequence, the duration of their larval development is widely influenced by host age at parasitization (Mellini, 1986). Other tachinid (and also sarcophagid) parasitoids do not show a developmental synchrony with their host and develop continuously until pupation. These species, especially those that form primary integumental respiratory funnels (*E. larvarum*), grow quickly following attack and kill the host rapidly, thereby behaving as zoonecrophages for most of their larval life. The degree of complexity of host–parasitoid interaction and the extent of host development following attack are related not only to developmental synchrony but also to respiration strategies. In fact, the tachinids that do not depend on host hormones, but form secondary funnels, grow continuously until pupation and keep the host alive longer than those building primary funnels. For *E. bryani* and *C. concinnata*, host–parasitoid interactions have also been found to be widely influenced by host age at parasitization (Coulibaly and Fanti, 1992; Caron et al., 2010).

Belshaw (1994) suggested that, similar to hymenopteran parasitoids, tachinids exhibiting a complex life history are generally less polyphagous than those that display rapid development. However, there are tachinids with a relatively narrow host range despite their independence from host hormones (*P. sylvestris*), whereas a high level of polyphagy is shown by a number of species displaying complicated host–parasitoid interactions (*C. concinnata*) (Godwin and Odell, 1984). In tachinids, polyphagy seems to be connected to the parasitoid's ability to avoid the host encapsulation response and/or indirect oviposition strategies (Askew and Shaw, 1986). In vivo production of species with a complex life history (whether based on respiration mode or hormonal interactions with the host) is not necessarily complicated, even on factitious hosts. For instance, the larval–pupal parasitoid *P. rufifrons* has successfully been cultured for many years in laboratory conditions on its factitious host *Galleria mellonella* (L.) (Mellini and Coulibaly, 1991). In contrast, host–parasitoid relationships may deeply influence the success of in vitro culture because tachinids exhibiting developmental synchrony with the host are known to be difficult to rear on artificial media (Dindo, 1998; Thompson, 1999). Independent of hormonal interactions, respiration adaptations of the parasitoid larvae have considerable impact on methods of in vitro rearing. Species that build primary funnels need to stay in contact with air from the first instar, which makes liquid media unsuitable for their culture. In vitro development may be more problematic for parasitoids that induce the formation of secondary, rather than primary, funnels (Dindo, 2011).

Aspects related to host–parasitoid interactions are known for only a few cryptochaetid species, such as *C. iceryae* (gregarious) and *C. grandicorne* (solitary). The larvae of these endoparasitoids do not build respiratory funnels but breathe through two caudal filaments containing tracheae, which become entangled with the host tracheae. Pupation may occur inside or outside the host scale carcass.

The host–parasitoid interactions displayed by *Pseudacteon* phorids with their fire ant hosts are rather complex and peculiar. The fly larvae develop and build respiratory structures in the head capsule of the host ant, which displays altered behavior and is finally decapitated. Larvae turn into pupae within the detached head capsule (Porter, 1998; Mathis and Philpott, 2012). Due to this manner of pupation and difficulties involved in the collection of fertilized eggs, in vitro rearing of *Pseudacteon* would be difficult to obtain on a large scale (Vogt et al., 2003). Conversely, sarcophagids attacking advanced larval stages or pupae of lepidopterans, such as *A. housei* and *Agria affinis* (Fallen), behave as zoosaprophages for most of their development and therefore show simple host–parasitoid interactions. Because of this characteristic, *A. housei* was successfully reared in vitro for many generations on different types of media (House and Traer, 1949).

4.4. PRODUCTION TECHNIQUES

4.4.1. In Vivo Production

Successful in vivo production depends on different factors, which are briefly illustrated in this section.

4.4.1.1. Larval Food: Natural Hosts versus Alternative Hosts

The classical method for production of entomophagous insects is the natural tritrophic system, which includes the plant (or other natural food), host insect, and parasitoid. By simplifying the system, we could make parasitoid production easier.

The first step consists of replacing the plant, or other natural food, with artificial diets to grow the insect hosts. There are many examples, especially with lepidopterous larvae used as hosts for parasitoid rearing (King and Leppla, 1984; Cohen, 2004). For instance, in Brazil, *L. minense* and *P. claripalpis* were reared on their natural host *Diatraea* spp., which was reared on an artificial diet to avoid the use of sugarcane (Rossi and Fowler, 2003). Systems based on rearing the natural host appear suitable for a number of tachinid species, including *C. concinnata*, *B. pratensis*, and other parasitoids of *L. dispar* (Odell and Godwin, 1979; Bourchier, 1991); *T. pennipes*, a parasitoid of *N. viridula* (Giangiuliani and Farinelli, 1995) and *L. franki*, a parasitoid of the weevil *Metamasius quadrilineatus* (Champion) (Suazo et al., 2006). The natural hosts of all these tachinids could be cultured without difficulties on either plants or artificial diets.

Natural hosts were also used to multiply nontachinid dipteran parasitoids, including the cryptochaetid. *Cryptochaetum iceryae*, which developed on *I. purchasi* growing on 17 plant species (Mendel and Blumberg, 1991), and the sarcophagid *A. affinis*, which was reared for many generations on *C. fumiferana* (House, 1967). Conversely, attempts of rearing the bombyliid fly *Exhyalanthrax* spp., which are parasitoids of the tsetse fly, on a natural host were unsuccessful (Markham, 1986). Vogt et al. (2003) described the mass production of the phorid *Pseudacteon curvatus* Borgmeier on its natural host, the red imported fire ant *Solenopsis invicta* Buren, which was fed a standard diet of live crickets, occasionally supplemented with other components.

The effect of host food, whether natural or artificial, should not be overlooked as it may influence parasitoid survival and quality (Thompson and Hagen, 1999). For example, Reitz and Trumble (1997) observed increased mortality of *A. marmoratus* when it was cultured on the natural host *Spodoptera exigua* (Hübner) fed three furanocoumarins derived from *Apium* sp. The effects of these allelochemicals on the tachinid were found to be mediated through their effects on the host larva.

Alternative hosts such as the lepidopterous larvae of *G. mellonella* are often easier to rear than natural hosts. The first studies with a parasitoid species should however start with rearing conditions similar to those observed in nature in order to obtain information on its basic biological, physiological, and behavioral characteristics. *Phryxe caudata* was first reared on its natural host *T. pityocampa* (Biliotti, 1956) and later on the factitious host *G. mellonella* (Delobel and Laviolette, 1969). The use of this factitious host was much more convenient and safe for laboratory rearing, especially taking into consideration the problems related to manipulation of the natural host, which bears dangerous urticating hair. Many physiological parameters were acquired thanks to the rearing of this tachinid on *G. mellonella* (Bonnot et al., 1976, 1984; Grenier et al., 1986). Similarly, *P. nigrolineata* (=*Pseudoperichaeta insidiosa* (Robineau-Desvoidy)), which parasitizes *Ostrinia nubilalis* Hübner in nature, was reared in the laboratory on *G. mellonella* (Grenier and Delobel, 1982); at least 21 other tachinid species were also reared on *G. mellonella* (18 are listed in Grenier, 1986). In Cuba, *L. diatraeae* was multiplied first on *Diatraea* spp. larvae, but later mass production occurred on *G. mellonella*. Indigenous to Cuba, *L. diatraeae* has been multiplied and released in sugarcane fields since 1945. An artificial diet for *G. mellonella* larvae was developed to improve the production system without reducing the parasitoid quality (Alemán et al., 1999). *Galleria mellonella* supplies several of the nutritional requirements for the development of many tachinid and hymenopterous parasitoids (Campadelli, 1988) and also for entomopathogenic nematodes (Ehlers and Shapiro-Ilan, 2005).Water-free artificial diets are a possibility for rearing *G. mellonella*. The absence of water results in reduced mold contamination, allowing the elimination of fungicides from the diet (Mellini and Coulibaly, 1991).

4.4.1.2. Infestation Mode

Extensive research concerning the rearing technique of *P. rufifrons* on *G. mellonella* revealed that the parasitization mode is crucial (Mellini and Coulibaly, 1991). In nature, *P. rufifrons* oviposits microtype eggs on foliage, and natural hosts (noctuid lepidopterans) become infested by eating "contaminated" leaves. In captivity, oviposition was obtained on wax substrates simulating leaves, and the parasitoid developed in *G. mellonella* larvae after they had fed on these artificial "leaves." The highest parasitization percentages were obtained when eight eggs per larvae were ingested.

Another tachinid displaying an indirect oviposition strategy (via planidia), *A. marmoratus*, was successfully reared on both natural hosts and *G. mellonella*. Parasitization was either performed by exposing host larvae to maggots larviposited on pieces of pleated filter paper (Coulibaly et al., 1993) or by using maggots that were mechanically extracted from fecund females. Female flies were placed in a 0.7% formalin–water solution and homogenized three times for 3 s each at 8500 rpm. Following this treatment, the uterus was ruptured and the maggots were released in the solution, which was then poured through a screen to exclude fly particles. Finally, the maggots were placed in a 1:3 hydroxyethylcellulose and water suspension and applied to corrugated cardboard disks with an aerosol sprayer. The disks were then placed close to host larvae (Gross and Johnson, 1985; Gross, 1994). For *L. diatraeae*, the infestation was done on *G. mellonella* larvae maintained in an appropriate-sized glass tube plugged with cotton on its extremities. This technique allowed control of the number of planidia introduced into the tube so that the effect of planidia numbers per host on the development of the parasitoid larvae could be determined (Grenier, 1981).

Galleria mellonella can be also used as a factitious host for tachinids displaying direct oviposition strategies, including *E. bryani* (Coulibaly and Fanti, 1992) and *E. larvarum*. *Exorista larvarum* has been maintained in culture in the laboratory of entomology of the University of Bologna since 1992 using methods described by Dindo et al. (1999). For both tachinids, parasitization occurs by exposing host larvae to flies for 30–60 min (three host larvae per female in the case of *E. larvarum*). The number of eggs per host proved to be very important for the success of parasitization. For *E. larvarum* and *E. japonica*, the best results were obtained when 4–6 or three eggs were deposited on last-instar hosts, respectively (Mellini and Campadelli, 1997; Nakamura, 1994). Parasitization of *G. mellonella* by *P. caudata* was done either manually by microinjection in the host of larvae obtained by dissecting the uterus of gravid females (Grenier et al., 1974) or by direct oviposition of the females onto host larvae maintained between two latticed frames (Grenier, 1977).

4.4.1.3. Influence of Host Age at Parasitization

Host age, another key factor for successful parasitization by tachinids (Mellini, 1986), had a significant effect on the development of *P. rufifrons* in

G. mellonella. The best parasitoid yields were obtained by infesting last instars (Mellini and Coulibaly, 1991). In the host–parasitoid system *H. zea*—*A. marmoratus*, early last instars were the most suitable hosts for parasitization, as lower percentages of late last instars were successfully parasitized (Bratti et al., 1992). The growth parameters (development and size) of the different instars of *P. caudata* varied greatly when parasitization occurred at the lowest weight versus the maximum weight of the last instar of *G. mellonella* (Grenier et al., 1974). The tachinid *P. nigrolineata* was able to parasitize any instar of its natural host *O. nubilalis* except the first one, which is probably due to an efficient synchronization with its host cycle (Rhamadane et al., 1987, 1988). Advanced host larval stages are the most suitable for *E. larvarum* development, but the complete life cycle of this parasitoid was also obtained in younger larvae, unless they were too small (Baronio et al., 2002).

4.4.1.4. Abiotic Conditions for Preimaginal Development

Temperature, relative humidity (RH), and photoperiod may all affect the success and duration of preimaginal development in tachinid parasitoids. For some of them, like *C. concinnata* in *L. dispar*, the development from first instar to adult occurred within a relatively wide range of temperature (15.6–29.4 °C) and humidity (20–60%) (Fusco et al., 1978). However, most species were reared at more restricted temperature and humidity levels, for instance 27±1 °C and 65±5% RH for *A. marmoratus* (Gross, 1994), or 26±1 °C and 75±5% RH for *E. larvarum* (Dindo et al., 1999). Some species, including *P. caudata* (in *G. mellonella*) and *L. franki* (in *M. quadrilineatus*), were successfully cultured at lower temperatures (around 21 °C) (Delobel and Laviolette, 1969; Suazo et al., 2006). High temperatures (over 30 °C) may be detrimental for tachinid parasitoid development. For example, in the system *G. mellonella*–*P. rufifrons*, no parasitoid pupae were obtained at 35 °C (Mellini and Coulibaly, 1991). A photoperiod of 16:8 (L:D) (Fusco et al., 1978; Mason et al., 1991; Mellini and Coulibaly, 1991; Giangiuliani and Farinelli, 1995; Dindo et al., 1999) or 14:10 (L:D) (Coulibaly and Fanti, 1992; Gross, 1994) was adopted for the rearing of the preimaginal stages of many tachinids. Long-day photoperiods could be necessary to prevent the host from entering diapause and inducing a developmental arrest of the parasitoid larva (e.g. *P. nigrolineata* in *O. nubilalis*) (Rhamadhane et al., 1988). The tachinid *I. aldrichi* exhibits an obligatory diapause that requires a period of 4 months at a temperature near 4 °C to be broken (Simões and Grenier, 1999).

A rearing system for the phorid fly *P. curvatus* was described by Vogt et al. (2003); parasitized host ants were held for 13 days at 27 °C and 60% humidity, and then moved to well-ventilated boxes ("attack boxes") placed at 28 °C, 80% RH, and a 12:12 (L:D) photoperiod. Emergence occurred about 10 days later. High RH was necessary for parasitization and fly survival, thus "attack boxes" were connected with a system producing steam in order to maintain the RH inside the box between 80% and 90%.

4.4.2. In Vitro Production

For all entomophagous insects, a main concern for their use in biological control is the production of quality organisms at a reasonable cost. In vitro rearing may be an effective tool to produce these natural enemies in the absence of hosts or prey. The availability of an artificial medium could be economically desirable by simplifying the production line. However, for dipteran parasitoids, it is more appropriate to speak of in vitro "rearing" on a relatively small scale instead of actual "production," as high adult yields have so far been obtained in only a few species on artificial media. Data on this topic are available for only Tachinidae and Sarcophagidae. The first step to succeed in the in vitro rearing is the evaluation of the nutritional requirements of the insect to be reared (see Chapter 7). Nevertheless, there are other needs and rearing conditions to satisfy, as it is briefly described in this section.

4.4.2.1. Nutritional Requirements and Other Needs of Dipteran Parasitoids

Various analytical approaches were employed to determine the nutritional requirements and define the medium composition: larval food analyses, whole-carcass analyses of the parasitoid, a nutritional balance sheet, or dietary supplementation and deletion studies, the latter being difficult to apply to endoparasitoids, due to the need to maintain stable pH and osmotic pressure.

The basic qualitative requirements for dipteran parasitoids are not very different from those of free-living insects, but some differences exist, particularly concerning aromatic amino acids and surfactants.

4.4.2.1.1. Nitrogen Sources

Nitrogen sources could be provided as proteins, peptides, or free amino acids. Dipteran parasitoids are carnivorous insects that exhibit very fast growth during larval stages (Grenier, 1980), especially when they have access to atmospheric oxygen through respiratory funnels. Thus, they need a protein-rich medium with a well-balanced composition to avoid the loss of time and energy in conversion metabolism. The 10 usual essential amino acids are required (see Chapter 7), but other ones are highly beneficial for normal growth. In Tachinidae, aromatic amino acids are required, especially at the end of larval development for cuticle tanning (Bonnot et al., 1976). Some tyrosine-rich dipeptides, such as β-alanyl-tyrosine, are present in mature tachinid larvae, for example in *P. caudata* (Delobel and Bonnot, 1976). To maintain acceptable levels of osmotic pressure, part of the essential amino acids must be provided in proteins or peptides, but some free amino acids may be required in some species. Casein, lactalbumine, ovalbumine, serumalbumine, soybean extract, and yeast are commonly used as crude proteins or hydrolysates. Supplying aromatic amino acids is complicated because some free amino acids may be toxic at high concentrations, like phenylalanine, or have low solubility, like tyrosine (Grenier et al., 1994;

Grenier, 2012). The best way to deliver aromatic amino acids is within peptides or proteins rich in tyrosine and/or phenylalanine. However, such types of polypeptides are not common in commercially available food sources and therefore are difficult to obtain in the marketplace.

4.4.2.1.2. Lipids

Lipids are mainly considered as a source of energy. Similarity of lipid composition between the parasitoid and host in some species suggests that host lipids may be absorbed with few modifications, especially neutral lipids (Delobel and Pageaux, 1981). The addition of lipids in diets improves the survival and fecundity of dipteran parasitoids. Some polyunsaturated fatty acids are necessary for development and reproduction. Phospholipids are important as key components in cell membranes and also play a role as molecular signals. Dietary sterols are required by all insects (House, 1961). Lipids may be supplied as free fatty acids or triglycerides, but their incorporation within aqueous base formulations requires surfactants. Free fatty acids are toxic for some species, but the degree of toxicity may depend on the emulsifying agent used (Thompson, 1977). Moreover, surfactants could have detrimental effects on larval respiration by modifying the cuticle characteristics near spiracles (Grenier et al., 1994).

4.4.2.1.3. Carbohydrates

They are usually considered as a source of energy, and hence a specific qualitative supply is not required. Trehalose, the most common nonreducing disaccharide in insects, plays an important role in the general metabolism (Thompson, 2003). To maintain the osmotic potential at an acceptable level, oligosaccharides (e.g. glucose) could be replaced by polysaccharides (e.g. glycogen). Sucrose may act as a feeding stimulant (Cohen, 2004).

4.4.2.1.4. Miscellaneous

Other nutrients, such as vitamins, organic acids, nucleic acids, and minerals, are required in addition to the three basic types of nutrients (proteins, lipids, and carbohydrates). These requirements are generally similar among insects, but they are poorly investigated in dipteran parasitoids. Nevertheless, inorganic cations, especially a proper ratio between Na and K, were found to be critical for some species.

4.4.2.1.5. Other Needs

Important physiological conditions need to be fulfilled besides nutrition to ensure complete in vitro development of dipteran parasitoids, including physicochemical factors, respiration, excretion, and hormonal titers. The pH has to be stabilized between 6.5 and 7.5, even if some species seem highly tolerant to extreme levels, such as *E. bryani* (Nettles, 1986). In different tachinid species, the optimum dietary osmolarity varies between 350 and 450 mOsm (Grenier, 2012).

The presentation of the food as a liquid medium, gelled or supported by absorbent cotton, is crucial for the fulfillment of larval respiration. The open functional gut in many tachinid larvae is an important factor to be taken into consideration. It becomes necessary to supply diet medium in large quantities or regularly renew it to reduce its degradation by excretory products. The life cycle of many Tachinidae is closely dependent upon the host hormonal levels, but this dependence may not be so strict in an in vitro system. The life cycle of the tachinid *P. caudata* is synchronized with its host cycle in nature, but it does not need hormones during the first and second molts when reared in vitro (Grenier et al., 1975). Nevertheless, the addition of 20-OH ecdysone to the media was necessary to trigger the first molt in *P. nigrolineata* and *P. rufifrons* (Grenier, 1988b; Fanti, 1991). Other tachinids, such as *L. diatraeae* and *E. bryani*, which develop independently of host hormones, can also develop from egg to adult in media devoid of hormones (Grenier et al., 1978; Nettles et al., 1980).

4.4.2.2. In Vitro Rearing

Early successful attempts to rear dipteran parasitoids on artificial media were obtained in sarcophagid flies attacking advanced host larval stages and involving simple host–parasitoid relationships. Research mostly focused on *A. housei*, which was reared from egg to adult with high adult yields in media based on raw materials (salmon and fresh liver) (House and Traer, 1949; Coppel et al., 1959) and on chemically defined media containing amino acids, RNA, dextrose, inorganic salts, B vitamins, choline, inositol, and agar as a gelling agent (House, 1954). Subsequently, House (1966) showed that inclusion of vitamin E was essential for reproduction and continuous in vitro culture of this parasitoid. Other successful attempts at rearing sarcophagid flies in vitro were attained in the 1950s (Arthur and Coppel, 1953; Smith, 1958).

All subsequent research on in vitro rearing of dipteran parasitoids has focused on tachinids, which have generally proven to be more difficult to be reared on artificial media than sarcophagids because of their more complex interactions with the host and more specific nutritional needs. Limited success was obtained with species displaying developmental synchrony with the host. For instance, *P. caudata* was reared in vitro only from the first to the third larval instar (Grenier et al., 1975). The medium composition was based on the body composition of the fly larvae and was possibly deficient in protein, but supplementing it with free amino acids, beyond a certain level, was detrimental for parasitoid development due to excessive osmotic pressure (Bonnot et al., 1976). *Pseudogonia rufifrons* was reared from first-instar larva to adult (with very low yields) on a subnatural medium based on *G. mellonella* pupal homogenate (Bratti and Monti, 1989) and on a medium containing bovine serum as the main ingredient, combined with trehalose, chicken egg yolk, and *G. mellonella* pupal hemolymph (Mellini et al., 1994). In both cases, host material (obtained from young pupae and thus rich in ecdysteroids) proved essential to promote the parasitoid molt from the first to the second instar. Subsequently,

Fanti (1991) showed that the first molt could be obtained by supplementing an artificial medium devoid of insect components with 20-OH ecdysone. *Archytas marmoratus*, another species exhibiting complex behavioral and physiological interactions with the host, was reared on a medium from Nettles et al. (1980) and on veal homogenate-based media, but parasitoid growth and development were poor despite the addition of lepidopterous pupal extracts (Bratti, 1994; Farneti et al., 1997). Limited results were also obtained with in vitro rearing efforts of *Lydella thompsoni* Herting, a larval parasitoid of *O. nubilalis*. Synchronization with the host life cycle in south France requires parasitization of a secondary host (*Archanara* spp., noctuids developing in reeds) for the first spring generation of *L. thompsoni* (Galichet et al., 1985; Plantevin and Grenier, 1990). In particular, first instars failed to molt to second instar on media based on pupal extract, either of *O. nubilalis* or of two nonhost lepidopterous species (Bratti, 1994). According to the author, this failure was possibly due to the fact that the extract was derived from an unsuitable host stage (the pupa) and/or nonpermissive hosts of *L. thompsoni*.

Complete in vitro development on an insect-free artificial medium has been obtained only for tachinids that do not display developmental synchrony with their hosts. For example, *L. diatraeae* was the first tachinid successfully reared in vitro from first instar to adult (although only a few flies were obtained). The rearing was performed on an artificial medium gelled with agarose and containing organic acids, inorganic salts, amino acids, B and fat-soluble vitamins, gelatin, protein hydrolysates, glycogen, adenosine triphosphate, lecithin, corn oil, and cholesterol (Grenier et al., 1978). Subsequently, *E. bryani* was reared from first-instar larva to adult on a medium with a similar composition combined with other ingredients (including soy flour and chicken egg yolk). Yields reached as high as 46% of the parasitoid maggots, but only after the latter had been allowed to develop in the host for the first 18–24 h before being transferred onto the artificial media. It was estimated that an average of 2000 flies per liter of medium could be obtained (Nettles et al., 1980; Bratti and Nettles, 1992). Another parasitoid of lepidopterous larvae, *P. laxa*, was reared from first-instar larva to adult on the same media, with low yields, whereas yields of about 80% were obtained on a subnatural substrate based on pupal homogenate of the host *H. zea* (Bratti and Nettles, 1995). Host material improved dramatically the in vitro growth and development of *P. laxa* (despite its nondependence on host hormones), possibly because it better satisfied the parasitoid nutritional needs as compared with the media devoid of insect components.

The best results with rearing a tachinid parasitoid on artificial media have been obtained with *E. larvarum*. Besides displaying a nonsynchronized development with the host, this parasitoid also builds primary integumental respiratory funnels and thus has a simpler life cycle than the species mentioned in this chapter, which all form secondary funnels. These characteristics, in addition to polyphagy and gregariousness, make this parasitoid particularly suitable for in vitro rearing (Dindo, 2007). The first artificial medium for *E. larvarum* was combined with *G. mellonella* pupal homogenate (Mellini et al., 1993), but host components were progressively deleted in subsequent studies. Different insect-free media were developed, based on skimmed milk (Mellini and Campadelli, 1996; Dindo et al., 2010), tissue culture media (TNM-FH) (Bratti et al., 1995), or veal homogenate (Dindo et al., 1999). The media also contained different amounts of yeast extract, chicken egg yolk, and other additives (sucrose, fetal bovine serum, or wheat germ), and were gelled with agar. Physical support is very important for successful in vitro rearing of *E. larvarum* because, in living hosts, its larvae breathe atmospheric oxygen from the beginning of their development through the primary funnels, as explained in the Section 4.3.3. Fecund adults were obtained on different media, with puparial yields (based on eggs) and adult emergence (based on puparia) reaching as high as 55% and 70%, respectively. These yields were comparable to those usually obtained from conventional in vivo rearing. On the tissue culture medium and a veal medium, *E. larvarum* was reared for five and three generations, respectively, with no drops in adult longevity and fecundity (Bratti et al., 1995; Dindo et al., 1999). Moreover, Mellini and Campadelli (1999) demonstrated that this tachinid can be mass produced in glass petri dishes (easier to manage) instead of multiwell plates. Further improvements in the in vitro rearing procedure included the use of absorbent cotton in lieu of more expensive agar (Dindo et al., 2003).

4.4.2.3. Continuous In Vitro Culture

Continuous in vitro culture of dipteran parasitoids with complete exclusion of the host, which implies direct oviposition on artificial substrate, has been achieved only for sarcophagid flies (e.g. *Sarcophaga aldrichi* Parker). This parasitoid was reared for many generations on pork liver and fish from larvae deposited on the medium by ovoviviparous females (Arthur and Coppel, 1953). Conversely, this goal has not been reached for any tachinid. The in vitro rearing of tachinid parasitoids is usually carried out by removing eggs or larvae from previously parasitized hosts or by dissecting gravid females in a physiological solution and placing them on the artificial media. Studies leading to eliminating the host from some steps of the parasitoid line of production have been performed on species showing indirect oviposition strategies, like *P. rufifrons* using artificial leaves made of bee wax (Mellini et al., 1980). Hatching, which normally occurs after the egg has been ingested, was induced outside of the host by centrifuging the eggs at 1000 rpm for 10–15 s. The newly hatched larvae were viable and could be used for in vitro rearing trials (Mellini and Campadelli, 1989).

In the case of *E. larvarum*, which oviposits directly on the host body, captive parasitoids oviposit most eggs on host larvae. However, whether hosts are available or not, many eggs are also indiscriminately oviposited on the cage surfaces and are usually lost. Starting from this observation, Dindo et al. (2007) used eggs oviposited on a plastic sheet to rear *E. larvarum* on a medium based on skimmed milk developed by Mellini and Campadelli (1996).These out-of-host eggs proved competitive with those removed from *G. mellonella* larvae to rear

E. larvarum in vitro, in terms of immature survival, adult production, and fly quality. This study showed that, at least for one generation, in vitro production of *E. larvarum* may be disengaged from dependence on a living host. The quality of in vitro-cultured tachinids (in terms of fecundity, egg viability, and adult survival) decreased over subsequent generations when plastic sheets were provided as oviposition substrates to parental or first-generation females (Marchetti et al., 2008). Moreover, in both studies cited in this section, the eggs oviposited on the plastic sheet were dramatically fewer than those oviposited on host larvae. Further research is needed on methods to stimulate oviposition by *E. larvarum* on artificial substrates and directly on artificial media. A better understanding of the chemical and physical cues involved in oviposition of this parasitoid could be valuable, as is shown for the congeneric species *E. japonica* where cylindrical shapes simulating hosts contributed to enhanced oviposition (Tanaka et al., 1999).

Table 4.2 provides examples of dipteran parasitoids reared on artificial media divided on the basis of the presence or absence of insect components, as suggested by Grenier and De Clercq (2003).

4.4.3. Adult Maintenance

Factors such as food, available space, and other abiotic conditions may affect adult survival, mating, fecundity, and success of parasitoid rearing. These aspects will be outlined in this section, especially with regard to tachinids.

4.4.3.1. Food and Water Supply

Although the nutritional needs of adult parasitoids are poorly understood, it is known that most of them feed on nectar or honeydew. These food sources, which are rich in carbohydrates, provide for their energetic requirements (Thompson and Hagen, 1999). According to Wäckers (2003), nectar might also contribute to egg maturation and other physiological processes despite the low content in amino acids, proteins, and lipids. In nature, some tachinids are able to feed on pollen (Campadelli, 1977). In captivity, adult tachinids are often provided with absorbent cotton soaked in honey- or sugar- water solutions and/or sugar cubes (Fusco et al., 1978; Mason et al., 1991; Mellini and Coulibaly, 1991; Coulibaly and Fanti, 1992; Kfir et al., 1989; Dindo et al., 1999, 2010; Sourakov and Mitchell, 2002). Honey solutions may also be gelled (Grenier, 1977; Quednau, 1993). Furthermore, the life span and fecundity of adult flies may be enhanced by providing raisins (Giangiuliani and Farinelli, 1995; Coombs, 1997) or mixtures of different carbohydrates and protein or yeast hydolysates (Campadelli, 1977). However, as emphasized by Wäckers (2003), little attention is usually given to the selection of the adult parasitoid food source based on the assumption that any sugar-rich substrate may be suitable. More attention should be focused on comparing different food sources with the aim to optimize the performance of adult parasitoids. However, adult feeding did not always prove to

TABLE 4.2 Examples of Dipteran Parasitoids Reared on Artificial Media with (W) or without (WO) Insect-Derived Material (See Text for Details)

Parasitoid		Type of Artificial Medium	Development Obtained
Family	Species		
Sarcophagidae	*Agria housei*	WO	First instar to fecund adult—many generations
	Sarcophaga aldrichi	WO	First instar (larviposited directly on medium) to fecund adult—many generations
Tachinidae	*Archytas marmoratus*	W	Second instar to pupa Third (= last) instar to adult
	Eucelatoria bryani	WO	First instar to adult
	Exorista larvarum	W WO	Egg to fecund adult Egg to fecund adult (3–5 generations)
	Lixophaga diatraeae	WO	First instar to adult
	Lydella thompsoni	W	Second instar to pupa
	Palexorista laxa	W or WO	First instar to adult
	Phryxe caudata	WO	First instar to third (= last) instar
	Pseudogonia rufifrons	W	First instar to adult

be indispensable. For example, although adults of the bombyliid *V. cana* refused to feed in captivity, females mated and oviposited in high numbers, but their survival was low (Du Merle, 1966).

Feeding on host hemolymph often occurs among hymenopteran parasitoids by stinging the host body with their piercing ovipositor (Jervis and Kidd, 1986), but host feeding is far less common in dipterans (Thompson and Hagen, 1999). This behavior was documented in *E. bryani*, which may perforate the host integument with a piercing organ (see Section 4.3.2) and feed on the hemolymph coming out from the puncture. Nettles (1987) demonstrated that the fecundity of this parasitoid was prolonged by host feeding, while a solution of free amino acids or bovine serum albumin was not a good substitute. Exposing host larvae to *E. bryani* in captivity may be beneficial not only for parasitization but also to allow host feeding.

Distilled water may be supplied to dipteran parasitoid adults by spraying over the cages (Quednau, 1993) or through moistened absorbent cotton in drinking troughs (Mason et al., 1991; Dindo et al., 2010). Collazo et al. (1997) described an automatic system to supply water to mass-produced *L. diatraeae* adults, which reduced production costs and improved adult quality. Purified tap water, such as reverse osmosis water, could be a good substitute for distilled water (J. Morales Ramos, personal communication, November 2012). Antimicrobials like chloramphenicol (Coulibaly and Fanti, 1992) may be added to the water source, but the use of nipagin, an antifungal frequently used in the food for phytophagous insects, should be avoided due to toxicity (Grenier, 1977).

4.4.3.2. Space Availability and Adult Management

Dipteran parasitoid adults are generally good flyers. The space available (cage volume) and adult density (number of individuals per cage) are two of the parameters to be considered for survival and mating success, and this is often a crucial phase for life cycle completion. Mating occurred easily when adults of *L. franki* were kept in large screen cages measuring 1.5×1.5×1.2m (with 10 males and 10 females) (Suazo et al., 2006). Larger cages measuring 3×4×2m were required for successful mating in the bombyliid *V. cana* (Du Merle, 1966). For other species like *E. larvarum*, mating was obtained in smaller Plexiglas cages measuring 40×30×30cm, each containing 70–80 flies with a 1:1 sex ratio (Dindo et al., 2007). Suitable small-sized (30.5×30.5×15.5cm) cardboard cages for laboratory rearing and field release of *A. marmoratus* were described by Gross (1990b). Other parameters that are crucial for mating are sex ratio, age, and light intensity. For example, in some cases, mating was best achieved when the adult sex ratio was male biased (Odell and Godwin, 1979; Zhang et al., 2003) or when newly emerged females were kept with older males (Ho et al., 2011).

For better management of parasitoid rearing, it could be helpful to know the age of the adults in relation to their reproductive physiology. A special automatic collector was designed by Grenier and Ogier (1978) for controlling the emergence of tachinid flies, allowing an accurate determination of egg maturation dynamics (Grenier et al., 1982). Estimations of the weights of *L. diatraeae* pupae may be obtained by measurements of their length and diameter (based on preestablished mathematical equations) (Grenier and Bonnot, 1983). The approach also allows evaluation of weight and fecundity of emerging adults because adult weight correlates with fecundity.

Oviposition or larviposition substrates (hosts for direct strategists or artificial devices for indirect strategists) have to be periodically exposed to the adults inside cages, as described in the Section 4.4.1. The addition of suitable soil and stones to mimic the natural environment was necessary for oviposition in *V. cana* (Du Merle, 1966).

4.4.3.3. Abiotic Conditions

Adult biological parameters (mating, preoviposition period, longevity, and parasitization capacity) are also affected by abiotic factors. Adults of *B. pratensis*

(Meigen) could be maintained at a relatively wide range of temperatures (15–26 °C) and humidities (50–90%) (Odell and Godwin, 1979). But for most dipterans, in particular Tachinidae, adults were usually maintained at a restricted range of temperature and RH (Coulibaly and Fanti, 1992; Dindo et al., 1999; Suazo et al., 2006). High RH levels (more than 80%) were required by the phorid *P. curvatus* (Vogt et al., 2003) and some tachinids, such as *Argyrophylax basifulva* (Bezzi) (Godfray, 1985).

The common photoperiod adopted for adult rearing is 16:8 or 14:10 (L:D), but natural light conditions of approximately 12:12 L:D were found suitable for *L. franki* (Suazo et al., 2006). Mellini (1991) reported that tachinid adults are usually active during the daytime, for instance *P. rufifrons* oviposits mostly between the sixth and the eighth hour of the photophase (Fanti, 1984), but phonotactic ormiine tachinids are rather active during the night (Fowler, 1987). Wineriter and Walker (1990) adopted a photoperiod of 16:8 to rear *O. ochracea* (Bigot), without specifying when the parasitoid adults were active.

Mating often requires high light intensity. For example, 8000–10,000 lux was necessary for *L. thompsoni* (Galichet et al., 1985). In some cases, mating also appeared to be influenced by wind conditions. A strong breeze provided by an electric fan was necessary for 50% mating success in *A. basifulva* (Godfray, 1985). However, some species like *E. larvarum* and *L. diatraeae* easily mate in captivity without special requirements of light intensity or wind speed (Grenier et al., 1982; Dindo et al., 2007). For this reason, such species are particularly suitable for mass production.

4.4.4. Quality Control

Quality parameters and assessment procedures of entomophagous insects have been widely discussed by van Lenteren (2003) and are further discussed in Chapter 9 of this book. As remarked by Grenier (2009), "quality" is not an absolute concept, and its definition depends on criteria that differ according to the production aims of each particular entomophagous arthropod (dipteran parasitoids in our case).

No guidelines for quality control have so far been developed for any dipteran parasitoid due to the relatively few species that have been mass produced. Research on quality control has focused on a limited number of tachinid species. The quality of in vivo-produced *L. diatraeae* has been the subject of different research works. Pintureau and Grenier (1992) showed that modifications of biological characteristics, such as development time and puparial size, were induced by long-term laboratory rearing of this tachinid on *G. mellonella*. Subsequently, Pintureau et al. (1995) compared the genetic variability of a number of characters (including puparial length in millimeters) between a *L. diatraeae* laboratory strain and a wild strain from Cuba. The laboratory strain showed a lower genetic variability, probably due to a genetic drift or bottleneck effect; reduced genetic variability is a concrete risk of the long-term rearing of beneficial insects that has to be taken into consideration. In the Cuban strain, two lines were successfully selected for puparial size (lowest and highest length), for eight generations, suggesting that such a selection could be used to improve the efficacy of this parasitoid as a biocontrol agent. Body size was found to be correlated with fitness and fecundity in *L. diatraeae* (King et al., 1976) and also in other tachinids like *E. bryani* (Reitz and Adler, 1995) and *C. concinnata* (Bourchier, 1991). In Cuba, attempts were made to maintain (and even improve) the quality of mass-reared *L. diatraeae* by modifying the composition of the *G. mellonella* diet, while taking into account production costs. A diet composed of Torula yeast, glycerine, corn, soybean flour, wheat, and sugar guaranteed the development, survival, and reproduction of both the host and the parasitoid. The *L. diatraeae* adult emergence, egg fertility, and flying ability were comparable to those obtained with conventional host diets, which included wax and honey (both expensive ingredients). The parasitism rate of the target host *D. saccharalis* did not differ significantly between laboratory-cultured and wild-collected *L. diatraeae* under greenhouse conditions (Alemán, 2000; Vidal et al., 2002). Both fecundity and flying ability (an important characteristic for augmentative release) were negatively affected in the laboratory population when compared to the wild population (Alemán et al., 2001).

Host weight is often correlated with the tachinid weight and indirectly influences fecundity or other parameters related to the parasitoid body size. Medium composition may also affect different characteristics of insect hosts and subsequently impact their parasitoids (Cohen, 2004). As shown for *L. diatraeae* and other tachinid species, host diet is very important for parasitoid quality. For instance, *A. marmoratus* puparial weight and other characters such as longevity were affected by the host diet (Gross et al., 1996). For gregarious species (*Eucelatoria* sp., *L. diatraeae*, and *E. larvarum*), puparial weight may also be affected by the number of parasitoids per host (Ziser et al., 1977; Grenier, 1981; Mellini and Campadelli, 1997).

Quality loss due to inbreeding depression (i.e. reduced fitness in a population as a result of breeding between related individuals) is a well-known risk of long-term insect rearing. Nakamura (1996) showed that, for laboratory-cultured *E. japonica*, the rates of adult emergence decreased in the F_2 generation under inbreeding conditions. He suggested adopting rotational breeding by pairing females with unrelated males in small and medium-scale laboratory rearing to prevent this problem. However, as emphasized by Grenier and De Clercq (2003), in the absence of inbreeding depression, crosses between different inbred strains of the same species could be a useful tool to select favorable characteristics or minimize the harmful effects of laboratory adaptation. The final objective remains the optimization of the efficiency of a parasitoid strain for biological control applications.

Quality control is even more important for in vitro production because there is a higher probability of poor parasitoid performance compared to in vivo rearing. Some parameters to be considered for comparison of artificially versus naturally reared entomophagous insects were suggested by Grenier and De Clercq (2003). Parasitism of the natural host *L. dispar* by *E. larvarum* reared on a skimmed milk-based medium was compared in the laboratory and in semi-field conditions with parasitoids reared on the factitious host *G. mellonella*. No significant difference in percentages of successfully parasitized *L. dispar* larvae was found between the in vitro and in vivo-produced *E. larvarum* in laboratory studies. The semifield trials were conducted in a cork forest in northern Sardinia, Italy, and each experimental unit consisted of a tree branch wrapped in a screen net. Five *L. dispar* mature larvae were released into the net with a couple of in vitro or in vivo-reared *E. larvarum*. The host larvae were removed after parasitoid flies died, which occurred within 48–72 h after release. In vitro-reared females oviposited fewer eggs and induced lower host mortality than the in vivo-reared ones (Dindo et al., 2002). Subsequently, Dindo et al. (2006) compared a number of biological and biochemical traits among *E. larvarum* reared on a milk medium, a veal homogenate-based medium (Dindo et al., 1999), and factitious host *G. mellonella*. Surprisingly higher puparial yields and weights were obtained in both artificial media than in host larvae; however, there was no significant difference in fecundity among the treatments. In another experiment with females emerged from puparia of similar weight, the host-reared flies produced significantly more eggs than the milk medium-reared ones. For in vitro-reared *E. larvarum*, therefore, puparial weight alone is not a reliable quality criterion to predict other traits like fecundity. Chemical analyses of *E. larvarum* mature larvae reared on both media showed lower amino acid content with deficiency of aromatic amino acids and excess of proline compared to the in vivo-reared parasitoids. The amino acid imbalance found in the in vitro-reared larvae could explain the lower number of eggs produced by the developing females, but this has not been experimentally proven.

4.4.5. Storage and Shipment Procedures

Quality control is also important during the storage and shipment processes of entomophagous insects. Leopold (2007) emphasized the importance of preserving entomophagous insects at low temperatures to extend their shelf life, and Coudron (2007) discussed the possibility of improving their cold storage in egg, immature, as well as adult stages. Each stage responds differently to short- and long-term storage at low temperature, and their ability to tolerate cold storage depends on the nutrient origin (artificial vs. natural food). In particular, it was better to feed adults of *Podisus maculiventris* (Say) on natural food, instead of artificial diet, for long-term storage at 10 °C. Conversely, it is more advantageous to rear this bug on artificial diet instead of natural food for short-term storage. This approach might also be used with dipteran parasitoids.

Storage of natural enemies is usually done by placing them, often at immature stages, at temperatures between 4 and 15 °C for short periods (van Lenteren and Tommasini, 2003). Data on storage of dipteran parasitoids are limited to tachinids. Fusco et al. (1978) showed that *C. concinnata* may be stored at 10–15.6 °C as developing maggots in *L. dispar* larvae for up to 2 weeks and as puparia for 2–4 weeks. Also *Sturmiopsis inferens* (Townsend) was stored in the pupal stage for 30 days at 15 °C without any adverse effects on fly emergence, female mating, and male longevity, whereas female fertility was slightly reduced (Easwaramoorthy et al., 2000). Leopold (1998) provides examples of tachinid storage as puparia for 8 months under diapausing conditions and suggests that dormancy could be exploited in developing cold-storage techniques to facilitate mass rearing. Gross and Johnson (1985) reported that *A. marmoratus* maggots extracted from females, stored for 14 and 21 days at 13 °C, yielded 89% and 51% parasitization of *H. zea* larvae, respectively. Extracted maggots could not be maintained at 13 °C for more than 3 days, as longer storage periods led to rapid decline of successful parasitization. Cryopreservation is a delicate technique, even for insect embryos, and is still in an early stage of research (Leopold, 1998).

Short-term storage (5 days) of in vitro-reared *E. larvarum* was possible at 15 or 20 °C. When the medium with eggs was restored at the standard rearing temperature of 26 °C, the eggs hatched and the parasitoids developed to the adult stage; however, fewer adults were obtained compared to controls, especially from eggs stored at 15 °C (Dindo et al., unpublished data).

Studies on shipment procedures for dipteran parasitoids are very scant. McInnes et al. (1976) described a method of shipping and handling tachinid puparia in individual gelatin capsules securely placed in holes in an expanded polystyrene block. On arrival, the blocks could be examined daily for adult emergence, which occurred inside the capsules. Only those containing well-formed adults were removed.

In recent years, a number of developments have occurred in insulation, energy storage, and humidity control that provide the opportunity for improving the conditions during transportation of insects. New vacuum panel insulation is up to 10 times more efficient than conventional expanded polystyrene, providing longer temperature control in smaller containers during shipping. Phase change materials, which absorb or release energy as they change from solid to liquid, are available with a wide range of transition temperatures that provide better options than the zero Celsius supplied by ice packs. Semipermeable membrane sachets of saturated salt solutions can maintain required humidity conditions inside sealed containers for extended periods. All these products are commercially available, allowing suitable combinations to be selected for the transport of most organisms for any desired temperature and humidity and for any required transport period while maintaining high standards of quality at reception (A. Parker, personal communication, July 2012).

4.5. PERSPECTIVES AND CONCLUDING REMARKS

Production of dipteran parasitoids is deeply related to the utilization of these beneficial insects in applied biological control. Future research should focus on the possibility of better exploiting dipteran parasitoids (in particular, but not exclusively, tachinids) in biocontrol strategies. The examples of success achieved in biological control discussed in this chapter should stimulate efforts to increase the use of species that have proven to be effective. For instance, applied biological control of lepidopterous sugarcane borers with *L. diatraeae*, currently performed successfully in the Americas, could be applied in other sugarcane-producing countries like China and other Asian countries. Moreover, the large-scale results obtained with tachinids, especially against sugarcane borers, should also stimulate the relaunch of research and application programs concerning *A. marmoratus*, *E. bryani*, and *P. laxa*, which appeared promising for applied biological control and mass production, even in vitro, in the 1980s and 1990s but were later neglected. In the literature, we have not found any information about the reasons why these experimental programs were neglected. As was evoked in the introduction, we may speculate that (1) the programs appeared to be expensive, mainly due to rearing costs—among other things, the bigger size of tachinids, compared to hymenopteran parasitoids, is unfavorable for mass production, as more food and space are necessary to multiply them; (2) the general image of flies was (and still is) unpopular, which may lead to a preference for other programs using other kinds of insects; and (3) some biological characteristics may be considered as negative for inundative biological control (e.g. adult flies are good fliers and may disperse too much from release points). The good results obtained, for instance with tachinids against sugarcane borers, may also stimulate the relaunch of programs with tachinids of similar biology and behavior, antagonists of maize lepidopterous pests. Moreover, in greenhouses, where the added value of the culture is high, some tachinids (such as *E. larvarum*) could be of interest for controlling caterpillars. Research aimed at improving the rearing technique and field release may make the use of tachinids more convenient and efficient. Future research could also be focused on the search for new dipteran parasitoid species that are suitable for biocontrol.

Increasing the use of dipteran parasitoids in applied biological control requires the improvement of their rearing techniques for mass production. Future research on these issues may deal with species that are recognized as promising biological control agents but are not yet exploited due to difficulties encountered in their rearing. The rearing techniques of the species currently mass produced such as *L. diatraeae* may be rendered more efficient through the development of in vitro rearing procedures. As argued by Grenier and De Clercq (2003), even if this technology cannot completely replace standard in vivo rearing methods, it may contribute to making the production process easier and more flexible.

Mass-rearing capability is one of the main factors that influence the selection of a parasitoid species for applied biological control. The potential of *E. larvarum* as a biocontrol agent could be exploited considering that this tachinid is one of the most promising species for in vivo and in vitro mass production. Additional laboratory and field studies aimed at assessing the efficacy of *E. larvarum* against selected target species could be performed, possibly involving insect producers and farmers. With regard to in vitro rearing in general, Grenier (2009) emphasized that the best research approach would be the constitution of "consortiums" among different institutions, either public or private, aimed at developing research programs and obtaining support from government or supranational commissions. A similar approach would benefit the production of dipteran parasitoids and their utilization in biological control. Aspects of dipteran parasitoid biology of relevance for rearing and issues related to in vivo and in vitro culture, including quality control, storage, and shipment procedures, may all be the subject of further research addressed with novel approaches. For example, nutrigenomics (genomics and proteomics applied to nutrition) may deliver helpful information on the way that nutrition affects the patterns of gene expression by using microarray techniques (see Chapter 7). The implicated genes could be recognized and typified for use as molecular markers for the characterization of the response (e.g. performance levels) of the insect to different nutritional sources (Coudron et al., 2006; Chang and Coudron, 2007). This approach, originally developed for heteropteran predators, might also be applied to dipteran parasitoids with the purpose of improving in vitro rearing and quality control techniques.

Besides nutritional considerations, the insect-rearing field may benefit from food technology principles as discussed by Cohen (2004), such as the extrusion process, flash sterilization, freeze-drying applications, and different approaches for nutrient processing. Also, more mechanization and automation of the production line have to be developed to reduce costs. As a general recommendation, it could be helpful to develop cooperation between biologists, physiologists, chemists, and physicists.

Finally, it should be remembered that insect-rearing technology is useful not only for large-scale field releases, but also on the small scale for entomological research concerning aspects of parasitoid biology, physiology, and behavior. The availability of efficient rearing techniques for tachinids or other dipterans may also be helpful in studies concerning pesticide effects on nontarget species (Marchetti et al., 2009, 2012). Aspects concerning dipteran parasitoid rearing are therefore important for biological control from different points of view, and all deserve to be exploited.

REFERENCES

Aldrich, J.R., Khrimian, A., Zhang, A., Shearer, P.W., 2006. Bug pheromones (Hemiptera, Heteroptera) and tachinid fly host-finding. Denisia 19, 1–17.

Alemán, J., 2000. Quality assessment in *Lixophaga diatraeae* Townsend (Diptera: Tachinidae) mass-rearing. Rev. Protección Veg. 15, 64–65.

Alemán, J., Muñoz, L., Plana, L., Llanes, G., Fernández, M., Vidal, M., García, G.A., 1999. Effect of artificial recycled diet on some indicators of quality in the *Galleria mellonella* Linneo (Lepidoptera: Pyralidae) and *Lixophaga diatraeae* Townsend (Diptera: Tachinidae). Rev. Protección Veg. 14, 161–166.

Alemán, J., Richards, M.L., Plana, L., Llanez, G., Fernández, M., Vidal, M., 2001. Comportamiento de indicadores de calidad en poblaciones salvajes y de laboratorio de *Lixophaga diatraeae* Townsend (Diptera: Tachinidae). Rev. Protección Veg. 16, 15–19.

Allen, G.E., Pape, T., 1996. Description of female and biology of *Blaesoxipha ragg* Pape (Diptera: Sarcophagidae), a parasitoid of *Sciarasaga quadrata* Rentz (Orthoptera: Tettigoniidae) in Western Australia. Aust. J. Entomol. 35, 135–145.

Arrigoni, E.B., 1992. Cana-de-açúcar: moscas e pequeñas vespas fazem o controle. Man. de Controle Biológico. Rio de Janeiro, Soc. Nac. Agr. 21–22.

Arthur, A.F., Coppel, H.C., 1953. Studies on dipterous parasites of the spruce budworm, *Choristoneura fumiferana* (Clemens) (Lepidoptera Tortricidae). I. *Sarcophaga aldrichi* Parker (Diptera: Sarcophagidae). Can. J. Zool. 31, 374–391.

Askew, R.R., Shaw, M.R., 1986. Parasitoid communities: their size, structure and development. In: Waage, J., Greathead, D. (Eds.), Insect Parasitoids. Academic Press, London, pp. 225–264.

Baronio, P., Campadelli, G., 1979. Ciclo biologico di *Gonia cinerascens* Rond. (Dipt. Tachinidae) allevata in ambiente condizionato sull'ospite di sostituzione *Galleria mellonella* L. (Lep., Galleriidae). Boll. Ist. Ent. Univ. Bologna 34, 35–54.

Baronio, P., Sehnal, F., 1980. Dependence of the parasitoid *Gonia cinerascens* on the hormones of its lepidopterous hosts. J. Insect Physiol. 26, 619–626.

Baronio, P., Dindo, M.L., Campadelli, G., Sighinolfi, L., 2002. Intraspecific weight variability in tachinid flies: response of *Pseudogonia rufifrons* to two host species with different size and of *Exorista larvarum* to variations in vital space. Bull. Insectol. 55, 55–61.

Belshaw, R., 1994. Life history characteristics of Tachinidae (Diptera) and their effect on polyphagy. In: Hawkins, B.A., Sheehan, W. (Eds.), Parasitoid Community Ecology, Oxford University Press, New York, pp. 145–162.

Biliotti, E., 1956. Biologie de *Phryxe caudata* Rondani (Dipt. Larvaevoridae) parasite de la chenille processionnaire du pin (*Thaumetopoea pityocampa* Schiff.). Rev. Pathol. Vég. Entomol. Agric. France 35, 50–65.

Biliotti, E., Démolin, G., du Merle, P., 1965. Parasitisme de la processionnaire du pin par *Villa quinquefasciata* Wied. apud Meig (Dipt. Bombyliidae). Importance du comportement de ponte du parasite. Ann. Epiphyt. 16, 279–288.

Blumenthal, E.M., Fusco, R.A., Reardon, R.C., 1979. Augmentative release of two established parasite species to suppress populations of the gypsy moth. J. Econ. Entomol. 72, 281–288.

Bonnot, G., Delobel, B., Grenier, S., 1976. Composition corporelle en acides aminés du parasitoïde *Phryxe caudata* Rond. (Diptera) au cours de sa croissance larvaire. J. Insect Physiol. 22, 505–514.

Bonnot, G., Delobel, B., Grenier, S., 1984. Elevage, croissance et développement de *Phryxe caudata* (Diptera, Tachinidae) sur son hôte de substitution *Galleria mellonella* (Lepidoptera, Pyralidae) et sur milieu artificiel. Bull. Soc. Linn. Lyon 53, 313–320.

Bourchier, R.S., 1991. Growth and development of *Compsilura concinnata* (Meigan) (Diptera: Tachinidae) parasitizing gypsy moth larvae feeding on tannin diets. Can. Entomol. 123, 1047–1055.

Bratti, A., 1994. In vitro rearing of *Lydella thompsoni* Herting and *Archytas marmoratus* (Towns.) (Dipt. Tachinidae) larval stages: preliminary results. Boll. Ist. Ent. "G. Grandi" Univ. Bologna 44, 93–100.

Bratti, A., Nettles Jr., W.C., 1992. In vitro rearing of *Eucelatoria bryani*: improvement and evaluation of factors affecting efficiency. Ent. Exp. Appl. 63, 213–219.

Bratti, A., Nettles Jr., W.C., 1995. Comparative growth and development in vitro of *Eucelatoria bryani* Sab. and *Palexorista laxa* (Curran) (Diptera Tachinidae) fed a meridic diet and a diet of *Helicoverpa zea* (Boddie) (Lepidoptera Noctuidae) pupae. Boll. Ist. Ent. "G. Grandi" Univ. Bologna 49, 119–129.

Bratti, A., Monti, M., 1989. Allevamento "in vitro" delle larve di *Pseudogonia rufifrons* Wied. (Dipt. Tachinidae) su omogeneizzato di crisalidi di *Galleria mellonella* L. (Lep. Galleriidae). Boll. Ist. Ent. "G. Grandi" Univ. Bologna 43, 115–126.

Bratti, A., Nettles, W.C., Fanti, P., 1992. Influence of *Helicoverpa zea* (Lepidoptera: Noctuidae) age during the last instar on rates of parasitization by the larval–pupal parasitoid, *Archytas marmoratus* (Diptera: Tachinidae). Environ. Entomol. 21, 1196–1201.

Bratti, A., Campadelli, G., Mariani, M., 1995. In vitro rearing of *Exorista larvarum* (L.) on diet without insect components. Boll. Ist. Ent. "G. Grandi" Univ. Bologna 49, 225–236.

Bustillo Pardey, A.E., 2009. Acciones para reducir las poblaciones de *Diatraea*. Carta Trimestral—CENICANA 31, 10–15.

Campadelli, G., 1977. Tecniche di allevamento per Ditteri Tachinidi con particolare riguardo agli adulti. Boll. Ist. Ent. Univ. Bologna 33, 215–240.

Campadelli, G., 1988. *Galleria mellonella* L. quale ospite di sostituzione per i parassitoidi. Boll. Ist. Ent. Univ. Bologna 42, 47–65.

Caron, V., Myers, J.H., Gillespie, D.R., 2010. The failure to discriminate: superparasitism of *Trichoplusia ni* Hübner by a generalist tachinid parasitoid. Bull. Entomol. Res. 100, 255–261.

Cerretti, P., Tschorsnig, H.P., 2010. Annotated host catalogue for the Tachinidae (Diptera) of Italy. Stutt. Beit. Natur. 3, 305–340.

Chang, C.L., Coudron, T.A., 2007. Improving fruit fly nutrition and performance through proteomics. In: van Lenteren, J.C., De Clercq, P., Johnson, M. (Eds.), Proceedings of the 11th Meeting of the Working Group Arthropod Mass Rearing and Quality Control—28/10/2007–1/11/2007, Canada Bulletin IOBC Global, Montreal, pp. 24–27 No. 3.

Clausen, C.P., 1962. Entomophagous Insects. Hafner, New York, USA.

Cohen, A.C., 2004. Insect Diets: Science and Technology. CRC Press, Boca Raton, USA.

Colazza, S., Giangiuliani, G., Bin, F., 1996. Fortuitous introduction and successful establishment of *Trichopoda pennipes* F.: adult parasitoid of *Nezara viridula* (L.). Biol. Control 6, 409–411.

Collazo, D., Castellanos, J.A., Verdese, Y.R., 1997. Sistema automatizado para el su ministro de agua a las crias masivas de *Lixophaga diatraeae* (Diptera: Tachinidae). Cienc. Tec. Agric. Protec. Plantas 10, 7–17.

Coombs, M., 1997. Influence of adult food deprivation and body size on fecundity and longevity of *Trichopoda giacomelli* a South American parasitoid of *Nezara viridula*. Biol. Control 8, 119–123.

Coombs, M., Sands, D.P.A., 2000. Establishment in Australia of *Trichopoda giacomelli* (Blanchard) (Diptera: Tachinidae), a biological control agent of *Nezara viridula* (L.). Aust. J. Entomol. 39, 219–222.

Cooper, T.M., Frank, J.H., Cave, R.D., Burton, M.S., Dawson, J.S., Smith, B.W., 2011. Release and monitoring of a potential biocontrol agent, *Lixadmontia franki*, to control an invasive bromeliad-eating weevil, *Metamasius callizona*, in Florida. Biol. Control 59, 319–325.

Coppel, H.C., House, H.L., Maw, M.G., 1959. Studies on dipterous parasites of the spruce budworm, *Choristoneura fumiferana* (Clemens) (Lepidoptera Tortricidae). VII. *Agria affinis* (Fall.) (Diptera: Sarcophagidae). Can. J. Zool. 37, 817–830.

Coudron, T.A., 2007. Improving cold storage of insects with dietary changes. In: van Lenteren, J.C., De Clercq, P., Johnson, M. (Eds.), Proceedings of the 11th Meeting of the Working Group Arthropod Mass Rearing and Quality Control—28/10/2007–1/11/2007, Canada Bulletin IOBC Global, Montreal, pp. 34–37 No. 3.

Coudron, T.A., Yocum, G.D., Brandt, S.L., 2006. Nutrigenomics: a case study in the measurement of insect response to nutritional quality. Ent. Exp. Appl. 121, 1–14.

Coulibaly, F., Fanti, P., 1992. *Eucelatoria bryani Sabr.* (Diptera Tachinidae) rearing on the factitious host *Galleria mellonella* L. (Lepidoptera Galleriidae): effect of host age at exposure to the parasitoid females. Boll. Ist. Emt. "G. Grandi" Univ. Bologna 46, 229–238.

Coulibaly, A.K., Bratti, A., Fanti, P., 1993. Allevamento di *Archytas marmoratus* (Town.) (Dipt. Tachinidae) su *Galleria mellonella* L. (Lep. Galleriidae): numero ottimale di planidi per la parassitizzazione e ritmo di larvideposizione del parassitoide. Boll. Ist. Ent. "G. Grandi" Univ. Bologna 47, 13–25.

Danyk, T., Mackauer, M., Johnson, D.L., 2005. The influence of host suitability and host range of grasshopper species utilized by *Blaesoxipha atlantis* (Diptera: Sarcophagidae) in the field. Bull. Entomol. Res. 95, 571–578.

Delobel, B., Bonnot, G., 1976. Présence de beta alanyltyrosine chez le Diptère Tachinaire *Phryxe caudata* Rond. Importance relative des acides aminés libres, peptidiques et protéiques. Ann. Zool. Ecol. Anim. 8, 493–497.

Delobel, B., Laviolette, P., 1969. Elevage de *Phryxe caudata* Rond. (Larvaevoridae) parasite de *Thaumetopoea pityocampa* Schiff. sur un hôte de remplacement *Galleria mellonella*. C. R. Acad. Sci. Paris 268, 2436–2438.

Delobel, B., Pageaux, J.F., 1981. Influence de l'alimentation sur la composition en acides gras totaux des Diptères Tachinaires. Ent. Exp. Appl. 29, 281–288.

Deng, Z.Y., He, W.Z., Tan, Y.M., Chen, Y.Z., Liang, C.X., 2008. Biological and ecological studies of *Lixophaga diatraeae* Townsend. Sugar Crops of China 4, 4–6.

Deng, Z.Y., Fang, F.X., Liu, H.B., Pan, X.H., Wui, J., Huang, D.F., Liu, X.J., 2010. Field demonstration on control effect of *Lixophaga diatraeae* and *Trichogramma chilonis* on sugar-cane borer. Sugar Crops of China 3, 9–11.

Depalo, L., Marchetti, E., Baronio, P., Martini, A., Dindo, M.L., 2010. Location, acceptance and suitability of *Spodoptera littoralis* and *Galleria mellonella* as hosts for the parasitoid *Exorista larvarum*. Bull. Insectol. 63, 65–69.

Depalo, L., Dindo, M.L., Eizaguirre, M., 2012. Host location and suitability of the armyworm larvae *Mythimna unipuncta* for the tachinid parasitoid *Exorista larvarum*. BioControl 57, 471–479.

Dindo, M.L., 1998. Allevamento di Ditteri Tachinidi su diete artificiali. Atti Accad. Naz. Ital. Entomol. Rend. 46, 105–133.

Dindo, M.L., 2007. Rearing the parasitoid *Exorista larvarum* (L.) (Diptera, Tachinidae) on artificial media. In: van Lenteren, J.C., De Clercq, P., Johnson, M. (Eds.), Proceedings of the 11th Meeting of the Working Group Arthropod Mass Rearing and Quality Control—28/10/2007–1/11/2007, Canada Bulletin IOBC Global, Montreal, pp. 43–46 No. 3.

Dindo, M.L., 2011. Tachinid parasitoids: are they to be considered as koinobionts? BioControl 56, 249–255.

Dindo, M.L., Farneti, R., Scapolatempo, M., Gardenghi, G., 1999. In vitro rearing of the parasitoid *Exorista larvarum* (L.) (Diptera: Tachinidae) on meat homogenate-based diets. Biol. Control 16, 258–266.

Dindo, M.L., Grenier, S., Sighinolfi, L., Baronio, P., 2006. Biological and biochemical differences between in vitro- and in vivo-reared *Exorista larvarum*. Ent. Exp. Appl. 120, 167–174.

Dindo, M.L., Marchetti, E., Baronio, P., 2007. In vitro rearing of the parasitoid *Exorista larvarum* (Diptera: Tachinidae) from eggs laid out of host. J. Econ. Entomol. 100, 26–30.

Dindo, M.L., Stangolini, L., Marchetti, E., 2010. A simplified artificial medium for the in vitro rearing of *Exorista larvarum* (Diptera: Tachinidae). Biocont. Sci. Tech. 20, 407–410.

Dindo, M.L., Marchetti, E., Galvagni, G., Baronio, P., 2003. Rearing of *Exorista larvarum* (Diptera Tachinidae): simplification of the in vitro technique. Bull. Insectol. 56, 253–257.

Dindo, M.L., Verdinelli, M., Baronio, P., Serra, G., 2002. Laboratory and field performance of in vitro and in vivo-reared *Exorista larvarum* (L.), a natural enemy of cork-oak defoliators. In: Villemant, C., Sousa, E. (Eds.), Integrated Protection in Oak Forests (Oeiras-Lisbonne-Portugal). 1–4 Ottobre 2001. IOBC Bull., 25, 147–150.

Disney, R.H.L., 1994. Scuttle Flies: The Phoridae. Chapman & Hall, London, UK.

Du Merle, P., 1966. Modèle de cage permettant d'obtenir la ponte d'un diptère Bombyliidae, *Villa quinquefasciata* Wied. AP. Meig. Entomophaga 11, 325–330.

Easwaramoorthy, S., Kurup, N.K., Santhalakshmi, G., Shanmugasundaram, M., 2000. Effect of low temperature storage on the viability of puparia of *Sturmiopsis inferens* Townsend (Diptera: Tachinidae) a larval parasitoid of sugar-cane moth borers. J. Biol. Control 14, 63–65.

Eggleton, P., Belshaw, R., 1992. Insect parasitoids: an evolutionary overview. Phil. Trans. R. Soc. London (Series B) 337, 1–20.

Ehlers, R.U., Shapiro-Ilan, D.I., 2005. Mass production. In: Grewal, P.S., Ehlers, R.U., Shapiro-Ilan, D.I. (Eds.), Nematodes as Biocontrol Agents, CABI Publishing, Wallingford, UK, pp. 65–78.

Fanti, P., 1984. Ovideposizione di *Gonia cinerascens* Rond. (Diptera: Tachinidae) in condizioni sperimentali: variazioni nel tempo e ritmi giornalieri. Boll. Ist. Ent. "G. Grandi" Univ. Bologna 38, 167–179.

Fanti, P., 1991. Fattori ormonali inducenti la prima muta larvale del parassitoide *Pseudogonia rufifrons* Wied. *(Diptera Tachinidae)* in substrati di crescita in vivo e in vitro. Boll. Ist. Ent. "G. Grandi" Univ. Bologna 45, 47–59.

Farneti, R., Dindo, M.L., Cristiani, G., 1997. In vitro rearing of the larval-pupal parasitoid *Archytas marmoratus* (Townsend) (Diptera: Tachinidae) on oligidic diets: preliminary results. Boll. Ist. Ent. "G. Grandi" Univ. Bologna 51, 53–61.

Feener, D.H., Brown, B.V., 1997. Diptera as parasitoids. Annu. Rev. Entomol. 42, 73–97.

Fowler, H.G., 1987. Field behavior of *Euphasiopterix depleta* (Diptera: Tachinidae): phonotactically orienting parasitoids of mole crickets (Orthoptera: Gryllotalpidae: *Scapteriscus*. J. N. Y. Entomol. Soc. 95, 474–480.

Frank, J.H., Walker, T.J., Bennett, F.D., 1996. The introduction, establishment and spread of *Ormia depleta* in Florida. Biol. Control 6, 368–377.

Frank, J.H., Parkman, J.P., 1999. Integrated pest management of pest mole crickets with emphasis on the southeastern USA. Int. Pest Manage. Rev. 4, 39–52.

Fuester, R.W., Kenis, M., Swan, K.S., Kingsley, P.C., Lopez Vaamonde, C., Hérard, F., 2001. Host range of *Aphantorhaphopsis samarensis* (Diptera: Tachinidae), a larval parasite of the gypsy moth (Lepidoptera: Lymantriidae). Environ. Entomol. 30, 605–611.

Fusco, R.A., Rhoads, L.D., Blumenthal, M., 1978. *Compsilura concinnata*: effect of temperature on laboratory propagation. Environ. Entomol. 7, 15–18.

Galichet, P.F., Riany, M., Agounke, D., 1985. Bioecology of *Lydella thompsoni* Herting (Dip., Tachinidae) within the Rhone delta in southern France. Entomophaga 30, 315–328.

Giangiuliani, G., Farinelli, D., 1995. Technique for the laboratory rearing for *Trichopoda pennipes* F. (Diptera: Tachinidae), an adult parasitoid of the southern green stink bug, *Nezara viridula* (L.) (Hemiptera: Pentatomidae). J. S. Afr. Soc. Hort. Sci. 5, 55–56.

Godfray, H.C.J., 1985. Mass rearing of the tachinid fly *Argyrophilax basifulva*, a parasitoid of the greater coconut spike moth (*Tirathaba* spp.) (Lep. Pyralidae). Entomophaga 30, 211–215.

Godfray, H.C.J., 1994. Parasitoids: Behavioral and Evolutionary Ecology. Princeton University Press, Princeton, USA.

Godwin, P.A., Odell, T.M., 1984. Laboratory study of competition between *Blepharipa pratensis* and *Parasetigena silvestris* (Diptera: Tachinidae) in *Lymantria dispar* (Lepidoptera: Lymantriidae). Environ. Entomol. 21, 1059–1063.

Greathead, D.J., 1980. Biological control of tsetse flies: an assessment of insect parasitoids as control agents. Biocontrol News Inf. 1, 111–123.

Greathead, D.J., 1986. Parasitoids in classical biological control. In: Waage, J., Greathead, D.J. (Eds.), Insect Parasitoids, Academic Press, London, pp. 289–318.

Grenier, S., 1977. Effets nocifs de la Nipagine M sur le parasitoïde *Phryxe caudata* (Dipt.: Tachinidae). Entomophaga 22, 223–236.

Grenier, S., 1980. Développement endoparasitaire et croissance pondérale larvaire du parasitoïde *Lixophaga diatraeae* (Dip.: Tachinidae) dans un hôte de substitution *Galleria mellonella* L. (Lep.: Pyralidae). Entomophaga 25, 17–26.

Grenier, S., 1981. Influence du superparasitisme sur la durée du développement larvaire et le poids du parasitoïde *Lixophaga diatraeae* élevé dans un hôte de substitution *Galleria mellonella*. Ent. Exp. Appl. 29, 69–75.

Grenier, S., 1986. Biologie et physiologie des relations hôtes-parasitoïdes chez 3 Tachinaires (Diptera, Tachinidae) d'intérêt agronomique. Développement en milieux artificiels. Lutte biologique. Thèse de Doctorat d'Etat ès Sci., Lyon UCB/INSA, n°IDE8604, 156 pp. + 61 pp. d'illustrations.

Grenier, S., 1988a. Applied biological control with tachinid flies (Diptera, Tachinidae): a review. Anz. Schadl. Pflanz. Umwelt. 61, 49–56.

Grenier, S., 1988b. Developmental relationships between the tachinid parasitoid *Pseudoperichaeta nigrolineata* and two host species—hormonal implications. In: Boulétreaeu, M., Bonnot, G. (Eds.), Parasitoid Insects, European Workshop, Lyon, 1987. Colloq. INRA, 48, pp. 87–89.

Grenier, S., 2009. In vitro rearing of entomophagous insects—past and future trends: a minireview. Bull. Insectol. 62, 1–6.

Grenier, S., 2012. Artificial rearing of entomophagous insects, with emphasis on nutrition and parasitoids—general outlines from personal experience. Karael. Sci. Eng. J. 2, 1–12.

Grenier, S., Bonnot, G., 1983. Evolution pondérale du parasitoïde *Lixophaga diatraeae* (Dipt. Tachinidae) de la sortie de l'hôte à l'émergence. Estimations des poids par mensurations de la pupe. Entomophaga 28, 259–270.

Grenier, S., De Clercq, P., 2003. Comparison of artificially vs. naturally reared natural enemies and their potential for use in biological control. In: van Lenteren, J.C. (Ed.), Quality Control and Production of Biological Control Agents—Theory and Testing Procedures, CABI Publishing, Oxon, UK, pp. 115–131.

Grenier, S., Delobel, B., 1982. *Pseudoperichaeta insidiosa*, un nouveau Tachinaire (Diptera) élevé dans *Galleria mellonella* (Lepidoptera). Entomophaga 27, 141–146.

Grenier, S., Liljestrhöm, G., 1991. Préférences parasitaires et particularités biologiques des Tachinaires (Diptera Tachinidae). Bull. Soc. Linn. Lyon 60, 128–141.

Grenier, S., Ogier, P., 1978. Description d'un appareil collecteur automatique pour contrôler l'émergence de mouches. Ann. Zool. Ecol. Anim. 10, 139–145.

Grenier, S., Barthelemy, S., Bonnot, G., 1982. Dynamique de la maturation des oeufs chez le parasitoïde *Lixophaga diatraeae* (Diptera, Tachinidae) élevé dans un hôte de substitution *Galleria mellonella* (Lepidoptera, Pyralidae). Reprod. Nutr. Dev. 22 (3), 523–535.

Grenier, S., Bonnot, G., Delobel, B., 1974. Définition et mise au point de milieux artificiels pour l'élevage in vitro de *Phryxe caudata* Rond. (Diptera, Tachinidae). I—survie du parasitoïde en milieux dont la composition est basée sur celle de l'hémolymphe de l'hôte. Ann. Zool. Ecol. Anim. 6, 511–520.

Grenier, S., Bonnot, G., Delobel, B., 1975. Définition et mise au point de milieux artificiels pour l'élevage in vitro de *Phryxe caudata* Rond. (Diptera, Tachinidae). II—croissance et mues larvaires du parasitoïde en milieux définis. Ann. Zool. Ecol. Anim. 7, 13–25.

Grenier, S., Bonnot, G., Delobel, B., 1986. Physiological considerations of importance to the success of in vitro culture: an overview. J. Insect Physiol. 32, 403–408.

Grenier, S., Bonnot, G., Delobel, B., Laviolette, P., 1978. Développement en milieu artificiel du parasitoïde *Lixophaga diatraeae*. (Towns.) (Diptera, Tachinidae). Obtention de l'imago à partir de l'oeuf. C. R. Acad. Sci. Paris, Série D 387, 535–538.

Grenier, S., Greany, P., Cohen, A.C., 1994. Potential for mass release of insect parasitoids and predators through development of artificial culture techniques. In: Rosen, D., Bennett, F.D., Capinera, J.L. (Eds.), Pest Management in the Subtropics: Biological Control—A Florida Perspective. Intercept. Pub., Andover, UK, pp. 181–205 Chapter. 10.

Gross, H.R., 1990a. Field release and evaluation of *Archytasmarmoratus* (Diptera: Tachinidae) against larvae of *Heliothis zea* (Lepidoptera: Noctuidae) in whorl-stage corn. Environ. Entomol. 19, 1122–1129.

Gross, H.R., 1990b. Disposable cage for rearing and releasing parasitoids. Fla. Entomol. 78, 513–515.

Gross, H.R., 1994. Mass propagation of *Archytas marmoratus* (Diptera: Tachinidae). Environ. Entomol. 23, 153–189.

Gross, H.R., Johnson, R., 1985. *Archytas marmoratus* (Diptera: Tachinidae): advances in large-scale rearing and associated biological studies. J. Econ. Entomol. 78, 183–189.

Gross, H.R., Rogers, C.E., Carpenter, J.E., 1996. Development of *Archytas marmoratus* (Diptera: Tachinidae) reared in *Galleria mellonella* larvae (Lepidoptera: Pyralidae) feeding on selected diets. Biol. Control 6, 158–163.

Haeselbarth, E., 1979. Zur Parasitierung der Puppen von Forleule (*Panolis flammea* [Schiff.]), Kiefernspanner (*Bupalus piniarius* [L.]) und Heidelbeerspanner (*Boarmia bistortana* [Goeze]) in bayerischen Kiefernwäldern. Zeitschr. Angew. Ent. 87, 186–202.

Ho, G.T., Ichiki, R.T., Nakamura, S., 2011. Reproduction biology of the microtype tachinid fly *Zenillia dolodsa* (Meigen) (Diptera: Tachinidae). Entomol. Sci. 14, 210–215.

Hoddle, M.S., 2006. Historical review of control programs for *Levuana iridescens* (Lepidoptera: Zygaenidae) in Fiji and examination of possible extinction of this moth by *Bessa remota* (Diptera: Tachinidae). Pac. Sci. 60, 439–453.

Horgan, F.G., Myers, J.H., Van Meel, R., 1999. *Cyzenis albicans* (Diptera: Tachinidae) does not prevent the outbreak of winter moth (Lepidoptera: Geometridae) in birch stands and blueberry plots on the lower mainland of British Columbia. Environ. Entomol. 28, 96–107.

House, H.L., 1954. Nutritional studies with *Pseudosarcophaga affinis* (Fall.), a dipterous parasite of the spruce budworm, *Choristoneura fumiferana* (Clemens). I. A chemically defined medium and aseptic-culture technique. Can. J. Zool. 32, 331–341.

House, H.L., 1961. Insect nutrition. Annu. Rev. Entomol. 6, 13–26.

House, H.L., 1966. Effects of vitamin E and A on growth and development and the necessity of vitamin E for reproduction in the parasitoid *Agria affinis* (Fallen) (Diptera: Sarcophagidae). J. Insect Physiol. 12, 409–417.

House, H.L., 1967. The decreasing occurrence of diapause in the fly *Pseudosarcophaga affinis* through laboratory-reared generations. Can. J. Zool. 45, 149–153.

House, H.L., Traer, M.G., 1949. An artificial food for rearing *Pseudosarcophaga affinis* (Fall.), a parasite of the spruce budworm *Choristoneura fumiferana* (Clem.). 79th Ann. Rep. Ent. Soc. Ontario, pp. 50–53.

Howarth, F.G., 1991. Environmental impacts of classical biological control. Annu. Rev. Entomol. 36, 485–509.

Hulme, M.A., Green, G.W., 1984. Biological control of forest insect pests in Canada 1969–1980: Retrospect and prospect. In: Kelleher, J.S., Hulme, M.A. (Eds.), Biological Control Programmes against Insects and Weeds in Canada 1969–1980, Commonwealth Agriculture Bureaux, Slough, UK, pp. 215–227.

Ichiki, R., Shima, H., 2003. Immature life of *Compsilura concinnata* (Meigen) (Diptera: Tachinidae). Ann. Entomol. Soc. Am. 96, 161–167.

Ichiki, R.T., Kainoh, Y., Kugimiya, S., Yamawaki, Y., Nakamura, S., 2011. The parasitoid fly *Exorista japonica* uses visual and olfactory cues to locate herbivore-infested plants. Ent. Exp. Appl. 138, 175–183.

Jervis, M.A., Kidd, A.C., 1986. Host-feeding strategies in hymenopteran parasitoids. Biol. Rev. 61, 395–434.

Kenis, M., Lopez Vaamonde, C., 1998. Classical biological control of the gypsy moth, *Lymantria dispar* (L.) in North America: prospects and new strategies. Proceedings: Population Dynamics, Impact and Integrated Management of Forest Defoliating Insects, U.S. Dep. Agric.For. Serv. Gen. Tech. Rep, pp. 213–221 NE-247.

Kfir, R., Graham, D.Y., Van Vuuren, R.V., 1989. An improved method for mass rearing *Paratheresia claripalpis* for biological control of lepidopteran stalk borers. Ent. Exp. Appl. 51, 37–40.

King, E.G., Coleman, R.J., 1989. Potential for biological control of *Heliothis* species. Annu. Rev. Entomol. 36, 485–509.

King, E.G., Leppla, N.C. (Eds.), 1984. Advances and Challenges in Insect Rearing, Government Printing Office, Washington, DC, USA.

King, E.G., Miles, L.R., Martin, D.F., 1976. Some effects of superparasitism by *Lixophaga diatraeae* on sugar-cane borer larvae in the laboratory. Ent. Exp. Appl. 20, 261–269.

King, E.G., Sanford, J., Smith, J.W., Martin, D.F., Hensley, S.D., 1981. Augmentative release of *Lixophaga diatraeae* (Dipt.: Tachinidae) for suppression of early season sugar-cane borer populations in Louisiana. Entomophaga 26, 59–69.

Lauzière, I., Legaspi, J., Legaspi, B., Saldana, R., 2001. Field release of *Lydella jalisco* Woodley (Diptera: Tachinidae) in sugar-cane and other gramineous crops for biological control of *Eoreuma loftini* (Dyar) (Lepidoptera: Pyralidae) in Texas. Subtrop. Plant Sci. 49, 53–64.

Leopold, R.A., 1998. Cold storage of insects for integrated pest management. In: Hallman, G.J., Denlinger, D.L. (Eds.), Temperature Sensitivity in Insects and Application in Integrated Pest Management, Westview, Boulder, CO, pp. 235–267.

Leopold, R.A., 2007. Colony maintenance and mass-rearing: using cold storage technology for extending the shelf-life of insects. In: Vreysen, M.J.B., Robinson, A.S., Hendrichs, J. (Eds.), Area-Wide Control of Insect Pests, from Research to Field Implementation, Springer, Dordrecht, The Netherlands, pp. 149–162.

Lewis, W.J., Tumlison, J.H., 1988. Host detection by chemically mediated associative learning in a parasitic wasp. Nature 331, 257–259.

Lopez, R., Ferro, D.N., Van Driesche, R.G., 1995. Two tachinid species discriminate between parasitized and non-parasitized hosts. Ent. Exp. Appl. 74, 37–45.

Mannion, C., Carpenter, J.C., Gross, H.R., 1995. Integration of inherited sterility and a parasitoid, *Archytas marmoratus* (Diptera: Tachinidae) for managing *Helicoverpa zea* (Lepidoptera: Noctuidae): acceptability and suitability. Environ. Entomol. 24, 1679–1684.

Marchetti, E., Baronio, P., Dindo, M.L., 2008. In vitro rearing of the tachinid parasitoid *Exorista larvarum* with exclusion of the host insect for more than one generation. Bull. Insectol. 61, 333–336.

Marchetti, E., Alberghini, S., Battisti, A., Squartini, A., Baronio, P., Dindo, M.L., 2009. Effects of conventional and transgenic *Bacillus thuringiensis galleriae* toxin on *Exorista larvarum* (Diptera: Tachinidae), a parasitoid of forest defoliating Lepidoptera. Biocontrol Sci. Tech. 19, 463–473.

Marchetti, E., Alberghini, S., Battisti, A., Squartini, A., Dindo, M.L., 2012. Susceptibility of adult *Exorista larvarum* to conventional and transgenic *Bacillus thuringiensis galleriae* toxin. Bull. Insectol. 65, 133–137.

Markham, R.H., 1986. Biological control of tsetse: prospects and progress in the use of pupal parasites. Insect Sci. Appl. 7, 1–4.

Mason, C.E., Jones, R.L., Thompson, M.M., 1991. Rearing *Lydella thompsoni* (Diptera: Tachinidae), a parasite of the European corn borer (Lepidoptera: Pyralidae). Ann. Entomol. Soc. Am. 84, 179–181.

Mason, P.G., Erlandson, M.A., 1994. Invitation paper (C. P. Alexander Fund): the potential of biological control for management of grasshoppers (Orthoptera: Acrididae) in Canada. Can. Entomol. 126, 1459–1491.

Massó Villalón, E., 2007. Producción y uso de entomófagos en Cuba. Fitosanidad 11, 67–73.

Mathis, K.A., Philpott, S.M., 2012. Current understanding and future prospects of host selection, acceptance, discrimination, and regulation of phorid fly parasitoids that attack ants. Psyche 2012, 9. http://dx.doi.org/10.1155/2012/895424, Article ID 895424.

McInnes, R.S., Albert, D.J., Alma, P.J., 1976. A new method of shipping and handling tachinid parasites of *Paropsini* (Col. Chrysomelidae). Entomophaga 21, 367–370.

McManus, M., Csóka, G., 2007. History and impact of gypsy moth in North America and comparison to recent outbreaks in Europe. Acta Silv. Lign. Hung. 3, 47–64.

Medina, H., 2002. Resultados del programa nacional de lucha biológica del MINAZ en sus 20 años de trabajo. Rev. Protección Veg. 17 (3), 187.

Mellini, E., 1986. Importanza dello stadio dell'ospite, al momento della parassitizzazione, per la biologia dei Ditteri Larvevoridi. Frust. Entomol. 7, 1–23.

Mellini, E., 1991. Sinossi di biologia dei Ditteri Larvevoridi (Studi sui Ditteri Larvevoridi. L contributo). Boll. Ist. Ent. "G. Grandi" Univ. Bologna 45, 1–38.

Mellini, E., 1997. Dalla predazione al parassitoidismo nell'ordine dei Ditteri. Boll. Ist. Ent. "G. Grandi" Univ. Bologna 51, 91–159.

Mellini, E., Campadelli, G., 1989. Prove di incubazione extrauterina e di schiusa delle uova microtipiche di *Pseudogonia rufifrons* Wied. (studi sui Ditteri Larvevoridi. XLVIII contributo). Boll. Ist. Ent. "G. Grandi" Univ. Bologna 43, 105–113.

Mellini, E., Campadelli, G., 1996. Formulas for "inexpensive" artificial diets for the parasitoid *Exorista larvarum* (L.). (Studies on Diptera Tachinidae. LXVIII contribution). Boll. Ist. Ent. "G. Grandi" Univ. Bologna 50, 95–106.

Mellini, E., Campadelli, G., 1997. Analisi del superparassitoidismo di *Exorista larvarum* (L.) nell'ospite di sostituzione *Galleria mellonella* L. (Studi sui Ditteri Tachinidi. LXXII contributo). Boll. Ist. Ent. "G. Grandi" Univ. Bologna 51, 1–11.

Mellini, E., Campadelli, G., 1999. Allevamento di parassitoidi su diete artificiali: l'apporto dell'Istituto di Entomologia "G. Grandi" (Bologna) sui Ditteri Tachinidi. Inf. Tore Fitopat. 49, 26–34.

Mellini, E., Coulibaly, A.K., 1991. Un decennio di sperimentazione sul sistema ospite—parassita *Galleria mellonella* L.—*Pseudogonia rufifrons* Wied.: sintesi dei risultati. Boll. Ist. Ent. "G. Grandi" Univ. Bologna 45, 191–249.

Mellini, E., Campadelli, G., Dindo, M.L., 1993. Artificial culture of the parasitoid *Exorista larvarum* L. (Dipt. Tachinidae) on bovine serum-based diets. Boll. Ist. Ent. "G. Grandi" Univ. Bologna 47, 221–229.

Mellini, E., Campadelli, G., Dindo, M.L., 1994. Possibile impiego di plasma bovino nell'allestimento di diete artificiali per le larve del parassitoide *Pseudogonia rufifrons* Wied. In: Viggiani, G. (Ed.), M.A.F.—Convegno Lotta Biologica, Acireale 1991, Sper. Pat. Veg., Roma, Italy, pp. 145–150.

Mellini, E., Malagoli, M., Ruggeri, L., 1980. Substrati artificiali per l'ovideposizione dell'entomoparassita *Gonia cinerascens* Rond. (Diptera Larvaevoridae) in cattività. Boll. Ist. Ent. "G. Grandi" Univ. Bologna 35, 127–156.

Mendel, Z., Blumberg, D., 1991. Colonization trials with *Cryptochaetum iceryae* and *Rodolia iceryae* for improved biological control of *Icerya purchasi* in Israel. Biol. Control 1, 68–74.

Mendel, Z., Assael, F., Zeidan, S., Zehavi, A., 1998. Classical biological control of *Palaeococcus fuscipennis* (Burmeister) (Homoptera: Margarodidae) in Israel. Biol. Control 12, 151–157.

Mes Démarches, 2005. Santé et protection des végétaux. http://mesdemarches.agriculture.gouv.fr/Santé-et-protection-des-vegetaux.

Monteith, L.G., 1963. Habituation and associative learning in *Dryno bohemica* Men. (Diptera: Tachinidae). Can. Entomol. 95, 418–426.

Montes, M., 2008. Producción masiva y aplicación de biorreguladores nativos en Cuba. Fitosanidad 12, 241–242.

Morrison, L.W., 2012. Biological control of *Solenopsis* fire ants by *Pseudacteon* parasitoids: theory and practice. Psyche, 11. Article ID 424817.

Muniz-Reyes, E., Lomell-Flores, J.R., Sanchez-Escudero, J., 2011. Parasitoides nativos de *Rhagoletis pomonella* Walsh (Diptera: Tephritidae) en tejocote *Crataegus* spp. en el centro de Mexico. Acta Zool. Mex. 27, 425–440.

Nakamura, S., 1994. Parasitization and life history parameters of *Exorista japonica* (Diptera: Tachinidae) using the common armyworm *Pseudaletia separata* (Lepidoptera: Noctuidae) as a host. Appl. Entomol. Zool. 29, 133–140.

Nakamura, S., 1996. Inbreeding and rotational breeding of the parasitoid fly, *Exorista japonica* (Diptera: Tachinidae), for successive rearing. Appl. Entomol. Zool. 31, 433–441.

Nettles, W.C., 1986. Effects of soy flour, bovine serum albumin, and three amino acid mixtures on growth and development of *Eucelatoria bryani* (Diptera: Tachinidae) reared on artificial diets. Environ. Entomol. 15, 1111–1115.

Nettles, W.C., 1987. *Eucelatoria bryani* (Diptera: Tachinidae): effect on fecundity of feeding on hosts. Environ. Entomol. 16, 437–440.

Nettles, W.C., Burks, M.L., 1975. A substance from *Heliothis virescens* larvae stimulating larviposition by females of the tachinid, *Archytas marmoratus*. J. Insect Physiol. 21, 965–978.

Nettles, W.C., Wilson, C.M., Ziser, S.W., 1980. A diet and methods for the in vitro rearing of the tachinid *Eucelatoria* spp. Ann. Entomol. Soc. Am. 73, 180–184.

Nicholls, I.C., Pérez, N., Vasquez, L., Altieri, M.A., 2002. The development and status of biologically based integrated pest management in Cuba. Int. Pest Manage. Rev. 7, 1–16.

Odell, T.M., Godwin, P.A., 1979. Laboratory techniques for rearing *Blepharipa pratensis*, a tachinid parasite of gypsy moth. Ann. Entomol. Soc. Am. 72, 632–635.

O'Hara, J.E., 2008. Tachinid flies (Diptera: Tachinidae). In: Capinera, J.L. (Ed.), Encyclopedia of Entomology, second ed. Springer, Dordrecht, pp. 3675–3686.

O'Hara, J.E., 2013. History of tachinid classification (Diptera, Tachinidae). ZooKeys 316, 1–34.

Orr, M.R., Seike, S.H., Benson, W.W., Gilbert, L.E., 1995. Flies suppress fire ants. Nature 373, 292–293.

Pape, T., 1987. The Sarcophagidae (Diptera) of Fennoscandia and Denmark. Brill/Scand. Sci., Leiden, The Netherlands.

Pape, T., 1990. Two new species of *Sarcophaga* Meigen from Madeira and mainland Portugal (Diptera: Sarcophagidae). Tijdschr. Entomol. 133, 39–42.

Pape, T., 1994. The world *Blaesoxipha* Loew, 1861 (Diptera: Sarcophagidae). Entomol. Scand. Suppl. 45, 1–247.

Paur, J.P., Gray, D.A., 2011. Individual consistency, learning and memory in a parasitoid fly, *Ormia ochracea*. Anim. Behav. 82, 825–830.

Pintureau, B., Grenier, S., 1992. Variability of morphological and biological characteristics in two strains of *Lixophaga diatraeae*. Biol. Control 2, 176–180.

Pintureau, B., Grenier, S., Paris, A., Ogier, C., 1995. Genetic variability of some biological and morphological characters in *Lixophaga diatraeae* (Diptera: Tachinidae). Biol. Control 5, 231–236.

Plantevin, G., Grenier, S., 1990. Ecophysiologie des relations hôtes-parasitoïdes: synchronization des cycles de développement. Bull. Soc. Ent. Fr. 95, 49–61.

Plowes, R.M., Folgarait, P.J., Gilbert, L.E., 2012. The introduction of the fire ant parasitoid *Pseudacteon nocens* in North America: challenges when establishing small populations. BioControl 57, 503–514.

Porter, S.D., 1998. Biology and behavior of *Pseudacteon* decapitating flies (Diptera: Phoridae) that parasitize *Solenopsis* fire ants (Hymenoptera: Formicidae). Fla. Entomol. 81, 292–309.

Porter, S.D., Pesquero, M.A., 2001. Illustrated key to the *Pseudacteon* decapitating flies (Diptera: Phoridae) that attack *Solenopsis saevissima* complex fire ants in South America. Fla. Entomol. 84, 802–808.

Proshold, F.I., Gross, H.R., Carpenter, J.E., 1998. Inundative release of *Archytas marmoratus* (Diptera: Tachinidae) against the corn earworm and fall armyworm (Lepidoptera: Noctuidae) in whorl-stage corn. J. Entomol. Sci. 33, 241–255.

Quednau, F.W., 1993. Reproductive biology and laboratory rearing of *Ceranthia samarensis* (Villeneuve) (Diptera: Tachinidae), a parasitoid of the gypsy moth, *Lymantria dispar*. Can. Entomol. 125, 749–759.

Quezada, J.R., DeBach, P., 1973. Bioecological and population studies of the cotton-cushion scale, *Icerya purchasi* Mask., and its natural enemies, *Rodolia cardinalis* Mul. and *Cryptochaetum iceryae* Will., in southern California. Hilgardia 41, 631–638.

Ramadhane, A., Grenier, S., Plantevin, G., 1987. Physiological interactions and development synchronizations between non-diapausing *Ostrinia nubilalis* larvae and the tachinid parasitoid *Pseudoperichaeta nigrolineata*. Ent. Exp. Appl. 45, 157–165.

Ramadhane, A., Grenier, S., Plantevin, G., 1988. Photoperiod, temperature and ecdysteroid influences on physiological interactions between diapausing *Ostrinia nubilalis* larvae and the tachinid parasitoid *Pseudoperichaeta nigrolineata*. Ent. Exp. Appl. 48, 275–282.

Reitz, S.R., Adler, P.H., 1995. Fecundity and oviposition of in vivo rearing of *Eucelatoria bryani*, a gregarious parasitoid of *Helicoverpa zea* and *Heliothis virescens*. Ent. Exp. Appl. 75, 175–181.

Reitz, S.R., Trumble, J.T., 1997. Effects of linear furanocoumarins on the herbivore *Spodoptera exigua* and the parasitoid *Archytas marmoratus*. Host quality and parasitoid success. Ent. Exp. Appl. 84, 9–16.

Rossi, M.N., Fowler, H.G., 2003. Temporal patterns of parasitism in *Diatraea saccharalis* Fabr. (Lep., Crambidae) populations at different spatial scales in sugar-cane fields in Brazil. J. Appl. Entomol. 127, 501–508.

Sabrosky, C.W., Reardon, R.C., 1976. Tachinid parasites of the gypsy moth, *Lymantria dispar*, with keys to adults and puparia. Misc. Publ. Entomol. Soc. Am. 10, 1–126.

Sampaio, M.V., Bueno, V.H.P., Silveira, L.C.P., Auad, A.M., 2009. Biological control of insect pests in the tropics. In: Del Claro, K., Oliveira, P.S., Rico-Gray, V. (Eds.), Tropical Biology and Conservation Management, vol. 3. UNESCO Publishing-Eolss Publishers, Oxford, UK, pp. 28–70.

Schniederkötter, K., Lakes-Harlan, R., 2004. Infection behavior of a parasitoid fly, *Emblemasoma auditrix*, and its host cicada *Okanagana rimosa*. J. Insect Sci. 4 (36), 1–7.

Sharma, K., Vander Meer, R.K., Fadamiro, H.Y., 2011. Phorid fly, *Pseudacteon tricuspis*, response to alkylpyrazine analogs of a fire ant, *Solenopsis invicta*, alarm pheromone. J. Insect Physiol. 57, 939–944.

Shoaib, M., Cagan, L., 2004. Natural enemies of slugs and snails recorded in Slovakia. Acta Fytsotech. Zootech. 7, 275–278.

Simões, A.M., Grenier, S., 1999. An investigation into the overwintering capability of *Istocheta aldrichi* (Mesnil) (Diptera: Tachinidae) a parasitoid of *Popillia japonica* beetle Newman (Coleoptera: Scarabaeidae) on Terceira Island, Azores. Arquipelago 17a, 23–36.

Simões, A.M., Dindo, M.L., Grenier, S., 2004. Development and yields of the tachinid *Exorista larvarum* in three common Noctuidae of Azores Archipelago and in a laboratory host. Bull. Insectol. 57, 145–150.

Smith, R.W., 1958. Parasites of nymphal and adult grasshoppers (Orthoptera: Acrididae) in Western Canada. Can. J. Zool. 36, 217–262.

Sourakov, A., Mitchell, E.R., 2002. Laboratory biology of *Chetogena scutellaris* (Diptera: Tachinidae), a parasitoid of Noctuidae, reared on fall armyworm and cabbage looper. Fla. Entomol. 85, 341–343.

Stireman, J.O., 2002. Host location and selection cues in a generalist tachinid parasitoid. Ent. Exp. Appl. 103, 23–34.

Stireman, J.O., O'Hara, J.E., Wood, D.M., 2006. Tachinidae: evolution, behavior, and ecology. Annu. Rev. Entomol. 51, 525–555.

Suazo, A., Arismendi, N., Frank, J.H., Cave, R.D., 2006. Method for continuously rearing *Lixadmontia franki* (Diptera: Tachinidae), a potential biological control agent of *Metamasius callizona* (Coleoptera: Dryophthoridae). Fla. Entomol. 89, 340–353.

Tanaka, C., Kainoh, Y., Honda, H., 1999. Comparison of oviposition on host larvae and rubber tubes by *Exorista japonica* Townsend (Diptera: Tachinidae). Biol. Control 14, 7–10.

Thompson, S.N., 1977. Lipid nutrition during larval development of the parasitic wasp, *Exeristes*. J. Insect Physiol. 23, 279–283.

Thompson, S.N., 1999. Nutrition and culture of entomophagous insects. Annu. Rev. Entomol. 44, 561–592.

Thompson, S.N., 2003. Trehalose—the insect 'blood' sugar. Adv. Insect Physiol. 31, 205–285.

Thompson, S.N., Hagen, K.S., 1999. Nutrition of entomophagous insects and other arthropods. In: Bellows, T.S., Fisher, T.W. (Eds.), Handbook of Biological Control. Academic Press, San Diego, pp. 594–652.

van Lenteren, J. (Ed.), 2003. Quality Control and Production of Biological Control Agents—Theory and Testing Procedures, CABI Publishing, Oxon, UK.

van Lenteren, J., 2012. The state of commercial augmentative biological control: plenty of natural enemies, but a frustrating lack of uptake. BioControl 57, 1–20.

van Lenteren, J.C., Bueno, V.H.P., 2003. Augmentative biological control of arthropods in Latin America. BioControl 48, 123–139.

van Lenteren, J.C., Tommasini, M.G., 2003. Mass production, storage, shipment and release of natural enemies. In: van Lenteren, J.C. (Ed.), Quality Control and Production of Biological Control Agents—Theory and Testing Procedures, CABI Publishing, Oxon, UK, pp. 181–189.

van Lenteren, J.C., Bale, J., Bigler, F., Hokkanen, H.M.T., Loomans, A.J.M., 2006. Assessing risks of releasing exotic biological control agents of arthropod pests. Annu. Rev. Entomol. 51, 609–634.

Vet, L.E.M., Lewis, W.J., Papaj, D.R., van Lenteren, J.C., 1990. A variable response model for parasitoid foraging behavior. J. Insect Behav. 3, 471–491.

Vidal, A.M., Dominguez, F., Alemán, J., 2002. Obtención de *Galleria mellonella* de alta calidad para la producción de diferentes medios biológicos (Entomófagos y Entomopatógenos). Rev. Protección Veg. 17, 153–154.

Vogt, J., Portes, S., Nordlund, D.A., Smith, R., 2003. A modified rearing system for production of *Pseudacteon curvatus* (Diptera: Phoridae), a parasitoid of imported fire ants. Biol. Control 28, 346–353.

Wäckers, F.L., 2003. The parasitoids' needs for sweets: sugar in mass rearing and biological control. In: van Lenteren, J.C. (Ed.), Quality Control and Production of Biological Control Agents—Theory and Testing Procedures, CABI Publishing, Oxon, UK, pp. 59–72.

Weir, E.H., Contreras, W., de Weir, K.G., 2007. Biological control of *Diatraea* spp. (Lepidoptera: Pyralidae) in sugar-cane crops in Venezuela. Rev. Biol. Trop. 55, 655–658.

Wineriter, S.A., Walker, T.J., 1990. Rearing phonotactic parasitoid flies (Diptera: Tachinidae, Ormiini, *Ormia* spp.). Entomophaga 35, 621–632.

Yeates, D.K., Greathead, D., 1997. The evolutionary pattern of host use in the Bombyliidae (Diptera): a diverse family of parasitoid flies. Biol. J. Linn. Soc. 60, 149–185.

Zhang, F., Toepfer, S., Kuhlmann, U., 2003. Basic biology and small-scale rearing of *Celatoria compressa* (Diptera: Tachinidae), a parasitoid of *Diabrotica virgifera virgifera* (Coleoptera: Chrysomelidae). Bull. Entomol. Res. 93, 569–575.

Ziser, S.W., Wojtowicz, J.A., Nettles, W.C., 1977. The effects of number of maggots per host on length of development, puparial weight and adult emergence of *Eucelatoria* sp. Ann. Entomol. Soc. Am. 70, 733–736.

This page intentionally left blank

Chapter 5

Mass Rearing *Bemisia* Parasitoids for Support of Classical and Augmentative Biological Control Programs

John A. Goolsby,[1] Matthew A. Ciomperlik,[2] Gregory S. Simmons,[3] Charles J. Pickett,[4] Juli A. Gould[5] and Kim A. Hoelmer[6]

[1]USDA-ARS, Knipling-Bushland Livestock Insects Research Laboratory, Cattle Fever Tick Research Laboratory, Edinburg, TX, USA, [2]US Department of Agriculture, Animal and Plant Health Inspection Service, Plant Protection and Quarantine, Center for Plant Health Science and Technology (USDA-APHIS-PPQ-CPHST), Edinburg, TX, USA, [3]USDA-APHIS-PPQ-CPHST, Salinas, CA, USA, [4]California Department of Food and Agriculture, Sacramento, CA, USA, [5]USDA-APHIS-PPQ-CPHST, Buzzards Bay, MA, USA, [6]US Department of Agriculture, Agricultural Research Service, European Biological Control Laboratory (USDA-ARS-EBCL), Montpellier, France

5.1. INTRODUCTION

The development of efficient mass-rearing systems for *Bemisia* parasitoids was crucial for the implementation of classical and augmentative biological control programs for this exotic pest (Gould et al., 2008). Early systems relied on adapting methods for the production of *Encarsia formosa* (Gahan), a parasitoid of the greenhouse whitefly, *Trialeurodes vaporariorum* (Westwood). Production for the Nearctic parasitoid *Eretmocerus eremicus* Rose and Zolnerowich was developed using this system, but new production techniques were needed for the exotic Palearctic parasitoids that were imported during the biological control program, such as *Eretmocerus mundus* Mercet, *Eretmocerus emiratus* (Zolnerowich and Rose), and *Eretmocerus hayati* (Zolnerowich and Rose), which are unable to develop successfully in *Trialeurodes* species (Goolsby et al., 1998) (Fig. 5.1).

In this chapter, we describe several rearing techniques developed in support of augmentative biological control demonstration projects in California, Arizona, and Texas to establish new species of *Eretmocerus* and for greenhouse biological control evaluations in several states (Goolsby and Ciomperlik, 1999; Pickett et al., 2004; Goolsby et al., 2004). The species reared on *Bemisia tabaci* (Gennadius)

Mass Production of Beneficial Organisms. http://dx.doi.org/10.1016/B978-0-12-391453-8.00005-4
2014 Published by Elsevier Inc.

biotype "B" using the systems described here include all of the *Eretmocerus* species that were collected in the worldwide effort to introduce more effective *Bemisia* parasitoids to the United States (Kirk et al., 2000; Legaspi et al., 1996; Goolsby et al., 1998, 2004; Goolsby et al., 2009). These include some of the most effective Palearctic and North American species: *E. emiratus*, *E. mundus*, *E. hayati*, and *E. eremicus* (Rose and Zolnerowich).

The initial rearing methods for production of *B. tabaci* and its imported natural enemies are described in this chapter. These methods were used at the US Department of Agriculture, Animal and Plant Health Inspection Service, Plant Protection and Quarantine (USDA-APHIS-PPQ) Mission Biological Control Laboratory (MBCL) in Mission, Texas, which is in Hidalgo County in the Lower Rio Grande Valley; the methods used were smaller in scale than those described in the second part of this chapter. Rearing systems at MBCL used both hibiscus and eggplant as host plants. The rearing operations were used to support small-scale colonization of parasitoids, colony maintenance, and quarantine evaluations (Goolsby et al., 1996, 1998) of new exotic *B. tabaci* silverleaf whitefly cultures collected from the worldwide foreign exploration effort. Parasitoids were reared in laboratory colonies from 1991 to 2000 to provide material for field cage evaluations in Arizona, California, and Texas; to provide material for greenhouse biological control evaluations in Colorado, Mississippi, and New York; to provide seed material for the rearing operations in California; and for direct field releases throughout the United States.

FIGURE 5.1 Field cage production system. (Top row) Left: *Eretmocerus hayati* male antennating emerging female in advance of mating; right: *Eretmocerus emiratus* female ovipositing into *Bemisia tabaci* second-instar nymph. (Second row) Left: *Eretmocerus mundus* pupae inside exuviae of whitefly (note empty areas on either side of the pupae, which allows the parasitized whitefly to float as compared to an unparasitized whitefly, which sinks; this feature is used for separation of parasitoid pupae); right: *Encarsia sophia* pupae inside exuviae (note the presence of dark meconia, which are characteristic of this parasitoid genus). (Third row) Left: *Bemisia tabaci* biotype B adults, eggs, and exuviae on the underside of an eggplant leaf; right: high density of *B. tabaci* adults on cucumber leaves in the field in the Lower Rio Grande Valley of Texas, 1994. (Fourth row) Left: vacuum collection of whitefly adults from the mother colony to use for infesting field cages; right: release of parasitoid adults onto infested plants in field cage. (Fifth row) Left: field cage full of infested eggplants which have been inoculated with parasitoids. Plants are mature and ready for harvest of leaves with parasitized whitefly; right: harvested leaves drying on racks for one day. (Sixth row) Left: harvested leaves inside Plexiglas emergence cages, which are used to collect adult parasitoids; right: emergence cage with black shroud to force adult parasitoids toward light and into petri dishes used for collection. (Seventh row) Left: greenhouse production system large eggplants infested with *B. tabaci* and shrouded to contain the adult parasitoids that have just been released; right: eggplants held in greenhouse for maturation of parasitoid pupae. (Eighth row) Left: funnel showing bulk unparasitized *B. tabaci* and parasitoid pupae that are floating in water and have just been removed from the eggplant leaves using a high-pressure flat fan sprayer; right: funnel showing separation of unparasitized whitefly that sink to bottom and top layer of floating parasitoid pupae. (Bottom row) Left: parasitoid pupae drying on nylon mesh cloth; right: parasitoid pupae being weighed to determine approximate numbers. (For color version of this figure, the reader is referred to the online version of this book.)

Fifty-eight populations and/or species of parasitoids were reared at MBCL from 1991 to 1999. Initially, all cultures in quarantine were colonized on whitefly-infested hibiscus, *Hibiscus rosa-sinensis* L. Following release from quarantine, parasitoid cultures were increased in the laboratory production facility on both whitefly-infested hibiscus and eggplant (*Solanum melongena* L.). Production in the laboratory facility allowed for stable year-round rearing, which could be used for inoculation of outdoor rearing cages. Outdoor cages were designed for economical rearing of multiple species for field release. Together, the use of these two systems resulted in the production of several million parasitoids from the 58 species and/or populations of natural enemies collected in the foreign exploration effort. A third, high-output greenhouse-based production system was developed for production of parasitoids for local augmentative release programs. This system was primarily used for production of *Eretmocerus* spp. for release in Arizona and California.

5.2. LABORATORY CULTURE

Plant Culture. All parasitoid and predator cultures were reared on whitefly-infested *H. rosa-sinensis* var. "Brilliant Red." This variety produced large, glossy leaves that were ideal for oviposition by *B. tabaci*. Plants were grown in 15.2cm (6in) wide pots from cuttings or liners purchased from commercial nurseries in Florida. Mature plants were ready for whitefly oviposition approximately 2months after potting. They were recycled twice and then discarded. More than 500 plants were held in rotation to supply adequate numbers of plants for the rearing colonies. Before using plants in the insect colonies, newly expanding young leaves were removed to force oviposition by the silverleaf whitefly onto only the large, mature leaves. The large, mature leaves were easier to handle in the rearing process. Plants were also completely hand wiped with towels to remove aphids, mites, and other pests. Plants were watered weekly in the environmental growth chambers.

Whitefly Oviposition. A population of *B. tabaci* collected from Mission, Texas, was used to initiate the "mother" whitefly colony. Approximately 10,000 adult whiteflies were used to start the mother colony. To ensure long-term viability of the colony, a new colony was started every 2years, always replacing the previous colony with field-collected whiteflies from the local agricultural area near Mission, Texas. Mature hibiscus plants were placed in an environmental growth chamber set at 27°C and a 14:10 light:dark (L:D) photoperiod. The lights were fluorescent with special ballasts, which had a flicker rate of more 750 cycles per second. A combination of Vitalitet® and Chromo-50® 40W bulbs produced a full spectrum of artificial light. Each chamber was illuminated from overhead with two each of the above fluorescent bulbs. Inside the chamber, plants were placed inside a 0.5×0.5×1m aluminum frame fitted with a white organza shroud to confine the emerging adult whitefly on the hibiscus plants. Approximately 100 mature clipped hibiscus leaves bearing fourth-instar whiteflies were stapled to paper towels and placed in the cage for emergence and subsequent oviposition onto new plants. The plants were kept in the whitefly-infested shroud cage for 2 days, at which point egg density reached approximately 50 eggs per cm^2. After oviposition, whiteflies were removed with a handheld vacuum. The short oviposition period produced whitefly populations with a strict cohort on each plant, which were then held at 27 °C for 7 days. After this period of time, the whitefly had developed to mostly second-instar nymphs suitable for parasitization. A portion of plants (1 per 15) were held for maturation of the whitefly to produce adults for the next oviposition cycle. Spider mite outbreaks (*Tetranychus* spp.) were avoided by quarterly inoculative releases of the phytoseiid predator mite, *Phytoseiulus persimilis* Athias-Henriot, applied at a rate of 500 mites per 25 plants.

Parasitoid Culture. Eretmocerus spp. (Hymenoptera: Aphelinidae) are solitary, internal parasites of whitefly that oviposit external to developing second- to third-instar whitefly nymphs, complete larval development in the whitefly nymphs, and pupate in mummified fourth-instar whitefly larvae. Adult *Eretmocerus* emerge from the mummified whitefly larva through an exit hole cut in the dorsum of the host. *Eretmocerus* spp. void their meconia (fecal pellets) outside the body of the host whitefly. *Encarsia* spp. have the same general biology, except that males develop as a hyperparasitoid of developing female *Encarsia* or *Eretmocerus* spp. In addition, *Encarsia* void their meconia as they pupate inside the body of the parasitized whitefly. Parasitoids were reared in environmental growth chambers operated under the same conditions as described in this chapter. Whitefly-infested plants were placed inside cages designed to prevent escape of the extremely small aphelinids. The whitefly-parasitoid cages were 0.25×0.25×0.5 m with three sides made of white translucent Plexiglas® and with top and front sides made of clear Plexiglas. A 10 cm hole on the top was covered in fine mesh organza for ventilation. The right side had a similar hole with a 20 cm organza sleeve, which allowed access for watering and wasp collection. The left side had a 30 cm oblong hole with a sock large enough to allow for addition of potted, 15.2 cm (6 in) wide hibiscus plants. The cages could hold a total of four hibiscus plants.

One hundred adult *Eretmocerus* parasitoids of mixed sex were aspirated into vials and released into each cage, which contained two to four hibiscus plants infested with mostly first- and second-instar whitefly. For *Encarsia* spp., plants with a larger proportion of third- and fourth-instar whitefly were used. For autoparasitic (autoparasitoids lay male eggs in the body of female larvae of their own species, or closely related taxa) *Encarsia*, like *Encarsia sophia* Viggiani, a third plant containing already parasitized whitefly was added to allow for production of male progeny. Plants with parasitized whitefly were held for 14 days until the majority of the parasitoids had pupated. At this point, leaves were clipped, scanned under a dissecting scope for contamination of aphids or autoparasitic *Encarsia* spp. in *Eretmocerus* cultures, stapled to paper towels, and hung up in emergence cages. If the whitefly nymphs on leaves from *Eretmocerus* colonies were contaminated with *Encarsia*, the unwanted *Encarsia* pupae were scraped off the leaf with a pair of forceps. As parasitoids emerged, they were collected with an aspirator and used to inoculate the next series of cages or removed for field releases or evaluation studies. The top right-hand corners of production and emergence cages were streaked with honey using an eyelash brush, since the addition of honey significantly increased the longevity and fecundity of the parasitoids.

Fifty-eight species or populations of parasitoids were reared using these methods. At the peak of production in 1995, six insect production workers were required to manage the parasitoid and whitefly colonies. Periodic samples of each parasitoid colony were checked for purity using randomly amplified polymorphic DNA polymerase chain reaction (RAPD-PCR). If contamination was detected, a new culture was initiated from each of 10 isofemale lines that matched the original parental RAPD-PCR banding pattern. To minimize contamination, each growth chamber contained no more than one *Eretmocerus* and one *Encarsia* species in combination. If congeneric species were reared in combination, morphological differences between adults and pupae of *Eretmocerus* and *Encarsia* allowed production workers to detect early stages of contamination. Since the *Eretmocerus* spp. from similar appearing species or different populations were difficult or impossible to distinguish using macromorphological characters, only one *Eretmocerus* spp. was reared per chamber. These meticulous anticontamination efforts carried out by the production and genetics teams resulted in no populations being lost due to contamination during 9 years of insect rearing.

5.3. OUTDOOR FIELD CAGE PRODUCTION

Plant Culture. All parasitoid cultures were reared on whitefly-infested eggplant, *S. melongena* var. "Ichiban." This variety of eggplant produced large leaves that were ideal for oviposition by *B. tabaci*. Kale, *Brassica oleracea* "Acephala." was used from November to February because of its tolerance to cool, winter temperatures. Plants were grown from seed in 15.2 cm (6 in) wide pots. Mature plants were ready for whitefly oviposition approximately 3 months after potting. More than 1000 6 in plants were kept for the production cycle, with new plants started each week to maintain the supply of young vigorous plants.

Whitefly Oviposition. As with the laboratory culture, a locally collected population of *B. tabaci* was used as the "mother" colony in the outdoor field cage production. The colony was replaced every 2 years with a new field-collected whitefly population. Mature plants were placed in $3.1 \times 3.1 \times 1.8$ m high field cages covered in 52×52 mesh Lumite Saran™ (Lumite Industries, Gainesville, GA, USA). Cages were pinned to the ground using large 25×1.23 cm ($10 \times 1/2$ in) soil auger screws. Cages were sealed along the bottom edges with weights made of sand-filled lengths of 4 in fire hose. Plants and whitefly were reared under ambient outdoor temperatures, except during the coldest months from November to February (the mean minimum and maximum temperatures for McAllen, Texas, in January are 7.6 and 20.1 °C, respectively). During the winter, cages were covered with clear greenhouse plastic, and a single electric 12.5 amp, 1300 W portable heater was placed in each cage, which increased the temperatures in the cages by about 5 °C above ambient temperatures. Eight outdoor cages with 40 eggplants were needed for a continuous supply of whitefly for inoculation of parasitoid production cages. All the plants were watered daily on an automated irrigation timer. Liquid fertilizer was added to the irrigation water monthly, or as needed, as per label rates (20–20–20, Peter's®, Allentown, PA, USA). The whitefly production facility was 1.6 km (1 m) distant from the parasitoid production facility to reduce the possibility of contamination from parasitoid colonies. As whitefly adults emerged, they were vacuumed off using a high-volume, low-speed ventilation box fan. The adults were collected in a fine-mesh cotton organza sock for transfer to the parasitoid production facility or to inoculate new whitefly production cages. Approximately 10,000 whiteflies were added to each cage. Plants were monitored until whitefly egg density reached 50 eggs per cm^2, at which point adult whitefly were removed by vacuuming. The short oviposition period produced a defined whitefly cohort on the plants, which was held for 7–21 days ("d"), at which point they contained mostly second-instar whitefly nymphs that were suitable for parasitization. Spider mites and cotton aphid (*Aphis gossypii*) outbreaks were avoided by quarterly inoculative releases of the phytoseiid predatory mite (*P. persimilis*) or the aphelinid parasitoid (*Aphelinus gossypii* Timberlake), respectively.

Parasitoid Culture. Parasitoids were reared in the same cages with the same environmental parameters as those described here. One hundred and twenty 15.2 cm (6 in) potted eggplants were placed in cages and allowed to mature to the six-leaf stage, which took approximately one month. The same number of kale plants was used during the winter months, but plants were held for 2 months prior to inoculation with whiteflies. Adult whiteflies were then added using the methods described here. Parasitoid cages were zippered with an additional Velcro® flap to prevent escape or contamination of parasitoids. An onsite weather station provided degree-day information, so that we could predict when parasitoids would be ready for harvest (von Arx et al., 1983).

Two to 6000 parasitoids of mixed sex were added to each cage, which contained plants infested with mostly first- and second-instar whiteflies for production of *Eretmocerus* spp.; third and fourth instars were used for production of *Encarsia* spp. Plants with parasitized whiteflies were held from 2 weeks to 2 months (depending on outdoor temperatures) until the majority of the unparasitized whiteflies had emerged and the parasitoids had reached the pupal stage. At this point, the leaves were clipped; scanned for contamination by *Encarsia* spp., aphids, and so on; stapled to paper towels; and hung up in indoor laboratory emergence cages maintained at 25 °C. Emergence cages (1 × 1.5 × 1 m) were covered with black shrouds, except for two round openings at the top facing the fluorescent lights. The round openings were designed for placement of large 10cm petri dishes (bottom halves). Emerging parasitoids gathered on the top half of the petri dish. As parasitoids emerged and gathered on the petri dishes, they were removed and closed with the other piece of the dish, which was streaked with honey. The numbers of parasitoids were counted using a subsample on a fixed grid. Petri dishes were sealed with parafilm and placed in a temperature-controlled cabinet at 16°C with a 14:10L:D cycle. Dishes containing adult parasitoids were shipped interstate to cooperators, held for inoculation of cages, or used for field release. Aside from plant production, one person was able to manage the daily tasks of the entire outdoor production facility.

The outdoor cage production procedures provided a low-cost, low-input method for multiplying parasitoids. However, production was somewhat unpredictable during cooler winter months. If the mean temperature in the cage was below 16°C, development of the parasitoids stopped. It was not unusual for the parasitoid generation time to take up to 2.5months during the winter. Toward the end of the project, heated greenhouses were used instead of the outdoor cages for mass production. Although the production from heated greenhouses was more predictable, the cost for the structures was considerably higher than for the outdoor cages. The outdoor cage method should be considered as an alternative method for multiplication of whitefly parasitoids during warmer months, especially if funds are limited.

5.4. LARGE-SCALE GREENHOUSE-BASED SYSTEM

A joint project by the California Department of Food and Agriculture (CDFA), USDA-APHIS-PPQ, the Imperial County Agricultural Commissioner's office, and private industry was launched in Imperial County, California, to mass rear the most effective exotic whitefly parasitoid species. The goal of this project was to increase biological control of *B. tabaci* biotype B in spring melons by rearing and releasing several species of whitefly parasitoids. Rates of parasitism by native parasites of whitefly (*E. eremicus* and *Encarsia* spp.) are generally low in the spring melon crop in desert production areas of California and Arizona, which is where whitefly populations first start to rapidly increase, leading to high regional populations subsequently infesting cotton and alfalfa after the melon harvest. In addition, the mass-rearing initiative supported the classical biological control program with the goal of establishing new species of parasitoids. Benefits of this program included the release of more than 60million exotic species of whitefly parasitoids into the Imperial Valley, which helped establish several new species; the transfer of rearing technology; providing new, more effective whitefly parasitoids to the beneficial insect industry; and improving the use of mechanized beneficial insect delivery systems in cooperation with industry.

Plant Production. In the Imperial Valley, plants were grown all year in a greenhouse without artificial lights. Suitable plants were produced in

6–8 weeks. Eggplant seeds were planted into one gallon plastic pots filled with peat-based soil (Redi-Earth™, Scotts, Marysville, OH, USA) mixed with coarse-grained sand (Play Sand, Quikrete™, Atlanta, GA, USA) to make a mixture of approximately 60:40 of peat moss and sand. Osmocote™ slow-release fertilizer (14–14–14, 3–4 month formulation, Scotts) or a similar generic brand was mixed into the soil at a rate of 0.177 ml/0.028 m^3. Before use, the soil mix was sterilized with an electric box sterilizer by bringing the temperature of the soil to 77 °C for one half-hour. The main eggplant varieties used were "Black Beauty" and "Whopper," although good results were also achieved with Asian varieties, such as "Sennari" and "Ping Tung." Plants were grown in a greenhouse equipped with both heating and cooling, so that plants were grown at temperatures ranging from 22 to 33 °C throughout the year. Plants were provided additional fertilizer through the irrigation system with a soluble fertilizer (20–20–20, Peter's™, Allentown, PA, USA) at a dilution ratio of 16:1 of water to fertilizer. Under these conditions, a plant usable for whitefly oviposition was produced within 5–8 weeks depending on the season. Plants that were ready for production were growing robustly with a height of 0.6–1.0 m and leaves as wide as 19 cm. Such plants had 25–40 large leaves that were suitable for production. Good plant health was the single most important factor for good parasitoid production.

Insect and Disease Control. Insect infestations caused problems in several ways. Eggplants infested with insect pests were weaker and less able to tolerate the high density of whitefly nymphs necessary for high parasitoid production. Infested plants grew slower, became deformed, and shed leaves in the parasitoid rearing before the parasitoids were mature. Pest infestations also reduced oviposition activity by both whitefly and parasitoids by making the plants less suitable for feeding and oviposition and also by simply taking up space on the leaf.

Infestations of *Bemisia* in the clean plant production greenhouse caused problems when they became established on a cohort of plants before plants were exposed to whitefly in the production system. Early whitefly infestation led to contamination by *Encarsia* spp. (especially *Encarsia hispida* De Santis), which had become pests in the greenhouses after they had been in production. Whiteflies that became established too early also emerged as adults in the parasitoid exposure cages, resulting in heavy feeding and excreting honeydew, which weakened plants and interfered with parasitoid oviposition.

For all of these reasons, it was important to maintain a pest-free environment in the plant production greenhouses as much as possible. To reduce pest entry, greenhouse vents and doors were screened with 52×52 mesh (Lumite Saran™, Lumite Industries). A combination of pesticides and natural-enemy releases was used, as needed, to keep plants as clean as possible. Pesticides with short or no residual activity were used so that treated plants could be placed into the parasitoid production system with little delay. Pesticides were frequently rotated to avoid problems with the buildup of resistance. Pesticides were tested for phytotoxicity to eggplant before use; in some cases, application rates had to be adjusted to levels lower than recommended by the label to avoid damage to eggplant leaves. Young seedlings were the most susceptible to damage from spraying, and, whenever possible, spraying was avoided in plants younger than 2–3 weeks. Most of the pesticide applications during plant production were made before their introduction into the whitefly and parasitoid colonies to avoid applying pesticides directly to the insect cultures. Treatment of pests in the insect colonies was predominantly limited to releases of natural enemies, especially spider mite predators and aphid parasitoids, although exceptions are noted in this chapter.

The pesticide treatments included pyrethrin–rotenone (Pyrellin EC™, 2.5 ml/l, Webb Wright Corp., Fort Myers, FL, USA), pyrethrin–piperonyl butoxide (Concern™, 2.5 ml/l, Scotts), insecticidal soap (M-Pede™, 19.2 ml/l, Mycogen Corp., San Diego, CA, USA), Cyfluthrin (Tempo 2 EC™, 0.3 ml/l, Bayer Corp., Leverkusen, Germany), and Azadirachtin (BioNeem™, 23 ml/l, Safer, Lititz, PA, USA). These pesticides were used alone at these rates or mixed with M-pede insecticidal soap for control of the following insect pests: green peach aphid, *Myzus persicae* (Sulzer); cotton aphid, *Aphis gossypii* Glover; *B. tabaci* (Gennadius); long-tailed mealybug, *Pseudococcus longispinus* (Targioni-Tozzetti); and western flower thrips, *Frankliniella occidentalis* (Pergande). Although all pesticide treatments helped in the control of the listed pests, most of the applications were made for control of *B. tabaci* in the production greenhouse for clean eggplant, as the other pests were occasional problems that could be controlled by releases of commercially available natural enemies for aphids, mealybugs, and thrips. Treatments for whitefly were made on average once a week from May to September. The most frequent treatments were insecticidal soap alternated with insecticidal soap mixed with a pyrethrin compound.

Aphids were controlled effectively by releases of *Aphidius colemani* Viereck at the rate of 2–5 per plant. Infestations of long-tailed mealybug were treated with releases of *Cryptolaemus montrouzieri* Mulsant, the mealybug destroyer, at the rate of 1–2 beetles per plant. Thrips infestations were treated with releases of the predatory mite, *Amblyseius cucumeris* Oudemans, at rates of 5–20 mites per plant. Thrips were also treated with releases of the minute pirate bug, *Orius insidiosus* (Say), at the rate of 1–5 bugs per plant. As minute pirate bugs are also whitefly predators, these were released only in the plant production greenhouse. Mite species that caused problems included the two-spotted spider mite, *Tetranychus urticae* Koch, and the broad mite, *Polyphagotarsonemus latus* (Banks). At the first sign of two-spotted spider mite infestation, plants were treated with mixed releases of mite predators: *P. persimilis*, *Mesoseiulus longipes* (Evans), and *Neoseiulus californicus* (McGregor). These were released at the rate of 10–20 mites per plant. Broad mite infestations and severe two-spotted mite infestations were treated with avermectin (Avid 0.15 ec™, 0.6 ml/l, Syngenta Inc., Basel, Switzerland).

Other pests included ants, especially argentine ants, *Linepithema humile* (Mayr), which tended aphids and preyed on parasitoid pupae, and fungus gnats,

Bradysia spp. Ants were treated with applications of hydramethylnon ant bait (Amdro Pro®, BASF, Ludwigshafen, Germany) by putting 10–15 g in a small petri dish placed in the corner of each whitefly or parasitoid cage. Fungus gnats were treated with soil drenches of *Bacillus thuringiensis* var. *israelensis* (Gnatrol™, 10.0 ml/l, Valent, Walnut Creek, CA) once a week.

Occasionally, the whitefly colony would become infested with undesirable parasitoids (especially *Encarsia* spp.), greenhouse whitefly, or spider mites. Initially, badly infested colonies were destroyed and new ones were started, but this method was a setback for production as the colonies could actively produce whitefly for 4–8 weeks. Because Imperial Valley populations of *B. tabaci* have high levels of acephate resistance (nearly 100%) (Steve Castle, USDA-ARS, Phoenix, AZ, USA, personal communication), while aphelinid parasitoids, greenhouse whitefly, and spider mites are completely susceptible, we treated infested whitefly colonies with acephate (Orthene 75 S, 0.77 ml/l, Valent) to eliminate everything except whitefly populations.

Within the whitefly colony and parasitoid rearing, plants were occasionally affected by root rot diseases, mainly *Pythium* spp. and *Phytophthora* spp. In general, these diseases were a problem only during the hottest months, when the greenhouse evaporative cooling system was less effective at keeping the greenhouse at optimal temperatures below 38 °C. The combination of heat stress and high densities of feeding whiteflies coupled with exposure to plant pathogens sometimes resulted in diseased plants, leading to premature leaf drop and loss of many developing parasitoids. These problems were primarily controlled by avoiding overwatering, which created favorable conditions for root rot diseases, and by applying preventive application of fungicides as needed. When disease conditions were prevalent, alternating weekly drench applications of metalaxyl (Subdue 2E, 0.31 ml/l, Syngenta) or fosetyl aluminum (Aliette 80WP, 1.0 ml/l, Bayer) were made.

Whitefly Colony. To start a new colony, whitefly adults were collected from the field and transferred to colony cages with eggplant as described in this chapter. Initial whitefly colonies were collected in the late winter from cole crops, such as cabbage, broccoli, and cauliflower, where large populations of whitefly infestations could be found. Whitefly adults were collected with a gas-powered vacuum (D-VAC Vacuum Insect Net Model 24, Rincon-Vitova Insectaries, Inc., Ventura, CA, USA) fitted with a nylon organdy bag over the end of a collection hose. Whitefly adults were collected for 5 min periods into the bags, which were then tied off with a piece of twine and transferred to an ice chest maintained at 13–15 °C for transport to the laboratory.

Whiteflies were released inside a 0.9 × 0.5 × 0.4 m high wooden cage with a slanted glass top. A fluorescent light source was placed over the cage to attract the whitefly to the top of the glass and aid in their separation from other insects present in the collection samples. Whitefly were collected with an aspirator constructed with 0.64 cm (¼ in) plastic tubing attached to 0.64 cm (¼ in) aluminum tubing fitted into a 9 dram vial with a rubber plug. The aspirator was connected to an electric vacuum pump (Model G582DX-S55NXMLD-6711, 1/3 HP, 5.5 A 115 V, Gast, Mfg. Corp., Benton Harbor, MI, USA) with plastic tubing of the same diameter. Whitefly adults were collected into the vials, inspected for other species, and stored at 13–15 °C. When other species were prevalent in the collection, repeated retransference of collected whiteflies to clean cages was necessary to obtain a pure colony. Whiteflies were transferred to the greenhouse in a 13–15 °C ice chest to release on eggplant in cages to start a new whitefly colony.

Plants that were 4–6 weeks old were washed with a fine spray of water to remove dust and pesticide residues, damaged and yellow leaves were removed, and plants were placed into 1.8 × 1.8 × 1.8 m whitefly oviposition cages inside a 6 × 12 m greenhouse. The greenhouse temperature was maintained within a range of 20–38 °C throughout the year. Production in the late winter and early spring for the first releases on spring melons was an important time of the year for rearing. Observations indicated that when temperatures fell below 20 °C, production levels of both whitefly and parasitoids were poor. Conversely, temperatures above 38 °C resulted in production decline, since eggplant appeared to be less tolerant to high levels of whitefly feeding at higher temperatures and many leaves became yellow and died before parasitoids could mature.

Cages were filled with 14–20 plants per cage depending on the size of the plants. The cages were made of Lumite Saran™ (52 × 52 mesh) material with doors secured by zippers with a flap of fabric secured by Velcro™ to keep small insects from entering around the zipper teeth. Plants were watered 2–3 times per week using a drip irrigation system, depending on the temperature and growing conditions. Plants were fertilized weekly with a soluble fertilizer (Peters™, 20–20–20) at a dilution ratio of 16:1 water to fertilizer.

Sufficient adult whitefly were added to the oviposition cage until a target density of 15 or more whitefly per cm^2 was reached. Plants were exposed to ovipositing adult whiteflies for 24–48 hours depending on temperature. Whitefly adults were added by collecting from the eggplant in the whitefly colony using a battery-powered electric vacuum device with a fine-mesh organdy bag fit over the collection nozzle (Modified CDC Backpack Aspirator Model 1412, John Hock Co., Gainesville, FL, USA). Whitefly were collected for approximately 2–3 min, then transferred to cages at the rate of 0.121 of adult whitefly per 20 plants.

After whiteflies were introduced, the plants were monitored over the next 1–3 days by collecting 10 2.54 cm^2 plugs per cage and counting the number of eggs with a dissecting microscope or by inspection of several intact leaves with a 10× hand lens. Once target density was reached, all adult whitefly were removed by vacuuming, as described here.

After adult whitefly removal, plants were monitored every 2–3 days to remove any remaining whitefly adults, remove any damaged or yellowing leaves, and wash off accumulated honeydew. Plants were monitored once a week for whitefly stage development by examining leaves of several plants in a cage with a 10× hand lens. Whitefly nymphs typically reached the second instar, preferred by *Eretmocerus* spp. (Headrick et al., 1996), in 7–8 days during warmer summer months and about 12–13 days during cooler periods of the year. Once it was observed that approximately 50% of the whitefly nymphs were in second instar or later stages, the cage was readied for parasitoid introduction.

Parasitoid Production. Parasitoids were released as adult wasps or as parasitoid pupae inside the cages. To protect parasitoids from heat during transit, they were carried to the rearing cages in Styrofoam ice chests cooled to 13–16 °C. Cooling the parasitoids also slowed their activity, which made them less likely to fly to the roof of the cage away from whitefly-infested plants. Adult parasitoids were placed on the soil at the base of the plants in 20 dram vials or in 100 cm petri dishes so that they would fly or crawl up to the leaves as they gradually warmed. Parasitoids were released at the rate of 667 adults per plant using vials or petri dishes that contained no more than about 1000 wasps per dish. Parasitoid pupae were released in 60 cm petri dishes, which were also placed on the soil at the base of the plant, at the rate of 1334 pupae per plant. More parasitoid pupae than adult wasps were released because emergence rates were less than 100%.

Cages were left undisturbed for one week after parasitoid introduction to avoid disrupting adult wasps searching for whitefly nymphs. After one week, several leaves from the upper and middle parts of the plants were inspected for adult parasitoid searching activity. If no parasitoids were observed searching, any adult whitefly that emerged from nymphs were removed by vacuuming. This was necessary to reduce feeding stress on the eggplant while parasitoids were developing. This procedure was continued on an average of 2–3 days until harvest. It was important to carefully balance the density of whiteflies with the rate of parasitism. When plants were oviposited too heavily by whiteflies, lower than expected rates of parasitism occurred, and prolonged adult whitefly feeding caused the plant to drop leaves before parasitoid development was completed. For the same reason, it was important to grow healthy and robust plants by providing adequate amounts of fertilizer, controlling pests, controlling high temperatures, and making sure plants were not stressed by either excessively dry or wet conditions.

Ten days after parasitoid introduction, plants were monitored for developing parasitoid pupae. Fourth-instar whitefly nymphs were monitored with a 10× hand lens for signs of advancing degrees of parasitism, such as displaced mycetomes, development of an amber-colored cuticle, parasitoid eye development, and/or the appearance of a margin on the developing parasitoid pupae within the whitefly (Roltsch, California Department of Food and Agriculture, http://www.cdfa.ca.gov/phpps/ipc/biocontrol/insects/19eretmocerus-lifestages.pdf). Older leaves, typically holding the oldest and most developed parasitoids, sometimes yellowed and dropped from the plant and were collected to begin the parasitoid harvest.

When the monitoring showed that >50% of the parasitized whitefly held mature parasitoid pupae on the remaining healthy leaves, harvesting started.

This was determined by sampling 10 2.54 cm^2 leaf plugs from leaves on the middle of the plant and counting pupae with a dissecting microscope. However, some of the insectary workers were able to accurately determine the state of harvest readiness by inspecting several plants per cage using only a hand lens.

If the parasitoids were not ready for harvest, the plants were monitored every 2 days until parasitoids were fully developed. During summer months, parasitoids were typically ready for harvest about 10–12 days after their introduction. During winter months, harvest commenced 14–20 days after parasitoid introduction. At harvest time, all leaves from the plants in the cage were stripped, stacked one layer deep on paper towels, placed in large brown paper bags (30.5 × 17.8 cm opening by 43.2 cm deep), and transported to the laboratory in a large ice chest maintained at about 13–16 °C. These bags were kept in the laboratory in an incubator maintained at 16 °C until processing.

Parasitoid Processing. Parasitoid pupae still inside the host were removed from eggplant leaves with pressurized water and separated from dead whitefly nymphs and exuvia using a water and air separation process. Leaves with parasitoid pupae were removed from the leaves with a spray gun with an adjustable cone-shaped nozzle (Triggerjet 22650 Spray Gun™, conejet spray tip #5500-ppb; TeeJet Spraying Systems Co., Wheaton, IL, USA) attached to a sink using a washing machine hose modified with a threaded hose attachment, which delivered a fine spray of water under pressure. Pupae were washed from leaves, one leaf at a time, by placing the leaf over a piece of nylon organdy stretched over a 30.5 cm (12 in) diameter metal kitchen colander and held tight with rubber bands. Parasitoid pupae separated from the washed unparasitized whitefly leaves were transferred to a 1 l glass beaker with approximately 0.5 l of water with ≈1.5 ml of liquid detergent (concentrated detergent, low phosphate, no perfume) to break the surface tension of the water, mixed well, and allowed to settle for ≈5 min. Because there is a buoyant cell of air within the whitefly exuvia around the developed parasitoid, parasitized whiteflies with fully developed parasitoid pupae float, while unparasitized whitefly pupae sink. Floating parasitoid pupae were removed from the top of the beaker with a small 6.4 cm (2.5 in) diameter steel kitchen strainer and transferred to a sheet of dry nylon organdy stretched over a steel frame and allowed to dry. This stirring and straining process was repeated 3–4 times until no more parasitoid pupae were found floating on the surface of the water.

Pupae were air-dried on the laboratory bench at room temperatures (20–24 °C). Parasitoids generally took ½ to 1 h to dry, and drying was speeded by providing extra air circulation with a table fan set at medium speed. Once the parasitoid material was dry, it was further processed to remove dry dead whitefly nymphs and exuvia that did not sink in the water separation process. Dried pupae were brushed into a 0.5 l paper container, adding enough pupae to cover no more than about 3–4 cm of the bottom of the container. Dead whitefly nymphs and exuvia (=scales) were much lighter than parasitoid pupae and could be removed by shaking the containers, allowing scales to float from the container, and then removing them by suction or blowing air. We used a 1.2m (4ft) wide power exhaust fume hood with the front sash set low enough to create enough suction to pull scales out of the container while being gently shaken. Other methods for removing dried scales could be used if an exhaust fume hood was not available, such as fans or vacuums to remove dried scales while shaking the parasitoid material. After processing, parasitoid pupae were stored until ready for use. These techniques resulted in very clean parasitoid material with very low numbers of whitefly pupae (see Table 5.1).

TABLE 5.1 Production Statistics for *Eretmocerus emiratus* Mass Rearing on Eggplant in 1997

Harvest Date	Generation	Input (1000s)	Number Produced/ Cage (1000s)	Increase Factor	% WF
3/24	4	24	206	8.6	1.6
3/31	5	84	299	3.6	0.5
3/31	6	50	196	3.9	3.0
4/2	7	29	154	5.3	4.0
4/11	9	60	65	1.1	3.6
4/8	10	35	113	3.2	8.0
4/14	11	20	371	18.5	1.6
4/17	12	95	81	0.8	1.5
4/21	13	20	490	24.6	1.2
4/21	14	31	331	10.7	2.3
4/24	15	35	129	3.6	–
4/24	16	44	172	3.9	–
4/29	17	75	114	1.5	2.4
4/29	18	60	133	2.2	2.0
5/1	19	35	66	1.9	1.5
5/5	24	76	499	6.6	1.7
5/6	25	110	580	5.3	3.2
5/7	26	25	347	13.9	6.3
7/7	45	25	186	7.5	11.9
Average			236.4	6.5	4.14

A sample of production data collected from 20 generations of *E. emiratus* reared between March and July 1997 is presented in Table 5.1. These data show that parasitoid production averaged 236,000 per cage, with a range of 65 to 580,000 per cage. For our production facility, this translated into 19,000 to 172,000 parasitoids per m^2 (1800 to 16,000 parasitoids produced per ft^2). Depending on the season, this level of production was achieved in 10–28 days. A greenhouse with better heating and cooling control would allow for uniform production toward the shorter end of this range. The increase rate of production averaged 6.5-fold the parasitoid inputs with a range of 0.8–24.6 (Table 5.1). Unparasitized whiteflies that remained in the cleaned product averaged 4.1%, with a range of 0.5–8% (Table 5.1). Factors that led to lower production included infesting plants with too few or too many whiteflies, late introduction of parasitoids (after whitefly nymphs had passed second instar), poor emergence of adult parasitoids, cool nighttime greenhouse temperatures (below 20 °C), prolonged high temperatures (above 35 °C), and poor-quality plants infested with aphids or spider mites or with root or foliar plant diseases. Controlling these factors was a complex process that required attention to maintenance processes of host plants, whitefly hosts, and parasitoid populations. With refinement of rearing techniques, and with gained worker experience, annual production increased. Production levels of exotic parasitoids for the years 1994–1998 were 9.3, 5.4, 8.3, 30.9, and 47.6 million, respectively, allowing large numbers to be released in augmentative and classical biological control programs.

Storage of Parasitoid Pupae. The cleaned and separated parasitoid pupae were stored in paper food containers or plastic vials in an incubator maintained at 15.5–16 °C before use in the field or shipping to cooperators. At these temperatures, parasitoid emergence was delayed safely and with little effect for 3–5 days. Longer storage times and lower temperatures decreased the emergence rates and longevity of adult parasitoids (J. Gould, USDA-APHIS, Buzzards Bay, MA, USA, unpublished data).

5.5. CONCLUSION

Three systems for production of *Bemisia* parasitoids in the genera *Eretmocerus* and *Encarsia* were described, one a smaller scale system for initial production and evaluation of the numerous cultures collected during the foreign exploration effort and two methods for larger scale production to conduct augmentative biological control demonstration projects in support of classical biological control by establishment of new species.

Efficient production systems depended on providing high-quality host plants that were free of pests, good environmental control, and careful control and monitoring of the whitefly host population. When these conditions were met, the greenhouse-based production systems could produce millions of parasitoids per week, with production levels as high as 172,000 parasitoids/m^2/generation.

Similar results could be expected for other aphelinid parasitoid species. The methods developed for the *Bemisia* parasitoids should be adaptable for other parasitoids of whitefly species targeted by classical and/or augmentative biological control programs.

The classical biological control program directed against *B. tabaci* biotype B (=silverleaf whitefly) in the 1990s was one the largest and most comprehensive programs in the history of biological control. The foreign exploration program for natural enemies of *B. tabaci* was comprehensive, covering 30 countries, and more than 130 shipments of natural enemies were sent to quarantine facilities in the United States between 1991 and 1998. Climate matching was used to match the affected areas in the United States with locations within the native distribution of *B. tabaci* and was used to prioritize foreign exploration. The ARS European Biological Control Laboratory in Montpellier, France, was extremely valuable to the biological control program; its staff engaged in nearly year-round exploration, which led to the discovery of many parasitoids for evaluation by U.S. researchers.

Mass-rearing facilities were established in Tucson, Arizona; Imperial and Sacramento, California; and Mission, Texas. Hundreds of millions of *Eretmocerus* and *Encarsia* species were mass reared for several years for release and evaluation in the areas affected by the silverleaf whitefly, which included the subtropical agricultural areas of the United States and Mexico. Mass-rearing techniques improved dramatically over the course of the program, beginning with laboratory rearing in environmental chambers on whitefly-infested hibiscus plants, then using heated, outdoor field cages with large pots of kale and eggplant, and finally using highly managed greenhouses that used large-leaf eggplants and mechanical removal of parasitoid pupae. The substantial number of parasitoids available for release enabled a large-scale field evaluation of biological control as an integrated component of management programs. In retrospect, the silverleaf whitefly biological control program clearly demonstrated the potential benefits of classical biological control in annual row-crop agriculture. The success of the program was made possible by a robust and productive mass-rearing effort of the key natural enemies. We hope that the information provided in this chapter will be useful to future researchers faced with mass rearing aphelinid parasitoids at the scale needed for a large, multistate biological control program.

5.6. USDA DISCLAIMER

Mention of specific pesticides is for informational purposes only and does not constitute a pest control recommendation by USDA. Information provided regarding commercial products does not constitute a recommendation or endorsement by USDA. USDA is an equal opportunity provider and employer.

REFERENCES

Goolsby, J.A., Legaspi, J.C., Legaspi Jr., B.C., 1996. Quarantine evaluation of exotic parasitoids of the sweet potato whitefly, *Bemisia tabaci* (Gennadius). Southwest. Entomol. 21, 13–21.

Goolsby, J.A., Ciomperlik, M.A., Legaspi Jr., B.C., Legaspi, J.C., Wendel, L.E., 1998. Laboratory and field evaluation of parasitoids of exotic parasitoids of *Bemisia tabaci* (Gennadius) (biotype "B") (Homoptera: Aleyrodidae) in the Lower Rio Grande Valley of Texas. Biol. Control 12, 127–135.

Goolsby, J.A., Ciomperlik, M.A., 1999. Development of parasitoid inoculated seedling transplants for augmentative biological control of silverleaf whitefly (Homoptera: Aleyrodidae). Fla. Entomol. 82, 1–14.

Goolsby, J.A., DeBarro, P.J., Kirk, A.A., Sutherst, R., Canas, L., Ciomperlik, M.A., Ellsworth, P., Gould, J., Hoelmer, K.A., Naranjo, S.J., Rose, M., Roltsch, W., Ruiz, R., Pickett, C., Vacek, D.P., 2004. Post-release evaluation of the biological control of *Bemisia tabaci* biotype "B" in the USA and the development of predictive tools to guide introductions for other countries. Biol. Control 32, 70–77.

Gould, J., Hoelmer, K.A., Goolsby, J.A. (Eds.), 2008. Classical Biological Control of *Bemisia Tabaci* in the United States, Springer, Amsterdam, p. 343.

Headrick, D.H., Bellows Jr., T.S., Perring, T.M., 1996. Behaviors of female *Eretmocerus* sp. nr. *californicus* (Hymenoptera: Aphelinidae) attacking *Bemisia argentifolii* (Homoptera: Aleyrodidae) on cotton, *Gossypium hirsutum* (Malvaceae) and melon, *Cucumis melo* (Cucurbitaceae). Biol. Control 6, 64–75.

Kirk, A.A., Lacey, L.L., Brown, J.K., Ciomperlik, M.A., Goolsby, J.A., Vacek, D.A., Wendel, L.E., Napometh, B., 2000. Variation in the *Bemisia tabaci* s. l. species complex (Hemiptera: Aleyrodidae) and its natural enemies leading to successful biological control of *Bemisia* biotype B in the USA. Bull. Entomol. Res. 90, 317–327.

Legaspi, J.C., Legaspi Jr., B.C., Carruthers, R.I., Goolsby, J.A., Jones, W.A., Kirk, A.A., Moomaw, C., Poprawski, T.J., Ruiz, R.A., Talekar, N.S., Vacek, D., 1996. Foreign exploration for natural enemies of *Bemisia tabaci* from Southeast Asia. Subtrop. Plant Sci. 48, 43–48.

Pickett, C.H., Simmons, G.S., Lozano, E., Goolsby, J.A., 2004. Augmentative biological control of whiteflies using transplants. Biol. Control 49, 665–688.

von Arx, R., Baumgärtner, J., Delucchi, V., 1983. Developmental biology of *Bemisia tabaci* (Gennadius) (Sternorrhyncha: Aleyrodidae) on cotton at constant temperatures. Mitt. Schweiz. Entomol. Ges. 56, 389–399.

FURTHER READING

Goolsby, J.A., Pfannenstiel, R.S., Evans, G.A., 2009. New state record for the silverleaf whitefly parasitoid *Encarsia sophia* in Texas. Southwest. Entomol. 34, 327–328.

Pickett, C.H., Brown, J., Simmons, G.S., Van Mantgem, L., Kumar, H., 1997. Release of Parasites into Citrus for Control of Silverleaf Whitefly Infesting Cotton in the San Joaquin Valley. http://www.cdfa.ca.gov/phpps/ipc/biocontrol/bc-annualreports.htm.

Chapter 6

Mass Rearing of the Stem-Galling Wasp *Tetramesa romana*, a Biological Control Agent of the Invasive Weed *Arundo donax*

Patrick J. Moran,[1] John A. Goolsby,[2] Alexis E. Racelis,[3] Allen C. Cohen,[4] Matthew A. Ciomperlik,[5] K. Rod Summy,[3] Don P.A. Sands[6,*] and Alan A. Kirk[7,*]

[1]US Department of Agriculture, Agricultural Research Service (USDA-ARS), Exotic and Invasive Weeds Research Unit, Albany, CA, USA, [2]USDA-ARS, Knipling-Bushland Livestock Insects Research Laboratory, Cattle Fever Tick Research Laboratory, Edinburg, TX, USA, [3]Department of Biology, University of Texas–Pan American, Edinburg, TX, USA, [4]Insect Rearing Education and Research Program, Department of Entomology, North Carolina State University, Raleigh, NC, USA, [5]US Department of Agriculture, Animal and Plant Health Inspection Service, Plant Protection and Quarantine, Center for Plant Health Science Technology (USDA-APHIS-PPQ-CPHST), Edinburg, TX, USA, [6]Commonwealth Science, Industry and Research Organization (CSIRO), Brisbane, Queensland, Australia, [7]USDA-ARS, European Biological Control Laboratory, Montpellier, France

6.1. INTRODUCTION

The ultimate goal of the mass rearing of beneficial organisms for biological control is to release the control agent in sufficient numbers to regulate the populations of targeted pests while overcoming, at least in the short term, negative population regulation by the agent's own natural enemies or climatic factors. The basic requirements for mass rearing are similar across all types of biological control agents: food sufficient in quality and quantity to allow immature development and adult reproduction, a physical and environmental habitat that maximizes expression of feeding and reproductive behaviors, and protection from natural enemies. In successful systems, the costs associated with rearing are more than balanced by the economic and environmental benefits of the release of large numbers of efficacious agents. Insects have similar basic nutritional

*Retired.

Mass Production of Beneficial Organisms. http://dx.doi.org/10.1016/B978-0-12-391453-8.00006-6
2014 Published by Elsevier Inc.

requirements regardless of the type of food consumed (Thompson and Hagen, 1999; Nation, 2002; see Chapter 7), but insects that feed on vegetative plant tissues (herbivorous or phytophagous) have food quality and abiotic environmental limitations that are unique to this food source. Mass rearing of predatory and parasitic insects (covered in Chapters 2, 3, 4, and 5) in some cases requires rearing of plant-feeding prey-hosts. However, this chapter is unique in addressing the mass rearing of an herbivorous insect as a beneficial organism for biological weed control. We therefore focus our discussion in the broader context of the need for and applicability of the biological weed control approach.

The objective of this chapter is to describe a mass rearing system for the shoot tip-galling wasp *Tetramesa romana* Walker (Hymenoptera: Eurytomidae). This wasp is a nonnative insect released to control the invasive perennial grass known as arundo, giant reed, or carrizo cane (*Arundo donax* L. (Poaceae)), which is one of the most damaging invasive weeds in the Lower Rio Grande Basin of Texas and Mexico, and throughout the southwestern United States (Tracy and DeLoach, 1999; Goolsby and Moran, 2009; Giessow et al., 2011). We provide background information on the biological approach to control of arundo and other invasive weeds, and an overview of past examples of mass rearing of insect agents for weed control. We then summarize biological information about the arundo wasp that is relevant to the development of mass rearing. We next describe a plant-based rearing system that has been implemented for releases of arundo wasps in the Lower Rio Grande Basin. We address challenges encountered in the development of this rearing system and solutions. An attempt to develop an artificial diet yielded novel insights about the feeding biology of immature plant-galling wasps. We provide a brief overview of the application of arundo wasps for biological control of giant reed, and conclude by providing insights about the utility of the rearing procedure for other weed biological control projects, in particular those targeting invasive grasses.

6.1.1. Critical Needs in a Mass Rearing Program for Biological Weed Control

Biological control using introduced insects is a safe, efficacious, and cost-effective weed control strategy (Crutwell McFadyen, 1998; Goeden and Andres, 1999; Forno and Julien, 2000; Syrett et al., 2000; Culliney, 2005; Van Driesche et al., 2010), when given enough time (5–20 years) to verify impact. Benefit-to-cost ratios range as high as 300:1 or more for field-established agents that impact plant survival, growth, reproduction, and/or population spread (Page and Lacey, 2006; van Wilgen and De Lange, 2011). Plant-feeding insects reared for weed control have the same basic requirements for food, habitat, and protection as do other types of mass reared organisms. However, the process of selecting plant substrate and habitat for feeding and reproduction is complex, even in situations in which only the host plant species is available. A myriad of abiotic factors, such as light, temperature, humidity, and the presence of nonhost odors, all influence substrate selection by females for oviposition and larviposition and the stimulation of immature feeding behavior (Bernays and Chapman, 1994; Chapman, 2003; Müller-Schärer and Schaffner, 2008). Physiological and nutritional qualities of the plant play critical roles before and after selection (Rohrfritsch, 1992; Waring and Cobb, 1992; Price, 2000; De Bruxelles and Roberts, 2001; Kerchev et al., 2012). Therefore, detailed information about plant and insect biology and behavior is required for mass rearing.

Insects that induce galls on plants have developed intricate associations with the biochemical and physiological processes associated with plant resource allocation (Raman et al., 2005). Most insect-induced galls, including the ones associated with the arundo wasp, are of the "histoid prosoplasmic" type, consisting of inner nutritive and storage regions and outer protective regions that together form an organ that is foreign to normal plant organogenesis, yet consists entirely of normal plant biomolecules (Rohrfritsch, 1992; Raman et al., 2005). Galling likely evolved as a way to obtain all three basic survival requirements—food, habitat, and protection—in one location (Stone and Schönrogge, 2003). Nevertheless, gall-inducing organisms (gallers) are attacked by parasites, predators, and inquilines (which live in the gall and may consume gall contents and/or modify the gall) (Wiebes-Rijks and Shorthouse, 1992; Veldtman et al., 2011).

Among 13,000 plant-galling insects, or 2% of all known insect species (Dreger-Jauffret and Shorthouse, 1992) feeding on 15,000 plant hosts (Raman et al., 2005), 30 species had been introduced and established in the field as weed biological control agents by 2004 (Muniappan and McFadyen, 2005), including three wasps in the superfamily Chalcidoidea that includes the arundo wasp (La Salle, 2005). Members of the order Diptera (flies) constitute the majority of established galling insects. In recent years, at least nine additional galling insects have been released, including the arundo wasp *T. romana* for biological control of giant reed (Goolsby and Moran, 2009; Moran and Goolsby, 2009; Racelis et al., 2010), one other wasp (Hymenoptera: Cynipidae), seven members of the dipteran family Cecidomyiidae (one adventive), and one beetle (Coleoptera: Brentidae) (Adair, 2005; Ostermeyer and Grace, 2007; Djamankulova et al., 2008; Impson et al., 2008; Adair et al., 2009; Gagné et al., 2009; Markin and Horning, 2010; Zachariades et al., 2011). Additional gallers under consideration include a tephritid fly for control of cape ivy (Balciunas and Smith, 2006; Balciunas et al., 2010). About one-half of galling insects released gall vegetative plant parts. Gall-forming insects offer high host plant species specificity for biological weed control and also host tissue specificity, for example to target flowering plant organs while not harming forestry uses of exotic acacia trees in South Africa (Muniappan and McFadyen, 2005). The negative effects of galling insects on host plants could be limited by evolutionary pressure on gallers to minimize damage in long-term symbiotic relationships (Raman et al., 2005) and high galler susceptibility to parasites (Wiebes-Rijks and Shorthouse, 1992; Julien and Griffiths, 1998). However, galling insects have been used successfully in biological weed control. Examples include flower and seed gallers against acacia (*Acacia* spp.) and mesquite (*Prosopis* spp.) in South Africa (Moran and

Hoffmann, 2012); reduction of *Centaurea* spp. (spotted and diffuse) knapweed growth and reproduction by flower head-galling tephritid flies (*Urophora* spp.) (Myers, 2007; Story et al., 2008) and the root-galling weevil *Cyphocleonus achates* Fahraeus (Myers, 2007; Knochel and Seastedt, 2010; Van Hezewijk and Bourchier, 2012); early evidence of similar impacts on Russian knapweed (*Acroptilon repens* (L.) DC) by a cecid midge, *Jaapiella ivannikovi* Fedotova, and the cynipid wasp, *Aulacidea acroptilonica* Tyurebaev (Djamankulova et al., 2008); and the impacts of galling on crofton weed (*Ageratina adenophora* (Spreng.) King & H. Rob) (Bucellato et al., 2012).

6.1.2. Mass Rearing of Weed Biological Control Agents: Past Successes

Mass rearing of insects as weed biological control agents may be necessary for releases of large numbers (thousands or more) of new "classical" agents (nonnative insects intentionally released to control nonnative weeds) at a limited number of field sites, or for releases of small populations at many sites. "Mass rearing" is here defined as amplification, specifically for field release, of an insect population for at least several generations under controlled or semicontrolled conditions at a site or facility that is not itself a field release site. Large inoculative releases of classical agents may be needed to achieve a minimum population size to complete a full generation and/or overwinter, due to the negative impacts of abiotic climatic factors and generalist predators (Goeden and Louda, 1976; Crawley, 1986; Memmott et al., 1998; Keane and Crawley, 2002). Mass rearing approaches for biological weed control agents are less developed than for entomophagous predators and parasitoids (covered in Chapters 2, 3, 4, and 5), although "ease of culturing" is one criterion applied to the selection of candidate weed control agents for laboratory host range testing (Goeden and Andres, 1999; Sheppard and Raghu, 2005). Small-scale quarantine laboratory rearing is required to determine life cycles of candidate agents and perform host range and efficacy studies (e.g. Price et al., 2003; Martin et al., 2004; Herrick et al., 2011). However, only rarely are attempts made to mass rear insects for weed control, as human-aided or natural agent dispersal among field sites is the predominant approach (Van Driesche et al., 2002, 2010). The first use of mass rearing was the spectacularly successful program against prickly pear cacti in Australia in the 1920s and 1930s (Dodd, 1940; Goeden and Andres, 1999) (Table 6.1). Two other early examples involved ragweed (*Ambrosia artemisiifolia* L.) (Reznik, 1991) and *Orobanche* spp. parasitic plants in the former Soviet Union and neighboring countries (Table 6.1); the *Orobanche* spp. project represented the only application of a mass reared insect for biological weed control in a crop setting (fruit groves) (Klein and Kroschel, 2002). Technology transfer and rearing are ongoing today for plant-based systems targeting the aquatic weeds hydrilla and giant salvinia (Table 6.1) (Center et al., 1997; Freedman et al., 2001; Flores and Carlson, 2006; Harms et al., 2009a, b). Plant-based systems for terrestrial weeds, including purple loosestrife (Blossey and Hunt, 1999), houndstongue

es of Successful Mass Rearing of Weed Biological Control Agents*

	Insect Agent	Host	Country or Region	Reference(s)
ɔuntia th)	*Cactoblastis cactorum* (Berg) (Lepidoptera: Pyralidae)	Plant	Australia, later Indonesia, South Africa, and Pacific islands	Dodd (1940), sumn in Goeden and Anc (1999)
spp.)	*Tarachidia candefacta* Hübner (Lepidoptera: Noctuidae) and *Zygogramma suturalis* (Fab.) (Coleoptera: Chrysomelidae)	Plant	Russia	Kovalev and Runev (1970), Kovalev anc Vechernin (1986) a Reznik (1991)
:he spp.)	*Phytomyza orobanchia* Kalt. (Diptera: Agromyzidae)	Plant	Russia	Klein and Kroschel
ticillata ıaritaceae)	*Hydrellia pakistanae* Deonier (Diptera: Ephydridae)	Plant	USA	Center et al. (1997) Freedman et al. (20 and Harms et al. (2
(*Matricaria* :aceae)	*Omphalapion hookeri* (Kirby) (Coleoptera: Apionidae)	Plant	Canada	McClay and De Cle Floate (1999)
hrum ıe)	*Galerucella calmariensis* L. and *G. pusilla* Duftschmidt (Coleoptera: Chrysomelidae)	Plant	USA	Blossey and Hunt (
	Hylobius transversovittatus Goeze (Colopetera: Curculionidae)	Meridic diet with plant component	USA	Blossey et al. (200C
	H. transversovittatus	Meridic diet	USA	Tomic-Carruthers (2

:amples of Successful Mass Rearing of Weed Biological Control Agents*—cont'd

·d	Insect Agent	Host	Country or Region	Reference(s)
laleuca (Cav.) S. F. Blake	*Oxyops vitiosa* Pascoe (Coleoptera: Curculionidae)	Meridic diet with plant component	USA	Wheeler and (2001)
opis spp.)	*Evippe* sp. #1 (Lepidoptera: Gelechiidae) and *Prosopidopsylla flava* Burckhardt (Homoptera: Psyllidae)	Plant	Australia	van Klinken e
Salvinia molesta iniaceae)	*Cyrtobagous salviniae* Calder (Coleoptera: Curculionidae)	Plant	USA	Flores and C and Harms e
fuse knapweeds *culosa* Lam. and steraceae)	*Cyphocleonus achates* Fahraeus (Coleoptera: Curculionidae)	Meridic diet	USA	Goodman et and Tomic-C (2009)
(*Cynoglossum* oraginaceae)	*Mogulones cruciger* Herbst (Coleoptera: Curculionidae)	Plant	Canada	Smith et al. (
per (*Macfadyena* A. H. Gentry	*Carvalhotingis visenda* (Boheman) (Hemiptera: Tingidae)	Plant	Australia	Dhileepan et
o donax L.,	*Tetramesa romana* Walker (Hymenoptera: Eurytomidae)	Plant	USA	This chapter
o donax L.	*Rhizaspidiotus donacis* (Leonardi) (Hemiptera: Diaspidiae	Plant (rhizomes)	USA	J. Goolsby, P. Kirk, D. Sand Racelis, unpu

weed biological control agents is defined as the process of increasing insect populations in confinement under controlled or semicont t is not itself a field release site. Success is defined as rearing, from egg or first-instar larvae to reproductive adult, of numbers of insects cess, under this definition, does not depend on the level of field establishment or biological control impact.

(Smith et al., 2009), scentless chamomile (McClay and De Clerck-Floate, 1999), cat's claw creeper (Dhileepan et al., 2010), and mesquite (van Klinken et al., 2003), have involved field cages or greenhouses at research facilities using potted or planted weeds. All of these mass reared weed control agents have established field populations and most are having an impact, indicating the value of the mass rearing approach. The rearing system described here for the arundo wasp represents the first use of mass rearing for biological control of a grass weed, and its first use for a shoot tip-galling insect.

Examples of successful mass rearing of weed biological control agents on artificial diets are limited to the purple loosestrife root-feeding weevil *Hylobius transversovittatus* Goeze reared on a meridic diet (consisting of a combination of both crude and purified components of known molecular structure) (Blossey et al., 2000; Tomic-Carruthers, 2009) (Table 6.1) and a root-galling weevil *C. achates* Fahraeus (released to control knapweeds in the United States) reared on the same diet (Goodman et al., 2006; Tomic-Carruthers, 2009) (Table 6.1). No systematic techniques are available for artificial diet-based rearing, due likely to the presence of feeding stimulants (Chapman, 2003) and nutritional profiles that are available only in the specific tissues on which the insect is specialized to feed within the target weed. In contrast, at least 41 predatory and 38 parasitic entomophagous insects have been reared on artificial diets, either meridic or oligidic (the latter composed entirely of crude ingredients) (Grenier et al., 1994; Thompson and Hagen, 1999; Grenier, 2009), although only two predators and three parasitoids have been reared on holidic diets (in which all components have known molecular structure). Coccinellid beetle and hemipteran true bug predatory insects can be reared on oligidic diets based on meat or other protein sources (Cohen, 1985; Rojas et al., 2000; Coudron et al., 2002) or on factitious prey (used in rearing because they are easy to rear themselves), but not actual field prey (Cocuzza et al., 1997). Meridic diets have been used to rear chrysopid (Neuroptera) predators (Thompson and Hagen, 1999). A few parasitoids have been reared on insect-free artificial diets (Grenier et al., 1994; Xie et al., 1997; Hu and Vinson, 1998), but most require tissue culture extract or some other insect derivative (Thompson and Hagen, 1999; Ferkovich et al., 2000). Most endoparasitic wasps and tachnid (Diptera) species can be reared only on whole-insect hosts, whether an intact field host (Johanowicz and Mitchell, 2000; Saleh et al., 2010; Van Nieuwenhove et al., 2012), irradiated field host (Pelanchar et al., 2009), or factitious host (Ozkan, 2006; Yokoyama et al., 2010).

6.1.3. The Critical Need for Mass Rearing Programs for Exotic Invasive Environmental Weeds

Nonnative, invasive weeds of noncrop areas (pasture, rangeland, and natural terrestrial and aquatic habitats) exert an impact of at least US$8 billion per year in control costs and damage (Pimental et al., 2005). The perennial giant grass *A. donax* occupies hundreds of river miles and tens of thousands of hectares along the Rio Grande and its tributaries in the United States and Mexico (Yang et al., 2011), and it causes $1–5 million per year in water loss damage (Seawright et al., 2009) by consuming three times as much water as native plants (Giessow et al., 2011; Watts and Moore, 2011). Mechanical and chemical control methods cannot prevent the spread of arundo and other major invasive weeds in noncrop areas throughout their invaded ranges. Biological control is a cost-effective and efficacious method for controlling weeds in rangelands and natural areas (Nechols et al., 1995; Tracy and DeLoach, 1999; Culliney, 2005; Van Driesche et al., 2010; Van Driesche, 2012), including riparian corridors that produce and convey ecologically and economically valuable water resources. A key example is saltcedar (*Tamarix* spp.), which has been controlled over wide areas using the biological method (Carruthers et al., 2008; Dudley and Bean, 2012). There is little commercial incentive to mass produce insects as weed biological control agents in noncrop areas. Mass production is nonetheless essential for biological control of weeds like arundo that are rapidly expanding their already-large distributions (15% increase since 2002 along the Rio Grande) (Yang et al., 2011) and that benefit from human-exacerbated disturbances such as flooding and fires (Coffman et al., 2010; Giessow et al., 2011), which could hinder insect agent establishment. It was in this context that mass-rearing technology for the arundo wasp was developed to support biological control of arundo by the US Department of Agriculture-Agricultural Research Service (USDA-ARS).

6.1.4. *Arundo donax* as an Invasive Weed

Arundo is a large (up to 15 m) perennial grass native from Mediterranean Europe to India, and invasive in North and South America, South Africa, and Australia (Perdue, 1958; Tracy and DeLoach, 1999). Arundo was introduced for use as roofing and fence material and planted for erosion control, and has since spread through clonal reproduction, developing dense monotypic stands (Decruyenaere and Holt, 2005; Boland, 2006). In North America, the ecological and economic impacts of arundo are most severe in the Lower Rio Grande Basin of Texas and Mexico (over 15,000 ha) (Yang et al., 2011), and in the agricultural Central Valley and coastal drainages in California (over 10,000 ha) (Giessow et al., 2011). Arundo spreads specifically in narrow riparian corridors, which are ecologically and economically critical habitats alongside creeks, canals, rivers, and lakes (Spencer et al., 2005; Goolsby and Moran, 2009; Giessow et al., 2011). In addition to removing valuable water resources in arid regions, arundo increases flood-related erosion and damage to infrastructure (Giessow et al., 2011), reduces the recreational value of waterways (Hughes and Mickey, 1993), and fuels wildfires (Coffman et al., 2010). In south Texas, cover provided by arundo increases survival of the cattle fever tick (Racelis et al., 2012a), a disease-vectoring pest along the US–Mexico border. Dense arundo populations are associated with reduced biodiversity of plants (Quinn and Holt, 2008; Racelis et al., 2012b), arthropods (Herrera and Dudley, 2003; Going and Dudley, 2008), fish (McGaugh et al., 2006), and some birds (Bell, 1997; Giessow et al., 2011). Control by chemical and mechanical methods is sometimes effective (Spencer et al., 2011), but shoots often regenerate from rhizomes, which contain more than 50% of total plant biomass (Thornby et al., 2007; Spencer et al., 2008). In Mediterranean Europe, arundo shoots are thin and brittle and stands are visibly less dense than in similar habitats in North America due to the presence of several insect herbivores in the native range (Kirk et al., 2003). Two insect agents from Europe have been released to date in the North American arundo biological control program: the arundo wasp (Goolsby and Moran, 2009; Goolsby et al., 2009a) and the arundo armored scale *Rhizaspidiotus donacis* Leonardi (Hemiptera: Diaspididae) (Goolsby et al., 2011).

6.2. BIOLOGICAL, ECOLOGICAL, AND BEHAVIORAL INFORMATION ABOUT THE ARUNDO WASP

6.2.1. Discovery and Characterization

6.2.1.1. Taxonomic Background

The arundo wasp is in the superfamily Chalcidoidea, family Eurytomidae (Hymenoptera), which has 1400–1500 species in 88 genera worldwide (Noyes, 2012), 250 species and 23 genera in the Nearctic (DiGaulio, 1998), and about 600 species in 31 genera in the Palearctic (Noyes, 2012), with possibly many more species as yet undescribed (Lotfalizadeh et al., 2007). Eurytomid biology varies greatly, with the majority of species feeding as ectoparasitoids or (less commonly) endoparasitoids of mostly phytophagous insects, including gallers (DiGaulio, 1998; Lotfalizadeh et al., 2007). Three phytophagous genera consist entirely of grass-feeding gallers or borers: *Cathilaria*, *Eurytomocharis*, and *Tetramesa* (DiGaulio, 1998). Several eurytomids have been released or considered for biological weed control, including one gall-inducing *Eurytoma* sp. as a candidate agent of strawberry guava in South Africa (Winkler et al., 1996); a leaf-borer, *Eurytoma bryophylli*, considered for control of mother-of-millions weed in Australia (Neser, 2008); *Eurytoma* sp. considered for wild asparagus in Australia (Kleinjan and Edwards, 2006); and a seed feeder, *Eurytoma attiva* Burks, released and having impact against black sage (*Cordia curassavica* (Jacq.) Roem and Schult.) in Malaysia (Simmonds, 1980).

The genus *Tetramesa* (in subfamily Eurytominae) has at least 200 species, including 125 in the Palearctic (Noyes, 2012), 112 in Europe (http://www.faunaeur.org), and 65 in North America north of Mexico (Phillips, 1936; DiGaulio, 1998). The genus consists entirely of grass-feeding stem gallers and borers, many host specific to one grass genus (Claridge, 1961; Al-Barrak et al., 2004). Three galling species are minor pests of grain crops in the United States and Canada (Holmes and Blakeley, 1971; Sterling, 1976; La Salle, 2005; Shanower and Waters, 2006) and Spain (Cantero-Martínez et al., 2003). A stem-boring *Tetramesa* sp. has been considered for biological control of dropseed (*Sporobolus* spp.) grasses, which are invasive in Australia (Witt and McConnachie, 2004). Another species, *Tetramesa phragmites* (Erdos), is present in adventive (accidentally introduced) populations on common reed (*Phragmites australis* (Cav.) Trin. ex Steud.) in North America and is considered monophagous (Tewksbury et al., 2002). The arundo project is the first to use a stem-galling eurytomid, *T. romana*, as a "classical" weed biological control agent imported from the native range, and the first to release nonnative insect agents of any type against a nonnative perennial grass weed.

6.2.1.2. Biology

Arundo stands in their native range in Mediterranean France and Spain were observed with dead, broken main and lateral stem tips with insect exit holes at stem break points, and swellings on live shoot tips indicative of gall formation. The arundo wasp was discovered as the causal agent of stem tip breakage and exit holes. The arundo wasp is native to the Mediterranean basin (Steffan, 1956; Claridge, 1961; Kirk et al., 2003), and is abundant on arundo in Mediterranean Spain, southern France, and Italy. Galls can be collected from lateral and main arundo canes in the spring, summer, and fall (year-round in southern Spain). In surveys in southern Europe, galls were found on *Arundo plinii* Turra, but never on *P. australis* or other grasses. Quarantine laboratory testing of 34 grass species besides *A. donax* and dicots with habitat associations with arundo demonstrated that the arundo wasp is able to induce gall formation and complete development only on the genus *Arundo* (Goolsby and Moran, 2009), which has no native members in North or South America.

Moran and Goolsby (2009) provide a description of the adult female of the arundo wasp (Fig. 6.1(a)) and life cycle information. Arundo wasps that emerge from European collections as well as from the US-based mass rearing colony are at least 90% female (Moran and Goolsby, 2009). Females are pro-ovigenic (Fig. 6.1(d)) (Vårdal et al., 2003), as they contain a full complement of eggs at emergence, and thus reproduce parthenogenetically. Adult females live an average of 3.7 days at 25–30 °C in colonies and produce an average of 21 adults (maximum of 66 observed among 182 productive wasps), with 80% or more of lifetime production occurring within the first 4 days. Females alight on arundo shoots and insert their ovipositor at a node close to the shoot tip (Fig. 6.1(b)). Eggs are deposited singly but in close proximity to one another at various positions around the apical meristem. Eggs hatch within 5 days, and galls initiate prior to hatch. Galls are visible as swellings on the stem within 7–14 days (Fig. 6.1(c)). Larvae feed in separate chambers within the gall. According to cynipid wasp terminology, the arundo wasp gall is of the polythalamous, plurilocular type (Askew, 1984). Under culturing conditions (temperatures between 25 and 30 °C and high humidity), larvae complete three stadia (Fig. 6.1(e)) within an average of 26 days after oviposition. The pupating larva excretes a meconium containing accumulated larval wastes and cuticle, and gall tissue hardens at this time. Pupation requires about 7 days (Fig. 6.1(f)). The life cycle requires 30–35 days from adult to adult at 27 °C. However, emergence requires up to 3 months under cool subtropical winter night

FIGURE 6.1 The arundo wasp *Tetramesa romana*. (a) Adult female. (b) Female ovipositing into an arundo shoot tip. (c) Gall produced in colony. (d) Ovaries of colony-produced newly emerged adult female showing parthenogenetically produced eggs. (e) Third-instar larva. (f) Advanced (sclerotized) pupa. *All images from Moran and Goolsby (2009)*. (For color version of this figure, the reader is referred to the online version of this book.)

conditions (5–15 °C), a delay put to advantage in mass rearing (see Section 6.3). Dry, galled stems with third-instar larvae and pupae can house live immatures for several months (P. Moran, J. Goolsby, and L. Gilbert, University of Texas–Austin, unpublished data), suggesting that drought-induced quiescence is a possible survival mechanism in the field. Under warm laboratory conditions (constant 32 °C) with high humidity (80% relative humidity (RH)), the arundo wasp completed its development in 40 days and reproduction by females was similar to that under normal culturing conditions (P. Moran, unpublished data).

6.2.1.3. Impact

Native range observations indicated that the arundo wasp caused damage to arundo stems (Kirk et al., 2003). Bird or small-mammal feeding on galls enhances damage in both the native range and the United States by causing stem breakage. In a quarantine greenhouse study conducted with 4-week-old potted arundo stems, infestation with six arundo wasp females per stem per week for 12 weeks decreased final stem length by 92% and leaf length by 44%, and induced production of short lateral stems suitable for the next generation of wasps (Goolsby et al., 2009c). Similar reductions in shoot heights occurred in a nonquarantine greenhouse test involving infestation of shoots with one or five wasps per shoot per week for 12 weeks (Fig. 6.2). Field impacts are being evaluated in the United States.

FIGURE 6.2 Effect of the arundo wasp on shoot height. Center cage shows plants allowed to grow for 12 weeks with no arundo wasp infestation. The cage on the left contains plants exposed to one wasp per shoot per week for 12 weeks, and, on right, five wasps per shoot per week. (For color version of this figure, the reader is referred to the online version of this book.)

6.2.1.4. Field Release

Laboratory host range, life cycle, and impact results led to the 2009 permitting of the arundo wasp for field release in North America by the USDA's Animal and Plant Health Inspection Service (APHIS). Prior to the start of mass rearing, isolated adventive populations were found in Laredo (Racelis et al., 2009) and Austin, Texas (Goolsby et al., 2009b), and in southern California (Dudley et al., 2006). The Laredo population was the first source of insects for the development of mass rearing procedures. Wasps produced from collections in Mediterranean France and Spain later became the predominant output of the rearing program.

6.2.2. Considerations for Mass Rearing of the Arundo Wasp

6.2.2.1. Conditions for Female Oviposition and Survival

Arundo wasp females deposit eggs only at nodes close to the shoot tip (Fig. 6.3(a)), consistent with egg placement close to apical meristem tissues by cynipid galling wasps (Bronner, 1992; Silva and Shorthouse, 2006). Because of the potential importance of plant vigor to stimulate oviposition, only well-watered and fertilized young main shoots were offered to adults. Probing behavior was not always indicative of oviposition, as counts of ovipositor probing

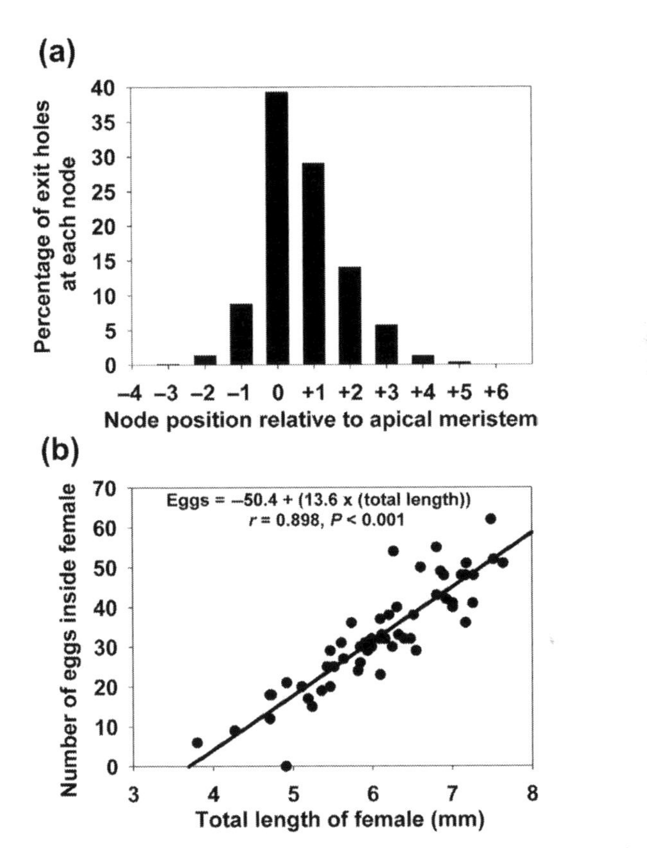

FIGURE 6.3 Importance of plant position and body size on oviposition by the arundo wasp. (a) Locations of exit holes (among 2075 counted) as percentage at each node below or above the shoot tip marked at the time of female release. *(Data resummarized from Moran and Goolsby (2009).)* (b) Relationship between female body length (including antennae) and egg load in dissected ovaries ($n=60$ females).

events on arundo stems were much higher than counts of offspring produced (Goolsby and Moran, 2009). Special care was taken to maintain high ambient humidity under warm greenhouse temperature regimes in oviposition cages, as this approach increased adult survival time and hence probing opportunities. Substantial variation in adult female size was observed, and was positively correlated with egg load (Fig. 6.3(b)). However, the relationship between female fecundity (eggs produced) and fertility (offspring produced) could not be determined due to the deposition of eggs inside the stem. Output of adult wasps was tracked over time to verify maintenance of optimal conditions for wasps.

6.2.2.2. *Condition of the Plant for Gall Development*

The plant vigor hypothesis (Price, 2000) suggests that herbivores favor and/or perform better on rapidly growing tissues with high nutrient and moisture content. Galls contain tissues with high nitrogen and/or sugars (Raman et al., 2005) and are considered potent nutrient resource sinks (Price, 1992; Dennill, 1988; Raman et al., 2005, 2006; Dorchin et al., 2006). Galling insects used in weed control are sensitive to plant vigor: a leaf-galling moth (*Epiblema strenuana* Walker (Lepidoptera: Tortricidae)) selected *Parthenium hysterophorus* L. (Asteraceae) plants that appeared vigorous in terms of height, number of branches, and flowers (Dhileepan, 2004). The root-galling weevil *C. achates* selected fertilized diffuse knapweed (Lejeune et al., 2005), although this was not the case for bud-galling tephritid flies (*Urophora* spp.). A stem-galling midge (*Rhopalomyia* sp.) (Diptera: Cecidomyiidae) produced more galls on fertilized compared to unfertilized scentless chamomile (Hinz and Muller-Scharer, 2000). Outside the realm of biological control, galling sawfly (Hymenoptera: Tenthredinidae) performance decreased under drought stress (Price, 1992) due to avoidance of stressed tissues (Craig et al., 1989). Gall production in natural systems is tied to production of new growth or fruit (Rohrfritsch, 1992). Observations in Spain indicated that active arundo wasp galls (containing larvae, pupae, or teneral adults) were most common on young main and lateral shoots. Across field sites in the Lower Rio Grande Basin, galls are most common on vigorous shoots occurring adjacent to permanent water sources, in the spring in moist soil, and at normally dry sites affected by significant rainfall (Racelis et al., 2009). Water was provided in sufficient or excess quantities during the gall development stage of mass rearing to maximize colony productivity. For example, frequent watering under hot, dry greenhouse conditions was necessary to prevent the development of high osmolality and/or low nutrient content in galled tissues and subsequent negative effects on adult wasp output. Galling chalcid wasps develop only on diets that fall within certain ranges of osmolality (osmotic pressure) (Thompson, 1983; Nation, 2002).

6.2.2.3. The Plant Physiological Basis of Gall Development

Galls are produced via continuous chemical communication between the galling insect and its plant host (Bronner, 1992; Rohrfritsch, 1992). Gall formers likely manipulate the genetic regulation of their host plants (Hartley, 1998; Schonrogge et al., 2000), but the chemical signals involved remain largely unknown (La Salle, 2005; Raman et al., 2005). Cynipid wasp females inject proteolytic enzymes along with the eggs, whereas chalcids do not (Bronner, 1992). Oviposition by the chalcid arundo wasp is, however, sufficient to initiate gall formation, with stimulation likely coming from the eggs prior to hatch. Gall development of the arundo wasp involves three stages: (1) initiation, completed before the arundo wasp larvae hatch; (2) growth and differentiation, completed during the first stadium; and (3) maturation, occurring during the second and third stadia, which are time frames of heavy feeding (Bronner, 1992; La Salle, 2005; Silva and Shorthouse, 2006). Wasp galls consist of an inner nutritive layer subdivided into nutritive cells (consumed by the larvae), a storage nutritive layer, and a sclerenchyma layer. Beyond these is a protective outer layer. Nutritive cells have high concentrations of glucose, di- and triacylglycerides, free amino acids, and both structural and enzymatic proteins (Raman et al., 2005). Lipids accumulate in nutritive cells, while starch accumulates in a gradient peaking at the sclerenchyma (Bronner, 1992). Larval feeding on nutritive cells is necessary to stimulate their continued production (Rohrfritsch, 1992; Raman et al., 2005; Silva and Shorthouse, 2006). The gall thus provides a small amount of high-nutrient food. Light, water, and nutrients were made abundantly available to the plants to promote assimilation of metabolites into galls.

6.2.2.4. Subspecific Genotype Adaptation

Insects or mites used in biological weed control agents may show host specificity at the subspecies level for the geographic region and/or genotype of the weed on which they evolved (Goolsby et al., 2006). Arundo wasps from several sites in Spain, France, and the adventive Texas population were reared on one common south Texas source of arundo plants. Microsatellite analyses revealed genotypic differences among the wasp accessions (D. Tarin et al., unpublished data). We found no differences in host range among these different accessions (Goolsby and Moran, 2009); however, Spanish wasps had somewhat (1.4-fold) higher productivity in quarantine rearing than did French wasps on shoots grown from south Texas arundo rhizomes (Moran and Goolsby, 2009), which were found to be genetically matched to arundo sites in Spain (D. Tarin et al., unpublished data). Releases were conducted with mixed wasp populations. Efforts to re-collect samples from multiple field sites with different climates could determine differential establishment and preadaptation among wasp accessions.

6.3. A PLANT-BASED MASS REARING SYSTEM FOR THE ARUNDO WASP

6.3.1. Description of the Mass Rearing System

6.3.1.1. Source of Insects

Galled *A. donax* stems were collected from riparian sites near rivers or creeks in Perpignan, France, and from Coloma and Granada, Spain. A separate colony was initiated using galled stems collected from an adventive population of *T. romana* in Laredo, Texas. Both main stems (length 20–40 cm) and lateral stems (length 5–20 cm) were collected. Galled stems were stripped of leaves and placed in Plexiglas® emergence boxes (60 cm long × 45 cm deep × 45 cm tall), clear on top and either clear or white on the bottom and sides. Boxes were maintained at 25–30 °C with a 14:10 h fluorescent light:dark regime, and were moistened daily with water mist applied with a handheld pump sprayer. Approximately 3 g of honey was applied as a linear streak inside each emergence box as adult food. Wasps were collected every 1–2 days with handheld insect aspirators. Males are sterile, and the female:male sex ratio was at least 9:1 (Moran and Goolsby, 2009).

6.3.1.2. Mass Rearing Cages

Cages were constructed by the USDA-APHIS Aircraft and Equipment Operations (AEO) Division, Moore Air Base, Edinburg, Texas. The cages measure 2.1 m tall by 0.8 m wide by 0.8 m deep, and consist of a lightweight aluminum metal frame (0.063 in sheet metal thickness) and screen material (Fig. 6.4(a) and (b)). The metal frame is held together with corner plates secured with soldered pin-type connectors rather than screws. The cages are equipped with a metal shelf on the bottom and at 1 m height and two hinged doors, creating two levels on which arundo wasps can be confined on plants for oviposition. Based on studies of light penetration under greenhouse conditions (see Section 6.4.2.2), the mass rearing protocol uses only the top level. Each door is secured with two clip-type latches. The inner surface of the door and the door contact surface on the cage frame are lined with foam gasket material (approximately 0.3 cm thick) to create a tight seal. The two shelves have holes on the rear side of the cage so that water emitters (DIG Corp., Vista, CA, USA) can be used. Emitters for both shelves enter the cage on the bottom through holes with rubber gaskets. Emitters for the upper shelf pass through the lower cage, and then through holes in the upper shelf. Cage screening is sealed to the cages using plastic square spline material inserted with the screening into grooves in the aluminum cage panels. Initially, the cage screen material consisted of black organza silk mesh (20 mesh size) (Los Dos Rios, McAllen, TX, USA). The screening was changed to black "one-way stretch" nylon (20 mesh size) (Los Dos Rios) to reduce moisture and algal buildup on cage surfaces and to reduce costs, as the nylon material was cheaper and resistant to deterioration. This change did not affect wasp output (Racelis and Moran, unpublished data).

Other cage designs were tested and found to be inferior for mass rearing of the arundo wasp in terms of wasp output per arundo shoot or per unit labor. These included 40 cm long, 12 cm diameter cylindrical cages, which were used to confine wasps on single shoot tips for biological and host range studies in quarantine (Moran and Goolsby, 2009); larger cylinders or "tomato cages", 20–30 cm in diameter and 0.8–1.2 m tall, with black silk organza screening, which were used on single pots containing one or more arundo stems; and square cages, 0.8 m on each side, placed directly on the floor of the greenhouse with 6–8 young arundo shoots inside.

6.3.1.3. Plant Culture

Arundo rhizomes are obtained from local populations in the Lower Rio Grande Valley of south Texas. Rhizomes are planted in commercial soil containing 80% peat moss (Sunshine Mix #1, Sun-Gro Horticulture, Bellevue, WA, USA) in 20 cm diameter pots (volume of 0.008 m^3 soil) supplemented with approximately 10 g (1 US tablespoon) pelletized urea per pot (Wilbur Ellis, Edinburg, TX, USA). Plants are maintained either onsite or at a local nursery, under greenhouse conditions in both cases, with ambient lighting, temperature between 25 and 35 °C, and water application every 1–2 days with emitters or overhead spray irrigation. After 4 weeks, plants with shoots 0.5–0.8 m tall, 3–6 nodes and leaves, and no lateral stems are moved into cages for infestation with arundo wasps.

FIGURE 6.4 Mass rearing of the arundo wasp. (a) Cage screened with one-way stretch nylon material containing plants and wasps released from collection jar. (b) Arrangement of wasp oviposition cages in greenhouse. (c) Arundo shoots in gall maturation greenhouse. (d) Harvested shoots prepared for progeny wasp emergence. (e) Wasp emergence barrels containing shoots and overhead light. (f) Collection jar to capture wasps emerging from barrel. (For color version of this figure, the reader is referred to the online version of this book.)

6.3.1.4. Arundo Wasp Infestation

Six to eight pots (each containing 1–4 shoots) are placed into the top shelf of the oviposition cages. Insects are released from plastic colony collection jars onto plants (Fig. 6.4(a)). Cages are positioned in the greenhouse in a pattern to maximize light exposure and air circulation (Fig. 6.4(b)). The oviposition greenhouse is 30 m long by 15 m wide and has a semicircular plastic roof with plastic shading material that reduces sunlight by 40% (P. Moran, unpublished data).

Temperature is maintained between 20 and 30 °C with built-in evaporative cooling fans. Humidity is enhanced with portable high-intensity mist blowers running for 5 min per hour between 0900 and 1900 h. Ambient lighting is sufficient between March 1 and November 1, when day lengths are at least 11.8 h in south Texas (NOAA, 2011). In late fall and winter (November to March), supplemental lighting (sodium halide lamps, 1000 W) is used from 1700 to 2000 h to maintain a 12 h day length. Between 70 and 100 wasps are released into each cage, with separate cages for each wasp accession–collection source. Wasps are allowed access to the plants for 7 days, well beyond the expected 4–5-day life span of the adult female (Moran and Goolsby, 2009). A small number of uncaged "sentinel" arundo plants in the greenhouse attract the few surviving wasps that escape cages when plants are removed.

6.3.1.5. *Arundo Wasp Gall Maturation*

When plants are removed from cages, they are fertilized with a second dose of 10 g urea per pot and placed in separate greenhouses (Fig. 6.4(c)) maintained at 30–35 °C to promote rapid gall development, with evaporative cooling and overhead irrigation. An outdoor screened area is also used for maturation; in the summer, conditions are largely similar to those in the greenhouses, while during winter low nighttime temperatures increase gall development time two- to threefold (J. Goolsby, unpublished data). This delay is suitable for accumulation of large numbers of progeny wasps for spring releases. Greenhouses and the screened outdoor area typically contain 500–2000 plants each due to weekly staggering of wasp infestations. A third dose of 10 g urea is added to each pot 28 days after cage removal. Under warm greenhouse or summer outdoor conditions, arundo shoots are harvested 35 days after removal from the infestation cages by cutting at the soil line.

6.3.1.6. *Collection of Progeny Wasps*

Developed leaves and the whorl of immature leaves at the shoot tip are removed from shoots (Fig. 6.4(d)). Cut shoot ends are dipped in paraffin wax to seal in moisture. Galls are typically found 2–3 nodes below the shoot tip at the time of harvest; the galled area was the apical meristem at the time of oviposition (Fig. 6.3(a)) (Moran and Goolsby, 2009). All shoots from each week of oviposition cage infestations are pooled by wasp accession and placed in emergence barrels (Fig. 6.4(e)) in an air-conditioned (25 °C) room. During weekends, the emergence room is darkened and chilled to 18 °C to slow emergence. The emergence barrels consist of a 0.6 m internal diameter, 1.2 m tall fiberboard cylinder with a metal frame (208 l internal volume) (FLR 10653, Pelican Sales, Greif Inc., Melbourne, FL, USA) and a 60°-angled spun aluminum funnel top (fabricated by the USDA-APHIS AEO Division at Moore Air Base), attached to the cylinder with a tension clamp that encircles the cone. The top of the cone has a threaded opening to accommodate an inverted 3.8 l plastic jar with a 12 cm diameter opening. Each barrel, conical top, and jar are positioned under a high-intensity LED light (PAR 38 NV-PPW12BKQ, 18W, 1000 lumens) (New View Lighting, Las Vegas, NV, USA) mounted onto an arm connected with bolts to the tension clamp. Emerging wasps show a strong phototropic response (Racelis et al., 2010) and fly into collection jars (Fig. 6.4(f)). The jars are ventilated with one-way screening, similar to that used in the oviposition cages, and glued to cover two oblong holes (approximately 8 cm long and 2 cm wide) cut into the jar directly below the threaded opening. Jars are collected and wasps counted daily. Shoots are discarded from barrels after one month, and total progeny wasp yield is determined.

6.3.2. Output of the Mass Rearing System

The mass rearing system was implemented in 2009. Figure 6.5 shows a running average of daily output between 2010 and 2012. Wasp accessions from Spanish Mediterranean locations comprised 70% of total production by July 2010 (Fig. 6.5) to match the geographic origin of released wasps to the genetically determined biotypes of invasive arundo present in the Lower Rio Grande Basin. Wasps are obtained throughout the year with seasonal variation in production, which is controlled to meet seasonal needs for large-scale releases. Plants are produced in large numbers (200–300 4-week-old pots per week with 1–4 stems per pot) in greenhouses between mid-January and April to supply up to 36 wasp oviposition cages per week (eight pots per cage). Galls are maturated on these stems in greenhouses and stems harvested between February and June. These shoots are combined with others on which wasps were allowed to oviposit between November of the previous year and January, and on which galls maturated slowly in the outdoor screen house under cool winter conditions. The result is a spring–early summer peak in wasp production (Fig. 6.5) for release at field sites to attack young arundo main and lateral

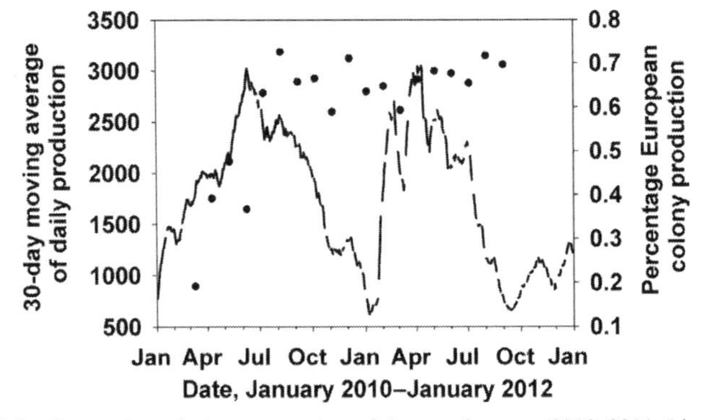

FIGURE 6.5 Seasonal trends in mass rearing of the arundo wasp, 2010–2011. Line (left axis) shows 30-day running average of daily production across all wasp accessions. Scatterplot (right axis) shows average monthly proportion of the daily yield consisting of European wasp accessions.

shoots, which develop in large numbers during the spring. To reduce cooling costs, gall maturation greenhouses are taken out of production during the summer, reducing gall production between July and October (Fig. 6.5). Most gall maturation takes place in the shade house during this time. Plant production increases in October to provide stems for the winter shade house wasp generation, which is initiated in November. Plant production is then kept to less than 200 pots per week for indoor winter wasp production, which serves mainly to maintain colonies.

Wasps have established populations at all of the sites at which they have been released along the Rio Grande (Racelis et al., 2010; A. Racelis and J. Goolsby, unpublished data), as a result of high production control (Leppla and Ashley, 1989) (defined as consistent high quality of wasps produced) and the effectiveness of the large-scale release approach. The arundo wasp system involves built-in variability in product control (output of agents) (Leppla and Ashley, 1989) to meet seasonal needs. The results of early experiments (see Section 6.4) were used to improve process control (the maintenance of constant values for key life history parameters such as adult survival and fecundity) (Leppla and Ashley, 1989) (see Chapter 9).

6.4. CHALLENGES ENCOUNTERED AND ADDRESSED IN THE DEVELOPMENT OF MASS REARING

6.4.1. Challenges to Efficient Mass Rearing

The arundo wasp mass rearing system is based on a thorough knowledge of the life history and behavior of the wasp, of plant physiological processes as they relate to galling (see Section 6.2), and of the high efficacy potential of galling insects (see Section 6.1). Logistical challenges were addressed to develop this system (see Section 6.3) with limited space and labor resources, as well as limitations in control over environmental conditions. Key challenges included the maintenance of plants that were favorable for galling, development of suitable environmental conditions for adult infestation of shoots, and development of a system for efficient collection of progeny wasps.

Arundo reproduces only through rhizome expansion and stem rooting. Weekly excavation and planting of rhizomes were significant labor and cost factors, but were necessary because replanting of rhizomes yielded small, thin shoots. Logistical challenges were encountered in moving hundreds of arundo shoots per week through the wasp rearing process. A peat-based commercial soil was used, rather than the sandy or clay soils that are typical of arundo field sites, to minimize the weight of pots. For plant infestation, variable adult wasp density, cage size, environmental conditions (light, temperature, and humidity), as well as shoot quality factors were tested and then controlled to promote probing behavior and successful oviposition on an abundant supply of suitable shoot tip tissue. To collect adults, adequate shoot moisture and lighting were necessary to allow adults to chew holes in galls and to fly into collection containers.

6.4.2. Key Challenges Addressed in Mass Rearing

6.4.2.1. Adult Survival

Tests indicated that high-intensity misting fans elevated relative greenhouse humidity to 80% or higher, which increased parent adult survival and improved colony output. Provisioning was considered to increase wasp survival. Wasps lived significantly longer (typically 5–10 days) when confined in vials with sucrose (0.3 or 1.0 M reagent-grade solution applied to paper tissue and placed in the vial cap) than in vials with water alone (log-rank test from survival analysis, $\chi^2=14.9$, $df=1$, $P=0.0001$, SAS PROC LIFETEST, SAS Institute, Cary, NC, USA). In behavioral observations of 50–100 wasps, at least one wasp was present on cotton balls containing 10% honey solution for 51% of the time, while balls soaked with sucrose hosted wasps for only 14–19% of the time, and water-only balls less than 1%. In separate studies, glucose, fructose, and sucrose were four- to 12-fold higher in honey-fed compared to unfed 3-day-old wasps (F. Wackers, Lancaster University, Lancaster, UK, unpublished data), indicating uptake. Wasp survival could thus be enhanced by feeding on a natural source of sugars such as honey; however, in our system, problems like mold buildup and ant invasions made feeding unpractical. Provisioning of adults could be useful for colony initiation starting with small numbers of parental wasps.

6.4.2.2. Adult Oviposition

Limited lighting was associated with low progeny production. Observations showed that probing behavior was higher in chambers equipped with sodium halide lamps compared to fluorescent lamps. In the infestation greenhouse, rooftop plastic heat screening reduced light intensity by 40% (1657 $\mu mol/m^2/s$ outside compared to 1030 inside, readings taken at 0900, 1200, and 1500 h). Light levels were threefold to 13-fold higher in top cages than bottom cages in the stacked cage design ($F=162$, $df=1, 94$, $P<0.0001$; ANOVA) (Fig. 6.6(a)). Within top cages, light levels were 2.6-fold higher at shoot tips than at the base of the shoot ($F=49.8$, $df=1, 47$, $P<0.0001$). Wasp output per cage was significantly (by 1.4-fold) higher in top than in bottom cages (Fig. 6.6(b)) ($F=6.6$, $df=1, 50$, $P=0.01$). The final rearing protocol thus used only top cages' output. Varying the density of wasps in oviposition cages (between 35 and 80 on eight pots, each containing 1–4 arundo shoots) did not affect the number of progeny produced ($F=0.82$, $df=3, 50$, $P=0.49$). The lowest dose (about one wasp per shoot) is thus adequate to induce sufficient galls for wasp production while maximizing the number of wasps available for field release.

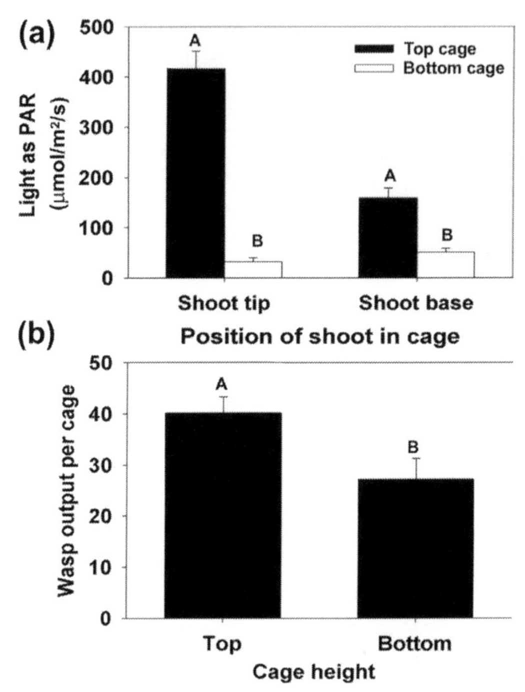

FIGURE 6.6 Influence of variable cage lighting on reproductive output of the arundo wasp in greenhouses. (a) Effect of top vs bottom cage position in double-stack cages on light levels at the shoot tip and at the base of the shoot. (b) Effect of cage position on output per cage.

6.4.2.3. Plant Condition at Time of Oviposition

Urea fertilization at planting promoted rapid growth of tender shoot tip tissue. Pest problems occasionally arose in plant production. A tip-boring fly, *Chaetopsis massyla* Walker (Diptera: Ulidiidae), damaged shoot tips. A bait device with spinosad was used to control this fly (Goolsby and Mangan, 2010). Wasp progeny yield was sixfold higher on main shoot tips than on 2–5 cm long lateral shoots (arising from pruned main shoots) in cages containing both shoot types (A. Racelis and P. Moran, unpublished data), indicating that wasps preferred main shoots.

6.4.2.4. Plant and Environmental Conditions during Maturation

Nitrogen fertilization increases soluble protein in grasses (Cuomo and Anderson, 1996; Wolf and Optiz Von Bobefeld, 2003), which may then be mobilized into gall tissue (Raman et al., 2005), explaining why urea was provided immediately after infestation and 4 weeks later, as galls neared maturity. Temperature extremes in greenhouses were avoided in mass rearing, but the use of an outdoor shade house in summer with temperatures ranging as high as 39 °C did not adversely affect colony production.

6.4.2.5. *Conditions during Emergence of Progeny Wasps*

Waxing cut arundo shoot ends allowed them to retain moisture so that teneral adults could chew their way out of galls, while avoiding the need for stem misting and subsequent mold problems. Conical emergence barrel tops with a 60° angle led to increased upward movement of wasps into lighted collection jars, relative to 45° angle cones used in early designs. Improvements in emergence were measured as decreased numbers of dead adults found trapped in dissected cut shoots or in barrels.

6.5. USE OF MASS REARED ARUNDO WASPS FOR BIOLOGICAL CONTROL OF ARUNDO

6.5.1. Use of Cold Inebriation to Prolong Wasp Life

Arundo wasps are stored overnight in some cases prior to field releases and are often transported long distances to remote sites along the Rio Grande. Wasps are routinely stored in emergence jars (see Section 6.3.1.6) or vials at 5 °C for up to 24 h with little or no mortality. In pilot tests, 60–70% of wasps confined in vials or petri dishes survived 3 days of storage in darkness at 5 °C (Racelis et al., 2010) or 10 °C (A. Racelis, unpublished data), with longer wasp recovery times required after removal from 5 °C than from 10 °C. Wasps survived (30%) up to 7 days under 5 °C conditions. Wasps that recovered from chilling increased their populations by 3.5- to 6-fold (Racelis et al., 2010). Based on these results, cold storage of wasps is now a routine part of the wasp transport and release process.

6.5.2. Packaging of Wasps for Field Release

Wasps are released in some cases by opening emergence jars at field sites. However, many wasps are packaged for an aerial release system at remote field sites. The objectives were to deliver the wasps with little or no mortality and to maximize the speed with which large numbers of wasps could be released (Racelis et al., 2010). The package consists of a cardboard box frame, 12 cm long by 6 cm wide and 6 cm deep, with a sealable 2 cm diameter hole in the top. The frames are wrapped in brown heavyweight paper with a semicircular cut to admit a funnel to load 100 wasps through this sealable hole. The aerial release method is described by Racelis et al. (2010). Cooler-stored boxes containing wasps are loaded into an uncooled chute or "magazine" mounted onto the floor of an airplane. Packages drop through a hole in the plane floor when a release pin is removed. Blades in the bottom of the release chute slice open one end of the box. Boxes land in arundo stands within a few seconds of release, and wasps emerge from the sliced end. In warm temperatures (25–30 °C), all wasps emerge within 5 min (A. Racelis, unpublished data), and at low field temperature (10 °C), 70% of wasps emerge within 1 h (Racelis et al., 2010), demonstrating the effectiveness of this release technique.

6.5.3. Status of the Arundo Wasp Mass Release Program

The arundo wasp *T. romana* has established reproductive populations along all lengths of the Rio Grande between Del Rio and Brownsville, Texas, that contain arundo, a river span of some 500 km covering about 25 release sites. This accomplishment is attributable to the success of mass rearing, a strategic release plan to match arundo seasonal phenology, and the ability of the wasps to disperse 15–25 km per year along waterways (Racelis et al., 2009). Wasps of three Spanish accessions as well as one French and one adventive (Laredo, Texas) accession have been released. Ongoing studies are examining differential establishment of these accessions across multiple field sites with differing microclimates using molecular methods. Releases have been made in Mexico along the Rio Grande and in the south-central region using the mass rearing protocol (M. Martínez Jiménez, Instituto Mexicano de Tecnologia del Agua (IMTA), Cuernavaca, Mexico, personal communication, November 2011). Additional releases are planned throughout the Lower Rio Grande Basin and other arid southwestern US areas invaded by arundo.

6.6. INVESTIGATIONS INTO AN ARTIFICIAL DIET FOR REARING OF THE ARUNDO WASP

6.6.1. Nutritional Needs and Feeding Constraints

The development of an artificial diet and subsequent rearing system for the arundo wasp would allow large-scale production in vitro. A prerequisite to diet development is an understanding of the basics of the feeding biology of the desired species (Cohen, 2003). The arundo wasp is highly specialized to feed on galled stem tip tissue, and its larval feeding biology is poorly understood. Ascertaining the physical and nutritional requirements of arundo wasp larvae and the feeding stimuli they receive from the plant is difficult due to the larvae's cryptic endophagous feeding behavior deep inside gall tissues, which might not be replicable in ex situ observations on diets. The importance of matching physical characteristics (such as solid vs liquid) and nutritional profiles (protein, lipid, carbohydrate, mineral, and water content) in diet development to the insect's feeding requirements is discussed by Cohen (2003). Gall tissue is not nutritionally unique in plants; other plant tissue such as seed endosperm, pollen grains, and nectar share its nutritional features (Bronner, 1992). However, nutritive gall tissue is available only in limited quantities for biochemical and textural analysis. For this reason, comparisons of artificial feeding substrates with the natural food, as recommended by Cohen (2003), are difficult for the arundo wasp. Despite these difficulties, we studied the mouthpart morphology and feeding behavior of arundo wasp larvae on gall tissue ex situ and used this information to formulate meridic artificial diets based on gels or slurries. We also examined arundo callus tissue as a feeding substrate. To our knowledge, the arundo wasp is the first stem-galling wasp for which artificial diet development has been attempted.

6.6.2. Novel Feeding Modality of the Arundo Wasp Larva

The gross morphology of the mouthparts of the arundo wasp larva is similar to that of parasitic wasp larvae (Fig. 6.7(a) and (b)) (Snodgrass, 1935), suggesting that the feeding modality is proximally mechanical but ultimately suctorial. First, the mandibles damage nutritive cells, then saliva- or gut-derived digestive enzymes are secreted, and finally imbibing of liquid food occurs through the modified prementum and postmentum or other mouthparts. This is one feeding model for cynipid galling wasp larvae (Bronner, 1992). The feeding behavior of arundo wasp larvae was observed live and in digital video recordings on an artificial diet and arundo callus, as well as inside partially dissected galls (Fig. 6.7(c) and (d)). We discovered that arundo wasp larvae use a type of extra-oral digestion (EOD) that is different from the cynipid model noted here, and very different from the EOD described by Cohen (1995) for arthropods in general. Arundo wasp larvae produced a droplet of clear fluid, presumably from salivary glands or filtered gut fluids; suspended the droplet on the prementum and postmentum for 5–30 s; and then applied the droplet to the diet or callus. This process of droplet dabbing was followed by biting or rasping of the softened or macerated diet material. The release of digestive fluid was continuous, suggesting that this process is also used within actual plant gall chambers. This mode of suctorial larval feeding, involving secretion of salivary enzymes first and then cell lysis, has been observed in some cynipid wasp larvae (Rohrfritsch, 1992) but has not been previously reported in a chalcidoid galling wasp larva. Chewing action after droplet secretion presumably helps dislodge liquid food from cellular debris for easier uptake. The digestive fluid of other insects that use EOD in combination with a tissue-macerating mechanism contains pectinase (Cohen and Wheeler, 1998). However, we could not capture adequate amounts of the arundo wasp larval digestive fluid for enzyme analyses. The presence of a "blind'' gut (no connection to anus) in these larvae was confirmed, consistent with cynipids (Bronner, 1992) and chalcid and most hymenopteran parasitoids (Gordh et al., 1999), in which a meconium containing larval wastes and exuviae is excreted at the time of pupation.

6.6.3. Diet Components Evaluated for the Arundo Wasp

6.6.3.1. Gelled Diets

The stimulation of specialized feeding by the arundo wasp larva could demand a specific host plant "signature" in diets, including alkaloid compounds such as arundavine (Khuzhaev et al., 2004). We produced extracts by freeze drying ungalled arundo stems and grinding tissue to 0.1–1.0 mm diameter particle size. We then either added whole arundo powder to diets or first extracted the powder with hexane, ethanol, or water washes to remove molecules that were soluble in each solvent. Extracts were added to diets in ranges from 0.1% to 10% of diet content on a wet weight basis. Recognizing the importance of diet texture

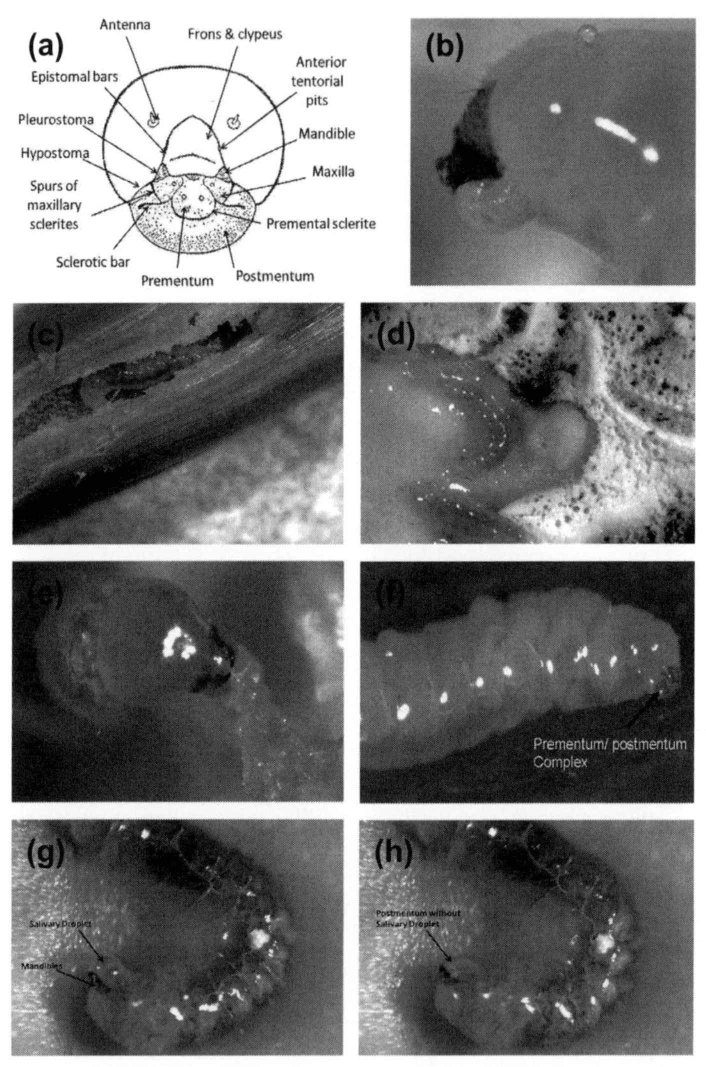

FIGURE 6.7 Larval feeding biology of the arundo wasp. (a) Generalized wasp larval head, anterior view (based on *Sphecophaga*, a parasitoid (Ichneumonidae)) *(Modified from Snodgrass (1935))*. (b) Larval head of the arundo wasp. The bulbous structure at the bottom of the head is the postmentum, and the structure above the left mandible (dark anteriorly-pointed structure) is the prementum. Methylene blue staining was used to highlight pre- and postmentum, frons, and clypeus. (c) Cutaway of gall showing third-instar larva feeding on gall tissue. (d) Arundo wasp larvae feeding on an artificial diet consisting of a cellulose slurry mixed with a liquid solution of casein and plant tissue culture. (e) Larva showing positive feeding response to callus by rubbing its postmentum on callus tissues. (f) Larva on callus dyed red to show movement of food bolus in larval gut by peristalsis and antiperistalsis. (g and h) Larva on cellulose slurry diet before (g) and after (h) depositing a drop of salivary fluid onto the diet substrate as part of the extra-oral digestive feeding strategy described in this chapter. (For color version of this figure, the reader is referred to the online version of this book.)

TABLE 6.2 Example of a Gel-Based Artificial Diet Tested with Second- and Third-Instar Larvae of the Arundo Wasp

Diet Component	Weight or Volume (per Liter Diet)
Agar solution (5%)	200 ml
Arundo extract (aqueous)	500 ml
Sucrose	4 g
Starch	1 g
Casein (Sigma sodium salt)	4 g
Sodium propionate	0.2 g
Potassium sorbate	0.2 g
Yeast extract	2 g
Wesson salt mixture	0.5 g
Vanderzant vitamins	2 g
Ascorbic acid	0.5 g

to maximize phagostimulation, gut motility, and nutrient protection (Cohen, 2003), we used various texture-inducing materials such as microparticulate cellulose, xanthan gum, locust bean gum, and arundo powder that had been boiled to remove soluble nutrients and extracted with hexane to remove lipids, resulting in an inert powder. We tested both agar- and carrageenan-based gel diets, often with pectin to enhance gelling effects and provide a pectin "flavor," to simulate the extracellular matrix found in terrestrial plants. Agar, carrageenan, and pectin are among the macromolecules that bind water and modify the texture (viscosity or gel capacity) of aqueous materials used in insect diets (Cohen, 2003). We observed secretory feeding behavior by larvae with no detectable differences in responses among these hydrocolloids, so we concluded that *T. romana* will feed on gelled diets. Table 6.2 shows one of the dozens of gel diet formulations tested. In this case, a casein-based diet (derived from Harley and Willson, 1968) was used with an aqueous extract of arundo. We also tested diets based on soy, rice, wheat germ, wheat bran, and flax seed meal as plant components with supplements of yeast extract, brewer's yeast, torula yeast, various types of sugars, starches, lipid sources (including various sterols and polar lipids), vitamin mixtures, and mineral mixtures as described by Cohen (2003). None of the gel-based diets supported larvae for more than a few days, and no molting occurred.

6.6.3.2. *Arundo Callus Diets*

We hypothesized that the arundo wasp requires living tissue to feed, and that arundo tissue culture would thus be an attractive food source. Tissue cultures were produced by Dr Sergei F. Krasnyanski of the Plant Transformation Laboratory in the Department of Horticulture at North Carolina State University. Callus was grown in tissue culture medium using the methods of Marton and Czako (2007). The feeding response is evident in Fig. 6.7(e), in which a larva is seen rubbing its postmentum against the arundo callus, and ingestion was verified using dyed callus (Fig. 6.7(f)). Arundo tissue culture sustained larvae for up to 2 weeks, with several second instars molting to third instars. However, mold outbreaks terminated the tests before larvae could pupate. The delicate balance of keeping tissue cultures free of microbial invasion was complicated by the introduction of larvae. In fact, a microbial intermediary may be involved in the galling process (Raman et al., 2005), making the use of antibiotics problematic.

6.6.3.3. *Slurry-Based Diets*

In an attempt to combine plant-derived stimuli with ingredients in a meridic diet, a cellulose slurry diet was made from a nutrient base of casein protein, carbohydrate (sugars and starch), lipids (such as lecithin), mineral and vitamin mixtures, and plant tissue culture media prescribed for banana, tobacco, and rose (obtained from Carolina Biological Company, Burlington, NC, USA). We hypothesized that hormones present in these slurries, such as auxins and gibberellins, would simulate cellular fluids in the nutritive layer of galls, which are also rich in hormones (Rohrfritsch, 1992; Raman et al., 2005). Solutions of diet components were filter sterilized with a 0.22 μM filter system, and the filtrate was mixed with autoclaved microparticulate cellulose (Sigma) to form the slurry (Fig. 6.7(d)). A slurry diet with plant tissue culture media as the base nutrient mixture elicited feeding activity (Fig. 6.7(g) and (h)) and growth, but no larvae pupated.

6.6.4. Conclusions: Artificial Diet Development

The previously undescribed feeding mechanism of the arundo wasp larva is likely reflective of all galling eurytomid larvae. Conventional gel-based diets, either with or without extracts from ungalled arundo stems, failed to stimulate sufficient feeding for growth and survival. Arundo tissue culture appeared to contain at least some of the necessary stimuli for feeding and sufficient nutrients for larval growth, but it was susceptible to microbial growth. Aseptic culture is used for host resistance and insecticide screening involving grape phylloxera (*Daktulosphaira vitifoliae* Fitch), which induces simple pouch galls on leaves and nodules on roots cultured from callus. Aseptic methods have not been used successfully for a galling insect with an endophagous larval stage (Beiderbeck, 2005) and present logistical challenges for mass rearing. Slurry formulations combined the practicality of meridic gel-based diets with plant (non-arundo)-based stimuli, and they sustained arundo wasp larvae for several days, with molts to the next instar in a few cases. A possible future direction would be to test arundo callus or cultured gall tissue extracts as stimulants in slurry-based diets.

6.7. CONCLUSIONS AND FUTURE DIRECTIONS

6.7.1. Success of the Mass Rearing System for the Arundo Wasp

The mass rearing system developed for the arundo wasp has been successful in amplifying the wasp population for field releases in the Rio Grande Basin, and the technology is suitable for transfer to other agencies and facilities. Pitfalls and challenges associated with variable environmental conditions, wasp behavior, and reproductive biology and logistics have been addressed. The mass-rearing protocol is adaptable to constraints in rearing facilities. For example, arundo shoots could be propagated from 1-year-old main shoots positioned horizontally in soil (Boland, 2006), rather than from rhizomes. Biochemical characterization of nutrient and stimulants in gall tissue could facilitate development of an artificial diet. Wasp impact is being measured initially in terms of increased gall density on arundo shoot tips, decreased shoot length (Goolsby et al., 2009c), and shoot tip death. However, decreases in new shoot recruitment and increased visibility through dense monotypic stands are the long-term indicators of the impact of the arundo wasp *T. romana* released in combination with the arundo armored scale *R. donacis* in the Lower Rio Grande Basin of Texas and Mexico.

6.7.2. Mass Rearing of Biological Control Agents and Invasive Grasses

This chapter demonstrates the feasibility of plant-based mass rearing to expedite field release and establishment of the arundo wasp. Mass rearing has been underutilized as an approach to biological weed control. The current regulatory requirements for the introduction of insect agents for weed control require comprehensive investigations of their life history, feeding and reproductive behavior, environmental requirements, as well as host range. The information obtained can provide a solid basis for mass-rearing protocols. Grasses and sedges dominate the list of the world's worst weeds in control costs and economic and environmental damage (Holm et al., 1977; Witt and McConnachie, 2004). Insects have an important though largely unexplored role as potential biological control agents for nonnative, invasive grasses, extending beyond riparian habitats into rangelands, for example guineagrass in Texas and Mexico (Mercadier et al., 2009). Insects that, like the arundo wasp, are of types that are new or uncommon as weed control agents will likely be required. With sufficient information, these novel agents will be amenable to mass rearing for biological control, leading to increased protection of water and other natural resources.

ACKNOWLEDGMENTS

Greenhouse facilities were provided by the USDA Animal and Plant Health Inspection Service (APHIS), Pest Detection, Diagnostics and Management Laboratory (PDDML), Moore Air Base, Edinburg, Texas. We thank the many students and other USDA-ARS staff who carried out the mass-rearing protocols. We thank Rupert Santos and Andrew Parker in particular for daily coordination of mass-rearing activities and Matthew Rector for assistance with field releases. Nadie Tomic-Carruthers, Link Smith, and the editors provided critical reviews. This work was supported in part by the US Department of Homeland Security, Science and Technology Directorate and Customs and Border Protection. Mention of trade names or commercial products in this article is solely for the purpose of providing specific information and does not imply recommendation or endorsement by the US Department of Agriculture. USDA is an equal opportunity provider and employer.

REFERENCES

Adair, R.J., 2005. The biology of *Dasineura dielsi* Rusaamen (Diptera: Cecidomyiidae) in relation to the biological control of *Acacia cyclops* (Mimosaceae) in South Africa. Aust. J. Entomol. 44, 446–456.

Adair, R.J., Burgess, T., Serdani, M., Barber, P., 2009. Fungal associations in *Asphondylia* (Diptera: Cecidomyiidae) galls from Australia and South Africa: implications for biological control of invasive acacias. Fungal Ecol. 2, 121–134.

Al-Barrak, M., Loxdale, H.D., Brookes, C.P., Dawah, H.A., Biron, D.G., Alsagair, O., 2004. Molecular evidence using enzyme and RAPD markers for sympatric evolution in British species of *Tetramesa* (Hymenoptera: Eurytomidae). Biol. J. Linn. Soc. 83, 509–525.

Askew, R.R., 1984. The biology of gall wasps. In: Ananthakrishnan, T.N. (Ed.), The Biology of Gall Insects, Oxford and IBH Press, New Delhi, pp. 223–271.

Balciunas, J., Smith, L., 2006. Prerelease efficacy assessment, in quarantine, of a tephritid gall fly being considered as a biological control agent for cape-ivy (*Delairea odorata*). Biol. Control 39, 516–524.

Balciunas, J., Mehelis, C., van der Westhuizen, L., Neser, S., 2010. Laboratory host range of *Parafruetreta regalis* (Diptera: Tephritidae), a candidate agent of biological control of cape-ivy. Environ. Entomol. 39, 841–848.

Beiderbeck, R., 2005. Dual aseptic culture of gall-inducing arthropods and their host plants. In: Raman, A., Schaefer, C.W., Withers, T.M. (Eds.), Biology, Ecology and Evolution of Gall-Inducing Arthropods, vol. 2. Science Publishers, Enfield, NH, USA, pp. 731–760.

Bell, G., 1997. Ecology and management of *Arundo donax*, and approaches to riparian habitat restoration in southern California. In: Brock, J.H., Wade, M., Pysek, P., Green, D. (Eds.), Plant Invasions: Studies from North America and Europe, Backhuys Publishers, Leiden, The Netherlands, pp. 103–113.

Bernays, E.A., Chapman, R.F., 1994. Behavior: the process of host selection. Host-Plant Selection by Phytophagous Insects, Chapman and Hall, New York, pp. 95–165.

Blossey, B., Eberts, D., Morrison, E., Hunt, T.R., 2000. Mass rearing the weevil *Hylobius transversovittatus* (Colopetera: Curculionidae), biological control agent of *Lythrum salicaria*, on semiartificial diet. J. Econ. Entomol. 93, 1644–1656.

Blossey, B., Hunt, T.R., 1999. Mass-rearing methods for *Galerucella calmariensis* and *G. pusilla* (Coleoptera: Chrysomelidae), biological control agents of *Lythrum salicaria* (Lythraceae). J. Econ. Entomol. 92, 325–334.

Boland, J.M., 2006. The importance of layering in the rapid spread of *Arundo donax* (giant reed). Madroño 53, 303–312.

Bronner, R., 1992. The role of nutritive cells in the nutrition of cynipids and cecidomyiids. In: Shorthouse, J.D., Rohrfritsch, O. (Eds.), Biology of Insect-Induced Galls. Oxford University Press, Oxford, UK, pp. 118–140.

Bucellato, L., Byrne, M.J., Witkowski, E.T.F., 2012. Interactions between a stem gall fly and a leaf-spot pathogen in the biological control of *Ageratina adenophora*. Biol. Control 61, 222–229.

Cantero-Martínez, C., Angas, P., Lampurlanés, J., 2003. Growth, yield, and water productivity of barley (*Hordeum vulgare* L.) affected by tillage and N fertilization in Mediterranean semiarid, rainfed conditions of Spain. Field Crops Res. 84, 341–357.

Carruthers, R.I., DeLoach, C.J., Herr, J.C., Anderson, G.L., Knutson, A.E., 2008. Saltcedar area-wide pest management in the western United States. In: Koul, O., Cuperus, G., Elliott, N. (Eds.), Areawide Pest Management: Theory and Implementation, CAB International, Wallingford, UK, pp. 252–279.

Center, T.D., Grodowitz, M.J., Cofrancesco, A.F., Jubinsky, G., Snoddy, E., Freedman, J.E., 1997. Establishment of *Hydrellia pakistanae* (Diptera: Ephydridae) for the control of the submersed aquatic plant *Hydrilla verticillata* (Hydrocharitaceae) in the southeastern United States. Biol. Control 8, 65–73.

Chapman, R.F., 2003. Contact chemoreception in feeding by phytophagous insects. Annu. Rev. Entomol. 48, 455–484.

Claridge, M.F., 1961. A contribution to the biology and taxonomy of some Palearctic species of *Tetramesa* Walker (=*Isosoma* Walk.; =*Harmolita* Motsch.), with particular reference to the British fauna. Trans. R. Entomol. Soc. Lond. 113, 175–216.

Cocuzza, G.E., Declercq, P., van de Veire, M., Decock, A., Degheele, D., Vacante, V., 1997. Reproduction of *Orius laevigatus* and *Orius albidipennis* on pollen and *Ephestia kuehniella* eggs. Entomol. Exp. Appl. 82, 101–104.

Coffman, G.C., Ambrose, R.F., Rundel, P.W., 2010. Wildfire promotes dominance of invasive giant reed (*Arundo donax*) in riparian ecosystems. Biol. Invasions 12, 2723–2734.

Cohen, A.C., 1985. Simple method for rearing the insect predator *Geocoris punctipes* (Heteroptera: Lygaeidae) on a meat diet. J. Econ. Entomol. 78, 1173–1175.

Cohen, A.C., 1995. Extra-oral digestion in predatory arthropods. Annu. Rev. Entomol. 40, 85–103.

Cohen, A.C., 2003. Insect Diets: Science and Technology. CRC Press, Boca Raton, FL, USA.

Cohen, A.C., Wheeler, A.G., 1998. Role of saliva in the highly destructive four-lined plant bug (Hemiptera: Miridae). Ann. Entomol. Soc. Am. 91, 94–100.

Coudron, T.A., Wittmeyer, J., Kim, Y., 2002. Life history and cost analysis for continuous rearing of *Podisus maculiventris* (Say) (Heteroptera: Pentatomidae) on a zoophytophagous artificial diet. J. Econ. Entomol. 95, 1159–1168.

Craig, T.P., Itami, J.K., Price, P.W., 1989. A strong relationship between oviposition preference and larval performance in a shoot-galling sawfly. Ecology 70, 1691–1699.

Crawley, M.J., 1986. The population biology of invaders. Philos. Trans. R. Soc. Lond. B 314, 711–731.

Crutwell McFadyen, R.E., 1998. Biological control of weeds. Annu. Rev. Entomol. 43, 369–393.

Culliney, T.W., 2005. Benefits of classical biological control for managing invasive plants. Crit. Rev. Plant Sci. 24, 131–150.

Cuomo, G.J., Anderson, B.E., 1996. Nitrogen fertilization and burning effects on rumen protein degradation and nutritive value of native grasses. Agron. J. 88, 439–442.

De Bruxelles, G.Z., Roberts, M.H., 2001. Signals regulating multiple responses to wounding and herbivores. Crit. Rev. Plant Sci. 20, 487–521.

Decruyenaere, J.G., Holt, J.S., 2005. Ramet demography of a clonal invader, *Arundo donax* (Poaceae), in southern California. Plant Soil 277, 41–52.

Dennill, G.B., 1988. Why a gall former can be a good biocontrol agent: the gall wasp *Trichilogaster acaciaelongifoliae* and the weed *Acacia longifolia*. Ecol. Entomol. 13, 1–9.

Dhileepan, K., 2004. The applicability of the plant vigor and resource regulation hypothesis in explaining *Epiblema* gall moth-*Parthenium* weed interactions. Entomol. Exp. Appl. 113, 63–70.

Dhileepan, K., Treviño, M., Bayliss, D., Saunders, M., Shortus, M., McCarthy, J., Snow, E.L., Walter, G.H., 2010. Introduction and establishment of *Carvalhotingis visenda* (Hemiptera: Tingidae) as a biological control agent for cat's claw creeper *Macfadyena unguis-cati* (Bignoniaceae) in Australia. Biol. Control 55, 58–62.

DiGaulio, 1998. Eurytomidae. In: Gibson, G.A.P., Huber, J.T., Woolley, J.B. (Eds.), Annotated Keys to the Genera of Nearctic Chalcidoidea (Hymenoptera), National Research Council (NRC) of Canada, Ottawa, pp. 477–495.

Djamankulova, G., Khamraev, A., Schaffner, U., 2008. Impact of two shoot-galling biological control candidates on Russian knapweed, *Acroptilon repens*. Biol. Control 46, 101–106.

Dodd, A.P., 1940. The Biological Campaign against Prickly Pear. Commonwealth Prickly Pear Board, Brisbane, Australia.

Dorchin, N., Cramer, M.D., Hoffmann, J.H., 2006. Photosynthesis and sink activity of wasp-induced galls in *Acacia pycnantha*. Ecology 87, 1781–1791.

Dreger-Jauffret, F., Shorthouse, J.D., 1992. Diversity of gall-inducing insects and their galls. In: Shorthouse, J.D., Rohrfritsch, O. (Eds.), Biology of Insect-Induced Galls, Oxford University Press, Oxford, UK, pp. 8–33.

Dudley, T.L., Lambert, A., Kirk, A., 2006. Augmentation biological control of *Arundo donax*. In: Hoddle, M.S., Johnson, M.W. (Eds.), Proceedings, Fifth California Conference on Biological Control, July 25–27, 2006, Riverside, California, USA, University of California, Riverside, pp. 141–144.

Dudley, T.L., Bean, D.W., 2012. Tamarisk biocontrol, endangered species risk and resolution of conflict through riparian restoration. Biol. Control 57, 331–347.

Ferkovich, S.M., Shapiro, J., Carpenter, J., 2000. Growth of a pupal ectoparasitoid, *Diapetimorpha introit*, on an artificial diet: stimulation of growth rate by a lipid extract from host pupae. Biol. Control 45, 401–413.

Flores, D., Carlson, J.W., 2006. Biological control of giant salvinia in East Texas waterways and effects on dissolved oxygen levels. J. Aquat. Plant Manag. 44, 115–121.

Forno, I.W., Julien, M.H., 2000. Success in biological control of aquatic weeds by arthropods. In: Gurr, G., Wratten, S. (Eds.), Biological Control: Measures of Success, Kluwer, Dordrecht, The Netherlands, pp. 159–187.

Freedman, J.E., Grodowitz, M.J., Cofrancesco, A.F., Bare, R., 2001. Mass-Rearing *Hydrellia pakistanae* Deonier, a Biological Control Agent of *Hydrilla verticillata* (L. f.) Royle, for Release and Establishment. US. Army Engineer Environmental Research and Development Center, Technical Report ERDC/EL TR-01-24 U.S. Army Engineer Research and Development Center, Vicksburg, MS, USA.

Gagné, R.J., Wright, S.A., Purcell, M.F., Brown, B.T., Pratt, P.D., Center, T.D., 2009. Description of the larva of *Lophodiplosis trifida*, an Australian gall midge (Diptera: Cecidomyiidae) and biocontrol agent of paperbark in Florida, USA. Fla. Entomol. 92, 593–597.

Giessow, J., Casanova, J., Leclerc, R., MacArthur, R., Fleming, G., Giessow, J., 2011. *Arundo donax*: Distribution and Impact Report. California Invasive Plant Council (Cal-IPC), Sacramento, CA, USA.

Goeden, R.D., Andres, L.A., 1999. Biological control of weeds in terrestrial and aquatic environments. In: Bellows, T.S., Fisher, T.W. (Eds.), Handbook of Biological Control, Academic Press, San Diego, CA, USA, pp. 871–890.

Goeden, R.D., Louda, S.M., 1976. Biotic interference with insects imported for weed control. Annu. Rev. Entomol. 21, 325–342.

Going, B.M., Dudley, T.L., 2008. Invasive riparian plant litter alters aquatic insect growth. Biol. Invasions 10, 1041–1051.

Goodman, C.L., Phipps, S.J., Wagner, R.W., Peters, P., Wright, M.K., Nabli, H., Saathoff, S., Vickers, B., Grasela, J.J., McIntosh, A.H., 2006. Growth and development of the knapweed root weevil, *Cyphocleonus achates*, on a meridic larval diet. Biol. Control 32, 238–246.

Goolsby, J.A., DeBarro, P.J., Makinson, P.R., Pemberton, R.W., Hartley, D.M., Frohlich, D.R., 2006. Matching the origin of an invasive weed for selection of a herbivore haplotype for a biological control programme. Mol. Ecol. 15, 287–297.

Goolsby, J.A., Moran, P.J., 2009. Host range of *Tetramesa romana* Walker (Hymenoptera: Eurytomidae), a potential biological control of giant reed, *Arundo donax* L. in North America. Biol. Control 49, 160–168.

Goolsby, J.A., Moran, P.J., Adamczyk, J.A., Kirk, A.A., Jones, W.A., Marcos, M.A., Cortés, E., 2009a. Host range of the European, rhizome-stem feeding scale *Rhizaspidiotus donacis* (Leonardi) (Hemiptera: Diaspididae), a candidate biological control agent for giant reed, *Arundo donax* L. (Poales: Poaceae) in North America. Biocontrol Sci. Technol. 19, 899–918.

Goolsby, J.A., Moran, P.J., Falk, J., Gilbert, L., 2009b. Distribution and spread of an adventive population of the biological control agent, *Tetramesa romana* in Austin, Texas. Southwest. Entomol. 34, 329–330.

Goolsby, J.A., Spencer, D.F., Whitehand, L., 2009c. Pre-release assessment of impact on *Arundo donax* by the candidate biological control agents, *Tetramesa romana* (Hymenoptera: Eurytomidae) and *Rhizaspidiotus donacis* (Homoptera: Diaspididae) under quarantine conditions. Southwest. Entomol. 34, 359–376.

Goolsby, J.A., Mangan, R., 2010. Use of spinosad bait (BF-120) for management of *Chaetopsis massyla* in shadehouse-grown *Arundo donax*. Southwest. Entomol. 35, 573–574.

Goolsby, J.A., Kirk, A.A., Moran, P.J., Racelis, A.E., Adamczyk, J.J., Cortés, E., Marcos García, M.A., Martinez Jimenez, M., Summy, K.R., Ciomperlik, M.A., Sands, D.P.A., 2011. Establishment of the armored scale *Rhizaspidiotus donacis*, a biological control agent of *Arundo donax*. Southwest. Entomol. 36, 373–374.

Gordh, G., Legner, E.F., Caltagirone, L.E., 1999. Biology of parasitic Hymenoptera. In: Bellows, T.S., Fisher, T.W. (Eds.), Handbook of Biological Control, Academic Press, New York, pp. 355–381.

Grenier, S., 2009. In vitro rearing of entomophagous insects-past and future trends: a minireview. Bull. Insectol. 62, 1–6.

Grenier, S., Greany, P.D., Cohen, A.C., 1994. Potential for mass release of insect parasitoids and predators through development of artificial culture techniques. In: Rosen, D., Bennett, F.D., Capinera, L. (Eds.), Pest Management in the Subtropics: Biological Control—A Florida Perspective. Intercept, Andover, MA, USA, pp. 181–205.

Harley, K.L.S., Willson, B.W.W., 1968. Propagation of a cerambycid borer on meridic diet. Can. J. Zool. 46, 1265–1266.

Harms, N., Grodowitz, M., Nachtrieb, J., 2009a. Mass-Rearing *Hydrellia pakistanae* Deonier and *H. balciunasi* Bock for the Management of *Hydrilla verticillata*. Publication ERDC/TN APCRP-BC-12 U.S. Army Engineer Research and Development Center, Vicksburg, MS, USA.

Harms, N., Grodowitz, M., Hachtrieb, J., 2009b. Mass-Rearing *Cyrtobagous salviniae* Calder and Sands for the Management of *Salvinia molesta* Mitchell. Publication ERDC/TN APCRP-BC-16 U.S. Army Engineer Research and Development Center, Vicksburg, MS, USA.

Hartley, S.E., 1998. The chemical composition of plant galls: are levels of nutrients and secondary compounds controlled by the gall-former? Oecologia 113, 492–501.

Herrera, A.M., Dudley, T.L., 2003. Reduction of riparian arthropod abundance and diversity as a consequence of giant reed (*Arundo donax*) invasion. Biol. Invasions 5, 167–177.

Herrick, N.J., Salom, S.M., Kok, L.T., McAvoy, T.J., 2011. Life history, development, and rearing of *Eucryptorrhynchus brandti* (Coleoptera: Curculionidae) in quarantine. Ann. Entomol. Soc. Am. 104, 718–725.

Hinz, H.L., Muller-Scharer, H., 2000. Influence of host condition on the performance of *Rhopalomyia* n sp. (Diptera: Cecidomyiidae), a biological control agent for scentless chamomile, *Tripleurospermum perforatum*. Biol. Control 18, 147–156.

Holm, L.G., Plucknett, D.L., Pancho, J.V., Herberger, J.P., 1977. The World's Worst Weeds, Distribution and Biology. University of Hawaii Press, Honolulu, HI.

Holmes, N.D., Blakeley, P.E., 1971. The rye jointworm (Hymenoptera: Eurytomidae), a new insect pest in western Canada. Can. Entomol. 103, 277–280.

Hu, J.S., Vinson, S.B., 1998. The in vitro development from egg to prepupa of *Campoletis sonorensis* (Hymenoptera: Ichneumonidae), in an artificial medium: importance of physical factors. J. Insect Physiol. 44, 455–462.

Hughes, B.G., Mickey, K.L., 1993. A Reanalysis of Recreational and Livestock Trespass Impacts on the Riparian Zone of the Rio Grande, Big Bend National Park, Texas, Final Report. US Department of the Interior, National Park Service, Southwest Region, Santa Fe, NM, USA.

Impson, F.A.C., Kleinjan, C.A., Hoffmann, J.H., Post, U.A., 2008. *Dasineura rubiformis* (Diptera: Cecidomyiidae), a new biological control agent for *Acacia mearnsii* in South Africa. S. Afr. J. Sci. 104, 247–249.

Johanowicz, D.L., Mitchell, E.R., 2000. A novel method to rear *Diadegma insulare* (Hymenoptera: Ichneumonidae), a parasitoid of the diamondback moth (Lepidoptera: Plutellidae). Fla. Entomol. 83, 377–379.

Julien, M.H., Griffiths, M.W., 1998. Biological Control of Weeds: A World Catalogue of Agents and Their Target Weeds, fourth ed. CABI Publishing, Oxford, UK.

Keane, R.M., Crawley, M.J., 2002. Exotic plant invasions and the enemy release hypothesis. Trends Ecol. Evol. 17, 164–170.

Kerchev, P.I., Fenton, B., Foyer, C.H., Hancock, R.D., 2012. Plant responses to insect herbivory: interactions between photosynthesis, reactive oxygen species and hormonal signaling pathways. Plant Cell Environ. 35, 441–453.

Khuzhaev, V.U., Zhalolov, I., Turgunov, K., Tashkhodzhaev, B., Levkovich, M.G., Aripova, S.F., Shashkov, A.S., 2004. Alkaloids from *Arundo donax*. XVI: structure of the new dimeric indole alkaloid arundavine. Chem. Nat. Comp. 40, 261–265.

Kirk, A.A., Widmer, T., Campobasso, G., Carruthers, R.A., Dudley, T.L., 2003. The potential contribution of natural enemies from Mediterranean Europe to the management of the invasive weed *Arundo donax* (Graminae: Arundinae) in the USA. Kings Beach, California, USA, 2–4 October, 2003 In: Pirosko, C. (Ed.), Proceedings of the California Invasive Plant Council Symposium, vol. 7. California Invasive Plant Council, Sacramento, CA, pp. 62–68.

Klein, O., Kroschel, J., 2002. Biological control of *Orobanche* spp. with *Phytomyza orobanchia*, a review. Biol. Control 47, 245–277.

Kleinjan, C.A., Edwards, P.B., 2006. Asparagus weeds in Australia: a South African perspective with emphasis on biological control prospects. Plant Prot. Q. 21, 63–68.

Knochel, D.G., Seastedt, T.R., 2010. Reconciling contradictory findings of herbivore impacts on spotted knapweed (*Centaurea stoebe*) growth and reproduction. Ecol. Appl. 20, 1903–1912.

Kovalev, O.V., Runeva, T.D., 1970. *Tarachidia candefacta* Hubn. (Lepidoptera: Noctuidae), an efficient phytophagous insect for biological control of weeds in the genus *Ambrosia*. Acad. Sci. USSR Entomol. Rev. 49, 23–36 (in Russian).

Kovalev, O.V., Vechernin, V.V., 1986. Description of a new wave process in populations with reference to introduction and spread of the leaf beetle *Zygogramma suturalis* F. (Coleoptera: Chrysomelidae). Entomol. Obozr. 65, 21–38 (in Russian).

La Salle, J., 2005. Biology of gall inducers and evolution of gall induction in Chalcidoidea (Hymenoptera: Eulophidae, Eurytomidae, Pteromalidae, Tanaostigmatidae, Torymidae). In: Raman, A., Schaefer, C.W., Withers, T.M. (Eds.), Biology, Ecology and Evolution of Gall-Inducing Arthropods, vol. 2. Science Publishers, Enfield, NH, USA, pp. 507–537.

Lejeune, K.D., Suding, K.N., Sturgis, S., Scott, A., Seastedt, T.R., 2005. Biological control insect use of fertilized and unfertilized diffuse knapweed in a Colorado grassland. Environ. Entomol. 34, 225–234.

Leppla, N.C., Ashley, T.R., 1989. Quality Control in Insect Mass Production—a Review and Model. Bulletin of the ESA, Ecological Society of Australia, Canberra, Australia.

Lotfalizadeh, H., Delvare, G., Rasplus, J.-Y., 2007. Phylogenetic analysis of Eurytominae (Chalcidoidea: Eurytomidae) based on morphological characters. Zool. J. Linn. Soc. 151, 441–510.

Markin, G.P., Horning, C.J., 2010. Discovery of a gall-forming midge, *Asphondylia pilosa* Kieffer (Diptera: Cecidomyiidae) on Scotch broom (*Cytisus scoparius* (L.) Link) (Fabaceae). J. Kans. Entomol. Soc. 83, 260–263.

Martin, C.G., Cuda, J.P., Awadzi, K.D., Medal, J.C., Habeck, D.H., Pedrosa-Macedo, J.H., 2004. Biology and laboratory rearing of *Episimus utilis* (Lepidoptera: Tortricidae), a candidate for classical biological control of Brazilian peppertree (Anacardiaceae) in Florida. Environ. Entomol. 33, 1351–1361.

Marton, L., Czako, M., 2007. Sustained totipotent culture of selected monocot genera. US Patent No. 7,303,916. US Patent Service, Washington, DC.

McClay, A., De Clerck-Floate, R., 1999. Establishment and early effects of *Omphalapion hookeri* (Kirby) (Coleoptera: Apionidae) as a biological control agent for scentless chamomile, *Matricaria perforata* Mérat (Asteraceae). Biol. Control 14, 85–95.

McGaugh, S., Hendrickson, D., Bell, G., Cabral, H., Lyons, K., McEachron, L., Munoz, O., 2006. Fighting an aggressive wetlands invader: a case study of giant reed, (*Arundo donax*) and its threat to Cuatro Ciénegas, Coahuila, Mexico. In: Lozano-Vilano, L., Contreas-Balderas, A.J. (Eds.), Studies of North American Desert Fishes in Honor of E.P. (Phil) Pister, Conservationist, Universidad Autónoma de Nuevo León, Facultad de Ciencas Biológicas, Monterrey, México, pp. 100–115 (in Spanish).

Memmott, J., Fowler, S.V., Hill, R.L., 1998. The effect of release size on the probability of establishment of biological control agents: gorse thrips (*Sericothrips staphylinus*) released against gorse (*Ulex europaeus*) in New Zealand. Biocontrol Sci. Technol. 8, 103–115.

Mercadier, G., Goolsby, J.A., Jones, W.A., Tamesse, J.L., 2009. Results of preliminary survey in Cameroon, Central Africa, for potential natural enemies of *Panicum maximum* Jacq. (Poales: Poaceae), guineagrass. Subtrop. Plant Sci. 61, 31–36.

Moran, P.J., Goolsby, J.A., 2009. Biology of the galling wasp *Tetramesa romana*, a biological control agent of giant reed. Biol. Control 49, 169–179.

Moran, V.C., Hoffman, J.H., 2012. Conservation of the fynbos biome in the Cape Floral Region: the role of biological control in the management of invasive alien trees. Biol. Control 57, 139–149.

Müller-Schärer, H., Schaffner, U., 2008. Classical biological control: exploiting enemy escape to manage plant invasions. Biol. Invasions 10, 859–874.

Muniappan, R., McFadyen, R.E., 2005. Gall inducing arthropods used in the biological control of weeds. In: Raman, A., Schaefer, C.W., Withers, T.M. (Eds.), Biology, Ecology and Evolution of Gall-Inducing Arthropods, vol. 2. Science Publishers, Enfield, NH, USA, pp. 709–730.

Myers, J., 2007. How many and what kind of agents for the biological control of weeds: a case study with diffuse knapweed. In: Vincent, C., Goettel, M.S., Lazarovits, G. (Eds.), Biological Control, a Global Perspective, CABI International, Wallingford, UK, pp. 70–79.

Nation, J.L., 2002. Insect Physiology and Biochemistry. CRC, Boca Raton, FL, USA.

Nechols, J.R., Andres, L.A., Beardsley, J.W., Goeden, R.D., Jackson, C.G., 1995. Biological Control in the Western United States, Accomplishments and Benefits of Regional Research Project W-84, 1964–1989, Publication 3361. University of California Division of Agriculture and Natural Resources, Oakland, CA, USA.

Neser, O.C., 2008. *Eurytoma bryophylli* sp. n. (Hymenoptera: Eurytomidae), a leaf borer of *Bryophyllum delagoense* (Crassulaceae) from Madagascar and a candidate for the biocontrol of the plant in Australia. Afr. Entomol. 16, 60–67.

NOAA, 2011. Cimatography of the United States No. 21, 1981–2010. Available from: http://www.ncdc.noaa.gov/oa/climate/normals/usnormals.html.

Noyes, J.S., 2012. Universal Chalcidoidea Database. Available at: http://www.nhm.ac.uk/chalcidoids.

Ostermeyer, N., Grace, B.C.S., 2007. Establishment, distribution and abundance of *Mimosa pigra* biological control agents in northern Australia: implications for biological control. Biol. Control 52, 703–720.

Ozkan, C., 2006. Laboratory rearing of the solitary egg larval parasitoid, *Chelonus oculator* Panzer (Hymenoptera: Braconidae) on a newly recorded factitious host *Plodia interpunctella* (Hübner) (Lepidoptera: Pyralidae). J. Pest Sci. 79, 27–29.

Page, A.R., Lacey, K.L., 2006. Economic Impact Assessment of Australian Weed Biological Control. CRC for Australian Weed Management Technical Series No 10, Cooperative Research Centre for Australian Weed Management, Glen Osmond, Australia.

Pelanchar, J., Holler, T., Moses-Rowley, A., McGovern, R., Sivinski, J., 2009. Evaluation of irradiated Caribbean fruit fly (Diptera: Tephrtitidae) larvae for laboratory rearing of *Doryctobracon areolatus* (Hymenoptera: Braconidae). Fla. Entomol. 92, 535–537.

Perdue Jr., R.E., 1958. *Arundo donax*—source of musical of musical reeds and industrial cellulose. Econ. Bot. 12, 368–404.

Phillips, W.J., 1936. A Second Revision of the Chalcid Flies of the Genus *Harmolita* (Isosoma) of America North of Mexico, with Descriptions of 20 New Species. Technical Bulletin No. 518 US Department of Agriculture, Washington, DC.

Pimental, D., Zuniga, R., Morrison, D., 2005. Update on the environmental and economic costs associated with alien-invasive species in the United States. Ecol. Econ. 52, 273–288.

Price, D.L., Hough-Goldstein, J., Smith, M.T., 2003. Biology, rearing, and preliminary evaluation of host range of two potential biological agents for mile-a-minute weed, *Polygonum perfoliatum* L. Environ. Entomol. 32, 229–236.

Price, P.W., 1992. Evolution and ecology of gall-inducing sawflies. In: Shorthouse, J.D., Rohrfritsch, O. (Eds.), Biology of Insect-Induced Galls, Oxford University Press, Oxford, UK, pp. 208–224.

Price, P.W., 2000. Host plant resource quality, insect herbivores, and biocontrol. In: Spencer, N.R. (Ed.), Proceedings of the X International Symposium on Biological Control of Weeds, 4–14 July, 1999, Bozeman, Montana, USA, US Department of Agriculture, Agricultural Research Service, Washington, DC, USA, pp. 583–590.

Quinn, L.D., Holt, J.S., 2008. Ecological correlates of invasion by *Arundo donax* in three southern California riparian habitats. Biol. Invasions 10, 591–601.

Racelis, A.E., Goolsby, J.A., Moran, P.J., 2009. Seasonality and movement of adventive populations of the arundo wasp *Tetramesa romana*, a biological control agent of giant reed (*Arundo donax* L.) in south Texas. Southwest. Entomol. 34, 347–357.

Racelis, A.E., Goolsby, J.A., Penk, R., Jones, W.K., Roland, T.J., 2010. The development of an inundative, aerial release technique for the arundo wasp, a biological control agent of the invasive *Arundo donax*. Southwest. Entomol. 35, 495–501.

Racelis, A.E., Davey, R.B., Goolsby, J.A., Pérez de León, A.A., Varner, K., Duhaime, R., 2012a. Facilitative ecological interactions between invasive species: *Arundo donax* (Poaceae) stands as favorable habitat for cattle ticks (Acari: Ixodidae) along the US-Mexico border. J. Med. Entomol. 49, 410–417.

Racelis, A.E., Rubio, A., Vaughan, T., Goolsby, J.A., 2012b. Passive restoration potential of riparian areas invaded by giant reed (*Arundo donax*) in Texas. Ecol. Restor. 30, 103–105.

Raman, A., Schaefer, C.W., Withers, T.M., 2005. Galls and gall-inducing arthropods: an overview of their biology, ecology, and evolution. In: Raman, A., Schaefer, C.W., Withers, T.M. (Eds.), Biology, Ecology and Evolution of Gall-Inducing Arthropods, vol. 1. Science Publishers, Enfield, NH, USA, pp. 1–33.

Raman, A., Madhavan, S., Florentine, S.K., Dhileepan, K., 2006. Metabolite mobilization in the stem galls of *Parthenium hysterophorus* induced by *Epiblema strenuana* inferred from the signatures of isotopic carbon and nitrogen and concentrations of total non-structural carbohydrates. Entomol. Exp. Appl. 119, 101–107.

Reznik, S.Y., 1991. The effects of feeding damage in ragweed *Ambrosia artemisiifolia* (Asteraceae) on populations of *Zygogramma suturalis* (Coleoptera: Chrysomelidae). Oecologia 88, 204–210.

Rohrfritsch, O., 1992. Patterns in gall development. In: Shorthouse, J.D., Rohrfritsch, O. (Eds.), Biology of Insect-Induced Galls, Oxford University Press, Oxford, UK, pp. 60–86.

Rojas, M.G., Morales-Ramos, J.A., King, E.G., 2000. Two meridic diets for *Perillus bioculatus* (Heteroptera: Pentatomidae), a predator of *Leptinotarsa decemlineata* (Coleoptera: Chrysomelidae). Biol. Control 17, 92–99.

Saleh, A., Allawi, T.F., Ghabeish, I., 2010. Mass rearing of *Neochrysocharis formosa* (Westwood) (Eulophidae: Hymenoptera), a parasitoid of leafminers (Agromyzidae: Diptera). J. Pest. Sci. 83, 59–67.

Schonrogge, K., Harper, L.J., Lichtenstein, C.P., 2000. The protein content of tissues in cynipid galls (Hymenoptera: Cynipidae): similarities between cynipid galls and seeds. Plant Cell Environ. 23, 215–222.

Seawright, E.K., Rister, M.E., Lacewell, R.D., McCorkle, D.A., Sturdivant, A.W., Yang, C., Goolsby, J.A., 2009. Economic implications for the biological control of *Arundo donax*: Rio Grande Basin. Southwest. Entomol. 34, 377–394.

Shanower, T.G., Waters, D.K., 2006. A survey of five stem-feeding insect pests of wheat in the northern Great Plains. J. Entomol. Sci. 41, 40–48.

Sheppard, A.W., Raghu, S., 2005. Working at the interface of art and science: how best to select an agent for classical biological control? Biol. Control 34, 223–235.

Silva, M.D., Shorthouse, J.D., 2006. Comparison of the development of stem galls induced by *Aulacidea hieracii* (Hymenoptera: Cynipidae) on hawkweed and by *Diplolepis spinosa* (Hymenoptera: Cynipidae) on rose. Can. J. Bot. 84, 1052–1074.

Simmonds, F.J., 1980. Biological control of *Cordia curassavica* (Boraginaceae) in Malaysia. Entomophaga 25, 363–364.

Smith, E.G., De Clerck-Floate, R.A., Van Hezewijk, B.H., Moyer, B.H., Pavlik, E., 2009. Costs of mass-producing the root weevil *Mogulones cruciger*, a biological control agent for houndstongue (*Cynoglossum officinale* L.). Biol. Control 48, 281–286.

Snodgrass, R.E., 1935. Principles of Insect Morphology. Cornell University Press, Ithaca, NY, USA.

Spencer, D.F., Ksander, G.G., Whitehand, L.C., 2005. Spatial and temporal variation in RGR and leaf quality of a clonal riparian plant, *Arundo donax*. Aquat. Bot. 81, 27–36.

Spencer, D.F., Stocker, R.K., Liow, P.-S., Whitehand, L.C., Ksander, G.G., Fox, A.M., Everitt, J.H., Quinn, L.D., 2008. Comparative growth of giant reed (*Arundo donax* L.) from Florida, Texas, and California. J. Aquat. Plant Manag. 46, 89–96.

Spencer, D.F., Ksander, G.G., Tan, W., Liow, P.-S., Whitehand, L.C., 2011. Influence of application timing on the impact of glyphosate on giant reed (*Arundo donax* L.). J. Aquat. Plant Manag. 49, 106–110.

Steffan, J.R., 1956. Notes sur la biologie d'*Harmolita romana* (Walk.) (Hym. Eurytomidae). Bull. Soc. Entomol. France 61, 34–35.

Sterling, J.D.E., 1976. Resistance to barley jointworm. Can. Agric. 21, 13–14.

Stone, G.N., Schönrogge, K., 2003. The adaptive significance of insect gall morphology. Trends Ecol. Evol. 18, 512–522.

Story, J.M., Smith, L., Corn, J.G., White, L.J., 2008. Influence of seed head-attacking biological control agents on spotted knapweed reproductive potential in western Montana over a 30-yr period. Environ. Entomol. 37, 510–519.

Syrett, P., Briese, D.T., Hoffmann, J.H., 2000. Success in biological control of terrestrial weeds by arthropods. In: Gurr, G., Wratten, S. (Eds.), Biological Control: Measures of Success, Kluwer, Dordrecht, The Netherlands, pp. 159–187.

Tewksbury, L., Casagrande, R., Blossey, B., Häfliger, P., Schwarzlander, M., 2002. Potential for biological control of *Phragmites australis* in North America. Biol. Control 23, 191–212.

Thompson, S.N., 1983. *Brachymeria lasus*: effects of nutrient level on in vitro growth of a chalcid insect parasite. Exp. Parasitol. 55, 312–319.

Thompson, S.N., Hagen, K.S., 1999. Nutrition of entomophagous insects and other arthropods. In: Bellows, T.S., Fisher, T.W. (Eds.), Handbook of Biological Control, Academic Press, San Diego, CA, USA, pp. 594–652.

Thornby, D., Spencer, D.F., Hanan, J., Sher, A., 2007. L-DONAX, a growth model of the invasive weed species, *Arundo donax*. L. Aquat. Bot. 87, 275–284.

Tomic-Carruthers, N., 2009. Rearing *Hylobius transversovittatus* and *Cyphocleonus achates* larvae on artificial diets (Coleoptera: Curculionidae). Fla. Entomol. 92, 656–657.

Tracy, J.L., DeLoach, C.J., 1999. Suitability of classical biological control for giant reed (*Arundo donax*) in the United States. In: Bell, C.E. (Ed.), Arundo and Saltcedar Management Workshop Proceedings. Ontario, California, 17 June 1998, University of California Cooperative Extension, Holtville, CA, USA, pp. 73–109.

Van Driesche, R.V., Blossey, B., Hoddle, M., Lyon, S., Reardon, R. (Eds.), 2002. Biological Control of Invasive Plants in the Eastern US, US Department of Agriculture-Forest Service (USFS), Forest Health Technology Enterprise Team, Morgantown, WV, USA.

Van Driesche, R.G., Carruthers, R.I., Canter, T., Hoddle, M.S., Hough-Goldstein, J., Morin, L., Smith, L.M., Wagner, D.L., et al., 2010. Classical biological control for the protection of natural ecosystems. Biol. Control 54 (Suppl. 1), S2–S33.

Van Driesche, R.G., 2012. The role of biological control in wildlands. Bio. Control 57, 131–137.

Van Hezewijk, B.H., Bourchier, R.S., 2012. Impact of *Cyphocleonus achates* on diffuse knapweed and its interaction with *Larinus minutus*. Biol. Control 62, 113–119.

van Klinken, R.D., Fichera, G., Cordo, H., 2003. Targeting biological control across diverse landscapes: the release, establishment, and early success of two insects on mesquite (*Prosopis* spp.) in Australian rangelands. Biol. Control 26, 8–20.

Van Nieuwenhove, G.A., Bezdjian, L.P., Ovruski, S.M., 2012. Effect of exposure time and ratio of hosts to female parasitoids on offspring production of *Diachasmimorpha longicaudata* (Hymenoptera: Braconidae) reared on *Anastrepha fraterculus* (Diptera: Tephritidae) larvae. Fla. Entomol. 95, 99–104.

van Wilgen, B.W., De Lange, W.J., 2011. The costs and benefits of biological control of invasive alien plants in South Africa. Afr. Entomol. 19, 504–511.

Vårdal, H., Sahlén, G., Ronquist, F., 2003. Morphology and evolution of the cynipoid egg (Hymenoptera). Zool. J. Linn. Soc. 139, 247–260.

Veldtman, R., Lado, T.F., Botes, A., Proches, S., Timm, A.E., Geertsema, H., Chown, S.L., 2011. Creating novel food webs on introduced Australian acacias: indirect effects of galling biological control agents. Divers. Distrib. 17, 958–967.

Waring, G.J., Cobb, N.S., 1992. The impact of plant stress on herbivore population dynamics. In: Bernays, E.A. (Ed.), Insect–Plant Interactions. CRC Press, Boca Raton, FL, USA, pp. 167–226.

Watts, D.A., Moore, G.W., 2011. Water-use dynamics of an invasive reed, *Arundo donax*, from leaf to stand. Wetlands 31, 725–734.

Wheeler, G.S., Zahniser, J., 2001. Artificial diets and rearing methods for the *Melaleuca quinquenervia* (Myrtales: Myrtaceae) biological control agent *Oxyops vitiosa* (Coleoptera: Curculionidae). Fla. Entomol. 84, 439–441.

Wiebes-Rijks, A.A., Shorthouse, J.D., 1992. Ecological relationships of insects inhabiting cynipid galls. In: Shorthouse, J.D., Rohrfritsch, O. (Eds.), Biology of Insect-Induced Galls, Oxford University Press, Oxford, UK, pp. 238–257.

Winkler, C., Smith, C.W., Pedrosa Macedo, J.H., 1996. The stem-gall wasp *Eurytoma* sp. (Hymenoptera: Eurytomidae): a potential biological control agent against *Psidium cattleianum*. In: Moran, V.C., Hoffmann, J.H. (Eds.), Proceedings of the 9th International Symposium on Biological Control of Weeds, Stellenbosch, South Africa, 19–26 January, 1996, University of Cape Town, Rondebosch, South Africa, pp. 219–221.

Witt, A.B.R., McConnachie, A.J., 2004. The potential for classical biological control of invasive grass species with special reference to invasive *Sporobolus* spp. (Poaceae) in Australia. In: Cullen, J.M., Briese, D.T., Kriticos, D.J., Lonsdale, W.M., Morin, L., Scott, J.K. (Eds.), Proceedings of the XI International Symposium on Biological Control of Weeds, 27 April–2 May 2003, Canberra, Australia, Commonwealth Science and Industrial Research Organization (CSIRO), Canberra, Australia, pp. 198–202.

Wolf, D., Optiz Von Boberfeld, W., 2003. Effects of nitrogen fertilization and date of utilization on the quality and yield of tall fescue in winter. J. Agron. Crop Sci. 189, 47–53.

Xie, Z.N., Wu, Z.X., Nettles Jr., W.C., Saldana, G., Nordlund, D.A., 1997. In vitro culture of *Trichogramma* spp. on artificial diets containing yeast extract and ultracentrifuged chicken egg yolk but devoid of insect components. Biol. Control 8, 107–110.

Yang, C., Everitt, J.H., Goolsby, J.A., 2011. Mapping giant reed infestations along the Texas–Mexico portion of the Rio Grande with aerial photography. Invasive Plant Sci. Manag. 4, 402–410.

Yokoyama, V.Y., Cáceres, C.E., Kuenen, L.P.S., Wang, X.G., Rendón, P.A., Johnson, M.W., Daane, K.M., 2010. Field performance and fitness of an olive fruit fly parasitoid, *Psyttalia humilis* (Hymenoptera: Braconidae), mass reared on irradiated Medfly. Biol. Control 54, 90–99.

Zachariades, C., Hoffmann, J.H., Roberts, A.P., 2011. Biological control of mesquite (*Prosopis* species) (Fabaceae) in South Africa. Afr. Entomol. 19, 402–415.

Chapter 7

Artificial Diet Development for Entomophagous Arthropods

Juan A. Morales-Ramos,[1] M. Guadalupe Rojas[1] and Thomas A. Coudron[2]
[1]National Biological Control Laboratory, US Department of Agriculture, Agricultural Research Service (USDA-ARS), Stoneville, MS, USA, [2]Biological Control of Insects Research Laboratory, USDA-ARS, Columbia, MO, USA

7.1. INTRODUCTION

Artificial diets were first developed as a tool for entomological research and quickly evolved as aids for bioassay of potential insecticides, entomopathogens, and plant resistance characteristics (Stone and Sims, 1992; Cohen, 2001; Grenier, 2009). Their potential was soon acknowledged as tools for insect control in sterile male release and augmentative biological control strategies (King et al., 1985; Knipling, 1992; Cohen, 2001, 2004; Grenier, 2009). The most successful mass-rearing programs based on artificial diets were achieved with herbivore insects utilized in sterile male releases, in entomopathogen mass production, or as host or prey for entomophagous insects (King and Leppla, 1984). Developing artificial diets for entomophagous arthropods has been a more challenging enterprise and will be the focus of this chapter.

Artificial diets have been stated to be critical to the advancement of mass production and eventual commercialization of entomophagous arthropods as biological control agents (Yazlovetsky, 1992). Rearing parasitoids and predators in vivo requires the rearing of multiple organisms. Mass rearing is a complex enterprise, and its complexity multiplies as more than one species must be reared. Artificial diets are intended to eliminate the need to rear host (or prey) species, thereby reducing the complexity of the system to manageable levels. King et al. (1985) considered the development of artificial diets a requirement for the success of augmentative biological control, and Cohen (1992) considered the lack of suitable artificial diets the greatest barrier to the mass production of entomophagous insects. However, decades later, few artificial diets have been utilized successfully and consistently in the commercial mass production of entomophagous arthropods (Grenier, 2009). Riddick (2009) concluded in his review that most existing artificial diets were inferior

Mass Production of Beneficial Organisms. http://dx.doi.org/10.1016/B978-0-12-391453-8.00007-8

to natural or factitious prey in producing high-quality predators. The lack of rigorous artificial diet evaluation has been considered an important factor for the low success of artificial diets to date (Grenier and De Clercq, 2003; Riddick, 2009).

Artificial diets have played a very important role in advancing the science of entomology. Cohen (2001) estimated that artificial diets were used in 45% of the articles that required some type of insect rearing and were published in journals of the *Entomological Society of America* (from 1998 to 2000). However, their promising potential in augmentative biological control remains mostly unfulfilled. In this chapter, we will focus on problems of commercial implementation of artificial diets for entomophagous arthropods and potential ways to solve them. We will not make an attempt to enumerate and describe the large number of artificial diets that have been produced in the last 50 years. Excellent reviews on artificial diets for entomophagous arthropods have been published in recent years (Grenier et al., 1994; Etzel and Legner, 1999; Thompson, 1999; Thompson and Hagen, 1999; Grenier, 2009; Riddick, 2009).

Artificial diets for entomophagous arthropods have not always been developed with the intent for use in commercial mass production but rather as means to facilitate biological research (Cohen, 2001). In most cases, published artificial diets for entomophagous arthropods have not been developed to the level required for commercial application and still may be seen as a work in progress. Successful diets are often difficult and expensive to produce in large quantities because they require the use of crude food ingredients and costly media for encapsulation. Most published artificial diets that have been effective for producing healthy arthropods are still unfinished when it comes to commercial use.

7.1.1. Levels of Development

Artificial diets are classified as holidic, meridic, and oligidic according to the level of chemical definition, which ranges from fully chemically defined, to partially chemically defined, to totally undefined (Cohen, 2004). Meridic artificial diets are also defined as lacking any insect components (Grenier and De Clercq, 2003). If artificial diets are to play an important role in the mass production of natural enemies, efforts in their development should have as final objectives the successful mass production of entomophagous arthropods at a cost suitable for commercial application and a process that can be automated. For conceptualization, we have organized the process of artificial diet development into six levels required to achieve these goals. These levels of artificial diet development focus on how easily an arthropod can be produced in artificial media and how effective they are in fulfilling their intended purpose. Efforts to develop artificial diets to the level of commercialization require much more than just perfecting the diet formulation. However, this chapter will focus exclusively on the efforts to improve and refine the artificial diets for entomophagous arthropods.

Implementation of rearing procedures, packing, diet preparation and dispensing, and mechanization will be left for future discussions.

Level 1: Arthropod completes development and reproduces in the artificial diet formulation. Often, at this level of development, artificial diets are typically chemically defined (i.e. holidic), and arthropods are reared in small containers or petri dishes. Only a few individuals are produced in artificial media at this level.

Level 2: Arthropods complete development and reproduce for at least five successive generations on an artificial diet. In order to maintain a colony for multiple continuous generations, a viable population must be maintained in vitro. A viable population represents a colony size of around 100 individuals at each life stage. Some of the rearing procedures must be simplified to facilitate this level of production. For example, artificial diets may remain holidic in their composition, but in many cases some chemically defined components are replaced by natural components for simplification.

Level 3: Arthropods are produced in artificial diet in sufficient numbers and reproducibility to make possible a statistically valid comparison of biological parameters against arthropods reared on natural hosts. The need to produce arthropods in vitro in substantial numbers (several hundred individuals) drives the development of more sophisticated rearing procedures, and means to obtain eggs in the absence of hosts or prey are often developed at this stage. Diets are mostly meridic in composition but may still include some chemically defined components.

Level 4: Arthropods are produced on an artificial diet in sufficient numbers to make experimental field releases and statistically valid field evaluations. Field releases demand the production of thousands of individuals, driving the development of sophisticated rearing procedures. Techniques for enhancing diet mixing and dispensing, oviposition, egg handling, immature feeding, adult emergence, and packing for release must be developed. At this level, meridic diets with provisions for microbial contamination and storage are common.

Level 5: Arthropods are produced on an artificial diet in sufficient numbers to be released in a commercial field and achieve at least substantial control of the target pest. Release of arthropods in a commercial field with the purpose of controlling a pest will demand the production of tens of thousands to hundreds of thousands of high-quality individuals. Some level of mechanization must occur to maintain that level of production. Rearing systems and diet must be entirely reliable, or the enterprise will fail. Diet composition tends to be oligidic at this level, with few chemically defined components, including mostly vitamins and some amino acids. The diet quality at this level must be of high enough quality to produce natural enemies as good as or closely comparable to those reared in natural host or prey. Given that the diet must be produced in mass, the formulation must be stable enough for temporary storage.

Level 6: Production levels of arthropods on an artificial diet are limited only by space, and production costs are suitable for commercialization. Production capabilities at this level should reach millions of individuals. Most of the rearing procedures must be either mechanized or extremely simplified in order to attain this level of production. Expensive materials and diet ingredients are substituted by less expensive alternatives. Diets at this level are typically oligidic in composition and stable. Mass preparation of diet involves mechanized procedures and sophisticated sterilization procedures like flash sterilization and/or extrusion. Reliable packing, shipping, and release procedures must be developed.

Developing artificial diets from levels 1 to 6 implicates the usage of various disciplines from insect nutrition and insect dietetics to food-processing sciences (Singh, 1977; Cohen, 2004). Cohen (2001, 2004) considered insect nutrition and insect dietetics as two extremes of a continuum of practicality. Developing artificial diets requires knowledge of chemical, physiological, biophysical, and physical processes that interact during the ingestion, digestion, absorption, and assimilation of food by insects.

7.1.2. Degrees of Difficulty

In developing artificial diets, all aspects of arthropod biology have relevance, but the nature of some species' biology adds difficulty to the process. For instance, parasitoid koinobionts require the host to be alive during their development (Askew and Shaw, 1986). Hormonal interactions among parasitoid and host play an important role in parasitoid development (Vinson, 1984; Beckage, 1985; Grenier et al., 1986). In contrast, idiobionts consume their host in a fashion more akin to that of a predator (Quicke, 1997), with the difference being that parasitoids complete their development by consuming one single host. These differences in biology are fundamental for determining the degree of difficulty in developing artificial diets. Reviews of artificial diets show that more success has been obtained when developing artificial diets for idiobionts than for koinobionts (Quicke, 1997; Grenier and De Clercq, 2003). Egg parasitoids, most of which are considered idiobionts, have special needs regarding osmotic balance because they develop in liquid media (Grenier et al., 1986, 1994). However, all parasitoids have a biological characteristic that reduces the degree of difficulty in developing artificial diets. Parasitoids are able to complete development in one single host that provides all their nutritional needs; therefore, in theory a complete biochemical analysis of the host composition should yield the basis for a functional artificial diet. This is not always true because many parasitoids are capable of inducing changes in their host chemistry by the action of venoms (Morales-Ramos et al., 1995; Quicke, 1997). Chemical analyses of parasitized or envenomized hosts could provide better information for developing a functioning diet (Rojas et al., 1996).

Predatory arthropods may feed not only on different prey species but also on different plant-derived foods (Jervis and Kidd, 1996; Dixon, 2000). Omnivore predators are known to feed on plant nectar, pollen, fruit, foliage, and fungi in addition to prey (Coll and Guershon, 2002; Lundgren, 2009a, b). Feeding on nonprey items can provide significant benefits for the development, survival, and fecundity of predatory arthropods (Eubanks and Styrsky, 2005; Lundgren, 2009b). Even some predatory arthropods considered selective, such as *Phytoseiulus persimilis* Athias-Henriot (Zhang, 2003), have been reported to feed on extrafloral nectar and sugary solutions that provided significant improvements on immature survival and adult fecundity in the presence of prey (Rojas and Morales-Ramos, 2008). Nonprey foods provide a great variety of nutrients, including sugars (in nectars), complex carbohydrates, protein, some free amino acids, lipids, sterols, vitamins, and minerals. Pollen protein contains high amounts of proline, and pollen is also a good source of phytosterols (Lundgren, 2009a, b). Fungus can provide carbohydrates, protein, essential lipids, and ergosterol (Lundgren, 2009a, b). Consumption of nonprey foods increases the complexity of the predators' diet and alters the profile of amino acids and fatty acids that are assimilated, which can be relevant to artificial diet development (Rojas et al., 1996; Cohen, 2004). In order to achieve an optimal artificial diet for these predators, it is necessary to identify their alternate food sources, determine their composition, and find the proportion at which each is consumed. The complexity of the nutritional ecology of predatory arthropods is why a chemical analysis of the prey does not always provide the basis for a good artificial diet formulation.

7.2. ARTHROPOD NUTRITION

Nutrition is the most basic subject relevant for artificial diet development. Diets must provide all essential nutrients to allow complete development and reproduction.

7.2.1. Carbohydrates

Carbohydrates are an important energy source. In addition, they are required to produce chitin, which is an amino polysaccharide found on the exoskeleton of arthropods (Chippendale, 1978). The most common sugars involved in arthropod nutrition are monosaccharides and disaccharides (Singh, 1977), some of which are arthropod specific (Cohen, 2004).

Typically, sugars refer to the monosaccharides (glucose, fructose, and galactose) and the disaccharides (maltose and sucrose). The requirements of each specific sugar vary depending upon the type of arthropod. Most insects are able to absorb and metabolize fructose and glucose, but some monosaccharides such as arabinose, ribose, xylose, and galactose, while readily absorbed, are not metabolized (Chippendale, 1978). Generalist feeders are able to digest disaccharides, such as sucrose and maltose, while specialist feeders are not. Among generalist feeders are herbivores and predators, and specialist feeders include parasitoids and some mites (Singh, 1977; Cohen, 2004). Predatory arthropods can obtain carbohydrates from prey in the form of the disaccharide trehalose and the polysaccharide glycogen, which are the two main forms of carbohydrate reserves in insects; however, insect amylase can also digest starch (Chippendale, 1978), which is present in many nonprey food sources consumed by omnivore predators.

7.2.2. Lipids

Lipids are essential structural components of the cell membrane. They also provide an efficient way to store and provide metabolic energy during sustained demands, and they provide a barrier for water conservation in the arthropod cuticle (Downer, 1978; Chapman, 1998).

7.2.2.1. Fatty Acids

In general, polyunsaturated fatty acids such as linoleic and linolenic acids are essential in insect nutrition. Insects are either unable to synthesize them altogether or incapable of synthesizing them in sufficient quantities. The inability of insects to synthesize polyunsaturated fatty acids has been confirmed in some species, and limited capacity has been observed in other species such as mosquitoes, aphids, and cockroaches (Downer, 1978; Chapman, 1998). Derivatives of polyunsaturated fatty acids, known as eicosanoids, stimulate oviposition in crickets and may be important for reproduction in all insects (Chapman, 1998).

7.2.2.2. Sterols

Insects are unable to synthesize sterols, and therefore they are essential nutrients throughout Insecta (House, 1961; Chapman, 1998). Sterols play a variety of important roles in insect physiology as components of subcellular membranes, precursors of hormones, constituents of surface wax of the cuticle, and constituents of lipoprotein carrier molecules (Downer, 1978; Chapman, 1998). Insects obtain sterols from cholesterol, but other important sources of sterols include plant phytosterols and ergosterol from fungi (Downer, 1978; Cohen, 2004).

7.2.2.3. Phospholipids

Insects synthesize phospholipids in the fat body (Downer, 1978). Phospholipids play an important role in lipid transfer and in the synthesis of vitellin and other lipoproteins (Agosin, 1978; Shapiro, 1988). Phospholipids are easily digested and absorbed by insects and are a good source of polyunsaturated fatty acids (Chapman, 1998).

7.2.3. Proteins

Proteins are classified according to their solubility and function as globulins, nucleoproteins, lipoproteins, and insoluble protein. Globulins include enzymes, antibodies, and protein hormones. Nucleoproteins are associated with nucleic acids and ribosomes. Lipoproteins often serve as transport proteins, and insoluble proteins are passive compounds often referred to as structural proteins (Agosin, 1978).

7.2.3.1. Amino Acid Profile

Arthropod diets should contain at least these 10 essential amino acids: leucine, isoleucine, valine, threonine, lysine, arginine, methionine, histidine, phenylalanine, and tryptophan (Chapman, 1998; Cohen, 2004; Lundgren, 2009a). These 10 amino acids are considered essential in insects because of insects' inability to synthesize them. Other amino acids that can be synthesized but only in insufficient quantities, or that require large consumption of energy for their synthesis, may also be required in the insect diet (House, 1961; Chapman, 1998; Lundgren, 2009a). Tyrosine is a major component of sclerotin and is required in large quantities during molting (Hopkins, 1992). Proline is important during flight initiation by elevating sugar metabolism (Wigglesworth, 1972; Carter et al., 2006; Lundgren, 2009a); serine, cysteine, glycine, aspartic acid, and glutamic acid are essential for silkworm growth (Chapman, 1998).

In recent years, assessment of the proportion of amino acids, lipids or fats, or glycogen in arthropods through a "carcass analysis" has provided useful information for comparing the nutritional value of food in relation to the arthropods consuming the food. For example, Lindig et al. (1981) conducted an analysis of essential and total amino acids in the carcasses of three insect pests (the pecan weevil *Curculio caryae* (Horn), southwestern corn borer *Diatraea grandiosella* Dyar, and tarnished plant bug *Lygus lineolaris* (Palisot de Beauvois)). A comparative amino acid analysis of host plants of these insect herbivores (i.e. pecan kernels, inner whorls of corn, and pods of green bean) was also conducted. The authors concluded that the percentage of essential amino acids in each insect was comparable to that in each host. Lindig et al. (1981) surmised that the analysis of the insect carcass could provide information on the protein types to incorporate into an artificial diet for healthy development of the insect species under investigation. A broader application of this approach to include other nutrients such as sugars and lipids may prove valuable.

7.2.4. Vitamins

Also named growth factors, vitamins are grouped according to their solubility in water and lipids. Vitamins are usually required for growth in insects, which in general are unable to synthesize them (Chapman, 1998).

Water-soluble vitamins include vitamin C and the B complex, which is thiamin (B_1), riboflavin (B_2), niacin=nicotinamide (B_3), pantothenic acid (B_5), pyridoxine (B_6), biotin (B_7), folic acid (B_9), and cobalamins (B_{12}). The B vitamins function as cofactors for enzymes and are required in the diet of all insects (Chapman, 1998), except for vitamin B_{12}, which is not universally required (House, 1961; Cohen, 2004). Predatory arthropods can obtain B complex vitamins from prey, but nonprey foods such as fungi also contain high quantities of B vitamins (Lundgren, 2009a). Vitamin C (ascorbic acid) is more important in herbivorous than in entomophagous insects, but it is known to play a role in the molting process (Chapman, 1998; Cohen, 2004). Vitamin C is also an antioxidant and may play an important role in the detoxification processes and as protection against microbial infection (Cohen, 2004). Inositol and choline are constituents of some phospholipids. Choline plays a role in spermatogenesis and oogenesis in addition to its structural role in phospholipids, and it is probably required in all insects. Inositol is known to be required in Coleoptera and plays a role in the nervous system (Chapman, 1998).

Lipid-soluble vitamins include retinol, carotenoids (A), tocoferols (E), calciferol (D), and phyloquinone (K). Only vitamins A and E are known to be required in insects, where they play a role in the synthesis of pigments and in reproduction, respectively (Chapman, 1998). Both vitamins A and E are also antioxidants (Cohen, 2004). Fat-soluble vitamins are usually found in high quantities in nonprey foods like pollen and grains (Lundgren, 2009a).

7.2.5. Minerals

While some elements like nitrogen, sulfur, iron, and phosphorous can be obtained from organic sources, other essential elements for growth and reproduction must be obtained from inorganic sources (minerals), which cannot be biosynthesized. Twenty-four elements are known to be essential for living matter, and they are listed here in the order of importance: hydrogen, oxygen, carbon, nitrogen, calcium, phosphorous, chlorine, potassium, sulfur, sodium, magnesium, iron, copper, zinc, silicone, iodine, cobalt, manganese, molybdenum, fluorine, tin, chromium, selenium, and vanadium (Hammond, 1996). Elements can be divided according to their ionic charge into cation (+) and anion (−). Cations include metals like iron, sodium, potassium, magnesium, manganese, calcium, zinc, and copper. Anions include chloride, sulfur, fluoride, phosphorous, and iodine. Minerals are compounds that consist of combinations of cations and anions. House (1961) mentions potassium, sodium, phosphorus, magnesium, manganese, zinc, and copper as important in insect growth. Iron is important in several enzyme pathways, including in the synthesis of DNA (Cohen, 2004). Calcium is required to a lesser extent in arthropods than in vertebrates, but it still plays an important role in muscular excitation (Cohen, 2004).

Rarely have nutrient or tissue levels of trace elements been determined for insects (Cohen, 2004; Nation, 2008). Hence, the concentrations of trace elements have seldom been evaluated for their impact on insect production. This is in sharp contrast to the focus on protein, lipid, and carbohydrate levels. Yet the nutritional and physiological importance of trace elements in insects has been well established (Dadd, 1985; Lavilla et al., 2010), along with their role in regulating gene expression via transcription factors (Zinke et al., 2002). The primary reason for this gap in the knowledge and assessment of trace elements has been the difficulty associated with the detection and accurate determination of concentrations. Recent advances with inductively coupled plasma mass spectrometry (ICPMS) have provided an excellent solution for the simultaneous determination of several trace elements over a range of low concentrations, making it an ideal method for working with insects.

Information on the dynamics between the levels of trace elements in food and levels retained within the insect reared on the food has been a starting point for the application of ICPMS in determining trace element levels in predatory heteropterans (Coudron et al., 2012). Their result clearly showed that the levels of trace elements in the food stream affect the levels within the predator *Podisus maculiventris*, and those levels likely affect the health of the beneficial insect. Specifically, they demonstrated that trace element concentrations in nymphs were different than in adults and that the content of trace elements in both nymphs and adults differed greatly from the content in the food. Although the trace element content in eggs differed from that in the food, there was no significant difference in the trace element content in eggs regardless of the food source for nine of the 10 trace elements measured, demonstrating homeostasis of trace elements in eggs independent of the food source. This showed the ability of *P. maculiventris* to actively and selectively regulate absorption, sequestration, and/or accumulation of each trace element, and showed that the regulation varies with the developmental stage. The authors reasoned that the results indicate we can compare the bioavailability of trace elements in different food sources using these methods, which would be another piece of information seldom recorded previously. Additionally, trace elements have a unique characteristic of not being formed or destroyed (which makes them different from proteins, lipids, and carbohydrates) but having an integral physiological role. These qualities may enable the use of trace elements to determine relationships between nutrition and fitness in ways that could make them excellent biomarkers as described here.

7.3. DETERMINING THE BASIC FORMULATION

7.3.1. Chemical Analysis

A prelude to developing efficient artificial diets for predatory insects may involve an assessment of the biochemical makeup of the predator, its prey, and the host plant. An examination of the biochemical makeup of natural enemies has been conducted in just a few studies. For example, Specty et al. (2003) did a carcass analysis of the predatory lady beetle *Harmonia axyridis* Pallas and its prey aphid *Acyrthosiphon pisum* (Harris) or factitious prey *Anagasta (=Ephestia) kuehniella* (Zeller) eggs. They discovered that the biochemical composition of the carcass of the lady beetle varied according to the food source. *Harmonia axyridis* reared on aphids had lower protein content in their body than those reared on *A. kuehniella* eggs. Similarly, the lipid content of *H. axyridis* was lower in individuals when reared on aphids than when reared on factitious food. In a companion study, Zapata et al. (2005) found that a biochemical analysis of carcasses of the predatory mirid bug *Dicyphus tamaninii* Wagner proved useful in reformulating an artificial diet for this species. Carcasses of individuals reared on an artificial diet showed deficiencies in the content of essential fatty acids and amino acids, as compared to individuals reared on their prey (Zapata et al., 2005).

A technique that determines the composition of key elements, such as carbon (C), nitrogen (N), and phosphorous (P), in the body of arthropods and their food has been termed ecological stoichiometry by some (Sterner and Elser, 2002) and nutrient stoichiometry by others (Schoo et al., 2009). According to stoichiometry theory, the relative ratios of C:N:P in the tissues of an organism should match those of its food source. Any apparent imbalances in any of the elements, especially N and P, in the natural enemy could suggest that its food source was inadequate in one or more of these elements. Model organisms have typically been plants, but now plant feeders (herbivores) and their natural enemies (omnivores and carnivores) have been incorporated into stoichiometric analyses (Fagan and Denno, 2004; Raubenheimer et al., 2007). There have been no studies using stoichiometry to assess the elemental composition of natural enemies used in biological control of crop pests. This technology should be useful in designing and refining artificial diets.

Artificial diets have been developed by using information from chemical analysis of the host or prey that was used to approximate the nutritional requirements of its parasitoid or predator (Rojas et al., 1996, 2000; Fercovich et al., 1999; Reeve et al., 2003). The most suitable natural food source for the targeted entomophagous arthropod was used to develop chemically defined artificial diet formulations. These formulations were typically designed to replicate the concentration and ratios of the major nutritional component of the host or prey. This was intended to eliminate the process of trial and error by simultaneously providing all major dietary components in adequate concentrations. Chemical analyses are performed with high-performance liquid chromatography (HPLC), gas chromatography (GC), and mass spectrometry (MS). These methods of analysis are specially directed to determine the content, concentration, and ratios of amino acids, carbohydrates, lipids, protein, and vitamins found in the bodies of the host or prey. Quantitative analysis of the major nutritional components can be done by comparing mathematical integrations of instrument outputs of samples with outputs generated with standardized solutions of the target components. This method allows determination of the specific instrument sensitivity for each component and provides a method to convert area units to parts per million (Rojas et al., 1996, 2000; Fercovich et al., 1999; Reeve et al., 2003).

Amino acids, salts, carbohydrates, water, and fat-soluble vitamins can be analyzed by HPLC (Dionex D500 microbore system, Thermo Scientific Dionex, Sunnyvale, CA, USA). Amino acids are extracted with a 0.1 N HCl aqueous solution at a 1:10 sample:solvent ratio. Fresh samples are transferred to a centrifuge tube containing traces of bezophenyl urea and centrifuged for 15 min at 14,000 rpm. The solution is filtered using a 22 μm nylon filter. Twenty microliter of the sample is injected into the HPLC equipped with an AminoPac-PA10 column as directed by the manufacturer (Thermo Scientific Dionex) (Table 7.1).

Mineral salt, carbohydrate, and water-soluble vitamin samples are prepared in the same manner but using deionized Milli-Q water (Millipore, Billerica, MA, USA) as the extraction solvent. These types of compounds can also be analyzed with the Dionex D500 system by using the IonPac-AS11-HC column for anions (fluoride, chloride, nitrite, bromide, nitrate, phosphate, and sulfate), IonPac-CS-12A for cations (lithium, sodium, ammonium, potassium, magnesium, and calcium), CarboPac-PA10 for carbohydrates (glucosamine, galactose, glucose, fructose, maltose, and sucrose), and Waters Symmetry C18 for

TABLE 7.1 Synopsis of the Analytical Methods for Detailed Chemical Analysis of Basic Nutrients Present in Bean Leaves

Chemical Groups	Solvent for Extraction	Analytical Instrument	Analytical Column
Amino acids	0.1 N HCl	ICMB	AminoPac-PA10
Anions	Water	ICMB	IonPac-AS11-HC
Cations	Water	ICMB	IonPac-CS-12A
Carbohydrate	Water	ICMB	CarboPac-PA10
Water-soluble vitamins	Water	HPLC	Acclaim-C18
Fat-soluble vitamins	1% citric acid in 8% ethanol–water	HPLC	Acclaim-120 C18 5 μm
Soluble protein	0.9% NaCl solution	SPM	
Fatty acids	Methanol–chloroform (2:1)	GC/MS	HP-5% MS
Sterols	Methanol–chloroform (1:2)	GC/MS	HP-5% MS

ICMB = ion chromatography, microbore; HPLC = high-performance liquid chromatography; SPM = spectrophotometry; GC/MS = gas chromatography and mass spectrometry; and water = reverse osmosis (e.g. Milli-Q water, Millipore) or distilled.

water-soluble vitamins (ascorbic acid, thiamine, pyridoxal chloride, niacinamide, PABA, folic acid, B_{12}, and riboflavin) (Table 7.1). Fat-soluble vitamins can be extracted from 350 mg of sample with 4 ml of 1% citric acid in 80% ethanol–water plus 2 ml hexane. The mixture is shaken and the vortex mixed for 1 min, then placed in an ultrasonic bath at 40 °C for 5 min. The sample is then centrifuged at 10 °C for 5 min to allow the hexane layer to be separated. The hexane portion is washed three times in this manner and then concentrated by evaporation under a stream of nitrogen to a 1 ml volume (Dong and Pace, 1996). The Dionex HPLC system can be used to analyze samples using the 120 C18 5 μm column (vitamin A acetate, vitamin D, and vitamin E) (Table 7.1).

7.3.1.1. *Soluble Protein*

Fresh material is extracted with 0.9% sodium chloride containing traces of sodium azide at a 1:10 ratio into a 2 ml centrifuge tube by maceration and then centrifugation at 14 rpm for 15 min. A sample of the liquid is then reacted according to the Bio-Rad (Detergent Compatible) protein assay protocol. A standard curve is calculated, also following the protocol procedure. The percentage of soluble protein is measured in a UV–vis auto-spectrophotometer at a 750 nm wave length.

7.3.1.2. *Fatty Acids and Sterols*

Samples are dried in a vacuum oven at 40 °C and 20 psi pressure for 24 h. Fatty acids are extracted with a methanol–chloroform (2:1) solution to a 1:10 ratio sample solvent. Samples are set at room temperature for 1 h, and liquid is then filtered through a glass pipet packed with glass wool. Samples are rinsed three times with equal amounts of extracting solvent, filtered and combined. The combined samples are dried under a stream of nitrogen on a hot plate. Dry samples are reacted to their methyl ester form following the Alltech GC Boron trifluoride–methanol procedure. Dry methyl esters are redissolved in 0.700 ml hexane for their analysis. Samples can be analyzed using an Agilent Technologies 6890N GC with a 5975 inert XL Mass Selective Detector (MSD), and an Agilent 19091S-433 HP-5MS 5% Phenyl Methyl Siloxane capillary column (Agilent, Santa Clara, CA, USA) (Table 7.1). Sterols are extracted with a methanol–chloroform (1:2) solution to a 1:10 ratio of sample to solvent for 1 h and filtered as described here. Residues must be washed three times with an equal volume of methanol–chloroform and distilled water, making sure that layer separation is distinct. The bottom layer of each rinse is collected, combined in a vial containing sodium sulfate, and dried on a hot plate under a stream of nitrogen. The samples are redissolved in extracting solvent and analyzed in a GCMS as described here.

7.3.2. Water Content

The optimal water content of an artificial diet is approximately the same as the water content in the body of the host or prey (Cohen, 2004). The body water content may differ among organisms and must be determined for each host or prey species during the process of artificial diet development. Wigglesworth (1972) mentioned that water content in insect bodies may range from 50 to 90% of total body weight. For instance, the water content in *Tenebrio molitor* L., *Zophobas morio* F., *Galleria mellonella* L., *Bombyx mori* L. larvae, and *Acheta domesticus* L. nymphs is 61.9, 57.9, 58.5, 82.7, and 77.1%, respectively (Finke, 2002). One simple way to determine water content is by the differential of wet weight minus dry weight. Water content is very important, especially in artificial diets for parasitoids because their water needs must be completely fulfilled by the formulation (Thompson, 1980, 1981; Rojas et al., 1996; Fercovich et al., 1999). In egg parasitoids, such as *Trichogramma* spp., water content also impacts osmotic pressure by affecting the survival of parasitoid larvae (Grenier, 1994). Because predator arthropods are able to drink water freely, the water content in predator artificial diets can be more flexible. Predators with extra-oral digestion such as phytoseiid mites, heteropterans, and chrysopids can process solid foods (Cohen, 1985, 1998; Cohen and Urias, 1986; De Clercq and Degheele, 1992; Cohen and Smith, 1998).

7.3.3. Basic Nutrient Ratios

The ratio of three basic nutrient types, lipid, protein, and carbohydrate, can be used as the basis for an artificial diet formulation. The ratios of lipid, protein, and carbohydrate from food items are calculated as $R_L = L/(L+P+C)$, $R_P = P/(L+P+C)$, and $R_C = C/(L+P+C)$, where L, P, and C are the contents in grams per 100 g of lipid, protein, and carbohydrate, respectively. These ratios change in different arthropod species and nonprey food sources, and can be characteristic of the basic nutrition of each species.

Detailed nutritional analyses of most commercially available insect species have been performed, including *T. molitor*, *Z. morio*, *G. mellonella*, *A. domesticus* (Finke, 2002), *B. mori* (Finke, 2002; Mishra et al., 2003), *Antheraea assamensis* Helfer (=*Antheraea assama*) (Mishra et al., 2003), *Antheraea pernyi* (Guérin-Méneville) (Zhou and Han, 2006), *Samia ricinii* (Boisduval) (Longvah et al., 2011), *Musca domestica* L. (Finke, 2013; Inaoka et al., 1999), *Hermetia illucens* L., *Chilecomadia moorei* Silva, *Blatta lateralis* Walker (Finke, 2013), and *A. kuehniella* (Specty et al., 2003). Lipid, protein and carbohydrate ratios of those commercially produced insects can be calculated from published data (Fig. 7.1) and utilized as the basis for a carnivore diet. The basic nutrient ratio can be used to set the basic limits for these nutrients during the artificial diet development process. However, this method has a more important application during diet-refining stages (Section 7.5.2) and for substitution of chemically defined ingredients.

7.4. PRESENTATION

Artificial diets can have a consistency ranging from liquid to solid, depending mostly on two factors: water content and molecular cohesion. The first artificial

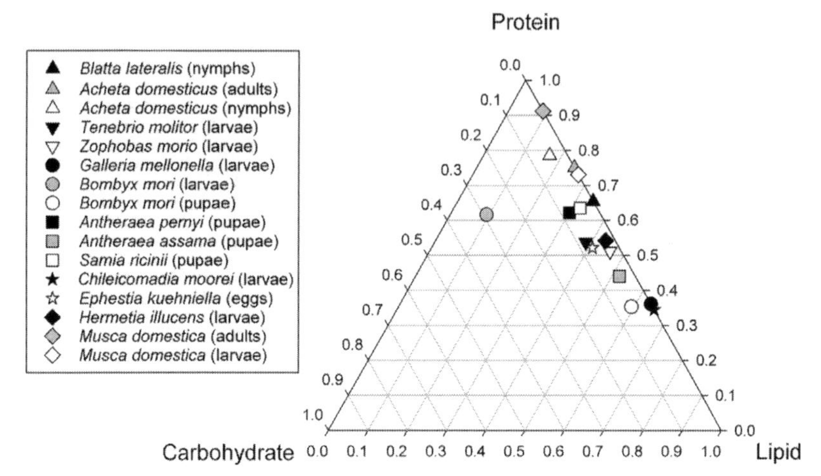

FIGURE 7.1 Basic nutrient ratios of lipid, protein and carbohydrate of commercially produced insects. *Data adapted from Inaoka et al. (1999), Finke (2002, 2013), Mishra et al. (2003), Specty et al. (2003), Zhou and Han (2006), and Longvah et al. (2011).*

diets have been liquid (Thompson, 1975; House, 1978). The major disadvantage of liquid diets is that all nutritional components must be in solution or in stable suspension. Providing adequate sources of protein and lipids in liquid diets becomes difficult for some arthropod species (Rojas et al., 2000). Protein must be provided in soluble form (globulin), and lipids must be water soluble by using the saponified forms. Because egg parasitoids typically develop in a semiliquid medium within the host egg, a successful artificial diet is likely to be presented as an encapsulated liquid (Grenier, 1994). The same often applies to endoparasitoids (koinobionts), which reside in the hemocoel of the living hosts (Quicke, 1997). Endoparasitoids have access to a continuous source of globulins and lipids in the form of phospholipids and glycoproteins inside the living host (Agosin, 1978; Chapman, 1998). Insect eggs contain vitellum, which is composed of glycoproteins and lipo-glycoproteins (vitellin). Lipo-glycoproteins consist of a combination of phospholypids, carbohydrates, and proteins (Chen, 1978). The complexity of such molecules allows the storage of large quantities of nutrients with little increase in osmotic pressure. Liquid diets for egg parasitoids must provide all nutrients in solution and at the same time maintain an optimal osmotic pressure to maintain parasitoid good health (Grenier, 1994).

7.4.1. Feeding Adaptations

Feeding adaptations determine the way that food is processed by insects before ingestion, and they greatly influence how artificial diets must be presented. Mouth parts in insects have been classified as mandibulate and hustellate (Borror et al., 1976). Most coleopteran and neuropteran adult predators have mandibulate mouth parts with well-developed chewing mandibles. Larvae of some Coleoptera and most larvae of Neuroptera possess modified mandibles with channels that enable them to suck digested liquids of their prey; this mandible type has been named grasping-sucking (Snodgrass, 1935). Hymenopteran parasitoids usually have a hymenopteriform larva with a poorly defined head capsule, but with well-defined functional mandibles (Gauld and Bolton, 1996). Heteropteran predators like the rest of the Heteroptera possess hustellate, highly modified mouth parts (Borror et al., 1976) specialized for sucking liquefied food following extra-oral digestion (Cohen, 1990). Mite predators, such as members of Chelicerata, have chelicerae as mouth parts used in extra-oral digestion. Chelicerae of phitoseiid mites can be articulated, chelate, and dentate pincer-like structures capable of cutting holes in the prey's body (Walter and Proctor, 1999).

7.4.2. Encapsulation of Liquid Diets

Liquid diets require some type of enclosure for containment and presentation, regardless of whether the entomophagous arthropod develops within or feeds through the containment. Liquid artificial diets have been encapsulated to mimic artificial eggs using different materials, such as paraffin, PVC, polyethylene, or polypropylene; *Trichogramma* species are perhaps the most successful example of rearing egg parasitoids by this method (Grenier, 1994). House (1978) used Parafilm® to encapsulate a semiliquid artificial diet for the endoparasitoid *Itoplectis conquisitor* (Say). Microliter volumes of liquid or semiliquid diets have been encapsulated in the wells formed with stretched Parafilm® to rear the predatory heteropteran *Perillus bioculatus* (F.) (Rojas et al., 2000), and domes formed with Mylar®–Parafilm® were used to contain a diet for *P. maculiventris* (Say) (Wittmeyer et al., 2001). Larger volume pouches were used to present a semiliquid diet for *Lygus hesperus* Knight (Patana, 1982) and *L. lineolaris* (Cohen, 2000). Because of its physical properties, Parafilm® is a difficult material to sterilize, form, or stretch in a mechanized system. Some absorbent materials like polyester padding (Rojas et al., 1996) or cotton (Dindo et al., 2003) have been used to contain liquid diets for the pteromalid ectoparasitoid *Catolaccus grandis* (Burks) and the tachinid endoparasitoid *Exorista larvarum* (L.), respectively. Such materials can be easily cut to size and sterilized in an autoclave before being impregnated with the diet formulation. Other fabric materials, made of synthetic fibers of high melting points such as nylon and polypropylene, provide equally viable media for diet containment and presentation. Diets with semiliquid to semisolid consistencies resembling gel or paste can be presented sandwiched between two sheets of screens made of high-fusion-point synthetic materials (Rojas, unpublished data). Screens provide a way for the arthropods to walk without getting trapped by the sticky diet, while at the same time providing easy access to the diet for feeding.

7.4.3. Gels and Carriers for Solid Formulations

Traditionally, solid diets have been developed for insects with well-developed chewing mouth parts. However, Cohen (1990, 1998) showed that arthropods with extra-oral digestion can feed on solid diets, even if they possess sucking mouth parts with no functional mandibles. Entomophagous arthropods with extra-oral digestion include spiders, phytoseiid mites, heteropterans, neuropterans, and some coleopteran larvae (Cohen, 1998). Larvae of the ectoparasitoid Hymenoptera also feed by extra-oral digestion (Cohen, 2004), and this was proven by successfully rearing *C. grandis* on solid diets (Rojas et al., 1996). Solid diet formulations for predatory insects have also been encapsulated using Parafilm® to create a barrier simulating the host cuticle; for instance, Cohen and Urias (1986) rolled cylindrical shapes of diet and wrapped them with Parafilm® to rear *Geocoris punctipes* (Say). De Clercq and Degheele (1992) used the same procedure and diet formulation to rear *Podisus sagitta* (F.) and *P. maculiventris*. A solid diet placed in cell culture wells sealed with Parafilm® was used to rear *Chrysoperla rufilabris* Burmeister (Cohen and Smith, 1998).

Solid diets have advantages over liquid and semiliquid diets, including non-stickiness and in many cases direct presentation without the need for encapsulation or containment. However, solid diets can have a large range of variation in water content; for instance, agar-based diets can contain over 90% water (Rojas et al., 1996; Cohen, 2004). Gelling agents are commonly used to make a diet solid. Cohen (2004) provided an excellent review of the most commonly used gelling agents and commented that these agents are usually the most expensive ingredient in diets. Gelling and thickening agents have long been utilized in the food industry. For instance, alginates, kappa and iota carrageenan, gelatin, gellan gum, pectins, methyl cellulose, and mixtures of xanthan and locust bean gums function as gelling agents; while gum Arabic, guar gum, carboxyl methyl cellulose, microcrystalline cellulose, and lambda carrageenan function as thickening agents (Fellows, 2009). Future directions on artificial diet development for entomophagous arthropods should include studies on the use of these agents to modify the consistency of the formulations.

7.5. DIET REFINING

Typically after an artificial diet formulation has proven to be sufficient to allow the target arthropod to complete development and reproduce for more than five generations (level 2), the job of refining the artificial diet begins. Artificial diets that achieve Level 2 criteria often produce arthropods of substantially lower quality than those produced in their natural host or prey. In most cases, first- or second-generation diet formulations (levels 1 and 2) tend to contain chemically defined components that are often too expensive to be used commercially. Additionally, successful artificial diet formulations are often susceptible to microbial contamination, which hinders their application in mass production systems.

Because refining an artificial diet requires extensive and precise methods of evaluation and comparison, only diets that have reached level 3 are ready for the refining process due to the large number of in vitro-produced specimens that are required by the evaluation procedures.

7.5.1. Improving Diet Quality

7.5.1.1. Diet Evaluation Techniques

The high complexity of an artificial diet formulation makes it difficult to analyze the impact of small modifications made for improvement. The change in concentration of one single component automatically affects the concentration of the rest of the diet components. One way to begin the process of refining the diet formulation is to create several versions of the formulation with different modifications. These versions could have different caloric ratios or different concentrations of a particular set of nutrients. The most important aspect of this process is the method of evaluating the results of these diet variations on the biology of the in vitro-reared arthropod.

The biological parameters that are relevant for mass production include immature survival, development time, adult size, fecundity, egg viability, progeny sex ratio (in parthenogenetic species), and doubling time (DT, defined as the time required for the population to double in numbers) (Cohen and Urias, 1986; De Clercq and Degheele, 1992; Carey, 1993; Rojas et al., 1996, 2000; Carpenter and Greany, 1998; Cohen and Smith, 1998; Morales-Ramos et al., 1998; Fercovich et al., 1999; Wittmeyer and Coudron, 2001). Parameters that are relevant for effectiveness as biological control agents include dispersal, host–prey finding, and host–prey handling capacities (Morales-Ramos et al., 1998). Evaluation of all the parameters mentioned here can be time-consuming, especially for species with relatively long life cycles. Some biological parameters can be correlated to one another, making the evaluations simpler to conduct (Grenier and De Clercq, 2003). For instance, Greenberg et al. (1995) reported that the pupal weight of *C. grandis* correlates with fecundity. However, that study was conducted in vivo and with only the natural host, *Anthonomus grandis* (Boheman), and no comparison was done among diet treatments (Greenberg et al., 1995). Changes in diet can have more profound effects on key physiological processes during the development of entomophagous arthropods impacting biological parameters independently of each other (Grenier and De Clercq, 2003). A good example of the independence of body weight and fecundity was mentioned by Grenier and De Clercq (2003), where *P. maculiventris* reared on artificial diet were smaller than those reared on *T. molitor* larvae, but they had similar fecundities (De Clercq et al., 1998); on the other hand, *P. bioculatus* reared on diet had similar size to those reared on *Leptinotarsa decemlineata* (Say) eggs, but their fecundity was only one-tenth (Rojas et al., 2000).

Life history analysis by the use of life tables provides a means to determine the impact of multiple biological parameters on population growth. Carey

(2001) categorizes life history analyses by three types of life table methods: (1) Lotka, (2) field, and (3) classical life tables. Classical and field life tables focus on the analysis of age-specific mortality, but Lotka life tables use age-specific birth and survivorship to compute statistics associated with the Lotka stable population model (Lotka, 1907, 1928; Dublin and Lotka, 1925), such as net reproductive rate (R_o) and intrinsic rate of increase (r_m) (Carey, 2001). Immature survival, progeny sex ratio, development time, adult fecundity, and longevity can be all summarized by the use of a Lotka life table in a single value defined as intrinsic rate of increase (r_m) (Carey, 1993). This method is the most powerful tool of evaluation available and the most accurate, and it provides a value (r_m) that is a measure of population fitness (Roff, 1992). Net reproductive rate (R_o) is defined as the average number of female offspring that would be born to a cohort of females, and it is calculated as follows:

$$R_o = \sum_{x=1}^{k} lx\ mx$$

where lx is the proportion surviving at age x, mx is female progeny produced per female of age x, and k is the oldest age. Intrinsic rate of increase (r_m) is the rate of natural increase of a population that has reached stable age distribution and is calculated by the Euler–Lotka equation (Carey, 1993) as

$$1 = \sum_{x=1}^{k} e^{-rmx} lx\ mx$$

The exact value of r_m can be obtained by this equation using recursive iteration while changing the value of r_m in the equation until the results yield the value of 1. DT can be calculated as

$$Ln\,(2)/r_m$$

Life history table analysis as a tool for evaluation will be covered in more detail in Chapter 8.

Another method to evaluate different versions of an artificial diet formulation is by comparing the efficiency of food utilization or the efficiency of food conversion into arthropod biomass. Analysis of consumption, assimilation, excretion, and conversion reveals how organisms respond to different foods and how food components affect growth (Scriber and Slansky, 1981). The most useful statistics are efficiency of food conversion of digested food (ECD=B/(I−F)) and efficiency of ingested food (ECI=B/I), where B=biomass gained, I=food ingested, and F=food excreted (frass) (Waldbauer, 1968; Scriber and Slansky, 1981; Wiegert and Petersen, 1983). Food utilization analysis requires dry weight measurements, but wet weight measurements can be utilized if the approximate moisture content is known. Wet weight estimates are useful if we need to maintain our specimens alive. Mean moisture content of arthropod species can be predetermined as well as that of the food provided. Food utilization can be used for evaluating the impact of different diet formulations on the growth of immature arthropods. However, this method is not suitable for evaluating the impact of diets on fecundity. Nevertheless, food utilization is a relatively quick way to eliminate the less advantageous formulations by reducing the number of diet variations to be compared using life table analysis.

The process of diet refining can be long and tedious due mostly to the time and labor required by the evaluation techniques available, which may require investing several years of research. These difficulties often have been real obstacles to develop artificial diets beyond level 3. Later in this chapter, we will discuss the possibility of substantially reducing the time and effort required to evaluate artificial diets by using proteomics and genomics as a method to determine gene expression. The use of "omics" technology to evaluate and compare artificial diet formulations requires the characterization of key molecular markers, which may be different in each species of arthropod.

7.5.1.2. Feeding Stimulants

Nutritional deficiency can result from the lack of some key nutrients in a formulation, but also can be the result of low levels of consumption. Even artificial diets that match the nutritional characteristics of the natural host or prey may be perceived by the entomophagous arthropod as unrecognized or "unpalatable." Some artificial diet formulations may be consumed only after the onset of starvation and as the last resort to survive. Such formulations may still produce reproductive adults, but their quality may be substantially compromised. It is generally accepted that phagostimulation by chemical compounds is required to produce an excitatory effect for continuous feeding (Simpson, 1995; Chapman, 1995, 1998). Some secondary plant chemicals with no nutritional value have been identified as feeding stimulants in some herbivore insects, and so have some nutrients like sucrose and inositol (Hanson, 1983). Compounds with no nutritional value that stimulate feeding have been termed token stimuli (Matthews and Matthews, 1978). Some organic acids resulting from the fungal decay of wood, such as oxalic acid (Espejo and Agosin, 1991; Green et al., 1995), stimulated feeding in the Formosan subterranean termite *Coptotermes formosanus* Shiraki (Morales-Ramos et al., 2009). In predatory insects, waxes present in the cuticle of prey may act as feeding stimulants (Greany and Hagen, 1981). Fatty acids present in *Tribolium castaneum* Jacquelin du Val induced feeding in the reduviid predator *Peregrinator biannulipes* Montrouzier & Signoret (Tebayashi et al., 2003). Some secondary plant chemicals are known to play a role in the prey-finding sequence (Hagen, 1987), but little information exists on the phagostimulatory properties of those chemicals. The value of feeding stimulants to improve artificial diets has not been fully explored, but their potential has been recognized. For instance, feeding was stimulated and resulted in increased weight gain in the mirid mite-predator *Hyaliodes vitripennis* (Say) when β-sistosterol was added to a meridic artificial diet

(Firlej et al., 2006). It is likely that feeding stimulants will play an important role in the improvement of artificial diets for entomophagous arthropods in the future.

7.5.2. From Chemically Defined to Economically Sound

Chemically defined ingredients such as amino acids, sterols, fatty acids, and vitamins add to the cost of artificial diet formulation. In order to develop an artificial diet formulation from level 3 to level 4, adequate substitutions must be found that make the formulation economically viable. For instance, the adequate content of fatty acids may be achieved using vegetable or fish oils (single or in combination) with a fatty acid profile similar to that of the chemically defined formulation. The fatty acid profile of the most common oils used as food is available online at the USDA nutritional database (USDA, 2010). The USDA database also provides detailed content on the most important nutrients present in common food items including protein (amino acid profile), carbohydrate, lipid (fatty acid profile), vitamins, minerals, water, and so on. Controlling the content of each of the most vital nutrients in a diet while making substitutions of chemically defined ingredients can be a difficult process. We propose a method consisting of six steps for diet refining: (1) determine the relative content of the three basic nutrient types: lipid, protein, and carbohydrate; (2) the amino acid profile of proteins (chemically defined additions may still be necessary in the case of amino acids); (3) the relative content of the three types of lipids: saturated, monounsaturated, and polyunsaturated; (4) the relative content of the three types of carbohydrates: sugars, digestible polysaccharides, and indigestible polysaccharides (fiber); (5) the final content of vital nutrients in the resulting diet formulation; and (6) the vitamin, mineral, and sterol deficiencies and water balance.

The relative content of lipid, protein, and carbohydrate provides a useful starting point for making substitution decisions. The content of these three basic nutrient groups can be determined relatively easily (as discussed in this chapter) from prey or host specimens and from artificial diet formulations. The ratios of lipid, protein, and carbohydrate from common foods are calculated as shown in Section 7.3.2. These ratios can be plotted in a ternary graph, as shown in Fig. 7.2 for the ratios contained in food ingredients commonly used in insect artificial diets. This graphic representation provides visual information on the similarities of undefined food ingredients with the prey or host. Commercially mass-produced insects could be used as ingredients in artificial diets (Fig. 7.1) without compromising the basic principle of rearing a single species. Many commercially produced insects are processed and sold as dry powders. Such flours can be used as artificial diet ingredients. As the commercialization of insects becomes more prevalent, a larger variety of species may become available for inclusion in artificial diets in the future.

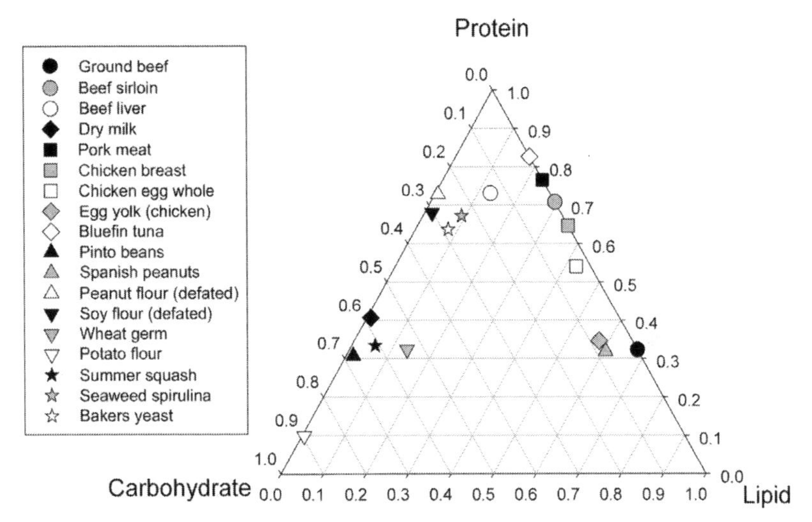

FIGURE 7.2 Basic nutrient ratios of lipid, protein, and carbohydrate of food sources commonly used in insect artificial diets. *Data adapted from USDA (2010).*

Amino acid profiles of protein can be calculated in a similar way as the basic nutrient ratios by converting amino acid content in grams per 100 g into ratios based on the total amino acid content.

$$RAai = Aai \div \sum_{i=1}^{n} Aai$$

Amino acid content converted into ratio profiles can be compared independently of the relative protein content, which is established during step 1. Comparisons and adjustments are important only for the essential and conditional essential amino acids. Figure 7.3 provides an example comparing the amino acid profiles of four commercially produced insects, four proteins of animal origin, and four proteins of vegetable origin. Some amino acids are present in higher proportion in insect protein than in vegetable or vertebrate protein, such as proline and tyrosine (Fig. 7.3(c)). Proline can be very important as a source of energy to initiate flight, and omnivore predators obtain it from pollen feeding (Wigglesworth, 1972; Carter et al., 2006; Lundgren, 2009a). Tyrosine plays an important role in the process of sclerotization as precursors of proteins associated with cuticle tanning (Hopkins, 1992) and is probably required in all arthropods in higher quantities than in vertebrates. For this reason, meridic artificial diets may have to be supplemented with proline and tyrosine. Other amino acids may have to be added for some predators with prey choices containing unusual amino acid profiles or omnivore predators feeding in nonprey food with a unique profile of amino acids.

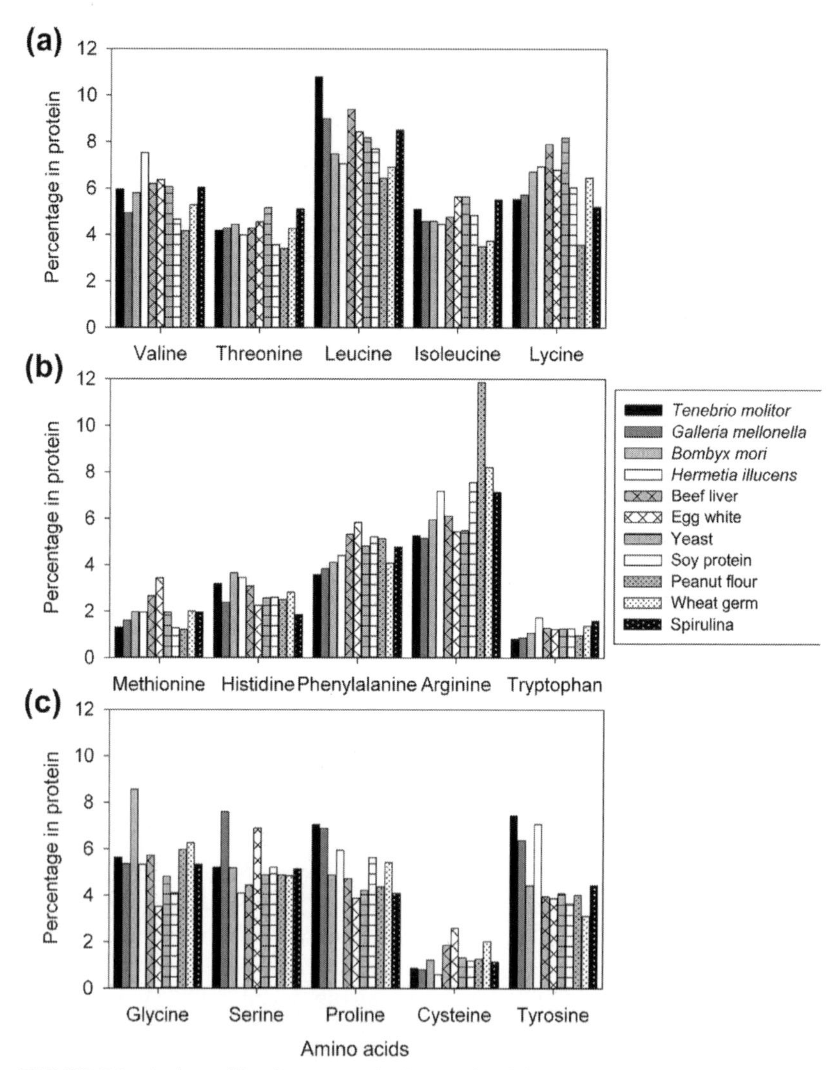

FIGURE 7.3 Amino acid ratios present in the protein of four commercially produced insects compared with two foods of animal origin and five foods of vegetable origin. (a) Simple-molecule essential amino acids, (b) complex-molecule essential amino acids, and (c) amino acids that are required in some species or hard to synthesize. *Insect data adapted from Finke (2002, 2013), food data adapted from USDA (2010).*

It is not necessary to create a full fatty acid profile for the substitution of lipids with chemically defined ingredients in step 3. Ratios of saturated, monounsaturated, and polyunsaturated fatty acids can be calculated using the same method used for the basic nutrient ratios. These ratios can also be plotted in a ternary graph for better visual comparison (Fig. 7.4). Additionally, the actual content of essential polyunsaturated fatty acids such as linoleic and linolenic

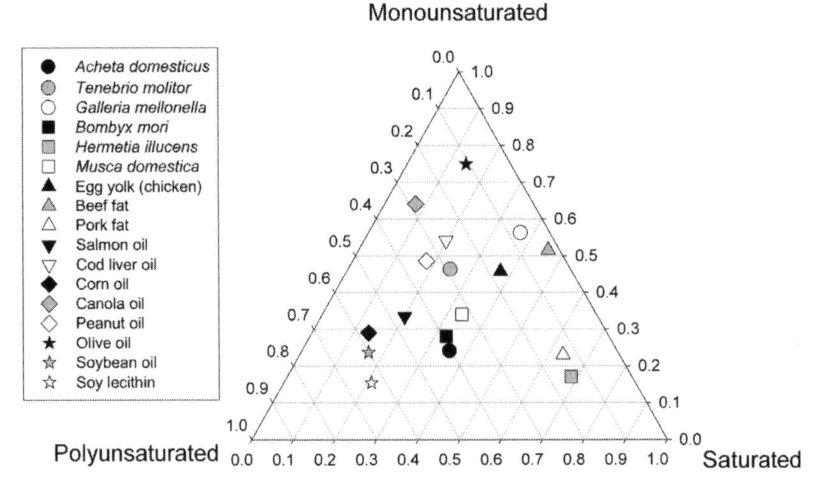

FIGURE 7.4 Ratios of three basic fatty acid types, saturated, monounsaturated, and polyunsaturated, present in the fat of six commercially produced insects compared to fats and oils of food products. *Insect data adapted from Finke (2002, 2013); and food data adapted from USDA (2010).*

acids can be compared in step 5. Some vegetable (corn and soybean) and fish (salmon) oils provide high quantities of polyunsaturated fatty acids. Olive and canola oils provide high quantities of monounsaturated fatty acids (Fig. 7.4). Commercially produced insects have different ratios of the three types of fatty acids, providing a diverse variety of choices that is useful for making decisions when choosing an insect as the protein source for the diet.

Carbohydrate substitution in step 4 is relatively simpler than protein and lipid substitutions. Because many chemically defined carbohydrates, such as sucrose, fructose, glucose, maltose, and starch, are commercially available, adjusting sugar content in artificial diets can be done entirely by adding chemically defined ingredients. However, some food product such as honey, potato, corn starch, and powder milk can provide sufficient amounts of carbohydrates. The fiber component consists mostly of chitin, but it can also include cellulose in omnivore predators, and can be substituted by cellulose powder. Fiber can be important in aiding the digestion of other food components even if the predators are unable to digest it.

Natural product substitutes for chemically defined components add many ancillary substances to the diet other than the intended components. It becomes necessary to keep track of these changes in order to maintain a correct balance of all nutrients as established by the original chemically defined artificial diet formulation. This can be accomplished in step 5 by a matrix operation (Fig. 7.5), where a vector **v** contains the proportions of each of the diet ingredients **i** (including those that are chemically defined). This vector is multiplied by a matrix **M** consisting of rows that represent the ingredients (eggs, powdered milk, beef liver, vegetable oil, yeast, etc.) and columns that represent

$$\mathbf{v} = (v_1, v_2, v_3, \ldots\ldots v_n),$$

Where elements are ingredient proportions with values between 0 and 1

$$\mathbf{M} = \begin{bmatrix} m_{11}\ m_{12}\ m_{13}\ m_{14} \ldots\ldots m_{1n} \\ m_{21}\ m_{22}\ m_{23}\ m_{24} \ldots\ldots m_{2n} \\ m_{31}\ m_{32}\ m_{33}\ m_{34} \ldots\ldots m_{3n} \\ m_{41}\ m_{42}\ m_{43}\ m_{44} \ldots\ldots m_{4n} \\ \ldots\ldots \\ \ldots\ldots \\ m_{n1}\ m_{n2}\ m_{n3}\ m_{n4} \ldots\ldots m_{nn} \end{bmatrix}$$

Where elements m_{ij} are gram per 100 g with values between 0 and 100, sub indexes i and j represent ingredients and nutrients, respectively.

$$\mathbf{vM} = \mathbf{d} = (d_1, d_2, d_3, d_4, \ldots\ldots\ldots d_n)$$

Where $d_j = \sum_{i=1}^{n} v_i m_{ij}$ for j = 1 to n with final values in grams per 100 g

FIGURE 7.5 Method for calculating the total nutrient content of artificial diet formulations by multiplying the diet ingredient vector **v** by the food matrix **M**. The diet ingredient vector consists of ingredient ratios (0–1). The food matrix consists of the nutrient content in grams per 100 g of key nutrients (columns) contained in the artificial diet ingredients (rows).

the nutrients to be considered (protein, lipid, sugar, polysaccharide, vitamins, sterol, water, etc.). Elements of matrix **M** represent the content of each nutrient **j** in grams per 100 g for each of the diet ingredients **i**. The nutrient content of the natural ingredients can usually be found in the USDA's nutritional database (USDA, 2010), as mentioned in this chapter. Matrix **M** needs to include only those nutrients that are critical for the development of the relevant entomophagous species, but it can include as many nutrients as desired. These could include individual ingredients such as starch, glucose, and proline, and nutrient groups like lipid, protein, and carbohydrate. Multiplying vector **v** times matrix **M** results in vector **d**, which contains the same number of elements as a matrix row (Pielou, 1984) and represents the amounts in grams per 100 g of each nutrient contained in the new modified artificial diet (Fig. 7.5). Values of chemically defined ingredients in matrix **M** will be 100 for their corresponding nutrient and 0 for the rest of the nutrients.

The results from this matrix operation provide an approximate content in grams per 100 g of all the critical ingredients in the diet formulation. This information can be utilized to determine and adjust the content of essential nutrients in the formulation (step 6). Because varying the content of one ingredient affects other ingredients, the diet vector **d** must be recalculated after every modification that is done to vector **v** and matrix **M**.

Water content is a critical part of this analysis because many raw natural ingredients contain large amounts of water. This potential problem can be overcome by using ingredients in their dry presentation. Some basic natural ingredients such as beef liver, potato, milk, egg yolk, and egg white are available as dry powders. Using dry ingredients provides much more flexibility when adjusting the water balance of artificial diet formulations, but even dry ingredients contain a small amount of water that should be included in the matrix operation. After steps 1 to 6 have been completed, the result is a meridic formulation that closely resembles the initial chemically defined formulation. At this point, the new diet is ready to be evaluated using the methods described in this chapter.

7.5.3. Diet Preservation

Artificial diets for entomophagous arthropods contain a balanced set of nutrients suitable for the growth of contaminant microbial agents such as bacteria and fungi. Even small levels of contamination can substantially compromise the nutritional value of the artificial diet. In addition, metabolic byproducts produced by the microbial contaminants are often toxic for the arthropods. Cohen (2004) listed 17 species of bacteria and nine species of fungi commonly found in insect artificial diets. In the same publication, Cohen (2004) presents an excellent list of chemicals that have been utilized to control microbial growth in insect artificial diets. While preservers like methylparaben, propionic acid, sodium benzoate, and formalin have been effective in herbivore diets (Debolt, 1982; Funke, 1983), entomophagous arthropods are often intolerant of those chemicals (Rojas, unpublished data).

7.6. FUTURE PERSPECTIVES

7.6.1. Nutri-Omics

Quality of food is broadly accepted as affecting the fitness and performance of beneficial insects, yet the association of nutrient quality with fitness (e.g. development and performance) is poorly understood. Most often, studies have been limited to the effect of diet on biological parameters (e.g. rate of development, weight and fecundity, and performance parameters like searching for and consuming food). The coarseness and delayed timing of those measurements make it difficult to correlate differences in development or performance with specific dietary ingredients. This has undoubtedly resulted in oversight of critical dietary components and an inability to optimize dietary formulations.

Reoccurring instances where empirically formulated diets required the addition of small amounts of natural food to achieve prey-fed performance but the ingredient within the natural food remained undefined are excellent examples of these phenomena. The delayed measurement of nutritional quality as expressed via biological and performance parameters can prevent early detection and correction of nutritional deficiencies during highly sensitive growth stages. While these concerns are easily stated, they have been difficult to remedy.

Missing have been sensitive and early indicators of food quality. Recognizing that biological and performance functions are controlled by physiological functions, efforts have been made to measure the response of cellular functions to dietary changes, in the hope of providing sensitive and early indicators of interest. Unfortunately, using biochemical measurements to detect cellular responses to dietary changes has had limited success. This is probably because many of the biochemical substances measured were likely to be under homeostatic control (e.g. vitellogenin). Hence, these substances demonstrate minimal changes or are slow to change in response to minor dietary alterations, and that change may be difficult to measure (Coudron et al., 2006). Several examples will be given in this section to suggest specific substances (like transcription factors) that may serve as more sensitive and early indicators.

Monitoring global gene expression (via RNA or protein) provides an opportunity to acquire information about the impact of nutrition on a wide range of biochemical and behavioral parameters. Potentially, a select set of those genes could serve as early and sensitive indicators that can be used to guide factitious host selection or diet formulation processes in ways that researchers have not been doing in the past (Coudron et al., 2006; Yocum et al., 2006). This approach is a more thorough assessment of the insect response to dietary changes. Additionally, this analysis is directed by the insect response rather than testing a predetermined substance for suitability as an indicator. That is to say, the results identify a marker rather than test a known substance for its suitability to serve as a marker. Admittedly, the challenge will be to "translate" those findings into meaningful marker-assisted selection of dietary ingredients. Yet, there is merit in exploring gene expression as a means to optimize dietary selections.

The use of marker-assisted selection of nutrients for insects is in its infancy. The concept was tested with artificial diets for the beneficial heteropteran *P. bioculatus*, the two-spotted stink bug (Coudron et al., 2006). Of approximately 6000 genes surveyed in a cDNA microarray, 47 genes displayed a greater than 1.5-fold difference between artificial diet- and prey-fed third-instar nymphs, giving an extensive set of genes for study. Of those, 31 genes were upregulated in the diet-fed nymphs. Additionally, 61% of the genes were distributed in the biological functional class of cellular processes, and 55% were categorized in the biological functional class of physiological processes (Coudron et al., 2006). From over 90 gene fragment sequences, four transcripts were selected for investigation that had repeatable and significant differences between diet- and prey-fed nymphs (Yocum et al., 2006). Two gene transcripts upregulated in diet-fed *P. bioculatus* nymphs compared to prey-fed nymphs had sequences similar to the tyrosine hydroxylase (TH) gene from *Apis mellifera* L. and the chitin-binding protein, *Gasp*, from *Drosophila melanogaster* Maigen. (The sequence of two transcripts that were upregulated in prey-fed nymphs did not match any gene of known function; this is an example of how a good biomarker may be of an unknown substance.) TH catalyzes the rate-limiting step in the biosynthesis of catecholamines (e.g. neurotransmitters) and impacts most biochemical processes, including ecdysis, behavior, immune response, and reproduction. The *Gasp* precursor is a peritrophin-like protein thought to be associated with the peritrophins, a family of chitin-binding proteins associated with the peritrophic matrix of insects.

The TH and *Gasp* genes are of particular interest because they meet several of the criteria of ideal diet biological markers (i.e. they indicate a response to nutritional quality, have a high level of expression for ease of detection, and occur early in development). A positive correlation was also found between levels of expression of these two genes and the number of generations the insects had been reared on the artificial diet (Yocum et al., 2006). Two of the diet-fed upregulated TH genes were tested for their responsiveness to nutrient alterations (Fig. 7.6). In one test, diet-fed individuals were switched to prey, which resulted in the disappearance of the upregulated TH genes, demonstrating that the artificial diet was the cause of the upregulation. In another test, liver was removed from the artificial diet, which again resulted in the disappearance of the upregulated TH genes, demonstrating that liver was the ingredient within the artificial

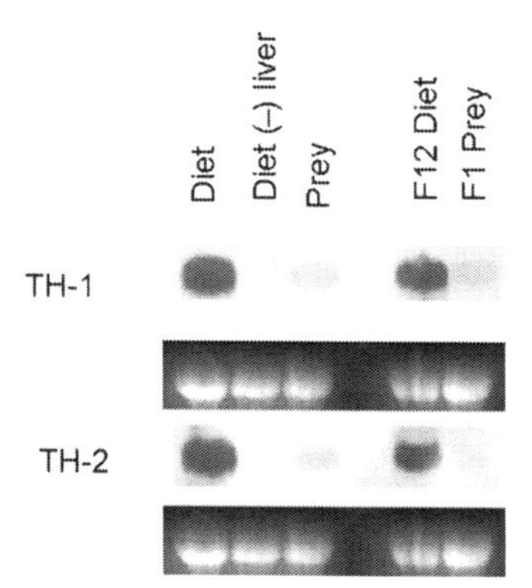

FIGURE 7.6 Expression of two *Perillus bioculatus* candidate biomarkers in response to dietary modifications. Northern blot analysis of two RNA transcripts (TH-1 and TH-2; with sequence homologies to tyrosine hydroxylase) isolated from fourth-instar nymphs fed an artificial diet (Diet), an artificial diet in which liver was removed (Diet (–) liver), or prey *Trichoplusia ni* (Prey). In another study, F12 generation of diet-fed fourth-instar nymphs (F12 Diet) was switched to a prey diet, and the resulting F1 generation of prey-fed fourth-instar nymphs was analyzed (F1 prey). Total RNA (7.5 μg) was separated on a 1.2% formaldehyde–agarose gel, transferred onto positively charged nylon membrane and probed with the clone indicated to the left of each panel. Equal loading was ensured by ethidium bromide–stained ribosomal RNA, shown below its respective Northern blot.

diet responsible for upregulating the TH genes. These tests indicate how gene or protein biomarkers can be used to evaluate nutrient components in diets.

As more examples of biomarkers emerge, one can anticipate a convergence to a set of genes and gene products that fit the criteria of good biomarkers and are indicative of specific dietary excesses or deficiencies. In many cases, these may be genes not previously thought of as predictors of nutrient quality. Currently, we are in the early stages of translating genomic and proteomic profiles into biomarkers that can be used to guide insect diet formulations. However, combinations of new technologies, such as RNA-sequence (which enables rapid sequencing of RNA) and expression profiling (which enables simultaneous comparison of levels of expressed RNA sequences from several samples), will accelerate global gene analysis and comparisons, which in turn will expedite the discovery of biomarkers that are useful for marker-assisted dietary choices.

7.6.2. Dietary Self-Selection

The ability of some insects to regulate their intake of nutrients has been well documented and has been referred to as dietary self-selection (Waldbauer and Friedman, 1991). The term dietary self-selection specifically refers to the consumption of different types of food items in an effort to achieve a specific balance of nutrients that is required by an organism (Waldbauer and Friedman, 1991; Behmer, 2009). Nutrient regulation by dietary self-selection has been demonstrated experimentally in a variety of insects, including *Supella longipalpa* F. (Cohen et al., 1987), *Locusta migratoria* L. (Raubenheimer and Simpson, 1993; Trumper and Simpson, 1993), *Spodoptera littoralis* (Boisduval) (Lee et al., 2002), *Spodoptera exempta* (Walker) (Lee et al., 2004), *Helicoverpa zea* (Boddie) (Waldbauer et al., 1984; Schiff et al., 1988), *Helicoverpa virescens* F. (Lee et al., 2006), *Anastrepha obliqua* (Macquart) (Cresoni-Pereira and Zucoloto, 2001), *Tribolium confusum* Jacquelin du Val (Waldbauer and Bhattacharya, 1973), and *T. molitor* (Morales-Ramos et al., 2011). Dietary self-selection has also been demonstrated in some predatory species, including *H. axyridis* (Coccinellidae) (Soares et al., 2004), *Agonum dorsale* (Pontoppidan) (Carabidae) (Mayntz et al., 2005), *Ectatomma ruidum* Roger (Formicidae) (Cook and Behmer, 2010), and the wolf spider *Pardosa prativaga* (Koch) (Mayntz et al., 2005).

Most dietary self-selection studies have shown that insects regulate their food choices so that a species-specific balance of digestible carbohydrate and protein can be attained. For instance, *T. confusum* feeding on a three-component diet mix developed faster than when feeding on pure diets of each of the individual components (Waldbauer and Bhattacharya, 1973). Last instars of *H. zea* self-selected an 80:20 protein–carbohydrate ratio when offered two diets of identical and complete nutritional content except that one lacked casein and the other lacked glucose. The self-selected ratio supported the best food utilization and food conversion when compared to each of the individual diets and a diet with a 1:1 casein–glucose ratio (Waldbauer et al., 1984).

In some cases, diets self-selected by immature stages can benefit the adult stages as well. For instance, self-selected diet ratios by *T. molitor* larvae significantly increased adult fecundity and longevity (Morales-Ramos et al., 2013). Self-selected diets can also be stage specific. For example, *S. longipalpa* developed better on a two-component diet of casein and glucose when allowed to self-select the proportions of a two-component diet rather than when on a diet of either component singularly or a 1:1 casein–glucose ratio. However, when forced to feed on a diet with the same component ratio as the self-selected diet, the development rate was slower than the 1:1 ratio diet. The consumption ratio of the two components changed from high carbohydrate at the beginning to an equal ratio with protein by the end of the stadium (Cohen et al., 1987).

Insects feeding on diverse diets also self-select to balance nutrients other than carbohydrate and protein (Waldbauer and Friedman, 1991; Behmer, 2009). Dietary self-selection for lipids and vitamins has been documented in *H. zea*. Schiff et al. (1988) reported that last-instar larvae of *H. zea* self-selected a superior mix of two incomplete diets. This experiment was similar to the one done by Waldbauer et al. (1984), except that the lipid and vitamin content were changed in the two diets. Trumper and Simpson (1993) demonstrated that salt intake was regulated by self-selection in nymphs of *L. migratoria* when using incomplete artificial diets. However, regulation of protein and carbohydrate was stronger since nymphs of *L. migratoria* were driven to consume a 12-fold suboptimal salt intake in order to balance protein and carbohydrate intake in a two-choice experiment with incomplete diets (Trumper and Simpson, 1993).

Self-selection is particularly important in omnivore species, which feed on a variety of food sources. Many of the most important and commercially produced arthropod predators are omnivores at least to some degree (Eubanks and Styrsky, 2005). Nonprey foods not only are a way to survive in the absence of prey; but also, for most omnivore predators, they are important sources of energy for flight and also play a role in reproduction and fecundity (Lundgren, 2009b). For instance, the syrphid predator *Rhingia campestris* Meigen consumes different ratios of pollen and nectar depending on the ovarial developmental stage (Haslett, 1989). Another important clue that indicates the importance of nonprey foods in reproduction is the fact that different sexes consume different proportions of nonprey foods (Lundgren, 2009b). One example is *Coleomegilla maculata* DeGeer, where females consume 10 times more corn pollen than males (Lundgren et al., 2005).

It is reasonable to assume that consumption of nonprey food not only is supplementary but also plays an important role in supplying all the nutritional requirements of omnivore predators. Consequently, omnivore predators must have the ability to adjust their nutritional needs by self-selection when confronted with a high diversity of food choices that have different nutritional characteristics. Soares et al. (2004) demonstrated that the coccinellid predator

H. axyridis self-selected optimal ratios of two aphid prey species, *Aphis fabae* Scopoli and *Myzus persicae* Sulzer, by showing that females allowed to self-select produced significantly more progeny than females fed with either of the two prey species alone. Mayntz et al. (2005) took a different approach to demonstrate self-selection capabilities in the carabid predator *A. dorsale* by offering adults a choice between lipid-rich and protein-rich diets composed of powdered locust. Another approach consisted of allowing the wolf spider *P. prativaga* to choose between live prey of fruit flies (*D. melanogaster*) fed with lipid- vs protein-rich diets (Mayntz et al., 2005). In both of the last two cases, predators showed the capacity to adjust their intake of protein and lipid by choosing among the two types of food (Mayntz et al., 2005). However, no studies have been reported on self-selection of prey vs nonprey foods in omnivore predators.

Dietary self-selection has not been extensively used as the basis for artificial diet development or refinement. However, Waldbauer et al. (1984) obtained a better diet formulation for *H. zea* after adjusting the casein-to-glucose ratios to those self-selected by the larvae. This achievement prompted Cohen (2004) to suggest dietary self-selection as a potential way to refine artificial diets. More recently, Morales-Ramos et al. (2013) were able to refine a dietary supplement for *T. molitor* using self-selected ratios of six components by groups of larvae. Adult *T. molitor* females fed with a supplement formulation derived from self-selected ratios by larvae produced significantly more progeny than females fed with other supplements or with no supplement (Morales-Ramos et al., 2013). In theory, self-selection should produce valuable information to adjust the ratios of some components in artificial diets for omnivore predators.

7.6.3. Endosymbionts

In Chapter 3, microorganisms associated with heteropteran predators are discussed, with a focus on endosymbiotic and pathogenic interactions. Here, we will address the probiotic gut microbiota as related to insect diets, although the gut microbiota represent all aspects of microbial relationships, from pathogenic to obligate mutualism. Ferreting out the complex roles of insect gut microbiota is in its initial stages and hence not well understood, but it is commonly accepted that gut microbiota have a significant role in insect nutrition via food digestion. Hence, to assist in diet development and optimization, some degree of gut microbiota manipulation may prove helpful. The two most common effects of artificial diet are the depletion of normal gut microbiota and the presence of foreign bacteria associated with contamination of the diet.

Probiotic bacteria provide valuable metabolic resources for many entomophagous arthropods through a symbiotic, multitrophic interaction between the insect and its food. Although awareness of gastric ceca containing bacteria in Heteroptera dates back nearly 100 years (Glasgow, 1914) for the predatory heteropterans, details of the relationship between the insect and its gut microbiota remain undefined. Genomic and proteomic methods (discussed in this chapter) have recently been used to accumulate information on the gut microbiome and have resulted in the characterization of indigenous and transient populations of bacteria communities in the insect alimentary canal, the detection of new species, and more frequent links to biological roles.

Gut bacterial communities adapt by the transfer of plasmids and other content between bacterial strains (Dillon and Dillon, 2004). We reason that this relationship and adaptation between the insect and its microbiota are critical to the production of insects optimized for fitness. However, using microbiota to optimize the production of insects will require a better understanding of the molecular relationships between the insect and its microbiome. The complexity and plasticity of the gut microbiota and the influence that rearing has on gut microbiota will challenge such studies.

It is assumed that gut microbiota are derived from the surrounding environment such as food, but also are passed from one generation to the next, most likely associated with the egg. Most of what we know today about heteropteran gut endosymbionts is from studies with phytophagous pentatomids (Prado and Almeida, 2009), although preliminary studies have begun to document the effects of rearing on gut endosymbionts in the heteroperan predator *P. maculiventris* (Coudron, unpublished data). There is much to learn before we will be able to effectively use endosymbionts to optimize the production of entomophagous arthropods. For example, in addition to identification of microbial species, the following would be important to know for production purposes: the degree of persistence of ingested strains, particular niches in the gut where specific strains colonize, which microbial species colonize the gut habitat, interactions among bacteria, the role of indigenous gut microbiota in preventing or suppressing infections, and the influence of food on the strains that constitute the gut microbiota.

7.7. CONCLUDING REMARKS

In this chapter, we have presented concepts of biochemistry, analytical chemistry, biophysics, mathematics, biology, physiology, demographics, behavior, genetics, genomics, dietetics, and more, all applied to the development and improvement of artificial diets for entomophagous arthropods. All the methods discussed in this chapter focus on advancing artificial diets from rearing to mass production. Advancing the technology of artificial diets to commercial application will require additional skills in engineering, artificial intelligence, robotics, food processing, and materials science. It is evident that advancing the development of artificial diets is a multidisciplinary enterprise, requiring the integration of multiple skills and the collaboration of teams to fulfill the goals of researchers and industry for the use of beneficial arthropods. The authors of this chapter wish to convey the importance of incorporating new ideas and technology in our efforts to meet those goals. The accelerated pace at which the field of artificial diets for arthropods is advancing makes us optimistic that the inherently complex challenge of diet development will be met in the next decade, and subsequently those goals achieved.

REFERENCES

Agosin, M., 1978. Functional role of proteins. In: Rockstein, M. (Ed.), Biochemistry of Insects. Academic Press, New York, NY, pp. 57–92.

Askew, R.R., Shaw, M.R., 1986. Parasitoid communities: their size, structure and development. In: Waage, J., Greathead, D. (Eds.), Insect Parasitoids, Academic Press, Orlando, FL, pp. 225–264.

Beckage, N.E., 1985. Endocrine interactions between endoparasitic insects and their hosts. Annu. Rev. Entomol. 30, 371–413.

Behmer, S., 2009. Insect herbivore nutrient regulation. Annu. Rev. Entomol. 54, 165–187.

Borror, D.J., DeLong, D.M., Triplehorn, C.A., 1976. An Introduction to the Study of Insects, fourth ed. Holt, Rinehart and Winston, New York, NY, pp. 852.

Carey, J.R., 1993. Applied Demography for Biologists with Special Emphasis on Insects. Oxford University Press, Oxford, UK, pp. 206.

Carey, J.R., 2001. Insect biodemography. Annu. Rev. Entomol. 46, 79–110.

Carpenter, J.E., Greany, P.D., 1998. Comparative development and performance of artificially reared versus host-reared *Dipetimorpha introita* (Cresson) (Hymenoptera: Ichneumonidae) wasps. Biol. Control 11, 203–208.

Carter, C., Shafir, S., Yehonatan, L., Palmer, R.G., Thornburg, R., 2006. A novel role for proline in plant floral nectars. Naturwissenschaften 93, 72–79.

Chapman, R.F., 1995. Chemosensory regulation of feeding. In: Chapman, R.F., de Boer, G. (Eds.), Regulatory Mechanisms in Insect Feeding, Chapman & Hall, New York, NY, pp. 101–136.

Chapman, R.F., 1998. The Insects, fourth ed. Cambridge University Press, Cambridge, UK, pp. 770.

Chen, P.S., 1978. Protein synthesis in relation to cellular activation and deactivation. In: Rockstein, M. (Ed.), Biochemistry of Insects, Academic Press, New York, NY, pp. 57–92.

Chippendale, G.M., 1978. The function of carbohydrates in insect life processes. In: Rockstein, M. (Ed.), Biochemistry of Insects, Academic Press, New York, NY, pp. 1–55.

Cohen, A.C., 1985. Simple method for rearing the insect predator *Geocoris punctipes* (Heteroptera: Lygaeidae) on a meat diet. J. Econ. Entomol. 78, 1173–1175.

Cohen, A.C., 1990. Feeding adaptations of some predaceous Hemiptera. Ann. Entomol. Soc. Am. 83, 1215–1223.

Cohen, A.C., 1992. Using a systematic approach to develop artificial diets for predators. In: Anderson, T.E., Leppla, N.C. (Eds.), Advances in Insect Rearing for Research & Pest Management, Westview Press, Boulder, CO, pp. 77–91.

Cohen, A.C., 1998. Solid-to-liquid feeding: the inside(s) story of extra-oral digestion in predaceous Arthropoda. Am. Entomol. 44, 103–116.

Cohen, A.C., 2000. New oligidic production diet for *Lygus hesperus* Knight and *L. lineolaris* (Palisot de Beauvois). J. Entomol. Sci. 35, 301–310.

Cohen, A.C., 2001. Formalizing insect rearing and artificial diet technology. Am. Entomol. 47, 198–206.

Cohen, A.C., 2004. Insect Diets Science and Technology. CRC Press, Boca Raton, FL.

Cohen, A.C., Smith, L.K., 1998. A new concept in artificial diets for *Chrysoperla rufilabris*: the efficacy of solid diets. Biol. Control 13, 49–54.

Cohen, A.C., Urias, N.M., 1986. Meat-based artificial diets for *Geocoris punctipes* (Say). Southwest. Entomol. 11, 171–176.

Cohen, R.W., Heydon, S.L., Waldbauer, G.P., Friedman, S., 1987. Nutrient self-selection by the omnivorous cockroach *Supella longipalpa*. J. Insect Physiol. 33, 77–82.

Coll, M., Guershon, M., 2002. Omnivory in terrestrial arthropods: mixing plant and prey diets. Annu. Rev. Entomol. 47, 267–297.

Cook, S.C., Behmer, S.T., 2010. Macronutrient regulation in the tropical ant *Ectatomma ruidum* (Formicidae): a field study in Costa Rica. Biotropica 42, 135–139.

Coudron, T.A., Mitchell, L.C., Sun, R., Robertson, J.D., Pham, N.V., Popham, H.J.R., 2012. Dietary composition affects levels of trace elements in the predator *Podisus maculiventris* (Say) (Heteroptera: Pentatomidae). Biol. Control 61, 141–146.

Coudron, T.A., Yocum, G.D., Brandt, S.L., 2006. Nutrigenomics: a case study in the measurement of insect response to nutritional quality. Entomol. Exp. Appl. 121, 1–14.

Cresoni-Pereira, C., Zucoloto, F.S., 2001. Dietary self-selection and discrimination threshold in wild *Anastrapha obliqua* females (Diptera, Tephritidae). J. Insect Physiol. 47, 1127–1132.

Dadd, R.H., 1985. Nutrition: organisms. In: Kerkut, G.A., Gilbert, L.I. (Eds.), Comprehensive Insect Physiology Biochemistry and Pharmacology, vol. 4. Pergamon Press, Oxford, pp. 313–390.

Debolt, J.W., 1982. Meridic diet for rearing successive generations *of Lygus hesperus*. Ann. Entomol. Soc. Am. 75, 119–122.

De Clercq, P., Degheele, D., 1992. A meat-based diet for rearing the predatory stinkbugs *Podisus maculiventris* and *Podisus sagitta* [Het.: Pentatomidae]. Entomphaga 37, 149–157.

De Clercq, P., Merlevede, F., Tirry, L., 1998. Unnatural prey and artificial diets for rearing *Podisus maculiventris* (Heteroptera: Pentatomidae). Biol. Control 12, 137–142.

Dillon, R.J., Dillon, V.M., 2004. The gut bacteria of insects: nonpathogenic interactions. Annu. Rev. Entomol. 49, 71–92.

Dindo, M.L., Marchetti, E., Galvani, G., Baronio, P., 2003. Rearing *Exorista larvarum* (Diptera: Tachinidae): simplification of the *in vitro* technique. Bull. Insectology 56, 253–257.

Dixon, A.F.G., 2000. Insect Predator–Prey Dynamics Lady Beetles & Biological Control. Cambridge University Press, Cambridge, UK, pp. 257.

Dong, M.W., Pace, J.L., 1996. A rapid HPLC method for multivitamins analysis. LC-GC 14 (9), 794–808.

Downer, R.G., 1978. Functional role of lipids in insects. In: Rockstein, M. (Ed.), Biochemistry of Insects, Academic Press, New York, NY, pp. 57–92.

Dublin, L.I., Lotka, A.J., 1925. On the true rate of natural increase. J. Am. Stat. Assoc. 20, 305–339.

Etzel, L.K., Legner, E.F., 1999. Culture and colonization. In: Bellows, T.S., Fisher, T.W. (Eds.), Handbook of Biological Control, Academic Press, San Diego, CA, pp. 125–197.

Espejo, E., Agosin, E., 1991. Production and degradation of oxalic acid by brown rot fungi. Appl. Environ. Microbiol. 57, 1980–1986.

Eubanks, M.D., Styrsky, J.D., 2005. Effects of plant feeding on the performance of omnivorous "predators". In: Wäckers, F.L., van Rijn, P.C.J., Bruin, J. (Eds.), Plant-Provided Food for Carnivorous Insects: A Protective Mutualism and Its Applications, Cambridge University Press, Cambridge, UK, pp. 148–177.

Fagan, W.F., Denno, R.F., 2004. Stoichiometry of actual vs. potential predator-prey interactions: insights into nitrogen limitation for arthropod predators. Ecol. Lett. 7, 876–883.

Fellows, P.J., 2009. Food Processing Technology Principles and Practice, third ed. CRC Press, Boca Raton, FL pp. 913.

Fercovich, S.M., Morales-Ramos, J.A., Rojas, M.G., Oberlander, H., Carpenter, J.E., Greany, P., 1999. Rearing the ectoparasitoid *Diapetimorpha introita* on an artificial diet: supplementation with insect cell line-derived factors. BioControl 44, 29–45.

Finke, M.D., 2002. Complete nutrient composition of commercially raised invertebrates used as food for insectivores. Zoo Biol. 21, 269–285.

Finke, M.D., 2013. Complete nutrient content of four species of feeder insects. Zoo Biol. 32, 27–36.

Firlej, A., Chouinard, G., Coderre, D., 2006. A meridic diet for the rearing of *Hyaliodes vitripennis* (Hemiptera: Miridae), a predator of mites in apple orchards. Biocontrol Sci. Tech. 16, 743–751.

Funke, B.R., 1983. Mold control for insect rearing media. Bull. Entomol. Soc. Am. 29, 41–44.

Gauld, I., Bolton, B., 1996. The Hymenoptera. Oxford University Press, Oxford, UK, pp. 332.

Glasgow, H., 1914. The gastric caeca and the caecal bacteria of the heteroptera. Biol. Bull. 26 (3), 101–171.

Greany, P.D., Hagen, K.S., 1981. Prey selection. In: Norlund, D.A., Jones, R.L., Lewis, W.J. (Eds.), Semiochemicals—Their Role in Pest Control, Wiley, New York, NY, pp. 121–135.

Green III, F., Clausen, C.A., Kuster, T.A., Highley, T.L., 1995. Induction of polygalacturonase and the formation of oxalic acid by pectin in brown-rot fungi. World J. Microbiol. Biotechnol. 11, 519–524.

Greenberg, S.M., Morales-Ramos, J.A., King, E.G., Summy, K.R., Rojas, M.G., 1995. Biological parameters for mass propagation of *Catolaccus grandis* (Hymenoptera: Pteromalidae). Environ. Entomol. 24, 1322–1327.

Grenier, S., 1994. Rearing of *Trichogramma* and other egg parasitoids on artificial diets. In: Wajnbergiand, E., Hassan, S.A. (Eds.), Biological Control with Egg Parasitoids, CAB International, Wallingford, UK, pp. 73–92.

Grenier, S., 2009. In vitro rearing of entomophagous insects—past and future trends: a minireview. Bull. Insectology 62, 1–6.

Grenier, S., De Clercq, P., 2003. Comparison of artificially vs. naturally reared natural enemies and their potential for use in biological control. In: van Lenteren, J.C. (Ed.), Quality Control and Production of Biological Control Agents Theory and Testing Procedures, CABI Publishing, Oxon, UK, pp. 115–131.

Grenier, S., Drlobel, B., Bonnot, G., 1986. Physiological considerations of importance to the success of in vitro culture: an overview. J. Insect Physiol. 32, 403–408.

Grenier, S., Greany, P.D., Cohen, A.C., 1994. Potential of mass release of insect parasitoids and predators through development of artificial culture techniques. In: Rosen, D., Bennett, F.D., Capinera, J.L. (Eds.), Pest Management in the Subtropics: Biological Control—A Florida Perspective, Intercept 10, Andover, UK, pp. 181–205.

Hagen, K.S., 1987. Nutritional ecology of terrestrial insect predators. In: Slansky Jr., F., Rodriguez, J.G. (Eds.), Nutritional Ecology of Insects, Mites, Spiders, and Related Invertebrates, John Wiley & Sons, New York, NY, pp. 533–577.

Hammond, C.R., 1996. The elements. In: Lide, D.R. (Ed.), CRC Handbook of Chemistry and Physics, 77th edition 1996–1997, CRC Press, Boca Raton, FL, pp. 1–34, section 4.

Hanson, F.E., 1983. The behavioral and neurophysiological basis of food plant selection by lepidopterous larvae. In: Ahmad, S. (Ed.), Herbivorous Insects, Academic Press, New York, NY, pp. 3–23.

Haslett, J.R., 1989. Adult feeding by holometabolous insects: pollen and nectar as complementary nutrient sources for *Rhingia campestris* (Diptera: Syrphidae). Oecologia 81, 361–363.

Hopkins, T.L., 1992. Insect cuticle sclerotization. Annu. Rev. Entomol. 37, 273–302.

House, H.L., 1961. Insect nutrition. Annu. Rev. Entomol. 6, 13–26.

House, H.L., 1978. An artificial host: encapsulated synthetic medium for in vitro oviposition and rearing the endoparasitoid *Itoplectis conquisitor* (Hymenoptera: Ichneumonidae). Can. Entomol. 110, 331–333.

Inaoka, T., Okubo, G., Yokota, M., Takemasa, M., 1999. Nutritive value of house fly larvae and pupae fed on chicken feces as food source for poultry. Jpn. Poult. Sci. 36, 174–180.

Jervis, M.A., Kidd, N.A.C., 1996. Phytophagy. In: Jervis, M.A., Kidd, N.A.C. (Eds.), Insect Natural Enemies Practical Approaches to Their Study and Evaluation, Chapman & Hall, London, UK, pp. 375–394.

King, E.G., Leppla, N.C., 1984. Advances and Challenges in Insect Rearing. United States Department of Agriculture, Agricultural Research Service, New Orleans, LA, pp. 306.

King, E.G., Hopper, K.R., Powell, J.E., 1985. Analysis of systems for biological control of crop arthropod pests in the U.S. by augmentation of predators and parasites. In: Hoy, M.A., Herzog, D.C. (Eds.), Biological Control in Agricultural IPM Systems, Academic Press, Orlando, FL, pp. 201–227.

Knipling, E.F., 1992. Principles of Insect Parasitism Analyzed from New Perspectives, Practical Implications for Regulating Insect Populations by Biological Means. Agricultural Handbook No. 693 USDA-ARS, Washington, DC, pp. 337.

Lavilla, I., Rodriguez-Liñares, G., Grrido, J., Bendicho, C., 2010. A biogeochemical approach to understanding the accumulation patterns of trace elements in three species of dragonfly larvae: evaluation as biomonitors. J. Environ. Monit. 12, 724–730.

Lee, K.P., Behmer, S.T., Simpson, S.J., Raubenheimer, D.R., 2002. A geometric analysis of nutrient regulation in the generalist caterpillar *Spodoptera littoralis* (Boisduval). J. Insect Physiol. 48, 655–665.

Lee, K.P., Simpson, S.J., Raubenheimer, D.R., 2004. A comparison of nutrient regulation between solitarious and gregarious phases of the specialist caterpillar *Spodoptera exempta* (Walker). J. Insect Physiol. 50, 1171–1180.

Lee, K.P., Behmer, S.T., Simpson, S.J., 2006. Nutrient regulation in relation to diet breadth: a comparison of *Heliothis* sister species and hybrid. J. Exp. Biol. 209, 2076–2084.

Lindig, O.H., Hedin, P.A., Poe, W.E., 1981. Amino acids in pecan weevil, southwestern corn borer and tarnished plant bug, and at their feeding sites. Comp. Biochem. Physiol. 68A, 261–263.

Longvah, T., Mangthya, K., Ramulu, P., 2011. Nutrient composition and protein quality evaluation of eri silkworm (*Samia ricinii*) prepupae and pupae. Food Chem. 128, 400–403.

Lotka, A.J., 1907. Relation between birth rates and death rates. Science 26, 21–22.

Lotka, A.J., 1928. The progeny of a population element. Am. J. Hyg. 8, 875–901.

Lundgren, J.G., 2009a. Nutritional aspects of non-prey foods in the life histories of predaceous Coccinellidae. Biol. Control 51, 294–305.

Lundgren, J.G., 2009b. Relationship of Natural Enemies and Non-prey Foods. Progress in Biological Control Series. vol. 7. Springer, pp. 453. www.springer.com.

Lundgren, J.G., Huber, A., Widenmann, R.N., 2005. Quantification of consumption of corn pollen by the predator *Coleomegilla maculata* (Coleoptera: Coccinellidae) during anthesis in an Illinois cornfield. Agric. For. Entomol. 7, 53–60.

Matthews, R.W., Matthews, J.R., 1978. Insect Behavior. John Wiley & Sons, New York, NY, pp. 507.

Mayntz, D., Raubenheimer, D., Salomon, M., Toft, S., Simpson, S.J., 2005. Nutrient-specific foraging in invertebrate predators. Science 307, 111.

Mishra, N., Hazarika, N.C., Narain, K., Mahanta, J., 2003. Nutritive value of non-mulberry and mulberry silkworm pupae and consumption pattern in Assam, India. Nutr. Res. 23, 1303–1311.

Morales-Ramos, J.A., Rojas, M.G., King, E.G., 1995. Venom of *Catolaccus grandis* (Hymenoptera: Pteromalidae) and its role in parasitoid development and host regulation. Ann. Entomol. Soc. Am. 88, 800–808.

Morales-Ramos, J.A., Rojas, M.G., Coleman, R.J., King, E.G., 1998. Potential use of in vitro-reared *Catolaccus grandis* (Hymenoptera: Pteromalidae) for biological control of the boll weevil (Coleoptera: Curculionidae). J. Econ. Entomol. 91, 101–109.

Morales-Ramos, J.A., Rojas, M.G., Nimocks III, D., 2009. Some organic acids acting as stimulants of recruitment and feeding for the Formosan subterranean termite (Isoptera: Rhinotermitidae). Sociobiology 54, 861–871.

Morales-Ramos, J.A., Rojas, M.G., Shapiro-Ilan, D.I., Tedders, W.L., 2011. Self-selection of two diet components by *Tenebrio molitor* (Coleoptera: Tenebrionidae) larvae and its impact on fitness. Environ. Entomol. 40, 1285–1294.

Morales-Ramos, J.A., Rojas, M.G., Shapiro Ilan, D.I., Tedders, W.L., 2013. Use of nutrient self-selection as a diet refining tool in *Tenebrio molitor* (Coleoptera: Tenebrionidae). J. Entomol. Sci. 48, 206–221.

Nation, J.L., 2008. Nutrition. In: Nation, J.L. (Ed.), Insect Physiology and Biochemistry, CRC Press, Boca Raton, FL, pp. 69–90.

Patana, R., 1982. Disposable diet packet for feeding and oviposition of *Lygus hesperus* (Hemiptera: Miridae). J. Econ. Entomol. 75, 668–669.

Pielou, E.C., 1984. The Interpretation of Ecological Data—A Primer on Classification and Ordination. John Wiley & Sons, New York, NY, pp. 263.

Prado, S.S., Almeida, R.P., 2009. Phylogenetic placement of pentatomid stink bug gut symbionts. Curr. Microbio. 58, 64–69.

Quicke, D.L.J., 1997. Parasitic Wasps. Chapman & Hall, London, UK.

Raubenheimer, D., Simpson, S.J., 1993. The geometry of compensatory feeding in the locust. Anim. Behav. 45, 953–964.

Raubenheimer, D., Mayntz, D., Simpson, S.J., Tøft, S., 2007. Nutrient-specific compensation following diapause in a predator: implications for intraguild predation. Ecology 88, 2598–2608.

Reeve, J.D., Rojas, M.G., Morales-Ramos, J.A., 2003. Artificial diet rearing methods for *Thanasimus dubius* (Coleoptera: Cleridae), a predator of bark beetles (Coleoptera: Scolytidae). Biol. Control 27, 315–322.

Riddick, E.W., 2009. Benefit and limitations of factitious prey and artificial diets on life parameters of predatory beetles, bugs, and lacewings: a mini-review. BioControl 54, 325–339.

Roff, D.A., 1992. The Evolution of Life Histories, Theory and Analysis. Chapman & Hall, New York, NY, pp. 535.

Rojas, M.G., Morales-Ramos, J.A., 2008. *Phytoseiulus persimilis* (Mesostigmata: Phytoseiidae) feeding on extrafloral nectar: reproductive impact of sugar sources in presence of prey. Biopestic. Int. 4, 1–5.

Rojas, M.G., Morales-Ramos, J.A., King, E.G., 1996. In vitro rearing of the boll weevil (Coleoptera: Curculionidae) ectoparasitoid *Catolaccus grandis* (Hymenoptera: Pteromalidae) on meridic diets. J. Econ. Entomol. 89, 1095–1104.

Rojas, M.G., Morales-Ramos, J.A., King, E.G., 2000. Two meridic diets for *Perillus bioculatus* (Heteroptera: Pentatomidae), a predator of *Leptinotarsa decemlineata* (Coleoptera: Chrysomelidae). Biol. Control 17, 92–99.

Schiff, N.M., Waldbauer, G.P., Friedman, S., 1988. Dietary self-selection for vitamins and lipids by larvae of the corn earworm *Heliothis zea*. Entomol. Exp. Appl. 46, 240–256.

Schoo, K.L., Aberle, N., Malzahn, A.M., Boersma, M., 2009. Does the nutrient stoichiometry of primary producers affect the secondary consumer *Pleurobrachia pileus*? Aquat. Ecol. http://dx.doi.org/10.1007/s10452-009-9265-4.

Scriber, J.M., Slansky Jr, F., 1981. The nutritional ecology of immature insects. Annu. Rev. Entomol. 26, 183–211.

Shapiro, J.P., 1988. Lipid transport in insects. Annu. Rev. Entomol. 33, 297–318.

Simpson, S.J., 1995. Regulation of meal: chewing insects. In: Chapman, R.F., de Boer, G. (Eds.), Regulatory Mechanisms in Insect Feeding, Chapman & Hall, New York, NY, pp. 137–156.

Singh, P., 1977. Artificial Diets for Insects, Mites, and Spiders. Plenum Press, New York, NY.

Snodgrass, R.E., 1935. Principles of Insect Morphology. McGraw-Hill Book Company Inc., New York, NY, pp. 667.

Soares, A.O., Coderre, D., Schanderl, H., 2004. Dietary self-selection behavior by the adults of the aphidophagous lady beetle *Harmonia axyridis* (Coleoptera: Coccinellidae). J. Anim. Ecol. 73, 478–486.

Specty, O., Febvay, G., Grenier, S., Delobel, B., Piotte, C., Pageaux, J.-F., Ferran, A., Guillaud, J., 2003. Nutritional plasticity of the predatory ladybeetle *Harmonia axyridis* (Coleoptera: Coccinellidae): comparison between natural and substitution prey. Arch. Insect Biochem. Physiol. 52, 81–91.

Sterner, R.W., Elser, J.J., 2002. Ecological Stoichiometry. Princeton University Press, Princeton, NJ.

Stone, T.B., Sims, S.R., 1992. Insect rearing and the development of bioengineered crops. In: Anderson, T.E., Leppla, N.C. (Eds.), Advances in Insect Rearing for Research & Pest Management, Westview Press, Boulder, CO, pp. 33–40.

Tebayashi, S., Kawahara, T., Kim, C., Nishi, A., Takahashi, K., Miyanoshita, A., Horiike, M., 2003. Feeding stimulants eliciting the probing behavior for *Peregrinator biannulipes* Montrouzier et Signore (Hemiptera: Ruduviidae) from *Tribolium confusum* (Jacquelin du Val). Z. Naturforsch. C J. Biosci. 58, 295–299.

Thompson, S.M., 1975. Defined meridic and holidic diets and aseptic feeding procedures for artificially rearing the ectoparasitoid *Exeristes roborator* (Fabricius). Ann. Entomol. Soc. Am. 68, 220–226.

Thompson, S.M., 1980. Artificial culture techniques for rearing larvae of the chalcidoid parasite, *Brachymeria intermedia*. Entomol. Exp. Appl. 27, 133–143.

Thompson, S.M., 1981. *Brachymeria lasus*: culture in vitro of a chalcid insect parasite. Exp. Parasitol. 52, 414–418.

Thompson, S.N., 1999. Nutrition and culture of entomophagous insects. Annu. Rev. Entomol. 44, 561–592.

Thompson, S.N., Hagen, K.S., 1999. Nutrition of entomophagous insects and other arthropods. In: Bellows, T.S., Fisher, T.W. (Eds.), Handbook of Biological Control, Academic Press, San Diego, CA, pp. 594–652.

Trumper, S., Simpson, S.J., 1993. Regulation of salt intake by nymphs of *Locusta migratoria*. J. Insect Physiol. 39, 857–864.

USDA, 2010. USDA National Nutrient Database for Standard Reference, Release 22. http://www.ars.usda.gov/Services/docs.htm?docid=8964.

Vinson, S.B., 1984. Parasitoid-host relationship. In: Bell, W.J., Cardé, R.T. (Eds.), Chemical Ecology of Insects, Sinauer Associates, Inc., Sunderland, MA, pp. 205–233.

Waldbauer, G.P., 1968. The consumption and utilization of food by insects. Adv. Insect Physiol. 5, 229–288.

Waldbauer, G.P., Bhattacharya, A.K., 1973. Self-selection of an optimum diet from a mixture of wheat fractions by the larvae of *Tribolium confusum*. J. Insect Physiol. 19, 407–418.

Waldbauer, G.P., Friedman, S., 1991. Self-selection of optimal diets by insects. Annu. Rev. Entomol. 36, 43–63.

Waldbauer, G.P., Cohen, R.W., Friedman, S., 1984. Self-selection of optimal nutrient mix from defined diets by larvae of the corn earworm *Heliothis zea*. Physiol. Zool. 57, 590–597.

Walter, D.E., Proctor, H.C., 1999. Mites Ecology, Evolution, and Behavior. CABI Publishing, Wallingford, Oxon, UK, pp. 322.

Wiegert, R.G., Petersen, C.E., 1983. Energy transfer in insects. Annu. Rev. Entomol. 28, 455–486.

Wigglesworth, V.B., 1972. The Principles of Insect Physiology, seventh ed. Chapman and Hall, London, UK, pp. 827.

Wittmeyer, J.L., Coudron, T.A., Adams, T., 2001. Reproductive development in the predacious spiner soldier bug, *Podisus maculiventris* (Say) (Heteroptera: Pentatomidae) on artificial diet. Invertebr. Reprod. Dev. 39, 9–20.

Wittmeyer, J.L., Coudron, T.A., 2001. Life table parameters, reproductive rate, intrinsic rate of increase, and estimated cost of rearing *Podisus maculiventris* (Heteroptera: Pentatomidae) on an artificial diet. J. Econ. Entomol. 94, 13–44.

Yocum, G.D., Coudron, T.A., Brandt, S.L., 2006. Differential gene expression in *Perillus bioculatus* nymphs fed a suboptimal artificial diet. J. Insect Physiol. 52, 586–592.

Yazlovetsky, I.G., 1992. Development of artificial diets for entomophagous insects by understanding their nutrition and digestion. In: Anderson, T.E., Leppla, N.C. (Eds.), Advances in Insect Rearing for Research & Pest Management, Westview Press, Boulder, CO, pp. 41–62.

Zapata, R., Specty, O., Grenier, S., Febvay, G., Pageaux, J.F., Delobel, B., Castane, C., 2005. Carcass analysis to improve a meat-based diet for the artificial rearing of the predatory mirid bug *Dicyphus tamaninii*. Arch. Insect Biochem. Physiol. 60, 84–92.

Zhang, Z.-Q., 2003. Mites in Greenhouses Identification, Biology and Control. CABI Publishing, Wallingford, Oxon, UK, pp. 244.

Zhou, J., Han, D., 2006. Safety evaluation of protein of silkworm (*Antheraea pernyi*) pupae. Food Chem. Toxicol. 44, 1123–1130.

Zinke, I., Schütz, C.S., Katzenberger, J.D., Bauer, M., Pankratz, M.J., 2002. Nutrient control of gene expression in *Drosophila*: microarray analysis of starvation and sugar-dependent response. EMBO J. 15, 6162–6173.

Chapter 8

Life Tables as Tools of Evaluation and Quality Control for Arthropod Mass Production

Maribel Portilla,[1] Juan A. Morales-Ramos,[1] M. Guadalupe Rojas[1] and Carlos A. Blanco[2]

[1]*National Biological Control Laboratory, USDA-Agricultural Research Service, Stoneville, MS, USA,* [2]*USDA Animal and Plant Health Inspection Service, Riverdale, MD, USA*

8.1. INTRODUCTION

Arthropods, due to their great impact on human activities, influence on the environment, abundant diversity, and relative ease to rear, have been ideal organisms to study. Arthropod pests, in general, cause economic damage to plants and animals during certain life stages, and the duration of these stages can have great variation in length depending on the species. These damaging stages usually correspond to portions of the life cycle when growth, development, or reproduction exerts great demands on feeding resources. An understanding of arthropod life cycles in relation to the length of the damaging developing periods, when arthropods can have the greatest demand for certain diets, is of the upmost importance for management approaches as well as for rearing methods.

Life tables, as a basic concept, are descriptions of survival potential at various ages or stages. A life table was first applied to study the demographic parameters of a forest caterpillar (Morris and Miller, 1954). Since then, innumerable improvements and applications have been published using arthropods as study organisms (Peterson et al., 2009). Understanding critical life stages of arthropod development and their influence on the population structure is of great importance for arthropod-rearing systems. Their birth rates, age structure at any given moment of time, death rates, and generation times are basic parameters that aid us in understanding arthropod populations and can be obtained by the construction of life tables (Carey et al., 2012). A clear advantage for the use of life tables in an arthropod-rearing facility is that the information obtained allows a complete life history analysis, which provides the possibility of controlling the population structure. Although it is a common practice to maintain a structure of overlapping generations of slightly different ages in a rearing facility,

Mass Production of Beneficial Organisms. http://dx.doi.org/10.1016/B978-0-12-391453-8.00008-X

arthropods born from a limited number of females are usually kept together in cohorts in order to control the population. These subpopulations are ideal for conducting life table studies and understanding which rearing factors have the greatest influence on their intrinsic rate of natural increase (r_m; see description in this chapter) (Eliopoulos, 2006).

Diets and environmental conditions for quality rearing of arthropods have been developed to maximize growth and reproduction by meeting or surpassing their minimum nutritional and environmental requirements. These diets and conditions are the products of intensive research, observations, and trial-and-error efforts. Altering rearing conditions can have potential negative, neutral, or beneficial effects on arthropod development, and these manipulations are usually done with the aim of improving the quality of arthropod-rearing systems and/or to reduce arthropod production costs.

Subtle changes in diets may have effects that are not easily perceived during arthropod rearing, and a few generations may pass before these are apparent. This is critically important, especially when a decline in arthropod production has been the outcome of manipulating rearing diets or conditions. Life table studies are a way of measuring the implications that small, not easily or immediately detectable changes may have on the reproductive capacity and growth or maintenance of an arthropod colony. Life tables are usually based on important parameters such as the capacity of females to produce females and the positive, neutral, or negative population growth index. Another important advantage of a life table is that they summarize multiple life history parameters like immature survival, progeny sex ratio, development time, adult fecundity, and longevity by a single value defined as the intrinsic rate of increase (r_m) (Carey, 1993). The life table provides an overall evaluation of the impact of environmental factors or nutrients in all life history parameters at once. This chapter reviews the basic methodology of life table constructions and analysis with a particular emphasis on real cases of insect studies.

8.2. LIFE TABLE

Extensive knowledge of an insect life cycle under laboratory conditions is an essential prerequisite to colony management (Singh and Moore, 1985). Note that any influence could affect some aspects of the life cycle of an organism, altering for example birth rates and age structure of the population. Estimations of the stability of age distribution and the maximal rate of increase under optimal conditions in arthropod colonies can be determined with a life table. Pressat (1985) defined a life table as a detailed description of the mortality of a population giving the probability of dying and various other statistics at each age. These measures are considered essential tools for ecological studies and also should be considered valuable for insect rearing management. The life span and length of the insect stages, number of insects alive at a given age, and egg production per day are parameters required to be recorded on a regular basis for arthropod quality control and assurance in rearing facilities. This type of information can be directly used in life table construction. Approaches to life table construction are available in the literature providing a general description, organization, and terminology and thereby making the life table an easy tool to use. Historically, insects (as a group) have received the greatest emphasis in examining population change and life table analysis; this is due to their ecological significance in ecosystems, their importance in agriculture and disease transition, and their suitability to be studied as an organism (Preston et al., 2001).

8.2.1. Life Table Construction

Basically, there are three methods to construct life tables: (1) the complete cohort life table (horizontal), which follows a group of same-aged individuals from birth throughout their lives (this type of life table is more commonly used for insects); (2) a static life table (vertical) that is made from data collected from all ages at one particular time, assuming a stable age distribution from generation to generation (this is a less accurate method because it compares population sizes from different cohorts across the entire range of ages); and (3) a life table constructed from mortality data collected from a specified time period (vertical), which also assumes a stable age distribution. The vertical tables can be a more useful technique for large mammals or can answer questions concerning human populations.

This chapter will introduce the basic concepts and methods of complete and abridged cohort life table analysis for insects only. A complete cohort life table is constructed by daily recordings of the number of deaths in an initial cohort of identical individuals at each point in time until all have died: in contrast, an abridged life table deals with age intervals greater than one day (Carey, 1993). The information for both life tables is recorded in nine columns representing the life history budgets of age-specific mortality, which are calculated based on the number of individuals alive at the beginning of each age interval and the number of deaths in the interval (Fig. 8.1). This type of life table was termed a budget and life table by Southwood (1978).

Figure 8.1 presents a hypothetical case of a budget and life table. The first column represents the age class (x), which stars at 0 at the time the egg is oviposited. The number of live individuals at the beginning of age x (N_x) is entered in the second column of the life table. The third column is the fraction living at age x (l_x) and is calculated as N_x/N_0, where N_0 is the number of individuals alive at the start age and N_x is the number of individuals alive at age x. Columns 4 and 5 represent the fraction living (p_x) and fraction dying (q_x), respectively, between age x and age $x+1$, where $p_x=N_{x+1}/N_x$, and $q_x=1-p_x$. The p_x and q_x values can also be interpreted as probabilities of surviving or dying from age x to age $x+1$, respectively (Carey, 1993). Column 6 represents the fraction dying within interval x and $x+1$ (d_x), and it is calculated as $d_x=l_x-l_{x+1}$. The value of d_x differs from that of q_x in that d_x is the proportion of individuals who die during age x, while

1	2	3	4	5	6	7	8	9
x	N_x	l_x	p_x	q_x	d_x	L_x	T_x	e_x
0	500	1	0.9	0.1	0.1	0.95	11.38	11.38
1	450	0.9	0.987	0.013	0.012	0.894	10.43	11.589
2	444	0.888	0.89	0.11	0.098	0.839	9.536	10.739
3	395	0.79	0.901	0.099	0.078	0.751	8.697	11.009
4	356	0.712	0.902	0.098	0.07	0.677	7.946	11.16
5	321	0.642	0.96	0.04	0.026	0.629	7.269	11.322
6	308	0.616	0.906	0.094	0.058	0.587	6.64	10.779
7	279	0.558	0.932	0.068	0.038	0.539	6.053	10.848
8	260	0.52	0.927	0.073	0.038	0.501	5.514	10.604
9	241	0.482	0.934	0.066	0.032	0.466	5.013	10.4
10	225	0.45	0.956	0.044	0.02	0.44	4.547	10.104
11	215	0.43	0.953	0.047	0.02	0.42	4.107	9.551
12	205	0.41	0.98	0.02	0.008	0.406	3.687	8.993
13	201	0.402	0.985	0.015	0.006	0.399	3.281	8.162
14	198	0.396	0.985	0.015	0.006	0.393	2.882	7.278
15	195	0.39	0.923	0.077	0.03	0.375	2.489	6.382
16	180	0.36	0.956	0.044	0.016	0.352	2.114	5.872
17	172	0.344	0.93	0.07	0.024	0.332	1.762	5.122
18	160	0.32	0.888	0.113	0.036	0.302	1.43	4.469
19	142	0.284	0.845	0.155	0.044	0.262	1.128	3.972
20	120	0.24	0.842	0.158	0.038	0.221	0.866	3.608
21	101	0.202	0.891	0.109	0.022	0.191	0.645	3.193
22	90	0.18	0.833	0.167	0.03	0.165	0.454	2.522
23	75	0.15	0.533	0.467	0.07	0.115	0.289	1.927
24	40	0.08	0.8	0.2	0.016	0.072	0.174	2.175
25	32	0.064	0.656	0.344	0.022	0.053	0.102	1.594
26	21	0.042	0.429	0.571	0.024	0.03	0.049	1.167
27	9	0.018	0.444	0.556	0.01	0.013	0.019	1.056
28	4	0.008	0.25	0.75	0.006	0.005	0.006	0.75
29	1	0.002	0	1	0.002	0.001	0.001	0.5
30	0	0						

FIGURE 8.1 Example of a hypothetical budget and life table. Columns: (1) age (x); (2) number alive at age x (N_x); (3) fraction living at age x (l_x); (4) fraction living between ages x and $x+1$ (p_x); (5) fraction dying between ages x and $x+1$ (q_x); (6) fraction dying within interval x (d_x); (7) number of days lived in interval x (L_x); (8) total days to be lived beyond age x (T_x); and (9) life expectancy (e_x). (For color version of this figure, the reader is referred to the online version of this book.)

q_x is the probability of an individual dying during age x (Carey, 1993). Column 7 is the number of days lived in an age interval (L_x); assuming that an individual who dies within an age interval does so at midpoint, then $L_x = (l_x + l_{x+1})/2$. Column 8 is the total days to be lived beyond age x by an individual of this age (T_x), and it is calculated as:

$$T_x = \sum_{y=x}^{w} Ly \tag{8.1}$$

where w is the oldest age (Carey, 1993). Column 9 is the life expectancy (e_x), which represents the expectation of life at age x, and it is calculated as $e_x = T_x/l_x$. Carey and Judge (2001) introduced a way to calculate e_x, directly eliminating columns 6 and 7, as:

$$e_x = \frac{1}{2} + \frac{\sum_{y}^{w} -{}_{w} Ly}{l_x} \tag{8.2}$$

Information on birth rates and progeny sex ratios (recorded in nine columns) is used to construct a reproductive budget or fecundity schedule. This type of life table is termed a life and fertility table (Southwood, 1978) (Fig. 8.2). Information on life and fertility tables is used to estimate the demographic parameters that represent population growth rates (Carey, 1993) (see Appendix III). The construction of life and fertility tables is similar to that of budget and life tables, but in this case the reproduction budget is emphasized. Life and fertility tables generally consist of nine columns, but the number of columns may change depending on specific applications. In this chapter, we will focus on the format of Carey (1993).

Figure 8.2 presents a hypothetical case of a life and fertility table. The first three columns of the life and fertility table are the same as those in the budget and life table: age class x, number of females alive at the middle of age x (N_x), and fraction alive at age x (l_x). The number of eggs oviposited by the cohort of age x (E_x) is entered in column 4. Column 5 is the gross fecundity (M_x), which is defined as the total progeny produced by a female alive at the middle of age x and is calculated as $M_x = E_x/N_x$.

Column 6 (Fig. 8.2) is net fecundity (m_x), defined as the female progeny produced by a female of age x. Net fecundity becomes relevant in species with arrhenotoky (Gauld and Bolton, 1996) or other types of parthenogenesis, in which the sex ratio of the progeny can change in response to environmental factors (Quicke, 1997). For instance, the progeny sex ratio of *Catolaccus grandis* (Burks) was significantly more male biased when reared in some artificial diet formulations (Rojas et al., 1996). The progeny sex ratio changes with female age in arrhenotokous parasitoids, especially in single-mating species. For instance, the boll weevil (*Anthonomus grandis* Boheman) parasitoids *C. grandis* and *Heterospilus megalopus* Marsh produce female-biased progeny at the onset of fecundity, but the progeny sex ratio becomes more biased toward males as females aged (Morales-Ramos and Cate, 1992, 1993a). Therefore, it is

1	2	3	4	5	6	7	8	9
x	N_x	l_x	E_x	M_x	m_x	l_xm_x	xl_xm_x	r_x
0	500	1	0	0	0	0	0	0
1	450	0.9	0	0	0	0	0	0
2	444	0.888	0	0	0	0	0	0
3	395	0.79	0	0	0	0	0	0
4	356	0.712	0	0	0	0	0	0
5	321	0.642	0	0	0	0	0	0
6	308	0.616	0	0	0	0	0	0
7	279	0.558	0	0	0	0	0	0
8	260	0.52	0	0	0	0	0	0
9	241	0.482	0	0	0	0	0	0
10	225	0.45	0	0	0	0	0	0
11	215	0.43	0	0	0	0	0	0
12	205	0.41	0	0	0	0	0	0
13	201	0.402	0	0	0	0	0	0
14	198	0.396	0	0	0	0	0	0
15	195	0.39	50	0.256	0.128	0.05	0.75	0.00382
16	180	0.36	1005	5.583	2.792	1.005	16.08	0.06463
17	172	0.344	2020	11.744	5.872	2.02	34.34	0.10943
18	160	0.32	3782	23.638	11.819	3.782	68.076	0.17259
19	142	0.284	4952	34.873	17.437	4.952	94.088	0.19036
20	120	0.24	5045	42.042	21.021	5.045	100.9	0.16337
21	101	0.202	4987	49.376	24.688	4.987	104.727	0.13604
22	90	0.18	3230	35.889	17.944	3.23	71.06	0.07423
23	75	0.15	2500	33.333	16.667	2.5	57.5	0.04840
24	40	0.08	1252	31.300	15.650	1.252	30.048	0.02042
25	32	0.064	854	26.688	13.344	0.854	21.35	0.01173
26	21	0.042	328	15.619	7.810	0.328	8.528	0.00380
27	9	0.018	99	11.000	5.500	0.099	2.673	0.00097
28	4	0.008	26	6.500	3.250	0.026	0.728	0.00021
29	1	0.002	3	3.000	1.500	0.003	0.087	0.00002
30	0	0						
					GRR	R_o	$\Sigma\, xl_xm_x$	
Sum					165.421	30.133	610.935	1

FIGURE 8.2 Example of a hypothetical life and fertility table. Columns: (1) age (x); (2) number alive at age x (N_x); (3) fraction living at age x (l_x); (4) total eggs produced by a cohort of age x (E_x); (5) gross fecundity = eggs produced per female of age x (M_x); (6) net fecundity = number of female progeny produced by a female of age x (m_x); (7) fecundity function (l_xm_x); (8) age and fecundity function (xl_xm_x); and (9) age-specific increase rate fraction (r_x). (For color version of this figure, the reader is referred to the online version of this book.)

recommended to determine and record the age-dependent progeny sex ratio in such species in order to calculate their accurate m_x values.

The fecundity function, calculated as the product of l_x and m_x (l_xm_x), is entered in column 7 of the life and fertility table (Fig. 8.2). The sum of l_xm_x across all ages is the net reproductive rate (R_o), which will be explained in Section 8.2.2. Column 8 (Fig. 8.2) contains the age and fecundity function calculated by multiplying l_x, m_x, and x (xl_xm_x). This age and fecundity function is used in column 9 to calculate the age-specific increase rate (r_x) fraction using Eqn (8.3).

$$r_x = e^{-r_m x} l_x m_x \tag{8.3}$$

The sum of these fractions must total 1, and this is used to calculate the intrinsic rate of increase (r_m). The values of r_x are dependent on the r_m value. The sum of this column is used to determine the true value of r_m by iteration, and this method will be fully explained in Section 8.2.2.

8.2.2. Demographic Parameters and Life History Analysis

Demographic parameters measure characteristics of the population, such as mortality and survival rates, population increase rates, and doubling time; and biological parameters measure characteristics of individuals, such as egg viability, development time, fecundity, and longevity (Krebs, 2001). The most useful demographic parameters for population comparisons include the (1) net reproductive rate (R_o), (2) mean generation time (T), (3) intrinsic rate of increase (r_m), (4) finite rate of increase (λ), and (5) doubling time (DT).

Net reproductive rate (R_o). This is defined as the mean female progeny produced by a female during its life span. The net reproductive rate is calculated by the sum of the maternity function over all age classes (Eqn (8.4)), which is the sum of column 6 in the life and fertility table (Fig. 8.2).

$$R_o = \sum_{x=0}^{w} l_x m_x \tag{8.4}$$

where w is the oldest age. The R_o value can be useful as an evaluation tool because it measures net fecundity (Krebs, 2001). Net fecundity is usually equivalent to 50% of the gross fecundity in sexually reproducing diploid species. But, as was discussed in Section 8.2.1, in parthenogenetic species the value of net fecundity relative to gross fecundity changes with age and can be impacted by environmental factors due to changes in the progeny sex ratio (Quicke, 1997). Then, in the case of parthenogenetic species, the R_o value may provide a quicker way to compare two populations exposed to different conditions of temperature, humidity, and so on, or feeding on different diets.

Mean generation time (T). This is defined as the mean age of reproduction (Carey, 1993), which characterizes T as the mean interval separating the births of one generation from those of the next generation (Pressat, 1985). Mean generation time is also defined as the mean period elapsing between the birth of parents and the birth of their offspring (Krebs, 2001). Mean generation time is calculated by dividing the sum of xl_xm_x by R_o (Eqn (8.5)), which is the sum of column 7 divided by the sum of column 6 of the life and fertility table (Fig. 8.2).

$$T = \frac{\sum_{x=0}^{w} x l_x m_x}{R_o} \tag{8.5}$$

The value of T is mostly impacted by development time, but the length of the preovipositional period may have some impact. For instance, the development time and preovipositional period of *Perillus bioculatus* (F.) females feeding on an artificial diet were both significantly longer than those of females feeding on eggs of *Leptinotarsa decemlineata* (Say) (Rojas et al., 2000). In the boll weevil parasitoid *C. grandis*, temperature impacted both development time (Morales-Ramos and Cate, 1993b) and preovipositional period, which resulted in significant differences in generation time (Morales-Ramos and Cate, 1992).

The intrinsic rate of increase (r_m). This is the rate of natural increase in a closed population that has been subject to a constant age-specific schedule of fertility and mortality for a long period, and has converged to be a stable population (Carey, 1993). A stable population is defined as a population with a stable age distribution (i.e. with stable proportions of all age classes) (Carey, 1993). Populations of organisms living in the wild do not often attain stable age distributions due to the constant changes in mortality factors and food availability (Southwood, 1978). Populations under mass production conditions of artificially controlled temperature, humidity, and photoperiod, and continuous provisioning of food, have a much greater probability of attaining a stable age distribution. This fact makes the demographic parameters more meaningful and useful for the evaluation of populations under controlled environments.

The intrinsic rate of increase is considered the most important of the demographic parameters by some authors. Roff (1992) considers r_m a measure of fitness of a population. Intrinsic rate of increase is basically a special case of a crude growth rate, and it was determined by analytical approximation using Lotka's (1907, 1913) equation, which is presented as Eqn (8.6) in this chapter.

$$\sum_{x=0}^{w} e^{-r_m x} l_x m_x = 1 \tag{8.6}$$

This equation cannot be solved explicitly for r_m (Krebs, 2001) and requires the use of iteration and trial and error. This is accomplished by modifying the r_m value and substituting it in Eqn (8.3); this is used in the calculation of column 9 of the life and fertility table presented in Section 8.2.1. The sum of this column, represented by Eqn (8.6), must be equal to 1 in order to find the true value of r_m. Iterations can be done manually using a spreadsheet template or may also be done automatically by an iteration routine using a programming language. Because the Lotka equation (Eqn (8.6)) is a function of x, l_x, and m_x, all of the main biological parameters, including development time, immature survival, adult longevity, fecundity, and the progeny sex ratio, are summarized in the r_m value. This fact makes r_m extremely important for evaluating artificial diets. Nutrition can impact each biological parameter differently, affecting some parameters favorably while detrimentally impacting others at the same time (J. A. Morales-Ramos, unpublished). Artificial diet evaluation becomes less complicated if all the biological parameters can be simultaneously evaluated with one single value.

Finite rate of increase (λ). This is the factor by which a population will increase each x period of time (Krebs, 1972). The finite rate of increase, as it is calculated from the life and fertility table, is dependent on the period of time represented by x (e.g. days or weeks), and it is a function of r_m (Eqn (8.7)).

$$\lambda = e^{r_m} \tag{8.7}$$

The finite rate of increase is useful for modeling population growth. The population size after one unit of time x is determined by $N_{x+1} = N_x \lambda$, assuming a stable age distribution. This model is largely dependent on the units used in age x within the life tables, and each iteration of the model will represent a change in population during a period of time of magnitude x (Krebs, 2001).

The doubling time (DT). This is defined as the period of time required to double the population in size (Carey, 1993). DT is expressed in the same time units as those used in the life and fertility table for x, because DT is a function of r_m (Eqn (8.8)).

$$\mathrm{DT} = \frac{\ln(2)}{r_m} \tag{8.8}$$

DT is essentially a transformation of r_m and provides a value that is easier to relate to mass production. Comparisons of DT among populations can provide a powerful tool for artificial diet evaluation or for determination of optimal rearing conditions.

8.3. CASE STUDIES

This section gives examples of cohort (horizontal) life table constructions for three different species of insects. Calculations were based on formulae and definitions from Keyfitz (1964), Roughgarden (1979), Pressat (1985), Roff (1992), Carey (1993), Carey and Judge (2001), and Krebs (1978, 2001). The examples presented in this section are based on real data. The data are available in MS Excel spreadsheets that are downloadable from the website (http://booksite.elsevier.com/9780123914538). The spreadsheets calculation presented in this chapter can be used as a model for any arthropod-rearing system, for either pests or biological control agents. A generic spreadsheet to calculate life and fertility tables using the equations presented in this chapter is also available on the website.

8.3.1. Complete Cohort Life Table of the Southern Green Stink Bug, *Nezara viridula* (L.) (Heteroptera: Pentatomidae)

Two artificial diets, a fresh chicken egg yolk-based diet and dry chicken egg yolk-based diet, were evaluated as alternative food sources for rearing the southern green stink bug (SGSB) (Portilla et al., unpublished data). Both fresh- and dry-yolk diets led to survival to adult stages. However, only the diet that produced females and males with longer longevity, higher fecundity, and larger egg mass sizes was used for this example.

Population size of initial cohort: In this example, the initial cohort was 61 eggs of an identical age (fertilized eggs) from a single SGSB female reared on an artificial diet. The total number of individuals that reached the beginning of each stage over the entire generation was determined, providing a longitudinal perspective of the mortality of the original cohort of 61 individuals from the moment of birth thorough consecutive ages until no individual remained alive. The population size at each stage and the number of eggs produced by the adults were daily recorded. In the online data example "Excel file case 1" (column 2, Life & Budget tab), Age Class (Days), x, is denoted as a given age group in units of time within the entire population lapse, from eggs until of all the individuals die. This might be expressed in days, weeks, or years depending on the life span of the organism, or it may be expressed as stages in the life cycle, such as in insects.

In Excel file case 1 (column 2, Life & Budget tab), "From 0 to 128 days" designates the exact age of *N. viridula* from birth to death. "Age Interval (Days)" is denoted as a given age group in units of time within the population, from eggs until all individuals died, including the interval from exact age x to exact age $x+1$. In this example from Excel file case 1 (column 3, Life & Budget tab):

$$X_0 \text{ to } X_0+1=0-1$$

$$X_{10} \text{ to } X_{10}+1=10-11$$

$$X_{100} \text{ to } X_{100}+1=100-111$$

Alive at beginning of interval (N_x): Number of individuals alive at the beginning of each age interval from the exact age x to exact age $x+1$. Alive female and male adult stages could be recorded separately, denoted as N_{fx} and N_{mx}; for example (from Excel file case 1, columns 3–5, Life & Budget tab):

$$N_0=61$$

$$N_{10}=59$$

$$N_{100}=21$$

$$N_{f0} = 0$$

$$N_{f10} = 0$$

$$N_{f100} = 10$$

$$N_{m0} = 0$$

$$N_{m10} = 0$$

$$N_{m100} = 11$$

Complete life table parameters. Survival l_x is designated as the proportion of a cohort surviving from birth to exact age x. The parameter can be found by dividing the number of individuals living at the beginning of each age (x) by the initial number of eggs. For example (Excel file case 1, column 6, Life & Budget tab):

$$l_0 = N_0\,(100)/N_0 = 61\,(100)/6 = 1$$

$$l_{10} = N_{10}\,(100)/N_0 = 59\,(100)/61 = 0.967$$

$$l_{100} = N_{100}\,(100)/N_0 = 21\,(100)/61 = 0.344$$

Survival in interval x (p_x). This is the fraction of a cohort surviving in the interval from age x to age $x+1$. For example (Excel file case 1, column 7, Life & Budget tab):

$$P_0 = l_1/l_0 = 1/1 = 1$$

$$P_{10} = l_{11}/l_{10} = 0.967/0.967 = 1$$

$$P_{100} = l_{101}/l_{100} = 0.344/0.344 = 1$$

Death (q_x). This is denoted as the proportion of the original cohort dying in the interval from age x to age $x+1$, and it is found by subtracting p_x from the value of 1. For example (Excel file case 1, column 8, Life & Budget tab):

$$q_0 = 1 - p_0 = 1 - 1 = 0$$

$$q_{10} = 1 - p_{10} = 1 - 1 = 0$$

$$q_{100} = 1 - p_{100} = 1 - 1 = 0$$

Death in interval x (d_x). This is denoted as the fraction of a cohort dying in the interval from age x to age $x+1$. It is found by subtracting l_{x+1} from l_x. For example (Excel file case 1, column 9, Life & Budget tab):

$$d_0 = l_0 - l_1 = 1.000 - 1.000 = 0$$

$$d_{10} = l_{10} - l_{11} = 0.967 - 0.967 = 0$$

$$d_{100} = l_{100} - l_{101} = 0.344 - 0.344 = 0$$

Per capita survival in interval x (L_x). This is denoted as the fraction of the interval that is alive in the interval from age x to age $x+1$. For example (Excel file case 1, column 10, Life & Budget tab):

$$L_0 = l_0 - (1/2)\,d_0 = 1 - (1/2)\,0 = 1$$

$$L_{10} = l_{10} - (1/2)\,d_{10} = 0.967 - (1/2)\,0 = 0.967$$

$$L_{100} = l_{100} - (1/2)\,d_{100} = 0.344 - (1/2)\,0 = 0.344$$

Total survival at age x (T_x). This is an essential calculation to estimate the life expectancy, which is the number of insect-days lived by the cohort at age x and beyond. By substituting in Eqn (8.1):

$$T_0 = \sum_{y=0}^{127} L_y$$

$$T_{10} = \sum_{y=10}^{127} L_y$$

$$T_{100} = \sum_{y=100}^{127} L_y$$

For example (Excel file case 1, column 11, Life & Budget tab):

$$T_0 = L_0 + L_1 + \cdots + L_{127} + L_{127} = 85.041$$

$$T_{10} = L_{10} + L_{11} + \cdots + L_{127} + L_{128} = 75.221$$

$$T_{100} = L_{100} + L_{101} + \cdots + L_{127} + L_{128} = 5.664$$

Life expectancy at age x (e_x). This is the average amount of time yet to be lived by the individuals surviving to a particular age. For example (Excel file case 1, column 12, Life & Budget tab):

$$e_0 = T_0/l_0 = 85.041/1 = 85.041$$

$$e_{10} = T_{10}/l_{10} = 75.221/0.967 = 77.771$$

$$e_{100} = T_{100}/l_{100} = 5.664/0.344 = 16.454$$

Demographic parameters calculated from life tables. The demographic parameters usually estimated from life tables are described in this chapter and after the calculation of each parameter is represented as a single value. To compare those demographic parameters, it is necessary to have information on the degree of uncertainty associated with its estimates, expressed as their variances. As mentioned by Maia et al. (2000), variance can be calculated by the Monte Carlo methods, such as bootstrap, jackknife, and randomization tests. Detailed information on those methods can be found in Manly (1997). Most of the estimates of associated life table parameters are reported in the literature without any measure of improbability, because the r_m estimation using the iterative method is a computer-intensive technique itself, and adding another computer-intensive technique to calculate variances makes these studies more complicated. Moreover, it is important to mention that methods such as the jackknife method are not appropriate for samples with small survival rates or when surviving individuals have little or no oviposition or only a few individuals with very high oviposition. Such cases tend to produce negative values of the jackknife estimates of r_m.

An easier and more practical way than using a jackknife method to calculate variances of demographic parameters that are associated with a complete cohort life table with high survival is by creating groups as a replicate. For example, three groups of nine females and males of *N. viridula* were created since 27 adult females and 28 males survived at the end of the cycle. Each group (replication) was calculated separately (see Excel file case 1).

Gross fecundity (M_x). This is defined as the average number of total offspring per female in the interval x. Substituting in the formula $M_x = E_x/N_x$ in Excel file case 1 (column 5, Rep 1–3 tabs), we obtain:

$$M_{50} = E_{50}/N_{50} = 97/9 = 10.778$$

$$M_{60} = E_{60}/N_{60} = 74/3 = 24.667$$

$$M_{70} = E_{70}/N_{70} = 50/2 = 25.0$$

Net fecundity (m_x). As explained in Section 8.2.1, net fecundity represents the number of female progeny produced by a female of age x, which is obtained by multiplying gross fecundity by the female proportion of sex ratio (PSR). In our example (Excel file case 1, column 6, Rep 1–3 tabs), the PSR is 0.48 (27 females:28 males). Substituting for the three replicates:

Replicate 1

$$m_{50} = 10.778\,(0.48) = 5.389$$

$$m_{60} = 24.667\,(0.48) = 11.84$$

$$m_{70} = 25.0\,(0.48) = 12.0$$

Replicate 2

$$m_{50} = 0\,(0.48) = 0$$

$$m_{60} = 9.111\,(0.48) = 4.373$$

$$m_{70} = 0\,(0.48) = 0$$

Replicate 3

$$m_{50} = 0\,(0.48) = 0$$

$$m_{60} = 8.778\,(0.48) = 4.213$$

$$m_{70} = 8.667\,(0.48) = 4.16$$

The maternity function is the product of net fecundity and survival. For example (Excel file case 1, column 7, Rep 1–3 tabs):

Replicate 1

$$l_0 m_0 = 4.759$$

$$l_{10} m_{10} = 3.631$$

$$l_{100} m_{100} = 2.453$$

Replicate 2

$$l_0 m_0 = 0$$

$$l_{10} m_{10} = 4.023$$

$$l_{100} m_{100} = 0$$

Replicate 3

$$l_0 m_0 = 0$$

$$l_{10} m_{10} = 3.876$$

$$l_{100} m_{100} = 3.827$$

Net reproductive rate (R_o). This is the sum of maternity function over all age classes, and it is defined as the average number of offspring left by each female during her lifetime. Substituting the data of column 7 into Eqn (8.4) for replicates 1, 2, and 3, we obtain:

$$R_o = 62.412\ (\text{replicate}\ \ 1)$$

$$R_o = 132.627\ (\text{replicate}\ \ 2)$$

$$R_o = 169.770\ (\text{replicate}\ \ 3)$$

The intrinsic rate of increase (r_m). The r_m values for the three replicates were obtained by iterating and changing the values of r_m in Eqn (8.3) in column 9 (Excel file case 1, Rep 1–3 tabs) until the sum produced the value of 1. When the sum of column 9 was equal to 1 (Eqn (8.6)), the value of r_m that was last used was identified as the true r_m value. These values were 0.0701, 0.0766, and 0.0773 for replicates 1, 2, and 3, respectively.

Finite rate of increase (λ). This is the factor by which a population will increase each day, and it is related to the life table schedule in the equation (Krebs, 1972). By substituting the r_m values obtained for the three replicas into Eqn (8.7), we obtain:

$$\lambda = e^{0.0701} = 1.072$$

$$\lambda = e^{0.0766} = 1.079$$

$$\lambda = e^{0.0773} = 1.080$$

Mean generation time (T) is the mean age of reproduction, which characterizes T as the mean interval separating the births of one generation from those of the next generation (Pressat, 1985). Substituting the sum of column 8 (Excel file case 1, Rep 1–3 tabs) and the obtained R_o values for replicates 1, 2, and 3 into Eqn (8.5), we obtain:

$$T = 3785.984/62.412 = 60.660$$

$$T = 9017.325/132.627 = 67.990$$

$$T = 12159.010/169.770 = 71.020$$

The doubling time. DT is calculated using the following formula (Krebs, 1972), which is the expression for geometrical increase. By substituting the r_m values obtained for replicates 1, 2, and 3 into Eqn (8.8), we obtain:

$$\text{DT} = (\log_e 2)/0.0701 = 9.88$$

$$\text{DT} = (\log_e 2)/0.0766 = 9.04$$

$$\text{DT} = (\log_e 2)/0.0773 = 8.96$$

Taking the values of the third replicate, the general interpretation of this calculation is that *N. viridula* reared in this specific diet has a net reproductive rate (R_o) of 169.770 females per newborn female per generation in a mean time (T) of 71.020 days. If the intrinsic rate of increase (r_m) is 0.077, then the finite rate of increase (λ) is 1.080 individuals per day per female. Therefore, for every female present this day, 1.080 individuals will be present on the next day, and the population will double in 8.96 days (i.e. the DT).

As we mentioned in this chapter, a variety of factors could affect the life cycle of any organism, altering the birth rates and age structure of the population. This alteration can also be observed among individuals of the same species under the same rearing conditions. For example, if we compared replicate 1 against replicate 3, the growth rates were lower in replicate 1, showing that *N. viridura* in group 1 has a net reproductive rate (R_0) of 62.412 females per newborn female per generation in a mean time (T) of 60.660 days, with an intrinsic rate of increase (r_m) of 0.0701 giving a finite rate of increase (λ) of 1.072 individuals per day per female and doubling its population in 9.88 days. Thus, if 1000 individuals of *N. viridula* were present in a colony this day, 1072 would be present on the next day in group 1 and 1080 in group 3. The alteration on birth rates in insect colonies could be more noticeable when physical conditions vary, such as temperature, relative humidity, diet components, or even texture of the diet. Studies on the effects on demographic parameters of arthropods reared under different conditions can be found in the literature. For example, the r_m of *Hypothenemus hampei* (Ferrari) (Coleoptera: Scolytidae) could vary from 0.018 to 0.071 depending on the moisture of the diet content (Portilla et al., 2000); the r_m value in *Helicoverpa* (Heliothis) *virescens* (F.) (Lepidoptera: Noctuidae) ranged from 0.100 to 0.206 when the concentration of protein in the diet varied from 2.15% to 2.51% (Blanco et al., 2009); and the r_m of the parasitoid *Cephalonomia stephanoderis* Betrem (Hymenoptera: Bethylidae) varied at different temperatures, from 0.089 at 23°C to 0.106 at 29°C (Portilla and Streett, 2008). Terán-Vargas et al. (1990) obtained r_m values that ranged from 0.0612 to 0.101 for *Anticarsia gemmatalis* Hübner (Lepidoptera: Noctuidae) using different varieties of soybeans. In general, subtle changes may have effects that are not easily perceived during arthropod rearing, and several generations may pass before these are apparent.

8.3.2. Abridged Cohort Life Table for the CBB, *Hypothenemus hampei*, at Different Temperatures Using an Artificial Diet

The abridged life table organizes age groupings into larger intervals, enabling users to summarize the information concisely while still retaining the basic life table format and concepts. Moreover, cohorts can be used independently for growth rate calculations. The results from each cohort can be incorporated in the parameter demographic estimation. The internal reproductive behavior of some insects, like in our example of the coffee berry borer (CBB), makes it difficult to monitor the daily mortality of individuals in a cohort over their entire life to determine the precise time of death in all stages; in this case, once the diet pellet was open to determine cohort mortality for the stages, the population housed inside the diet pellet was not usable for further observations. Therefore, a practical solution for insects with internal reproduction is using destructive cohorts. In our example, the population sizes of each initial synthetic cohort were different depending on availability and the purpose of the study. Each evaluated diet pellet was cut open and discarded.

Life span of insect from birth to adult and adult preoviposition. This study was conducted under laboratory conditions using four constant temperatures: 23, 25, 27, and 29°C ± 1. Eggs oviposited recently by CBB were collected from colonies maintained on a Cenibroca diet under laboratory conditions (Portilla, 1999a). To obtain the life span of CBB from birth to adult, 500 pellets of diet per treatment were individually inoculated with recently laid CBB egg and held inside the dark environment chambers at 85% relative humidity (RH) at each selected temperature until the last adult was obtained and initiated its first oviposition. Four infested diet pellets were observed and discarded every day at each temperature. The developmental times for eggs, larval, and pupal stages of CBB were calculated.

GLM analysis showed significant differences in developmental time between temperatures (Table 8.1). The developmental time from egg to adult of CBB at 23°C was significantly longer than at 25°C, 27°C, and 29°C. The overall GLM analysis showed that temperature significantly affected the developmental time in each CBB immature stage. Eggs hatched significantly earlier at 29°C, which was significantly different from the hatching times at 27° C and 25°C ($p<0.05$) and at 23°C ($p<0.01$). The developmental times for the larval, prepupal, and pupal stages also varied significantly between treatments, and they developed faster at high temperatures. The median duration of preoviposition time varied from 8.6 + SD = 0.9 at 23°C to 3.6 + 0.58 at 29°C (Table 8.1).

Gross fecundity and sex ratio. Newly emerged adult females of CBB obtained from first-generation colonies maintained in the laboratory were used to infest the diet. Two thousand four hundred diet pellets (600 per treatment) were evaluated. These were placed at a constant temperature in

TABLE 8.1 Mean Developmental Times of the Immature Stages of *Hypotenemus hampei* at Four Temperatures

	Developmental Times (Days) (Means ± Standard Deviation (SD))			
Life Stages	**23°C**	**25°C**	**27°C**	**29°C**
Egg	6.16 ± 1.07a	5.76 ± 0.74ab	5.46 ± 1.03b	3.30 ± 0.64c
Larva	15.41 ± 1.59a	14.3 ± 1.68b	12.18 ± 1.03c	10.70 ± 2.15d
Prepupa	1.95 ± 0.39a	1.48 ± 0.50b	1.23 ± 0.42c	1.00 ± 0d
Pupa	7.36 ± 0.99a	6.41 ± 0.56b	5.76 ± 0.83c	5.11 ± 0.69d
Total time to adult	30.8 ± 2.41a	27.98 ± 1.97b	24.65 ± 1.86c	20.11 ± 2.55d
Preoviposition period	8.6 ± 0.9a	5.9 ± 0.76b	5.3 ± 1.09b	3.6 ± 0.58c

Means ± SD followed by the same letter in each row are not significantly different ($p<0.05$ Tukey test).

environmental chambers set at 23, 25, 27, and 29°C ± 1 and 85% RH during 40 days under total darkness. Every 2 days, 25 pellets were randomly selected, cut open, and discarded from each treatment, and their contents counted and recorded.

The highest values of gross fecundity M_x were found at 25°C and 27°C (see Case 2 Excel file) and were due mainly to 30% of their population being eggs at this time, which were probably oviposited by the second generation. This means that at 29°C, CBB might reach an equal or higher population than at 25°C and 27°C; however, at 29°C, the loss of diet moisture content was faster than at the lower temperatures. Thus, the number of CBB ovipositing under these conditions was low. The sex ratio of adult progeny was about 10 females to 1 male. The GLM analysis was not significantly affected at any temperature in the mean number of adult females and males found per CBB female and male, respectively. The number of eggs per clutch was similar for each temperature, and in most samples only one clutch per CBB was found, explaining the results about the lack of significant difference in the mean number of adults collected. Adult mortality was not considered in this study as it was very low or zero in all samples, suggesting a survival (l_x) of 1 for all temperatures.

Fertility and mortality of immature stages. Diet colonization was started by introducing 1400 recently laid CBB eggs obtained from the first generation. Thirty-five infested diet pellets with 10 eggs each were observed every day, and these samples were evaluated until the last first instar was obtained. The ratio of total collected eggs divided by total hatched eggs indicated the level of fertility. Samples of 100-unit size were used to calculate the mortality of immature stages. Ten groups (10 stages per group) of each immature stage with identical ages (first instar, second instar, prepupae, and pupae) were held at each temperature and observed until each stage changed to the next stage. The number of dead stages found per group in each treatment gave the percentage of mortality for the immature stages.

Based on appearance, total mortality from eggs to adult was not affected by temperature: 46.4, 45.5, 47.8, and 45.4 at 23, 25, 27, and 29°C, respectively. The highest mortality was observed during the first stadium, with rates of 17, 13, 14, and 16% at 23, 25, 27, and 29°C, respectively; and the lowest mortality was observed during the egg and prepupae stages. Fertility also was not affected by the temperature, with rates of 94.6%, 93.5%, 93.2%, and 94.6% at 23, 25, 27, and 29°C, respectively. These values were calculated from the ratio of the total number of eggs at the beginning of the cohort divided by the number of viable eggs.

Demographic parameters. Calculation of demographic parameters was based on the formulae and procedures described here. Adult survival (l_x) was 1 for all temperatures. Data on developmental stage mortality are noted in Table 8.1, and immature mortality was taken for estimating the survival at age x from adult emergence to death. From the number of eggs counted per sample, the total oviposition was calculated. From the numbers of eggs counted in the second sample (two after infestation), the number of eggs counted in the previous sample was subtracted, and the difference was assumed as m_x. This parameter was estimated assuming a sex ratio of CBB of 10 females to 1 male.

Survival (l_x). Adult survival of CBB was 1 for all temperatures because no mortality was recorded; however, immature mortality was incorporated. For example (Excel file case 2, column 3):

$$(23°C)\ l_{30} = 25\ (0.611)/25 = 0.611$$

$$(25°C)\ l_{28} = 25\ (0.619)/25 = 0.619$$

$$(27°C)\ l_{24} = 25\ (0.603)/25 = 0.603$$

$$(29°C)\ l_{20} = 25\ (0.619)/25 = 0.619$$

Gross fecundity (M_x). Eggs per female of CBB that were counted in the second sample were subtracted from the number of eggs counted in the previous sample.

$$M_x = E_x/N_x$$

Example (Excel file case 2, column 5):

$$(23°C)\ M_{50} = 11.68$$

$$(25°C)\ M_{50} = 2.88$$

$$(27°C)\ M_{50} = 1.72$$

$$(29°C)\ M_{50} = 3.64$$

Net fecundity (m_x). This is the number of CBB females/females at age x, and it was obtained by multiplying the gross fecundity by the female proportion of the sex ratio obtained in Table 8.4 (10 female and 1 male). In our example, 0.9 is for all temperatures.

$$m_x = M_x\ (0.9)$$

For example (Excel file case 2, column 6):

$$(23°C)\ m_{50} = 10.51$$

$$(25°C)\ m_{50} = 2.59$$

$$(27°C)\ m_{50} = 1.54$$

$$(29°C)\ m_{50} = 3.27$$

Substituting the fecundity function $l_x m_x$ from the example in Excel file case 2 (column 7):

$$(23°C)\, l_{50}m_{50} = (0.611)(10.51 = 6.42$$

$$(25°C)\, l_{50}m_{50} = (0.619)(2.592 = 1.60$$

$$(27°C)\, l_{50}m_{50} = (0.603)(1.548 = 0.93$$

$$(29°C)\, l_{50}m_{50} = (0.619)(3.276 = 2.03$$

Net reproductive rate (R_o). Substituting the data of column 7 (Excel file case 2) in Eqn (8.4):

$$R_{o,23} = \sum_{x=0}^{70} l_x m_x = 13.97$$

$$R_{o,25} = \sum_{x=0}^{68} l_x m_x = 22.82$$

$$R_{o,27} = \sum_{x=0}^{64} l_x m_x = 21.38$$

$$R_{o,29} = \sum_{x=0}^{60} l_x m_x = 15.24$$

The intrinsic rate of increase (r_m). The r_m values for the four temperatures were obtained by iterating and changing the values of r_m in Eqn (8.3) using the sum of column 9 (Appendix IV) until the sum produced the value of 1. As explained in this chapter, when the sum of column 9 was equal to 1 (Eqn (8.6)), the value of r_m last used was identified as the true r_m value. These values were 0.051, 0.063, 0.071, and 0.078 for temperatures of 23, 25, 27, and 29°C, respectively.

Finite rate of increase (λ). By substituting the values obtained for r_m into Eqn (8.7), we obtain:

$$\lambda\,(23°C) = e^{0.051} = 1.049$$

$$\lambda\,(23°C) = e^{0.063} = 1.060$$

$$\lambda\,(23°C) = e^{0.071} = 1.070$$

$$\lambda\,(23°C) = e^{0.078} = 1.081$$

Mean generation time (T). Substituting the sums of columns 7 and 8 (Excel file case 2) into Eqn (8.5) for each of the four temperature treatments, we obtain:

$$T(23°C) = 759.79/13.97 = 47.66$$

$$T(25°C) = 1213.32/22.82 = 45.93$$

$$T(27°C) = 1007.16/21.38 = 40.79$$

$$T(29°C) = 578.81/15.24 = 37.97$$

Doubling time (DT). And, finally, substituting the r_m values obtained for each temperature treatment into Eqn (8.8), we obtain:

$$DT = (\log_e 2)/0.0701 = 9.88\ (23°C)$$

$$DT = (\log_e 2)/0.0766 = 9.04\ (25°C)$$

$$DT = (\log_e 2)/0.0773 = 8.96\ (27°C)$$

$$DT = (\log_e 2)/0.0773 = 8.96\ (29°C)$$

The general interpretation for this calculation of CBB reared at different temperatures is as follows. Although the intrinsic and finite rates of increase of CBB directly rose due to temperature, the R_0 values (females/female/generation) were higher at 25 and 27°C. However, the generation time of CBB at 27°C was 1.12-fold lower than at 25°C. Thus, 27°C was considered the most promising temperature for mass propagation of *H. hampei*. The results of this study were used to develop a mass-rearing system for three species of African parasitoids (*C. stephanoderis*, *Prorops nasuta* (Waterson), and *Phymastichus coffea* (Lasalle)) of CBB, which included colony establishment, maintenance, production, and quality control (Portilla and Streett, 2008).

8.3.3. Life Tables for *Cephalonomia stephanoderis* and *Prorops nasuta* (Hymenoptera: Bethylidae) Ectoparasitoids of *Hypothenemus hampei*, Using an Artificial Diet

The parasitic Hymenoptera are arguably one of the most important insect groups. Many species play valuable roles in pest control as well as maintain the diversity of natural communities (Quicke, 1997). Mass production of parasitoids is linked to host production, and life tables for both parasitoid and pest are important to get a better understanding of how to maintain equilibrium without a negative impact on both populations. Complete and abridged life tables can be used for parasitoids depending on their oviposition behavior. One of the most conspicuous features of an insect parasitoid is whether or not its larvae develop as endoparasitoids or as ectoparasitoids (Carey, 1993). However, this explains only some aspects of parasitic wasp biology. Askew and Shaw (1986) drew attention to another aspect of their biology: those parasitoids whose hosts do not develop further after parasitization are referred to as idiobionts, and those whose hosts carry on their development for at least a while after parasitization are called koinobionts. In most cases, a parasitoid attacks only a single host stage (larva, prepupa, pupa, or adult), but there are a number of exceptions like *C. stephanoderis* and *P. nasuta* that attack larvae, prepupae, pupae, and sometimes pharate adults (Portilla, 1999a). *Cephalonomia stephanoderis*, and *P. nasuta* are idiobiont ectoparasitoids that allow the construction of a complete cohort life table. In this study, an abridged cohort life table for both parasitoids was constructed for growth rate estimations.

For this example, growth rate estimations were calculated based on formulae and procedures used in case studies 1 and 2. Results are presented with tables and figures, with a detailed description of all biological factors that can be obtained by daily observation when constructing a life table.

Development time and immature survival. Cohorts of 291 hosts recently parasitized by *C. stephanoderis* and 547 hosts recently parasitized by *P. nasuta* were used for this study. The cohort size depended on the availability within the rearing facilities. Each cohort was divided into smaller cohorts, placed into culture rearing boxes (20–30 parasitized hosts/box), and held inside the environmental chambers until the last adult was obtained. Parasitized hosts were observed every day, and the developmental times of the eggs, larvae, prepupae, and pupae stages of *C. stephanoderis* and *P. nasuta* were measured (Table 8.2). Dead wasps of various stages were counted, and the immature survival rate was calculated. From the ratio of unhatched eggs to total initial stages, wasp fertility was estimated.

TABLE 8.2 Mean Developmental Times of the Immature Stages of *Cephalonomia stephanoderis* and *Prorops nasuta*

	Developmental Times (Days) (Means ± Standard Deviation (SD)) $n = 180$	
Life Stages	***C. stephanoderis***	***P. nasuta***
Egg (days)	3.00 ± 0.64a	4.48 ± 0.51b
Larva (days)	4.16 ± 0.94a	4.96 ± 0.73b
Cocoon (days)	3.84 ± 0.68a	2.84 ± 0.55b
Prepupa	4.08 ± 0.70a	4.12 ± 0.60a
Pupa (days)	13.60 ± 1.15a	13.32 ± 0.90a
Total time to adult (days)	28.68 ± 1.88a	29.72 ± 1.20b

Means ± SD followed by the same letter in each row are not significantly different ($p < 0.05$ Tukey test).

The immature survival rate was calculated by using a synthetic cohort for each parasitoid. The fraction of death was calculated by finding the sum of all dead in each developmental stage and dividing by the initial numbers for each stage (synthetic cohort). This value was used for incorporation into the adult survival parameter (l_x). Data were compared using GLM analysis, which indicated that these parasitoids have a marked difference among them regarding survival.

Adult reproduction and survival. Fifteen females of *C. stephanoderis* and 15 females of *P. nasuta* were used to evaluate the fecundity and growth rates of increase. Recently emerged and fertilized females were selected and individually confined in plastic petri dishes (25 × 10 mm), which contained 30 second-instar larvae, prepupae, and pupae stages each (10 individuals per stage), plus a few eggs and first-larval stages for feeding. These females were held inside an environmental chamber (25°C, 85% RH, 12:12 (D:L) photoperiod). The number of eggs oviposited every day by each parasitoid of both bethylids was counted and recorded until the last wasp died. Petri dishes were cleaned daily by removing any dead or parasitized hosts and replacing them with new ones. Means of number of eggs oviposited per female, longevity, and preoviposition, oviposition, and postoviposition periods of *C. stephanoderis* and *P. nasuta*, and their host preference for ovipositing, were calculated. Parasitized hosts of each parasitoid were placed inside a petri dish to develop under the same environmental conditions as the adult female. These parasitized stages were kept until the last adult was obtained and the progeny sex ratio was determined. This information was used to estimate the mean number of female progeny produced per female.

Figure 8.3 shows the percentages of parasitized hosts for both parasitoid species that were offered equal numbers of CBB second instars, prepupae, and pupae. *Cephalonomia stephanoderis* had percentages of parasitism in prepupae, pupae, and second-instar larvae of 61%, 37%, and 3%, respectively. *Prorops nasuta* showed host preferences for prepupae, second-instar larvae, and pupae of 50%, 45%, and 5%, respectively. Significant differences were found for percentages of parasitism in second-instar larvae, prepupae, and pupae. Preoviposition periods and longevity were longer for *P. nasuta* than *C. stephanoderis*. The oviposition and postoviposition periods were not significantly different among parasitoids. These results showed that *C. stephanoderis* and *P. nasuta* continued to survive after females had finished ovipositing, and they continued to feed on CBB hosts during the postreproductive period.

Oviposition began on day 4 for *C. stephanoderis* and day 5 for *P. nasuta* (Fig. 8.4(a)). Mean daily egg production (m_x values) for *C. stephanoderis* increased during the first week and remained at higher levels until the end of the second week. Subsequently, the egg production by this parasitoid declined gradually over a 5–6 week period. In contrast, *P. nasuta* showed stable mean daily egg production, with the highest level observed from 19 days to 45 days.

Adult survival values (l_x) are summarized for both parasitic wasps in Fig. 8.4(b). A similar trend in adult survival was found for both parasitoid species. However, *C. stephanoderis* survival began to decline by the end of the third week, whereas

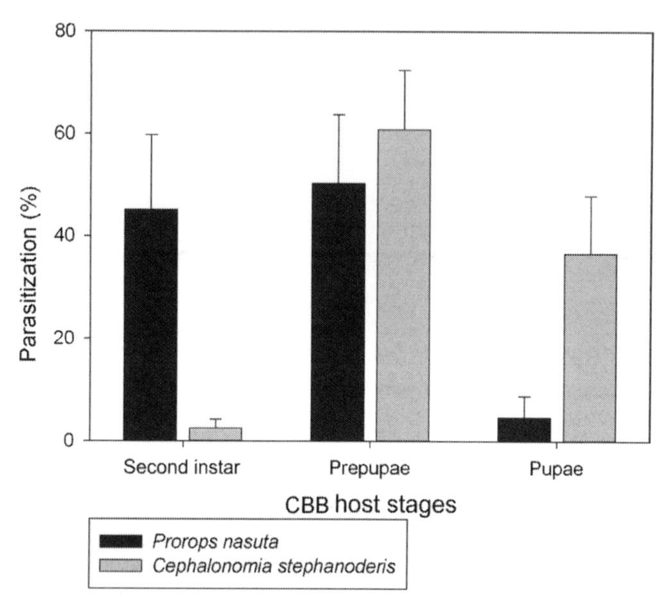

FIGURE 8.3 Percentage of hosts parasitized by *Cephalonomis stephanoderis* and *Prorops nasuta*; 95% confidence limits of the mean (Tukey test α=0.05).

P. nasuta survival began to decline in the fifth week. Adult survival for both parasitoid species continued to gradually decline until *C. stephanoderis* and *P. nasuta* female survival ended at 68 and 87 days, respectively.

Reproductive potential and rates of increase. Demographic parameters are presented in Table 8.3. Calculation of these parameters for both parasitoid species can be found in Excel file case 3. Significant differences were found in the reproductive potential of *C. stephanoderis* and *P. nasuta*. A higher gross fecundity was obtained for *C. stephanoderis* with 73.1±26.2 female and male eggs/wasp, versus 35.9±24.7 female and male eggs/wasp obtained for *P. nasuta*. The highest finite rate of increase was found in *C. stephanoderis*. The net reproductive rate for *C. stephanoderis* was 46.15 female wasps per newborn female in a mean generation time of 47.4 days. The daily rate of increase was 1.09 daughter females per female, with a doubling time of 7.2 days. The net reproductive rate for *P. nasuta* was 18.33 daughter females per female in a mean time of 58.6 days, with a daily rate of increase of 1.05 females per female and a doubling time of 13.12 days. Figure 8.4(c) shows the differences among these increasing population parameter values, although both parasitoids had a similar logarithmic regression model for their reproductive phase.

In general, the life table shown here assumes a hypothetical cohort subjected throughout its lifetime to the age-specific mortality rates prevailing for the actual population over a specific period. Most aspects of the life history of *C. stephanoderis* and *P. nasuta* have been collected from studies carried out in captivity using a hypothetical cohort because, under normal circumstances, its complete life cycle occurs inside coffee berries. The life table for *P. nasuta*

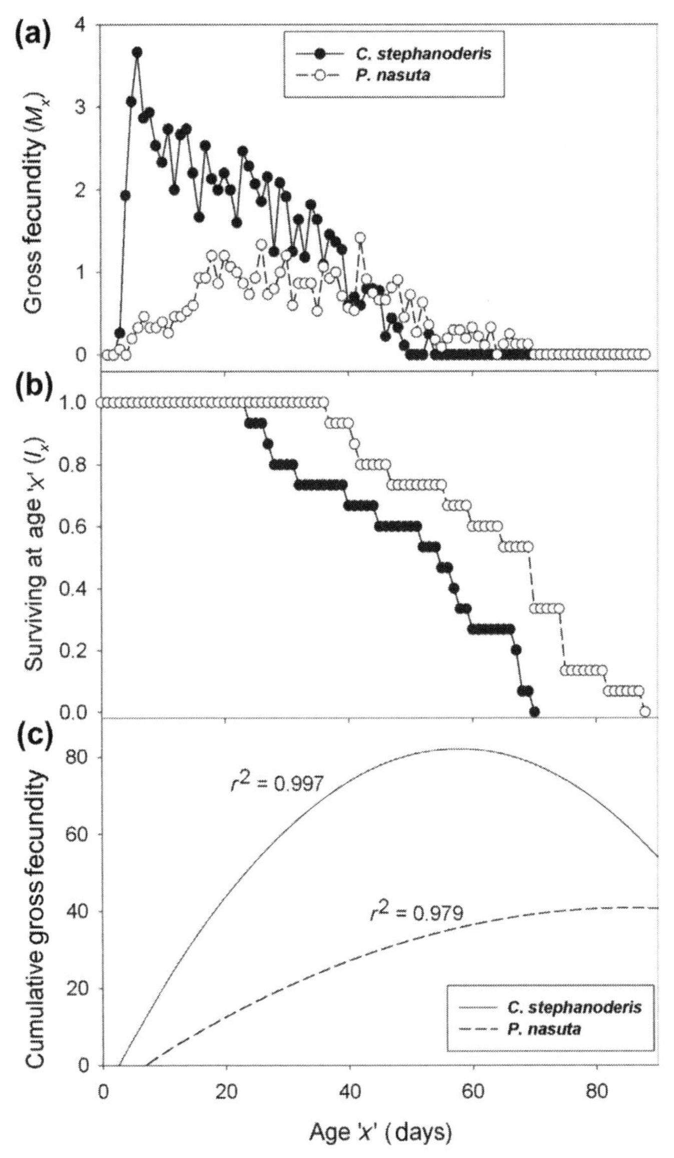

FIGURE 8.4 Life table data of *Cephalonomia stephanoderis* and *Prorops nasuta* restricted to the adult stage. (a) Gross fecundity (M_x); (b) survival (l_x); and (c) regression of cumulative gross fecundity.

showed higher mortality values for every immature stage when compared to *C. stephanoderis*. The lower mortality values observed in *C. stephanoderis* was probably due to the females selecting higher quality hosts (big and healthy) for oviposition. *Prorops nasuta* were found laying a higher percentage of eggs on host larvae before paralysis had occurred, which led to higher mortality. The capacity of predation by adults of *P. nasuta* was another characteristic that could

TABLE 8.3 Life Table Statistics for *Cephalonomia stephanoderis* and *Prorops nasuta* Reared on *Hypothenemus hampei* Developed on a Cenibroca Artificial Diet

	Parasitoids	
Parameters and Units	*C. stephanoderis*	*P. nasuta*
Gross fecundity (M_x)*	80.49	39.44
Net fecundity (m_x)**	67.29	31.09
Net reproductive rate (R_o)†	46.02	18.47
Mean generation time (T)‡	46.39	58.59
Doubling time (DT)§	7.60	12.85
Intrinsic rate of increase (r_m)§§	0.091	0.054
Finite rate of increase (λ)¶	1.095	1.055

*Total offspring/female.
**Females/females at age x.
†Daughters/newborn females.
‡Mean age of reproduction (days).
§Time required to (λ) double the number.
§§Rate of natural increase (daughters/female/day).
¶Individuals/female/day.

affect the mortality of immature parasitoids. Females of *P. nasuta* often oviposit on the injured host, and these hosts that are injured by predation do not provide all of the necessary resources for the survival of the parasitoid progeny. All of these behavioral features of *P. nasuta* provide a better understanding of the reasons why this parasitoid has been difficult to rear under laboratory conditions.

According to these results, *C. stephanoderis* produced more females per female than *P. nasuta* (R_o), but both parasitoids need a minimum of 40 suitable host stages to assure progeny. The speed at which the colony increased (r_m) is the most important parameter, and *C. stephanoderis* obtained the high intrinsic rate of increase values. The mass production of both parasitoids depends on the host's reproductive potential. Therefore, all of these results give some insight as to why the parasitoids have been so unsuccessful in the field, because to proliferate they evidently need a range of host stages. Thus, if a wasp enters an infected berry too early, she will find only eggs and young larvae, which may consume and thereby kill the CBB female too, but she won't have any offspring (Baker, 1999). On the other hand, if the wasp arrives too late, she would find mostly adult CBB and would have to wait until they start breeding, by which time harvest may interfere. All of this suggests that *C. stephanoderis* and *P. nasuta* will be incompatible in efficient commercial production when using CBB natural hosts for reproduction. An artificial diet used to rear CBB hosts provided an economical and feasible system for mass rearing *C. stephanoderis* and *P. nasuta*. This rearing system enabled us to adjust the number of hosts provided to each parasitoid. Also, it was possible to offer the preferred stages for the preoviposition period, feeding, and oviposition (Portilla and Streett, 2008).

8.3.4. Comparison of Demographic Parameters of *Perillus bioculatus* Feeding on Factitious Prey (*Anthonomus grandis*) Larvae and Natural Prey (*Leptinotarsa decemlineata*) Eggs

Perillus bioculatus prey primarily on the Colorado potato beetle, *L. decemlineata*. This beetle is the most important pest of potatoes and tomatoes in North America and Europe (Schalk and Stoner, 1976, 1979; Hare, 1980; Ferro et al., 1983; Hare and Moore, 1988; Senanayake and Holliday, 1990). The Colorado potato beetle threatens crops by its ability to not only cause economic damage but also develop resistance to most available types of pesticides (Harris and Svec, 1981; Boiteau et al., 1987; Heim et al., 1990; Tisler and Zehnder, 1990; Ioannidis et al., 1991; Grafius, 1995). Therefore, developing biological alternatives to control this pest is critical.

Releases of *P. bioculatus* have been effective in controlling first and second generations of *L. decemlineata* in the field, resulting in significant decreases in crop damage (Biever and Chauvin, 1992; Hough-Goldstein and Whalen, 1993; Cloutier and Bauduin, 1995). However, it has been estimated that a release density of 1–3 *P. bioculatus* nymphs per plant is required to achieve control of *L. decemlineata* (Tamaki and Butt, 1978; Biever and Chauvin, 1992). Such release densities would require thousands of nymphs per hectare, creating the need for an economical method to mass produce this predator. Without a good artificial formulation available, the use of factitious prey may be a viable way to mass produce *P. bioculatus* economically. Saint-Cyr and Cloutier (1996) presented some potential factitious prey choices to mass produce *P. bioculatus*. They concluded that mealworms, *Tenebrio molitor* (L.), and house crickets, *Acheta domesticus* (L.), are viable prey for *P. bioculatus* and do not alter the predator's preference for its natural prey. During the mid-1980s, mechanized mass production of the boll weevil was developed (Roberson and Wright, 1984), and by the 1990s capabilities to mass produce *A. grandis* reached millions of individuals. The objective of this study was to determine if the boll weevil could be a good alternative prey for mass producing *P. bioculatus*. Although the mass production of *A. grandis* was a government-funded project, at the time this study was completed (1998), the technology for mass producing *A. grandis* was well established and had a good potential to be transferred. In subsequent years, the boll weevil eradication program was established in all of the eastern US cotton-producing states and eventually forced the shutdown of boll weevil mass production. As a consequence, the potential for practical application of the results of this study diminished, publication priorities shifted, and this research remained unpublished until now.

Experimental procedures. The *P. bioculatus* used in this study were obtained from the USDA Animal and Plant Health Inspection Service's Plant

Protection and Quarantine Biological Control Center in Mission, Texas. This colony has been reared with the cabbage lopper, *Trichoplusia ni* (Hübner), larvae; however, the *P. bioculatus* eggs used in this study were produced by females reared on *L. decemlineata* eggs in the same facility. The Mission, Texas, facility also provided shipments of *L. decemlineata* eggs for the comparisons in these experiments. Boll weevil larvae used in this study were produced on an artificial diet at the USDA-ARS's Biological Control and Mass Rearing Research Unit at Mississippi State University. *Perillus bioculatus* was reared at a constant 26±1°C, 50±10% RH, and a 16:8 (L:D) hour photoperiod, and the experiments were performed under these conditions.

A total of 143 *P. bioculatus* eggs were assigned to the control group to be fed with *L. decemlineata* eggs. A total of 202 *P. bioculatus* eggs were assigned to the factitious prey group; this was approximately 40% more than in the control group, in anticipation of higher mortality based on preliminary observations. The reason to increase the starting numbers in the factitious prey group was to obtain a minimum of 40 adult females; any increase in this number improves the resolution of the estimated parameters. The groups of *P. bioculatus* eggs were placed in a plastic petri dish (140mm diameter) and allowed to develop at the conditions described here. Six days later, the first instars hatched and were provided with reverse-osmosis (RO) water until they molted to the second stadium. The second instars were collected using a fine brush and individually placed in six-well tissue culture dishes. The covers of the tissue culture dishes were modified with six screened windows, one on top of each well. Nymphs from the control group were provided daily with three *L. decemlineata* eggs, and those in the factitious prey group were provided with a single *A. grandis* third instar. All nymphs were also provided daily with RO water. Nymphs were observed daily to record any mortality and change of stadium, evidenced by the presence of exuviae. The amount of food provided was duplicated for each increasingly older stadium, so third, fourth, and fifth instars received 6, 12, and 24 *L. decemlineata* eggs or 2, 4, and 8 *A. grandis* third instars, respectively. The amount of prey provided was in accordance with the consumption rates reported by Tamaki and Butt (1978). Mortality and stadia length were recorded for treatment and factitious prey groups.

Adults completing development from each group were paired (male and female), placed in square petri dishes (95×95mm) lined with a piece of tissue paper, and daily provided with food consisting of 48 *L. decemlineata* eggs or 16 *A. grandis* third instars for the control and factitious prey groups, respectively. Both groups also were provided with RO water in dental cotton wicks. The pairs were placed in an environmental chamber with the same conditions described in this section, and they were inspected daily for oviposition and mortality. The number of eggs oviposited by each female was recorded until all of the females died.

Demographic parameters. Data consisting of number alive at each age from egg to adult (N_x) and mean number of eggs oviposited at every age (E_x) were used to calculate a life and fertility table as in Fig. 8.2 (see Excel file case 4). The gross fecundity (M_x) and net fecundity (m_x) were calculated assuming a 1:1 sex ratio in both groups. This assumption is reasonable considering that *P. bioculatus* is a sexually reproducing diploid species in which the sex is determined by homozygosis or heterozygosis of the sex alleles. The l_x values are plotted in Fig. 8.5(a), and the $l_x m_x$ is presented in Fig. 8.5(b). Net

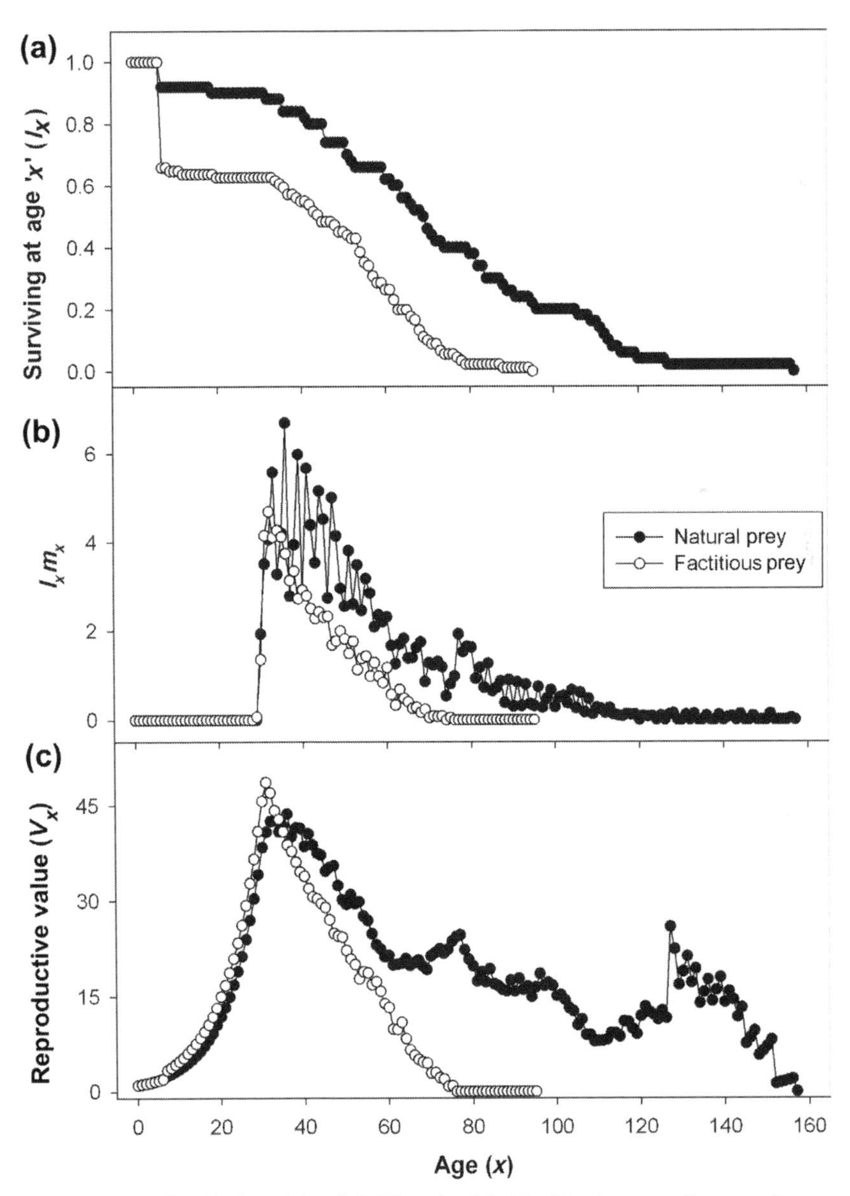

FIGURE 8.5 Life table data of the whole life cycle of *Perillus bioculatus* reared on natural versus factitious prey. (a) Survival at age x (l_x); (b) fecundity function ($l_x m_x$); and (c) reproductive value (V_x).

reproductive rate was calculated using Eqn (8.4) as the sum of column 8 in Excel file case 4 as:

$$R_{o,c} = \sum_{x=0}^{156} l_x m_x = 160.47$$

$$R_{o,f} = \sum_{x=0}^{94} l_x m_x = 76.76$$

where $R_{o,c}$ and $R_{o,f}$ are the net reproductive rates of the control and factitious prey groups, respectively. Generation time for the control group (T_c) and the factitious prey group (T_f) was calculated using Eqn (8.5) as:

$$T_c = \sum_{x=0}^{156} x l_x m_x / R_{o,c} = \frac{8811.86}{160.47} = 54.91$$

$$T_f = \sum_{x=0}^{94} x l_x m_x / R_{o,f} = \frac{3266.18}{76.76} = 42.55$$

Although the reproductive output of the control group was higher than that of the factitious prey group, the generation time was over 12 days longer in the control group. As a result, little difference among the groups was observed in the true r_m, which was 0.118 and 0.112 for the control and factitious prey groups, respectively. Substituting in Eqn (8.7), we obtain the value of the finite rate of increase for the control (λ_c) and the factitious (λ_f) prey groups, respectively, as:

$$\lambda_c = e^{r_{m,c}} = e^{0.118} = 1.125$$

$$\lambda_f = e^{r_{m,f}} = e^{0.112} = 1.118$$

Doubling time for the control (DT_c) and factitious prey (DT_f) groups was obtained by substituting into Eqn (8.8) as:

$$DT_c = \frac{\ln(2)}{r_{m,c}} = \frac{0.693}{0.118} = 5.87$$

$$DT_f = \frac{\ln(2)}{r_{m,f}} = \frac{0.693}{0.112} = 6.2$$

The values of the demographic parameters were close enough among the two groups that their differences may not be significant. The downside of the demographic parameters calculated from the life table is that no statistical test is yet available. The only exception to this is the R_o parameter, which can be analyzed as individual sums of total female progeny produced by each of the females included in the life table plus the estimated number of females that did not complete development, which would have a 0 value for their individual R_0. These numbers can then be analyzed using a regular ANOVA or GLM analysis. Efron (1983) proposed Tukey's jackknife method as a way to generate variation on single-parameter estimation. The jackknife method is a way to generate several estimations of the demographic parameters so that a statistical analysis can be done (Maia et al., 2000). The jackknife method works by systematically calculating demographic parameters after the deletion of a single observation, replacing it, deleting the next observation, and so on until a number of parameter estimates is obtained that is equal to the number of data points (Efron, 1983; Roff, 1992). When applying the jackknife method to life table data, it is necessary to correct the l_x for each deletion. Similarly, an estimate of the number of females that did not reach adulthood is necessary. For this example, the numbers of estimated females not reaching adulthood was five and 34 for the control and factitious prey groups, respectively. The total numbers of generated estimates of the parameters were 50 and 91 for the control and factitious prey groups, respectively. Statistical analysis of generated data showed that the values of all of the parameters were significantly different between the groups (Table 8.4). The control group had a significantly higher r_m value, resulting in higher λ and shorter DT. However, the DT in the control group was just over half a day shorter than in the factitious prey group. So if we can wait half a day to double the population of a *P. bioculatus* colony, then the use of *A. grandis* as factitious prey produces satisfactory results.

TABLE 8.4 Statistical Comparison of Demographic Parameters of *Perillus bioculatus* Feeding on Natural versus Factitious Prey Using the Jackknife Method to Generate Multiple Estimates of the Parameters

	Prey Provided	
Demographic Parameters	**Natural***	**Factitious****
Net reproductive rate (R_0)	159.19±2.07a	76.76±0.92b
Generation time (T)	54.24±0.24a	42.55±0.07b
Intrinsic rate of increase (r_m)	0.1181±0.0002a	0.1119±0.0003b
Doubling time (DT)	5.87±0.012b	6.2±0.015a
Finite rate of increase (λ)	1.125±0.0003a	1.118±0.0003b

Means±standard deviation. Means with the same letter within rows are not significantly different after Student's T test at $\alpha=0.05$.
**L. decemlineata* eggs (=control), $n=50$.*
***A. grandis* third instars, $n=91$.*

One interesting effect of consuming *A. grandis* was the shortening of the generation time (T) of *P. bioculatus* by over a week. The impact of this may be minimal, but a good way to visualize it is by calculating reproductive values (V_x). The reproductive value is defined as the contribution that an individual of age x will make to the future generation relative to the contribution in population number that one newborn individual will make over the remaining life of the female (Roughgarden, 1979; Roff, 1992; Carey, 1993). In some cases, the reproductive value is an age-dependent measurement of fitness (Roff, 1992) and can provide information on optimal ages for inoculative releases or for colony growth. The reproductive value can be calculated as an additional column in the life and fertility table by using Eqn (8.9).

$$V_x = \frac{e^{xr_m}}{l_x} \sum_{y=x}^{w} e^{-r_m y} l_y m_y \tag{8.9}$$

The second multiplier of Eqn (8.9) is essentially a partial sum of column 10 (r_x) of the life and fertility table (Excel file case 4) and can be easily calculated after the true value of r_m has been determined (see Excel file case 4, column 11). A plot of V_x of *P. bioculatus* reared on natural and factitious prey is presented in Fig. 8.5(c). *Perillus bioculatus* reared on natural prey present a peak of reproductive potential at age 36 and then an extended period with an almost constant reproductive potential lasting until age 140. When reared on *A. grandis* larvae, *P. bioculatus* present a sharp peak of reproductive potential at age 31 followed by a quick constant decline (Fig. 8.5(c)). It seems that the factitious prey is sufficiently adequate for *P. bioculatus* development and adult maturation, but it is inadequate to maintain adults for an extended period of time. Studying the differences in the nutrient content between *A. grandis* third instars and *L. decemlineata* eggs could provide clues for the improvement of current artificial diets by identifying key nutrients affecting adult longevity and fecundity.

8.4. CONCLUDING REMARKS

The objective of this chapter was to provide a better understanding of complete life table constructions and how essential they are in rearing systems to determine growth rates of insect colonies. Krebs (2001) said that demographic parameters cannot predict the future; but they are still useful to make projections of what will happen if specific assumptions are fulfilled. Collecting data required to calculate life tables can be extremely time-consuming, especially if the life cycle of the studied species is longer than a month. However, if such data are available, life table calculation is a simple task due to the availability of computational software (although software especially designed for calculating life tables is commercially unavailable). Development of such software can be easily accomplished by someone versed in computer programming by following the instructions provided in this chapter. Developing new methods for statistical analysis of demographic parameters is a more difficult task. Available methods are still inadequate and extremely time-consuming (such as the jackknife method). A teaming enterprise among population biologists and statisticians is required to achieve this goal. A practical method to analyze demographic parameters statistically could provide new tools for life table comparisons, enhancing the potential applications of life table analysis.

REFERENCES

Askew, R., Shaw, M., 1986. Parasitoid communities: their size, structure and development. In: Waage, J., Greathead, D. (Eds.), Insect Parasitoids, Academic Press, London, pp. 225–264.

Baker, P., 1999. La broca del café en Colombia: Informe final del proyecto MIP para el café DFID. Cenicafe—CABI Bioscience, Ascot, UK, p. 93.

Biever, K.D., Chauvin, R.L., 1992. Suppression of the Colorado potato beetle (Coleoptera: Chrysomelidae) with augmentative releases of predaceous stinkbugs (Hemiptera: Pentatomidae). J. Econ. Entomol. 85, 720–726.

Blanco, C.A., Portilla, M., Abel, C.A., Winters, H., Ford, R., Streett, D., 2009. Soybean flour and wheat germ proportions in artificial diet and their effect on the growth rates of the tobacco budworm, *Heliothis virescens*. J. Insect Sci. 9, Article 59.

Boiteau, G., Parry, R.H., Harris, C.R., 1987. Insecticide resistance in New Brunswick populations of the Colorado potato beetle (Coleoptera: Chrysomelidae). Can. Entomol. 119, 459–463.

Carey, F.G., 1993. Applied Demographic for Biologists with Special Emphasis on Insects. Oxford University Press, Oxford, UK, p. 206.

Carey, J.R., Judge, D.S., 2001. Principles of biodemography with especial reference to human longevity. Popul. Engl. Sel. 13, 9–40.

Carey, J.R., Papadopoulos, N.T., Papanastasiou, S., Diamantidis, A., Nakas, C.T., 2012. Estimating changes in mean population age using the death distributions of live-captured medflies. Ecol. Entomol. 37, 359–369.

Cloutier, C., Bauduin, F., 1995. Biological control of the Colorado potato beetle *Leptinotarsa decemlineata* (Coleoptera: Chrysomelidae) in Quebec by augmentative releases of the two-spotted stinkbug *Perillus bioculatus* (Hemiptera: Pentatomidae). Can. Entomol. 127, 195–212.

Efron, B., 1983. The Jackknife, the Bootstrap and Other Resampling Plans. Monograph 38 Society for Industrial and Applied Mathematics, Philadelphia, PA.

Eliopoulos, P.A., 2006. Life tables of *Venturia canescens* (Hymenoptera: Ichneumonidae) parasitizing the Mediterranean flour moth (Lepidoptera: Pyralidae). J. Econ. Entomol. 99, 237–243.

Ferro, D.N., Morzuch, B.J., Margolies, D., 1983. Crop loss assessment of the Colorado potato beetle (Coleoptera: Chrysomelidae) on potatoes in western Massachusetts. J. Econ. Entomol. 76, 349–356.

Gauld, I., Bolton, B., 1996. The Hymenoptera. Oxford University Press, Oxford, UK, p. 332.

Grafius, E.J., 1995. Is local selection followed by dispersal a mechanism for rapid development of multiple insecticide resistance in the Colorado potato beetle? Am. Entomol. 41, 104–109.

Hare, J.D., 1980. Impact of defoliation by the Colorado potato beetle on potato yields. J. Econ. Entomol. 73, 369–373.

Hare, J.D., Moore, R.E.B., 1988. Impact and management of late-season populations of the Colorado potato beetle (Coleoptera: Chrysomelidae) on potato in Connecticut. J. Econ. Entomol. 81, 914–921.

Harris, C.R., Svec, H.J., 1981. Colorado potato beetle resistance to carbofuran and several other insecticides in Quebec. J. Econ. Entomol. 74, 421–424.

Heim, D.C., Kennedy, G.G., Van Duyn, J.W., 1990. Survey of insecticide resistance among North Carolina Colorado potato beetle (Coleoptera: Chrysomelidae) populations. J. Econ. Entomol. 83, 1229–1235.

Hough-Goldstein, J., Whalen, J., 1993. Inundative release of predatory stink bugs for control of Colorado potato beetle. Biol. Control 3, 343–347.

Ioannidis, P.M., Grafius, E., Whalon, M.E., 1991. Patterns of insecticide resistance to azinphos-methyl, carbofuran, and permethrin in the Colorado potato beetle (Coleoptera: Chrysomelidae). J. Econ. Entomol. 84, 1417–1423.

Keyfitz, N., 1964. The intrinsic rate, of natural increase and the dominant root of the projection matrix. Popul. Stud. 18, 293–308.

Krebs, C.J., 1972. Ecology: The Experimental Analysis of Distribution and Abundance. Harper & Row, New York.

Krebs, C.J., 1978. Aggression, dispersal, and cycling changes in population of small rodents. In: Kramer, L., Pliner, P., Alloway, T. (Eds.), Aggression, Dominance, and Individual Spacing, Plenum Publishing Corporation, pp. 49–60.

Krebs, C.J., 2001. Ecology: The Experimental Analysis of Distribution and Abundance, fifth ed. Wesley Longman, San Francisco, CA, p. 695.

Lotka, A.J., 1907. Relation between birth rates and death rates. Science 26, 21–22.

Lotka, A.J., 1913. A natural population norm. II. J. Wash. Acad. Sci. 3, 289–293.

Maia, A. de H.N., Luiz, A.J.B., Campanhola, C., 2000. Statistical inference on associated fertility life table parameters using jackknife technique: computational aspects. J. Econ. Entomol. 93, 511–518.

Manly, B.F.J., 1997. Randomization, Bootstrap and Monte Carlo Methods in Biology, second ed. Chapman & Hall, London.

Morales-Ramos, J.A., Cate, J.R., 1992. Laboratory determination of age-dependent fecundity, development, and rate of increase of *Catolaccus grandis* (Burks) (Hymenoptera: Pteromalidae). Ann. Entomol. Soc. Am. 85, 469–476.

Morales-Ramos, J.A., Cate, J.R., 1993a. Reproductive biology of *Heterospilus megalopus* (Hymenoptera: Braconidae), a parasitoid of *Anthonomus grandis*. Ann. Entomol. Soc. Am. 86, 734–739.

Morales-Ramos, J.A., Cate, J.R., 1993b. Temperature-dependent developmental rates of *Catolaccus grandis* (Hymenoptera: Pteromalidae). Environ. Entomol. 22, 226–233.

Morris, R.F., Miller, C.A., 1954. The development of life tables for the spruce budworm. Can. J. Zool. 32, 283–301.

Peterson, R.K.D., Davis, R.S., Higley, L.G., Fernandes, O.A., 2009. Mortality risk in insects. Environ. Entomol. 38, 2–10.

Portilla, M., 1999a. Mass rearing technique for *Cephalonomia stephanoderis* (Hymenoptera: Bethylidae) on *Hypothenemus hampei* (Coleoptera: Scolytidae) developed using Cenibroca artificial diet. Rev. Colomb. Entomol. 25, 57–66.

Portilla, M., Mumford, J., Baker, P., 2000. Reproductive potential response of continuous rearing of *Hypothenemus hampei* (Coleoptera: Scolytidae) developed using Cenibroca artificial diet. Rev. Colomb. Entomol. 26, 99–105.

Portilla, M., Streett, D., 2008. Avances investigativos en la producción masiva automatizada de la broca del café *Hypothenemus hampei* (Coleoptera: Scolytidae) y de sus parasitoides sobre dietas artificiales. Sis. Agroeco. Mod. Biomatematic. 1 (1), 9–12.

Pressat, R., 1985. The Dictionary of Demography. Bell and Bain, Ltd, Glasgow, UK, p. 243.

Preston, S.H., Heuveline, P., Guillot, M., 2001. Demography: Measuring and Modeling Population Processes: Malden Blackwell Publishers, Oxford, UK, p. 312.

Quicke, D., 1997. Parasitic Wasps. Chapman and Hall, London, UK, p. 470.

Roberson, J.L., Wright, J.E., 1984. Production of boll weevils, *Anthonomus grandis* grandis. In: King, E.G., Leppla, N.C. (Eds.), Advances and Challenges in Insect Rearing. Agricultural Research Service, U.S. Department of Agriculture, New Orleans, LA, pp. 188–192.

Roff, D.A., 1992. The Evolution of Life Histories, Theory and Analysis. Chapman and Hall, New York, NY, p. 535.

Rojas, M.G., Morales-Ramos, J.A., King, E.G., 1996. In vitro rearing of the boll weevil (Coleoptera: Curculionidae) ectoparasitoid *Catolaccus grandis* (Hymenoptera: Pteromalidae) on meridic diets. J. Econ. Entomol. 89, 1095–1104.

Rojas, M.G., Morales-Ramos, J.A., King, E.G., 2000. Two meridic diets for *Perillus bioculatus* (Heteroptera: Pentatomidae), a predator of *Leptinotarsa decemlineata* (Coleoptera: Chrysomelidae). Biol. Cont 17, 92–99.

Roughgarden, J., 1979. Theory of Population Genetics and Evolutionary Ecology: An Introduction. MacMillan Publishing Co. Inc., New York, NY, p. 612.

Saint-Cyr, J.F., Cloutier, C., 1996. Prey preference by the stinkbug *Perillus bioculatus*, a predator of the Colorado potato beetle. Biol. Control 7, 251–258.

Senanayake, D.G., Holliday, N.J., 1990. Economic injury levels for Colorado potato beetle (Coleoptera: Chrysomelidae) on 'Norland' potatoes in Manitoba. J. Econ. Entomol. 83, 2058–2064.

Schalk, J.M., Stoner, A.K., 1976. Colorado potato beetle populations and their effect on tomato yield in Maryland. HortScience 11, 213–214.

Schalk, J.M., Stoner, A.K., 1979. Tomato production in Maryland: effects of different densities of larvae and adults of the Colorado potato beetle. J. Econ. Entomol. 72, 826–829.

Singh, P., Moore, R.F., 1985. Handbook of Insect Rearing. Elsevier Science, New York vol. 1.

Southwood, T.R.E., 1978. Ecological Methods, with Particular Reference to the Study of Insect Populations. Chapman and Hall, London, UK, p. 524.

Tamaki, G., Butt, B.A., 1978. Impact of *Perillus bioculatus* on the Colorado Potato Beetle and Plant Damage. USDA Technical Bulletin No. 1581. Washington, DC.

Terán-Vargas, A.P., Vera-Graziano, J., Carrillo-Sánchez, J.L., Alatorre-Rosas, R., 1990. Efectos de genotipos de soya en la longevidad, mortalidad y reproducción de *Anticarsia gemmatalis* (Lepidoptera: Noctuidae). Agrociencia 1, 57–69.

Tisler, A.M., Zehnder, G.W., 1990. Insecticide resistance in the Colorado potato beetle (Coleoptera: Chrysomelidae) on the eastern shore of Virginia. J. Econ. Entomol. 83, 666–671.

This page intentionally left blank

Chapter 9

Concepts and Methods of Quality Assurance for Mass-Reared Parasitoids and Predators

Norman C. Leppla
Entomology and Nematology Department, University of Florida, Institute of Food and Agricultural Sciences, Gainesville, FL, USA

9.1. INTRODUCTION

Quality assurance for mass-reared arthropods expands concepts previously incorporated into a total quality control system for insect production (Leppla and Fisher, 1989; Leppla, 2003, 2009) and increases emphasis on the postproduction use of parasitoids and predators in integrated pest management (Fig. 9.1). Quality assurance includes all of the factors that comprise development, production, delivery, application, and evaluation of the products, as well as feedback on their effectiveness (Juran and Godfrey, 1998; Feigenbaum, 1983; Webb, 1984). These factors can be measured and managed to assure the performance of mass-produced arthropods and to detect changes that could compromise their effectiveness in the field. Parasitoids and predators must be healthy enough to survive for a specified period after being released and effective in locating and killing or significantly injuring prey or hosts, as well as to possibly reproduce. Often, they must withstand handling, packaging, storage, shipment over long distances, transport to the field, and mechanical application (van Lenteren and Tommasini, 2003; Rull et al., 2012). Their postproduction quality can therefore be determined by characteristics such as the number of live organisms per package and their age, weight, size, tolerance to a wide range of temperatures and humidity levels, resistance to environmental hazards, mobility, longevity, fecundity, and impact on target pests. However, quality assurance in the mass production of beneficial arthropods is not limited to measurements and tests of organisms; rather, it is the design and management of systems that produce and deliver products that satisfy customer expectations (Deming, 1986; Burt, 2002).

Quality assurance for mass-produced arthropods was created to improve the reliability of producing and using products that meet required specifications and

Mass Production of Beneficial Organisms. http://dx.doi.org/10.1016/B978-0-12-391453-8.00009-1

standards. High-quality arthropods were not always available when their widespread use in pest management was a new concept (Smith, 1966; Knipling, 1966, 1979; King and Leppla, 1984; Singh and Moore, 1985). Reliable arthropod production was achieved only after industrial quality control procedures designed to consistently produce nonliving products were adapted for mass rearing insects (Boller and Chambers, 1977a, b; Chambers, 1975, 1977; ANSI/ASQC, 1987),

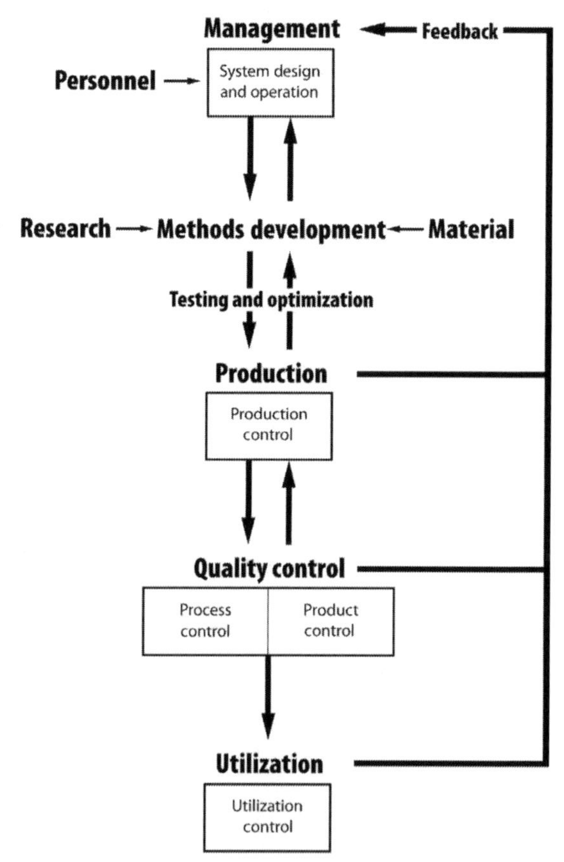

FIGURE 9.1 A quality assurance system for mass-reared arthropods expanded from total quality control for insect production (Leppla, 2009), with increased emphasis on utilization, customer support, and feedback. Management provides administrative support for designing and operating the entire system. The personnel division reports directly to management. The research and material divisions provide input to the methods development department, which is responsible for testing and optimizing rearing facilities, equipment, materials, and methods. The production division performs standard operating procedures for mass rearing the arthropods and controls required production inputs and processes. Rearing processes and products are monitored by the quality control division to assure that the arthropods meet established specifications and standards. The utilization division assesses product quality along the supply chain and assures that customer expectations are met. Management receives feedback on product quality from the production, quality control, and utilization divisions, enabling decisions to be made about adjusting the system.

keeping in mind the physiology, ethology, and ecology of the production and target populations. From the beginning to the present, a primary goal has been to minimize genetic divergence of the founder and laboratory populations, particularly if behavioral traits must be preserved (Boller, 1972; Mackauer, 1976; Huettel, 1976; Leppla and Ashley, 1989). Much of the evolution of quality assurance for mass-produced arthropods can be obtained from conference proceedings (Aeschlimann, 1996) and the 12 International Organization for Biological Control (IOBC) Global Working Group on Quality Control of Mass-Reared Arthropods (AMRQC) workshops (1981 to present, most available at http://www.AMRQC.org.). Boller and Leppla (2006) traced the history of the AMRQC, formerly the Global Working Group on Quality Control of Mass-Reared Arthropods and currently the Global Working Group on Mass Rearing and Quality Assurance.

Mass rearing was defined by Chambers (1977) as the "production of insects competent to achieve program goals with an acceptable cost/benefit ratio and in numbers per generation exceeding ten thousand to one million times the mean productivity of the native population female." Moreover, mass production, in contrast to small-scale rearing by an individual person, is accomplished as an industrial activity, usually involving a single species to support biologically based pest management, in factory-like facilities with controlled environments, artificial diets and oviposition substrates, mechanized equipment, and a succession of essential processes performed by separate work units (Nakamori et al., 1992; Lacasse et al., 2006; Leppla, 2009). Mackauer (1972) defined mass production of entomophagous insects as the rearing per generation cycle of one million times the mean number of offspring per female. Although this chapter describes quality assurance for mass-produced arthropods, the principles can be applied to any magnitude of rearing, especially parasitoids, predators, and herbivorous natural enemies.

The nucleus of a quality assurance program for mass-reared arthropods is the production capability and associated production, process, and product quality control (Leppla and Fisher, 1989; Bruzzone et al., 1993). A few species of arthropods have been mass reared in "biofactories" that employ several hundred workers during three shifts per 24-hour day, seven days per week (Leppla and Ashley, 1989). The functions performed in these biofactories are separated in time and space, so mass rearing systems must be highly coordinated. Efficient production requires that the arthropods have high levels of fertility and fecundity, exhibit rapid and synchronous development, are easy to harvest, and sustain high yields (Smith and Nordlund, 2000). Less industrialized rearing and associated quality control programs have been used successfully to produce beneficial arthropods, such as *Trichogramma* spp. (Morrison, 1985; Qiu et al., 1992; Pavlik, 1993; Wuehrer and Hassan, 1993; Bigler, 1994; Cerutti and Bigler, 1995; Bourchier et al., 2000; Liu and Smith, 2000; Prezotti et al., 2002; Kolliker-Ott et al., 2004). However, the largest production is of the Mediterranean fruit fly, *Ceratitis capitata* (Wiedemann), and screwworm fly, *Cochliomyia hominivorax* (Coquerel), for sterile male release. Production of the Mediterranean fruit fly has relied on a well-developed quality control program to reach several billion per week (Rendón et al., 2005). Quality control programs also have been adopted widely for producing the codling moth, *Cydia pomonella* (Linnaeus), (Hathaway et al., 1973; Bloem et al., 1998), mosquitoes (Dame, 1989), tropical fruit flies (Leppla, 1989; Caceres et al., 2007) and many other arthropod species (Moore et al., 1985; Zlotin and Chepurnaya, 1994).

The purpose of this chapter is to describe the principles of quality assurance for mass-produced arthropods, especially as they apply to parasitoids and predators. Quality assurance encompasses every aspect of arthropod mass production for augmentation biological control. Its application can accelerate the use of commercial natural enemies (Penn et al., 1998; van Lenteren, 2011) and enable increased growth of the industry (Hoy et al., 1991). Markets for mass-produced natural enemies are increasing, but potential customers are seeking assurance that the products are reliable and effective. For this chapter, customers are consumers or clientele who do not necessarily pay for the products (e.g. government biological control programs), as well as paying customers. Regardless, it is essential for them to be involved in assuring the quality of products they obtain because products must meet their specifications and standards. Customer satisfaction ultimately depends on how well the beneficial arthropods suppress arthropod and weed pests of agriculture, communities, and environmental areas, such as wetlands and rangelands.

This chapter provides a general framework for building a complete quality assurance system, followed by examples of product quality control for natural enemies and associated quality assessment and control data acquisition and analysis. Quality assurance systems can be reviewed periodically according to a basic guide that is included. The chapter concludes with a description of research priorities for mass-reared natural enemies. Excluded are alternative rearing methods and studies of parasitoid and predator field effectiveness.

9.2. QUALITY ASSURANCE IN THE MARKETPLACE

As in any business, including commercial biological control, successful products and services are defined by their usefulness and profitability (Penn et al., 1998). Natural enemy product effectiveness, however, often is difficult to quantify because of widely varying pest damage, naturally occurring biological control, natural enemy dispersal, and many other variables. Consequently, standards have been developed to specify the quality of commercial natural enemies. The Association of Natural Biocontrol Producers developed standards for selected commercial natural enemies using a system developed by the American Society for Testing and Materials International (http://www.astm.org). Moreover, at least one company, Bio-Bee Biological Systems (http://bio-bee.com), uses International Organization for Standardization (ISO) 9001 (http://www.iso.org/iso/home.html) and IOBC international standards for mass producing insects (van Lenteren et al., 2003) (http://users.ugent.be/~padclerc/AMRQC/images/guidelines.pdf). ISO International defines a standard as "a document that provides requirements, specifications, guidelines or characteristics that can be used consistently to ensure that materials, products, processes and services are fit for their purpose." Specifications are essential characteristics and tolerances, limiting values and other definitions for materials, products, services, processes, systems, or persons contained within the provisions of a standard (Juran et al. 1979).

Quality assurance systems for commercial natural enemies include all aspects of production, including completion of standard operating procedures, the results of monitoring rearing processes, and attainment of the performance standards for resulting products. Also included are contractual agreements, such as on-time delivery of the required number of living parasitoids or predators. Typical production control characteristics are ranges of temperature and relative humidity, as well as the yield per rearing container for process control. For product control, standards usually describe the following measurable characteristics: species or strain identity and purity, age, size, weight, motility, survival, host location, and level of parasitism (Smith, 1996). These kinds of characteristics can be used to identify products and their sources and compare products over time to detect changes. Requirements for postproduction quality assurance of mass-produced arthropods include packaging, shipping and handling, reliability of delivery, product care, application, and effectiveness (Brazzel et al., 1986; Bloem et al., 2002; Anonymous, 2003; FAO/IAEA/USDA, 2003; Enkerlin and Quinlan, 2004; Enkerlin, 2007; Blomefield et al., 2011). Cost and payment terms also affect customer satisfaction. For *Encarsia formosa* Gahan, the cost per functional insect was proposed as a useful quality guideline (Vasquez et al., 2004).

The acceptance of natural enemy products in the marketplace results not only from meeting written standards but also is indicated by the use of these products by satisfied customers. There are at least 30 large (>10 employees) biological control companies in the world that produce and sell about 200 arthropod species of natural enemies (van Lenteren, 2011). Currently, global sales are conservatively estimated to be at least $200 million per year, based on an anticipated 15–20% annual growth rate from $50 million in 2001 (van Lenteren, 2003a). Mass-produced parasitoids and predators are provided to farmers, ranchers, nursery operators, grower associations, crop advisors, pest managers, homeowners, and others for use in field and glasshouse crops, orchards, outdoor landscapes, interior plantscapes, pastures, livestock operations, forests, waterways, horse farms, and a few other appropriate situations. Accordingly, the acceptable quality of commercial parasitoids and predators is determined by these markets. Natural enemies will continue to be mass produced and used in greater quantities if they are affordable and suppress pests to levels below acceptable thresholds.

9.3. CUSTOMER INVOLVEMENT IN QUALITY ASSURANCE

The quality of mass-produced parasitoids and predators is not only intrinsic but also is related to the needs and expectations of the customers. Therefore, to maintain sales, producers and suppliers must provide high-quality products and also make sure that the customers are satisfied (Penn et al., 1998; Bolckmans, 1999). Customers are more likely to be satisfied if they are involved and educated in the release and evaluation of the natural enemy products they purchase (Bolckmans, 1999). Based on experience, the most helpful suppliers develop application guidelines and provide the required number of natural enemies to match the level of a pest infestation. Especially critical and easy to measure are the numbers of living and dead natural enemies in a package. Every package should be subsampled prior to shipment to ensure that it contains at least the number ordered; the normal target is approximately 15% overpacking. During shipment, the natural enemies are exposed to abnormal circadian rhythms, high densities, reduced oxygen concentrations, condensation, and possibly temperature and humidity extremes. Therefore, the natural enemies should be removed from the shipping container immediately when received and assessed by the customer for viability. Customers should be given quick and simple assessment methods that enable them to evaluate the products and expeditiously return feedback about quality to the suppliers, who in turn inform the producers. For some products, such as predatory mites or beetles, this is a short-term evaluation with motion indicating survival and some level of fitness. A longer period may be required to confirm the emergence and movement of parasitic wasps and midges. To facilitate these kinds of evaluations and maximize the effectiveness of natural enemy products, customers need better information and training (Bolckmans, 2003).

Customer feedback is an essential driver of product improvement. For example, a vegetable grower organization in British Columbia, Canada provided feedback to four natural enemy producers on the quality of *Phytoseiulus persimilis* Athias-Henriot (Acari: Phytoseiidae) and *E. formosa* (Hymenoptera: Aphelinidae) supplied for use in their glasshouses (Glenister et al., 2003). The producers were identified only by code and the grower's quality consultant notified them within a few weeks of the quantity of *E. formosa* in their shipments. During a 3-year trial, from 1997 to 1999, the quantity of *E. formosa* from one company improved from less than 5% of the number ordered to consistently 100%. Products from the second company followed the same pattern, achieving 100% in 1999. In 1997, only about 50% of the *E. formosa* shipments from the third company contained the expected number but more than 80% was achieved in 1998 and 1999. For the fourth company, the average number of *E. formosa* actually declined from about 50% the first year to below 20% by the end of the study. Emergence also was measured and exhibited the same general pattern as quantity—that is, packages with low numbers of insects had low levels of emergence. Unfortunately, the quality consultant was not able to measure the relative fecundity of the *E. formosa*. The performance of these companies probably was inconsistent during the study because there were changes in personnel, rearing capabilities, feedstock sources or quality, weather, packaging, routing, and quality assurance programs. In response to rapid feedback, it should be possible for producers to control the variables that determine product quality and reliably ship specified numbers of high-quality natural enemies. Ideally, the most successful producers and suppliers provide technical support to help growers assure that the natural enemies they purchase are effective and take rapid corrective action if necessary.

9.4. BUILDING A COMPLETE QUALITY ASSURANCE SYSTEM

Quality assurance in the mass production of arthropods, with minimal emphasis on product utilization, has been referred to as "total quality control in insect mass production" (Leppla and Fisher, 1989; Bernon and Leppla, 1994; Leppla, 2003). This organizational structure was adapted from manufacturing industries and similarly involves designing and operating systems that satisfy customer requirements for product performance (Fig. 9.1). The production capability is developed and optimized by continuously testing new methods and materials, monitoring standard operating procedures, evaluating the resulting products, determining product effectiveness, and providing feedback to make operational adjustments. In the spirit of W.E. Deming, J.M. Juran, A.V. Feiningbaum, and other pioneers of industrial quality assurance, every detail of an arthropod mass production and utilization system should be designed and managed to maximize product quality and effectiveness (Feigenbaum, 1983). Juran et al. (1979) defined quality assurance as "the activity of providing, to all concerned, the evidence needed to establish confidence that the quality function is being adequately performed." Accordingly, the quality function for arthropod mass production and utilization encompasses all of the following divisions of the system, regardless of how simple or complex (Leppla and Fisher, 1989) (Fig. 9.2):

9.4.1. Management

The first step in establishing an arthropod quality assurance system is to reach written agreement on exactly what customers need and expect in terms of product specifications and quality standards (Anonymous, 2006), delivery and evaluation, customer support remotely and in the field, and responsiveness to making adjustments (Leppla, 2003). Management must fully participate in planning the organization, establishing policies for its operation, and making necessary changes based on feedback from the production, quality control, and utilization divisions. Standard operating procedures for production are written cooperatively with the facility managers and key employees, including a sequence of detailed rearing processes, associated assignments, and schedules in the form

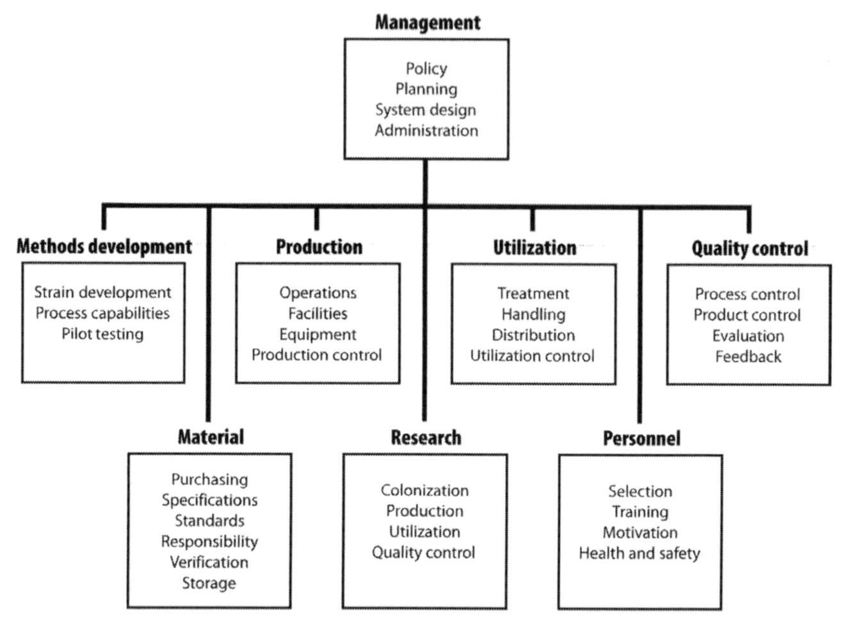

FIGURE 9.2 Divisions and associated functions of a quality assurance system for arthropod mass production described initially in total quality control for insect production *Leppla and Fisher, 1989.*

of instructions and check sheets (Minno and Holler, 1987; Fisher, 2009b). Schedules are arranged to provide coordination between processes accomplished at different times or locations. Considerable effort is devoted to defining competencies for rearing quality insects and selecting, training, evaluating, and compensating appropriate personnel (Fisher, 1984a; Singh and Ashby, 1985). This kind of organization and planning extends from production through all of the postproduction activities. Management provides ongoing administrative support for the entire quality assurance system.

Management can analyze the market and establish the quality assurance system by developing a comprehensive business plan. An example for producing sterile insects is the "Model Business Plan for a Sterile Insect Production Facility" (Anonymous, 2008). It describes commercial issues, such as ownership, costs and financing, organizational structure, site selection, intellectual property, liability insurance, and markets and pricing for the primary and waste products. Additional considerations for arthropod mass production include accessibility of supplies, facility design and operation (Fisher, 1984a), environmental requirements, health hazards, labor availability and employment conditions, and economics (Nasreen et al., 2011). The business plan should include a clear description of how the system will be managed and provided with adequate resources (Singh and Ashby, 1985). It will define quality assurance for delivering high-quality products, improving product effectiveness, and increasing associated customer satisfaction.

9.4.2. Methods Development

An arthropod mass rearing program typically has a distinct methods development division that conducts process capability studies to acquire, establish, maintain, and evaluate the colonies, as well as improve associated materials, facilities, equipment, and procedures (Goodenough, 1984; Griffin, 1984b; Harrell and Gantt, 1984; Edwards et al., 1996). Founders for colonies may be collected from several populations of the same species and propagated, with the best being selected for mass rearing or to replace the one in production. Methods development also can support utilization by helping to invent new ways of treating, handling, packaging, distributing, and evaluating the natural enemy products. Potential advancements for arthropod production, delivery, and evaluation, including those from research, are pilot tested within the existing quality assurance system before being adopted. The methods development division does not have a troubleshooting responsibility to correct immediate shortcomings in implementing established protocols. On the contrary, it seeks to progressively increase product quality and reduce costs by optimizing the system. Because suitable facilities are essential and expensive, methods development gives special consideration to their design, construction, maintenance, and, if necessary, modification (Griffin, 1984a; Owens, 1984; Goodenough and Parnell, 1985; Fisher and Leppla, 1985; Anonymous, 2004; Fisher, 2009a).

9.4.3. Materials

Materials and supplies must be selected, purchased, received, inventoried, stored, and used according to the quality assurance system (e.g. anticipated production levels, reliability of suppliers, distance from sources, required lead times). Specifications and standards for materials and supplies are described in detail within written purchasing contracts. Suppliers are responsible for the quality of their products and not allowed to unilaterally decide to substitute alternatives. Every shipment is labeled with received and "use by" dates. Rather than being assumed, required quality specifications and standards should be verified for all materials on receipt. Even paper products and water can be contaminated with substances, such as pesticides and other toxicants, that can reduce the quality of mass-reared arthropods. It is particularly difficult to control the quality and purity of dietary ingredients (Brewer and Lindig, 1984; ODell et al., 1997; Cohen, 2004; Inglis and Sikorowski, 2009). Detailed records are kept on all materials and supplies used in arthropod production and utilization (ODell, 1992).

9.4.4. Production

Mass production of beneficial arthropods often is unreliable and the resulting products are highly variable in quantity and quality (Knipling, 1979; Leppla, 2007; Vasquez and Morse, 2012). To increase reliability and reduce variability, standard operating procedures must be performed diligently and problems corrected rapidly (Anonymous, 2006). Production control data sheets are particularly useful for identifying the source of a problem because most production deficiencies are caused by failure to perform standard operating procedures. However, when production problems occur beyond environmental or equipment failures and procedural mistakes, management requests diagnostic assistance from other divisions (e.g. quality control, methods development and research) to quickly identify the causes. Production facilities are designed to be durable, efficient, and easy to clean. Every environment is kept within the specified range of temperature, relative humidity, photoperiod, and air flow and purity (Fisher, 2009a). Rooms with different rearing processes should have separate air handling equipment and be arranged in a linear sequence from ultra-clean to less clean with traffic patterns that do not allow employees to backtrack. Ultra-clean areas require high-efficiency particulate air (HEPA) air filtration (Goodenough and Parnell, 1985) or laboratory benches with HEPA-filtered laminar airflow. Linear workflow traffic patterns and sanitation are essential for arthropod natural enemy production (van Frankenhuyzen et al., 2004; Ingles and Sikorowski, 2009), especially tritrophic systems that incorporate plants and phytophagous hosts.

9.4.5. Research

Applied research that supports arthropod mass production and quality assurance is conducted to improve colonization, production, quality control, and utilization. More effective natural enemy species or strains are sought, along with better colonization procedures (Hoffman et al., 1984; Leppla, 1989; Nunney, 2002; Benedict et al., 2009). The goals of research in support of production generally are to adapt materials, invent equipment, and develop methods that improve the system and cost less (Vacari et al., 2012). Typical examples are plants or diets that produce more natural enemies or sometimes suitable alternative hosts (Bernal et al., 1999; Blossey and Hunt, 1999; Karamaouna and Copland, 2000; Serrano and Lapointe, 2002; Wheeler, 2003, 2006; Geden and Kaufman, 2007; Giang and Ueno, 2007; Van Hezewijk et al., 2008; Grenier, 2009; Henry et al., 2010; Wyckhuys et al., 2011). Production, process, and product control procedures can be improved most effectively by conducting research within mass rearing facilities (Couillien and Gregoire, 1994; Miyatake, 2011). Without supporting research, improvements in the production and use of mass-reared arthropods will be incremental based on customer feedback. The research division may have overlapping responsibilities with methods development, or they may be combined depending on the quality assurance system.

9.4.6. Utilization

Mass-produced arthropods must be competent to achieve program goals (Chambers, 1977), as prescribed by established specifications and standards. Quality assurance is more than determining if products meet production requirements; it also involves the maintenance of postproduction quality (Dominiak et al., 2011). Uncomplicated and rapid product control evaluations that are indicative of natural enemy performance should be used to assess product quality prior to shipment and at key points along the supply chain to application in cropping systems. Product quality can be affected by treatments, such as feeding adults, cold storage, and preconditioning (Leopold, 2000, 2007; Niijima et al., 2002; Tezze and Botto, 2004; Chen and Leopold, 2007; Mahmoud and Lim, 2007; Luczynski et al., 2008; Cancino et al., 2009; Chen et al., 2011; Ghazy et al., 2012; Maes et al., 2012). Automated counting, handling, and packaging equipment are required to efficiently collect and ship large numbers of arthropods. Additionally, for natural enemies, it often is necessary to differentiate predators from prey and parasitized from nonparasitized hosts. Differentiation of males from females may be required for arrhenotokous parasitoids because host quality can affect sex ratios. Natural enemy products should at least be checked for identity, purity, and quantity before being shipped. During transport, containers must both preserve product quality and match the requirements of the shipper. Due to their short time of effectiveness, natural enemies usually are shipped via air carriers. Effectiveness of parasitoids and predators can be assured further by optimizing the number applied and the rate and timing of applications, using mechanized application equipment or other methods that ensure complete coverage, and monitoring natural enemy/pest ratios.

9.4.7. Personnel

Quality assurance for arthropod mass production and utilization depends on selecting, training, and motivating appropriate personnel and protecting their health and safety (Fisher, 2009b; Reinecke, 2009). Arthropod mass rearing is highly repetitive and often tedious, requiring attention to detail and consistent accuracy. It is somewhat analogous to other monotonous industrial processes, particularly those involving living organisms, such as potting plants or tending confined animals. Therefore, employees who do not easily tolerate repetitive tasks are rarely successful in the material, production, personnel, and quality control divisions but may fit in management, methods development, research, and utilization. A potential employee should be asked questions designed to determine personality fit (Fisher, 2009b). Some employees enjoy routine tasks, such as those involved in rearing arthropods, and work well together in small groups. In these cases, it can be extremely disruptive to add an employee who does not fit. Regardless of personality type, however, employees must be competent and diligent in performing their assigned tasks. Exposure to respiratory irritants, such as moth scales, and other health risks must be minimized through protocols and use of protective equipment (Wolf, 1984, 1985; Kfir, 1994; Suarthana et al., 2012).

Arthropod mass rearing has become a profession (Cohen, 2001), especially to support areawide pest management (Robinson and Hendrichs, 2007; Leppla et al., 2009). Competence is gained mostly on the job by working with experienced employees, although training can be accelerated considerably by observing other quality assurance systems, studying technical publications, networking with arthropod rearing specialists, and participating in formal classes, such as the Mississippi State University Insect Rearing Workshop and courses at institutions of higher education. A matrix of competencies can be developed with the subject matter of training in the left column and a hierarchy of positions across the top row (Juran et al., 1979). Applied to arthropod mass production, the competencies column includes adaptability, customer focus, independence, initiative, problem solving, quality of work, teamwork, and understanding of rearing principles (Fisher, 1984b, 2009b). The competencies for each position can be described in the internal squares of the matrix. The highest level position, manager, should master the entire production system, including facility design and management, nutrition and diets, microbial management, quality control, applied population genetics, environmental biology, and health and safety (Schneider, 2009). It has been suggested that a certification for each position could be earned by mastering all of the competencies at the corresponding level in the matrix. This kind of training and certification could provide motivation and justify increased advancement and compensation.

9.4.8. Quality Control

The goal of quality control is to maximize yields of high-quality arthropods by rapidly detecting and correcting deviations from standard operating procedures and minimizing variability in production (Bigler, 1989; Sagarra et al., 2000; Bueno et al., 2006). It is accomplished by monitoring production materials and procedures, periodically sampling arthropod stages during development to ensure process control, and evaluating living organisms or associated end products through product control (Leppla, 2009). Monitoring based on efficient and accurate sampling and evaluation throughout production is required to maintain key characteristics of natural enemies within specified ranges. The frequency and extent of quality control evaluations depends on the stability of the system. Sampling, analysis, record keeping, and reporting should be efficient and limited to the amount of information needed to maintain product effectiveness. The quality control division is responsible for rapid data acquisition, analysis, and feedback in appropriate formats. Extending beyond quality control, quality assurance incorporates utilization control and comprehensive feedback from customers about the effectiveness of the product (Table 9.1).

9.5. QUALITY ASSESSMENTS OF MASS-REARED NATURAL ENEMIES

At least minimal product quality control guidelines have been developed for many commercially produced natural enemies (Bigler, 1992; van Lenteren et al., 1993a,b; Leppla and Larson, 2004) (Table 9.2); however, they must be

TABLE 9.1 Quality Assurance Program for Natural Enemy Producers and Suppliers

	Frequency of Actions		
Quality Assurance Actions	**Routinely**	**Periodically**	**Possibly**
Production and Process Control			
Have up-to-date standard operating procedures		X	
Use check sheets to monitor rearing processes*	X		
Record daily production data	X		
Track changes in production**	X		
Review production with staff	X		
Improve product quality based on staff feedback		X	
Product Control			
Conduct product quality control assessments	X		
Have established minimum thresholds of quality	X		
Use standard shipping and handling procedures	X		
Customer Service			
Add customer feedback forms to shipments	X		
Add product use instructions to shipments	X		
Have an established customer service program†	X		
Improve product quality based on customer feedback		X	
Research			
Maintain internal methods improvement‡			X
Conduct publishable research			X
Communicate research needs to outside researchers		X	
Collaborate with researchers without funding projects			X
Contract for research and provide funding			X

Continued

TABLE 9.1 Quality Assurance Program for Natural Enemy Producers and Suppliers—cont'd

Quality Assurance Actions	Frequency of Actions		
	Routinely	Periodically	Possibly
Outreach			
Provide materials for training activities[§]		X	
Provide products for training activities		X	
Participate in training activities[¶]		X	
Describe quality assurance program		X	

**Rearing processes include materials, environments, equipment, and procedures.*
***Tracking by quality control charts, basic statistics, or at least graphs.*
[†]Customer service program with data on the problems and solutions.
[‡]Internal staff members are assigned to improve methods.
[§]Materials typically include product descriptions, application procedures, and information.
[¶]Training activities include trade shows, field days, demonstrations, classes, etc.

TABLE 9.2 Selected Natural Enemies that Have Been Produced Commercially and Consequently Have at Least Minimal Quality Standards

Natural Enemy	Family: Order
Amblyseius californicus (McGregor)	Acarina: Phytoseiidae
Amblyseius cucumeris (Oudemans)	Acarina: Phytoseiidae
Amblyseius (Neoseiulus) degenerans (Berlese)	Acarina: Phytoseiidae
Amblyseius fallacis (Garman)	Acarina: Phytoseiidae
Amblyseius swirskii (Athias-Henriot)	Acarina: Phytoseiidae
Anthocoris nemoralis (Fabricius)	Hemiptera: Anthocoridae
Aphelinus abdominalis (Dalman)	Hymenoptera: Aphelinidae
Aphidius colemani (Viereck)	Hymenoptera: Braconidae
Aphidius ervi (Haliday)	Hymenoptera: Braconidae
Aphidius matricariae (Haliday)	Hymenoptera: Braconidae
Aphidoletes aphidimyza (Rondani)	Diptera: Cecidomyiidae
Aphytis lingnanensis (Compere)	Hymenoptera: Aphelinidae
Aphytis melinus (DeBach)	Hymenoptera: Aphelinidae
Carcinops pumilio (Erichson)	Coleoptera: Histeridae
Chrysoperla carnea (Steph.)	Neuroptera: Chrysopidae

TABLE 9.2 Selected Natural Enemies that Have Been Produced Commercially and Consequently Have at Least Minimal Quality Standards—cont'd

Natural Enemy	Family: Order
Chrysoperla rufilabris (Burmeister)	Neuroptera: Chrysopidae
Cotesia plutellae (Kurdjumov)	Hymenoptera: Braconidae
Cryptolaemus montrouzieri (Mulsant)	Coleoptera: Coccinellidae
Cybocephalus nipponicus (Endrody-Younga)	Coleoptera: Cybocephalidae
Dacnusa sibirica (Telenga)	Hymenoptera: Braconidae
Dalotia coriaria (Kraatz)	Coleoptera: Staphylinidae
Delphastus catalinae (Horn)	Coleoptera: Coccinellidae
Dicyphus hesperus (Wagner)	Hemiptera: Miridae
Diglyphus isaea (Walker)	Hymenoptera: Eulophidae
Encarsia formosa (Gahan)	Hymenoptera: Aphelinidae
Eretmocerus eremicus (Rose)	Hymenoptera: Aphelinidae
Eretmocerus mundus (Mercet)	Hymenoptera: Aphelinidae
Feltiella acarisuga (Vallot)	Diptera: Cecidomyiidae
Galendromus helveolus (Chant)	Acarina: Phytoseiidae
Galendromus occidentalis (Nesbitt)	Acarina: Phytoseiidae
Goniozus legneri (Gordh)	Hymenoptera: Bethylidae
Heterorhabditis bacteriophora (Poinar)	Rhabditida: Heterorhabditidae
Heterorhabditis megidis (Poinar, Jackson & Klein)	Rhabditida: Heterorhabditidae
Hippodamia convergens (Guérin-Méneville)	Coleoptera: Coccinellidae
Hypoaspis miles (Berlese)	Acarina: Laelapidae
Leptomastix dactylopii (Howard)	Hymenoptera: Encyritidae
Macrolophus caliginosus (Wagner)	Hemiptera: Miridae
Mesoseiulus longipes (Evans)	Acarina: Phytoseiidae
Muscidifurax raptor (Girault & Sanders)	Hymenoptera: Pteromalidae
Muscidifurax raptorellus (Kogan & Legner)	Hymenoptera: Pteromalidae
Muscidifurax zaraptor (Kogan & Legner)	Hymenoptera: Pteromalidae
Nasonia vitripennis (Walker)	Hymenoptera: Pteromalidae
Neoseiulus californicus (McGregor)	Acarina: Phytoseiidae

Continued

TABLE 9.2 Selected Natural Enemies that Have Been Produced Commercially and Consequently Have at Least Minimal Quality Standards—cont'd

Natural Enemy	Family: Order
Neoseiulus cucumeris (Oudemans)	Acarina: Phytoseiidae
Orius spp. (*O. insidiosus, O. laevigatus*, etc.)	Hemiptera: Anthocoridae
Pediobius foveolatus (Crawford)	Hymenoptera: Eulophidae
Pentalitomastix plethorica (Caltagirone)	Hymenoptera: Encyrtidae
Phytoseiulus persimilis (Athias-Henriot)	Acarina: Phytoseiidae
Podisus maculiventris (Say)	Hemiptera: Pentatomidae
Rhyzobius lophanthae (Blaisdell)	Coleoptera: Coccinellidae
Scolothrips sexmaculatus (Pergande)	Thysanoptera: Thripidae
Spalangia cameroni (Perkins)	Hymenoptera: Pteromalidae
Spalangia endius (Walker)	Hymenoptera: Pteromalidae
Spalangia nigroaenea (Curtis)	Hymenoptera: Pteromalidae
Trichogrammatoidea bactrae (Nagaraja)	Hymenoptera: Trichogrammatidae
Trichogramma brassicae (Bezd. (=*T. maidis*))	Hymenoptera: Trichogrammatidae
Trichogramma cacoeciae (Marchal)	Hymenoptera: Trichogrammatidae
Trichogramma dendrolimi (Matsumura)	Hymenoptera: Trichogrammatidae
Trichogramma minutum (Riley)	Hymenoptera: Trichogrammatidae
Trichogramma platneri (Nagarkatti)	Hymenoptera: Trichogrammatidae
Trichogramma pretiosum (Riley)	Hymenoptera: Trichogrammatidae
Thripobius semiluteus (Boucek)	Hymenoptera: Eulophidae
Steinernema carpocapsae (Weiser)	Rhabditida: Steinernematidae
Steinernema feltiae (Filipjev)	Rhabditida: Steinernematidae
Steinernema scapterisci (Nguyen & Smart)	Rhabditida: Steinernematidae

Source: Information from Van Lenteren (2003b); Leppla and Larson (2004); Leppla and Johnson (2010).

improved and adapted for use in specific situations. The guidelines specify a limited number of parameters that can be measured to indicate consistent quality for a species. Life history and morphological characteristics are easier to measure than behavior (Noldus, 1989; Lux, 1991) and can be good indicators of quality, such as number of adults, rate of emergence, proportion of live insects, size or weight, sex ratio, and longevity. Highly variable characteristics that are hard to measure, such as fertility and fecundity, are used only if necessary to monitor production and if the natural enemies are expected to reproduce in the field. The life history and behavior of mass-produced natural enemies will be consistent in reasonably stable production systems (Gandolfi et al., 2003), yielding quality control data that can be used to detect deviations from historical values.

The quality of mass-reared parasitoids and predators has been assessed partially by conducting postshipment product quality control evaluations. In an unprecedented 2.5-year study, three unidentified surrogate customers ordered the following species from four different companies: *E. formosa*, *Trichogramma pretiosum* Riley (Hymenoptera: Trichogrammatidae), *Chrysoperla carnea* (Stephens) (Neuroptera: Chrysopidae), and *Hippodamia convergens* Guerin (Coleoptera: Coccinellidae) (O'Neil et al., 1998). The number, identity, and stage of insects received were evaluated. Depending on species, assessments were made of emergence, sex ratio, survival, reproduction, and hyperparasitism. Initially, few adult parasitoids had emerged during shipment, regardless of species. For *E. formosa*, the number of pupae received was equal to the number ordered or considerably greater, emergence averaged 41.3%, and survival of adults after two days was 89–95%. Shipments of *T. pretiosum* periodically contained fewer insects than were ordered and the maximum average level of survival was 86%. Considering the average of orders for three companies and three receiving locations, only four of nine had a sufficient number of *H. convergens*. About 80% of the beetles were alive and survival after 48 h averaged 82–96%, although 9–22% were parasitized. All of the presumptive *C. carnea* adults received from the three companies actually were *Chrysoperla rufilabis* (Burmeister). Because there often was considerable variability in the products, the study concluded that "working relations between producers, evaluators, and customers would benefit all." Further, "defining realistic expectations for the quality of commercially available natural enemies will serve to expand their use in augmentative biological control and integrated pest management programs."

Encarsia formosa was included in a subsequent quality assessment of whitefly parasitoids, along with *Aphidius colemani* Viereck (Hymenoptera: Braconidae) and *Aphidoletes aphidimyza* (Rondani) (Diptera: Cecidomyiidae) (Vasquez et al., 2004). Tests from IOBC international standards used to assess product quantity were total adult emergence and sex ratio (van Lenteren, 2003a), in addition to recording the number of insects that flew. The length of the tibia on a hind leg of *E. formosa* was measured as an indicator of body size. Arrival of the *E. formosa*-parasitized whitefly pupae at the expected time was highly variable, depending on the company, and only three of the six companies provided detailed information on how to handle and use their products. Parameters that varied significantly between companies were adult emergence on arrival, parasitized whitefly pupae in the container, and insects that could fly; however, total emergence, sex ratio and tibia length were not significantly different. Some shipments of *E. formosa* contained *Eretmocerus* spp., also a whitefly parasitoid. Of the 1500 to 5000 *E. formosa* ordered, 17–91% functional adults were received (number emerged on arrival plus number unable to fly subtracted from the total that emerged). Thus, it appears that the parameters most indicative of product quality in *E. formosa* are quantity received, purity, emergence on arrival, total emergence, and flight capability. Similar results were obtained for *A. colemani*, i.e. highly variable product quality from individual sources and between sources, high levels of emergence on arrival in some cases, and emergence of 61–69% with just above 50% females. Thus, in this study, the six companies delivered an inconsistent number of functional wasps.

Additional quality control studies have been conducted on mass-reared natural enemies, both commercial and governmental, including several *Trichogramma* spp. (Cerutti and Bigler, 1991; Losey and Calvin, 1995; Hassan and Zhang, 2001), *Splangia cameroni* Perkins (Hymenoptera: Pteromalidae) (Tormos et al., 2011), *Orius insidiosus* (Say) (Heteroptera: Anthocoridae) (Shapiro and Ferkovich, 2002) and *Epidinocarsis lopezi* (De Santis) (Hymenoptera: Encyrtidae) (Neuenschwander et al., 1989). In a study of postshipment product quality control for *A. colemani,* species identification was inaccurate, hyperparasitoids were present in most shipments, an excessive number of wasps emerged before receipt, and a large number of wasps did not parasitize aphids (Fernandez and Nentwig, 1997). The results of quality control tests prior to shipment were much more consistent for a congener, *Aphidius matricariae* (Haliday) (Enkegaard and Reitzel, 1991). The number of wasps produced averaged 1663.1 ± 17.3 (SEM) with 71.9 ± 3.6% emerging and 88.9% flying. Predictable production and synchronized emergence enabled packaging and shipping of an accurate number of wasps that did not emerge prior to receipt. In another study, *Aphytis melinus* DeBach (Hymenoptera: Aphelinidae) was evaluated in cooperation with five insectaries to determine the variability of wasp quality (Vasquez and Morse, 2012). All of the insectaries were located in California and the natural enemies were shipped every 2 months overnight to the University of California, Riverside. Quality assessments included percentage of live parasitoids 1–28 days after receipt, sex ratio, and size of female wasps. Statistically significant variation occurred in all three parameters between insectaries and throughout the year, even though the wasps were reared in controlled environments. The mean proportion of females was 0.436–0.591 between insectaries and 0.393–0.644 for all insectaries during the approximately 2-year period.

9.6. QUALITY ASSURANCE AND CONTROL DATA ACQUISITION AND ANALYSIS

A minimum amount of data must be obtained routinely to maintain production, process, and product control, as well as to monitor product utilization and consumer satisfaction. Typical examples of production control data are temperature and relative humidity for all of the rearing environments

(Chen et al., 2012). Yield of a developmental stage is typically used for process control, including the number and proportion of viable eggs, the number of larvae per rearing container, and the proportion of containers with microbial contamination. Production and postproduction utilization control parameters for natural enemies are quantity, purity, emergence (viability), adult size, sex ratio, flight, fecundity, longevity, and parasitism or predation. For augmentation biological control, customers are satisfied only if an acceptable level of pest population suppression is achieved (Bolckmans, 2003). Therefore, quality assurance data include both natural enemy characteristics that are indicative of quality and sensitive to change and feedback from customers. Routine monitoring is expensive and the information is used to make important decisions, so the set of criteria and sampling structure must be limited but adequate. More intense sampling and complicated statistical analyses are needed for supportive research.

A sampling scheme that produces consistent means and standard deviations for each parameter must be determined for every site-specific rearing system. Variation is minimized by having the same person use standardized techniques to obtain a consistent number of random samples at a prescribed time of day and week. Daily production control data is recorded on data sheets and preharvest process control is accomplished by periodically sampling immature insects (Akey et al., 1984). The rearing container is the sampling unit, not the batch or worker shift, and the number of containers sampled at a time will depend on their variation. Increased variation among rearing containers can indicate a biological problem, whereas variation in batches results from inconsistent production control. The most efficient indicators of product quality at production and receipt by the consumer are quantity and purity of natural enemies received, perhaps percentage emergence on arrival, total emergence, and flight capability. Customers expect shipments to arrive consistently on time, containing the species and number of natural enemies ordered in good condition and at the correct stage of development. The data are analyzed using simple statistical methods such as JMP (SAS Institute, 2004), and recorded on quality control charts (Feigenbaum, 1983; Chambers and Ashley, 1984; Wajnberg, 1991, 2003; Leppla, 2009) (Fig. 9.3). The mean and standard deviation for each product characteristic should be plotted over time and stay within the historical range (Chambers and Ashley, 1984), probably within 1–2 standard deviations. This kind of data management also can be applied to feedback from consumers, including both product control and effectiveness parameters. Informed decisions can thereby be made about reducing sources of variation in production and the supply chain.

9.7. QUALITY ASSURANCE SYSTEM REVIEW

Every function of a quality assurance system for mass-reared arthropods should be reviewed to determine if it conforms to established specifications and

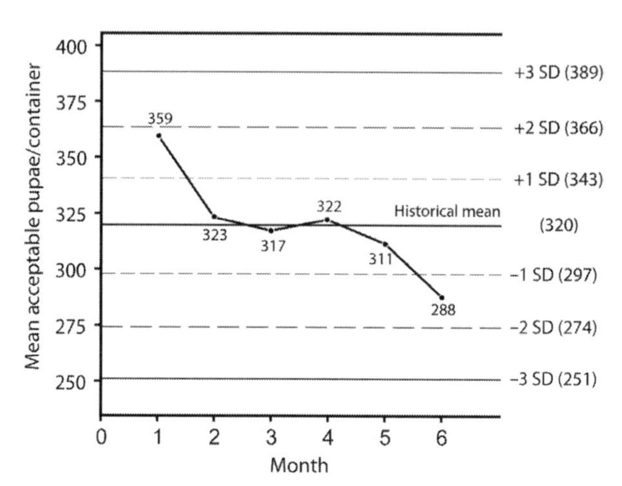

FIGURE 9.3 A quality control chart for the mean monthly number of acceptable fall armyworm, *Spodoptera frugiperda* (J.E. Smith) (Lepidoptera: Noctuidae) pupae per rearing container (*data from Leppla and Ashley, 1989*). The historical mean was 323.3 (rounded to 320) pupae per container and, using the monthly data, the standard deviation was 22.9. Mean and range charts can be used to monitor any measurable quality assurance parameter.

standards of performance. Depending on the function, reviews can be performed by personnel who accomplish the work, division managers, quality control or some other division, internal professional auditors, upper management, or outside experts. Generally, the complexity, rigor, independence, value, and cost of a review increase along this continuum. It has been argued that reviews within work units are biased or even dishonest, but employees who perform the work are most familiar with its nuances. These workers and their division managers will know when "corners are cut," such as circumventing traffic patterns, substituting substandard materials, and manipulating check sheets. Typical examples for traffic patterns in arthropod mass rearing facilities are using doors to avoid inefficient pass-throughs or back tracking through contaminated areas to retrieve a piece of equipment. Reviews by members of other divisions or shifts within the same division can be highly biased and overly critical, particularly if any of the reviewers once were assigned to the division or shift being reviewed. It can be enlightening, however, to compare the success of shifts that raise the same arthropods. To the present time, even the largest arthropod mass rearing facilities have not had internal professional auditors. Upper management often performs reviews, but the efforts tend to be superficial unless the managers have remained actively engaged in the quality assurance system. Managers who understand the inner workings of the system tend to have a positive influence on performance of the functions, whereas inexperienced managers may be fed misinformation and not be taken seriously. Consequently, the least biased and most comprehensive reviews of quality assurance programs for mass-reared arthropods probably can be performed by outside experts.

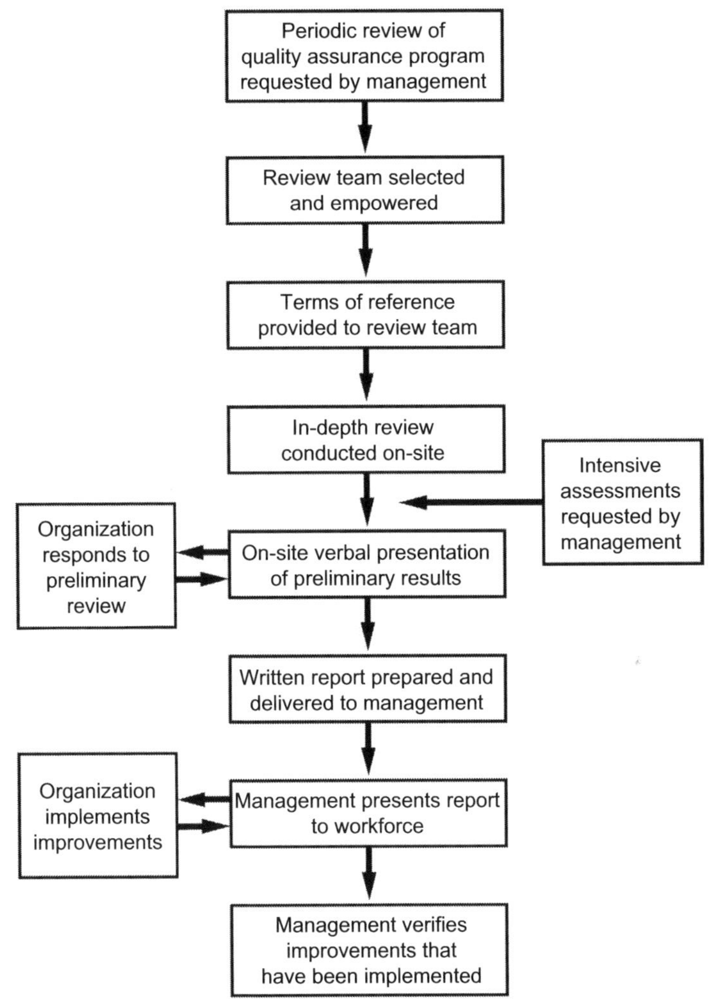

FIGURE 9.4 Process for reviewing a quality assurance system for arthropod mass production based on experience of the author and Juran et al. (1979).

A periodic review by outside experts begins when it is requested by upper management (Fig. 9.4). The entire quality assurance system is subject to the review and, if possible, should be performed when the system is stable and not experiencing a crisis. In practice, however, the resources for a thorough outside review typically become available when there is a major decrease in production. Reviews conducted during a crisis usually are rushed, lack careful planning, and can be stressful for everyone involved. This makes it difficult for review team members to develop rapport with the employees and earn their trust, avoid unproductive fault-finding, and gain an understanding of possible sources of system failure. Upper management must select review team members with the requisite knowledge, experience, stature, energy, diplomacy, willingness, and sometimes courage to conduct a successful review. Experts in arthropod mass production and utilization who have these attributes are uncommon and may not be available within the required timeframe. Once identified, suitable review team members are empowered to evaluate the system according to explicit written terms of reference. It is not appropriate to expect a review team to unilaterally identify the details of the review, although they often participate with upper management in refining the goals. The terms are typically compiled in a document that includes the mission and essential information about every function of the system, available human and physical resources, specific objectives for the review, reporting requirements, the timeframe, and associated expectations. The entire system likely will be evaluated for ways to make it more reliable and cost-effective, but this charge is too general. A more specific goal, for example, would be to evaluate every step in the product supply chain to identify causes for delays, so that high quality 2-day-old natural enemies can be delivered to customers consistently. In consultation with management and selected employees, the review team could develop actions intended to achieve this kind of goal for the review.

While the in-depth quality assessment system review is underway, the review team members should delve into the nuances of each function and periodically consult with management to refine the process. Managers may request that intensive evaluations be made concerning particular aspects of the system. They may have a special interest or perceive that a process requires an especially high level of scrutiny. This often occurs when a new piece of equipment is installed to automate a rearing procedure, such as a form-fill-seal machine to automate packaging and infestation of a diet. Adjustments invariably must be made as the workers and arthropods adapt to the change. As the review progresses, preliminary results are assembled and analyzed by the review team members; if gaps exist, additional information must be obtained. The time required to reflect and complete a comprehensive review is another reason for not conducting it when the system is failing and requires rapid restoration. A verbal report is presented onsite before the review is completed to explain preliminary findings and obtain feedback from the workforce. Care must be taken to identify strengths and opportunities while avoiding the impression that there are major deficiencies and a need for individual employees to improve. The final written report should be delivered to management within a few weeks, enabling managers to present it to the workforce expeditiously. Considerable discussion and detailed analysis of the findings by the managers and employees should take place before deciding to implement specific recommendations for improving the system. Feedback is then provided to the review team on each point in the report, including plans to implement changes as well as findings for which there is disagreement. The changes can be made either to restore or enhance specific aspects of the quality assurance system. Finally, management should verify that the improvements have been made and may ask a particularly insightful review team to remain available for consultation or a subsequent review.

The following outline for reviewing a quality assurance system for mass rearing arthropods and reporting the results was adapted from a World Wildlife Fund project audit framework (Anonymous, 2005):

- Approach
  - Define purpose and scope of the quality assurance system review
  - State reason for the quality assurance review
  - Review team members, qualifications, interest in review, and roles
  - Determine system functions being reviewed for performance
  - Define expectations and objectives of the review
  - Review methodology and timeline, data collection and analysis, ethics, and fairness
- Review of functions (successes and failures)
  - System design, quality, and effectiveness
  - Production, process, and product control data
  - Utilization control data
  - Communication and feedback mechanisms
  - Problems encountered that require urgent attention
- Conclusions
  - Reasons for function successes and failures
  - Insights into the findings
  - Innovations for increasing quality assurance
- Recommendations (based on evidence and insights)
- Attachments to report
  - Terms of reference for the review
  - Short biographies of the evaluators
  - Structure and timetable for the review
  - List of individuals interviewed
  - List of supporting documentation reviewed
  - Research instruments, questionnaires, and interview guides
  - Specific data acquired and summary tables

The World Wildlife Fund audit framework has more detail, along with a template for the terms of reference that includes the project background and context, purpose and objectives of the review, audience for the review, review issues and key questions (review matrix), quality and relevance of design, effectiveness, efficiency of planning and implementation, impact, potential for sustainability, replication and magnification, methodology, profile of the review team, evaluation timetable, cost, and logistical support.

9.8. RESEARCH ON QUALITY ASSESSMENT FOR MASS-REARED PARASITOIDS AND PREDATORS

An enormous increase in research and development must be made before use of mass-reared parasitoids and predators can become a major technology in integrated pest management. Research in both the laboratory and field, including demonstrations of effectiveness, is needed to assure the peak performance of natural enemies (Glenister and Hoffman, 1998; van Lenteren and Tommasini, 1999; Rendon et al., 2006). Virtually every aspect of augmentative biological control technology can be improved, including the feasibility of developing and marketing new natural enemy products, producing them reliably and efficiently, applying and evaluating them, and modifying the environment in which they are released (Table 9.3). A product is developed only if research has determined

TABLE 9.3 Research to Advance Quality Assurance in the Use of Mass-Produced Natural Enemies

Type of Research	Research Priorities
Feasibility/market analysis	Natural enemy effectiveness in controlling target pests Product development Quality assurance systems
Mass production	Rearing facilities Mechanized equipment Rearing materials Natural and artificial diets Rearing and harvesting techniques Production quality control Process quality control Product quality control
Utilization	Product handling and storage Automated counting technology Improved packaging and shipping Extended use period Efficient scouting techniques Optimal release numbers, rates and timing Mechanized application technologies Efficient product evaluation procedures Utilization quality control
Application environment	Site-specific pest prevention Systems to maximize the effectiveness of natural enemies Natural enemy use with pesticides Systems with multiple products and pests Effectiveness in seasonal and perennial crops

that the natural enemy is potentially effective, there is an identified market or need for the product, it can be produced and delivered to meet the demand, and an associated quality assurance system can be designed and implemented. Feasibility studies are conducted by entomologists in cooperation with managers who have considerable experience in analyzing potential markets and uses and determining cost/benefit.

Researchers should collaborate more frequently with biological control producers, suppliers, and customers to significantly improve the quality and effectiveness of mass-produced natural enemies. This is because the commercial biological control industry lacks the resources and often expertise to make significant improvements in rearing, packaging, storing, shipping, applying, and evaluating their products. Researchers could assist the industry by conducting quality assurance studies extending from feasibility through mass production and utilization to the application environment, thus assuring that their discoveries are used and evaluated. Researchers should assess product quality collaboratively and report the results to the supplier immediately so that problems can be identified quickly and corrected in the production and distribution chain. Product control assessments conducted by the recipient can indicate that a problem exists but not identify the cause (Webb et al., 1981; Steward et al., 1996; Thomson and Hoffmann, 2002). Possible causes include changes in hosts or diets (Rodriguez et al., 2002; Lopez et al., 2009), rearing materials and environments, personnel and their training, handling and packaging, transportation and routing, treatment on receipt, and testing procedures. Researchers and suppliers should jointly establish protocols for quality control assessments that specify taxon, sampling scheme, replication, materials, procedures, data analysis, and requirements for reporting. Once a standardized evaluation protocol is established for a product, it can be used by both trained suppliers and customers.

Typically, entomologists and agricultural engineers jointly design the rearing facilities and equipment, select materials, and establish the arthropod production line (Harrell and Gantt, 1984). They also collaboratively conduct postproduction research on handling, release, and evaluation of natural enemy products that could significantly increase their acceptance in the marketplace. The cost of natural enemy products could be reduced by engineering automated counting, packaging, storage, transportation, and application equipment (Pearson et al., 2002). Storage and stockpiling of natural enemies would enable suppliers and customers to better time applications, although prolonged storage usually is detrimental. For *A. aphidimyza*, stored females had low fecundity and therefore did not produce the number of progeny needed for successful biological control (Fernandez and Nentwig, 1997). Females that had not been stored were extremely useful in controlling a wide range of aphid species, whereas those that had been stored for months had little impact (Luczynski et al., 2007). If customers were able to hold *A. aphidimyza* pupae in a transparent package until adults emerge, viability of the product could be observed and emergence enhanced in the high humidity environment. Improvements through research in packaging, such as insulation requirements and monitoring equipment, would increase the effectiveness of natural enemies shipped over long distances. Well-designed packaging could protect pupae from ants and other predators, increase mating due to the confined space, and enable customers to determine the best time to apply the adults. Some natural enemy products are packaged in substrates that facilitate mechanical application, and these carriers could be improved.

Entomological research is needed to improve arthropod mass rearing systems, including natural and artificial diets (Reis et al., 2003; Wackers and van Lenteren, 2003; Arijs and De Clercq, 2004; Nathan et al., 2006; Vandekerkhove et al., 2006; Bonte and De Clercq, 2011; Bonte et al., 2011) and alternative rearing techniques (Bonte and De Clercq, 2010). For example, a parasitoid typically has higher levels of oviposition and survival on a natural host, whereas a factitious host can be much less expensive and more practical to use under conditions of mass production (Watson et al., 2000; Ramalho and Dias, 2003; Bonte and De Clercq, 2008; Tuncbilek et al., 2009). Artificial diets for natural enemies have been formulated to incorporate various gelling agents because agar-based diets are relatively expensive (Neuenschwander et al., 1989). These kinds of tradeoffs that affect colonization and colony maintenance can result in genetic bottlenecks that reduce survival and heterozygosity (McMurtry and Scriven, 1975; Jones et al., 1978; Bartlett, 1984, 1985; Joslyn, 1984; Mangan, 1992; Mohaghegh et al., 1998; Ode and Heinz, 2002; Hegazi and Khafagi, 2005; He and Wang, 2006; Gonzalez et al., 2007; Sarvary et al., 2008; Joyce et al., 2010). The goal is to produce natural enemies with characteristics that approximate those of the colony founders and can be maintained indefinitely (Messing et al., 1993; Prezotti et al., 2004). Customized production, process, and product quality control procedures are needed for each arthropod mass rearing system, along with appropriate statistical procedures (Pashley and Proverbs, 1981; Nordlund et al., 1997). Moreover, postproduction research could provide methods for customers to efficiently evaluate products at receipt and in the field (van Schelt and Mulder, 2000). Efficient scouting techniques could be developed by entomologists to optimize natural enemy release numbers, rates, and timing.

Mass-reared natural enemies are most effective if applied in site-specific pest prevention systems that conserve beneficial organisms and other natural controls. However, natural enemies often are evaluated and applied inappropriately in cropping systems and other situations designed to optimize the use of pesticides for maintaining pests below acceptable thresholds. In these systems, natural enemies generally are not compatible with pesticides and their use is precluded. After being released, natural enemies must survive and hopefully reproduce. Research must be continued on the effectiveness of natural enemies and the economic and ecological limits of their use in biological control (Collier and Van Steenwyk, 2004). Allied research is needed to determine the immediate and residual toxicity of chemical pesticides to natural enemies so that incompatibilities can be predicted and avoided (Pree et al., 1989). More systems must be designed to maximize the effectiveness of parasitoids and predators in concert with compatible pesticides, such as those used in organic agriculture. In these kinds of systems, most pesticides are not acceptable because of their nontarget effects, frequency of application, and long-term higher cost. Research also must be conducted to determine and demonstrate how natural enemies can be used in cropping systems based on biological and mechanical controls, with scouting and reasonable damage thresholds. Pest management can be complicated further when multiple parasitoids and predators are released to manage several pest species. Ultimately, reliance on natural enemies results from research that demonstrates their effectiveness in significantly reducing pest populations. According to Penn et al. (1998), "If biologically intensive integrated pest management is to become a predominant practice, and if a substantial increase in the use of mass-reared arthropods and other biologically based products is to be realized, the entire agricultural community (academia, government, growers and the pest control industry) must work together to develop comprehensive quality assurance programs and to specifically design pest management systems to accommodate these products."

9.9. CONCLUSION

This chapter on concepts and methods of quality assurance for mass-reared parasitoids and predators extends a previous book chapter, "The Basics of Quality Control for Insect Rearing" (Leppla, 2009). Here, emphasis is placed on mass rearing arthropod natural enemies, and the concept of total quality control in insect mass production has been expanded to emphasize customer involvement and satisfaction. This expansion combines all aspects of developing, producing, delivering, using, and evaluating natural enemy products. Mass production of high-quality arthropods requires a well-designed and managed system that is efficient and reliable. To maintain reliability, every function of the system should be reviewed periodically so that it continues to conform to established standard operating procedures. Key determinants of quality are assessed to ensure the performance of mass-produced arthropods and to detect changes that could compromise their effectiveness in the field.

The purpose for mass rearing beneficial arthropods is to satisfy customer needs and make a profit if the products are marketed. Therefore, a product must be effective, marketable, deliverable to meet demand, and produced and used according to an associated quality assurance system. Before investments are made in developing an arthropod product, a comprehensive business plan should be developed that includes these provisions and feasibility of the integrated pest management program that requires high-quality natural enemies or a market analysis for commercial ventures. It is essential to have customers involved in determining their needs and expectations and drafting associated product specifications and standards. Producers and suppliers must not only make sure that the customers are satisfied but also deliver high-quality products based on appropriate assessments. This can be challenging if there is a lack of consistent rearing and predictable performance of natural enemy products. In the absence of customer feedback that identifies the cause of unacceptable product quality, producers and suppliers typically resupply the purchased product and provide technical support in an attempt to assure that it ultimately is effective. Increased quantity does not compensate for poor quality, however. Natural enemies will continue to be produced and used in greater quantities if they are cost-effective and suppress pest populations to levels acceptable to the customers.

An arthropod mass production and utilization system designed and operated to maximize product quality and effectiveness includes the following divisions: management, personnel, methods development, research, material, production, quality control, and utilization. Management must fully participate in planning the organization, instituting policies for its operation, establishing standard operating procedures and schedules, assuring the competency of employees, and providing ongoing administrative support for the entire system. Employees must be appropriate, well trained, and conscientious in performing their assigned tasks, as well as periodically evaluated and rewarded. The methods development division is responsible for increasing product quality and reducing costs by optimizing the system, often by pilot testing potential advancements from research. Applied research is conducted to improve colonization, production, quality control, and utilization. The material division verifies that specifications and standards are met for all materials and supplies purchased from vendors. Arthropod mass production requires diligent adherence to standard operating procedures and immediate correction of deviations through production control. The quality control division monitors rearing processes and products, providing rapid feedback to production and management. Feedback directly to the production division enables immediate problems to be corrected. Management may use quality control feedback to make long-term changes to the system. The utilization division conducts product quality control assessments at key points along the supply chain to maintain product effectiveness and assure customer satisfaction.

Product quality standards have been developed for many of the more commonly produced predators and parasitoids. A set of assessments can be selected from these standards and adapted for use in specific situations to measure characteristics indicative of product quality and consistency. Depending on the species, preshipment and postshipment product quality control assessments could include the number, identity, and stage of insects supplied and their emergence, sex ratio, survival, and reproduction. Movement and hyperparasitism also are determined for some species. The data can be analyzed using simple statistics and recorded as means and standard deviations on quality control charts. Species studied in detail include *E. formosa*, *T. pretiosum*, *C. carnea*, *H. convergens*, *A. colemani*, *A. aphidimyza*, *Eretmocerus* spp., *Trichogramma* spp., *S. cameroni*, *O. insidiosus*, *E. lopezi* and *A. matricariae*. In these studies, the quality of natural enemies was highly variable between insectaries and throughout the year, probably because of limited employee experience and turnover. To reduce this variability, producers, suppliers, and customers should routinely conduct the same set of accurate assessments and rapidly share the results.

A minimum amount of monitoring is required to maintain production, process, product, and utilization control.

The quality of mass-produced natural enemies is improving incrementally due to advances in associated research and technology. Considerably more research is needed, however, to rapidly establish quality assurance systems that extend from feasibility through mass production to application and evaluation. Field studies to assure the peak performance of natural enemies are a high priority. Molecular techniques can be used to rapidly identify species and strains, screen colonies for pathogens and hyperparasites, and eventually genetically engineer natural enemies (Hoy, 1979). Research is required to produce natural enemies on alternative hosts and design and build reliable field insectaries. Entomologists, engineers, food technologists, and others could formulate, prepare, and package better diets; evaluate rearing materials and environments; increase automation of rearing processes; test alternative transportation options; and perhaps help train personnel. Postproduction research on the handling, release, and evaluation of natural enemy products is crucial and researchers, suppliers, and customers could jointly develop efficient product quality control assessments. Mass-reared natural enemies must be applied in integrated pest management systems designed to maximize their effectiveness, so research should be conducted on compatible practices. New selective pesticide chemistries could provide opportunities to combine biological and chemical control, especially when multiple pests are present or a backup is needed for the natural enemies. Because there are no pesticide-like labels for predators and parasitoids, it would help to have general guidelines for using each species that would minimize errors in handling, application rates, and timing. Thus, research is needed on almost every aspect of quality assurance for mass-produced arthropods to design and operate systems that produce and deliver products that satisfy customer expectations.

ACKNOWLEDGMENTS

Several of the quality assurance concepts presented in this chapter resulted from information provided by members of the Association of Natural Biocontrol Producers, especially Carol Glenister, IPM Laboratories, Inc. Carol Glenister; John Schneider, Mississippi State University, Department of Entomology and Plant Pathology; and Jim Nation, University of Florida, Entomology and Nematology Department provided very helpful reviews of the manuscript. I also thank Jane Medley, University of Florida, Entomology and Nematology Department, for preparing the figures.

REFERENCES

Aeschlimann, J.P., 1996. Proceedings of the IOBC Montpellier conference. Entomophaga 41, 307–531.

Akey, D.H., Jones, R.H., Walton, T.E., 1984. Systems analysis and automated data processing in insect rearing: a system for the biting gnat *Culicoides variipennis* and mosquitoes. In: King, E.G., Leppla, N.C. (Eds.), Advances and Challenges in Insect Rearing, Agricultural Research Service, United States Department of Agriculture, New Orleans, LA, pp. 269–291.

ANSI/ASQC, 1987. Quality Systems Terminology. American National Standards Institute/American Society for Quality Control, ANSI/ASQC A3-1987, Milwaukee, WI.

Anonymous, 2003. FAO/IAEA/USDA Manual for Product Quality Control and Shipping Procedures for Sterile Mass-Reared Tephritid Fruit Flies, Version 5. http://www-naweb.iaea.org/nafa/ipc/public/ipc-mass-reared-tephritid.html.

Anonymous, 2004. FAO/IAEA Generic Design, Technical Guidelines and Optimal Location of Tsetse Fly Mass-Rearing Facilities. IAEA-314-D4-CT09393. http://www-naweb.iaea.org/nafa/ipc/public/GenericDesign_2004.pdf.

Anonymous, 2005. WWF Template for Terms of Reference for Project and Program Evaluations. www.panda.org/standards/evaluations_terms_of_reference/.

Anonymous, 2006. FAO/IAEA Standard Operating Procedures for Mass-Rearing Tsetse Flies. http://www-naweb.iaea.org/nafa/ipc/public/Tsetse_Rearing_SOP_web.pdf.

Anonymous, 2008. FAO/IAEA Model Business Plan for a Sterile Insect Production Facility. IAEA-MBP ISBN: 978-92-0-110007 http://www-naweb.iaea.org/nafa/ipc/public/IAEA-MBP_SP_facility.pdf.

Arijs, Y., De Clercq, P., 2004. Liver-based artificial diets for the production of *Orius laevigatus*. Biocontrol (Dordrecht) 49, 505–516.

Bartlett, A.C., 1984. Genetic changes during insect domestication. In: King, E.G., Leppla, N.C. (Eds.), Advances and Challenges in Insect Rearing, Agricultural Research Service, United States Department of Agriculture, New Orleans, LA, pp. 2–8.

Bartlett, A.C., 1985. Guidelines for genetic diversity in laboratory colony establishment and maintenance. In: Singh, P., Moore, R.F. (Eds.), Handbook of Insect Rearing, vol. 1. Elsevier, Amsterdam, The Netherlands, pp. 7–17.

Benedict, M.Q., Knols, B.G.J., Bossin, H.C., Howell, P.I., Mialhe, E., Caceres, C., Robinson, A.S., 2009. Colonisation and mass rearing: learning from others. Malar. J. 8 (Suppl. 2), Art. No. S4 http://www.malariajournal.com/content/8/S2/S4.

Bernal, J.S., Luck, R.F., Morse, J.G., 1999. Host influences on sex ratio, longevity, and egg load of two *Metaphycus* species parasitic on soft scales: implications for insectary rearing. Entomol. Exp. Appl. 92, 191–204.

Bernon, G.L., Leppla, N.C., 1994. Nutrition and quality control in mass rearing phytophagous insects. In: Ochieng'-Odera, J.P.R. (Ed.), Techniques of Insect Rearing for the Development of Integrated Pest and Vector Management Strategies, Proc. Int. Group Training Course on Techniques of Insect Rearing for the Development of Integrated Pest and Vector Management Strategies, vol. I. ICIPE Science Press, Nairobi, Kenya, pp. 211–220.

Bigler, F., 1989. Quality assessment and control in entomophagous insects used for biological control. J. Appl. Entomol. 108, 390–400.

Bigler, F., 1992. Report [definitions of quality control]. In: Bigler, F. (Ed.), 6th Workshop of the IOBC Global Working Group "Quality Control of Mass-Reared Arthropods." Horsholm, Denmark, pp. 3–9.

Bigler, F., 1994. Quality control in *Trichogramma* production. In: Wajnberg, E., Hassan, S.A. (Eds.), Biological Control with Egg Parasitoids, CABI, Wallingford, UK, pp. 93–114.

Bloem, S., Bloem, K.A., Knight, A.L., 1998. Assessing quality of mass-reared codling moths (Lepidoptera: Tortricidae) using field release-recapture tests. J. Econ. Entomol. 91, 1122–1130.

Bloem, K.A., Fielding, L.S., Bloem, S., 2002. Handling: a forgotten factor in quality control. In: Leppla, N.C., Bloem, K.A., Luck, R.F. (Eds.), Proceedings Quality Control for Mass-Reared Arthropods, 8th and 9th Workshops of the IOBC Working Group on Quality Control of Mass-Reared Arthropods, p. 139. http://AMRQC.org.

Blomefield, T., Carpenter, J.E., Vreysen, M.J.B., 2011. Quality of mass-reared codling moth (Lepidoptera: Tortricidae) after long-distance transportation: 1. Logistics of shipping procedures and quality parameters as measured in the laboratory. J. Econ. Entomol. 104, 814–822.

Blossey, B., Hunt, T.R., 1999. Mass rearing methods for *Galerucella calmariensis* and *G-pusilla* (Coleoptera: Chrysomelidae), biological control agents of *Lythrum salicaria* (Lythraceae). J. Econ. Entomol. 92, 325–334.

Bolckmans, K.J.F., 1999. Commercial aspects of biological pest control. In: Albajes, R., Gullino, M.L., van Lenteren, J.C., Elad, Y. (Eds.), Integrated Pest and Disease Management in Greenhouse Crops, Kluwer Academic Publishing, Dordrecht, The Netherlands, pp. 310–338.

Bolckmans, K.J.F., 2003. State of affairs and future directions of product quality assurance in Europe. In: van Lenteren, J.C. (Ed.), Quality Control and Production of Biological Control Agents, Theory and Testing Procedures, CABI Publishing, Cambridge, MA, pp. 215–224.

Boller, E.F., 1972. Behavioral aspects of mass-rearing of insects. Entomophaga 17, 9–25.

Boller, E.F., Chambers, D.L., 1977a. Concepts and approaches. In: Boller, E.F., Chambers, D.L. (Eds.), Quality Control, an Idea Book for Fruit Fly Workers, IOBC, SROP, WPRS, pp. 4–131977/5.

Boller, E.F., Chambers, D.L., 1977b. Quality aspects of mass-reared insects. In: Ridgway, R.L., Vinson, S.B. (Eds.), Biological Control by Augmentation of Natural Enemies, Insect and Mite Control with Parasites and Predators, Plenum, New York, pp. 219–235.

Boller, E.F., Leppla, N.C., 2006. Global working group on arthropod mass rearing and quality control (1968–present). In: Boller, E.F., van Lenteren, J.C., Delucchi, V. (Eds.), International Organization for Biological Control of Noxious Animals and Plants, History of the First 50 Years (1956–2006), pp. 129–134.

Bonte, M., De Clercq, P., 2008. Developmental and reproductive fitness of *Orius laevigatus* (Hemiptera: Anthocoridae) reared on factitious and artificial diets. J. Econ. Entomol. 101, 1127–1133.

Bonte, M., De Clercq, P., 2010. Impact of artificial rearing systems on the developmental and reproductive fitness of the predatory bug, *Orius laevigatus*. J. Insect Sci. 10, 104. http://insectscience.org/10.104.

Bonte, M., De Clercq, P., 2011. Influence of predator density, diet and living substrate on developmental fitness of *Orius laevigatus*. J. Appl. Entomol. 135, 343–350.

Bonte, J., Vangansbeke, D., Maes, S., Bonte, M., Conlong, D., De Clercq, P., 2011. Moisture source and diet affect development and reproduction of *Orius thripoborus* and *Orius naivashae*, two predatory anthocorids from southern Africa. J. Insect Sci. 12, Art. No. 1.

Bourchier, B.H., Van Hezewijk, R.S., Smith, S.M., 2000. Searching speed of *Trichogramma minutum* and its potential as a measure of parasitoid quality. Biol. Control 17, 139–146.

Brazzel, J.R., Calkins, C.O., Chambers, D.L., Gates, D.B., 1986. Required Quality Control Tests, Specifications, and Shipping Procedures for Laboratory Produced Mediterranean Fruit Flies for Sterile Insect Control Programs. United States Department of Agriculture, Animal and Plant Health Inspection Service, Plant Protection and Quarantine, APHIS, pp. 81–51.

Brewer, F.D., Lindig, O., 1984. Ingredients for insect diets, quality assurance, sources, and storage and handling. In: King, E.G., Leppla, N.C. (Eds.), Advances and Challenges in Insect Rearing, Agricultural Research Service, United States Department of Agriculture, New Orleans, LA, pp. 45–50.

Bruzzone, N., Cáceres, C., Andrade, L., Guzman, N., Calderon, J., Rendon, P., 1993. Process control for medfly mass production at San Miguel Petapa, Guatemala: a system approach. In: Aluja, M., Liedo, P. (Eds.), Fruit Flies: Biology and Management, Springer, New York, pp. 289–294.

Bueno, V.H.P., Mendes, S.M., Mendes, C.L., 2006. Evaluation of a rearing-method for the predator *Orius insidiosus*. Bull. Insectol. 59, 1–6.

Burt, M., 2002. TQM (total quality management): managing quality, not just controlling or measuring it. In: Leppla, N.C., Bloem, K.A., Luck, R.F. (Eds.), Proceedings Quality Control for Mass-Reared Arthropods, 8th and 9th Workshops of the IOBC Working Group on Quality Control of Mass-Reared Arthropods, pp. 29–31. http://AMRQC.org.

Caceres, C., McInnis, D., Shelly, T., Jang, E., Robinson, A., Hendrichs, J., 2007. Quality management systems for fruit fly (Diptera: Tephritidae) sterile insect technique. Fla. Entomol. 90, 1–9.

Cancino, J., Ruiz, L., Lopez, P., Sivinski, J., 2009. The suitability of *Anastrepha* spp. and *Ceratitis capitata* larvae as hosts of *Diachasmimorpha longicaudata* and *Diachasmimorpha tryoni*: effects of host age and radiation dose and implications for quality control in mass rearing. Biocontrol Sci. Technol. 19, 81–94.

Cerutti, F., Bigler, F., 1995. Quality assessment of *Trichogramma brassicae* in the laboratory. Entomol. Exp. Appl. 75, 19–26.

Chambers, D.L., 1975. Quality in mass-produced insects, definition and evaluation. Controlling Fruit Flies by the Sterile-Insect Technique, IAEA-PL-582/3, Vienna, Austria, pp. 19–32.

Chambers, D.L., 1977. Quality control in mass rearing. Annu. Rev. Entomol. 22, 289–308.

Chambers, D.L., Ashley, T.R., 1984. Putting the control in quality control in insect rearing. In: King, E.G., Leppla, N.C. (Eds.), Advances and Challenges in Insect Rearing, Agricultural Research Service, United States Department of Agriculture, New Orleans, LA, pp. 256–260.

Chen, W., Leopold, R.A., 2007. Progeny quality of *Gonatocerus ashmeadi* (Hymenoptera: Mymaridae) reared on stored eggs of *Homalodisca coagulata* (Hemiptera: Cicadellidae). J. Econ. Entomol. 100, 685–694.

Chen, H., Opit, G.P., Sheng, P., Zhang, H., 2011. Maternal and progeny quality of *Habrobracon hebetor* Say (Hymenoptera: Braconidae) after cold storage. Biol. Control 58, 255–261.

Chen, H., Zhang, H., Zhu, K.Y., Throne, J.E., 2012. Induction of reproductive diapause in *Habrobracon hebetor* (Hymenoptera: Braconidae) when reared at different photoperiods at low temperatures. Environ. Entomol. 41, 697–705.

Cohen, A.C., Winter 2001. Formalizing insect rearing and artificial diet technology. Am. Entomol., 198–206.

Cohen, A.C., 2004. Insect Diets: Science and Technology. CRC Press, Boca Raton, FL.

Collier, T., Van Steenwyk, R., 2004. A critical evaluation of augmentative biological control. Biol. Control 13, 245–256.

Couillien, D., Gregoire, J.C., 1994. Take-off capacity as a criterion for quality control in mass-produced predators, *Rhizophagus grandis* (Coleoptera: Rhizophagidae) for the biocontrol of bark beetles, *Dendroctonus micans* (Coleoptera: Scolytidae). Entomophaga 39, 385–395.

Dame, D.A., 1989. The relationship of research to total quality control with special reference to sterile insect technique. J. Appl. Entomol. 108, 476–482.

Deming, W.E., 1986. Out of the Crisis. Massachusetts Institute of Technology Center for Advanced Engineering Study, Cambridge, MA.

Dominiak, B., Sundaralingam, S., Jiang, L., Nicol, H., 2011. Effect of conditions in sealed plastic bags on eclosion of mass-reared Queensland fruit fly, *Bactrocera tryoni*. Entomol. Exp. Appl. 141, 123–128.

Edwards, R.H., Miller, E., Becker, R., Mossman, A.P., Irving, D.W., 1996. Twin screw extrusion processing of diet for mass rearing the pink bollworm. Trans. ASAE 39, 1789–1797.

Enkegaard, A., Reitzel, J., 1991. A simple model for quality control of *Aphidoletes aphidimyza*, *Aphidius matricariae* and *Encarsia formosa*. In: Bigler, F. (Ed.), Proceedings 5th Workshop of the IOBC Working Group on Quality Control of Mass-Reared Arthropods, pp. 201–211, Wageningen, The Netherlands. http://AMRQC.org.

Enkerlin, W.R., 2007. Guidance for Packing, Shipping, Holding and Release of Sterile Flies in Area-Wide Fruit Fly Control Programmes. FAO Plant Production and Protection Paper 190. FAO, Rome, Italy.

Enkerlin, W.R., Quinlan, M.M., 2004. Development of an international standard to facilitate the transboundary shipment of sterile insects. In: Barnes, B.N. (Ed.), Proceedings 6th International Symposium on Fruit Flies of Economic Importance, Stellenbosch, South Africa, pp. 203–212.

FAO/IAEA/USDA, 2003. Product Quality Control and Shipping Procedures for Sterile Mass-Reared Tephritid Fruit Flies. (Manual, Version 5.0) IAEA, Vienna, Austria. http://www-naweb.iaea.org/nafa/ipc/public/d4_pbl_5_1.html.

Feigenbaum, A.V., 1983. Total Quality Control, third ed. McGraw-Hill, New York.

Fernandez, C., Nentwig, W., 1997. Quality control of the parasitoid *Aphidius colemani* (Hymenoptera: Aphidiidae) used for biological control in greenhouses. J. Appl. Entomol. 121, 447–456.

Fisher, W.R., 1984a. Production of insects for industry, the Dow Chemical rearing program. In: King, E.G., Leppla, N.C. (Eds.), Advances and Challenges in Insect Rearing, Agricultural Research Service, United States Department of Agriculture, New Orleans, LA, pp. 234–239.

Fisher, W.R., 1984b. The insectary manager. In: King, E.G., Leppla, N.C. (Eds.), Advances and Challenges in Insect Rearing, Agricultural Research Service, United States Department of Agriculture, New Orleans, LA, pp. 295–299.

Fisher, W.R., 2009a. Insectary design and construction. In: Schneider, J.C. (Ed.), Principles and Procedures for Rearing High Quality Insects, Mississippi State University, Mississippi State, pp. 9–41.

Fisher, W.R., 2009b. The insectary manager. In: Schneider, J.C. (Ed.), Principles and Procedures for Rearing High Quality Insects, Mississippi State University, Mississippi State, pp. 43–69.

Fisher, W.R., Leppla, N.C., 1985. Insectary design and operation. In: Singh, P., Moore, R.F. (Eds.), Handbook of Insect Rearing, vol. 1. Elsevier, Amsterdam, The Netherlands, pp. 167–183.

Gandolfi, M., Mattiacci, L., Dorn, S., 2003. Mechanisms of behavioral alterations of parasitoids reared in artificial systems. J. Chem. Ecol. 29, 1871–1887.

Geden, C.J., Kaufman, P.E., 2007. Development of *Spalangia cameroni* and *Muscidifurax raptor* (Hymenoptera: Pteromalidae) on live house fly (Diptera: Muscidae) pupae and pupae killed by heat shock, irradiation, and cold. Environ. Entomol. 36, 34–39.

Ghazy, N.A., Suzuki, T., Shah, M., Amano, H., Ohyama, K., 2012. Effect of long-term cold storage of the predatory mite *Neoseiulus californicus* at high relative humidity on post-storage biological traits. Biocontrol (Dordrecht) 57, 635–641.

Giang, H.T.T., Ueno, T., 2007. Improving parasitoid performance by improving adult food quality: a case study for the leafminer parasitoid *Hemiptarsenus varicornis* (Hymenoptera: Eulophidae). J. Fac. Agric. Kyushu Univ. 52, 57–61.

Glenister, C.S., Hale, A., Luczynski, A., 2003. Quality assurance in North America: merging customer and producer needs. In: van Lenteren, J.C. (Ed.), Quality Control and Production of Biological Control Agents, Theory and Testing Procedures, CABI Publishing, Wallingford, pp. 205–214.

Glenister, C.S., Hoffman, M.P., 1998. Mass-reared natural enemies: scientific, technological, and informational needs and considerations. In: Ridgway, R.L., Hoffmann, M.P., Inscoe, M.N., Glenister, C.S. (Eds.), Mass-Reared Natural Enemies: Application, Regulation and Needs, Proceedings Thomas Say Publications in Entomology, Entomological Society America, Lanham, MD, pp. 242–267.

Gonzalez, P.I., Montoya, P., Perez-Lachaud, G., Cancino, J., Liedo, P., 2007. Superparasitism in mass reared *Diachasmimorpha longicaudata* (Ashmead) (Hymenoptera: Braconidae), a parasitoid of fruit flies (Diptera: Tephritidae). Biol. Control 40, 320–326.

Goodenough, J.L., 1984. Materials handling in insect rearing. In: King, E.G., Leppla, N.C. (Eds.), Advances and Challenges in Insect Rearing, Agricultural Research Service, United States Department of Agriculture, New Orleans, LA, pp. 77–86.

Goodenough, J.L., Parnell, C.B., 1985. Basic engineering design requirements for ventilation, heating, cooling, and humidification of insect rearing facilities. In: Singh, P., Moore, R.F. (Eds.), Handbook of Insect Rearing, vol. 1. Elsevier, Amsterdam, The Netherlands, pp. 137–155.

Grenier, S., 2009. *In vitro* rearing of entomophagous insects—past and future trends: a minireview. Bull. Insectol. 62, 1–6.

Griffin, J.G., 1984a. General requirements for facilities that mass-rear insects. In: King, E.G., Leppla, N.C. (Eds.), Advances and Challenges in Insect Rearing, Agricultural Research Service, United States Department of Agriculture, New Orleans, LA, pp. 70–73.

Griffin, J.G., 1984b. Facility and production equipment. In: Sikorowski, P.P., Griffin, J.G., Roberson, J., Lindig, O.H. (Eds.), Boll Weevil Mass Rearing Technology, University Press of Mississippi, Jackson, MS, pp. 11–52.

Harrell, E.A., Gantt, C.W., 1984. Automation in insect rearing. In: King, E.G., Leppla, N.C. (Eds.), Advances and Challenges in Insect Rearing, Agricultural Research Service, United States Department of Agriculture, New Orleans, LA, pp. 74–76.

Hassan, S.A., Zhang, W.Q., 2001. Variability in quality of *Trichogramma brassicae* (Hymenoptera: Trichogrammatidae) from commercial suppliers in Germany. Biol. Control 22, 115–121.

Hathaway, D.O., Lydin, L.V., Butt, B.A., Morton, L.J., 1973. Monitoring mass rearing of the codling moth. J. Econ. Entomol. 66, 390–393.

He, X.Z., Wang, Q., 2006. Asymmetric size effect of sexes on reproductive fitness in an aphid parasitoid *Aphidius ervi* (Hymenoptera: Aphidiidae). Biol. Control 36, 293–298.

Hegazi, E., Khafagi, W., 2005. Developmental interaction between suboptimal instars of *Spodoptera littoralis* (Lepidoptera: Noctuidae) and its parasitoid *Microplitis rufiventris* (Hymenoptera: Braconidae). Arch. Insect Biochem. Physiol. 60, 172–184.

Henry, L.M., May, N., Acheampong, S., Gillespie, D.R., Roitberg, B.D., 2010. Host-adapted parasitoids in biological control: does source matter? Ecol. Appl. 20, 242–250.

Hoffman, J.D., Ignoffo, C.M., Peters, P., Dickerson, W.A., 1984. Fractional colony propagation. A new insect-rearing system. In: King, E.G., Leppla, N.C. (Eds.), Advances and Challenges in Insect Rearing, Agricultural Research Service, United States Department of Agriculture, New Orleans, LA, pp. 232–233.

Hoy, M.A., 1979. The potential for genetic improvement of predators for use in pest management programs. In: Hoy, M.A., McKelvey Jr., J.J. (Eds.), The Use of Genetics in Insect Management, Rockefeller Foundation, New York, pp. 106–115.

Hoy, M.A., Nowierski, R.M., Johnson, M.W., Flexner, J.L., 1991. Issues and ethics in commercial releases of arthropod natural enemies. Am. Entomol. 37, 74–75.

Huettel, M.D., 1976. Monitoring the quality of laboratory-reared insects: a biological and behavioral perspective. Environ. Entomol. 5, 807–814.

Inglis, D.G., Sikorowski, P.P., 2009. Microbial contamination and insect rearing. In: Schneider, J.C. (Ed.), Principles and Procedures for Rearing High Quality Insects. Mississippi State University, Mississippi, pp. 150–222.

Jones, S.L., Kinzer, R.E., Bull, D.L., Ables, J.R., Ridgway, R.L., 1978. Deterioration of *Chrysopa carnea* in mass culture. Ann. Entomol. Soc. Am. 71, 160–162.

Joslyn, D.J., 1984. Maintenance of genetic variability in reared insects. In: King, E.G., Leppla, N.C. (Eds.), Advances and Challenges in Insect Rearing, Agricultural Research Service, United States Department of Agriculture, New Orleans, LA, pp. 20–29.

Joyce, A.L., Aluja, M., Sivinski, J., Vinson, S.B., Ramirez-Ramirez, R., Bernal, J.S., Guillen, L., 2010. Effect of continuous rearing on courtship acoustics of five braconid parasitoids, candidates for augmentative biological control of *Anastrepha* species. Biocontrol (Dordrecht) 55, 573–582.

Juran, J.M., Godfrey, A.B., 1998. Juran's Quality Handbook, fifth ed. McGraw-Hill, New York. http://www.pqm-online.com/assets/files/lib/juran.pdf.

Juran, J.M., Gryna, F.M., Bingham, R.S., 1979. Quality Control Handbook, third ed. McGraw-Hill, New York.

Karamaouna, F., Copland, M.J., 2000. Host suitability, quality and host size preference of *Leptomastix epona* and *Pseudaphycus flavidulus*, two endoparasitoids of the mealybug *Pseudococcus viburni*, and host size effect on parasitoid sex ratio and clutch size. Entomol. Exp. Appl. 96, 149–158.

Kfir, R., 1994. Insect rearing and inhalant allergies. In: Ochieng'-Odero, J.P.R. (Ed.), Proc. Techniques of Insect Rearing for the Development of Integrated Pest and Vector Management Strategies, vol. 1. ICIPE Science Press, Nairobi, Kenya, pp. 277–284.

King, E.G., Leppla, N.C., 1984. Advances and Challenges in Insect Rearing. Agricultural Research Service, United States Department of Agriculture, New Orleans, LA.

Knipling, E.F., 1966. Introduction. In: Smith, C.N. (Ed.), Insect Colonization and Mass Production, Academic Press, NY, pp. 1–12.

Knipling, E.F., 1979. The Basic Principles of Insect Population Suppression and Management. Agriculture Handbook. vol. 512. United States Department of Agriculture.

Kolliker-Ott, U.M., Bigler, F., Hoffmann, A.A., 2004. Field dispersal and host location of *Trichogramma brassicae* is influenced by wing size but not wing shape. Biol. Control 31, 1–10.

Lacasse, B., Panneton, B., Fournier, F., 2006. Terminal speed and size sorting of insect eggs. Biocontrol Sci. Technol. 16, 953–969.

Leopold, R.A., 2000. Insect cold storage: using cryopreservation and dormancy as aids to mass rearing. In: Tan, K.H. (Ed.), Proceedings Area-Wide Control of Fruit Flies and Other Insect Pests, International Conference on Area-Wide Control of Insect Pests, and the 5th International Symposium on Fruit Flies of Economic Importance, Penang, Malaysia, pp. 315–324.

Leopold, R.A., 2007. Colony maintenance and mass-rearing: using cold storage technology for extending the shelf-life of insects. In: Vreysen, M.J.B., Robinson, A.S., Hendrichs, J. (Eds.), Area-Wide Control of Insect Pests, pp. 149–162.

Leppla, N.C., 1989. Laboratory colonization of fruit flies. In: Robinson, A.S., Hooper, G. (Eds.), World Crop Pests, Fruit Flies, Their Biology, Natural Enemies and Control, vol. 3B. Elsevier, Amsterdam, pp. 91–103.

Leppla, N.C., 2003. Aspects of total quality control for the production of natural enemies. In: van Lenteren, J.C. (Ed.), Quality Control and Production of Biological Control Agents, Theory and Testing Procedures, CABI Publishing, Cambridge, MA, pp. 19–24.

Leppla, N.C., 2007. Building partnerships to support research on commercial natural enemies. In: van Lenteren, J.C. DeClercq, P., Johnson, M.W. (Eds.), Proc. 11th Meeting of the IOBC Working Group, Arthropod Mass Rearing and Quality Control. Bulletin IOBC Global No. 3, Montreal, Canada, pp. 82–85.

Leppla, N.C., 2009. The basics of quality control for insect rearing. In: Schneider, J.C. (Ed.), Principles and Procedures for Rearing High Quality Insects, Mississippi State University, Mississippi State, pp. 289–306.

Leppla, N.C., Ashley, T.R., Winter 1989. Quality control in insect mass production: a review and model. Bull. Entomol. Soc. Am. 33–44.

Leppla, N.C., Davis, F.M., Schneider, J.C., 2009. Introduction. In: Schneider, J.C. (Ed.), Principles and Procedures for Rearing High Quality Insects, Mississippi State University, Mississippi State, pp. 3–8.

Leppla, N.C., Fisher, W.R., 1989. Total quality control in insect mass production for insect pest management. J. Appl. Entomol. 108, 452–461.

Leppla, N.C., Larson, B.C., 2004. Quality control methods for the production of natural enemies. Encycl. Pest Manage. Marcel Dekker.

Liu, F., Smith, S.M., 2000. Measurement and selection of parasitoid quality for mass-reared *Trichogramma minutum* Riley used in inundative release. Biocontrol Sci. Technol. 10, 3–13.

Lopez, O.P., Henaut, Y., Cancino, J., Lambin, M., Cruz-Lopez, L., Rojas, J.C., 2009. Is host size an indicator of quality in the mass-reared parasitoid *Diachasmimorpha longicaudata* (Hymenoptera: Braconidae)? Fla. Entomol. 92, 441–449.

Losey, J.E., Calvin, D.D., 1995. Quality assessment of four commercially available species of *Trichogramma* (Hymenoptera: Trichogrammatidae). J. Econ. Entomol. 88, 1243–1250.

Luczynski, A., Nyrop, J.P., Shi, A., 2007. Influence of cold storage on pupal development and mortality during storage and on post-storage performance of *Encarsia formosa* and *Eretmocerus eremicus* (Hymenoptera: Aphelinidae). Biol. Control 40, 107–117.

Luczynski, A., Nyrop, J.P., Shi, A., 2008. Pattern of female reproductive age classes in mass-reared populations of *Phytoseiulus persimilis* (Acari: Phytoseiidae) and its influence on population characteristics and quality of predators following cold storage. Biol. Control. 47, 159–166.

Lux, S., 1991. Diagnosis of behaviour as a tool for quality control of mass reared arthropods. In: Bigler, F. (Ed.), Proceedings 5th Workshop of the IOBC Global Working Group on Quality Control of Mass Reared Arthropods. Wageningen, The Netherlands, pp. 66–79.

Mackauer, M., 1972. Genetic aspects of insect production. Entomophaga 17, 27–48.

Mackauer, M., 1976. Genetic problems in the production of biological control agents. Annu. Rev. Entomol. 21, 369–385.

Maes, S., Machtelinckx, T., Moens, M., Grégoire, J., De Clercq, P., 2012. The influence of acclimation, endosymbionts and diet on the supercooling capacity of the predatory bug *Macrolophus pygmaeus*. Biocontrol (Dordrecht). http://dx.doi.org/10.1007/s10526-012-9446-2.

Mahmoud, A.M.A., Lim, U.T., 2007. Evaluation of cold-stored eggs of *Dolycoris baccarum* (Hemiptera: Pentatomidae) for parasitization by *Trissolcus nigripedius* (Hymenoptera: Scelionidae). Biol. Control 43, 287–293.

Mangan, R.L., 1992. Evaluating the role of genetic change in insect colonies maintained for pest management. In: Anderson, T.E., Leppla, N.C. (Eds.), Advances in Insect Rearing for Research and Pest Management, Westview Press, Boulder, CO, pp. 269–288.

McMurtry, J.A., Scriven, G.T., 1975. Population increase of *Phytoseiulus persimilis* on different insectary feeding programs. J. Econ. Entomol. 68, 319–321.

Messing, R.H., Klungness, L.M., Purcell, M., Wong, T.T.Y., 1993. Quality control parameters of mass-reared opiine parasitoids used in augmentative biological control of tephritid fruit flies in Hawaii. Biol. Control 3, 140–147.

Minno, M.C., Holler, T.C., 1987. Procedures Manual for Mass-Rearing the Caribbean Fruit Fly, *Anastrepha suspensa* (Loew) (Diptera: Tephritidae). Florida Department of Agriculture and Consumer Services, Division of Plant Industry, Gainesville, FL.

Miyatake, T., 2011. Insect quality control: synchronized sex, mating system, and biological rhythm. Appl. Entomol. Zool. (Jpn.) 46, 3–14.

Mohaghegh, J., De Clercq, P., Tirry, L., 1998. Effects of maternal age and egg weight on developmental time and body weight of offspring of *Podisus maculiventris* (Heteroptera: Pentatomidae). Ann. Entomol. Soc. Am. 91, 315–322.

Moore, R.F., Odell, T.M., Calkins, C.O., 1985. Quality assessment in laboratory reared insects. In: Singh, P., Moore, R.F. (Eds.), Handbook of Insect Rearing, vol. 1. Elsevier, Amsterdam, The Netherlands, pp. 107–135.

Morrison, R.K., 1985. *Trichogramma* spp. In: Singh, P., Moore, R.F. (Eds.), Handbook of Insect Rearing, vol. 1. Elsevier, Amsterdam, The Netherlands, pp. 413–417.

Nakamori, H., Kakinohana, H., Yamagish, M., 1992. Automated mass production system for fruit flies based on the melon fly, *Dacus cucurbitae* Coquillett (Diptera: Tephritidae). In: Anderson, T.E., Leppla, N.C. (Eds.), Advances in Insect Rearing for Research and Pest Management, Westview Press, Boulder, CO, pp. 441–454.

Nasreen, A., Gillespie, D.R., Mustafa, G., 2011. Graphical marginal analysis of the economics of natural enemy production: an example using a pilot mass rearing system for green lacewing. Biol. Control 57, 44–49.

Nathan, S.S., Kalaivani, K., Mankin, R.W., 2006. Effects of millet, wheat, rice, and sorghum diets on development of *Corcyra cephalonica* (Stainton) (Lepidoptera: Galleriidae) and its suitability as a host for *Trichogramma chilonis* Ishii (Hymenoptera: Trichogrammatidae). Environ. Entomol. 35, 784–788.

Neuenschwander, P., Haug, T., Ajounu, O., Davis, H., Akinwumi, B., Madojemu, E., 1989. Quality requirements in natural enemies used for inoculative release: practical experience from a successful biological control programme. J. Appl. Entomol. 108, 409–420.

Niijima, K., Yoshida, T., Akimoto, T., 2002. Egg collection and cold storage of adults in several Japanese chrysopids as biological control agents for mass rearing. Bull. Fac. Agric. Tamagawa Univ. 42, 15–29.

Noldus, L.P.J.J., 1989. Semiochemicals, foraging behavior and quality of entomophagous insects for biological control. J. Appl. Entomol. 108, 425–451.

Nordlund, D.A., Wu, Z.X., Greenberg, S.M., 1997. *In vitro* rearing of *Trichogramma minutum* Riley (Hymenoptera: Trichogrammatidae) for ten generations, with quality assessment comparisons of *in vitro* and *in vivo* reared adults. Biol. Control 9, 201–207.

Nunney, L., 2002. The population genetics of mass-rearing. In: Leppla, N.C., Bloem, K.A., Luck, R.F. (Eds.), Proceedings Quality Control for Mass-Reared Arthropods, 8th and 9th Workshops of the IOBC Working Group on Quality Control of Mass-Reared Arthropods, pp. 43–49. http://AMRQC.org.

Ode, P.J., Heinz, K.M., 2002. Host-size-dependent sex ratio theory and improving mass-reared parasitoid sex ratios. Biol. Control 24, 31–41.

ODell, T.M., 1992. Straggling in gypsy moth production strains: a problem analysis for developing research priorities. In: Anderson, T.E., Leppla, N.C. (Eds.), Advances in Insect Rearing for Research and Pest Management, Westview Press, Boulder, CO, pp. 325–350.

ODell, T.M., Keena, M.A., Willis, R.B., 1997. Dietary influence of iron formulation on the development of gypsy moth (Lepidoptera: Lymantriidae) in laboratory colonies. Ann. Entomol. Soc. Am. 90, 149–154.

O'Neil, R.J., Giles, K.L., Obrycki, J.J., Mahr, D.L., Legaspi, J.C., Katovich, K., 1998. Evaluation of the quality of four commercially available natural enemies. Biol. Control 11, 1–8.

Owens, C.D., 1984. Controlled environments for insects and personnel in insect rearing facilities. In: King, E.G., Leppla, N.C. (Eds.), Advances and Challenges in Insect Rearing, Agricultural Research Service, United States Department of Agriculture, New Orleans, LA, pp. 58–63.

Pashley, D.P., Proverbs, M.D., 1981. Quality control by electrophoretic monitoring in a laboratory colony of codling moths. Ann. Entomol. Soc. Am. 74, 20–23.

Pavlik, J., 1993. The size of the female and quality assessment of mass-reared *Trichogramma* spp. Entomol. Exp. Appl. 66, 171–177.

Pearson, T.C., Edwards, R.H., Mossman, A.P., Wood, D.F., Yu, P.C., Miller, E.L., 2002. Insect egg counting on mass rearing oviposition pads by image analysis. Appl. Eng. Agric. 18, 129–135.

Penn, S.L., Ridgway, R.L., Scriven, G.T., Inscoe, M.N., 1998. Quality assurance by the commercial producer of arthropod natural enemies. In: Ridgway, R.L., Hoffmann, M.P., Inscoe, M.N., Glenister, C.S. (Eds.), Mass-Reared Natural Enemies: Application, Regulation and Needs, Proceedings Thomas Say Publications in Entomology, Entomological Society America, Lanham, MD, pp. 202–230.

Pree, D.J., Archibald, D.E., Morrison, R.K., 1989. Resistance to insecticides in the common green lacewing *Chrysoperla carnea* (Neuroptera: Chrysopidae) in southern Ontario, Canada. J. Econ. Entomol. 82, 29–34.

Prezotti, L., Parra, J.R.P., Vencovsky, R., Coelho, A.S.G., Cruz, I., 2004. Effect of the size of the founder population on the quality of sexual populations of *Trichogramma pretiosum*, in laboratory. Biol. Control 30, 174–180.

Prezotti, L., Parra, J.R.P., Vencovsky, R., Dias, C.T.D.S., Cruz, I., Chagas, M.C.M., 2002. Flight test as evaluation criterion for the quality of *Trichogramma pretiosum* Riley (Hymenoptera: Trichogrammatidae): adaptation of the methodology. Neotrop. Entomol. 31, 411.

Qiu, H.G., Bigler, F., Van Bergeijk, K.E., Bosshart, S., Waldburger, M., 1992. Quality comparison of *Trichogramma maidis* emerged on subsequent days. Acta Entomol. Sin. 35, 449–455.

Ramalho, F.S., Dias, J.M., 2003. Effects of factitious hosts on biology of *Catolaccus grandis* (Burks) (Hymenoptera: Pteromalidae), a parasitoid of *Anthonomus grandis* Boheman (Coleoptera: Curculionidae). Neotrop. Entomol. 32, 305–310.

Reinecke, J.P., 2009. Health and safety issues in rearing arthropods. In: Schneider, J.C. (Ed.), Principles and Procedures for Rearing High Quality Insects, Mississippi State University, Mississippi State, pp. 9–41.

Reis, J., Oliveira, L., Garcia, P., 2003. Effects of the larval diet of *Pseudaletia unipuncta* (Lepidoptera: Noctuidae) on the performance of the parasitoid *Glyptapanteles militaris* (Hymenoptera: Braconidae). Environ. Entomol. 32, 180–186.

Rendón, P., Pessarozzi, C., Tween, G., 2005. The largest fruit fly mass-rearing facility in the world: lessons learned in management and R&D. Book of Extended Synopses. AO/IAEA International Conference on Area-Wide Control of Insect Pests: Integrating the Sterile Insect and Related Nuclear and Other Techniques, IAEA-CN-131/150, IAEA, Vienna, Austria, p. 36.

Rendon, P., Sivinski, J., Holler, T., Bloem, K., Lopez, M., Martinez, A., Aluja, M., 2006. The effects of sterile males and two braconid parasitoids, *Fopius arisanus* (Sonan) and *Diachasmimorpha krausii* (Fullaway) (Hymenoptera), on caged populations of Mediterranean fruit flies, *Ceratitis capitata* (Wied.) (Diptera: Tephritidae) at various sites in Guatemala. Biol. Control 36, 224–2006.

Robinson, A.S., Hendrichs, J., 2007. Area-Wide Control of Insect Pests, from Research to Field Implementation. Springer, Dordrecht, The Netherlands.

Rodriguez, L.E., Gomez, T.V., Barcenas, O.N.M., Vazquez, L.L.J., 2002. Effect of different factors on the culture of *Callosobruchus maculatus* (Coleoptera: Bruchidae) for the production of *Catolaccus* spp. (Hymenoptera: Pteromalidae). Acta Zool. Mex. Nueva 86, 87–101.

Rull, J., Birke, A., Ortega, R., Montoya, P., Lopez, L., 2012. Quantity and safety vs. quality and performance: conflicting interests during mass rearing and transport affect the efficiency of sterile insect technique programs. Entomol. Exp. Appl. 142, 78–86.

Sagarra, L.A., Vincent, C., Stewart, R.K., 2000. Mutual interference among female *Anagyrus kamali* Moursi (Hymenoptera: Encyrtidae) and its impact on fecundity, progeny production and sex ratio. Biocontrol (Dordrecht) 10, 239–244.

Sarvary, M.A., Hight, S.D., Carpenter, J.E., Bloem, S., Bloem, K.A., Dorn, S., 2008. Identification of factors influencing flight performance of field-collected and laboratory-reared, overwintered, and nonoverwintered cactus moths fed with field-collected host plants. Environ. Entomol. 37, 1291–1299.

SAS Institute, 2004. The SAS System Release 8.2 (TS2MO) for Windows, Cary NC. http://SAS.com.

Schneider, J.C., 2009. Principles and Procedures for Rearing High Quality Insects. Mississippi State University Press, Mississippi State, MS.

Serrano, M.S., Lapointe, S.L., 2002. Evaluation of host plants and a meridic diet for rearing *Maconellicoccus hirsutus* (Hemiptera: Pseudococcidae) and its parasitoid *Anagyrus kamali* (Hymenoptera: Encyrtidae). Fla. Entomol. 85, 417–425.

Shapiro, J.P., Ferkovich, S.M., 2002. Yolk protein immunoassays (YP-ELISA) to assess diet and reproductive quality of mass-reared *Orius insidiosus* (Heteroptera: Anthocoridae). J. Econ. Entomol. 95, 927–935.

Singh, P., Ashby, M.D., 1985. Insect rearing management. In: Singh, P., Moore, R.F. (Eds.), Handbook of Insect Rearing, vol. 1. Elsevier, Amsterdam, The Netherlands, pp. 185–215.

Singh, P., Moore, R.R., 1985. Handbook of Insect Rearing. vols 1 and 2. Elsevier, Amsterdam, The Netherlands.

Smith, C.N., 1966. Insect Colonization and Mass Production. Academic Press, New York.

Smith, R.A., Nordlund, D.A., 2000. Mass rearing technology for biological control agents of *Lygus* spp. Southwest. Entomol. (Suppl. 23), 121–127.

Smith, S.M., 1996. Biological control with *Trichogramma*: advances, successes, and potential of their use. Annu. Rev. Entomol. 41, 375–406.

Steward, B.V., Kintz, J.L., Horner, T.A., 1996. Evaluation of biological control agent shipments from three United States suppliers. Horttechnology 6, 233–237.

Suarthana, E., Shen, A., Henneberger, P.K., Kreiss, K., Leppla, N.C., Bueller, D., Lewis, D.M., Bledsoe, T.A., Janotka, E., Petsonk, E.L., 2012. Post-hire asthma among insect-rearing workers. J. Occup. Environ. Med. 54, 310–317.

Tezze, A.A., Botto, E.N., 2004. Effect of cold storage on the quality of *Trichogramma nerudai* (Hymenoptera: Trichogrammatidae). Biol. Control 30, 11–16.

Thomson, L.J., Hoffmann, A.A., 2002. Laboratory fecundity as predictor of field success in *Trichogramma carverae* (Hymenoptera: Trichogrammatidae). J. Econ. Entomol. 95, 912–917.

Tormos, J., Asis, J., Sabater-Munoz, B., Banos, L., Gayubo, S.F., Beitia, F., 2011. Superparasitism in laboratory rearing of *Spalangia cameroni* (Hymenoptera: Pteromalidae), a parasitoid of medfly (Diptera: Tephritidae). Bull. Entomol. Res. 102, 51–61.

Tuncbilek, A.S., Canpolat, U., Sumer, F., 2009. Suitability of irradiated and cold-stored eggs of *Ephestia kuehniella* (Pyralidae: Lepidoptera) and *Sitotroga cerealella* (Gelechidae: Lepidoptera) for stockpiling the egg-parasitoid *Trichogramma evanescens* (Trichogrammatidae: Hymenoptera) in diapause. Biocontrol Sci. Technol. 19, 127–138.

Vacari, A.M., De Bortoli, S.A., Dionísio, F., Borba, D.F., Martins, M.I.E.G., 2012. Quality of *Cotesia flavipes* (Hymenoptera: Braconidae) reared at different host densities and the estimated cost of its commercial production. Biol. Control 58, 127–132.

Vandekerkhove, B., Van Baal, E., Bolckmans, K., De Clercq, P., 2006. Effect of diet and mating status on ovarian development and oviposition in the polyphagous predator *Macrolophus caliginosus* (Heteroptera: Miridae). Biol. Control 39, 532–538.

van Frankenhuyzen, K., Ebling, P., McCron, R., Ladd, T., Gauthier, D., Vossbrinck, C., 2004. Occurrence of *Cystosporogenes* sp. (Protozoa, Microsporidia) in a multi-species insect production facility and its elimination from a colony of the eastern spruce budworm, *Choristoneura fumiferana* (Clem.) (Lepidoptera: Tortricidae). J. Invertebr. Pathol. 87, 16–28.

Van Hezewijk, B.H., De Clerck-Floate, R.A., Moyer, J.R., 2008. Effect of nitrogen on the preference and performance of a biological control agent for an invasive plant. Biol. Control 46, 332–340.

van Lenteren, J.C., 2003a. Need for quality control of mass-produced biological control agents. In: van Lenteren, J.C. (Ed.), Quality Control and Production of Biological Control Agents, Theory and Testing Procedures, CABI Publishing, Cambridge, MA, pp. 215–224.

van Lenteren, J.C., 2003b. Commercial availability of biological control agents. In: van Lenteren, J.C. (Ed.), Quality Control and Production of Biological Control Agents, Theory and Testing Procedures, CABI Publishing, Cambridge, MA, pp. 167–179.

van Lenteren, J.C., 2011. The state of commercial augmentative biological control: plenty of natural enemies, but a frustrating lack of uptake. Biocontrol 57, 1–20.

van Lenteren, J.C., Bigler, F., Waddington, C., 1993a. Quality control guidelines for natural enemies. In: Nicoli, G., Benuzzi, M., Leppla, N.C. (Eds.), Proceedings, 7th Workshop of the IOBC Working Group on Quality Control of Mass-Reared Arthropods, Rimini, Italy, pp. 222–230. http://AMRQC.org.

van Lenteren, J.C., Hale, A., Klapwijk, J.N., van Schelt, J., Steinberg, S., 1993b. Guidelines for quality control of commercially produced natural enemies. In: van Lenteren, J.C. (Ed.), Quality Control and Production of Biological Control Agents, Theory and Testing Procedures, CABI Publishing, Cambridge, MA, pp. 265–303.

van Lenteren, J.C., Tommasini, M.G., 1999. Mass production, storage, shipment and quality control of natural enemies. In: Albajes, R., Gullino, M.L., van Lenteren, J.C., Elad, Y. (Eds.), Integrated Pest and Disease Management in Greenhouse Crops, Kluwer, Dordrecht, The Netherlands, pp. 276–294.

van Lenteren, J.C., Tommasini, M.G., 2003. Mass production, storage, shipment and release of natural enemies. In: van Lenteren, J.C. (Ed.), Quality Control and Production of Biological Control Agents, Theory and Testing Procedures, CABI Publishing, Cambridge, MA, pp. 181–189.

van Schelt, J., Mulder, S., 2000. Improved methods of testing and release of *Aphidoletes aphidimyza* (Diptera: Cecidomyiidae) for aphid control in glasshouses. Eur. J. Entomol. 97, 511–515.

Vasquez, C.J., Morse, J.G., 2012. Fitness components of *Aphytis melinus* (Hymenoptera: Aphelinidae) reared in five California insectaries. Environ. Entomol. 41, 51–58.

Vasquez, G.M., Orr, D.B., Baker, J.R., 2004. Quality assessment of selected commercially available whitefly and aphid biological control agents in the United States. J. Econ. Entomol. 97, 781–788.

Wackers, F.L., van Lenteren, J.C., 2003. The parasitoid's need for sweets: sugars in mass rearing and biological control. In: van Lenteren, J.C. (Ed.), Quality Control and Production of Biological Control Agents, Theory and Testing Procedures, CABI Publishing, Cambridge, pp. 59–72.

Wajnberg, E., 1991. Quality control of mass-reared arthropods: a genetical and statistical approach. In: Bigler, F. (Ed.), Proceedings 5th Workshop of the IOBC Global Working Group on Quality Control of Mass Reared Arthropods, Wageningen, The Netherlands, pp. 15–25.

Wajnberg, E., 2003. Basic statistical methods for quality control workers. In: van Lenteren, J.C. (Ed.), Quality Control and Production of Biological Control Agents, Theory and Testing Procedures, CABI Publishing, Cambridge, MA, pp. 305–314.

Watson, D.M., Du, T.Y., Li, M., Xiong, J.J., Liu, D.G., Huang, M.D., Rae, D.J., Beattie, G.A.C., 2000. The effect of two prey species, *Chrysomphalus aonidum* and *Corcyra cephalonica*, on the quality of the predatory thrips, *Aleurodothrips fasciapennis*, reared in the laboratory. Biocontrol (Dordrecht) 45, 45–61.

Webb, J.C., 1984. The closed-loop system of quality control in insect rearing. In: King, E.G., Leppla, N.C. (Eds.), Advances and Challenges in Insect Rearing, Agricultural Research Service, United States Department of Agriculture, New Orleans, LA, pp. 87–89.

Webb, J.C., Agee, H.R., Leppla, N.C., Calkins, C.O., 1981. Monitoring insect quality. Tran. ASAE 24, 476–479.

Wheeler, G.S., 2003. Minimal increase in larval and adult performance of the biological control agent *Oxyops vitiosa* when fed *Melaleuca quinquenervia* leaves of different nitrogen levels. Biol. Control 26, 109–116.

Wheeler, G.S., 2006. Chemotype variation of the weed *Melaleuca quinquenervia* influences the biomass and fecundity of the biological control agent *Oxyops vitiosa*. Biol. Control 36, 121–128.

Wolf, W.W., 1984. Controlling respiratory hazards in insectaries. In: King, E.G., Leppla, N.C. (Eds.), Advances and Challenges in Insect Rearing, Agricultural Research Service, United States Department of Agriculture, New Orleans, LA, pp. 64–69.

Wolf, W.W., 1985. Recognition and prevention of health hazards associated with insect rearing. In: Singh, P., Moore, R.F. (Eds.), Handbook of Insect Rearing, vol. 1. Elsevier, Amsterdam, The Netherlands, pp. 157–165.

Wuehrer, B.G., Hassan, S.A., 1993. Selection of effective species/strains of *Trichogramma* (Hymenoptera: Trichogrammatidae) to control the diamondback moth *Plutella xylostella* L. (Lepidoptera: Plutellidae). J. Appl. Entomol. 116, 80–89.

Wyckhuys, K.A.G., Acosta, F.L., Garcia, J., Jimenez, J., 2011. Host exploitation and contest behavior in a generalist parasitoid partially reflect quality of distinct host species. Biocontrol Sci. Technol. 21, 953–968.

Zlotin, A.Z., Chepurnaya, N.P., 1994. General principles of quality control of the insect culture. Entomol. Rev. 73, 161–166.

Section II

This page intentionally left blank

Chapter 10

Production of Entomopathogenic Nematodes

David I. Shapiro-Ilan,[1] Richou Han[2] and Xuehong Qiu[2]
[1]USDA-ARS, SAA, SE Fruit and Tree Nut Research Unit, Byron, GA, USA, [2]Guangdong Entomological Institute, Guangzhou, China

10.1. INTRODUCTION

The objective of this chapter is the review and analysis of methodology for production of entomopathogenic nematodes (EPNs). Nematodes are round worms (also known as eelworms or threadworms) of the phylum Nematoda. EPNs have been defined as parasitic nematodes that are mutualistically associated with bacterial symbionts; all life stages of the nematode, except for the dauer stage (see below), are found exclusively inside the insect host (Grewal et al., 2005; Onstad et al., 2006). Historically, this definition of EPNs has referred to the families Steinernematidae and Heterorhabditidae. Recently, the definition of EPNs was expanded to include other nematodes, such as certain species of the genus *Oscheius* (Dillman et al., 2012a). Nonetheless, this chapter will focus exclusively on the genera *Heterorhabditis* Poinar and *Steinernema* Travassos because they are the only EPNs for which mass production methods have been developed, and they are the only ones sold commercially for biocontrol purposes.

More than 90 species of steinernematids and heterorhabditids have been described to date (at least 75 steinernematids and 18 heterorhabditids). The relationship between these nematodes and their bacterial symbionts is highly specific; bacteria from the *Xenorhabdus* spp. and *Photorhabdus* spp. are associated with *Steinernema* and *Heterorhabditis*, respectively (Poinar, 1990; Griffin et al., 2005; Lewis and Clarke, 2012). The bacteria are the primary agents responsible for killing the insect host and providing the nematodes with a source of nutrition. The nematodes are also critical to the symbiotic relationship because the bacteria *Xenorhabdus* spp. and *Photorhabdus* spp. (that are associated with nematodes used in biocontrol) depend entirely on the nematode to vector them from one insect host to another; additionally, the nematodes assist in suppressing the host immune system (Dowds and Peters, 2002; Lewis and Clarke, 2012).

Mass Production of Beneficial Organisms. http://dx.doi.org/10.1016/B978-0-12-391453-8.00010-8

A generalized life cycle of EPNs is depicted in Fig. 10.1. The infective juvenile nematode (IJ), or "dauer" stage (Fig. 10.2), which is the only free-living stage, enters the host via natural openings (i.e. mouth, anus, spiracles, or occasionally through the insect cuticle); in heterorhabditids, penetration through the cuticle may be facilitated by a specialized "tooth" structure (Dowds and Peters, 2002). After entering the insect's hemocoel, IJs release their symbiotic bacteria and the host dies usually within 24–72 h. Once inside the host, the developmentally arrested IJs reinitiate their life cycle and molt to fourth-stage juveniles (the reinitiation of the life cycle is referred to as "recovery"). The nematodes feed on multiplying bacteria and host tissues while completing their development (having six life stages: egg, four juvenile stages, and adult); the nematodes may carry out one to three generations within the host (Lewis and Clarke, 2012). When the nutritive value in the host becomes depleted, IJs exit from the insect cadaver, searching for new hosts (Lewis and Clarke, 2012). The heterorhabditid lifecycle differs from that of the steinernematid in that the first generation of heterorhabditids are exclusively hermaphrodites and subsequent generations include males, females, and hermaphrodites (Strauch et al., 1994; Koltai et al., 1995). In contrast, all steinernematid species possess only amphimictic forms (males and females), except one species that (similar to heterorhabditids) was found to also possess hermaphrodites in the first adult generation (i.e. *Steinernema hermaphroditum* Stock, Griffin, and Chaerani; Stock et al., 2004).

EPNs possess many positive attributes as biological control agents for suppression of insect pests (Shapiro-Ilan and Grewal, 2008; Shapiro-Ilan et al., 2012a). They are safe to humans and are generally safe to other nontarget organisms and the environment (Akhurst and Smith, 2002; Ehlers, 2005). The level of safety associated with EPNs has led to a lack of pesticide registration requirements in many countries such as the United States and nations in the European Union (Ehlers, 2005). Furthermore, EPNs have a wide host range (with few exceptions) and kill the target host relatively rapidly (Lewis and Clarke, 2012; Shapiro-Ilan et al., 2012a).

Based on the attractive attributes described above, EPNs have been developed as biocontrol agents on a commercial level. They are currently being produced by more than 10 companies in Asia, Europe, and North America, and, to date, at least 13 different species have reached commercial development: *Heterorhabditis bacteriophora* Poinar, *Heterorhabditis indica* Poinar, Karunakar and David, *Heterorhabditis marelatus* Liu and Berry, *Heterorhabditis megidis* Poinar, Jackson and Klein, *Heterorhabditis zealandica* Poinar, *Steinernema carpocapsae* (Weiser), *Steinernema feltiae* (Filipjev), *Steinernema glaseri* (Steiner), *Steinernema kushidai* Mamiya, *Steinernema kraussei* (Steiner), *Steinernema longicaudum* Shen and Wang, *Steinernema riobrave* Cabanillas, Poinar, and Raulston, and *Steinernema scapterisci* Nguyen and Smart (Lacey et al., 2001; Kaya et al., 2006; Georgis et al., 2006: unpublished). EPNs can suppress a wide variety of economically important pests, many of which are targeted

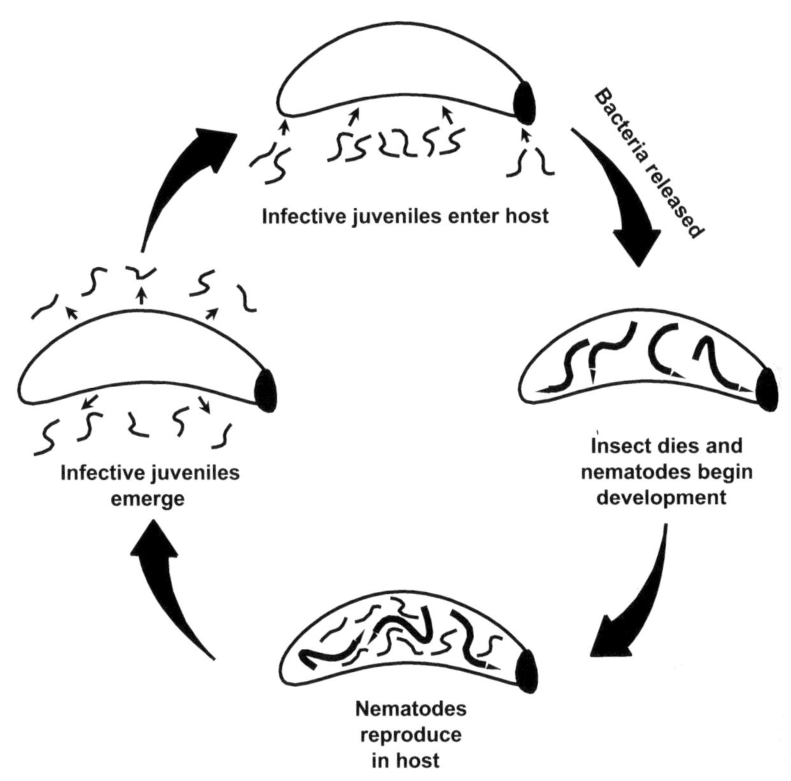

FIGURE 10.1 A generalized life cycle of entomopathogenic nematodes.

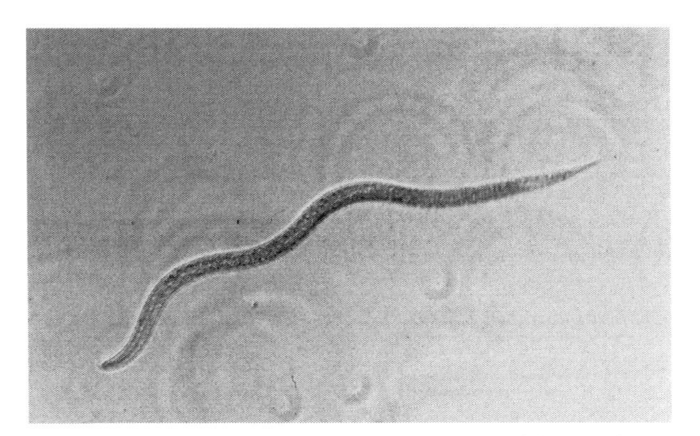

FIGURE 10.2 An infective juvenile steinernematid nematode.

commercially (Table 10.1, also see Shapiro-Ilan and Gaugler, 2002; Grewal et al., 2005; Georgis et al., 2006; Lacey and Georgis, 2012).

The efficacy of EPNs in suppressing insects depends on selecting the appropriate nematode for the target pest (Shapiro-Ilan et al., 2002b). A suitable

TABLE 10.1 Insects Pests Targeted for Control Using Entomopathogenic Nematodes*

Common Name	Scientific Name	Nematode(s)**
Artichoke plume moth	*Platyptilia carduidactyla* (Riley)	Sf, Sc
Banana moth	*Opogona sachari* (Bojer)	Hb, Sc
Banana root borer	*Cosmopolites sordidus* (Gemar)	Sc, Sf, Sg
Billbug	*Sphenophorus* spp.	Sc
Black cutworm	*Agrotis ipsilon* (Hufnagel)	Sc, Hb
Black vine weevil	*Otiorhynchus sulcatus* (F.)	Hb, Hmeg, Sk
Blue green weevils	*Pachneus* spp.	Sr, Hb
Borers	*Synanthedon* spp. (Lepidoptera: Sesiidae)	Sc, Hb, Sf
Cabbage maggot	*Delia radicum* (L.)	Sf
Cat flea	Ctenocephalides felis (Bouché)	Sc, Hb
Chinch bug	*Blissus leucopterus* (Say)	Sc
Codling moth	*Cydia pomonella* (L.)	Sf, Sc
Corn rootworm	*Diabrotica* spp.	Hb, Sf
Cranberry girdler	*Chrysoteuchia topiaria* (Zeller)	Sc
Diamondback moth	*Plutella xylostella* (L.)	Sc
Diaprepes root weevil	*Diaprepes abbreviatus* (L.)	Sr, Hb, Hi
Fungus gnats	Diptera: Sciaridae	Sf, Hb
Large pine weevil	*Hylobius abietis* (L.)	Hd, Sc, Sf
Leafminer	*Liriomyza* spp. (Diptera: Agromyzidae)	Sf, Sc
Mole crickets	*Scapteriscus* spp. (Orthoptera: Gryllotalpidae)	Ss, Sr, Sc
Navel orangeworm	*Amyelois transitella* (Walker)	Sc
Pecan weevil	*Curculio caryae* (Horn)	Sc
Plum curculio	*Conotrachelus nenuphar* (Herbst)	Sr
Small hive beetle	*Aethina tumida* (Murray)	Sr, Hi
Strawberry root weevil	*Otiorhynchus ovatus* (L.)	Hb, Hmar
Sweetpotato weevil	*Cylas formicarius* (F.)	Hb, Hi
Western flower thrips	*Frankliniella occidentalis* (Pergande)	Sf
White grubs	Coleoptera: Scarabaeidae	Hb, Hmeg, Hmar, Sg

**This table includes major target pests and the nematodes used to control them based on industry recommendations and refereed scientific articles indicating high levels of efficacy; the table is not meant to be an exhaustive list (see also Grewal et al., 2005; Georgis et al., 2006; Lacey and Shapiro-Ilan, 2008).*

***Hb = Heterorhabditis bacteriophora, Hd = H. downesi (Stock, Burnell and Griffin), Hi = H. indica, Hmar = H. marelatus, Hmeg = H. marelatus, Sc = Steinernema carpocapsae, Sf = S. feltiae, Sg = S. glaseri, Sk = S. kraussei, Sr = S. riobrave, Ss = S. scapterisci.*

candidate must possess a high level of virulence (disease-causing power) toward the host. The nematode must be able to evade host defenses and initiate pathogenesis. The ability of EPNs to persist in the environment may also contribute to host suitability and biocontrol efficacy. Indeed, in some cases enhanced persistence may even compensate for lower virulence. However, high levels of efficacy following EPN applications generally persist only for 2–6 weeks. Thus, EPNs are usually applied in an inundative or "pesticidal" manner where little or no recycling is expected; reapplication is required annually or seasonally. Nevertheless, there have also been some reports of prolonged pest control over several seasons or years (Shapiro-Ilan et al., 2002b).

Nematode persistence depends on the nematode species or strain, host density, and most importantly environmental factors (e.g. temperature, moisture; Shapiro-Ilan et al., 2006). EPNs are highly sensitive to desiccation and ultraviolet light. Therefore, applications made to soil or other cryptic habitats (and made during the early morning or evening) tend to be most successful. Temperature extremes (e.g. $<1\,°C$ and $>35\,°C$) can be detrimental to nematodes. The optimum temperature for maximum efficacy depends on nematode species or strain; some nematodes are relatively more heat tolerant, such as *H. indica*, and *S. riobrave*, whereas others are relatively more cold tolerant, such as *H. megidis* and *S. feltiae*.

The suitability of an EPN to control a particular target pest may also depend on the nematode's foraging strategy (Shapiro-Ilan et al., 2002b; Lewis and Clarke, 2012). Foraging strategies exhibited by EPNs exist along a continuum from ambushers to cruisers. Ambushers tend to use a sit-and-wait strategy: they usually stand on their tails (nictating) and wait until a host comes close. Cruisers generally actively seek out their hosts and cue into certain target volatiles prior to contacting the host. Examples of nematodes that exhibit foraging behavior characteristic of ambushers include *S. carpocapsae* and *S. scapterisci*. Those exhibiting behavior typical of cruisers include *H. bacteriophora*, *H. megidis*, and *S. glaseri*; those with intermediate search behaviors include *S. feltiae* and *S. riobrave*. Ambushers may be most successful at infecting mobile insects on or near the soil surface, whereas cruisers can be most successful at infecting sessile insects below the soil surface. Despite these generalities, some ambusher nematodes have been quite effective in suppressing below-ground pests (Shapiro-Ilan et al., 2009a; Dillon et al., 2007).

Research in the field of entomopathogenic nematology has advanced substantially in recent years in a variety of subject areas, including biological control (Grewal et al., 2005; Shapiro-Ilan et al., 2006; Lacey and Georgis, 2012; Dolinski et al., 2012), ecology (Lewis et al., 2006; Campos-Herrera et al., 2012), bioactive metabolites (Bode, 2009), symbiosis (Goodrich-Blair and Clarke, 2007), and genetics (Ciche, 2007; Dillman et al., 2012b). Research on improving mass production has also advanced (Shapiro-Ilan et al., 2012b). EPNs are currently produced using in vivo or in vitro (solid and liquid culture) methods (Friedman, 1990; Ehlers and Shapiro-Ilan, 2005; Shapiro-Ilan et al., 2012b). Each approach has its advantages and disadvantages in terms of production cost, technical expertise, economy of scale, and quality of the end product. Following production, a variety of formulations are available to facilitate application (Grewal, 2002). We provide a summary of EPN production methods, factors that affect efficiency, and methods for improvement, as well as future research directions needed.

10.2. IN VIVO PRODUCTION

10.2.1. Basic Method

Methods for in vivo production of EPNs have been reported by various authors (Dutky et al., 1964; Poinar, 1979; Woodring and Kaya, 1988; Lindegren et al., 1993; Flanders et al., 1996; Kaya and Stock, 1997; Shapiro-Ilan et al., 2002a). Generally, the approach is based on a two-dimensional system that relies on production in trays and shelves (Friedman, 1990; Shapiro-Ilan and Gaugler, 2002; Ehlers and Shapiro-Ilan, 2005; Shapiro-Ilan et al., 2012b). These systems usually revolve around the concept of a White trap (White, 1927), which is a device used for harvesting IJs that takes advantage of the progeny IJ's natural migration away from the host cadaver upon emergence. The White trap consists of a dish or tray on which the cadavers rest; the dish is surrounded by water, which is contained by a larger arena (Fig. 10.3).

The in vivo approach consists of inoculation, harvest (via White trap), concentration, and (if necessary) decontamination. Inoculation is achieved by applying IJs on a dish or tray containing the host insects and lined with absorbent paper or another material conducive to nematode infection such as soil, cheesecloth, or plaster of Paris. At approximately 2–5 days postinoculation, infected

FIGURE 10.3 A White trap. The area around the smaller Petri dish is filled with water below the level of dish. Nematode progeny crawl over the small dish lid into the water trap. (For color version of this figure, the reader is referred to the online version of this book.)

insects are transferred to the White traps. It is important not to allow the infections to progress too long before transfer; otherwise, the chance of the cadavers rupturing or harm to reproductive nematode stages will increase (Shapiro-Ilan et al., 2001). The infective juveniles are harvested after they emerge and migrate into the surrounding water trap. The scale of the White trap in size and number can be expanded to commercial levels; for instance, trays as large as 1 m^2 are conceivable as long as mechanisms for the IJs to enter the water trap are available. Following harvest, concentration of nematodes can be accomplished by gravity settling, such as in large cone-shaped vessels and/or vacuum filtration (Lindegren et al., 1993). Gravity settling is simple and straightforward, but one potential drawback is that prolonged lack of aeration can be detrimental to the nematodes. Centrifugation is also feasible (Kaya and Stock, 1997), but, for commercial in vivo operations, the capital outlay for a centrifuge of sufficient capacity may be unwarranted.

10.2.2. Factors Affecting Efficiency

Efficiency of production is derived from nematode yield relative to cost outputs. The primary expenses for in vivo production include the costs of insect hosts and labor. Methods to reduce these costs will be discussed in the subsequent section. Factors affecting nematode yield in in vivo production include the nematode and host species, inoculation parameters, and environmental conditions.

To some extent, variation in yield among nematode species is inversely proportional to IJ size (see Grewal et al., 1994; Hominick et al., 1997), although some species simply have innately high reproductive capacities (such as *H. indica* and *S. riobrave*). For example, yields of *S. riobrave* (average body length of IJ = 622 mm) may exceed 300,000 IJs per insect in *Galleria mellonella* L., whereas for *S. glaseri* (average body length of IJ = 1133 mm) yields do not exceed 50,000 IJs in the same host (Grewal et al., 1994; Stock and Hunt, 2005). Thus, choice of nematode species must be taken into account when developing in vivo production ventures. However, other considerations may also be critical in choosing which nematode(s) to produce, such as matching the appropriate nematode species to the selected target pests, environmental conditions of the target site, and marketing considerations (Shapiro-Ilan et al., 2002b).

The choice of insect host is also important to maximize yields and efficiency in production. Due to its high susceptibility to most nematodes, ease in rearing, wide availability, and ability to produce high yields, the last instar of the greater wax moth, *G. mellonella*, is the most common insect host used for in vivo laboratory and commercial EPN production (Woodring and Kaya, 1988). Only a few EPNs exhibit relatively poor reproduction in *G. mellonella* (due to extremes in host specificity), such as *S. kushidai* and *Steinernema scarabaei* Stock and Koppenhöfer appear to be especially adapted to hosts in the family Scarabaeidae (Order = Coleoptera) and *S. scapterisci* appears to be especially adapted to the order Orthoptera (Mamiya, 1989; Nguyen and Smart, 1990; Kaya and

Stock, 1997; Grewal et al., 1999; Koppenhöfer and Fuzy, 2003). Significant research and commercial application has also been achieved for production of EPNs in *Tenebrio molitor* L. (Blinova and Ivanova, 1987; Shapiro-Ilan et al., 2002a). Various other hosts have been studied for in vivo nematode production, including the navel orangeworm, *Amyelois transitella* (Walker), tobacco budworm, *Heliothis virescens* (F.), cabbage looper, *Trichoplusia ni* (Hübner), pink bollworm, *Pectinophora gossypiella* (Saunders), beet armyworm, *Spodoptera exigua* (Hübner), corn earworm, *Helicoverpa zea* (Boddie), gypsy moth, *Lymantria dispar* (L.), house cricket, *Acheta domesticus* (L.), and various beetles (Coleoptera) (Lindegren et al., 1979; Blinova and Ivanova, 1987; Cabanillas and Raulston, 1994; Grewal et al., 1999; Elawad et al., 2001).

Nematode yield is generally proportional to insect host size (Blinova and Ivanova, 1987; Flanders et al., 1996). However, IJ yield per milligram of insect (within host species) and susceptibility to infection is usually inversely proportional to host size or age (Blinova and Ivanova, 1987; Shapiro et al., 1999; Dolinski et al., 2007; Dias et al., 2008). In addition to yield, ease of culture and susceptibility to IJs are important factors when choosing a host (Blinova and Ivanova, 1987; Shapiro-Ilan and Gaugler, 2002). Ultimately, the choice of host species and nematode for in vivo production should rest on nematode yield per cost of insect biomass and the suitability of the nematode for the target pest (Blinova and Ivanova, 1987; Shapiro-Ilan et al., 2002a). Comparisons of cost efficiencies among different host species have rarely been studied. Blinova and Ivanova (1987) reported *T. molitor* to be superior in cost efficiency compared with *G. mellonella* and *Trichoplusia ni* for producing *S. carpocapsae*. In another study, Shapiro-Ilan et al. (2002a) observed that *H. bacteriophora* production in *T. molitor* was slightly advantageous relative to production in *G. mellonella,* whereas production efficiency for *S. carpocapsae* appeared to be equal in the two hosts (yield was higher in *G. mellonella*, but *T. molitor* was less expensive to produce). Finally, in addition to cost efficiency, nematode quality may also need to be considered when choosing a host because nematodes reared on various hosts may differ in quality (Abu Hatab et al., 1998), and nematodes can become adapted to the host they are reared on (Stuart and Gaugler, 1996), which can have negative implications for biocontrol efficacy.

Inoculation parameters including dosage, host density, and inoculation method can have a profound effect on the efficiency of production. For each particular nematode and host species, optimization of host density and inoculation rate is recommended for achieving maximum yields (Shapiro-Ilan et al., 2002a). In general, the number of hosts exhibiting patent signs of nematode infection increases with nematode concentration and decreases with host density per unit area (Shapiro-Ilan et al., 2002a). A dosage that is too low results in low host mortality. In contrast, a dosage that is too high has been suggested to potentially result in failed infections due to competition with secondary invaders (Woodring and Kaya, 1988). Thus, intermediate dosages can be used to maximize yield (Boff et al., 2000). For example, rates of approximately 25–200 IJs per insect are usually sufficient (depending on nematode species and method of inoculation) for infecting *G. mellonella*, whereas higher rates are generally needed to infect *T. molitor* (e.g. 100 to 600 IJs per insect). Crowding of hosts can lead to oxygen deprivation or buildup of ammonia, which suppresses nematode yield (Shapiro et al., 2000; Shapiro-Ilan et al., 2002a).

Inoculation method can also affect efficiency and yield. In vivo inoculation can be accomplished by pipetting or spraying nematodes onto a substrate, immersing insects in a nematode suspension, or applying the nematodes to the insect's food. Immersion of hosts is generally more time efficient but requires more nematodes than other procedures; for example, Shapiro-Ilan et al. (2002a) observed immersion of *G. mellonella* to be approximately four times more efficient in time than pipetting inoculum onto the hosts. However, some insects are more sensitive to being immersed than others. Relative to other methods, Blinova and Ivanova (1987) reported that infection of *T. molitor* by *S. carpocapsae* was increased using a feeding method. However, the feeding approach requires the additional step of removing infected cadavers from food remnants. Additionally, the food may introduce contaminants to the system. Therefore, inoculation procedures must be included in a cost efficiency analysis before deciding on a method.

Several environmental factors including optimum temperature, adequate aeration, and moisture can affect yield (Burman and Pye, 1980; Woodring and Kaya, 1988; Friedman, 1990; Grewal et al., 1994; Shapiro-Ilan et al., 2002a; Dolinski et al., 2007). Temperature is critical during the rearing process as it affects both yield and life cycle duration (time to emergence) (Grewal et al., 1994). Grewal et al. (1994) determined the optimum rearing temperature in *G. mellonella* for 12 species and strains of EPNs; optimum temperatures varied from 18 to 28 °C. For example, optimum rearing temperatures for certain strains of *S. feltiae* (a cold-tolerant species), *S. riobrave* (a heat-tolerant species), and *S. carpocapsae* (intermediate) were 18.5 °C, 28 °C, and 25 °C, respectively. In addition to appropriate temperatures, adequate aeration (Burman and Pye, 1980; Friedman, 1990) and high levels of humidity are important environmental factors that must be maintained throughout the production cycle (Woodring and Kaya, 1988). An optimum balance between aeration (to avoid build up ammonia or other harmful gases) and humidity should be reached, such as by using high-efficiency particulate air filters and a humidifying system.

10.2.3. Recent Advances and Future Directions

As indicated above, the efficiency of in vivo production is limited by the costs of insects and labor. Therefore, the economics of in vivo production can be improved substantially by producing the insect hosts "in-house" and mechanizing the process (thereby reducing labor). Various steps in the in vivo production can be mechanized (e.g. inoculation, harvest, concentration). Indeed, mechanization of the entire process from insect production through harvest and packaging has been proposed (Morales-Ramos et al., 2012; Shapiro-Ilan et al., 2012c).

Several approaches to mechanization of nematode inoculation and harvest have been developed or proposed. For example, improved inoculation devices have been developed to allow for mass inoculation on multiple shelves (Carr and Kolodny-Hirsch, 1993; Shapiro-Ilan et al., 2009b). A mechanized harvest device, LOTEK, allows for automated collection of IJs from stacked trays; the IJs are pumped to a central collection tank; unlike the White trap approach, harvesting does not require nematode migration to a water reservoir (Gaugler et al., 2002). Shapiro-Ilan et al. (2011) introduced a single-tray design where infected host cadavers are placed on a substrate intended as the formulation carrier (e.g. a gel) within the final package; once IJs emerge, the cadavers are removed and the final nematode product is ready for shipment or storage. In this approach, the process is streamlined substantially because nematodes emerge directly into their formulation and package; thus, additional concentration and packaging steps are removed from the process.

Advances have also been made in improving insect host production for in vivo EPN production. The host's reproductive capacity can be enhanced by optimizing host density and life cycle aspects of the rearing process (Morales-Ramos et al., 2012). Furthermore, Morales-Ramos et al. (2011a) developed a mechanized sifter to automatically separate insect hosts (e.g. *T. molitor*) of different sizes, thereby facilitating selection for nematode inoculation and continued culturing, among others. Optimization of insect diets can also lead to improved efficiency in insect host production (Morales-Ramos et al., 2011b). Morales-Ramos et al. (2011b) used a self-selection regimen to identify optimum ratios of carbohydrates (bran and potato) in the diet of *T. molitor*; recently, optimum levels of lipid and protein supplements were also determined using the same technique (Morales-Ramos, Rojas, Shapiro-Ilan, and Tedders, unpublished data). Based on tritrophic interactions, development of the host diet can also be used to improve the quality and production of nematodes. For example, a program has been initiated to simultaneously improve the diet of *T. molitor* (including carbohydrate, protein and lipid content) for optimal host production and nematode fitness (Shapiro-Ilan et al., 2008a, 2012c).

An alternate approach to in vivo production is culture and delivery of EPNs in their infected host cadavers (Jansson et al., 1993; Shapiro and Glazer, 1996; Del Valle et al., 2008). In this approach, nematodes are applied to the target site in their infected hosts, and pest suppression is subsequently achieved by progeny IJs that emerge from the cadavers. Production and application of nematodes in infected hosts may be more efficient than other in vivo production methods because several labor-intensive steps are removed from the process—harvest and concentration. Furthermore, laboratory studies have indicated that nematodes applied using the cadaver approach exhibited superior dispersal (Shapiro and Glazer, 1996), infectivity (Shapiro and Lewis, 1999), and survival (Perez et al., 2003) compared with nematodes applied using the standard approach in aqueous suspension. Pest control efficacy when using the cadaver application approach was also reported to be superior to standard application in aqueous suspension (Shapiro-Ilan et al., 2003a). The cadaver application approach has shown high levels of efficacy in suppression of various pests including the black vine weevil, *Otiorhynchus sulcatus* (F.), citrus weevil, *Diaprepes abbreviatus* (L.), guava weevil, *Conotrachelus psidii* Marshall, large pine weevil, *Hylobius abietis* (L.), and small hive beetle, *Aethina tumida* (Murray) (Shapiro-Ilan et al., 2003a, 2010; Dillon et al., 2007; Del Valle et al., 2008).

One potential drawback to applying EPNs in infected hosts is that the cadavers may be fragile or stick together, particularly when a soft-bodied insect host is used, such as *G. mellonella*. However, application of cadavers may be facilitated through formulations that have been developed to protect cadavers from rupture and improve ease of handling, such as a starch matrix with clay coating or an encapsulation process (Shapiro-Ilan et al., 2001, 2010; Del Valle et al., 2009). Alternatively, rupture and sticking together can be avoided by using a hard-bodied insect as the host, such as *T. molitor* (Shapiro-Ilan et al., 2008b). The use of hard-bodied insects can be further enhanced using a mechanized device that automatically wraps the cadavers between two pieces of masking tape; the IJs emerge between two rolls of tape without any loss in yield (Morales-Ramos et al., 2010; Shapiro-Ilan et al., 2010). Another barrier to the cadaver application method is devising procedures to distribute the infected hosts on a large scale; to this end, mechanized equipment for field distribution has been developed (Zhu et al., 2011). In yet another novel approach to using EPN-infected hosts, recently, nematodes applied in host cadavers were effective and highly persistent when added to bags of potting media for subsequent distribution to target pest sites (Deol et al., 2011).

10.3. IN VITRO PRODUCTION: SOLID CULTURE

10.3.1. Basic Method

The first attempt at EPN in vitro production was carried out on a solid medium axenically, but efforts to continuously produce EPNs failed because the required presence of the nematode's symbiotic bacterium was not yet known (Glaser, 1931). However, much later when the bacterial symbionts were discovered and their importance for EPN reproduction was recognized, a foundation was laid down for in vitro mass production of EPNs (Poinar and Thomas, 1966). Today, in vitro production of EPNs is based on introducing nematodes to a pure culture of their symbiont in a nutritive medium.

A prerequisite to in vitro production (using solid or liquid media) is the establishment of a monoxenic culture. Steps toward creating monoxenic cultures include isolation of the symbiotic bacteria and establishment of bacteria-free nematodes. The symbiotic bacteria can easily be isolated from nematode-infected insect larvae or surface sterilized IJs. The bacteria can occur in two phase variants, primary and secondary, which differ in several characteristics, including dye absorption, response to biochemical tests, and antibiotic production. Although EPNs have been reported to grow on secondary variant symbionts, the primary variant is most conducive to growth and infective juveniles tend to only retain the primary variant (Han and Ehlers, 2001). Single phase I colonies can be selected on indicator plates of nutrient bromothymol blue agar (NBTA) (per 1 l distilled water: peptone 5.0 g, beef extract 3.0 g, agar 15 g, bromothymol blue 0.025 g and 2,3,5-triphenyl tetrazolium chloride 0.04 g) and MacConkey (per 1 l distilled water: peptone 17 g, proteose peptone 3 g, lactose 10 g, bile salts 1.5 g, sodium chloride 5 g, neutral red 0.03 g, agar 13.5 g) (Akhurst, 1980) and cultured in lysogeny broth (LB) (per liter distilled water: NaCl 5 g, tryptone 10 g, yeast Extract 5 g) or yeast salts (YS) (per liter distilled water: $NH_4H_2PO_4$ 0.5 g, K_2HPO_4 0.5 g, $MgSO_4 \cdot 7H_2O$ 0.2 g, NaCl, yeast extract 5.0 g) broth. Stock cultures of bacteria are mixed with glycerol at 15% (v/v), and aliquots are frozen at −80 °C.

More laborious is the establishment of axenic nematodes (for subsequent introduction to the pure bacteria culture and initiation of production). In earlier work, to create monoxenic cultures, surface sterilized nematodes were added to a lawn of the bacterial symbionts (Akhurst, 1980; Wouts, 1981). However, Lunau et al. (1993) suggested that surface sterilization of IJs is insufficient to establish monxenicity because contaminating bacteria survive beneath the nematode's cuticle. Thus, an improved method was developed where axenic nematode eggs are obtained by rupturing gravid nematode females in an alkaline solution (Lunau et al., 1993; Han and Ehlers, 1998). Specifically, in this method, fertile nematode eggs from gravid females (*Steinernema* spp.) or hermaphrodites (*Heterorhabditis* spp.) are collected in Ringer's solution (per 1 l distilled water: NaCl 9.0 g, KCl 0.42 g, $CaCl_2$ 0.48 g, $NaHCO_3$ 0.2 g) and surface-sterilized in a solution of 2.5 ml 4 M NaOH, 0.5 ml 12% NaOCl, and 21.5 ml distilled water. After rinsing twice in sterile Ringer's solution, the eggs are transferred to sterile liquid media in which the J1 stage hatches. After 2 days (as long as the medium is without growth of contaminants), the J1s are combined with their symbiotic bacteria growing on the fortified lipid agar (1.6% nutrient broth, 1.2% agar, and 1% corn oil; Wouts, 1981). Once the monoxenic cultures are established, the nematode culture can be scaled-up in solid phase or liquid phase.

The in vitro solid technique involves the following four steps: preparation of solid medium, inoculation with bacteria, inoculation with nematodes, and harvest. Solid culture was first accomplished in two-dimensional arenas (e.g. Petri dishes) using various agar media such as those based on dog food, pork kidney, cattle blood, and other animal products (House et al., 1965; Hara et al., 1981; Wouts, 1981). Wouts (1981) developed an improved medium (less expensive and more consistent from batch to batch) that included yeast extract, nutrient broth, vegetable oil, and soy flour. Subsequently, in vitro solid culture was advanced considerably by the invention of a three-dimensional rearing system involving nematode culture on crumbled polyether polyurethane foam (Bedding, 1981).

This was a major breakthrough and opened the door for economic feasibility in commercial mass production.

A number of companies (Biotech Australia, De Groene Vlieg and Koppert Biological Systems, both from The Netherlands, and Bionema, Sweden) started to produce and sell EPNs for soil pests, mainly in greenhouses (Ehlers, 2001a). In the three-dimensional solid media method, a liquid medium is mixed with foam, autoclaved, and then inoculated with bacteria followed by the nematodes. Nematodes are then harvested within 2–5 weeks (Bedding, 1981, 1984) by placing the foam onto sieves immersed in water. Media for this approach was initially based on animal products such as pork kidney or chicken offal, but it was later improved and may include various ingredients including peptone, yeast extract, eggs, soy flour, and lard (Han et al., 1992, 1993). The approach was later expanded to autoclavable bags with filtered air being pumped in (Bedding, 1984) and culture vessels comprising a tray with side walls and overlapping lids that allow gas exchange through a layer of foam. Large-scale production was advanced further through various mechanisms, including using bags with gas permeable Tyvac® strips for ventilation, automated mixing and autoclaving, simultaneous inoculation of nematodes and bacteria, sterile room technology, and automated harvest through centrifugation (Gaugler and Han, 2002).

10.3.2. Factors Affecting Efficiency

The impact of nematode inoculum rate (IJs per unit of media) on yield varies among nematode strains (Han et al., 1992, 1993; Wang and Bedding, 1998). For example, *S. carpocapsae* (Agriotos strain) produced optimum yields at an intermediate inoculum size (2000 infective juveniles per g medium) (Han et al., 1993), whereas *S. carpocapsae* (CB2B strain) and *H. bacteriophora* (H06) were not affected by inoculum size (Han et al., 1992). Bacterial inoculum size does not appear to be important in yield determination (Han et al., 1992, 1993).

Temperature plays a crucial role in nematode development in solid culture (Han et al., 1992, 1993). A temperature of 27 °C supported optimal production for *Steinernema* sp. CB2B and 25 °C for *H. bacteriophora* H06 (Han et al., 1992). However, some species require lower or higher temperatures for growth and reproduction. Dunphy and Webster (1989) reported the optimum temperature of 30 °C for *H. bacteriophora* (=*heliothidis*) on lipid agar. When cultured in soy flour medium, maximum yields of Pakistani strains of *Steinernema pakistanense* Shanina, Anis, Reid, Rowe, and Maqbool; *Steinernema asiaticum* Anis, Shahina, Reid, and Rowe; *Steinernema abbasi* Elawad, Ahmad, and Reid; *Steinernema siamkayai* Stock, Somsook, and Kaya; *Steinernema carpocapsae*; *H. indica*; and *H. bacteriophora* were obtained at 32 ± 2 °C, whereas *S. feltiae* produced the highest yields at 20 ± 2 °C (Salma and Shahina, 2012). Thus, temperature adaptation of nematodes for maximum production in solid culture is related to the specific species or strains.

Culture time, being inversely related to temperature, should be optimized for maximum yield according to the needs of each nematode species or strain (Dunphy and Webster, 1989; Han et al., 1992, 1993). Increasing inoculum size can increase nematode growth and decrease culture time (Han et al., 1992). Longer culture times can provide higher yields, yet nematode mortality may also increase with time (Han et al., 1992, 1993) and culture time must be weighed against the cost of space and diminishing returns.

Media composition is a major factor that affects yield in solid culture. Increasing the lipid quantity and quality leads to increases in nematode yield (Dunphy and Webster, 1989; Han et al., 1992). The most suitable array of lipid components reflects the nematode's natural host composition (Abu Hatab et al., 1998; Abu Hatab and Gaugler, 2001). Nematode yields were significantly affected by the addition of water, yeast extract, and egg yolk for *S. carpocapsae* and by addition of water and lard for *H. bacteriophora*, which indicated the importance of selecting suitable medium components and combinations for these nematode species (Han et al., 1995). Soy flour is one of the media components both in solid culture and liquid culture (Ehlers and Shapiro-Ilan, 2005). Media containing soy flour yielded the highest population compared to media without soy flour in in vitro solid production of eight Pakistani strains (Salma and Shahina, 2012); the species tested included *S. pakistanense*, *S. asiaticum*, *S. abbasi*, *S. siamkayai*, *S. carpocapsae*, *S. feltiae*, *H. indica*, and *H. bacteriophora*. Other media ingredients that may have an effect on nematode yield include salts and proteins (Dunphy and Webster, 1989).

Yields in solid culture also depend upon the innate reproductive capacities of different nematode species and strains. For example, *H. bacteriophora* strain HbNJ yielded an average of 1.4×10^9/bag (Gaugler and Han, 2002), whereas an *S. scapterisci* strain yielded an average of 4.8×10^8/bag (Bonifassi et al., 1999). On same culture media and under the same culture conditions, *H. indica* strain LN2 yielded much more than *H. bacteriophora* strain H06 (9.3×10^5 IJs/g medium vs. 5.1×10^5 IJs/g medium) (Han and Qiu, unpublished data).

10.3.3. Recent Advances and Future Directions

Recently, advances have been made in expanding in vitro solid production of EPNs. For example, in China, a pilot factory was established for solid production of several EPN species based on the lower labor cost available in that country and an improved media and mechanization process. The factors influencing production efficiency were explored and optimized, including medium development, the culture parameters, recovery of the IJ inocula, formation of the IJs, extraction, and harvest (Han et al., 1995, 1997). A company called Guangzhou Greenfine Biotechnology, under the guidance of Guangdong Entomological Institute, is currently in commercial production; products from the solid culture system include *S. carpocapsae*, *S. feltiae*, *S. longicaudum*,

S. scapterisci, *H. bacteriophora*, and *H. indica*, which are provided for field trials in China and for internal and international markets.

10.4. IN VITRO PRODUCTION: LIQUID CULTURE

10.4.1. Basic Method

EPN in vitro liquid culture was first attempted by Stoll (1952) using raw liver extract in shaken flasks and was developed further by several researchers (Ehlers, 2001b; Shapiro-Ilan et al., 2012b). It is a complex rearing process that demands medium development and optimization, understanding of the biology of the nematode and their bacteria, and bioreactor development including understanding and controlling the process parameters. Collaboration between researchers in the public and private sectors contributed largely to the establishment of in vitro liquid culture.

Successful implementation of the liquid culture process faces the opposing challenges of supplying enough oxygen while preventing excessive shearing of nematodes (Pace et al., 1986; Buecher and Popiel, 1989; Friedman et al., 1989; Friedman, 1990). Initially, the issue was addressed using various approaches, such as relying on bubbling with a downward sparger coupled with limited agitation (Pace et al., 1986) or using an airlift fermenter coupled with a variable agitation regimen (Friedman et al., 1989). Innovations in mixing and aeration have been subsequently introduced, including internal (Strauch and Ehlers, 2000) and external (Neves et al., 2001) bioreactors. Internal loop vessels have baffles placed inside the single vessel, which create the channels required for circulation, whereas in external loop vessels circulation takes place through separate conduits. Another problem observed during agitation is foaming, which can be reduced by changes in the bioreactor design (Strauch and Ehlers, 2000) and use of antifoam or defoaming agents (Gaugler and Han, 2002).

The first attempt to use bioreactors was described by Pace et al. (1986) and the first commercial application of the liquid culture technology was made by the company Biosys (Palo Alto, California). The company was incorporated in 1987 and soon started to produce liquid culture nematode products. In 1992, large-scale production of *S. carpocapsae* began and was scaled-up to volumes of 80,000 l. Today, commercial EPNs resulting from liquid culture are produced by several companies.

In liquid culture, symbiotic bacteria are generally first introduced followed by the nematodes (Buecher and Popiel, 1989; Surrey and Davies, 1996; Strauch and Ehlers, 2000; Johnigk et al., 2004; Shapiro-Ilan et al., 2012b). A variety of ingredients for liquid culture media have been reported, including soy flour, yeast extract, canola oil, corn oil, thistle oil, egg yolk, casein peptone, milk powder, liver extract, and cholesterol (Surrey and Davies, 1996; Ehlers et al., 2000; Yoo et al., 2000). Media and species effects impact culture times in liquid culture, which may be as long as 3 weeks (Surrey and Davies, 1996; Chavarria-Hernandez and de la Torre, 2001), although many species can reach maximum IJ production in 2 weeks or less (Friedman, 1990; Ehlers et al., 2000; Strauch and Ehlers, 2000; Yoo et al., 2000; Neves et al., 2001).

EPN cultures in liquid are particularly vulnerable to contamination due to the even distribution of the fluids and organisms obtained through the mixing of liquid in bioreactors and based on the long process time (Ehlers and Shapiro-Ilan, 2005). Nematode yields can be severely reduced by the presence of nonsymbiotic microbes; such contamination can prevent further scale-up. Given that a nematode production process can last up to 3 weeks, maintenance of sterile conditions can be extremely challenging. The monoxenicity of the cultures must be ensured and verified from the onset of inoculum production. Monoxenic cultures to be used as inocula can be stored on shakers at 20 rpm and low temperature (generally 4 °C, but for some nematode species it should be higher) for several months until they are introduced into the bioreactor. Strain collections of nematodes can be kept in liquid nitrogen (Popiel and Vasquez, 1991). Cultures are always preincubated for 24–36 h with the specific symbiont bacteria before IJs are inoculated. The inoculum density for the symbiotic bacteria is between 0.5% and 1% of the culture volume. The inoculum of the nematodes varies depending on species and media composition. An optimal number of adults per milliliter can be calculated, which is defined by the percentage of IJs likely to recover. Generally, the nematode inoculum is between 5% and 10% of the culture volume. The culture medium should have a pH between 5.5 and 7.0 when the culture is started. The pH appears to be well regulated by the organisms themselves. Oxygen supply must be maintained at approximately 30% saturation, which may help to prevent the bacteria from shifting to the secondary phase. An important parameter is the aeration rate. Increasing the aeration rate often increases foaming. The addition of agents such as silicon oil usually prevents foaming. However, it should be used carefully because higher concentrations can be detrimental to the nematodes.

Owing to the potential of *Xenorhabdus* and *Photorhabdus* spp. to metabolize almost every kind of protein-rich medium, the selection of appropriate culture media for EPN production can largely follow economic aspects (Ehlers and Shapiro-Ilan, 2005). A standard medium should contain a carbon source (e.g. glucose or glycerol), a variety of proteins of animal and plant origin, yeast extract, and lipids of animal or plant origin (Pace et al., 1986; Friedman et al., 1989; Han et al., 1993; Surrey and Davies, 1996; Ehlers et al., 1998). The osmotic strength of the medium must not surpass 600 mmol/kg. Improvements of the medium and adaptation to the requirements of different species are feasible approaches to increase yields (Ehlers, 2001a). Essential amino acid requirements have only been defined for *S. glaseri* (Jackson, 1973). Nematodes have nutritional demands for sterols, but they can metabolize the necessary sterols from a variety of steroid sources (Ritter, 1988) that are provided through the addition of lipids of animal or plant origin. In general, *S. carpocapsae* requires proteins of animal origin (Yang et al., 1997), and it is unable to reproduce without the addition of lipid sources to the medium. *H. bacteriophora* produces offspring in a liquid medium without the addition of lipids (Han and Ehlers, 2001). *Photorhabdus luminescens* provides or metabolizes essential lipids; however, lipids should always be added to increase the total IJ fat content. The lipid composition of the medium has an effect on the fatty acid composition of the bacteria and IJs (Abu Hatab et al., 1998); low fatty content of IJ can reduce efficacy (Patel et al., 1997a, b).

10.4.2. Factors Affecting Efficiency

Although both steinernematids and heterorhabditids share the requirements of adequate aeration (without shearing), the approaches for maximizing yield of the two genera in liquid culture differ due to their life cycles and reproductive biology. Given that steinernematids (except one species) occur only as males and females and are capable of mating in liquid culture (Strauch et al., 1994), maximization of mating is paramount and can be achieved through bioreactor design and regulation of aeration (Neves et al., 2001). However, maximization of mating is not applicable for heterorhabditid production in liquid culture because the first generation is exclusively hermaphrodites and, although subsequent generations contain amphimictic forms, they cannot mate in liquid culture (Strauch et al., 1994). Thus, maximizing heterorhabditids yields in liquid culture depends on the degree of recovery (the developmental step when IJs molts to initiate completion of their life cycle).

Although levels of heterorhabditid recovery in vivo tend to be 100%, recovery in liquid culture can range from 0 to 85% (Ehlers et al., 2000; Jessen et al., 2000; Yoo et al., 2000). Recovery may be affected by various factors including nutrition, aeration, CO_2, lipid content, and temperature (Strauch and Ehlers, 2000; Ehlers et al., 2000; Jessen et al., 2000; Yoo et al., 2000). Recovery can be influenced during the bacterial preculture phase, prior to nematode introduction. A higher bacterial density leads to a higher food signal concentration, which in turn produces a higher recovery rate (Hirao and Ehlers, 2009). Nematodes should therefore be inoculated during the late stationary growth phase (Johnigk et al., 2004). The period starts with the increase of the pH after its distinct minimum at pH less than 8.0 (Johnigk et al., 2004). Glucose fed-batch reactors can increase bacterial density (Jeffke et al., 2000) and enhance food signal production, and thus they may be used to increase IJ recovery (Ehlers and Shapiro-Ilan, 2005). Jessen et al. (2000) reported that increasing the CO_2 concentration in the medium enhanced IJ recovery. However, the influence of decreasing pH caused by the CO_2 concentration was excluded. A pH less than 6.5 significantly reduced the IJ recovery. A protein source such as soybean can improve symbiotic bacterial growth and nematode recovery, which thereby promotes nematode production (Cho et al., 2011). Recovery in secondary-phase bacteria is lower than in primary-phase bacteria (Han and Ehlers, 2001; Hirao and Ehlers, 2009). No significant influence of inoculum density on IJ recovery of *S. carpocapsae* and *S. feltiae* has been detected, but a significant correlation between recovery and the age of IJ inoculum was detected for *S. carpocapsae* (Hirao and Ehlers, 2010).

Yield from liquid culture may also be affected by other factors, including media, inoculum parameters, nematode species, culture temperature, and culture time (Han, 1996; Ehlers et al., 2000). The central component of the liquid culture media is the lipid source and quantity (Abu Hatab et al., 1998; Yoo et al., 2000). Other nutrients that have been reported to affect yield positively include the glucose content (Jeffke et al., 2000), yeast extract content (Chavarria-Hernandez and de la Torre, 2001), and soybean protein (Cho et al., 2011). Similar to the other production approaches, nematode yield in liquid culture is generally inversely proportional to the size of the species (Ehlers et al., 2000). Some of the maximum average yields reported include 300,000 and 320,000 IJs per milliliter for *H. bacteriophora* and *S. carpocapsae*, respectively (Han, 1996), 138,000 IJs per milliliter for *H. megidis* (Strauch and Ehlers, 2000), 71,470 IJs per milliliter for *S. feltiae* (Chavarria-Hernandez and de la Torre, 2001) and 450,000 IJs per milliliter for *H. indica* (Ehlers et al., 2000).

Temperature has an influence on the time scale of the nematode development. For example, Ehlers et al. (2000) noted that recovery is influenced by culture temperature. Optimal culture temperature varies from species to species; for example, for *S. carpocapsae* and *S. feltiae*, the optimal culture temperature is 25 °C. The equipment and type of bioreactors used will also impact yield. For example, although flat-blade impellers, bubble columns, airlift, and internal loop bioreactors have all been successfully tested (Pace et al., 1986; Surrey and Davies, 1996; Ehlers et al., 1998), a direct comparison with flat-blade impeller-stirred tanks or airlift bioreactors internal loop bioreactors always yielded higher IJ concentrations (Ehlers and Shapiro-Ilan, 2005).

10.4.3. Recent Advances and Future Directions

Some recent advancements in liquid culture technology serve to increase quality and efficiency of production through various processes, such as optimizing media and bioprocess kinetics through modeling (Chavarria-Hernandez et al., 2006, 2010), as well as improvements in inoculum and bacterial cell density (Hirao and Ehlers, 2010), timing of inoculation (Johnigk et al., 2004), improving beneficial traits (such as heat and desiccation tolerance) in *Heterorhabditis* spp. (Mukuka et al., 2010; Anbesse et al., 2013), and downstream processing (Young et al., 2002). Future research and development in liquid culture, focusing on media optimization and bioreactor design, are expected to lead to additional benefits, such as higher yields and reduced costs.

10.5. ANALYSIS AND CONCLUSION

10.5.1. Comparison of Production Methods

A summary of advantages and disadvantages of each EPN production approach is presented in Table 10.2. Issues of concern include capital outlay, technical expertise, nematode quality, labor, and cost efficiency. In assessing the merits of the approaches among these issues, in vivo production and in vitro liquid culture are the extremes and in vitro solid culture is intermediate between them. For example, the level of capital outlay and expertise required is lowest for in vivo production, highest for in vitro liquid production, and intermediate for in vitro solid production (Friedman, 1990; Gaugler and Han, 2002; Shapiro-Ilan et al., 2012b). Another advantage to in vivo production is the ease of adapting the process to new/different nematode species; generally, the process can remain the same except for some slight modifications (e.g. temperature regimes). In contrast, in vitro techniques may require substantial modification when adapting to new nematode species based on media requirements, optimization of fermentation parameters, and downstream processing. Based on the versatility and low levels of technology and expertise required, in vivo production may be most appropriate for small startup companies or in lesser developed countries or regions (where labor may be less expensive).

On the other hand, a major disadvantage of in vivo production is the labor costs, which are highest for in vivo production and lowest for in vitro liquid production (in vitro solid production again being in between). Heretofore, due to the high labor costs, space requirements, and cost of insects, economy of scale (cost efficiency) tends to be lowest for in vivo production

TABLE 10.2 A Comparison of Production Approaches for Entomopathogenic Nematode

	Production Approach		
Issue	In Vivo	In Vitro—Solid	In Vitro—Liquid
Capital outlay	Low	Intermediate	High
Required expertise	Nominal	Intermediate	Extensive
Ease of achieving quality	Easy	Difficult	Difficult
Labor required	High	Intermediate	Low
Economy of scale	Low	Intermediate	High
Ease of adaptation to new nematode species	Easy	Difficult	Difficult

and highest for liquid production. Thus, on a global scale, by far the vast majority of nematode production is currently accomplished in in vitro liquid production.

Despite the reduced economy of scale relative to other production methods, in vivo production has managed to sustain itself as a cottage industry throughout the evolution of commercial in vitro enterprises (Shapiro-Ilan and Gaugler, 2002). In fact, in some markets, in vivo produced nematodes have remained competitive with, or even outcompeted, in vitro produced nematodes due to weakened consumer or distributor confidence in the quality of the in vitro product. Possibly, in vivo production will continue to expand based upon advancements in mechanization as described above.

In vitro solid production has several advantages. It does not require high technology inputs and large investments. The effect of phase variation on the yields is less than in liquid culture (Han and Ehlers, 2001). When it comes to large-scale production, the disadvantages can be overwhelming. In vitro solid production is labor intensive, vulnerable to contamination during upstream and downstream processing, and is difficult to monitor online. The uneven distribution of the nematodes in the medium prevents systematic sampling and thus improvement of the technique is required. However, today, a few companies still produce EPNs in solid culture, including Bionema (www.bionema.se), Andermatt Biocontrol AG (www.biocontrol.ch), BioLogic USA (www.bio-logic.us), and Guangzhou Greenfine Biotechnology (China). Recently, the free-living nematode *Panagrellus redivivus* (L.), a promising food source for first-feeding fish and crustaceans, was successfully cultured on *Xenorhabdus* bacteria in a solid in vitro system (Cao et al., 2008). In developing countries, the in vitro solid culturing system is still superior to liquid culture technology (Ehlers et al., 2000). Similar to in vivo production, efficiency can be increased through mechanization (labor reduction) and media enhancement (Gaugler and Han, 2002; Shapiro-Ilan and Gaugler, 2002).

An issue that has been in contention is the relative quality of nematodes produced using the various methods. Various authors and practitioners contend or have found evidence that in vitro produced nematodes can be inferior in quality compared with in vivo produced nematodes. For example, several studies indicate quality of nematodes produced using solid culture to be similar to nematodes produced by in vivo methods (Abu Hatab et al., 1998; Abu Hatab and Gaugler, 1999; Gaugler and Georgis, 1991), yet Yang et al. (1997) reported reduced fitness in *S. carpocapsae* produced in solid culture compared with in vivo culture. Furthermore, a number of reports also indicated reduced quality and field efficacy in in vitro liquid produced EPNs relative to those produced in vivo (Gaugler and Georgis, 1991; Cottrell et al., 2011), whereas others did not detect any differences (Gaugler and Georgis, 1991; Shapiro and McCoy, 2000a). In contrast, to our knowledge there have no reports in which in vitro produced nematodes caused superior field efficacy compared with in vivo produced nematodes.

The differences between production methods can vary by EPN species. For example, Gaugler and Georgis (1991) reported that the production method (in vivo, solid culture, or liquid culture) did not impact the efficacy of *S. carpocapsae* for control of Japanese beetle, *Popillia japonica* Newman, whereas the efficacy of *H. bacteriophora* produced in liquid culture was inferior to *H. bacteriophora* produced using either of the other methods.

The quality of in vitro production can vary from batch to batch. For example, in one field trial, Cottrell et al. (2011) observed reduced field efficacy in liquid produced *S. carpocapsae* relative to in vivo produced nematodes, yet there was no difference in another trial (the nematodes were from the same commercial source but trials were conducted using different batches). The quality of in vivo produced nematodes may also vary based on production sources (Gaugler et al., 2000). Therefore, it is conceivable that at least in some (if not many) cases, differences observed in nematode quality stem from batch variation rather than innate differences in production technique. Indeed, despite a few reports otherwise, there are vast numbers of examples where liquid produced EPNs have produced high levels of efficacy in a consistent manner. Additionally, with improvements in bioreactor design, media, and other parameters, the quality of in vitro liquid produced nematodes continues to advance (Chavarría-Hernández et al., 2010; Hirao and Ehlers, 2010).

10.5.2. Strain Selection, Improvement and Stability

One aspect that will improve EPN production regardless of culture method is the use of superior nematode strains. Strains that provide higher yield will obviously increase cost efficiency. Furthermore, strains that possess superior biocontrol traits (e.g. virulence, environmental tolerance, host-finding ability) will lead to improved cost efficiency because treatment of the target site will require fewer nematodes per unit area.

Selecting the most suitable EPN strain from a variety of candidates can be addressed simply by screening existing species and strains that may possess superior levels of desired traits. Additionally, surveys can be implemented to discover new strains, which can then be screened in comparison to existing strains; such surveys have been conducted extensively for EPNs (e.g. Shapiro-Ilan et al., 2003b; Bruck, 2004; Campos-Herrera et al., 2008; Malan et al., 2011). The screening process is often accomplished by first narrowing down the number of candidates in laboratory comparisons. The laboratory-based comparisons should address reproductive capacity in the culture system, as well as testing various desired traits, including efficacy against desired target pests under simulated field conditions.

Once a candidate EPN strain is selected, the nematode's reproductive capacity should be tested under full-scale production conditions (rather than just pilot scale). Furthermore, verification of pest control efficacy in the field is critical and should be addressed using the nematode in its final formulated product.

An entomopathogen that shows high virulence in the controlled laboratory environment could fail to suppress the target pest in the field due to various biotic or abiotic factors that may render the organism incompatible. For example, some laboratory strain selections that later proved to be successful in the field include *S. riobrave* (355 strain) and *H. indica* (HOM1 strain) for control of the citrus weevil, *D. abbreviatus* (Duncan and McCoy, 1996; Shapiro et al., 1999; Shapiro and McCoy, 2000b), and *S. riobrave* (355) for control of *Conotrachelus nenuphar* (Shapiro-Ilan et al., 2002c, 2004; Pereault et al., 2009). In contrast, *S. feltiae* was highly virulent to *C. nenuphar* in the laboratory but failed to control the pest under field conditions in Georgia peach orchards (Shapiro-Ilan et al., 2004). Indeed, strain selection that focuses primarily on virulence or mass production and ignores habitat preferences of the pathogen has often been unsuccessful (Shapiro-Ilan et al., 2012a).

If existing or newly discovered entomopathogen strains are not sufficient to reach expected production yields or cannot achieve desired levels of biological control efficacy, another option is to improve selected candidate strains through genetic approaches. Genetic improvement approaches can be directed toward enhancement of single or various beneficial traits, such as reproductive capacity, suitability to production or formulation regimens, virulence, and environmental tolerance. Methods for improvement may include nonmolecular or molecular approaches. One nonmolecular method entails directed selection for desired traits. Selection for improved virulence can be obtained by passing an entomopathogen through a susceptible host (Steinhaus, 1949; Daoust and Roberts, 1982). Some examples of genetic selection for other traits in EPNs include improvements of host finding (Gaugler et al., 1989) and nematicide resistance (Glazer et al., 1997). However, directed selection for a particular trait can inadvertently select for an inferior level of another trait (Gaugler, 1987). For example, a loss in storage capacity was observed in EPNs that had been selected for improved host finding (Gaugler et al., 1990).

Hybridization is another nonmolecular approach to strain improvement. In this approach, the transfer of beneficial traits from one strain to another is accomplished through controlled breeding. The use of hybridization was first demonstrated in EPNs by Shapiro et al. (1997); heat tolerance was transferred from one *H. bacteriophora* strain to another. Given that heterorhabditids produce both hermaphroditic and amphimictic forms, extra care must be taken to ensure that nematode progeny in controlled crosses arise from intended mating regimes rather than self-fertilization. Shapiro et al. (1997) accomplished this by using marker mutations. Hybridization in steinernematids is more straightforward in that only amphimictic forms exist. An example of hybridization in steinernematids is illustrated in the study of Shapiro-Ilan et al. (2005), which used hybridization of *S. carpocapsae* strains to develop superior environmental tolerance and virulence to the pecan weevil, *Curculio caryae* (Horn) (Shapiro-Ilan et al., 2005). The two nonmolecular approaches (selection and hybridization) have also been combined for development of superior EPN strains (Mukuka et al., 2010).

Progress has been made toward molecular approaches for improving EPNs; these procedures entail direct genetic modification of the organisms. For example, a strain of *H. bacteriophora* was improved for heat tolerance via transformation using a heat shock protein originating from *Caenorhabditis elegans* (Maupas) (Gaugler et al., 1997). A risk assessment study concluded that the transgenic organism was an unlikely environmental risk and thus the approach to improving biocontrol was considered viable (Gaugler, et al., 1997). The sequencing of entire genomes of EPNs and their symbionts (Duchaud et al., 2003; Bai and Grewal, 2007; Ciche, 2007; Bai et al., 2009; Schwartz et al., 2011) is expected to continue expanding and will undoubtedly enhance the potential for genetic strain improvement programs using molecular or nonmolecular approaches.

Once a desirable strain is chosen for mass production (based on existing cultures or stemming from an improvement program), it is imperative to ensure the stability of that strain. Regardless of the culture method, production efficiency and biocontrol efficacy can be jeopardized by attenuation of beneficial traits, which can result from repeated subculturing. This phenomenon of trait deterioration can be due to genetic factors (e.g. inbreeding, drift, inadvertent selection) or nongenetic factors (e.g. disease or nutrition) (Tanada and Kaya, 1993; Hopper et al., 1993; Chaston et al., 2011). Trait deterioration has been observed during laboratory culturing of EPNs; relatively rapid loss of various traits were reported, including virulence, environmental tolerance, reproductive capacity, and host finding (Shapiro et al., 1996; Wang and Grewal, 2002; Bai et al., 2005; Bilgrami et al., 2006). Both the nematodes and their bacterial symbionts are subject to trait loss (Bilgrami et al., 2006; Wang et al., 2007), and the cause was reported to be (at least in part) genetically based, with inbreeding depression being more prominent an issue than inadvertent selection (Bai et al., 2005; Adhikari et al., 2009; Chaston et al., 2011).

Therefore, manufacturers must implement precautions against strain deterioration. One approach to reduce trait deterioration is to maintain genetic diversity by storing isolates in liquid nitrogen (Nugent et al., 1996) or minimization of subculturing (Roush, 1990). Nematodes can be placed in liquid nitrogen for long-term storage; subculturing can be minimized by using the frozen material as seed cultures. There are, however, drawbacks to this approach. First, subculturing is inevitable because seed cultures eventually get used up and because mass culture or experiments routinely require many nematode generations; for instance, the number of generations needed for scale-up in liquid culture from shake flask to final bioreactor is extensive. Furthermore, because only small quantities of nematodes can be stored in each cryovial and mortality during storage can be high (Nugent et al., 1996), founder effects may be pronounced when nematodes are stored in nitrogen. Indeed, Wang and Grewal (2002) observed a reduction in desiccation tolerance of *H. bacteriophora* during storage in liquid nitrogen. One option to overcome founder effects (at least to some extent) is to optimize cryopreservation techniques to increase the number of surviving IJs (Bai et al., 2004). Nonetheless, few EPN researchers or commercial producers use liquid nitrogen because strains vary in their adaptability to cryogenic storage (Nugent et al., 1996), it is expensive, and mechanical failure, human error, or neglect can result in complete loss of genetic material. Another approach is to supplement genetic diversity in the laboratory by recollecting fresh nematode isolates from their source or mixing strains. Collection of fresh material, however, takes time, may introduce disease, and may be unreliable because it depends on the isolate remaining where it was last found. Mixing strains may reduce frequencies of desired alleles and, like the other methods, is only a temporary solution because subculturing is still inevitable.

Another approach to deterring trait deterioration in EPN strains that has recently been developed is the creation of homozygous inbred lines (Bai et al., 2005; Shapiro-Ilan et al., 2012b). Homozygous inbred lines in biocontrol agents were hypothesized to be impervious to certain genetic processes that impact trait stability (Hopper et al., 1993). The approach was first tested and validated in EPNs by Bai et al. (2005); selected inbred lines of *H. bacteriophora* remained stable during serial culture, whereas their wild-type parent strains deteriorated in various traits (virulence, environmental tolerance, and host finding). Thus, Bai et al. (2005) proposed that manufacturers create numerous inbred lines from promising candidate strains and select the lines that display high levels of desirable traits. Following further testing and validation of these selected lines (in full-scale production operations and biocontrol efficacy tests under field conditions), the inbred populations can then be used for commercial production. The approach provides strains that are both superior in production and biocontrol traits and stable during culture conditions.

In a variation of the inbred line approach developed by Bai et al. (2005), Anbesse et al. (2013) recommends creation of multiple heterorhabditid inbred lines in liquid culture (e.g. within a single tank). Heterorhabditids cannot mate in liquid culture, so all progeny are produced by self-crossing (via hermaphrodites) (Ehlers and Shapiro-Ilan, 2005); this process automatically creates multiple inbred lines. Similar to the study of Bai et al. (2005), in which inbred lines were generated in vitro on agar plates, the inbred lines produced during liquid culture deter trait deterioration (Anbesse et al., 2013). Anbesse et al. (2013) argued that the multiple inbred line approach (in liquid culture) is superior to the selection of only one or a few inbred lines (as in Bai et al., 2005) because genetic diversity is greater in the former. However, as long as a single (or hybridized) inbred line has been proven to have high biocontrol abilities (as is recommended prior to commercial production), genetic variation is not necessary; indeed, true-bred lines are used ubiquitously in crop production and animal husbandry. In fact, a drawback in allowing broad genetic diversity is that once the nematode population leaves the fermentation tank, the multitude of inbred lines in the nematode population (from that production batch) will be subject to trait deterioration in a similar manner as the wild-type population. On the other hand, the likelihood that postapplication deterioration would occur is minimal because EPNs are usually applied in an inundative manner with little or no recycling expected (Shapiro-Ilan et al., 2012a), and thus both approaches (multiple inbred lines and single or hybrid lines) may be viable. One drawback to the liquid culture approach that cannot be circumvented is that it only works for heterorhabditids. Creation of inbred lines for steinernematids must be done outside of liquid culture to avoid free-mating, such as via controlled sibling mating.

10.5.3. Conclusion

EPNs are commercially produced biocontrol agents that have become well established in several markets. Application of EPNs continues to increase on several continents. All three production approaches have played a role in the success of EPNs in the global market place, and all three will continue to contribute to EPNs expanded success in the future. In vitro liquid production is the most economical approach and is likely to continue to dominate the quantity of EPN production worldwide. Yet in some cases (where labor is less expensive), in vitro solid production may remain competitive. Although in vivo production is the least economical approach, it will likely continue to be appropriate for certain niche markets or for certain small or startup companies; improvements to in vivo production may increase cost efficiency, yet it is unlikely the economy of scale will ever reach that of in vitro approaches.

Despite the success of EPNs in achieving commercial level control for a variety of pests, EPN products remain cost prohibitive in many markets. Thus, additional advances to increase production efficiency and to reduce costs are required. For in vitro production, further advances in fermentation approaches and parameters will lead to reduced costs. In vivo production and in vitro solid production will continue to benefit from streamlining of processes (reducing labor). Additionally, higher efficiency for in vivo production can be achieved through improvements in insect culture methods or discovery of improved hosts for increased yields. Improvements in downstream processing (concentration and formulation) will also contribute substantially to increased cost efficiency for all production methods. As market demand for EPNs increases, additional scale-up of production will provide major benefits in reducing costs. Strain improvement programs may also lead to reduction in EPN production or application costs. We recommend developing some improvement programs that are specifically focused on enhancing traits required for higher EPN production levels (e.g. reproductive capacity) as the majority of improvement endeavors thus far have focused on enhancing traits required in postproduction application. An alternative approach to EPN production that may contribute to market expansion would be developing grower-based or cooperative level "do-it-yourself" ventures (this could be accomplished using any of the three production methods); this concept has been discussed by a number of researchers but thus far has not been attempted in earnest.

To achieve expanded success in biocontrol, as production of EPNs increases, application of the EPN products will need to fit into existing or emerging integrated pest managment strategies. Therefore, opportunities that facilitate incorporation into crop management systems should be seized, such as developing new (compatible) application methods or leveraging synergies between EPNs and other pest management tactics (Koppenhöfer and Grewal, 2005; Shapiro-Ilan et al., 2012a). As the use of broad-spectrum chemical insecticides decreases due to environmental and regulatory concerns and progress on EPN production and application advances, we anticipate that the importance of EPNs as an integral component of sustainable pest management systems will evolve into a more prominent role.

REFERENCES

Abu Hatab, M., Gaugler, R., 1999. Lipids of in vivo and in vitro cultured *Heterorhabditis bacteriophora*. Biol. Control 15, 113–118.

Abu Hatab, M., Gaugler, R., 2001. Diet composition and lipids of in vitro-produced *Heterorhabditis bacteriophora*. Biol. Control 20, 1–7.

Abu Hatab, M., Gaugler, R., Ehlers, R.U., 1998. Influence of culture method on *Steinernema glaseri* lipids. J. Parasitol. 84, 215–221.

Adhikari, B.N., Chin-Yo, L., Xiaodong, B., Ciche, T.A., Grewal, P.S., Dillman, A.R., Chaston, J.M., Shapiro-Ilan, D.I., Bilgrami, A.L., Gaugler, R., Sternberg, P.W., Adams, B.J., 2009. Transcriptional profiling of trait deterioration in the insect pathogenic nematode *Heterorhabditis bacteriophora*. BMC Genomics 10, 609.

Akhurst, R.J., 1980. Morphological and functional dimorphism in *Xenorhabdus* spp, bacteria symbiotically associated with insect pathogenic nematodes *Neoaplectana* and *Heterorhabditis*. J. Gen. Microbiol. 121, 303–309.

Akhurst, R., Smith, K., 2002. Regulation and safety. In: Gaugler, R. (Ed.), Entomopathogenic Nematology. CABI, New York, pp. 311–332.

Anbesse, S., Sumaya, N.H., Dörfler, A.V., Strauch, S., Ehlers, R.U., 2013. Stabilisation of heat tolerance traits in *Heterorhabditis bacteriophora* through selective breeding and creation of inbred lines in liquid culture. BioControl 58, 85–93.

Bai, C., Shapiro-Ilan, D.I., Gaugler, R., Yi, S., 2004. Effect of entomopathogenic nematode concentration on survival during cryopreservation in liquid nitrogen. J. Nematol. 36, 281–284.

Bai, C., Shapiro-Ilan, D.I., Gaugler, R., Hopper, K.R., 2005. Stabilization of beneficial traits in *Heterorhabditis bacteriophora* through creation of inbred lines. Biol. Control 32, 220–227.

Bai, X., Grewal, P.S., 2007. Identification of two down-regulated genes in entomopathogenic nematode *Heterorhabditis bacteriophora* infective juveniles upon contact with insect hemolymph. Mol. Biochem. Parasitol. 156, 162–166.

Bai, X., Adams, B.J., Ciche, T.A., Clifton, S., Gaugler, R., Hogenhout, S.A., Spieth, J., Sternberg, P.W., Wilson, R.K., Grewal, P.S., 2009. Transcriptomic analysis of the entomopathogenic nematode *Heterorhabditis bacteriophora* TTO1. BMC Genomics 10, Art. No. 205.

Bedding, R.A., 1981. Low cost in vitro mass production of *Neoaplectana* and *Heterorhabditis* species (Nematoda) for field control of insect pests. Nematologica 27, 109–114.

Bedding, R.A., 1984. Large scale production, storage, and transport of the insect-parasitic nematode *Neoaplectana* spp. and *Heterorhabditis* spp. Ann. Appl. Biol. 104, 117–120.

Buecher, E.J., Popiel, I., 1989. Liquid culture of the entomogenous nematode *Steinernema feltiae* with its bacterial symbiont. J. Nematol. 21, 500–504.

Bilgrami, A.L., Gaugler, R., Shapiro-Ilan, D.I., Adams, B.J., 2006. Source of trait deterioration in entomopathogenic nematodes *Heterorhabditis bacteriophora* and *Steinernema carpocapsae* during in vivo culture. Nematology 8, 397–409.

Blinova, S.L., Ivanova, E.S., 1987. Culturing the nematode-bacterial complex of *Neoaplectana carpocapsae* in insects. In: Sonin, M.D. (Ed.), Helminths of Insects, American Publishing Co., New Delhi, pp. 22–26.

Bode, H.B., 2009. Entomopathogenic bacteria as a source of secondary metabolites. Curr. Opin. Chem. Biol. 13, 224–230.

Boff, M., Wiegers, G.L., Gerritsen, L.J.M., Smits, P.H., 2000. Development of the entomopathogenic nematode *Heterorhabditis megidis* strain NLH-E 87.3 in *Galleria mellonella*. Nematology 2, 303–308.

Bonifassi, E., Fischer-Le Saux, M., Boemare, N., Lanois, A., Laumond, C., Smart, G., 1999. Gnotobiological study of infective juveniles and symbionts of *Steinernema scapterisci*: a model to clarify the concept of the natural occurrence of monoxenic associations in entomopathogenic nematodes. J. Invertebr. Pathol. 74, 164–172.

Bruck, D.J., 2004. Natural occurrence of entomopathogens in Pacific Northwest nursery soils and their virulence to the black vine weevil, *Otiorhynchus sulcatus* (F.) (Coleoptera: Curculionidae). Environ. Entomol. 33, 1335–1343.

Burman, M., Pye, A.E., 1980. *Neoaplectana carpocapsae*: respiration of infective juveniles. Nematologica 26, 214–219.

Cabanillas, H.E., Raulston, J.R., 1994. Pathogenicity of *Steinernema riobravis* against corn earworm, *Helicoverpa zea* (Boddie). Fundam. Appl. Nematol. 17, 219–223.

Campos-Herrera, R., Gómez-Ros, J.M., Escuer, M., Cuadra, L., Barrios, L., Gutiérrez, C., 2008. Diversity, occurrence, and life characteristics of natural entomopathogenic nematode populations from La Rioja (northern Spain) under different agricultural management and their relationships with soil factors. Soil Biol. Biochem. 40, 1474–1484.

Campos-Herrera, R., Barbercheck, M., Hoy, C.W., Stock, S.P., 2012. Entomopathogenic nematodes as a model system for advancing the frontiers of ecology. J. Nematol. 44, 162–176.

Cao, L., Qiu, X.H., Liu, X.F., Liu, X.L., Han, R.C., 2008. Nutrient potential of various *Xenorhabdus* and *Photorhabdus* bacteria for a free-living nematode *Panagrellus redivivus*. Nematology 10, 79–85.

Carr, C.W., Kolodny-Hirsch, D.M., 1993. Method and Apparatus for Mass Producing Insects, Entomopathogens and Entomoparasites. US Patent No. 5,178,094.

Chaston, J.M., Dillman, A.R., Shapiro-Ilan, D.I., Bilgrami, A.L., Gaugler, R., Hopper, K.R., Adams, B.J., 2011. Outcrossing and crossbreeding recovers deteriorated traits in laboratory cultured *Steinernema carpocapsae* nematodes. Int. J. Parasitol. 41, 801–809.

Chavarria-Hernandez, N., de la Torre, M., 2001. Population growth kinetics of the nematode, *Steinernema feltiae*, in submerged monoxenic culture. Biotechnol. Lett. 23, 311–315.

Chavarria-Hernandez, N., Espino-Garcia, J.J., Sanjuan-Galindo, R., Rodriguez-Hernandez, A.I., 2006. Monoxenic liquid culture of the entomopathogenic nematode *Steinernema carpocapsae* using a culture medium containing whey kinetics and modeling. J. Biotechnol. 125, 75–84.

Chavarría-Hernández, N., Ortega-Morales, E., Vargas-Torres, A., Chavarría-Hernández, J., Rodríguez-Hernández, A., 2010. Submerged monoxenic culture of the entomopathogenic nematode, *Steinernema carpocapsae* CABA01, in a mechanically agitated bioreactor: evolution of the hydrodynamic and mass transfer conditions. Biotechnol. Bioprocess. Eng. 15, 580–589.

Cho, C.H., Whang, K.S., Gaugler, R., Yoo, S.K., 2011. Submerged monoxenic culture medium development for *Heterorhabditis bacteriophora* and its symbiotic bacterium *Photorhabdus luminescens*: protein sources. J. Microbiol. Biotechnol. 21, 869–873.

Ciche, T.A., 2007. The biology and genome of *Heterorhabditis bacteriophora*. WormBook; the Online Review of C. *elegans* Biology, WormBook, http://dx.doi.org/10.1895/wormbook.1.135.1, http://www.wormbook.org.

Cottrell, T.E., Shapiro-Ilan, D.I., Horton, D.L., Mizell III, R.F., 2011. Laboratory virulence and orchard efficacy of entomopathogenic nematodes toward the lesser peachtree borer (Lepidoptera: Sesiidae). Environ. Entomol. 104, 47–53.

Daoust, R.A., Roberts, D.W., 1982. Virulence of natural and insect-passaged strains of *Metarhizium anisopliae* to mosquito larvae. J. Invertebr. Pathol. 40, 107–117.

Del Valle, E.E., Dolinski, C., Barreto, E.L.S., Souza, R.M., Samuels, R.I., 2008. Efficacy of *Heterorhabditis baujardi* LPP7 (Nematoda: Rhabditida) applied in *Galleria mellonella* (Lepidoptera: Pyralidae) insect cadavers to *Conotrachelus psidii*, (Coleoptera: Curculionidae) larvae. Biocontrol Sci. Technol. 18, 33–41.

Del Valle, E.E., Dolinksi, C., Barreto, E.L.S., Souza, R.M., 2009. Effect of cadaver coatings on emergence and infectivity of the entomopathogenic nematode *Heterorhabditis baujardi* LPP7 (Rhabditida: Heterorhabditidae) and the removal of cadavers by ants. Biol. Control 50, 21–24.

Deol, Y.S., Jagdale, G.B., Cañas, L., Grewal, P.S., 2011. Delivery of entomopathogenic nematodes directly through commercial growing media via the inclusion of infected host cadavers: a novel approach. Biol. Control 58, 60–67.

Dias, P.V.C., Dolinski, C., Molina, J.P.A., 2008. Influence of infective juvenile doses and *Galleria mellonella* (Lepidoptera: Pyralidae) larvae weight in the in vivo production of *Heterorhabditis baujardi* LPP7 (Rhabditida: Heterorhabditidae). Nematologia Brasileira 32, 317–321.

Dillman, A.R., Chaston, J.M., Adams, B.J., Ciche, T.A., Goodrich-Blair, H., Stock, S.P., Sternberg, P.W., 2012a. An entomopathogenic nematode by any other name. PLoS Pathog. 8, e1002527. http://dx.doi.org/10.1371/journal.ppat.1002527.

Dillman, A.R., Mortazavi, A., Sternberg, P.W., 2012b. Incorporating genomics into the toolkit of nematology. J. Nematol. 44, 191–205.

Dillon, A.B., Downes, M.J., Ward, D., Griffin, C.T., 2007. Optimizing application of entomopathogenic nematodes to manage large pine weevil, *Hylobius abietis* L. (Coleoptera: Curculionidae) populations developing in pine stumps, *Pinus sylvestris*. Biol. Control 40, 253–263.

Dolinski, C., Del Valle, E.E., Burla, R.S., Machado, I.R., 2007. Biological traits of two native Brazilian entomopathogenic nematodes (Heterorhabditidae: Rhabditida). Nematologia Brasileira 31, 180–185.

Dolinski, C., Choo, H.Y., Duncan, L.W., 2012. Grower acceptance of entomopathogenic nematodes: case studies on three continents. J. Nematol. 44, 226–235.

Dowds, B.C.A., Peters, A., 2002. Virulence mechanisms. In: Gaugler, R. (Ed.), Entomopathogenic Nematology, CABI, New York, pp. 79–98.

Duchaud, E., Rusniok, C., Frangeul, L., Buchrieser, C., Givaudan, A., Taourit, S., Bocs, S., Boursaux-Eude, C., Chandler, M., Charles, J.F., Dassa, E., Derose, R., Derzelle, S., Freyssinet, G., Gaudriault, S., Medigue, C., Lanois, A., Powell, K., Siguier, P., Vincent, R., Wingate, V., Zouine, M., Glaser, P., Boemare, N., Danchin, A., Kunst, F., 2003. The genome sequence of the entomopathogenic bacterium *Photorhabdus luminescens*. Nat. Biotechnol. 21, 1307–1313.

Duncan, L.W., McCoy, C.W., 1996. Vertical distribution in soil, persistence, and efficacy against citrus root weevil (Coleoptera: Curculionidae) of two species of entomogenous nematodes (Rhabditida: Steinernematidae; Heterorhabditidae). Environ. Entomol. 25, 174–178.

Dunphy, G.B., Webster, J.M., 1989. The monoxenic culture of *Neoaplectana carpocapsae* DD-136 and *Heterorhabditis heliothidis*. Revue. Nematol. 2, 113–123.

Dutky, S.R., Thompson, J.V., Cantwell, G.E., 1964. A technique for the mass propagation of the DD-136 nematode. J. Invertebr. Pathol. 6, 417–422.

Ehlers, R.U., 2001a. Mass production of entomopathogenic nematodes for plant protection. Appl. Microbiol. Biotechnol. 56, 623–633.

Ehlers, R.U., 2001b. Achievements in research of EPN mass production. In: Griffin, C.T., Burrell, A.M., Downes, M.J., Mulder, R. (Eds.), Developments in Entomopathogenic Nematodes/Bacterial Research, European Community Press, Luxembourg, EUR 19696-COST 819, pp. 68–77.

Ehlers, R.U., 2005. Forum on safety and regulation. In: Grewal, P.S., Ehlers, R.U., Shapiro-Ilan, D.I. (Eds.), Nematodes as Biological Control Agents, CABI, Wallingford, pp. 107–114.

Ehlers, R.U., Shapiro-Ilan, D.I., 2005. Mass production. In: Grewal, P.S., Ehlers, R.U., Shapiro-Ilan, D.I. (Eds.), Nematodes as Biological Control Agents, CABI, Wallingford, pp. 65–79.

Ehlers, R.U., Lunau, S., Krasomil-Osterfeld, K.C., Osterfeld, K.H., 1998. Liquid culture of the entomopathogenic nematodebacterium complex *Heterorhabditis megidis/Photorhabdus luminescens*. BioControl 43, 77–86.

Ehlers, R.U., Niemann, I., Hollmer, S., Strauch, O., Jende, D., Shanmugasundaram, M., Mehta, U.K., Easwaramoorthy, S.K., Burnell, A., 2000. Mass production potential of the bacto-helminthic biocontrol complex *Heterorhabditis indica–Photorhabdus luminescens*. Biocontrol Sci. Technol. 10, 607–616.

Elawad, S.A., Gowen, S.R., Hague, N.G.M., 2001. Progeny production of *Steinernema abbasi* in lepidopterous larvae. Int. J. Pest Manage. 47, 17–21.

Flanders, K.L., Miller, J.M., Shields, E.J., 1996. In vivo production of *Heterorhabditis bacteriophora* 'Oswego' (Rhabditida: Heterorhabditidae), a potential biological control agent for soil inhabiting insects in temperate regions. J. Econ. Entomol. 89, 373–380.

Friedman, M.J., Langston, S.L., Pollit, S., 1989. Mass Production in Liquid Culture of Insect-Killing Nematodes. Int Patent No. WO89/04602.

Friedman, M.J., 1990. Commercial production and development. In: Gaugler, R., Kaya, H.K. (Eds.), Entomopathogenic Nematodes in Biological Control, CRC Press, Boca Raton, pp. 153–172.

Gaugler, R., 1987. Entomogenous nematodes and their prospects for genetic improvement. In: Maramorosch, K. (Ed.), Biotechnology in Invertebrate Pathology and Cell Culture. Academic Press, San Diego, pp. 457–484.

Gaugler, R., Georgis, R., 1991. Culture method and efficacy of entomopathogenic nematodes (Rhabditida: Steinernematidae and Heterorhabditidae). Biol. Control 1, 269–274.

Gaugler, R., Han, R., 2002. Production technology. In: Gaugler, R. (Ed.), Entomopathogenic Nematology, CABI, New York, pp. 289–310.

Gaugler, R., Campbell, J.F., McGuire, T.R., 1989. Selection for host-finding in *Steinernema feltiae*. J. Invertebr. Pathol. 54, 363–372.

Gaugler, R., Campbell, J.F., McGuire, T.R., 1990. Fitness of a genetically improved entomopathogenic nematode. J. Invertebr. Pathol. 56, 106–116.

Gaugler, R., Wilson, M., Shearer, P., 1997. Field release and environmental fate of a transgenic entomopathogenic nematode. Biol. Control 9, 75–80.

Gaugler, R., Grewal, P., Kaya, H.K., Smith-Fiola, D., 2000. Quality assessment of commercially produced entomopathogenic nematodes. Biol. Control 17, 100–109.

Gaugler, R., Brown, I., Shapiro-Ilan, D.I., Atwa, A., 2002. Automated technology for in vivo mass production of entomopathogenic nematodes. Biol. Control 24, 199–206.

Georgis, R., Koppenhöfer, A.M., Lacey, L.A., Bélair, G., Duncan, L.W., Grewal, P.S., Samish, M., Tan, L., Torr, P., van Tol, R.W.H.M., 2006. Successes and failures in the use of parasitic nematodes for pest control. Biol. Control 38, 103–123.

Glaser, R.W., 1931. The cultivation of a nematode parasite of an insect. Science 73, 614–615.

Glazer, I., Salame, L., Segal, D., 1997. Genetic enhancement of nematicide resistance in entomopathogenic nematodes. Biocontrol Sci. Technol. 7, 499–512.

Goodrich-Blair, H., Clarke, D.J., 2007. Mutualism and pathogenesis in *Xenorhabdus* and *Photorhabdus*: two roads to the same destination. Mol. Microbiol. 64, 260–268.

Grewal, P.S., 2002. Formulation and application technology. In: Gaugler, R. (Ed.), Entomopathogenic Nematology, CABI, New York, pp. 265–287.

Grewal, P.S., Selvan, S., Gaugler, R., 1994. Thermal adaptation of entomopathogenic nematodes—niche breadth for infection, establishment and reproduction. J. Therm. Biol. 19, 245–253.

Grewal, P.S., Converse, V., Georgis, R., 1999. Influence of production and bioassay methods on infectivity of two ambush foragers (Nematoda: Steinernematidae). J. Invertebr. Pathol. 73, 40–44.

Grewal, P.S., Ehlers, R.-U., Shapiro-Ilan, D.I. (Eds.), 2005. Nematodes as Biological Control Agents, CABI, Wallingford.

Griffin, C.T., Boemare, N.E., Lewis, E.E., 2005. In: Grewal, P.S., Ehlers, R.-U., Shapiro-Ilan, D.I. (Eds.), Nematodes as Biological Control Agents, CABI, Wallingford, pp. 47–64.

Han, R.C., 1996. The effects of inoculum size on yield of *Steinernema carpocapsae* and *Heterorhabditis bacteriophora* in liquid culture. Nematologica 42, 546–553.

Han, R.C., Ehlers, R.U., 1998. Cultivation of axenic *Heterorhabditis* spp. dauer juveniles and their response to non-specific *Photorhabdus luminescens* food signals. Nematologica 44, 425–435.

Han, R.C., Ehlers, R.U., 2001. Effect of *Photorhabdus luminescens* phase variants on the in vivo and in vitro development and reproduction of the entomopathogenic nematodes *Heterorhabditis bacteriophora* and *Steinernema carpocapsae*. FEMS Microbiol. Ecol. 35, 239–247.

Han, R., Cao, L., Liu, X., 1992. Relationship between medium composition, inoculum size, temperature and culture time in the yields of *Steinernema* and *Heterorhabditis* nematodes. Fundam. Appl. Nematol. 15, 223–229.

Han, R., Cao, L., Liu, X., 1993. Effects of inoculum size, temperature, and time on in vitro production of *Steinernema carpocapsae* Agriotos. Nematologica 39, 366–375.

Han, R., Pang, X., Li, L., 1995. Optimization of the medium components for the solid culture of entomopathogenic *Steinernema* and *Heterorhabditis* nematodes. Nat. Enemies Insects 17, 153–164.

Han, R., Li, L., Pang, X., 1997. Modeling of the culture parameters for production of *Steinernema carpocapsae* and *Heterorhabditis bacteriophora* in solid culture. Nat. Enemies Insects 19, 75–83.

Hara, A.H., Lindegren, J.E., Kaya, H.K., 1981. Monoxenic mass production of the entomogenous nematode *Neoplectana carpocapsae* Weiser on dog food/agar medium. USDA Adv. Agric. W-16, 8 pp.

Hirao, A., Ehlers, R.U., 2009. Influence of cell density and phase variants of bacterial symbionts (*Xenorhabdus* spp.) on dauer juvenile recovery and development of biocontrol nematodes *Steinernema carpocapsae* and *S. feltiae* (Nematoda: Rhabditida). Appl. Microbiol. Biotechnol. 84, 77–85.

Hirao, A., Ehlers, R.-U., 2010. Influence of inoculum density on population dynamics and dauer juvenile yields in liquid culture of biocontrol nematodes *Steinernema carpocapsae* and *S. feltiae* (Nematoda: Rhabditida). Appl. Microbiol. Biotechnol. 85, 507–515.

Hominick, W.M., Briscoe, B.R., del Pino, F.G., Heng, J., Hunt, D.J., Kozodoy, E., Mracek, Z., Nguyen, K.B., Reid, A.P., Spiridonov, S., Stock, P., Sturhan, D., Waturu, C., Yoshida, M., 1997. Biosystematics of entomopathogenic nematodes: current status, protocols, and definitions. J. Helminthol. 71, 271–298.

Hopper, K.R., Roush, R.T., Powell, W., 1993. Management of genetics of biological-control introductions. Annu. Rev. Entomol. 38, 27–51.

House, H.L., Welch, H.E., Cleugh, T.R., 1965. Food medium of prepared dog biscuit for the mass-production of the nematode DD-136 (Nematoda: Steinernematidae). Nature 206, 847.

Jansson, R.K., Lecrone, S.H., Gaugler, R., 1993. Field efficacy and persistence of entomopathogenic nematodes (Rhabditida: Steinernematidae, Heterorhabditidae) for control of sweetpotato weevil (Coleoptera: Apionidae) in southern Florida. J. Econ. Entomol. 86, 1055–1063.

Jackson, G.J., 1973. *Neoaplectana glaseri*: essential amino acids. Exp. Parasitol. 34, 111–114.

Jeffke, T., Jende, D., Mätje, C., Ehlers, R.U., Berthe-Corti, L., 2000. Growth of *Photorhabdus luminescens* in batch and glucose fed-batch culture. Appl. Microbiol. Biotechnol. 54, 326–330.

Jessen, P., Strauch, O., Wyss, U., Luttmann, R., Ehlers, R.U., 2000. Carbon dioxide triggers dauer juvenile recovery of entomopathogenic nematodes (*Heterorhabditis* spp). Nematology 2, 319–324.

Johnigk, S.A., Ecke, F., Poehling, M., Ehlers, R.U., 2004. Liquid culture mass production of biocontrol nematodes, *Heterorhabditis bacteriophora* (Nematoda: Rhabditida): Improved timing of dauer juvenile inoculation. Appl. Microbiol. Biotechnol. 64, 651–658.

Kaya, H.K., Stock, S.P., 1997. Techniques in insect nematology. In: Lacey, L.A. (Ed.), Manual of Techniques in Insect Pathology. Academic Press, San Diego, pp. 281–324.

Kaya, H.K., Aguillera, M.M., Alumai, A., Choo, H.Y., de la Torre, M., Fodor, A., Ganguly, S., Hazir, S., Lakatos, T., Pye, A., Wilson, M., Yamanaka, S., Yang, H., Ehlers, R.-U., 2006. Status of entomopathogenic nematodes and their symbiotic bacteria from selected countries or regions of the world. Biol. Control 38, 134–155.

Koltai, H., Glazer, I., Segal, D., 1995. Reproduction of the entomopathogenic nematode *Heterorhabditis bacteriophora* Poinar. Fundam. Appl. Nematol. 18, 55–61.

Koppenhöfer, A.M., Fuzy, E.M., 2003. Ecological characterization of *Steinernema scarabaei*, a scarab-adapted entomopathogenic nematode from New Jersey. J. Invertebr. Pathol. 83, 139–148.

Koppenhöfer, A.M., Grewal, P.S., 2005. Compatibility and interactions with agrochemicals and other biocontrol agents. In: Grewal, P.S., Ehlers, R.U., Shapiro-Ilan, D.I. (Eds.), Nematodes as Biological Control Agents. CABI, Wallingford, pp. 363–381.

Lacey, L.A., Georgis, R., 2012. Entomopathogenic nematodes for control of insect pests above and below ground with comments on commercial production. J. Nematol. 44, 218–225.

Lacey, L.A., Shapiro-Ilan, D.I., 2008. Microbiol control of insect pests in temperate orchard systems: potential for incorporation into IPM. Annu. Rev. Entomol. 53, 121–144.

Lacey, L.A., Frutos, R., Kaya, H.K., Vail, P., 2001. Insect pathogens as biological control agents: do they have a future? Biol. Control 21, 230–248.

Lewis, E.E., Clarke, D.J., 2012. Nematode parasites and entomopathogens. In: Vega, F.E., Kaya, H.K. (Eds.), Insect Pathology, second ed. Elsevier, Amsterdam, pp. 395–424.

Lewis, E.E., Campbell, J., Griffin, C., Kaya, H., Peters, A., 2006. Behavioral ecology of entomopathogenic nematodes. Biol. Control 38, 66–79.

Lindegren, J.E., Hoffman, D.F., Collier, S.S., Fries, R.D., 1979. Propagation and storage of *Neoaplectana carpocapsae* Weiser using *Amyelois transitella* (Walker) adults. USDA Adv. Agric. 3, 1–5.

Lindegren, J.E., Valero, K.A., Mackey, B.E., 1993. Simple in vivo production and storage methods for *Steinernema carpocapsae* infective juveniles. J. Nematol. 25, 193–197.

Lunau, S., Stoessel, S., Schmidt-Peisker, A.J., Ehlers, R.U., 1993. Establishment of monoxenic inocula for scaling up in vitro cultures of the entomopathogenic nematodes *Steinernema* spp. and *Heterorhabditis* spp. Nematologica 39, 385–399.

Malan, A.P., Knoetze, R., Moore, S.D., 2011. Isolation and identification of entomopathogenic nematodes from citrus orchards in South Africa and their biocontrol potential against false codling moth. J. Invetebr. Pathol. 108, 115–125.

Mamiya, Y., 1989. Comparison of infectivity of *Steinernema kushidai* (Nematoda: Steinernematidae) and other steinernematid and heterorhabditid nematodes for three different insects. Appl. Entomol. Zool. 24, 302–308.

Morales-Ramos, J.A.M., Tedders, W.L., Dean, B., Shapiro-Ilan, D.I., Rojas, M.G., 2010. Apparatus for Packaging Arthropods Infected with Entomopathogenic Nematodes. US Patent Application No. 12/953, 719, DN 0140.07.

Morales-Ramos, J.A., Rojas, M.G., Shapiro-Ilan, D.I., Tedders, W.L., 2011a. Automated Insect Separation System. US Patent No. 8025027.

Morales-Ramos, J.A., Rojas, M.G., Shapiro-Ilan, D.I., Tedders, W.L., 2011b. Nutrient regulation in *Tenebrio molitor* (Coleoptera: Tenebrionidae): SELF-selection of two diet components by larvae and impact on fitness. Environ. Entomol. 40, 1285–1294.

Morales-Ramos, J.A., Rojas, M.G., Shapiro-Ilan, D.I., Kay, S., Tedders, W.L., 2012. Impact of adult weight, density, and age on reproduction of *Tenebrio molitor* (Coleoptera: Tenebrionidae). J. Entomol. Sci. 47, 208–220.

Mukuka, J., Strauch, O., Hoppe, C., Ehlers, R.-U., 2010. Improvement of heat and desiccation tolerance in *Heterorhabditis bacteriophora* through cross-breeding of tolerant strains and successive genetic selection. BioControl 55, 511–521.

Neves, J.M., Teixeira, J.A., Simoes, N., Mota, M., 2001. Effect of airflow rate on yield of *Steinernema carpocapsae* Az 20 in liquid culture in an external-loop airlift bioreactor. Biotechnol. Bioeng. 72, 369–373.

Nguyen, K.B., Smart Jr, G.C., 1990. *Steinernema scapterisci* n. sp. (Rhabditida: Steinernematidae). J. Nematol. 22, 187–199.

Nugent, M.J., O'Leary, S.A., Burnell, A.M., 1996. Optimised procedures for the cryopreservation of different species of *Heterorhabditis*. Fundam. Appl. Nematol. 19, 1–6.

Onstad, D.W., Fuxa, J.R., Humber, R.A., Oestergaard, J., Shapiro-Ilan, D.I., Gouli, V.V., Anderson, R.S., Andreadis, T.G., Lacey, L.A., 2006. An Abridged Glossary of Terms Used in Invertebrate Pathology, third ed. Society for Invertebrate Pathology. http://www.sipweb.org/glossary.

Pace, W.G., Grote, W., Pitt, D.E., Pitt, J.M., 1986. Liquid Culture of Nematodes. Int Patent No. WO 86/01074.

Patel, M.N., Perry, R.N., Wright, D.J., 1997a. Desiccation survival and water contents of entomopathogenic nematodes, *Steinernema* spp. (Rhabditida: Steinernematidae). Int. J. Parasitol. 27, 61–70.

Patel, M.N., Stolinski, M., Wright, D.J., 1997b. Neutral lipids and the assessment of infectivity in entomopathogenic nematodes: observations on four *Steinernema* species. Parasitology 114, 489–496.

Pereault, R.J., Whalon, M.E., Alston, D.G., 2009. Field efficacy of entomopathogenic fungi and nematodes targeting caged last-instar plum curculio (Coleoptera: Curculionidae) in Michigan cherry and apple orchards. Environ. Entomol. 38, 1126–1134.

Perez, E.E., Lewis, E.E., Shapiro-Ilan, D.I., 2003. Impact of host cadaver on survival and infectivity of entomopathogenic nematodes (Rhabditida: Steinernematidae and Heterorhabditidae) under desiccating conditions. J. Invertebr. Pathol. 82, 111–118.

Poinar, G.O., 1979. Nematodes for Biological Control of Insects. CRC Press, Boca Raton 277 pp.

Poinar, G.O., 1990. Biology and taxonomy of Steinernematidae and Heterorhabditidae. In: Gaugler, R., Kaya, H.K. (Eds.), Entomopathogenic Nematodes in Biological Control. CRC Press, Boca Raton, pp. 23–62.

Poinar Jr., G.O., Thomas, G.M., 1966. Significance of *Achromobacter nematophilus* Poinar and Thomas (Achromobacteriaceae: Eubacteriales) in the development of the nematode, DD-136 (*Neoaplectana* sp., Steinernematidae). Parasitology 56, 385–390.

Popiel, I., Vasquez, E.M., 1991. Cryopreservation of *Steinernema carpocapsae* and *Heterorhabditis bacteriophora*. J. Nematol. 23, 432–437.

Ritter, K.S., 1988. *Steinernema feltiae* (=*Neoaplectana carpocapsae*): effect of sterols and hypolipidemic agents on development. Exp. Parasitol. 67, 257–267.

Roush, R.T., 1990. Genetic considerations in the propagation of entomophagous species. In: Baker, R.R., Dunn, P.E. (Eds.), Critical Issues in Biological Control, Liss, New York, pp. 373–387.

Salma, J., Shahina, F., 2012. Mass production of eight Pakistani strains of entomopathogenic nematodes (Steinernematidae and Heterorhabditidae). Pak. J. Nematol. 30, 1–20.

Schwartz, H.T., Antoshechkin, I., Sternberg, P.W., 2011. Applications of high-throughput sequencing to symbiotic nematodes of the genus *Heterorhabditis*. Symbiosis 55, 111–118.

Shapiro-Ilan, D.I., Gaugler, R., 2002. Production technology for entomopathogenic nematodes and their bacterial symbionts. J. Ind. Microbiol. Biotechnol. 28, 137–146.

Shapiro, D.I., Glazer, I., 1996. Comparison of entomopathogenic nematode dispersal from infected hosts versus aqueous suspension. Environ. Entomol. 25, 1455–1461.

Shapiro-Ilan, D.I., Grewal, P.S., 2008. Entomopathogenic nematodes and insect management. In: Capinera, J.L. (Ed.), Encyclopedia of Entomology, second ed. Springer, Dordrecht, pp. 1336–1340.

Shapiro, D.I., Lewis, E.E., 1999. Comparison of entomopathogenic nematode infectivity from infected hosts versus aqueous suspension. Environ. Entomol. 28, 907–911.

Shapiro, D.I., McCoy, C.W., 2000a. Effect of culture method and formulation on the virulence of *Steinernema riobrave* (Rhabditida: Steinernematidae) to *Diaprepes abbreviatus* (Curculionidae). J. Nematol. 32, 281–288.

Shapiro, D.I., McCoy, C.W., 2000b. Virulence of entomopathogenic nematodes to *Diaprepes abbreviatus* (Coleoptera: Curculionidae) in the laboratory. J. Econ. Entomol. 93, 1090–1095.

Shapiro, D.I., Glazer, I., Segal, D., 1996. Trait stability in and fitness of the heat tolerant entomopathogenic nematode *Heterorhabditis bacteriophora* IS5 strain. Biol. Control 6, 238–244.

Shapiro, D.I., Glazer, I., Segal, D., 1997. Genetic improvement of heat tolerance in *Heterorhabditis bacteriophora* through hybridization. Biol. Control 8, 153–159.

Shapiro, D.I., Cate, J.R., Pena, J., Hunsberger, A., McCoy, C.W., 1999. Effects of temperature and host age on suppression of *Diaprepes abbreviatus* (Coleoptera: Curculionidae) by entomopathogenic nematodes. J. Econ. Entomol. 92, 1086–1092.

Shapiro, D.I., Lewis, E.E., Paramasivam, S., McCoy, C.W., 2000. Nitrogen partitioning in *Heterorhabditis bacteriophora*-infected hosts and the effects of nitrogen on attraction/repulsion. J. Invertebr. Pathol. 76, 43–48.

Shapiro-Ilan, D.I., Lewis, E.E., Behle, R.W., McGuire, M.R., 2001. Formulation of entomopathogenic nematode-infected-cadavers. J. Invertebr. Pathol. 78, 17–23.

Shapiro-Ilan, D.I., Gaugler, R., Tedders, W.L., Brown, I., Lewis, E.E., 2002a. Optimization of inoculation for in vivo production of entomopathogenic nematodes. J. Nematol. 34, 343–350.

Shapiro-Ilan, D.I., Gouge, D.H., Koppenhöfer, A.M., 2002b. Factors affecting commercial success: case studies in cotton, turf and citrus. In: Gaugler, R. (Ed.), Entomopathogenic Nematology, CABI, Wallingford, pp. 333–355.

Shapiro-Ilan, D.I., Mizell III, R.F., Campbell, J.F., 2002c. Susceptibility of the plum curculio, *Conotrachelus nenuphar*, to entomopathogenic nematodes. J. Nematol. 34, 246–249.

Shapiro-Ilan, D.I., Lewis, E.E., Tedders, W.L., Son, Y., 2003a. Superior efficacy observed in entomopathogenic nematodes applied in infected-host cadavers compared with application in aqueous suspension. J. Invertebr. Pathol. 83, 270–272.

Shapiro-Ilan, D.I., Gardner, W.A., Fuxa, J.R., Wood, B.W., Nguyen, K.B., Adams, B.J., Humber, R.A., Hall, M.J., 2003b. Survey of entomopathogenic nematodes and fungi endemic to pecan orchards of the Southeastern United States and their virulence to the pecan weevil (Coleoptera: Curculionidae). Environ. Entomol. 32, 187–195.

Shapiro-Ilan, D.I., Mizell III, R.F., Cottrell, T.E., Horton, D.L., 2004. Measuring field efficacy of *Steinernema feltiae* and *Steinernema riobrave* for suppression of plum curculio, *Conotrachelus nenuphar*, larvae. Biol. Control 30, 496–503.

Shapiro-Ilan, D.I., Stuart, R.J., McCoy, C.W., 2005. Targeted improvement of *Steinernema carpocapsae* for control of the pecan weevil, *Curculio caryae* (Horn) (Coleoptera: Curculionidae) through hybridization and bacterial transfer. Biol. Control 34, 215–221.

Shapiro-Ilan, D.I., Gouge, G.H., Piggott, S.J., Patterson Fife, J., 2006. Application technology and environmental considerations for use of entomopathogenic nematodes in biological control. Biol. Control 38, 124–133.

Shapiro-Ilan, D.I., Rojas, M.G., Morales-Ramos, J.A., Lewis, E.E., Tedders, W.L., 2008a. Effects of host nutrition on virulence and fitness of entomopathogenic nematodes: lipid and protein based supplements in *Tenebrio molitor* diets. J. Nematol. 40, 13–19.

Shapiro-Ilan, D.I., Tedders, W.L., Lewis, E.E., 2008b. Application of Entomopathogenic Nematode-Infected Cadavers From Hard-Bodied Arthropods for Insect Suppression. US Patent No. 7,374,773.

Shapiro-Ilan, D.I., Cottrell, T.E., Mizell III, R.F., Horton, D.L., Davis, J., 2009a. A novel approach to biological control with entomopathogenic nematodes: prophylactic control of the peachtree borer, *Synanthedon exitiosa*. Biol. Control 48, 259–263.

Shapiro-Ilan, D.I., Tedders, W.L., Morales-Ramos, J.A., Rojas, M.G., 2009b. Insect Inoculation System and Method. US Patent Application No. 12636245, DN. 0107.04.

Shapiro-Ilan, D.I., Morales, Ramos, J.A., Rojas, M.G., Tedders, W.L., 2010. Effects of a novel entomopathogenic nematode–infected host formulation on cadaver integrity, nematode yield, and suppression of *Diaprepes abbreviatus* and *Aethina tumida* under controlled conditions. J. Invertebr. Pathol. 103, 103–108.

Shapiro-Ilan, D.I., Tedders, W.L., Morales-Ramos, J.A., Rojas, M.G., 2011. System and Method for Producing Beneficial Parasites. US Patent Application No. 1 13/217,956, DN 172.07.

Shapiro-Ilan, D.I., Bruck, D.J., Lacey, L.A., 2012a. Principles of epizootiology and microbial control. In: Vega, F.E., Kaya, H.K. (Eds.), Insect Pathology, second ed. Elsevier, Amsterdam, pp. 29–72.

Shapiro-Ilan, D.I., Han, R., Dolinski, C., 2012b. Entomopathogenic nematode production and application technology. J. Nematol. 44, 206–217.

Shapiro-Ilan, D.I., Rojas, G., Morales-Ramos, J.A., Tedders, W.L., 2012c. Optimization of a host diet for in vivo production of entomopathogenic nematodes. J. Nematol. 44, 264–273.

Strauch, O., Ehlers, R.U., 2000. Influence of the aeration rate on yields of the biocontrol nematodes *Heterorhabditis megidis* in monoxenic liquid cultures. Appl. Microbiol. Biotechnol. 54, 9–13.

Strauch, O., Stoessel, S., Ehlers, R.U., 1994. Culture conditions define automictic or amphimictic reproduction in entomopathogenic rhabditid nematodes of the genus *Heterorhabditis*. Fundam. Appl. Nematol. 17, 575–582.

Steinhaus, E.A., 1949. Principles of Insect Pathology. McGraw-Hill, New York.

Stock, S.P., Hunt, D.J., 2005. Morphology and systematics of nematodes used in biocontrol. In: Grewal, P.S., Ehlers, R.U., Shapiro-Ilan, D.I. (Eds.), Nematodes as Biocontrol Agents, CABI, Wallingford, pp. 3–43.

Stock, S.P., Griffin, C.T., Chaerani, R., 2004. Morphological and molecular characterization of *Steinernema hermaphroditum* n. sp. (Nematoda: Steinernematidae), an entomopathogenic nematode from Indonesia, and its phylogenetic relationships with other members of the genus. Nematology 6, 401–412.

Stoll, N.R., 1952. Axenic cultivation of the parasitic nematode, *Neoaplectana glaseri,* in a fluid medium containing raw liver extract. J. Parasitol. 39, 422–444.

Stuart, R.J., Gaugler, R., 1996. Genetic adaptation and founder effect in laboratory populations of the entomopathogenic nematode *Steinernema glaseri*. Can. J. Zool. 74, 164–170.

Surrey, M.R., Davies, R.J., 1996. Pilot scale liquid culture and harvesting of an entomopathogenic nematode, *Heterorhabditis*. Fundam. Appl. Nematol. 17, 575–582.

Tanada, Y., Kaya, H.K., 1993. Insect Pathology. Academic Press, San Diego.

Wang, J.X., Bedding, R.A., 1998. Population dynamics of *Heterorhabditis bacteriophora* and *Steinernema carpocapsae* in in vitro solid culture. Fundam. Appl. Nematol. 21, 165–171.

Wang, X., Grewal, P.S., 2002. Rapid genetic deterioration of environmental tolerance and reproductive potential of an entomopathogenic nematode during laboratory maintenance. Biol. Control 23, 71–78.

Wang, Y., Bilgrami, A.L., Shapiro-Ilan, D., Gaugler, R., 2007. Stability of entomopathogenic bacteria, *Xenorhabdus nematophila* and *Photorhabdus luminescens*, during in vitro culture. J. Ind. Microbiol. Biotechnol. 34, 73–81.

White, G.F., 1927. A method for obtaining infective nematode larvae from cultures. Science 66, 302–303.

Woodring, J.L., Kaya, H.K., 1988. Steinernematid and Heterorhabditid Nematodes: A Handbook of Biology and Techniques. Southern Cooperative Series Bulletin 331 Arkansas Agricultural Experiment Station, Fayetteville, AR.

Wouts, W.M., 1981. Mass production of the entomogenous nematode *Heterorhabditis heliothidis* (Nematoda: Heterorhabditidae) on artificial media. J. Nematol. 13, 467–469.

Yang, H., Jian, H., Zhang, S., Zhang, G., 1997. Quality of the entomopathogenic nematode *Steinernema carpocapsae* produced on different media. Biol. Control 10, 193–198.

Yoo, S.K., Brown, I., Gaugler, R., 2000. Liquid media development for *Heterorhabditis bacteriophora*: lipid source and concentration. Appl. Microbiol. Biotechnol. 54, 759–763.

Young, J.M., Dunnill, P., Pearce, J.D., 2002. Separation characteristics of liquid nematode cultures and the design of recovery operations. Biotechnol. Prog. 18, 29–35.

Zhu, H., Grewal, P.S., Reding, M.E., 2011. Development of a desiccated cadaver delivery system to apply entomopathogenic nematodes for control of soil pests. Appl. Eng. Agric. 27, 317–324.

Chapter 11

Mass Production of Entomopathogenic Fungi: State of the Art

Stefan T. Jaronski
US Department of Agriculture, Agricultural Research Service, Sidney, MT, USA

Mention of trade names or commercial products in this chapter is solely for the purpose of providing specific information and does not imply recommendation or endorsement by the U.S. Department of Agriculture.

11.1. INTRODUCTION

The potential to control insects with fungi dates back to Augostino Bassi's 1835 demonstration that a fungus could cause a deliberately transmissible disease in silkworm (Steinhaus, 1956; Lord, 2007). In the late 1870s Metschnikoff observed a high proportion of *Metarhizium*-killed sugarbeet curculio *Cleonus punctiventris* Germar and proposed the concept of controlling this insect with conidia artificially produced on sterile brewer's mash (Metchnikoff, 1880; Steinhaus, 1975). His work was extended by Krassilstschik, who established a production facility using beer mash to produce a considerable amount conidia for distribution (Krassilstschik, 1888). In the United States, Lugger (1888) suggested the use of another fungus, now known as *Beauveria bassiana* (Balsamo) Vuillemin, to control the chinch bug (*Blissus leucopterus* (Say); Kansas, Nebraska, and neighboring states produced the fungus in vivo and attempted to augment natural populations with cadavers having sporulating fungus in the 1880s and 1890s (Billing and Glenn, 1911). During the early and middle twentieth century, citrus growers in Florida observed a series of fungi to be significant mortality factors for citrus pests. The Florida State Experiment Station produced and sold one fungus, *Aschersonia aleyrodis* Webber, to growers, while fungi attacking scale insects were often obtained and produced by the growers themselves (McCoy et al., 1988). With the advent of chemical insecticides, interest in all biological agents waned in the United States and Western Europe. It was not until the 1970s and 1980s that interest in microbial

Mass Production of Beneficial Organisms. http://dx.doi.org/10.1016/B978-0-12-391453-8.00011-X

TABLE 11.1 Commercial Mycoinsecticide Products Available in 2007

Species	Number of Products	Percent
Beauveria bassiana	45	37.2%
B. brongniartii	5	4.1%
Metarhizium anisopliae s.l.	44	36.4%
M. acridum	3	2.5%
Isaria fumosorosea	7	5.8%
I. farinosa	1	0.8%
Lecanicillium longisporium	2	1.7%
Lecanicillium muscarium	3	2.5%
Lecanicillium sp.	10	8.3%
Hirsutella thompsonii	1	0.8%
Total	121	

Source: Adapted from Faria and Wraight (2007).

agents resumed as the adverse environmental effects of chemical pesticides were better understood and alternatives began to be sought. Commercialization of fungi in the United States was limited, but efforts increased in a number of countries—most notably Brazil, Cuba, the former Czechoslovakia, the former U.S.S.R, and China—for economic as well as environmental reasons. Bartlett and Jaronski (1988) described many of these efforts.

Today, there are more than 100 commercial products based on entomopathogenic fungi (Table 11.1). De Faria and Wraight (2007) conducted a survey in 2006 and identified 129 active mycoinsecticide products; another 42 had been developed since the 1970s but were not commercially available at the time of the survey. In the United States, there are nine mycoinsecticides currently registered by U.S. Environmental Protection Agency; in the European Union (EU), 21 different fungi are registered in the Organisation for Economic Cooperation and Development (Agriculture and Agri-Food Canada, 2012).

A common characteristic of the entomopathogenic fungi is that, with the exception of the Microsporidia, they infect their hosts percutaneously, not perorally. With all but the Lagenidiales and Peronosporomycetes, the infectious stage is passively dispersed; the former two groups have motile zoospores that actively seek out their aquatic hosts. The life cycle, schematized in Fig. 11.1 and based on the Ascomycetes, begins when the spore contacts the arthropod cuticle, attaching initially by van der Waals forces, but then adheres more firmly and germinates within a few hours. A penetration hypha is produced, as well as

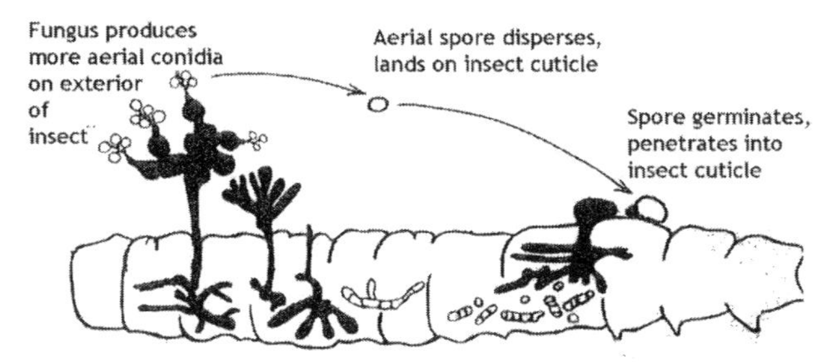

FIGURE 11.1 Schematic life cycle of the entomopathogenic fungi, exemplified by *Beauveria bassiana*.

FIGURE 11.2 *Beauveria bassiana* emerging from intersegmental areas of its host a Mormon cricket (*Anabrus simplex* Haldeman), and conidiating on the surface of the cuticle.

an appressorium or anchoring structure in some species. The hypha penetrates the arthropod cuticle by means of several enzymes and mechanical pressure. Once in the hemocoel, the fungus proliferates by means of yeast-like bodies (hyphal bodies in Entomophthorales, blastospores and mycelium in Ascomycetes, mycelium in Oomycetes). As the host dies, the fungus rapidly transforms into mycelium and, under ideal conditions (particularly an extended period of high humidity), emerges to conidiate on the exterior of the insect (Fig. 11.2).

Mass production methods for the entomopathogenic fungi up to 1983 were reviewed in detail by Bartlett and Jaronski (1988); the reader is referred to that publication. Since then, there have been a few general review publications, namely Feng et al. (1994) and Jenkins and Goettel (1997), as well as many specific studies about some aspect of mycoinsecticide mass production. The present chapter seeks to update the older literature, delineate important aspects, and identify areas of needed research.

11.2. PRODUCTION METHODS FOR THE IMPORTANT INSECT PATHOGENIC FUNGI

11.2.1. *Lagenidium giganteum* (Schenk)

Lagenidium giganteum (Oomycota: Lagenidiales) is the principal species studied for development, with larval mosquitoes as its principal target. A key aspect about this fungus is that the infectious stage is a motile zoospore that actively seeks out hosts and is produced by either mycelium or from germinated resting spores. Elution of nutrients from the mycelium and any carrier matrix (e.g. agar medium) is required for zoospore production from competent mycelium. Oospores—sexual resting bodies—can also produce zoospores upon rehydration, although the cues for relatively synchronous zoosporogenesis are unclear. Thus, efficient production of shelf-stable formulations is difficult. For a review of *L. giganteum* biology, see Kerwin (2007).

Exogenous sterols—specifically cholesterol, ergosterol, and campesterol—are essential nutrients for the production of both zoosporogenic mycelium (Domnas et al., 1977) and oospores (Kerwin and Washino, 1983). Attenuation of the zoosporogenic capability follows culture without these sterols (Lord and Roberts, 1986). Initially, production of zoosporogenic mycelium used agar-based media, primarily for experimental use (Jaronski et al., 1983). Subsequently, May and Vander Gheynst (2001) developed a solid substrate medium. This medium consisted of wheat bran supplemented with cholesterol, peptone, autolyzed yeast extract, glucose, corn oil, and lecithin, which was inoculated with a liquid preculture. Simpler media were subsequently developed by Maldonado-Blanco et al. (2011), who identified soybean meal and sunflower meal as suitable sources of the required sterols.

On a commercial scale, *L. giganteum* mycelium and oospores were produced in liquid fermentation using media consisting of crude carbon and nitrogen sources, with vegetable or fish oils providing the required sterols and unsaturated fatty acids (Kerwin and Washino, 1986); the unsaturated fatty acids, primarily triglycerides, were thought to help solubilize the sterols to optimize uptake and provide a higher percentage of fatty acids, thus increasing zoospore production (Kerwin and Washino, 1986). Another critical component is $CaCl_2 \cdot 2H_2O$. Fermentation output at pilot-scale level was 1–5 l fermenter volume per hectare of mosquito habitat, with a production cycle of 3–4 days (Kerwin and Washino, 1987). Commercial production used this method, with harvesting of the fungus from medium and storage in refrigerated containers; effectiveness of such material lasted 1–3 weeks (Vander Gheynst et al., 2000). Kerwin (2007) still considered economical scale-up of liquid fermentation a serious challenge. Although mycelium and oospore formulations of the fungus were registered and commercialized by Agraquest Inc. in the United States, the company abandoned the continued sale of the fungus. As an alternative, *L. giganteum* could be produced using wheat bran as a solid fermentation substrate (Vander Gheynst et al., 2000). The fungus retained its efficacy for 4 weeks. Glucose and wheat germ oil could increase the shelf life of the fungus and the whole culture could be efficacious against larval mosquitoes, at least in laboratory assays.

11.2.2. *Leptolegnia chapmani*

Leptolegnia chapmani Seymour (Straminipila: Peronosporomycetes) has been under study for a number of years as an alternative mosquito control agent (Zattau and McInnis, 1987). Much less work has been done regarding production of *L. chapmani*, primarily because the fungus remains in an experimental mode. Pelizza et al. (2011) evaluated a series of agar-based media for zoospore production. Although they observed that most media supported mycelial growth equally, zoospore production in an agar medium that contained 10% Fortisip (Nutricia, Wiltshire, UK), a complex human nutritional supplement, was 10-fold greater than any of the other media. The sterol requirement of *L. giganteum* does not seem to exist for *L. chapmani*.

11.2.3. *Coelomomyces* spp. Keilin

This group of more than 70 species, within the Phylum Chytridiomycota, Order Blastocladiales, are aquatic, obligatory pathogens of aquatic Diptera, including Culicidae, Psychodidae, Chironomidae, Simuliidae, and Tabanidae (Chapman, 1985; Whisler, 1985). These fungi are also notable for alternating sexual and asexual stages and involving copepods or ostracods as alternate hosts. Even though the *Coelomomyces* species have been observed to be very effective control agents of mosquitoes, the dependence on in vivo production has greatly limited their usefulness as inundative biocontrol agents (Scholte et al., 2004).

11.2.4. Entomophthorales

Previously classified within the Zygomycetes, Entomophthorales have been placed in a new subphylum, Entomophthoromycotina, representing a monophyletic lineage distinct from all the other fungi (Humber, 2012). Most species within this group are insect pathogens, with a few species also attacking nematodes and mites. A notable characteristic for most species is a degree of high host specificity. Many species have never been cultured in vitro because of their nutritional fastidiousness.

A major impediment in developing Entomophthorales—even for efficient inoculative biocontrol efforts, much less inundative applications—has been their biology. The conidia are short-lived after in vivo production and extremely difficult to produce efficiently in vitro. In vivo, hyphal bodies or protoplasts are the basic vegetative stages of these fungi. Resting spores (azygospores, and zygospores in some species), however, have been deemed to be the more promising infectious propagule for biocontrol efforts. In *Zoophthora radicans* (Brefeld) Batko and other species, the azygospores are formed in vivo by the proliferative hyphal bodies in the insect's hemolymph. These azygospores are generally dormant after production and are typically in the overwintering stage in temperate climates. Under specific environmental conditions, the resting spores germinate to produce primary conidia, which then give rise to infectious conidia, which then continue the fungus life cycle by infecting insect hosts. In limited field introductions, azygospores have been typically produced in vivo, being collected from laboratory-infected cadavers or from a soil matrix in which the infected insects have died. Neither process is efficient nor amenable for large-scale introduction (Hajek, 1997). Laboratory methods for isolating and culturing these fungi have been thoroughly discussed by Hajek et al. (2012).

Bartlett and Jaronski (1988) reviewed the status of entomophthoralean production up to that time. Some limited research into mass production of Entomophthorales had been conducted prior to 1984, namely efforts by Latge and collegues (Latge, 1975, 1981; Latge et al., 1978). Since then, there has been some effort beyond laboratory-level media to better study a particular species, but only a few studies attempted to mass produce an entomophthoralean fungus. A review of the literature reveals that only the species listed in Table 11.2 have been mass produced in vitro.

Culturing entomophthoralean fungi in vitro varies widely in difficulty, depending on the species and even the isolate. In general, *Neozygites* Witlaczil species are among the more difficult species to grow, and *Conidiobolus* (Costantin) Batko are the easiest. The earliest advances in the in vitro production of the Entomophthorales were the use of Grace's insect cell culture medium supplemented with fetal bovine serum (5% v/v), which simulated the insect hemolymph (Dunphy et al., 1978). This approach is still being used; for example, Kogan and Hajek (2000) produced azygospores of *Entomophaga miamaiga* Humber, Shimazu, and Soper with this medium. Alternatively, some species could be grown on Sabouraud dextrose agar supplemented with egg yolk and milk or coagulated egg–yolk–milk medium (Hajek et al., 2012). However, neither approach is practical—much less economical—for any level of mass production. Beauvais and Latge (1988) advanced the art with a medium of glucose, yeast extract, lactalbumin hydrolysate, NaCl, and 10% fetal bovine serum (GLEN medium). GLEN and modified Grace's medium have also been used to grow *Entomophthora thripidum* Samson, Ramakers, and Oswald in the laboratory, but not on a large scale (Freimoser et al., 2003).

Nolan (1988) initially devised a defined medium, which allowed growth in both stationary and shaken cultures in the absence of fetal calf serum; it was developed for protoplasts of the fungus *Entomophaga aulicae* (Reichardt)

TABLE 11.2 **Entomophthorales Species that Have Been Mass Produced In Vitro**

Species	Reference
Pandora nouryi	Zhou and Feng, 2010
Pandora (Erynia) neoaphidis	Hua and Feng, 2003; Uziel and Kenneth, 1998; Shah et al., 2007
Zoophthora radicans	Senthikumar et al., 2011; Pell and Wilding, 1992
Pandora delphasis	Uziel and Kenneth, 1998
Entomophaga aulicae	Nolan, 1988
Entomophthora thripidum	Freimoser et al., 2003
Entomophthora maimaiga	Hajek and Plymale, 2010
Batkoa sp.	Leite et al., 2005
Furia sp.	Leite et al., 2005
Neozygites floridana	Leite et al. 2000, 2005; Delalibera et al., 2003
Entomophthora (Conidiobolus) coronata	Wolf, 1951
Entomphthora obscura	Latge, 1981
Neozygites tanajoae	Delalibera et al., 2003
Entomophaga grylli pathotype 1	Sanchez Pena, 2005
Neozygites parvispora	Grundschober et al., 2001

Humber. The protein "requirement" was obviated and growth was enhanced by the addition of hematin (0.5 pg/ml) and oleic acid (1.0 pg/ml). Nolan (1993) developed a low-cost liquid medium for production of *E. aulicae* hyphal bodies competent to form conidia. The medium consisted of a basal medium of a series of salts, amino acids, sucrose, glucose, and buffer, plus 0.8% tryptic soy broth and 0.4% calcium caseinate; it supported the growth of several isolates. Inoculant protoplasts were grown on "traditional" Grace's tissue culture medium supplemented with 5% fetal bovine serum.

Hyphal bodies of *Neozygites floridana* (Weiser and Muma) Remaudier and Keller were produced in vitro using Grace's cell culture medium plus 0.33% lactalbumen hydrolysate and 0.33% yeastolate (Leite et al., 2000). Yields were over 10^6/ml. These hyphal bodies would produce primary conidia from which the infectious capilloconidia could be generated. Leite et al. (2005) refined the medium for *N. floridana* further, using glucose, skim milk, yeast extract, peptone, and trace salts, but the complexity of preparing this medium may preclude its use in large volume.

In seeking a mass production medium for *Z. radicans* azygospores, Senthikumar et al. (2011) tested a series of concentrations and ratios of sunflower oil or dextrose as a carbon source and a yeast extract or peptone as nitrogen source in liquid medium; they discovered the optimal ratio was 4:8 yeast extract to sunflower oil medium. Alternatively, *Z. radicans* could be produced as mycelium in a liquid culture medium of yeast extract, dextrose, and sunflower oil; mycelium was then harvested by filtration, washed, and prepared as a thin slurry. The slurry was treated with 10% maltose, matured at 4 °C, then air dried (McCabe and Soper, 1985; Wraight et al., 2003). This dry, marcescent mycelium was then prepared as a granular formulation, which when rehydrated and exposed to the appropriate environmental conditions would generate large numbers of the infectious conidia. The dry mycelium formulation produced equivalent or greater numbers of conidia than fungus on leafhopper cadavers. However, Li et al. (1993) discovered that dried mycelium preparations of *Z. radicans* and *Erynia (Pandora) neoaphidis* Remaudier and Hennebert did not survive milling or freezing.

The production process is complex and imposes serious constraints for mass production, especially for inundative releases of this fungus. The process could be feasible for inoculative release to establish epizootics, but it has not yet been capitalized upon. One impediment—at least in the United States, Canada, and the EU—is that any use of a microbial to control an insect is subject to registration with the respective regulatory authorities, and inoculative release be an insufficient commercial incentive.

Latge et al. (2004) described the best media for *Entomphthora virulenta* Hall and Dunn zygospores, consisting of dextrose and corn syrup as carbon sources and yeast extract, soybean flour, or cottonseed flour as the best nitrogen sources. The resulting zygospores had a 70% germination rate. A liquid culture medium for *E. neopahidis* was determined by Gray and Markham (1997) to consist of glucose, yeast extract, mycological peptone, KH_2PO_4, Na_2HPO_4, and 0.01% oleic acid in 1.5-l fermentation volumes. They obtained considerable mycelial biomass in batch but not continuous fermentations. A disadvantage was that large inoculum volumes were necessary to detoxify an essential nutrient, oleic acid.

Media for *Batkoa* (Keller) and *Furia* (Humber) species were devised by Leite et al. (2005). For *Furia*, the best medium consisted of 0.33% each of yeast extract, beef extract, and skim milk in a basal medium of 2.66% glucose and trace salts. Although several other combinations of nitrogen sources gave good yields, skim milk was associated with the best media. For *Batkoa*, in contrast, yeast extract was the best nitrogen source, whereas skim milk and peptone did not seem to be important nitrogen sources for this fungus. Both fungi produced mycelium in these media, which presumably could be processed and preserved using some modification of the McCabe and Soper (1985) process.

The Leite et al. (2005) study illustrates the differences in optimal media that can exist among species of Entomophthorales, especially differences in response to different concentrations of complex nitrogen sources, as well as to the sources themselves.

The most successful simple in vitro mass production of an entomophthoralean fungi using a simpler medium has been with *Pandora neoaphidis* (Remaud and Hennebert) Humber, *Pandora nouryi* (Remaud and Hennebert) Humber, and *Z. radicans* on broom corn millet (Hua and Feng, 2003, 2005; Zhou and Feng, 2010). Autoclaved broomcorn millet (with 36% moisture content) supplemented with Sabouraud dextrose broth and inoculated with the fungus evidently colonized the grains, which then served as surrogate aphid cadavers, allowing subsequent sporulation that was 2–3 times more abundant and with a greater duration than in aphid cadavers.

There have been several efforts of in vivo production. Mullens (1986) devised a method to infect large numbers of houseflies with *Entomophthora muscae* (Cohn) Fresenius. By this method, several hundred adult flies were exposed to conidial showers from infected insects in a small container, ensuring good dose transfer. Serial repetitions of these exposures could greatly increase the number of infected insects, which could then be released alive. Carruthers et al. (1997) described using *Entomophaga grylli* (Fresenius) Batko, produced in vivo, in inoculative releases against grasshoppers. The insects were infected by injecting protoplast cultures and were then released live into natural grasshopper populations.

The potential for using dried infected hosts (*Plutella xylostella* L. with *Z. radicans* resting spores) was presented by Pell and Wilding (1992). Second instar larvae were exposed to sporulating plate cultures, reared until death but before fungal sporulation occurred, and then dried at ambient temperature and 40% relative humidity. Steinkraus and Boys (2005) harvested large numbers of cotton aphids (*Aphis gossypii* Glover) infected with *Neozygites frezenii* (Nowak) Remaud and Keller from natural epizootics and preserved them by drying the aphid-infested leaves over silica gel. The dried cadavers had a very high level of sporulation when rehydrated, even 6 months later. In this manner, they were able to collect 25,572 infected aphids from 189 leaves. The key for success with such a method, however, is to find an appropriate epizootic, which may not always be possible.

11.2.5. Microsporidia

Microsporidia have been historically associated with the Protozoa, but in the last decade have been reclassified within the kingdom Fungi, as the Phylum Microsporidia (Keeling, 2009). Nevertheless, their biology and morphology are unique and distinctly different from the other fungi. A spore is the infectious agent, and the infection process most commonly involves the host ingesting the spores, explosive spore germination in the host gut, with insertion of the nucleus and associated cytoplasm into a gut epithelial cell through a thin, hollow, polar filament rapidly everted from the spore. The microsporidian then develops within the host cell. Vertical transmission via transovarial and transovum routes is also common among the microsporidia, as are intermediate hosts. Hosts of microsporidia span a wide range, from Protista to warm-blooded vertebrates, but in general each species is somewhat host specific. See Solter et al. (2012b) for a review of microsporidian biology. One species, *Paranosema (Nosema) locustae* (Canning), has been registered as a microbial pesticide and commercialized in the United States for the control of grasshoppers (USEPA, 2000). In general, however, biocontrol efforts with microsporidians have focused more on inoculative or augmentative approaches, for which smaller amounts of infective spores are needed than in inundative use.

A salient feature of microsporidia is their obligate intracellular development. Thus, these organisms can be mass produced only in vivo, in their hosts (Henry, 1985) or in invertebrate cell culture (Visvervara et al., 1999). For example, *P. locustae* is produced by perorally infecting grasshoppers with a dose that is optimal for spore production, allowing the pathogens to multiply within the host, fragmenting the cadavers in a mill and suspending the macerate in water. After the insect parts are removed by filtration, the resulting spore suspension is further purified and formulated with wheat bran carrier (Henry, 1985). Spores of *P. locustae* can be stored in refrigerated sterile water, lyophilized, or vacuum dried before formulation. Infected cadavers can also be dried and stored until further processing. The reader is referred to Solter et al. (2012a) and Solter et al. (2012b) for further information about mass production methods for microsporidia.

11.2.6. Ascomycete Hypocreales

Most commercial development of entomopathogenic fungi has been directed towards this group of fungi. These particular species, once placed in the Deuteromycete (Imperfect) fungi, have now been assigned to the family Clavicipitaceae within the Ascomycete order of Hypocreales, based on their molecular association with teleomorph stages. These fungi include the genera *Beauveria*, *Metarhizium*, *Isaria* (formerly *Paecilomyces*, *Lecanicillium* (formerly *Verticillium)*, and the species *Hirsutella thompsonii* Fisher, *Nomuraea rileyi* (Farlow) Samson, *Aschersonia aleyrodis* Webber, *Culicinomyces clavosporus* Romney and Rao, and *Tolypocladium cylindrosporum* (Gams). For recent reviews about the three most important genera, see Zimmermann (2007a) for *Beauveria*, Zimmermann (2007b) for *Metarhizium*, and Zimmermann (2008) for *Isaria*. Each of these references contains considerable information about the biology of the respective genera.

In general, these fungi have four major propagule types that can be used. In nature, the aerial conidium is the primary infectious propagule. Conidia are the spores that are produced on the exterior of fungus-killed insects. Blastospores are the proliferative stages within the insect for many of these fungi and can also be produced in liquid fermentation. (This ability is not surprising if one considers an insect as merely a six-legged flask of culture medium.) Under certain liquid fermentation conditions, mainly substitution of inorganic for organic nitrogen, *Beauveria* and *Metarhizium* can produce "microcycle" conidia (Thomas et al., 1987; Bosch and Yantoro, 1999; Zhang et al., 2009). These conidia are not true conidia and are produced on the ends of hyphal strands. Lastly, mycelium, the major form of fungal vegetative proliferation, or mycelial derivatives such as the microsclerotia, can be used. But in this last case, the mycelium or its derivatives are true producers of conidia—a way to deliver conidia to the insect.

Because of the significance of *Beauveria* and *Metarhizium* among the entomopathogenic Ascomycetes, emphasis on developments for their mass production follows. Specific production methods developed for *Isaria, Hirsutella, Aschersonia, Nomuraea, Lecanicillium,* and *Culicinomyces* are subsequently reviewed.

11.2.6.1. Solid Substrate Fermentation

The fundamentals of solid substrate fermentation to produce entomopathogenic fungi were detailed by Bartlett and Jaronski (1988); the reader is referred to that publication. Here, I will attempt to address process development since then. An overview of solid substrate fermentation systems has been presented by Krishna (2005). A description of practical small scale methods, yielding decagrams, even hectograms of conidia per kilogram of substrate may be found in Jaronski and Jackson (2012). Other descriptions of solid substrate production systems are presented in Aquino et al. (1977); Cherry et al. (1999); Perez-Guerra et al. (2003); and Bateman (2007).

11.2.6.1.1. The End Products of Solid Substrate Fermentation

Solid substrate fermentation, mimicking the natural conidiation processes, yields aerial conidia as the final product (Figs 11.3 and 11.4). Aerial conidia

FIGURE 11.3 Solid substrate (flaked barley) with well-sporulated *Beauveria bassiana*. Note the abundant conidia in the form of white powder on the substrate.

FIGURE 11.4 Commercial harvest of *Beauveria bassiana* conidia produced in a 1995 Mycotech Corporation pilot-scale, solid-substrate, fermentation system. This batch of conidia totaled 56 kg of conidial powder having a titer of 1.6×10^{11} conidia/g.

are the natural infectious bodies for these Ascomycetes. In a few isolated cases, the fermentation is terminated after the fungus has penetrated the nutritive substrate but before conidiation has begun (Kirchmair et al., 2007; Skinner et al., 2012). This process yields dried grain kernels colonized by *Beauveria* or, in particular, *Metarhizium anisopliae* (Metschnikoff) Sorokin, which remain competent for regrowth and sporulation upon rehydration. This rehydration occurs, for example, after such granules are applied into soil by a sowing machine in combination with a rotary harrow (Kirchmair et al., 2007) or mixed into horticultural soil (Skinner et al., 2012), creating focii of conidia within the habitat of the target insect.

11.2.6.1.2. Substrates and Media

A wide variety of organic materials have been evaluated as substrates for the Ascomycetes. Although rice and barley seem to be the major substrates used in the tropics and the Northern Hemisphere, respectively, there has been considerable effort to identify low-cost agricultural materials, especially byproducts and waste products, as suitable substrates. Such efforts have been particularly frequent in India, Pakistan, and China. Table 11.3 lists some of these alternative substrates. Comparisons, especially among different reports, are very difficult because of different fungus species and strains, different fermentation conditions, and different supplementations. A few studies have quantitatively examined certain basic nutritional aspects, such as C:N ratio (Ortiz-Urquiza

TABLE 11.3 Solid Substrates Evaluated for Production of the Principal Entomopathogenic Ascomycetes

Substrate	Fungus
Agricultural byproducts	*Metarhizium*
Bagasse ± 2% dextrose	*Beauveria*
Barley	*Beauveria*
Beetroot	*Beauveria*
Broken rice	*Metarhizium*
Broken rice + $CaCl_2$	*Beauveria, Metarhizium*
Carrot tubers	*Beauveria*
Cassava chips	*Metarhizium*
Chickpea	*Metarhizium*
Coconut cake	*Beauveria*
Cottonseed cake	*Beauveria*
Finger millet	*Beauveria*
Grade 3 unpolished rice	*Metarhizium*
Groundnut cake	*Beauveria*
Kodo millet	*Metarhizium*
Maize	*Beauveria, Metarhizium*
Maize bran ± 2% dextrose	*Beauveria*
Mijo grains + organic nitrogen	*Nomuraea*
Millet	*Beauveria, Metarhizium*
Neem cake	*Beauveria*
Pearl millet	*Beauveria*
Potato tubers	*Beauveria*
Prawn waste	*Beauveria*
Press mud ± 2% dextrose	*Beauveria*
Rice	*Metarhizium*
Rice + saccharomyces	*Beauveria, Metarhizium*
Rice bran ± 2% dextrose	*Beauveria, Metarhizium*
Rice flour	*Metarhizium*

Continued

TABLE 11.3 Solid Substrates Evaluated for Production of the Principal Entomopathogenic Ascomycetes—cont'd

Substrate	Fungus
Rice hulls, sawdust:rice bran ± 2% dextrose	*Beauveria*
Rice husk ± 2% dextrose	*Beauveria*
Sesame cake	*Beauveria*
Sorghum	*Beauveria, Metarhizium*
Soyabean	*Beauveria, Metarhizium*
Sugarcane bagasse ± yeast, molasses	*Beauveria, Metarhizium*
Sugarcane press mud	*Beauveria*
Tapioca rind	*Beauveria*
Tapioca tubers	*Beauveria*
Wheat	*Beauveria, Metarhizium*
Wheat bran + $Al(SO_4)_3$	*Beauveria*
Wheat bran + organic nitrogen	*Nomuraea*
Wheat bran ± 2% dextrose	*Beauveria*

et al., 2010) or controlled levels of carbohydrates (Domenech et al., 1998). Nevertheless, rice remains the substrate of choice where it is readily and cheaply available. In North America and Europe, barley—particularly dry, flaked barley—seems to be preferred because of its superior handling characteristics and cost relative to rice (Jaronski and Jackson, 2012).

Several inorganic substrates have been identified in recent years. These include granules of calcined diatomaceous earth (diatomite); (Jaronski and Jackson, 2012; Crangle, 2011; Wikipedia Contributors, 2012), and open-pored clay granules (e.g. Seramis®; Seramis GmbH, 2012). Use of clay granules to produce *B. bassiana* has been described in detail by Desgranges et al. (1993) and Guillon (1997). Diatomite can be obtained in a range of sizes from several millimeter in diameter to a coarse powder; it has a high surface to volume ratio and can absorb aqueous liquids up to 110–140% of its weight. Seramis is a processed, particulate clay composed of kaolinite, illite, and quartz, produced in the Westerwald region of Germany; it is used primarily in hydroponics, plant bedding, and interior landscaping industries in Europe. It also is open-pored and highly absorbent (Seramis, 2012).

Unlike cereal grains or other organic materials, mineral carriers have the advantages of allowing flexible control of nutrients tailored for each fungus species and strain, as well as the ability to be recycled after washing and sterilization. This ability for recycling avoids disposal issues, which can be considerable, as is faced by one North American company having 7000-kg production runs. In the 1980s, I and colleagues developed a pilot-scale (100-kg batch size) process using Celetom granules with a liquid medium to produce $5–7 \times 10^{12}$ *B. bassiana* ARSEF252 conidia per kilogram of substrate (Jaronski, unpublished data). If harvesting was performed aseptically, it was possible to obtain a second round of conidiation after the first was removed and the substrate reincubated. A third flush became contaminated, however, with *Aspergillus*, and in general such practice is not recommended for a quality product.

Niedermayr et al. (2012) noted that an unidentified, open-pored clay granule produced only 10% of the spore yield of *Metarhizium brunneum* BIPESCO5 (also known as Met52) with several grains. Their comparison was flawed, however, because they used a packed bed column fermenter for the clay granules and plastic bags for the grains. Further, they did not optimize the liquid medium for the clay granules. The grain in bags yielded more typical yields for this fungus. Hemp-based animal litter (Hemparade; HempFlax, The Netherlands) has also been used as an inert carrier, impregnated with a nutrient medium (Breukelen et al., 2011). The hemp evidently afforded more than adequate porosity for good gas and heat transfer in a packed-bed column fermenter. Conidia could be removed from crushed, dried whole culture by simple mechanical classification. Although yields were excellent, comparable to those on barley in vented mushroom spawn bags (Jaronski, unpublished data), the cost of the substrate and nutrients was not revealed. In addition, the hemp does not seem to be recyclable. Another novel substrate is Amberlite IRA-900 ion exchange resin impregnated with nutrient medium, which has been used for *Aspergillus niger* van Tieghem but has not yet been assessed for any of the entomopathogenic Ascomycetes (Auria et al., 1990).

11.2.6.1.3. Equipment for Fermentation

By and large, most academic and industrial systems have used traditional solid substrate fermentation, either manually intensive or highly mechanized, for producing these Ascomycetes, with a few systems employing submerged liquid fermentation for specific species. In situations where labor costs are low, allowing labor-intensive approaches, polypropylene bags are used. These bags include specialized, vented mushroom spawn bags (e.g. Unicorn spawn bags; Unicorn Industries, Plano, TX, USA), $SacO_2$® microsacs (Combiness, Belgium), zipper-lock bags, and simple shopping bags (Jaronski and Jackson, 2012). A unifying characteristic is that the different plastic bags are autoclavable; unused, unopened, zipper lock bags are often sterile inside. Certain production operations in Africa use plastic bags for the initial mycelial colonization and growth, then transfer the cultures to open, nonsterile, plastic laundry hampers or tubs for the sporulation phase within a controlled environment (Cherry et al., 1999; Bateman, 2007). *Metarhizium* production in Brazil since the 1970s has

FIGURE 11.5 Typical plastic bag-based mass production of *Beauveria bassiana* as is practiced in many countries. *Photograph courtesy Itaforte Bioprodutos, Koppert and Miguel Rinçon Najera.*

involved solid substrate in plastic bags (Aquino et al., 1975; Mendonca, 1991) as has *Metarhizium* production in Nicaragua (Grimm, 2001). Figure 11.5 illustrates *B. bassiana* mass production in Latin America. Open trays of inoculated substrate have also been employed (Alves and Pereira, 1989; Claro, 2006).

Another direction of research pertains to the use of trays within controlled environment chambers. The technology underlying such systems is the Koji tray fermentation, whereby the inoculated substrate is a shallow bed in a tray with perforated bottom and open or mesh cover. Ye et al. (2006) described an upright incubation chamber containing 25 mesh-bottom, open trays, each with capacity of 2 kg of rice. The chamber had a substrate volume of 0.72 m^3 and occupied a surface area of 0.36 m^2. Although temperature could not be regulated, humidity could be controlled during the fermentation. With such a chamber, fully loaded, they obtained 2.4×10^{12} *B. bassiana* conidia per kilogram of substrate.

Several groups have further explored packed bed fermentation technology based on early research. Since the review by Bartlett and Jaronski (1988), there has been considerable development in such systems. Comprehensive reviews of these systems have been presented by Durand (1998, 2003), Durand et al. (1996), Raimbault (1998), and Krishna (2005), and will not be discussed further here. The most recent equipment development is the Prophyta packed bed fermenter (Luth and Eiben, 2003; Eiben and Luth, 2006; Prophyta, 2012). Problems with packed, aerated beds include uneven air flow through the substrate creating pockets of overheated, anaerobic conditions and pockets of substrate with moisture that are suboptimal for fungal growth, even with water-saturated air supply. With many of the Ascomycetes, the solid substrate culture should be broken up for optimal conidiation. This is not readily possible with most packed bed systems, but it is with plastic bags.

A major advance in large-scale solid substrate fermentation has been in the development of mushroom spawn production, using a steam-sterilized, double-cone blender holding several hundred kilograms of substrate (Maul et al., 1980). An example of this blender is illustrated in Bateman (2006). The device allows in situ steam sterilization of substrate, controlled cooling, and subsequent thorough mixing of liquid inoculum with the substrate. A chute at the pointed end of the cone allows for aseptic filling of spawn bags of other containers in an efficient, rapid manner. This system is in use not only by a large North American mushroom spawn producer but also by several other entities to produce atoxigenic *Aspergillus flavus* Heinrich and Link and mycoinsecticides. It overcomes a major impediment to very large-scale production—efficient preparation and sterilization of a large amount of substrate.

11.2.6.1.4. Fermentation Parameters: Moisture, Temperature, Gas Environment, and pH

Strictly speaking, solid substrate fermentation is really biphasic, with an initial step being a liquid fermentation to produce inoculum, the most common practice, although some researchers use an aqueous suspension of conidia to initiate the solid substrate phase. There are obvious advantages to liquid fermentation in producing inoculum for the solid phase. Conidia require 24 h to fully germinate and begin colonization of the substrate, while a liquid fermentation (blastospores and mycelia) begins immediately. Further, a liquid fermentation phase greatly multiplies the inoculum potential compared to conidia, because an inoculum should contain 10^7–10^8 propagules per milliliter with 60–100 ml inoculum per kilogram of grain substrate. Inoculum concentration affects the duration of the solid substrate phase to reach maximum conidial yield. Nuñez-Gaona et al. (2010) developed a model for *B. bassiana* grown on wheat bran substrate amended with sugarcane bagasse to predict the effect of inoculum concentration. The time to produce 1×10^{10} conidia per gram of substrate could be halved to 148 h by increasing the conidial concentration from 1×10^6 to 5×10^7 conidia per gram of dry substrate.

The ideal liquid fermentation should produce mostly, or entirely, blastospores and short hyphae for optimal dispersion of inoculum through the substrate. Abundant mycelial production, especially in balls or clusters, prevents good dispersion and even colonization of the solid substrate. A wide range of media have been used for the liquid fermentation phase. The simplest recipe is dextrose/sucrose as the carbon source and yeast extract as source of nitrogenous compounds and vitamins; trace salts are not necessary (Cherry et al. 1999; Bateman, 2007; Jaronski and Jackson, 2012). A medium that works very well with a wide range of *Beauveria* and *Metarhizium* isolates consists of 20–30 g/l glucose or sucrose and 15 g/l yeast extract, supplemented with 17 g/l liquid corn steep or 8 g/l corn steep liquor powder (e.g. Solulys®, Roquette Chemical and Bioindustries, Lestrem, France; Jaronski and Jackson, 2012). Corn steep evidently stimulates vigorous blastospore production of both fungi, with minimal mycelial formation. The liquid fermentation requires vigorous aeration, which can be achieved using 200–250 rpm for flask production or an air-lift bubbler fermentation vessel for larger quantities. A typical liquid fermentation cycle takes 72 h with conidia as inoculant or 48 h if a blastospore preculture is used.

In all cases, the solid substrate, whether organic or inert, must be hydrated and sterilized. The role of water in solid substrate fermentation of fungi was discussed in detail by Gervais and Molin (2003). In situations of potential contamination, the substrate can be hydrated with 0.04% H_2SO_4 and 0.097% KH_2PO_4 to inhibit bacterial growth (Jaronski, unpublished data). In some cases, published protocols using rice substrate have used nutritional additives in hydrating the substrate to increase conidial production: 2% dextrose (Mazumder et al., 1995), cane molasses and torula yeast (Calderon et al., 1995), urea, $(NH_4)_2SO_4$, and yeast extract (Domenech et al., 1998), yeast and molasses (Calderon et al., 1990), and sugar molasses (Sene et al., 2010). No supplements are needed for barley, oats, or wheat, presumably because these have more nitrogen and micronutrients than polished rice (Jaronski, unpublished data).

The typical duration of the solid substrate phase is 7–14 days. Depending on the nature of the inoculum (conidia or blastospores/mycelia), the substrate is colonized by fungus within the first 24–48 h, after which there is active mycelial proliferation through the substrate. Then, the culture needs to be dried in most cases. During the initial drying process, as water activity (a_w) begins to decrease through 0.99, there can be a burst of additional conidiation.

Fermentation parameters—particularly in packed bed approaches—have been dealt with in detail elsewhere (Raimbault, 1998; Krishna, 2005) and the reader is referred to these works. An extremely good discussion of the role of moisture in solid substrate fermentation may be found in Gervais and Molin (2003). A critical moisture level is needed for optimal fungal growth and sporulation. In terms of water activity, the critical level for *Beauveria* and *Metarhizium* is a_w 0.97–0.98 (Humphreys et al., 1989; Hallsworth and Magan, 1999; Nunez-Gaona et al., 2010). Significant interspecific differences in tolerances to a_w exist (Hallsworth and Magan, 1999). They reported optimal a_w ranging between 0.99 and 0.97 for *M. anisopliae* and *Isaria farinosa* (Holmskiold; formally *Paecilomyces farinosus* Holmskiold) and 0.998 for *B. bassiana*. Tarocco et al. (2005) reported that the greatest conidia yield of a *B. bassiana* on rice was obtained at an initial a_w of 0.99.

The amount of water to achieve this level will vary based on the nature of the substrate. For flaked barley substrate, optimal moisture is 52–56% (w/w) before autoclaving (Jaronski and Jackson, 2012). For rice and sorghum, 22–30% and ~75–76% moisture, respectively, seem to be optimum (Prakash et al., 2008). A researcher investigating production of a new isolate of either fungus should experiment to determine optimum moisture levels because there are differences among isolates of any species (Jaronski, unpublished data).

Because the solid substrate phase is an active fermentation, not only does oxygen need to be readily available to all parts of the substrate, but carbon dioxide and heat must be drawn off. In vented mushroom spawn bags O_2 falls to ~10% and CO_2 rises to 11% in the headspace of the bag within 24 h of inoculation (Jaronski, unpublished data). In bag fermentation, the substrate mass must be less than 7–10cm thick to allow proper heat dissipation and gas exchange. Additives to create more headspace within the grain substrate have included aluminum silicate (Saroha and Mohan, 2006), sugarcane bagasse (Nuñez-Gaona et al., 2010), rice bran/husk (Dorta et al., 1996), and wheat bran husk (Arcas et al., 1999). Forced-air packed bed fermentation, although potentially overcoming heat dissipation and gas exchange problems of still fermentations, often results in zones of poor growth and sporulation, especially if larger masses are employed. This aspect was investigated thoroughly by Underkofler et al. (1947). However, as Arzumanov et al. (2005) observed, forced aeration of packed beds may not be absolutely necessary, at least with certain Ascomycetes, as long as the bed geometry allows sufficient gas exchange and heat dissipation.

A recent advance in the empirical identification of environmental variables optimal for spore production with economy of effort has been the use of response surface methodology (Prakash et al., 2008; Dhar and Kaur, 2011; Deng et al., 2011; Qiu et al., 2013). Response surface methodology, which has been used in industry but rarely in academic studies, combines mathematical and statistical techniques to design experiments and identify optimal conditions using a reduced number of experiments; this approach eliminates the limitations and avoids the laboriousness of single-factor optimization in fermentation.

Conidial yields can vary among strains of each fungus species. For example, Arcas et al. (1999) determined that one strain of *B. bassiana* produced three times as many spores as a second under identical fermentation conditions. Conidial production of 15 *B. bassiana* isolates ranged from 1.11×10^{11} to 2.25×10^{13} conidia per gram of initial dry substrate when grown under identical solid substrate fermentation conditions (Jaronski, unpublished data). It has been my observation that, for the most part, conidial yield is genetically determined in both *M. anisopliae* and *B. bassiana* and can be inversely proportional to virulence for insects.

Kuźniar (2011) and Zhang et al. (2009) observed that exposure to light enhanced *B. bassiana* growth and conidiation; however, this is contradicted by the very high conidial yields obtained by Bradley et al. (1992) and subsequent commercial production of several *B. bassiana* strains in a completely dark fermentation environment. Similarly, although Onofre et al. (2001) stated that continuous illumination gave 2.5–5 times increase in spore production of *Metarhizium flavoviride*, this has not been my experience. Thus, a light requirement may be strain specific rather than a general phenomenon. On a practical level, a light requirement may be a considerable challenge in even shallow-packed bed fermentations. In bag fermentation, where transparent plastic is used to enclose the fermentation, light is more easily supplied but the substrate mass must be thin enough to allow light penetration through the entire mass.

11.2.6.1.5. Downstream Processing

For all purposes except immediate use, the conidia produced on solid substrate must be dried down to a moisture content <9% w/w or $a_w \leq 0.3$ (Bateman, 2007; Jaronski and Jackson, 2012). This low moisture is necessary for optimal shelf life regardless of whether conidia are formulated or not. Moore et al. (1996) observed that even trace amounts of moisture in the conidia result in foreshortened shelf life. A relatively short shelf life of a few weeks to a few months, depending on the fungus, can be obtained by refrigerating whole sporulated solid substrate within their original fermentation bags.

There are a number of published drying methods: simple opening of plastic fermentation bags; transfer of sporulated substrate to open trays (Bateman, 2007; Claro, 2006) or table tops (D.W. Roberts, personal communication; transfer to Kraft paper sacks (Jaronski and Jackson, 2012); or use of air-lift devices. In general, *Beauveria* conidia can be dried relatively quickly (within 2–3 days) without loss in viability, whereas *Metarhizium* conidia require slower drying (5–9 days; Hong et al., 2000; Jaronski and Jackson, 2012). There is only one study carefully examining the effect of drying temperature and duration of an entomopathogenic Ascomycete (*B. bassiana*) (Li et al., 2008). In that study, the viability of conidia was affected by different drying temperatures and speed of drying; 5 h at 35 °C had no effect on conidial viability. However, the effects of drying temperature and speed on the shelf life of formulated and unformulated conidia, an important commercial aspect, are not known.

Moisture endpoint is best measured using a water activity meter (see Jaronski and Jackson, 2012), but gravimetric moisture analysis is satisfactory. Water activity (a_w) is a measure of the biologically relevant moisture content of an object. The relationship between water activity and moisture content was graphically and algebraically presented for both absorption and desorption isotherms by Faria et al. (2009); the interested reader should refer to this report for more details.

Extreme desiccation of conidia ($a_w < 0.1$), however, can lead to two problems: one affects proper quality assurance testing and the second is a significant problem in operational use of such dry conidia. This situation stems from imbibitional damage of the plasma membranes within the conidia during improper dehydration. Problems with conidial rehydration were first observed by Moore et al. (1996) and subsequently studied in more detail by Faria et al. (2010). In a series of experiments, Faria et al. (2009) discovered that extreme desiccation of *Metarhizium* ($a_w < 0.3$) conidia requires careful rehydration before suspension in an aqueous medium. Use of cold water (≤15 °C) with dry conidia resulted in very low germination. Conidia of *B. bassiana* were more resistant to imbibitional damage, with conidia as dry as a_w of 0.02 having germination rates at normal levels, except when water was 0 °C. Proper rehydration of dry conidia can be by exposure of the conidia to moisture-saturated atmosphere for a minimum of 30 min, as described by Moore et al. (1996), or by use of warm (33–34 °C) water as described by Faria et al. (2009). Use of warm water has to be carefully controlled, however; as Xavier-Santos et al. (2011) observed, different isolates of *Metarhizium* responded differently to immersion at 31 °C and all isolates tested were adversely affected by either 45 °C for 60 min or prolonged exposure.

In operational conditions involving more than a few decagrams of conidia, the need for careful rehydration of very dry conidia can pose logistical challenges, especially with *Metarhizium*. Of course, one can carefully observe the drying process and terminate it when the conidia have reached a water activity of 0.3. However, emulsifiable oil formulations seem to confer protection from imbibitional damage (Xavier-Santos et al., 2011).

Mechanical separation of conidia from the dried substrate is the predominate method in commercial use. There are two ways to do this: sieving (Fig. 11.6) and mechanical agitation of the substrate with air collection using a cyclone dust collector (Fig. 11.7) (Jaronski and Jackson, 2012; Bateman, 2007). Graded sieves, ideally on a vibratory rather than rotary shaker, can separate out the conidia from substrate with 60–80% efficiency (Jaronski, unpublished data). The second method can be mechanically simple with spore collection using an appropriate vacuum cleaner system (Jaronski and Jackson, 2012) or sophisticated cyclone dust collector. One device using the latter technology has been specifically developed for harvesting fungal conidia (Bateman, 2012). In both cases, the conidia are physically dislodged from the substrate in a rotating drum agitator.

Washing the conidia off in dry or semidry substrate has not been pursued to any great extent, at least in the published literature. The *Beauveria* production

FIGURE 11.6 Harvesting conidia of *Metarhizium anisopliae* by mechanical classification using sieves on a vibratory shaker.

FIGURE 11.7 A cyclone dust collector used to concentrate conidia mechanically dislodged from solid substrate. The conidia are collected in the barrel beneath the conical collector. There is a high-efficiency particulate air (HEPA) filter on the exhaust to prevent escape of conidia into the environment.

system that I developed in the 1980s employed washing conidia from moist substrate using cold 0.1 M NaCl, removing nonconidial material with self-cleaning filters, adding a diatomaceous earth filter aid, and then concentrating the conidia with a continuous flow centrifuge or, alternatively, a ribbon filter. The resulting semidry paste was then rapidly air-dried as a thin layer or lyophilized (Jaronski, unpublished data). Cold saline solution was necessary to keep the conidia from germinating, and the filter aid allowed creation of a fine friable powder at the end of the process.

11.2.6.1.6. Major Technical Problems/Solutions in Solid Substrate Fermentation

The greatest problem in solid substrate fermentation is the scale-up to large capacity at commercial levels. For a mycoinsecticide to succeed commercially, a very large number of conidia must be produced as cheaply and efficiently as possible to compete with chemical insecticides. There are two directions in the production of entomopathogenic fungi by solid substrate fermentation for inundative applications, whether it be uniphasic or biphasic. The first is a low-input, manually intensive, simple technology (e.g. in plastic bags). In practice, especially in emerging rural economies, this system is appropriate and often targets limited, local use. The other is a high-technology, high-input, industrial approach, as can be seen in companies in North America and the EU.

The industrial approach is mandated by the considerable costs of development and registration and expensive wage rates. The typical bag fermentation is not amenable to such large-scale needs, particularly in the developed world where salaries are considerable. The high-technology alternative is capital intensive, often requiring specialized equipment (Ravensberger, 2011), although sparing manpower costs.

Applied to large-scale solid substrate fermentation, the need for large conidial numbers mandates efficient sterilization of very large amounts of solid substrate. For example, for one U.S. company, the fermentation batch size is 10,000 kg. Traditionally, moist heat (steam) is the method used. The V-cone blender sterilizer mentioned in Section 11.2.6.1.3 is one solution. There are alternative methods (e.g. radiation sterilization) already in commercial use in other industries that may be adaptable to fungal mass production. Of course, with plastic bags, the unit volumes are small (0.1–2 kg) and so it is relatively easy to steam sterilize. Use of recyclable solid substrate is highly desirable, yet the current state of the art largely relies on a grain, causing considerable waste disposal issues. In countries with emerging economies, many different alternative materials have been identified, but few seem to have been adopted by commercial enterprises.

Jenkins (1995), in reviewing yields of conidia for 13 fungi produced on solid substrate at that time, found only one instance where yields reached 1×10^{13} conidia per kilogram of dry substrate. The product, based on *B. bassiana*, which is sold by Laverlam International (previously known as Mycotech and Emerald BioAgriculture), was reported to have an operational yield of 2.6×10^{13} conidia per kilogram of substrate (Bradley et al., 1992).

There is also the demand for sufficient, yet efficient, fermentation space, although the mushroom and soy sauce industries have dealt with such problems and offer potentially useful technology. Harvesting problems are largely solved with the advent of cyclone dust collector technology, as discussed in Section 11.2.6.1.5.

11.2.6.2. Submerged Fermentation

Submerged liquid fermentation to produce mycoinsecticides has been a goal of fermentation microbiologists for many years. The technology lends itself to massive scale-up, using existing commercial equipment, and allows closer control of environmental variables and shorter process times (i.e., hours rather than days as for solid substrate fermentation). Submerged fermentation is currently used to produce commercial mycoinsecticides of *Isaria (Paecilomyces) fumosorosea* (Wize) and *Lecanicillium* spp.

11.2.6.2.1. The End Products of Submerged Fermentation

Submerged fermentation generally yields different propagules than solid substrate fermentation: blastospores, submerged, "microcycle" conidia, stabilized

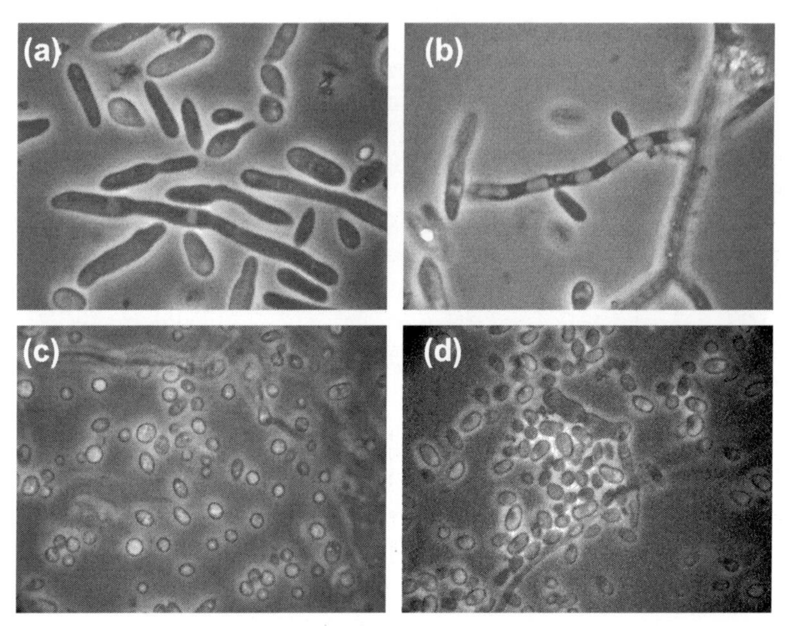

FIGURE 11.8 Ascomycete propagules obtained in submerged liquid fermentation: (a) blastospores of *Metarhizium anisopliae*; (b) hyphae of *Metarhizium robertsii* DWR346 giving rise to microcycle conidia; (c) blastospores of *Beauveria bassiana*; (d) microcycle conidia of *Metarhizium acridum*.

mycelial products, and microsclerotia. Blastospores (Fig 11.8(a), (c)) are vegetative cells by which the Ascomycetes proliferate through the body of an infected insect (Vestergaard et al., 1999), which could be considered a six-legged liquid fermentation vessel. Their growth is yeast-like, but it is not the true budding of yeasts. *Beauveria*, *Metarhizium*, *Nomuraea* and *Isaria* will grow as blastospores under the appropriate liquid fermentation conditions. The blastospore is a more environmentally fragile propagule than the aerial conidium, and special approaches are necessary to make it desiccation tolerant.

Submerged (or microcycle) conidia are produced by *Beauveria* and *Metarhizium* (Fig 11.8(b), (d)) (Thomas et al., 1987; Zhang, 2001). Microcycle conidiation has been defined as the production of conidia directly by a spore without the intervention of hyphal growth (Anderson and Smith, 1971). These conidia are morphologically and ultrastructurally different from true, aerial conidia; they lack one layer in the spore wall and have some different physical properties (Hegedus et al., 1990). They also germinate at a rate intermediate between blastospores and aerial conidia. The microcycle conidia of *Metarhizium acridum* (Driver and Milner) can evidently also be produced on solid medium (Zhang et al., 2009).

Submerged fermentation can be used to produce mycelial masses, which are then dried with preservatives to make granular formulations, such as was described in Section 11.2.4 for the Entomophthorales (Rombach et al., 1988).

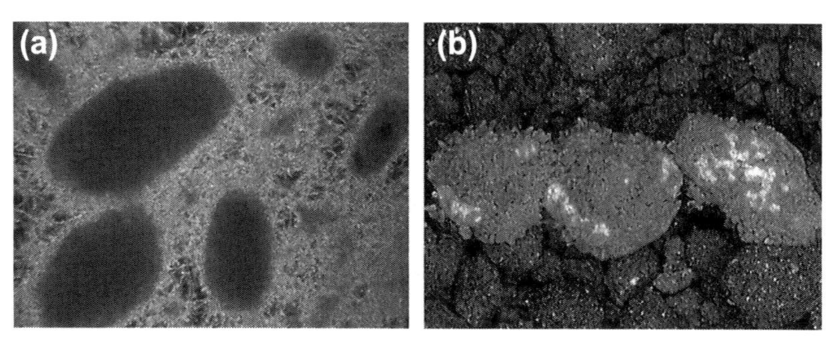

FIGURE 11.9 (a) Microsclerotia of *Metarhizium brunneum*; (b) granules formed from microsclerotia that have conidiated upon rehydration on moist soil.

When rehydrated, the mycelium generates large numbers of conidia. For a time in the 1990s, a *Metarhizium*-based, dried mycelial pellet, which would conidiate when rehydrated upon application to soil, was commercialized by Bayer AG as BIO1020. These pellets were produced in submerged fermentation using specific media and additives, which caused the mycelium to aggregate as large pellets. The pellets could then be harvested and dried to form 0.5–1 mm diameter granules (Andersch et al., 1993). This product and process were subsequently abandoned.

Fungi within the genus *Metarhizium*, except for *M. acridum*, can also produce compact melanized bodies, 50–200 μ in size, termed "microsclerotia" because of their similarity to such forms in plant pathogenic fungi (Fig. 11.8(d)). These microsclerotia are generated under certain nutrient and aeration conditions (Jackson and Jaronski, 2009a). So far, all strains within the nonacridum species of *Metarhizium* produce microsclerotia, although the numbers seem to vary by strain (Jackson and Jaronski, unpublished data). Experiments with multiple strains of *B. bassiana* and *Nomuraea rileyi* have failed to yield microsclerotia (Jackson and Jaronski, unpublished data). Microsclerotial fermentations can be simply filtered and air dried; the process has been scaled up to 100-l fermenters (Jackson and Jaronski, 2012). When microsclerotia are rehydrated, they rapidly conidiate and can thus be made into conidiogenic granules for use (Fig. 11.9). A noted feature is that microsclerotia can be simply air dried and have considerable shelf-life stability; in addition, the resulting granules can be sized as needed for particular applications. There are a number of considerations when choosing the most appropriate propagule for deployment against an insect. The reader is referred to Jackson et al. (2010) for an exploration of this topic; the discussion is not restricted to *Beauveria* and *Metarhizium*, but also includes relevant data about *Isaria fumosorosea*, which is otherwise discussed in Section 11.2.6.4.1.

11.2.6.2.2. Media

A variety of standard fermentation media can be and have been used. Carbon sources are typically glucose, while nitrogen sources can vary. Yeast extract, peptone, and inorganic nitrogen compounds have been used in laboratory experiments, while commercial scale fermentations use cottonseed flour (Pharmamedia®, Archer Daniels Midland Company, Decatur, IL, USA), or casamino acids (e.g. Solulys, mentioned earlier).

The nature of the nitrogen source affects the predominant propagule obtained. Microcycle conidium production by *B. bassiana* requires inorganic nitrogen and very high carbohydrate levels (Thomas et al., 1987), whereas blastospore production is favored by organic nitrogen, especially corn steep liquor or corn steep solids. Thomas et al. (1987) first described production of microcycle conidia by *B. bassiana*. Their medium consisted of 5% glucose and 1% KNO_3. This medium preferentially produced high numbers of submerged conidia, whereas other carbon sources strongly favored blastospore production. In work refining the production of *B. bassiana* blastospores and microcycle conidia, Chong-Rodríguez et al. (2011) discovered that a medium of glucose plus casamino acids or one containing glucose, peptone, and KNO_3 produced high numbers of blastospores, whereas a glucose-KNO_3-corn steep liquor medium produced mostly microcycle conidia. Slight variations in the ratio of carbon and nitrogen, or the nature of the carbon source, will also affect blastospore yields. For example, with 5% sucrose as the carbon source, *B. bassiana* produced 4.6×10^8 blastospores and microcycle conidia per milliliter with a 9:1 ratio of the two. When the media was supplemented with 5% sugarbeet molasses, *B. bassiana* generated 2×10^9 spores per milliliter, but only 64% were microcycle conidia with the remainder being blastospores (Jakobs-Schönwandt et al., 2011). Strain and species differences increase the complexity in defining an optimal medium for either blastospores or microcycle conidia (Vega et al., 2003).

Organic nitrogen, preferably as brewer's yeast, and high levels of sucrose seem to be essential for submerged conidia production by *M. acridum* (Jenkins and Prior, 1993; Leland et al., 2005a, 2005b), which will produce blastospores in a medium of glucose and C:N ratio > 30:1, using casamino acids as the nitrogen source (Jaronski, unpublished data). Issaly et al. (2005) further refined medium and fermentation parameters, affirming the use of sucrose and yeast extract and identifying an optimal C:N ratio of 1:6 for blastospores. Zhang et al. (2009) described production of microcycle conidia by one strain of *M. acridum* on an agar medium consisting of 3% sucrose, 0.5% yeast extract, 0.3% $NaNO_3$, and trace salts. The same fungus conidiated normally on quarter-strength Sabouraud dextrose agar (1% dextrose, 0.25% neopeptone, 0.025% yeast extract). Whether other *M. acridum* strains do the same remains to be determined. A high-osmolarity medium (using polyethylene glycol) seems to have allowed formation of a spore intermediate between blastospores and microcycle conidia in morphology and physical-chemical properties. These spores germinated more quickly than submerged conidia at a rate similar to blastospores and were more pathogenic, than either submerged conidia or aerial conidia (Leland et al., 2005b). Fargues et al. (2002) added 0.4% polyoxyethylene sorbitan monooleate. Microsclerotia of *Metarhizium* spp. can be produced using glucose-rich media having a C:N ratio > 30:1 (Jackson and Jaronski, 2009). The preferred nitrogen source can be casamino acids, such as HyCase M® (Kerry Group Services, United Kingdom) or Pharmamedia® (Archer Daniels Midland Company; Jaronski and Jackson, 2009b).

For production of marcescent mycelium, similar methods as for the Entomophthorales have been employed (Rombach et al., 1988). Magalhaes et al. (1994) identified 4% sucrose and 1% yeast extract as the best medium of several assessed for an isolate of *M. anisopliae*, yielding 9 g biomass per liter in 72 h and 1.6×10^{10} conidia per gram of dry mycelium after rehydration. This process has had very little implementation, however, and is best suited for small-scale applications. The Bayer mycelial pellet mentioned in Section 11.2.6.2.1 was produced in a glucose-yeast autolysate-K_2HPO_4 medium with $MgCl_2$ and trace salts. The medium was seeded at a very low rate (10^6 conidia/ml for small volumes; if a preculture was used, then it was added at 3% of final volume) to induce pellet formation (Andersch et al., 1993).

11.2.6.2.3. Fermentation Parameters: Air, Agitation, pH, Batch and Fed-batch Culture

A high degree of aeration for submerged fermentation is essential. The difficulty lies in the method of aeration. On a small, flask-scale level, rotary agitation at 300–400 rpm, ideally with baffled flasks, should be used. For *I. fumosorosea*, 20% dissolved oxygen is necessary for good blastospore production (Jackson, 2012). Machado et al. (1994) stated that the best air flow rate was 1.5 volumes of air per volume of medium per minute for mycelium production by *M. anisopliae*. De la Torre and Cardenas-Cota (1996) used air flow and agitation to achieve 20% saturation for optimal production of *I. fumosorosea*. The traditional stirring paddle method of agitation in larger volume fermenters can lead to mycelium and blastospore damage through the shear forces present at required speeds. Sparging fermenters overcome this problem but may not provide sufficient oxygen efficiently in larger volumes.

Recommended temperatures are those at which the particular fungus grows best, typically 24–28 °C, although there is one report that heat induction of *I. fumosorosea* inoculant for submerged fermentation strongly increased microcycle conidiation (De La Torre and Cardenas-Cota, 1996). In their study, they exposed the inoculum, a conidial suspension, to an initial 24-h incubation at 37 °C followed by 96 h at 30 °C. The germinating conidia immediately went into blastospore rather than mycelial growth and quickly gave rise to microcycle conidia. Anderson et al. (1978) saw a parallel heat induction of microcycle conidiation in *Paecilomyces varioti* Samson. Whether this phenomenon extends to *Beauveria* spp. or *Metarhizium* spp. remains to be determined.

11.2.6.2.4. Equipment for Submerged Fermentation

Discussion of equipment used in submerged fermentation—especially in large-scale, industrial production—is beyond the scope of this review. Typically, the same fermentation equipment as is used for other microorganisms is used for the entomopathogenic Ascomycetes. The literature about submerged liquid fermentation to produce microorganisms is vast and the reader is referred elsewhere for more information (e.g. Stansbury and Whitaker, 1993; Rao, 2010).

11.2.6.2.5. Downstream Processing

The product of liquid culture is typically harvested by filtration, then processed further to stabilize the fungal material. Industrial-scale filtration uses a ribbon filter or continuous flow centrifugation. Often, diatomaceous earth is added to aid in filtration because fungal culture will often clog filters.

If the fungus material (blastospores, microcycle conidia) is to be used within a short time after harvest, it can be preserved by refrigeration. Otherwise, proper preservation of the propagules is necessary for acceptable shelf life and optimal handling characteristics in formulations. Conversion of the fungal propagules into a dry powder can be accomplished by simple air drying, freeze drying, or spray drying, but preservation of the viability of the propagules, especially blastospores, has been the greatest challenge. Most of the research regarding blastospore preservation has been directed towards *I. fumosorosea* rather than *Beauveria* or *Metarhizium*, but it is discussed here because of its relevance to the latter fungi. Typically, blastospores lose considerable viability upon drying (Inch et al., 1986; Fargues et al., 1994). Microcycle conidia are less prone to this phenomenon but do suffer viability loss.

In simple air drying, relative humidity of the drying air significantly affects desiccation tolerance of *I. fumosorosea* blastospores (Jackson and Payne, 2007). A humidity level >40% allowed significantly higher rates of initial blastospore survival after drying versus drying with lower humidity air; air with a relative humidity (RH) >50% improved the shelf-life of the air-dried blastospore preparations. Fargues et al. (1979) discovered good blastospore preservation by lyophilization using powdered milk supplemented with glycerol as cryoprotectants. Lyophilization, however, is not economically practical on a large scale.

Spray drying is a common technique used in many microbial fermentations. Stephan and Zimmermann (1998) obtained 90% blastospore/microcycle conidium viability after spray drying *B. bassiana*, *M. anisopliae*, *M. acridum* and *I. fumosorosea* when the blastospores were suspended in 20% skimmed milk powder and 2.5% sugar-beet syrup before spray drying; unprotected submerged spores were killed. Yeast extract, soluble starch, hydroxyethyl-starch and bentonite clay failed to protect the propagules. Spray dryer inlet and outlet temperature, as well as the flow rate, are also important to minimize damage to blastospores. Stephan and Zimmermann (1998) documented that 64 °C and 48 °C were the respective critical temperatures for inlet and outlet temperatures.

Of relevance to this discussion is the work of Horaczek and Viernstein (2004), who reported a detailed comparison of lyophilization, spray-drying, and fluid-bed drying for processing of aerial conidia of *Beauveria brongniartii* (Saccardo) and *M. anisopliae*, harvested with water from agar substrate. They found that *M. anisopliae* was slightly more heat resistant than *B. brongniartii* (2 min at 50 °C). In addition, conidia of both fungi suffered considerable mortality (~65%) after spray drying, even with a 60 °C/40 °C inlet/outlet temperatures. *B. brongniartii* was best preserved with lyophilization, whereas the drying phase caused considerable mortality to *Metarhizium*. *M. anisopliae* was better preserved with either of two possible preservatives, skim milk and polyvinylpyrrolidone. Finally, fluid bed drying killed both fungi, even when the inlet temperature was lowered to 60 °C.

Chen et al. (2002), in examining several drying methods for *I. fumosorosea* conidia, observed that 20- to 24-h low-vacuum (0.1 MPa), low-heat (30 °C) drying was the best, yielding only a slight loss in conidial viability. High-vacuum freeze drying, high-vacuum room temperature drying, and heating-drying at 35 °C were all deleterious for conidial viability. Thus, desiccation and temperature intolerance are the most important constraints in processing conidia, and likely even more so for blastospores and microcycle conidia.

Effort has also been concentrated on media manipulations to produce more desiccation-tolerant blastospores. For example, Cliquet and Jackson (1999) evaluated the impact of amino acids, carbohydrates, trace metals, and vitamins on the freeze-drying tolerance of *I. fumosorosea* blastospores as well as hyphal growth and sporulation. Sodium citrate or galactose as the sole carbohydrate produced more desiccation-tolerant spores but yielded lower blastospore concentrations. Media containing glucose concentrations greater than 20 g/l and 13.2–40 g casamino acid per liter supported maximal production of desiccation-tolerant blastospores of *I. fumosorosea* using air drying but not necessarily the greatest yields per se (Jackson et al., 1997). Thus, there seems to be tradeoff between blastospore yield and desiccation tolerance. In the case of the BIO1020 mycelial pellets, they were simply harvested using a 0.1-mm pore sieve plate, washed, and dried in a fluidized bed granulator (Andersch et al., 1993) or freeze dried.

11.2.6.2.6. Major Technical Problems/Solutions in Submerged Fermentation

There are several major problems with blastospore or microcycle conidia production using submerged fermentation. Rarely are the desired propagules produced in high purity. Blastospores are "contaminated" with mycelium, whereas blastospores and mycelium are copresent with microcycle conidia. For example, a mixture of cell types dilutes out the desired type and necessitates larger fermentation volumes for a specified number of the desired spore. Such mixtures make harvesting and purification difficult or render the physical nature of the harvested and stabilized product incompatible with formulation. The hydrophilic nature of blastospores and microcycle conidia requires different liquid formulations than do the hydrophobic aerial conidia. Also, with few exceptions, the shelf life of dry blastospore or microcycle preparations is shorter than that of dry aerial conidia and often commercially unacceptable without special measures, such as vacuum packaging, low-temperature storage, or some sort of encapsulation. All of these measures can add considerable cost to a commercial mycoinsecticide.

Microsclerotia production with *Metarhizium*, on the other hand, offers a viable alternative, in which sporulating granules (in soil, under a plant canopy, or in other humid, protected microhabitats) are practical. Microsclerotia seem to be considerably stable as a granular formulation and produce abundant conidia when rehydrated. Unfortunately, microsclerotia production seems to be restricted to the genus *Metarhizium*. A number of isolates of *B. bassiana* failed to produce microsclerotia under any conditions tested. (Jaronski and Jackson, unpublished data). Microsclerotia production by the genera *Isaria, Verticillium*, or *Nomuraea* has not yet been investigated.

11.2.6.3. Other Novel Methods

In addition to several atypical methods for fungus production described by Bartlett and Jaronski (1988), a novel technology using media-impregnated nonwoven fabric bands was developed and commercialized in Japan in the mid-1990s as Biolisakamakiri, using *B. brongniartii* (Higuchi et al., 1997). A U.S. patent subsequently described the methodology and specifics for mechanization and scale-up (Higuchi et al., 1996).

Wood pulp fabric (and in later modifications rayon-polypropylene or rayon polyester), typically 0.5–2 mm in thickness and sometimes laminated, is first impregnated with a liquid culture medium, then is heat dried, with the drying process sterilizing and stabilizing the media. This fabric can be stored dry. The patent describes a machine that can produce the dry, medium-impregnated fabric in long rolls. A hydrophilic polymer added to the fabric improves medium absorption and fungal conidiation. To cultivate a fungus for conidial production, the fungus is first grown in a liquid medium, which is then applied to the dry fabric. With suitable humid incubation (>80% RH, ~25 °C.) for 3–14 days, the fungus abundantly colonizes and conidiates on the surface and even within the fabric, which is then partially or completely dried and refrigerated until use. Levels of $1–2 \times 10^8$ conidia/cm^2 were obtained in this manner. The method was developed for deployment of *B. brongniartii* against tree-dwelling Chrysomelidae beetles, who have a habit of moving up and down tree trunks. A conidiated band is wrapped around the tree trunk, and the insects pass beneath or over it in their movements, coming into contact with the conidia. If the semidry band is rewetted, for instance by rain, there is another flush of conidiation. This fabric carrier was also used to obtain *Lecanicillium* conidia at similar levels (Higuchi et al., 1996). U.S. workers have adapted the fabric methodology for *Metarhizium* in an effort to evaluate its potential to combat the Asian longhorned beetle (*Anoplophora glabripennis* [Motschulsky]; Dubois et al., 2004; Hajek et al., 2006; Shanley et al., 2009). The approach was also evaluated and showed promise versus the pecan weevil (Shapiro-Ilan et al., 2009).

In parallel to the above development, Jenkins and Lomer (1994) developed a similar method for mass production of *M. acridum* conidia on cellulose cloths. In their case, however, the strategy was to remove the conidia off the cloths by washing or by mechanical separation and SIR classification. The method was abandoned in favor of grain-based solid substrate fermentation.

Bringing the liquid fermentation facility to the side of the field or in the form of a self-contained, portable fermenter was another novel concept developed in the 2000s (Jackson et al., 2004). By the late 1990s, production of beneficial bacteria on site for use as soil amendments in managed turf had been demonstrated, and that technology was adapted to entomopathogenic fungi. Jackson and his associates developed and validated a self-contained system for producing *I. fumosorosea* blastospores on site for immediate application. Prepared media concentrates were diluted in a chemically disinfected (but not completely sterilized), portable, 100-l fermenter that could be placed beside the field or glasshouse intended for application. A rehydrated, dry blastospore preparation was used as inoculum. Fermentation variables were monitored and controlled by a computer attached to the unit. Yields of 8×10^{11}/l blastospores were obtained after 48 h of fermentation starting with an inoculum level of 5×10^{9} blastospores per liter of medium. Bacterial contamination was a periodic problem, however, despite use of low pH (4). The system was designed to accommodate limited user expertise and the greatest self-containment possible. This technology is readily adaptable to *Beauveria, Metarhizium,* and the other Ascomycetes. Commercial development ensued, initially targeting high-value turf situations, but was eventually ended due to cost.

11.2.6.4. Other Ascomycetes

11.2.6.4.1. *Isaria* (*Paecilomyces*)

The Ascomycete *I. fumosorosea* has generated considerable interest and commercial exploitation for the control of several glasshouse pests in recent years. De Faria and Wraight (2007) identified seven commercial products containing *I. fumosorosea* at that time; in the United States and EU, there is currently one registered strain, Apopka 97, and a second undergoing registration review, FE9901 (U.S. Environmental Protection Agency, 2010, 2011). A second species, *I. farinosa* (Holmsk.) Fr., has been occasionally identified as an insect pathogen of potential use, but there has been little effort directed towards its mass production.

Although *I. fumosorosea* can be produced using solid substrate fermentation with substrates and methods developed for *Beauveria* and *Metarhizium*, many strains have a far blue-near ultraviolet (UV) light requirement for good conidiation (Sakamoto et al., 1985; Sanchez-Murillo et al., 2004; Kuźniar, 2011). In continuous darkness, conidiation seems to be reduced to continued, robust, vegetative growth. This situation may be the cause of the relatively low yields in at least one assessment of different grains for producing *I. fumosorosea* conidia

(Kuźniar and Krysa, 2011). A need for light:dark cycles may also be present in some *I. fumosorosea* strains (De la Torre and Cardenas-Cota, 1996). However, not all strains of *Isaria fumosorosea* display this behavior. In screening many isolates of this species for potential commercialization, Mycotech Corporation identified four to five isolates that conidiated abundantly on agar or on solid substrate in continuous darkness (Jaronski, unpublished data).

Current commercial production seems to be based on liquid fermentation. Eyal et al. (1994) described the liquid fermentation production of Apopka 97 *I. fumosorosea* using molasses, cottonseed flour, and corn steep liquor to produce blastospores and mycelium that are then encapsulated in an alginate matrix to yield prill having the potential of producing conidia upon rehydration. This system formed the basis of commercial production of PFR97®, at least in initial years of commercialization. Current processes are proprietary. FE9901 is produced by its company in submerged fermentation and consists of blastospores. Considerations in producing *Isaria* blastospores by submerged fermentation are discussed in Section 11.2.6.2.

11.2.6.4.2. *Lecanicillium*

Lecanicillium muscarium Zare and Gams and *L. longisporum* Zare and Gams (both formerly classified as *Verticillium lecanii* [Zimmerman]) have attracted some attention as biocontrol agents of Homoptera and spider mites. In their 2007 survey, De Faria and Wraight (2007) noted 16 commercial products in existence. Two have been in commercial use in Europe since the 1980s. This genus is notable in that conidia are borne in slime balls and rarely in dry chains, unlike the other Hyprocreales fungi, which produce hydrophobic conidia.

Both submerged liquid and solid substrate fermentation have been used to produce these two fungi, using methods similar to those for the genera *Beauveria* and *Metarhizium*. Both aerial conidia and submerged conidia are produced in the respective fermentation systems. There are differences in morphology, germination, and growth patterns between the two types, but there is no clear indication in the literature about their relative desiccation tolerance.

Derakhshan et al. (2008) reported that molasses-yeast broth was the best liquid medium while rice yielded the highest conidial production. A wheat bran-sugarbeet pulp mixture (9:1 w/w) has also been touted as an excellent solid substrate by Grajek (1994). Feng et al. (2000) identified rice bran to be the best medium, better than cooked rice; a bran:husk ratio of 1:1 was almost as good. Of note is that spore production in liquid media, at least shake flask culture, decreased significantly after 9 days, indicating a dynamic situation in that system. Several agricultural waste products have also been examined with a view of lowering production costs (Lopez-Llorca and Carbonell, 1998; Lopez-Llorca et al., 1999; Shi et al., 2009). Optimal temperatures for growth and sporulation in both systems are typically 20–25 °C. There is virtually no public information about harvesting and subsequent processing of *Lecanicillium*. Presumably, the aerial conidia can be washed off solid substrate because they are not hydrophobic; submerged conidia can be harvested by methods described for other Hypocreales.

11.2.6.4.3. *Hirsutella thompsonii*

Hirsutella thompsonii Fisher has undergone several commercialization efforts, primarily in the United States in the 1970s and Cuba in the 1980s. Only one product made from this fungus was identified by De Faria and Wraight (2007) in their survey. Initial data (Van Winklehof and McCoy, 1984) indicated that microcycle conidia could be produced in submerged fermentation, but by only 1 of 15 strains evaluated, and conidial germination was poor. The earliest program used liquid fermentation to produce mycelial mats, which were refrigerated then formulated on the day of application by creating an aqueous slurry for spraying (McCoy et al., 1975). The medium consisted of dextrose, yeast extract, peptone, and essential mineral salts. Subsequently, solid substrate fermentation using wheat bran was substituted by the company commercializing the fungus (Jaronski, unpublished data). The fungus was marketed for the control of eriophyid mites in citrus, but sales were terminated in the 1980s for a number of technical reasons. In 1988, Latge et al. (1988) were also able to obtain microcycle conidia from submerged fermentation, but from a strain unique in this regard.

A low level of interest has continued, primarily in South Asia, but there do not seem to be any commercial products at present. Ground maize coated with molasses was the best solid medium for conidial production evaluated by Maimala et al. (1999). Sreerama Kumar et al. (2005) focused on submerged liquid fermentation and observed that low concentrations of polyethylene glycol in dextrose peptone medium allowed production of mycelial pellets that were competent to produce conidia. A powder formulation containing both mycelia and conidia was developed in India during the 2000s, followed by two liquid formulations, but efforts to commercialize were discontinued (Sreerama Kumar, 2010).

11.2.6.4.4. Nomuraea

Nomuraea rileyi (Farl.) Samson, a pathogen of certain Lepidoptera, has undergone relatively little development and commercialization. De Faria and Wraight (2007) identified only one product, but it is no longer being marketed. The conidia were produced on erect conidiophores much like in *Beauveria*, *Isaria*, and *Metarhizium*.

Both solid substrate and submerged liquid fermentation have been explored for mass production of this fungus, primarily in Latin America, South Asia, and China. Rice has been identified as an excellent substrate, although the best conidial yields require precooking in boiling water versus simple autoclaving (Mendez et al., 2010); the boiling presumably gelatinizes the starch in the rice, making it more available to the fungus. The liquid phase of this biphasic system used molasses and yeast extract. In India, Vimala Devi (2000), Kulkarni and Lingappa (2002), and Lalitha et al. (2008) identified sorghum and rice grains as the solid substrates of choice. The second group also supplemented their grain with 1% yeast extract for better conidial production. Vimala Devi et al. (2000) determined that 2% barley extract and 1% soybean extract provided cheap carbon and nitrogen sources for the liquid production phase, whereas sorghum seeds with 0.5% yeast extract served as the best solid substrate. They also noted that good aeration of the solid substrate was essential and closed plastic bags prevented conidiation.

In evaluating a variety of agricultural products and byproducts, Tincilley et al. (2004) identified sugarcane spent wash liquid medium (still culture), rice, finger millet, and groundnut cake as suitable solid substrates. Thakre et al. (2011) continued examination of alternative substrates; rice, sorghum, and refuse raw bananas yielded the greatest conidial numbers. A liquid medium for submerged fermentation, consisting of molasses and yeast extract, was developed either for still culture or to produce inoculum for rice gruel semisolid substrate (Ramegowda et al., 2007). In Colombia, a mass production system was devised using plastic bags with autoclaved mijo grains, inoculated with 8-day-old fragments of agar grown with the sporulated fungus (Villamizar et al., 2004). A recent Chinese study indicated that illumination was essential for conidiation in at least one strain of *N. rileyi* (JianWen et al., 2009), yet Bell (1975) stated that light had no effect on growth or sporulation. This discrepancy may be due to strain differences.

11.2.6.4.5. *Aschersonia* spp.

Aschersonia spp. are specific to Aleyrodidae (whiteflies) and Coccoidea (scale insects). Because of the growing impact of the former insect, the genus—particularly the species *A. aleyrodis*—has received periodic attention by researchers (Ramakers and Samson, 1984; Fransen and van Lenteren 1993). It frequently causes severe epizootics in these insects in the tropics and subtropics. It also was one of the first fungi used in inoculative biocontrol; in the early 1900s, U.S. citrus growers introduced infected insects into their orchards. One European company briefly considered its commercialization but did not do so. An overview of the biology of *A. aleyrodis* is given by Fransen (1990). This fungus differs from the other Ascomycetes because the aerial conidia are produced within pycnidia rather than on exposed structures.

Very little work on mass production of this fungus has been published, and most research has focused on *A. aleyrodis* rather than the other species in the genus. Ibrahim et al. (1993) observed that semisolid rather than liquid media were better for growth and sporulation, with macerated pumpkin being the best of the media they tested. They also observed that the fungus sporulated well on the surface of still liquid culture, a process developed by Czech scientists for *Beauveria* (Bartlett and Jaronski, 1988). Zhu et al. (2008) studied the nutritional requirements of one strain of *A. aleyrodis* in liquid culture (for mycelial biomass) and on agar media (for conidial production). They thereby identified a semisynthetic liquid medium of soluble starch, tryptone, Ca^{2+}, and folacin, and a solid medium of lactose, tryptone, Fe^{2+}, and Vitamin B_1. Use of an orthogonal matrix method allowed Zhu and coworkers to define the concentrations of each medium component. Very recently, optimal solid substrate medium for *Achersonia placenta* Berk. was identified by response surface method analysis to contain millet, KH_2PO_4 $MgSO_4$, albeit in an agar medium (Qiu et al., 2013).

11.2.6.4.6. Culicinomyces

Culicinomyces clavisporus Couch was investigated beginning in the 1980s as a biocontrol agent for control of larval mosquitoes (Sweeney, 1985). Since then, however, interest in this fungus seem to have waned, probably because of the commercial success of *Bacillus thuringiensis israelensis* Berliner (Bti) and *B. sphaericus* Meyer and Neide for mosquito control. One exception was a small resurgence regarding its potential to control biting midges (*Culicoides* Latreille), reported by Unkles et al. (2004), but there is little published literature about the topic. Conidia have been produced on wheat-bran solid substrate or in liquid media (corn meal extract, corn steep liquor, or standard nutrient broths), but yields were very low in comparison to efficacious field rates. The fungus has been experimentally grown in 750–1000 l fermenters in Australia and harvested by filtration or centrifugation (A. Sweeney, personal communication), but no details about the process are known to the author.

Additional experimental production using the marcescent process as described earlier for Entomophthorales was evaluated by Roberts et al. (1987) and Goettel et al. (1984). Mycelia were produced in liquid peptone-yeast extract-glucose medium, harvested by filtration treated with 10% sucrose, air dried to 13% moisture, then granulated. Although freshly dried and granulated marcescent mycelium produced abundant conidia that could be stored at room temperature or 4 °C, the product lost viability within 2 weeks. Mycelium stored at −20 °C did retain viability at least for 63 days. Given the biology and ecology of *Culicinomyces,* this fungus has potential as a persistent biological control agent for mosquitoes and perhaps *Culicoides* spp., but considerable technical advances in mass production are still necessary to achieve success.

11.3. PROCESS AND QUALITY CONTROL IN MASS PRODUCTION

A quality mycoinsecticide or fungal biocontrol agent is critical for successful use. A quality mycoinsecticide demands a thorough quality assurance component to any production effort. Despite what ostensibly seems to be fine control of fermentation variables, the production process (be it liquid or solid substrate) is only semicontrolled, especially when on a large commercial scale (>1000-l liquid fermenters, 10,000 kg solid substrate fermentation batches). For example, during the initial commercial production of *B. bassiana* GHA by Mycotech in the mid-1990s, meaningful differences were observed in the shelf life of conidial powders from 16 full-scale production runs, with the time to loss of 50% of the original conidial viability (LT_{50}s) at 25 °C ranging from 180 to 700 days. By 1999, after a new production facility had been established, the half lives of conidial powders ranged from 280 to 450 days (25 °C). These powders were produced under closely controlled environmental conditions and dried to 5–7% moisture. To ensure that a quality product will be produced over the long term, it behooves the mycoinsecticide enterprise to constantly monitor agent viability, physical specifications (e.g. moisture), contamination levels, virulence, and shelf life. Vigilance is paramount. See Bateman (2007) and Jenkins et al. (1998) for discussion of specific quality control parameters and methods.

There is another problem in mass production efforts—genetic changes (degeneration) in the fungus. Butt et al. (2006) expressed the situation succinctly, "Entomogenous fungi will degenerate when continuously cultured on nutrient-rich media." Genetic changes can encompass virulence determinants, changes in colony color and gross morphology, or decline in conidial or metabolite production. Some of the fungi can give rise to morphologically different sectors in radial colonies. I have found this particularly true of *Metarhizium* spp., less so with *B. bassiana*. Butt et al. (2006) presented a detailed discussion of this topic and the interested reader should refer to this work.

Changes in virulence with repeated subculture vary widely, depending on the fungus species and strain. Butt et al. (2006) summarized all the reports up to 2006. Since then, additional studies have been performed by Hussain et al. (2010), Shah et al. (2007), Ansari and Butt (2011), Rajanikanth et al. (2011), and Safavi (2011, 2012). Under commercial conditions, bioassays of 35 standard production runs of *B. bassiana* GHA, spanning 2 years, did not reveal any significant changes in virulence (LD_{50}) for the migratory grasshopper, *Melanoplus sanguinipes* (F.), nor did eight production batches for nymphal *Bemisia tabaci* (Gennadius) (Jaronski, unpublished data).

It is possible to restore, at least partially, lost attributes by in vivo passage through an insect (Shah et al., 2005). Other examples are given in Butt et al. (2006). In the course of selecting for improved heat tolerance of a *M. anisopliae* strain via continuous growth under selective conditions, the ability to sporulate was lost in two heat-tolerant clones but regained after one passage through a grasshopper (de Crecy et al., 2009); virulence was only partially restored.

Nevertheless, care must be taken to minimize the number of in vitro conidiation cycles. Typically in an industrial situation, multiple "mother" cultures are prepared from the second or third in vitro passage from an insect and preserved by freezing with cryoprotectants to −80 °C or lyophilized. From this mother culture, enough subcultures are prepared for 6 months or a year of production cycles, and they are frozen until used. When these are exhausted, another mother culture is thawed and the cycle repeated.

There is another aspect to potential genetic changes—the number of mitotic divisions and propagules generated during a typical production cycle. Consider that in one commercial production process, an initial inoculum of 1×10^9 conidia results in 4.5×10^{16} conidia at the end of a biphasic production cycle, a 4.5×10^7-fold multiplication. Yet, no changes in the fungal restriction fragment length polymorphism pattern were noted, indicating considerable genetic stability (Jaronski, unpublished data).

11.4. CURRENT KNOWLEDGE ABOUT EFFECT OF CULTURAL CONDITIONS ON PROPAGULE ATTRIBUTES

Fungal spores (blastospores and conidia) have a proscribed environmental range that affects their infectivity (efficacy). Spores require high humidity for germination, although this requirement can be obviated by the microhabitat of the insect cuticle or phylloplane boundary layer or be modified with formulations. In addition, spores can be subjected to desiccation in the target habitat following application in an aqueous spray. Conidial germination, as well as vegetative growth within the insect, is limited in both speed and extent by high (>32 °C) and low (<15 °C) temperatures. The speed of conidial germination, even at optimal temperatures, can be critical with frequently molting insects such as aphids. UV-A (315–406 nm) and UV-B (280–315 nm) are lethal to conidia, greatly shortening their persistence in the foliar habitat and limiting efficacy when the target insects are more likely to acquire spores from treated habitat rather than direct spray. See Jaronski (2010) for a full discussion of ecological considerations in the inundative use of entomopathogenic fungi.

The different fungus species and isolates within each species vary in their tolerances to these environmental factors (Devi et al., 2005), as well as in virulence and mass production potential. Typical development programs therefore seek to identify the best fungus for a particular use and target, and formulations are devised to accommodate fungal deficiencies to a great or lesser degree of success. In recent years, there have been attempts to improve efficacy through genetic modifications. A developmental program can therefore expend considerable effort in screening candidate fungi; emphasizing virulence, spore production, and shelf life; and optimizing fermentation variables to maximize spore production.

However, there may be another approach, at least to improve an already acceptable fungal agent: manipulation of fermentation conditions to improve environmental stress tolerance. Magan (2001) posed four key questions on this subject: "(1) can one manipulate the physiology of non-xerophilic/tolerant fungi to accumulate useful endogenous reserves into inocula for improved environmental stress tolerance?; (2) would this result in improved germination/growth under environmental stress?; (3) can this improve the establishment of inocula and conserved biocontrol potential in the field? and (4) does ecophysiological manipulation have a role in improving the production and quality of inocula?" Magan (2001) pointed out that there are xerotolerant/xerophilic fungi able to tolerate a range of water availability that inhibits the entomopathogenic species, which are inhibited by water activities below 0.95–0.93 a_w. The ability to tolerate water stress conditions in these fungi is associated with compatible solutes within the spores. The potential for such manipulations was identified by Matewele et al. (1994), who observed that mutants of *M. anisopliae* and *I. farinosa*, which were able to germinate and grow at lower water activities than parental strains and which were subsequently grown on low water activity medium (a_w=0·969), were more virulent against green leafhopper than the parental strains. Also, *B. bassiana* conidia, having an increased glycerol and erythritol content, germinated in low a_w (0.935) media but conidia having low glycerol and erythritol failed to germinate below 0.951 a_w (Hallsworth and Magan, 1995). In addition, Magan (2001) noted that *M. anisopliae* conidia from an insect had significantly greater polyol content and different sugar/polyol ratios than conidia produced on an agar medium. These data strongly suggest that endogenous solutes can be manipulated. Can manipulation affect stress tolerance and improve efficacy? Although there is not an abundance of data on this subject, what does exist poses intriguing possibilities.

11.4.1. Age of Conidia

Hall et al. (1994) observed that young conidia produced by all the isolates of *M. anisopliae* and *Lecanicillium* tested germinated faster than older ones, whereas *B. bassiana* conidia germinated at the same rate, regardless of culture age. The impact of culture age on conidial germination appeared to be species- and strain-dependent. More recently, Smith and Edgington (2011) showed that the capacity of conidia to withstand water stress developed by low water activity was related to the age of fungal cultures, implying that prolonged production cycle (conidial ageing) may also improve the fitness of conidia.

"Old" spores (collected after 14 days of fungal growth) of *B. bassiana* and *M. anisopliae* produced on agars composed of whey permeate or millet were more thermotolerant than conidia from quarter-strength Sabouraud dextrose yeast agar, although the differences were inconsistent among the two isolates of each species (Kim et al., 2010c). A mixed message was described by Rajanikanth et al. (2011), whereby conidia of six strains of *B. bassiana* from 14-day-old cultures had greater virulence for larval *Spodoptera litura* F. than conidia from 7-, 21-, and 28-day-old cultures.

A clue to what may be going on with age-related differences in conidia is offered by Kim et al. (2010c), who observed that two isolates each of *B. bassiana* and *M. anisopliae* had conidia of two types with different degrees of hydrophobicity, termed young (7-day-old) versus old (mature, 14-day-old) conidia. Thermotolerance seems to have been directly associated with hydrophobicity (age) of the conidia, although there were differences among the isolates and media used to generate the conidia. Quarter-strength Sabouraud dextrose yeast agar had the most pronounced difference between the two spore ages.

11.4.2. Conidia Produced under Certain Nutrient Conditions or under Osmotic Stress

Conidia of *M. anisopliae* and *I. farinosa*, produced on agar media adjusted to an a_w of 0.96, were significantly more infective for *Galleria mellonella* (L.) larvae at 86% and 78% (Hallsworth and Magan, 1994b). The series of studies by Hallsworth and Magan (1994a, 1994b, 1994c, 1995, 1996) were significant in that they lay the foundation for manipulation of fermentation media to change solute content of conidia and thereby improve desiccation tolerance. There are inter- and intraspecies differences in the response to particular growth conditions (Hallsworth and Magan, 1999). Overall, however, lowering a_w to less than 0.98 reduces mycelial growth and conidial yield, so there is a fine line between obtaining xerotolerance and not seriously affecting spore yield. Also, it should be kept in mind that these studies were conducted with agar media; it may be difficult to control a_w within fine tolerances on a solid substrate, especially when in large quantities under commercial conditions.

Lane et al. (1991a, 1991b) examined the effect of C:N ratios on blastospore production by *B. bassiana*. They observed that inorganic nitrogen-limited blastospores had a longer shelf-life than blastospores from carbon-limited media, which was accompanied by differences in morphology and endogenous reserves. In addition, although there were no differences in the LC_{50}s of carbon- and nitrogen-limited blastospores for a leafhopper, the virulence (LT_{50}) of nitrogen-limited blastospores was greater than their carbon-limited counterparts and the former adhered more firmly to insect cuticle than the latter.

Virulence of *M. anisopliae* conidia seems to be slightly affected by the C:N ratio of the (agar) medium, with the most virulent conidia having an endogenous C:N ratio <5.2:1, per the claims of Shah et al. (2005). The conidial C:N ratios were affected by the agar medium used, with a glucose:peptone medium having a C:N ratio of >35:1, or an osmotically stressful medium, via KCl addition, causing the greatest endogenous C:N ratio. However, virulence for *Tenebrio molitor* L. was really unaffected from a biological perspective, varying from 3.5 to 4.1 days for one strain and 3.5–3.9 days for a second strain. A problem with interpreting these data is that virulence, as time to 50% mortality, is an expression of mycelial development rather than conidial fitness per se.

In a related study with *B. bassiana*, Safavi et al. (2007) did not observe the effect of endogenous C:N ratio as did Shah et al. (2005) with *M. anisopliae*. Although the endogenous C:N ratio was affected by the different media, Safavi and his associates could not see a clear relationship among C:N ratio, germination rate, conidial PR1 activity, and virulence in their *B. bassiana* isolates. Osmotic stress caused the lowest conidial production, however.

Rangel et al. (2004) noted that conidia of two isolates of *M. anisopliae* obtained from insect cadavers were more sensitive to UV-B irradiation than those produced on a rich artificial medium, and they also germinated more slowly. The artificial medium also had an effect with conidia from rice substrate or two agar media having more UV-B tolerance than conidia on potato dextrose yeast agar. Rangel et al. (2006) subsequently observed that UV-B tolerance of *Metarhizium robertsii* (= *anisopliae*) conidia was increased at least twofold when the fungus was grown on agar medium containing nonpreferred carbon sources, such as fructose, galactose, or lactose versus a preferred carbon source, such as dextrose. With lactose, endogenous trehalose and mannitol accumulated to a significantly greater level. However, conidial yields were reduced, which could be a considerable disadvantage.

In follow-up work, Rangel et al. (2008a, 2008b) examined the effect of growing *M. robertsii* under different stress conditions. Conidia produced on a minimal agar medium and minimal medium supplemented with lactose had the faster germination and greater bioassay virulence for *T. molitor* than conidia from a rich medium (potato dextrose agar with yeast extract). These researchers also observed that conidia produced under conditions of carbon or nitrogen starvation possessed significantly greater heat and UV-B tolerance than conidia produced on nutrient-rich potato dextrose agar plus yeast extract (PDAY). The greater tolerances were associated with the greatest accumulation of trehalose and mannitol within the conidia. Similar results were seen with an osmotically stressful agar medium (0.8 M KCl or NaCl, a_w unknown). Conidial yield, however, could be severely affected by stressful conditions during mycelial growth.

Thermotolerance (conidial viability after exposure to 48 °C for 30 min) seems also to have been significantly increased when glucose (4% optimal) or starch (1%) were the carbon sources in agar media (Ying and Feng, 2006). When ≤50 µg/ml of Mn^{+2} was added, thermotolerance was extremely good and was greatly increased with sucrose and Fe^{+3} (Ying and Feng, 2006). The pH had some effect; thermotolerance was greatest when the fungus was grown at pH 5–6; osmotic stress, in the form of KCl, had an adverse effect. Amendments, such as Mn^{+2} or carbon supplementation with a grain-based solid substrate, are feasible, but again, whether this phenomenon extends to other strains or species remains to be elucidated.

An insight into the mechanisms of thermotolerance may be obtained from Leng et al. (2011). Using RNAi, they demonstrated that trehalose levels in *M. acridum* conidia were closely associated with tolerance to heat stress. There also seems to be a relationship between thermotolerance and formic-acid-extractable proteins in aerial conidia of *B. bassiana* and *I. fumosorosea* (Ying and Feng, 2004). The formic-acid-extractable proteins are affected by the nature of the carbon source, with glucose and sucrose causing significantly better thermotolerance than starch (in an agar medium).

Investigations about a_w effects can also be extended to solid substrate media. The a_w of rice substrate (modified by use of different normalities and amounts of HCl) affected the polyol content of *B. bassiana* conidia and the relative proportions of mannitol, arabitol, glycerol, and erythritol (Tarocco et al., 2005). The last two are the most biologically important: high intraconidial concentrations of glycerol and erythritol were associated with ability of conidia to germinate at much lower a_w than otherwise (Hallsworth and Magan, 1995). The optimal a_w for glycerol/erythritol accumulation was 0.980 +/− 0.005. Conidial production, however, was severely reduced by 95–99.8% (optimal a_w for conidial production was 0.999). A compromise a_w yielding acceptable polyol levels still resulted in a 72% reduction in yield.

Kim et al. (2011) discovered that millet grain as a solid substrate produced *B. bassiana* and *M. anisopliae* conidia that were more thermotolerant (to 43–47 °C) than conidia from agar-based media. Their data indicate that certain substances in the millet contributed to the observed thermotolerance. Use of millet supplemented with plant-derived oils, especially corn oil, further induced greater heat tolerance of *B. bassiana* conidia (45 °C for 90 min) (Kim et al., 2010a).

In a subsequent paper, Kim et al. (2010b) demonstrated that, for *I. fumosorosea*, ground corn solid substrate produced conidia with the most heat tolerance, compared to rice, soybean of red kidney bean substrates, and corn oil supplementation of the corn meal increased that heat tolerance (Kim et al., 2010c). They hypothesized that unsaturated fatty acids, such as linoleic acid and oleic acid, in the corn oil possibly explained the improved thermotolerance.

The addition of 1–4 mM salicylic acid, a plant cell-signaling metabolite, to an agar medium yielded conidia of one *M. robertsii* isolate with a doubling in heat tolerance, but not UV-B tolerance (Rangel et al., 2012). It should be noted that the amendment did reduce conidial yield somewhat and a defined agar base was used in the study. Whether the same phenomenon would result in either solid substrate or submerged liquid fermentation—or with other *Metarhizium* strains, species, and *B. bassiana*—remains to be determined. Many of the fermentation manipulations caused a significant decrease in spore production, to the extent that the manipulations were self-defeating. Nevertheless, further research may find a satisfactory compromise between stress tolerance and spore production.

11.4.3. Conidia Produced after Photoirradiation

Conidia produced from mycelium of *M. robertsii*, irradiated with UV-A while growing on PDAY, had somewhat elevated virulence levels similar to that of conidia produced on nutritionally minimal medium, but their germination rate was not increased, nor were UV-B resistance and thermotolerance improved (Rangel et al., 2008a, 2008b). This approach needs further investigation to determine the applicability of the observations to other strains and other species. Nevertheless, extension of this approach to solid substrate fermentation seems limited, although it may be possible when plastic bags are used.

11.5. THE CHALLENGE IN MASS PRODUCTION OF ENTOMOPATHOGENIC FUNGI

The fundamental consideration in potential commercialization of any of these fungi is whether an efficacious product can be developed and produced cheaply enough to compete with existing controls, especially chemicals. To return an acceptable profit on the research and development investment, low-cost mass production of an entomogenous fungus is only one of a number of technical constraints. Yet, mass production is basic to commercialization.

Bartlett and Jaronski (1988) examined the capacities of the different technologies at the time and concluded that high-efficiency biphasic solid substrate, such as that now practiced by at least one company in the United States, was the only commercially feasible system for very large mycoinsecticide production needs. For example, let us look at a potential U.S. market, European corn borer, *Ostrinia nubilaris* (Hübner), in maize, ignoring for the moment the dominance of Bt-maize. In 2012, there were 26 million hectares of maize subject to attack by European corn borer and corn earworm, *Helicoverpa zea* (Boddie) (U.S. Department of Agriculture, Economic Research Service, 2012; U.S. Department of Agriculture, 2012). Assuming 5% of the corn market is open to mycoinsecticide use, that is 1.3 million hectares. Much of the published literature on efficacy indicates that a rate of $1–2.5 \times 10^{13}$ conidia per hectare is needed for fungal products (De Faria and Wraight, 2007); however, *M. acridum* has a use rate of 5×10^{12} per hectare for locust control. If the use rate of mycoinsecticide is the more optimistic 5×10^{12} conidia per hectare, then a company would need to produce 6.5×10^{18} conidia for one growing season. With the public Mycotech/Laverlam yields of 2.6×10^{13} conidia/kg, that need would require 250,000 kg of substrate (25 production runs at Laverlam). Lower yields, such as those reported in the literature, or higher efficacious rates of fungus would require much larger amounts (Table 11.4).

A comparison of the economics and efficiencies of the two processes for producing *L. muscarium* was presented by Ravensberger (2011). For the manufacturer, liquid submerged fermentation yielded 2.1×10^{13} spores/l within 7 days in 1000-l fermenters, whereas solid substrate fermentation in specialized equipment of 100-kg substrate capacity yielded 8.1×10^{12} conidia/kg. Ravensberger (2011) estimated that new solid substrate equipment would cost €750,000 in 2007 versus €300,000 for a 1000-l liquid fermenter that had four times the production capacity. Ignoring quality of spores, their shelf life, losses to desiccation, and production of unwanted metabolites in submerged fermentation, that process may be more practical.

In contrast, for smaller-scale production serving local needs, solid substrate fermentation (e.g. in plastic bags as is practiced in Cuba, Brazil, and other Latin American countries) may be practical. About $5–10 \times 10^{16}$ spores would be needed for 5000 ha. Using the data in Table 11.4, 16,000–33,000 kg substrate (with a yield of 1.5×10^{12} per kilogram) would meet that need. If the highest yield obtained on a laboratory scale, 2×10^{13} per kilogram rice (Dorta et al., 1996), was operationally possible, then the substrate need would decrease to 1200–2400 kg, which is much more feasible.

For inoculative release of fungi in a biocontrol campaign (e.g. with Entomophthorales), such considerations are avoided; only small numbers of spores

TABLE 11.4 Estimates of Production Capacity Needed to Supply Enough Fungus Propagules to Treat 1.3×10^6 ha of Maize in the United States (5% of Total Potential Maize Market) Based on Documented Production Yields. The Total Spore Need Is 6.5×10^{18} spores, Based on a Use Rate of 5×10^{12} Spores/ha

Fermentation System	Yield per Unit	Fungus (Source)	Production Need
High-efficiency biphasic solid substrate fermentation	2.5×10^{13} conidia/kg substrate	*B. bassiana* GHA (Bradley et al., 1992)	2.6×10^5 kg
Low-efficiency (bag) solid substrate fermentation	1.5×10^{12} conidia/kg	*M. acridum* IMI330189 (Jenkins et al., 1998)	4.3×10^6 kg
Submerged fermentation	1×10^{12} blastospores/l	*I. fumosorosea* (Jackson et al., 1997)	6.5×10^6 l
Liquid surface culture	1×10^{14} conidia/m^2 of surface	*B. bassiana* (Bartlett and Jaronski, 1988)	6.5×10^4 m^2

are needed because the fungus will reproduce and proliferate, ideally causing an epizootic. In inundative applications, the fundamental challenge is to reduce fungus use rates. Although the typical application rate is on the order of $1–2 \times 10^{13}$ spores per hectare, lower application rates have been obtained by the selection of a more efficacious (yet still productive) strain. The official use rate for *M. acridum* against locusts is 5×10^{12} per hectare, which is already a lower rate than the generally accepted level for a mycoinsecticide, and good efficacy has been obtained with rates as low as 1.25×10^{12} per hectare. One-fourth the fungus per acre has a major impact on production needs.

In the past few years, efforts have been made towards creating fungal strains that are genetically modified for greater virulence (reviewed by St. Leger and Wang, 2010). Some of these transformations have greatly increased the efficacy of the fungus; for example, the incorporation of scorpion neurotoxin increased toxicity for hornworm 22-fold (Wang and St. Leger, 2007). If regulatory agencies will allow the use of such transformed fungi, application rates could be considerably lowered. Use rates per unit area have also been lowered by concentrating the spores into a narrow, targeted zone by modification of application equipment and methods to deliver more spores exactly where needed (Jaronski, 2010). If applied in a broadcast spray, 5×10^{12} conidia per hectare would result in a level of 5×10^4 conidia/cm^2 of surface area. With a 12.5-cm banded application of each row of plants, the conidial levels become 2.4×10^5/cm^2—a fivefold increase at the same rate per hectare, confined to the actual habitat of the target insect, or a potential fivefold reduction in the rate of fungus per hectare if the 5×10^4 conidia/cm^2 of surface area was efficacious. For example, Wraight and Ramos (2002) were able to increase the conidial deposition on leaf undersides 6- to 30-fold by modifying the arrangement of spray nozzles. Similarly, spores can be placed in the path of insects, such as on fiber bands wrapped around tree trunks to control Asian longhorned beetle (Shanley et al., 2009) or in compact spray bands in front of migrating locusts, affording considerable economies. Therefore, the situation with mass production of entomopathogenic fungi is not a dismal one. Considerable potential exists for fungi to be significant tools in managing insect pest populations.

REFERENCES

Agriculture and Agri-Food Canada (AAFC), 2012. Directory of Biopesticides in Organization for Economic Cooperation and Development Countries. http://www4.agr.gc.ca/AAFC-AAC/display-afficher.do?id=1315941819381&lang=eng.

Alves, S.B., Pereira, R.M., 1989. Production of *Metarhizium anisopliae* (Metsch.) Sorok and *Beauveria bassiana* (Bals.) Vuill in plastic trays. Ecossistema 14, 188–192.

Andersch, W., Hartwig, J., Homeyer, B., Stenzel, K., 1993. Self-supporting carrier-free cell granulates for combating pests and treating plants. U.S. Patent 5,418,164.

Anderson, J.G., Smith, J.E., 1971. The production of conidiophores and conidia by newly germinated conidia of *Aspergillus niger* (microcycle conidiation). J. Gen. Microbiol. 69, 185–197.

Anderson, J.G., Aryee, V., Smith, J.E., 1978. Microcycle conidiation in *Paecilomyces varioti*. FEMS Microbiol. Lett. 3 (1), 57–60.

Ansari, M.A., Butt, T.M., 2011. Effects of successive subculturing on stability, virulence, conidial yield, germination and shelf-life of entomopathogenic fungi. J. Appl. Microbiol. 110 (6), 1460–1469.

Aquino, M., Cavalcanti, V., Sena, R., Queiroz, G., 1975. Novel technology for the production of the fungus *Metarhizium anisopliae*. Bol. Tec. CODECAP 4, 11–31.

Aquino de, L.M.N., Vital, A.F., Cavalcanti, V.L.B., Nascimento, M.G., 1977. Culture of *Metarhizium anisopliae* (Metchn.) Sorokin in polypropylene bags. Bol. Tec. CODECAP 5, 7–11.

Arcas, J.A., Díaz, B.M., Lecuona, R.E., 1999. Bioinsecticidal activity of conidia and dry mycelium preparations of two isolates of *Beauveria bassiana* against the sugarcane borer *Diatraea saccharalis*. J. Biotechnol. 67 (2–3), 151–158.

Arzumanov, T., Jenkins, N., Roussos, S., 2005. Effect of aeration and substrate moisture content on sporulation of *Metarhizium anisopliae var. acridum*. Proc. Biochem. 40 (3/4), 1037–1042.

Auria, R., Hernandez, S., Raimbault, M., Revah, S., 1990. Ion exchange resin: a model support for solid state growth fermentation of *Aspergillus niger*. Biotechnol. Tech. 4 (6), 391–396.

Bartlett, M.C., Jaronski, S.T., 1988. Mass production of entomogenous fungi for biological control of insects. In: Burge, M.N. (Ed.), Fungi in Biological Control Systems. Manchester University Press, Manchester, UK, pp. 61–85.

Bateman, R., 2006. Transferring Mycopesticides from the Laboratory into the Field: The Importance of Enabling Technologies. Downloaded from http://www.mycoharvester.info/Enabling_technology.PDF, January 2, 2013.

Bateman, R., 2007. Production Processes for Anamorphic Fungi. IPARC, Silwood Park, Ascot, Berks. http://www.dropdata.net/mycoharvester/Mass_production.PDF (accessed 24.11.12.).

Bateman, R., 2012. Large-Scale Spore Extraction Unit (MycoHarvester Mk 3). Downloaded from http://www.mycoharvester.info/MH3_2012.pdf December 1, 2012.

Beauvais, A., Latge, J.P., 1988. A simple medium for growing entomophthoralean protoplasts. J. Invertebr. Pathol. 51, 175–178.

Bell, J.V., 1975. Production and pathogenicity of the fungus *Spicaria rileyi* from solid and liquid media. J. Invertebr. Pathol. 26 (1), 129–130.

Billings, F.H., Glenn, P.A., 1911. Results of the artificial use of the white-fungus disease in Kansas. USDA Bur. Entomol. Bull. 107, 58.

Bosch, A., Yantorno, O., 1999. Microcycle conidiation in the entomopathogenic fungus *Beauveria bassiana* Bals. (Vuill.). Process Biochem. 34, 707–716.

Bradley, C.A., Black, W.E., Kearns, R., Wood, P., 1992. Role of production technology in mycoinsecticide development. In: Leatham, G.E. (Ed.), Frontiers in Industrial Microbiology, Chapman & Hall, New York, pp. 160–173.

Breukelen, F.R. van, Haemers, S., Wijffels, R.H., Rinzema, A., 2011. Bioreactor and substrate selection for solid-state cultivation of the malaria mosquito control agent *Metarhizium anisopliae*. Proc. Biochem. 46 (3), 751–757.

Butt, T.M., Wang, C.S., Shah, F.A., Hall, R., 2006. Degeneration of entomogenous fungi. In: Eilenberg, J., Hokkanen, H.M.T. (Eds.), An Ecological and Societal Approach to Biological Control, Springer, Dordrecht, pp. 213–226.

Calderón, A., Fraga, M., Lujan, M., Sanchez, E., Pavis, C., Kermarrec, A., 1990. Reproduction of *Beauveria bassiana* (Bals.) Vuill. and *Metarhizium anisopliae* (Metsch.) Sor. on industrial by-products. Caribbean meetings on biological control. Guadeloupe (Antilles françaises), Novembre 5–7, 1990, 325–328.

Calderón, A., Fraga, M., Carreras, B., 1995. Production of *Beauveria bassiana* by solid state fermentation (SSF). Rev. Prot. Veg. 10 (3), 269–273.

Carruthers, RI., Walsh, S.R., Ramos, M.E., Larkin, T.S., Hostetter, D.L., Soper, R.S., 1997. The *Entomphaga grylli* (Fresenius) Batko species complex: its biology, ecology, and use for biological control of pest grasshoppers. Mem. Can. Entomol. Soc. 171, 329–353.

Chapman, H.C., 1985. Ecology and use of *Coelomomyces* species in biological control: a review. In: Couch, J.N., Bland, C.E. (Eds.), The Genus Coelomomyces, Academic Press, Orlando, pp. 82–91.

Chen, Y.T., Feng, M.-G., 2002. Optional drying procedures for mass production of *Paecilomyces fumosoroseus* conidia for insect control. Mycosystema 21 (4), 565–572.

Cherry, A., Jenkins, N., Heviefo, G., Bateman, R., Lomer, C., 1999. A West African pilot scale production plant for aerial conidia of Metarhizium sp. for use as a mycoinsecticide against locusts and grasshoppers. Biocontrol Sci. Technol. 9, 35–51.

Chong-Rodríguez, M.J., Maldonado-Blanco, M.G., Hernández-Escareño, J.J., Galán-Wong, L.J., Sandoval-Coronado, C.F., 2011. Study of *Beauveria bassiana* growth, blastospore yield, desiccation-tolerance, viability and toxic activity using different liquid media. Afr. J. Biotechnol. 10 (30), 5736–5742.

Claro, O.E., 2006. Low Technology Systems for Mass Production of Biopesticides Based on Entomopathogenic Fungi and Fungal Antagonists. Report, Instituto de Investigaciones de Sanidad Vegetal (INISAV), Havana, Cuba, Downloaded, from http://www.inisav.cu/OtrasPub/METODOS%20ARTESANALES%20DE%20PRODUCCI%C3%93N%20DE%20BIOPLAGUICIDAS.pdf, December 1, 2012.

Cliquet, S., Jackson, M.A., 2005. Impact of carbon and nitrogen nutrition on the quality, yield, and composition of blastospores of the bioinsecticidal fungus *Paecilomyces fumosoroseus*. J. Indian Microbiol. Biotechnol. 32, 204–10.

Crangle, R.D., 2011. Diatomite. 2011 Minerals Yearbook. U.S. Geological Survey, Washington DC. http://minerals.usgs.gov/minerals/pubs/commodity/diatomite/myb1-2011-diato.pdf. (accessed 04.12.12.).

de Crecy, E., Jaronski, S.T., Lyons, B., Lyons, T.J., Keyhani, N.O., 2009. Directed evolution of a filamentous fungus to thermotolerance. BMC Biotechnol. 9, 74. http://dx.doi.org/10.1186/1472-6750-9-74.

De Faria, M., Wraight, S., 2007. Mycoinsecticides and Mycoacaricides: a comprehensive list with worldwide coverage and international classification of formulation types. Biol. Control 43, 237–256.

Delalibera Jr. I., Hajek, A.E., Humber, R.A., 2003. Use of cell culture media for cultivation of the mite pathogenic fungi *Neozygites tanajoae* and *Neozygites floridana*. J. Invertebrate Pathol. 84, 119–127.

De la Torre, M., Cardenas-Cota, H.M., 1996. Production of *Paecilomyces fumosoroseus* conidia in submerged culture. Entomophaga 41 (3/4), 443–453.

Deng, X.J., Zhou, G.Y., Liu, J.A., Yan, R.K., Chen, C.L., Yang, Z.L., 2011. Optimization of solid-state fermentation conditions for *Beauveria bassiana* Bb III 22. Acta Agric. Univ. Jiangxiensis 33 (6), 1228–1235.

Derakhshan, A., Rabindra, R.J., Ramanujam, B., Rahimi, M., 2008. Evaluation of different media and methods of cultivation on the production and viability of entomopathogenic fungi, *Verticillium lecanii* (Zimm.) Viegas. Pak. J. Biol. Sci. 11 (11), 1506–1509.

Desgranges, C., Vergoignan, C., Lereec, A., Riba, G., Durand, A., 1993. Use of solid state fermentation to produce *Beauveria bassiana* for the biological control of European corn borer. Biotechnol. Adv. 11, 577–587.

Devi, K.U., Sridevi, V., Mohan, C.M., Padmavathi, J., 2005. Effect of high temperature and water stress on in vitro germination and growth in isolates of the entomopathogenic fungus *Beauveria bassiana* (Bals.) Vuillemin. J. Invertebr. Pathol. 88 (3), 181–189.

Dhar, P., Kaur, G., 2011. Response surface methodology for optimizing process parameters for the mass production of *Beauveria bassiana* conidiospores. Afr. J. Microbiol. Res. 5 (17), 2399–2406.

Domenech, F., Leon, M., Rodriguez-Leon, J.A., Rodriguez, D.E., Alfonso, A., Sanchez, D., 1998. Obtaining a medium for spore production of *Metarhizium anisopliae* in solid state fermentation. Rev. ICIDCA Sobre los Derivados de la Caña de Azúcar 32 (1), 15–22.

Domnas, A.J., Srebro, J.P., Hicks, B.F., 1977. Sterol requirement for zoospore formation in the mosquito-parasitizing fungus, *Lagenidium giganteum*. Mycologia 69, 875–886.

Dorta, B., Ertola, R.J., Arcas, J., 1996. Characterization of growth and sporulation of *Metarhizium anisopliae* in solid-substrate fermentation. Enzyme Microb. Technol. 19 (6), 434–439.

Dubois, T., Li, Z., Jiafu, H., Hajek, A.E., 2004. Efficacy of fiber bands impregnated with *Beauveria brongniartii* cultures against the Asian longhorned beetle, *Anoplophora glabripennis* (Coleoptera: Cerambycidae). Biol. Control 31 (3), 320–328.

Dunphy, G.B., Nolan, R.A., MacLeod, D.M., 1978. Comparative growth and development of two protoplast isolates of *Entomophthora egressa*. J. Invertebr. Pathol. 31, 267–269.

Durand, A., 1998. Solid state fermentation. Biofutur 181, 41–43.

Durand, A., 2003. Bioreactor designs for solid state fermentation. Biochem. Eng. J. 13 (2/3), 113–125.

Durand, A., Renaud, R., Maratray, J., Almanza, S., Diez, M., 1996. INRA-Dijon reactors for solid state fermentation: designs and applications. J. Sci. Ind. Res. 55 (5–6), 317–332.

Eiben, U., Lüth, P., 2006. Development of Novel Fungal Biocontrol Agents. orgprints.org/7717/1/short_paper,_prophyta,14.03.06.PDF, Downloaded December 27, 2012.

Eyal, J., Walter, J.F., Osborne, L., Landa, Z., 1994. Method for production and use of pathogenic fungal preparation for pest control. U.S. Patent 5,360,6067.

Fargues, J., Robert, P.H., Reisinger, O., 1979. Formulation des productions de masse de l'hyphomycete entomopathogene *Beauveria* en vue des applications phytosanitaires. Ann. Zool. Ecol. Anim. 11, 247–257.

Fargues, J., Maniania, N.K., Delmas, J.C., 1994. Infectivity of propagules of *Paecilomyces fumosoroseus* during in vitro development to *Spodoptera frugiperda*. J. Invertebr. Pathol. 64, 173–178.

Fargues, J., Smits, N., Vidal, C., Vey, A., Vega, F., Mercadier, G., Quimby, P., 2002. Effect of liquid culture media on morphology, growth, propagule production, and pathogenic activity of the Hyphomycete, *Metarhizium flavoviride*. Mycopathologia 154 (3), 127–138.

Faria, M., Hajek, A.E., Wraight, S.P., 2009. Imbibitional damage in conidia of the entomopathogenic fungi *Beauveria bassiana*, *Metarhizium acridum*, and *Metarhizium anisopliae*. Biol. Control 51 (3), 346–354.

Faria, M., Hotchkiss, J.H., Hajek, A.E., Wraight, S.P., 2010. Debilitation in conidia of the entomopathogenic fungi *Beauveria bassiana* and *Metarhizium anisopliae* and implication with respect to viability determinations and mycopesticide quality assessments. J. Invertebr. Pathol. 105 (1), 74–83.

Feng, M.G., Poprawski, T.J., Khachatourians, G.G., 1994. Production, formulation and application of the entomopathogenic fungus *Beauveria bassiana* for insect control: current status. Biocontrol Sci. Technol. 4 (1), 3–34.

Feng, K.C., Liu, L., Tzeng, M.Y., 2000. *Verticilliun lecanii* spore production in solid-state and liquid-state fermentations. Bioprocess Eng. 23 (1), 25–29.

Fransen, J.J., 1990. Natural enemies of whiteflies: fungi. In: Gerling, D. (Ed.), Whiteflies: Their Bionomics, Pest Status and Management, Intercept, Andover, UK, pp. 187–210.

Fransen, J.J., van Lenteren, J.C., 1993. Host selection and survival of the parasitoid *Encarsia formosa* on greenhouse whitefly, Trialeurodes vaporariorum, in the presence of hosts infected with the fungus *Aschersonia aleyrodis*. Entomol. Exp. Appl. 69 (3), 239–249.

Freimoser, F.M., Grundschober, A., Tuor, U., Aebi, M., 2003. Regulation of hyphal growth and sporulation of the insect pathogenic fungus *Entomophthora thripidum* in vitro. FEMS Microbiol. Lett. 222 (2), 281–287.

Gervais, P., Molin, P., 2003. The role of water in solid-state fermentation. Biochem. Eng. J. 13 (2/3), 85–101.

Goettel, M.S., Sigler, L., Carmichael, J.W., 1984. Studies on the mosquito pathogenic Hyphomycete *Culicinomyces clavisporus*. Mycologia 76 (4), 614–625.

Grajek, W., 1994. Sporogenesis of the entomopathogenic fungus *Verticillium lecanii* in solid-state cultures. Folia Microbiol. 39 (1), 29–32.

Gray, S.N., Markham, P.A., 1997. Model to explain the growth kinetics of the aphid-pathogenic fungus *Erynia neoaphidis* in liquid culture. Mycol. Res. 101, 1475–1483.

Grimm, C., 2001. Economic feasibility of a small scale production plant for entomopathogenic fungi in Nicaragua. Crop Prot. 20, 623–630.

Grundschober, A., Freimoser, F.M., Tuor, U., Aebi, M., 2001. In vitro spore formation and completion of the asexual life cycle of *Neozygites parvispora*, and obligate biotrophic pathogen of thrips. Microbiol. Res. 156, 247–257.

Guillon, M., 1997. Production of biopesticides: scale up and quality assurance. Microbial insecticides novelty or necessity? Proc. 1997 Brit. Crop Prot. Council Symp. No. 68, pp. 151–162.

Hajek, A.E., 1997. Ecology of terrestrial fungal entomopathogens. Adv. Microbiol. Ecol. 15, 193–249.

Hajek, A.E., Huang, B., Dubois, T., Smith, M.T., Li, Z., 2006. Field studies of control of *Anoplophora glabripennis* (Coleoptera, Cerambycidae) using fiber bands containing the entomopathogenic fungi *Metarhizium anisopliae* and *Beauveria* brongniartii. Biocontrol Sci. Technol. 16, 329–343.

Hajek, A., Papierok, B., Eilenberg, J., 2012. Research methodology for the Entomophthorales. In: Lacey, L.A. (Ed.), Manual of Techniques in Invertebrate Pathology, second ed. Academic Press, New York, pp. 285–316.

Hajek, A.E., Plymale, R.C., 2010. Variability in azygospore production among *Entomophaga maimaiga* isolates. J. Invertebr. Pathol. 104,157–159.

Hall, R.A., Peterkin, D.D., Ali, B., Lopez, V., 1994. Influence of culture age on rate of conidiospore germination in four deuteromycetous entomogenous fungi. Mycol. Res. 98, 763–768.

Hallsworth, J.E., Magan, N., 1994a. Effect of carbohydrate type and concentration on polyhydroxy alcohol and trehalose content of conidia of three entomopathogenic fungi. Microbiology 140 (10), 2705–2713.

Hallsworth, J.E., Magan, N., 1994b. Improved biological control by changing polyols/trehalose in conidia of entomopathogens. Proc. Brighton Crop Prot. Conf.—Pests and Diseases, pp. 1091–1096.

Hallsworth, J.E., Magan, N., 1994c. Effects of KCl concentration on accumulation of acyclic sugar alcohols and trehalose in conidia of three entomopathogenic fungi. Lett. Appl. Microbiol. 18 (1), 8–11.

Hallsworth, J.E., Magan, N., 1995. Manipulation of intracellular glycerol and erythritol enhances germination of conidia at low water availability. Microbiology 141 (5), 1109–1115.

Hallsworth, J.E., Magan, N., 1996. Culture age, temperature, and pH affect the polyol and trehalose contents of fungal propagules. Appl. Environ. Microbiol. 62 (7), 2435–2442.

Hallsworth, J.E., Magan, N., 1999. Water and temperature relations of growth of the entomogenous fungi *Beauveria bassiana, Metarhizium anisopliae*, and *Paecilomyces farinosus*. J. Invertebr. Pathol. 74 (3), 261–266.

Hegedus, D.D., Bidochka, M.J., Khachatourians, G.G., 1990. *Beauveria bassiana* submerged conidia production in a defined medium containing chitin, two hexosamines or glucose. Appl. Microbiol. Biotechnol. 33 (6), 641–647.

Henry, J.E., 1985. Effect of grasshopper species, cage density, light intensity, and method of inoculation on mass production of *Nosema locustae* (Microsporida: Nosematidae). J. Econ. Entomol. 78 (6), 1245–1250.

Higuchi, T., Fukushima, Y., Furumori, K., Yamamoto, K., Ouchi, M., 1996. Vermin exterminating element and vermin exterminating method. U.S. Patent 5,589,390.

Higuchi, T., Takeshi, S., Shuji, S., Mizobata, T., Kawata, Y., Nagai, J., 1997. Development of biorational pest control formulation against longicorn beetles using a fungus, *Beauveria brongniartii* (Sacc.) Petch. J. Ferment. Bioeng. 84, 236–243.

Hong, T.D., Jenkins, N.E., Ellis, R.H., 2000. The effects of duration of development and drying regime on the longevity of conidia of *Metarhizium flavoviride*. Mycol. Res. 104 (6), 662–665.

Horaczek, A., Viernstein, H., 2004. Comparison of three commonly used drying technologies with respect to activity and longevity of aerial conidia of *Beauveria brongniartii* and *Metarhizium anisopliae*. Biol. Control 31 (1), 65–71.

Hua, L., Feng, M.G., 2003. New use of broomcorn millets for the production of granular cultures of aphidpathogenic fungus *Pandora neoaphidis* for high sporulation potential and infectivity to *Myzus persicae*. FEMS Microbiol. Lett. 227 (2), 311–317.

Hua, L., Feng, M.G., 2005. Broom corn millet grain cultures of the entomophthoralean fungus *Zoophthora radicans*: sporulation capacity and infectivity to Plutella xylostella. Mycol. Res. 109, 319–325.

Humber, R.A., 2012. Entomophthoromycota: a new phylum and reclassification for entomophthoroid fungi. Mycotaxon 120, 477–492.

Humphreys, A.M., Matawele, P., Trinci, A.P.J., Gillespie, A.T., 1989. Effects of water activity on morphology, growth, and blastospore production of *Metarhizium anisopliae* and *Beauveria bassiana* in batch and fed-batch culture. Mycol. Res. 92, 257–264.

Hussain, A., Tian, M.Y., He, Y.R., Ruan, L., Ahmed, S., 2010. In vitro and in vivo culturing impacts on the virulence characteristics of serially passed entomopathogenic fungi. J. Food Agric. Environ. 8 (3–4 part 1), 481–487.

Ibrahim, Y.B., Lim, T.K., Tang, M.K., Teng, H.M., 1993. Influence of temperature, pH and selected growth media on germination, growth and sporulation of *Aschersonia placenta* and *Hypocrella raciborskii*. Biocontrol Sci. Technol. 3, 55–61.

Inch, J.M.M., Humphreys, A.M., Trinci, A.P.J., Gillespie, A.T., 1986. Growth and blastospore formation by *Paecilomyces fumosoroseus*, a pathogen of the brown planthopper (*Nilaparvata lugens*). Trans. Br. Mycol. Soc. 87, 215–222.

Issaly, N., Chauveau, H., Aglevor, F., Fargues, J., Durand, A., 2005. Influence of nutrient, pH and dissolved oxygen on the production of *Metarhizium flavoviride* Mf189 blastospores in submerged batch culture. Proc. Biochem. 40 (3–4), 1425–1431.

Jackson, M.A., 2012. Dissolved oxygen levels affect dimorphic growth by the entomopathogenic fungus *Isaria fumosorosea*. Biocontrol Sci. Technol. 22 (1), 67–79.

Jackson MA, Dunlap C, Jaronski ST. Ecological considerations in producing and formulating fungal entomopathogens for use in insect biocontrol. In: Roy HE, Vega FE, Chandler D, Goettel M, Pell JK, Wajnberg E, editors. Ecology of Fungal Entomopathogens. Dordrecht, The Netherlans: Springer; 2010. p. 129–46.

Jackson, M.A., Jaronski, S.T., 2009. Production of microsclerotia of the fungal entomopathogen *Metarhizium anisopliae* and their use as a biocontrol agent for soil-inhabiting insects. Mycol. Res. 113 (8), 842–850.

Jackson, M.A., Jaronski, S.T., 2012. Development of pilotscale fermentation and stabilisation processes for the production of microsclerotia of the entomopathogenic fungus *Metarhizium brunneum* strain F52. Biocontrol Sci. Technol. 22 (8), 915–930.

Jackson, M.A., Payne, A.R., 2007. Evaluation of the desiccation tolerance of blastospores of *Paecilomyces fumosoroseus* (Deuteromycotina: Hyphomycetes) using a lab-scale, air-drying chamber with controlled relative humidity. Biocontrol Sci. Technol. 17 (7), 709–719.

Jackson, M.A., McGuire, M.R., Lacey, L.A., Wraight, S.P., 1997. Liquid culture production of desiccation tolerant blastospores of the bioinsecticidal fungus *Paecilomyces fumosoroseus*. Mycol. Res. 101 (1), 35–41.

Jackson, M.A., Payne, A.R., Odelson, D.A., 2004. Liquid-culture production of blastospores of the bioinsecticidal fungus *Paecilomyces fumosoroseus* using portable fermentation equipment. J. Ind. Microbiol. Biotechnol. 31, 149–154.

Jakobs-Schönwandt, D., Lohse, R., Patel, A.V., 2011. Encapsulation materials and methods applied to biological insect control. IOBC/WPRS Bull. 66, 135–138.

Jaronski, S.T., 2010. Role of fungal ecology in the innundative use of entomopathogenic fungi. BioControl 55, 159–185.

Jaronski, S.T., Jackson, M.A., 2009. Further progress with Metarhizium microsclerotial production. IOBC/WPRS Bull. 45, 275–278.

Jaronski, S.T., Jackson, M.A., 2012. Mass production of entomopathogenic Hypocreales. In: Lacey, L.A. (Ed.), Manual of Techniques in Invertebrate Pathology, second ed. Academic Press, New York, pp. 255–284.

Jaronski, S.T., Axtell, R.C., Fagan, S.M., Domnas, A.J., 1983. In vitro production of zoospores by *Lagenidium giganteum* (Oomycetes: Lagenidiales) from solid media. J. Invertebr. Pathol. 41, 305–309.

Jenkins, N.E., 1995. Studies on Mass Production and Field Efficacy of *Metarhizium Flavoviride* for Biological Control of Locusts and Grasshoppers. PhD thesis, Cranfield University, UK.

Jenkins, N.E., Goettel, M.S., 1997. Methods for mass production of microbial control agents of grasshoppers and locusts. Mem. Entomol. Soc. Can. 171, 37–48.

Jenkins NE, Lomer CJ. Development of a new procedure for the mass production of conidia of *Metarhizium flavoviride*. Bull. OILB/SROP 1994;17:181–184.

Jenkins, N.E., Prior, C., 1993. Growth and formation of true conidia by *Metarhizium flavoviride* in a simple liquid medium. Mycol. Res. 97, 1489–1494.

Jenkins, N.E., Heviefo, G., Langewald, J., Cherry, A.J., Lomer, C.J., 1998. Development of mass production technology for aerial conidia for use as mycopesticides. Biocontrol News Inform. 19 (1), 21N–31N.

JianWen, L., ZhiBing, L., XingDan, L., SuHong, Z., Yan, P., YongJun, Z., 2009. Screen of solid media and optimization of fermentation condition of *Nomuraea rileyi*. Acta Phytophylactica Sin. 36 (5), 437–442.

Keeling, P., 2009. Five questions about microsporidia. PLoS Pathog. 5 (9), e1000489. http://dx.doi.org/10.1371/journal.ppat.1000489.

Kerwin, J.L., 2007. Oomycetes: *Lagenidium giganteum*. J. Am. Mosq. Control Assoc. 23 (sp2), 50–57.

Kerwin, J.L., Washino, R.K., 1983. Sterol induction of sexual reproduction in *Lagenidium giganteum*. Exp. Mycol. 7, 109–115.

Kerwin, J.L., Washino, R.K., 1986. Regulation of oosporogenesis by *Lagenidium giganteum*: promotion of sexual reproduction by unsaturated fatty acids and sterol availability. Can. J. Microbiol. 32, 294–300.

Kerwin, J.L., Washino, R.K., 1987. Ground and aerial application of the asexual stage of *Lagenidium giganteum* (Oomycetes: Lagenidiales) for the control of mosquitoes associated with rice culture in the Central Valley of California. J. Am. Mosq. Control Assoc. 3, 59–64.

Kim, J.S., Skinner, M., Parker, B.L., 2010a. Plant oils for improving thermotolerance of *Beauveria bassiana*. J. Microbiol. Biotechnol. 20 (9), 1348–1350.

Kim, J.S., Skinner, M., Hata, T., Parker, B.L., 2010b. Effects of culture media on hydrophobicity and thermotolerance of *Beauveria bassiana* and *Metarhizium anisopliae* conidia, with description of a novel surfactant based hydrophobicity assay. J. Invertebr. Pathol. 105 (3), 322–328.

Kim, J.S., Je, Y.H., Roh, J.Y., 2010c. Production of thermotolerant entomopathogenic *Isaria fumosorosea* SFP-198 conidia in corn–corn oil mixture. J. Ind. Microbiol. Biotechnol. 37 (4), 419–423.

Kim, J.S., Kassa, A., Skinner, M., Hata, T., Parker, B.L., 2011. Production of thermotolerant entomopathogenic fungal conidia on millet grain. J. Ind. Microbiol. Biotechnol. 38 (6), 697–704.

Kirchmair, M., Hoffmann, M., Neuhauser, S., Strasser, H., Huber, L., 2007. Persistence of GRANMET®, a *Metarhizium anisopliae* based product, in grape phylloxera-infested vineyards. IOBC/WPRS Bull. 30 (7), 137–142.

Kogan, P.H., Hajek, A.E., 2000. In vitro formation of resting spores by the insect pathogenic fungus *Entomophaga maimaiga*. J. Invertebr. Pathol. 75 (3), 193–201.

Krassilstschik, I.M., 1888. La production industrielle des parasites vegetaux pour la destruction des insectes nuisibles. Bull. Sci. Fr. Belg. 19, 461–472.

Krishna, C., 2005. Solid-state fermentation systems—an overview. Crit. Rev. Biotechnol. 25, 1–30.

Kulkarni, N.S., Lingappa, S., 2002. Evaluation of food grains for mass production of entomopathogenic fungus *Nomuraea rileyi* (Farlow) Samson. Karnataka J. Agric. Sci. 15 (2), 288–292.

Kuźniar, T., 2011. The influence of white light exposition on the growth, sporulation and pathogenicity of entomopathogenic fungi *Isaria fumosorosea* and *Beauveria bassiana*. IOBC/WPRS Bull. 66, 199–203.

Kuźniar, T., Krysa, A., 2011. Evaluation of crop plants grains for mass production of entomopathogenic fungus *Isaria fumosorosea*. Progr. Plant Prot. 51 (2), 776–780.

Lalitha, C., Manjula, K., Srinivasan, S., 2008. Suitability of agricultural products for mass production of *Nomuraea rileyi*. Indian J. Plant Prot. 36 (2), 315–316.

Lane, B.S., Trinci, A.P.J., Gillespie, A.T., 1991a. Endogenous reserves and survival of blastospores of *Beauveria bassiana* harvested from carbon- and nitrogen-limited batch cultures. Mycol. Res. 95 (7), 821–828.

Latge, J.P., 1975. Growth and sporulation of six *Entomophthora* species 2. Influence of various nitrogen sources. Mycopathologia 57, 53–57.

Latge, J.P., 1981. Comparaison des exigences nutritionnelles des Entomophthorales. Ann. Microbiol. 132B, 299–306.

Latge, J.P., Remaudiere, G., Diaquin, M., 1978. A new culture medium for growth and sporulation of Entompohthorales species pathogens of aphids. Ann. Microbiol. 120, 463–476.

Latge, J.P., Cabrera Cabrera, M.C., Prevost, M., 1988. Microcycle conidiation in *Hirsutella thompsonii*. Can. J. Microbiol. 34, 625–630.

Latge, J.P., Soper, R.S., Madore, C.D., 2004. Media suitable for industrial production of *Entomophthora virulenta* zygospores. Biotechnol. Bioeng. 19 (9), 1269–1284.

Leite, L.G., Alves, S.B., Filho, A.B., Roberts, D.W., 2005. Simple, inexpensive media for mass production of three entomophthoralean fungi. Mycol. Res. 109, 326–334.

Leland, J.E., Mullins, D.E., Vaughan, L.J., Warren, H.L., 2005a. Effects of media composition on submerged culture spores of the entomopathogenic fungus, *Metarhizium anisopliae var. acridum*, Part 1: comparison of cell wall characteristics and drying stability among three spore types. Biocontrol Sci. Technol. 15 (4), 379–392.

Leland, J.E., Mullins, D.E., Vaughan, L.J., Warren, H.L., 2005b. Effects of media composition on submerged culture spores of the entomopathogenic fungus, *Metarhizium anisopliae var. acridum* Part 2: effects of media osmolality on cell wall characteristics, carbohydrate concentrations, drying stability, and pathogenicity. Biocontrol Sci. Technol. 15 (4), 393–409.

Leng, Y., Peng, G., Cao, Y., Xia, Y., 2011. Genetically altering the expression of neutral trehalase gene affects conidiospore thermotolerance of the entomopathogenic fungus *Metarhizium acridum*. BMC Microbiol. 11 (32). http://dx.doi.org/10.1186/1471-2180-11-32.

Li, Z., Butt, T.M., Beckett, A., Wilding, N., 1993. The structure of dry mycelia of the entomomophthoralean fungi *Zoophthora radicans* and *Erynia neoaphidis* following different preparatory treatments. Mycol. Res. 97 (11), 1315–1323.

Li, Z., Lin, J., Ma, J., Wu, D., Zhang, Y., 2008. Influence of different drying temperatures for solid substrate after fermentation on conidia characteristics of the entomopathogenic fungus *Beauveria bassiana*. Wei Sheng Wu Xue Bao 48 (7), 887–892.

Leite, L.G., Smith, L., Moraes, G.J., Roberts, D.W., 2000. In vitro production of hyphal bodies of the mite pathogenic fungus *Neozygites floridana*. Mycologia 92 (2), 201–207.

Lopez-Llorca, L.V., Carbonell, T., 1998. Use of almond mesocarp for production of the entomopathogenic fungus *Verticillium lecanii*. Can. J. Microbiol. 44 (9), 886–895.

Lopez-Llorca, L.V., Carbonell, T., Salinas, J., 1999. Colonization of plant waste substrate by entomopathogenic and mycoparasitic fungi—a SEM study. Micron 30 (4), 325–333.

Lord, J.C., 2007. From Metchnikoff to Monsanto and beyond: the path of microbial control. J. Invertebr. Pathol. 89 (1), 19–29.

Lord, J.C., Roberts, D.W., 1986. The effects of culture medium quality and host passage on zoosporogenesis, oosporogenesis, and infectivity of *Lagenidium giganteum* (Oomycetes: Lagenidiales). J. Invertebr. Pathol. 48, 355–361.

Lugger, O., 1888. Fungi which kill insects. Univ. Minn. Coll. Ag. Bull. 4, 26–41.

Luth, P., Eiben, U., 2003. Solid-state fermenter and method for solid-state fermentation. U.S. Patent 6,620,614.

Machado, S.S., Magalhães, B.P., Dias, J.M.C.S., 1994. The effects of aeration on mycelium production by the entomopathogenic fungus *Metarhizium anisopliae* (Metsch.) Sor. in liquid fermentation. An. Soc. Entomol. Bras. 23 (3), 565–569.

Magalhães, B.P., Cabral de Sousa Dias, J.M., Ferreira, C.M., 1994. Mycelial production of *Metarhizium anisopliae* in liquid culture using different sources of carbon and nitrogen. Rev. Microbiol. 25 (3), 181–187.

Magan, N., 2001. Physiological approaches to improving the ecological fitness of fungal biocontrol agents. In: Butt, T.M., Jackson, C., Magan, N. (Eds.), Fungi as Biocontrol Agents: Progress, Problems and Potential, CAB International Publishing, Wallingford, pp. 239–252.

Maimala, S., Chandrapatya, A., Chamswarng, C., McCoy, C.W., 1999. Preliminary testing of biomass production of *Hirsutella thompsonii var. synnematosa*. Thai J. Agric. Sci. 32 (2), 229–239.

Maldonado-Blanco, M.G., Leal-López, E.Y., Ochoa-Salazar, O.A., Elías-Santos, M., Galán-Wong, L.J., Quiroz-Martínez, H., 2011. Effects of culture medium and formulation on the larvicidal activity of the mosquito pathogen *Lagenidium giganteum* (Oomycetes: Lagenidiales) against Aedes aegypti. Acta Trop. 117 (2), 114–118.

Matewele, P., Trinci, A.P.J., Gillespie, A.T., 1994. Mutants of entomopathogenic fungi that germinate and grow at reduced water activities and reduced relative humidities are more virulent to *Nephotettix virescens* (green leafhopper) than the parental strains. Mycol. Res. 98 (11), 1329–1333.

Maul, S.B., Lemke, P.A., Gerner, W.L., Yoder, J.B., 1980. Method and apparatus for sterile cultivation of cells on solid substrates. U.S. Patent 4,204,364.

May, B.A., Vander Gheynst, J.S., 2001. A predictor variable for efficacy of *Lagenidium giganteum* produced in solid state cultivation. J. Ind. Microbiol. Biotechnol. 27, 203–207.

Mazumder, D., Puzari K.C., Hazarika, L.K., 1995. Mass production of *Beauveria bassiana* and its potentiality on rice hispa. Ind. Phytopathol. 48,275–8.

McCabe, D., Soper, R.S., 1985. Preparation of an entomopathogenic fungal insect control agent. U.S. Patent 4,530,834.

McCoy, C.W., Hill, A.J., Kanavel, R.F., 1975. Large-scale production of the fungal pathogen *Hirsutella thompsonii* in submerged culture and its formulation for application in the field. Entomophaga 20 (3), 229–240.

McCoy, C.W., Samson, R.A., Boucias, D.G. Entomogenous fungi. In: Ignoffa CM, Mandava NB, editors. Handbook of Natural Pesticides Part A. Entomogenous Protozoa and Fungi, vol. V, Boca Raton, Florida: CRC Press; 1988. p. 151–236.

Méndez, A., Pozo, E. del, García, I., González, A., 2010. Evaluation of solid substrates for mass production of *Nomuraea rileyi* (Farlow) Samson. Rev. Prot. Veg. 25 (2), 108–112.

Mendonca, A.F., 1991. Mass production, application and formulation of *Metarhizium anisopliae* for control of sugarcane froghopper, *Mahanarva posticata* in Brazil. In: Lomer, C.J., Prior, C. (Eds.), Biological Control of Locusts and Grasshoppers. CAB International, Oxon, UK, pp. 239–244.

Metchnikoff, E.A., 1880. Zur Lehre über Insektenkrankenheiten. Zool. Anz. 3, 44–47.

Moore, D., Douro-Kpindou, O.K., Jenkins, N.E., Lomer, C.J., 1996. Effects of moisture content and temperature on storage of *Metarhizium flavoviride* conidia. Biocontrol Sci. Technol. 6 (1), 51–61.

Moore, D., Langewald, J., Obognon, F., 1997. Effects of rehydration on the conidial viability of *Metarhizium flavoviride* mycopesticide formulations. Biocontrol Sci. Technol. 7,87–94.

Mullens, B.A., 1986. A method for infecting large numbers of *Musca domestica* (Diptera: Muscidae) with *Entomophthora muscae* (Entomophthorales: Entomophthoraceae). J. Med. Entomol. 23, 457–458.

Niedermayr, V., Strasser, H., Pertot, I., Elad, Y., Gessler, C., Cini, A., 2012. Mass production of ecologically competent *Metarhizium anisopliae* spores for pest management. IOBC/WPRS Bull. 78, 103–106.

Nolan, R.A., 1988. A simplified defined medium for growth of *Entomophaga aulicae* protoplasts. Cancer J. Microbiol. 34,45–51.

Nolan, R.A., 1993. An inexpensive medium for mass fermentation production of *Entomophaga aulicae* hyphal bodies competent to form conidia. Can. J. Microbiol. 39 (6), 588–593.

Nuñez-Gaona, O., Saucedo-Castañeda, G., Alatorre-Rosas, R., 2010. Effect of moisture content and inoculum on the growth and conidia production by *Beauveria bassiana* on wheat bran. Braz. Arch. Biol. Technol. 53 (4), 771–777.

Onofre, S.B., Miniuk, C.M., de Barros, M., Azevedo, J.L., 2001. Growth and sporulation of *Metarhizium flavoviride var.* flavoviride on culture media and lighting regimes. Sci. Agric. 58 (3), 613–616.

Ortiz-Urquiza, A., Vergara-Ortiz, A., Santiago-Alvarez, C., Quesada-Moraga, E., 2010. J. Appl. Entomol. 134, 581–591.

Pelizza, S.A., Cabello, M.N., Tranchida, M.C., Scorsetti, A.C., Bisaro, V., 2011. Screening for a culture medium yielding optimal colony growth, zoospore yield and infectivity of different isolates of *Leptolegnia chapmanii* (Straminipila: Peronosporomycetes). Ann. Microbiol. 61, 991–997.

Pell, J.K., Wilding, N., 1992. The survival of *Zoophthora radicans* (Zygomycetes: Entomophthorales) isolates as hyphal bodies in mummified larvae of *Plutella xylostella* (Lep.: Yponomeutidae). Entomophaga 37 (4), 649–654.

Pérez-Guerra, N., Torrado-Agrasar, A., López-Macias, C., Pastrana, L., 2003. Main characteristics and applications of solid substrate fermentation. Electron. J. Environ. Agric. Food Chem. 2 (3), 343–350.

Prakash, G.V.S.B., Padmaja, V., Kiran, R.R.S., 2008. Statistical optimization of process variables for the large-scale production of *Metarhizium anisopliae* conidiospores in solid-state fermentation. Bioresour. Technol. 99 (6), 1530–1537.

Prophyta GmbH, 2012. Solid State Fermenter Technology from Downloaded. http://www.prophyta.de/wp-content/uploads/2012/06/Folder-Technology.pdf, December 27, 2012.

Qiu, J., Song, F., Qiu, Y., Li, X., Guan, X., 2013. Optimization of the medium composition of a biphasic production system for mycelial growth and spore production of *Aschersonia placenta* using response surface methodology. J. Invertebr. Pathol. 112 (2), 108–115.

Raimbault, M., 1998. General and microbiological aspects of solid substrate fermentation. Electron. J. Biotechnol. 1 (3), 174–188.

Rajanikanth, P., Subbaratnam, G.V., Rahaman, S.J., 2011. Effect of age and frequency of subculturing of different strains/isolates of *Beauveria bassiana* Vuillemin for their pathogenicity against *Spodoptera litura* Fabricius. Biochem. Cell. Arch. 11 (1), 79–84.

Ramakers, P.M.J., Samson, R.A., 1984. *Aschersonia aleyrodis*, a fungal pathogen of whitefly II. Application as a biological insecticide in glasshouses. J. Appl. Entomol. 97 (1–5), 1–8.

Ramegowda, G.K., Lingappa, S., Patil, R.K., 2007. Production of aerial conidia of *Nomuraea rileyi* (Farlow) Samson on liquid media. J. Entomol. Res. 31 (3), 225–227.

Rangel, D.E.N., Braga, G.U.L., Flint, S.D., Anderson, A.J., Roberts, D.W., 2004. Variations in UV-B tolerance and germination speed of *M. anisopliae* conidia produced on artificial and natural substrates. J. Invertebr. Pathol. 87, 77–83.

Rangel, D.E.N., Alston, D.G., Roberts, D.W., 2008a. Effects of physical and nutritional stress conditions during mycelial growth on conidial germination speed, adhesion to host cuticle, and virulence of *Metarhizium anisopliae*, an entomopathogenic fungus. Mycol. Res. 112, 1355–1361.

Rangel, D.E.N., Anderson, A.J., Roberts, D.W., 2008b. Evaluating physical and nutritional stress during mycelial growth as inducers of tolerance to heat and UV-B radiation in *Metarhizium anisopliae* conidia. Mycol. Res. 112, 1362–1372.

Rangel, D.E.N., Fernandes, É.K.K., Anderson, A.J., Roberts, D.W., 2012. Culture of *Metarhizium robertsii* on salicylic-acid supplemented medium induces increased conidial thermotolerance. Fungal Biol. 116 (3), 438–442.

Rao, D.G., 2010. Introduction to Biochemical Engineering. Tata McGraw Hill, New Delhi.

Ravensberger, W.R., 2011. A Roadmap to the Successful Development and Commercialization of Microbial Pest Control Products for Control of Arthropods. Progress in Biocontrol. vol. 10. Springer, Dordrecht, pp. 80–81.

Roberts, D.W., Dunn, H.M., Ramsay, G., 1987. A procedure for preservation of the mosquito pathogen *Culicinomyces clavisporus*. Appl. Microbiol. Biotechnol. 26 (2), 186–188.

Rombach MC, Aguda RM, Roberts DW. Production of *Beauveria bassiana* (Deuteromycotina: Hyphomycetes) in different liquid media and subsequent conidiation of dry mycelium. Entomophaga 1988;33:315–24.

Safavi, S.A., 2011. Successive subculturing alters spore-bound Pr1 activity, germination and virulence of the entomopathogenic fungus, *Beauveria bassiana*. Biocontrol Sci. Technol. 21 (8), 883–890.

Safavi, S.A., 2012. Attenuation of the entomopathogenic fungus *Beauveria bassiana* following serial in vitro transfers. Biologia 67 (6), 1062–1068.

Safavi, S.A., Shah, F.A., Pakdel, A.K., Reza Rasoulian, G., Bandani, A.R., Butt, T.M., 2007. Effect of nutrition on growth and virulence of the entomopathogenic fungus *Beauveria bassiana*. FEMS Microbiol. Lett. 270 (1), 116–123.

Sakamoto, N., Inoue, Y., Aoki, J., 1985. Effect of Light on the conidiation of *Paecilomyces fumosoroseus*. Trans. Mycol. Soc. Jpn. 26 (4), 499–509.

Sanchez-Murillo, R.I., Torre-Martinez, M., Aguirre-Linares, J., Herrera-Estrella, I., 2004. Light regulated asexual reproduction in *Paecilomyces fumosoroseus*. Microbiology 150, 311–319.

Sanchez Pena SR. In vitro production of hyphae of the grasshopper pathogen *Entomophaga grylli* (Zygomycota: Entomophthorales): Potential for production of conidia. Florida Entomology 2005;88:332–4.

Saroha, P., Mohan, J., 2006. Mass production of conidia of two isolates of *Beauveria bassiana* for the control of potato beetle. Plant Arch. 6 (1), 289–291.

Scholte, E.J., Knols, B.G.J., Samson, R.A., Takken, W., 2004. Entomopathogenic fungi for mosquito control: a review. J. Insect Sci. 4, 19.

Sene, L., Alves, L.F.A., Lobrigatte, M.F.P., Thomazoni, D., 2010. Production of conidia of *Metarhizium anisopliae* in solid media based on agroindustrial residues. Arq. Instit. Biol. (São Paulo) 77 (3), 449–456.

Senthilkumar, M., Nizam, M., Narayanasamy, P., 2011. Development of a semi-synthetic medium for production of azygospores of *Zoophthora radicans* (Brefeld) Batko, a pathogen of rice leaf folder. J. Biopestic. 4 (1), 43–47.

Seramis GmbH, 2012. Seramis Special Growing Granules. Downloaded from http://www.seramisuk.co.uk/technicaldata.html, December 23, 2012.

Shah, F.A., Wang, C.S., Butt, T.M., 2005. Nutrition influences growth and virulence of the insect-pathogenic fungus *Metarhizium anisopliae*. FEMS Microbiol. Lett. 251 (2), 259–266.

Shah, F.A., Allen, N., Wright, C.J., Butt, T.M., 2007. Repeated in vitro subculturing alters spore surface properties and virulence of *Metarhizium anisopliae*. FEMS Microbiol. Lett. 276 (1), 60–66.

Shanley, R.P., Keena, M., Wheeler, M.M., Leland, J., Hajek, A.E., 2009. Evaluating the virulence and longevity of non-woven fiber bands impregnated with *Metarhizium anisopliae* against the Asian longhorned beetle, *Anoplophora glabripennis* (Coleoptera: Cerambycidae). Biol. Control 50 (2), 94–102.

Shapiro-Ilan, D.I., Cottrell, T.E., Gardner, W.A., Leland, J., Behle, R.W., 2009. Laboratory mortality and mycosis of adult *Curculio caryae* (Coleoptera: Curculionidae) following application of *Metarhizium anisopliae* in the laboratory or field. J. Entomol. Sci. 44, 24–36.

Shi, Y., Xu, X., Zhu, Y., 2009. Optimization of *Verticillium lecanii* spore production in solid-state fermentation on sugarcane bagasse. Appl. Microbiol. Biotechnol. 82 (5), 921–927.

Skinner, M., Gouli, S., Frank, C.E., Parker, B.L., Kim, J.S., 2012. Management of *Frankliniella occidentalis* (Thysanoptera: Thripidae) with granular formulations of entomopathogenic fungi. Biol. Control 63 (3), 246–252.

Smith, C., Edgington, S., 2011. Germination at different water activities of similarly aged *Metarhizium* conidia harvested from ageing cultures. J. Stored Prod. Res. 47, 157–160.

Solter, L.F., Becnel, J.J., Vavra, J., 2012a. Research methods for entomopathogenic microsporidia and other protists. In: Lacey, L.A. (Ed.), Manual of Techniques in Invertebrate Pathology, Elsevier Ltd, Amsterdam, pp. 329–371.

Solter, L.F., Becnel, J.J., Oi, D.H., 2012b. Microsporidian entomopathogens. In: Vega, F.E., Kaya, H.K. (Eds.), Insect Pathology, Elsevier Ltd, Amsterdam, pp. 221–263.

Sreerama Kumar, P., 2010. *Hirsutella thompsonii* as a mycoacracidie for *Aceria guerreronis* on cocnut in India: research, development and other aspects. In: Sabelis, M.W., Bruin, J. (Eds.), Trends in Acarology, Proceedings of the 12th International Congress, Springer, Dordrecht, pp. 441–444.

Sreerama Kumar, P., Singh, L., Tabassum, H., 2005. Potential use of polyethylene glycol in the mass production of nonsynnematous and synnematous strains of *Hirsutella thompsonii* Fisher in submerged culture. J. Biol. Control 19 (2), 105–113.

St. Leger, R., Wang, C., 2010. Genetic engineering of fungal biocontrol agents to achieve greater efficacy against insect pests. Appl. Microbiol. Biotechnol. 85, 901–907.

Stanbury, P.F., Whitaker, A., 1993. Principles of Fermentation Technology. Pergamon Press, Oxford.

Steinhaus, E.A., 1956. Microbial control—the emergence of an idea. A brief history of insect pathology through the nineteenth century. Hilgardia 26, 107–160.

Steinhaus, E.A., 1975. Disease in a Minor Chord. Ohio State University Press, Columbus.

Steinkraus, D.C., Boys, G.O., 2005. Mass harvesting of the entomopathogenic fungus, *Neozygites fresenii*, from natural field epizootics in the cotton aphid, *Aphis gossypii*. J. Invertebr. Pathol. 88 (3), 212–217.

Stephan, D., Zimmermann, G., 1998. Development of a spray-drying technique for submerged spores of entomopathogenic fungi. Biocontrol Sci. Technol. 8 (1), 3–11.

Sweeney, A., 1985. The potential of the fungus *Culicinomyces clavisporus* as a biocontrol agent for medically important Diptera. In: Laird, M., Miles, J.W. (Eds.), Integrated Mosquito Control Strategies, Academic Press, London, pp. 269–284.

Tarocco, F., Lecuona, R.E., Couto, A.S., Arcas, J.A., 2005. Optimization of erythritol and glycerol accumulation in conidia of *Beauveria bassiana* by solid-state fermentation, using response surface methodology. Appl. Microbiol. Biotechnol. 68 (4), 481–488.

Thakre, M., Thakur, M., Malik, S., Ganger, S., 2011. Mass scale cultivation of entomopathogenic fungus *Nomuraea rileyi* using agricultural products and agro wastes. J. Biopestic. 4 (2), 176–179.

Thomas, K.C., Khachatourians, G.G., Ingledew, W.M., 1987. Production and properties of *Beauveria bassiana* conidia cultivated in submerged culture. Can. J. Microbiol. 33, 12–20.

Tincilley, A., Easwaramoorthy, S., Santhalakshmi, G., 2004. Attempts on mass production of *Nomuraea rileyi* on various agricultural products and byproducts. J. Biol. Control (India) 18 (1), 35–40.

Underkofler, L.A., Severson, G.M., Goering, K.J., Christensen, L.M., 1947. Commercial production and use of mold bran. Cereal Chem. 24, 1–22.

Unkles, S.E., Marriott, C., Kinghorn, J.R., Panter, C., Blackwell, A., 2004. Efficacy of the entomopathogenic fungus, *Culicinomyces clavisporus* against larvae of the biting midge, *Culicoides nubeculosus* (Diptera: Ceratopogonidae). Biocontrol Sci. Technol. 14 (4), 397–401.

U.S. Department of Agriculture, Economic Research Service (USDA-ERS), 2012. Adoption of Genetically Engineered Crops in the U.S. http://www.ers.usda.gov/data-products/adoption-of-genetically-engineered-crops-in-the-us.aspx Downloaded January 5, 2013.

U.S. Department of Agriculture, National Agricultural Statistics Service, 2012. Acreage. http://usda.mannlib.cornell.edu/usda/current/Acre/Acre-06-29-2012.pdf. Downloaded January 5, 2013.

U.S. Environmental Protection Administration (USEPA), 2000. *Nosema locustae* (117001) Fact Sheet. Downloaded from http://www.epa.gov/pesticides/chem_search/reg_actions/registration/fs_PC-117001_01-Oct-00.pdf, January 29, 2013.

U.S. Environmental Protection Agency (USEPA), 2010. Biopesticides Registration Action Document: *Paecilomyces fumosoroseus strain* FE 9901, pp. 16 http://www.regulations.gov/#!documentDetail;D=EPA-HQ-OPP-2010-0093 Downloaded November 22, 2012.

U.S. Environmental Protection Agency (USEPA), 2011. Biopesticides Registration Action Document: *Isaria fumosorosea* (Formerly *Paecilomyces fumosoroseus*) Apopka Strain 97, pp. 24. http://www.regulations.gov/#!documentDetail;D=EPA-HQ-OPP-2010-0088-0008 Downloaded November 22, 2012.

U.S. Environmental Protection Agency (USEPA), 2012. Biopesticides. http://www.epa.gov/pesticides/biopesticides/.

Uziel A, Kenneth RG. Influence of commercially derived lipids and a surfactant on the mode of germination and process of germ-tube formation in primary conidia of two species of *Erynia* subgenus *Neopandora* (Zygomycotina: Entomophthoralis). Mycopathologia 1998;144:153–63.

Van Winkelhof, A.J., McCoy, C.W., 1984. Conidiation of *Hirsutella thompsonii var. synnematosa* in submerged culture. J. Invertebr. Pathol. 43, 59–68.

Vander Gheynst, J.S., May, B.A., Karagosian, M., 2000. The effect of cultivation methods on the growth rate and shelf life of *Lagenidium giganteum*. ASAE Ann. Int. Mtg. 5.

Vega, F.E., Jackson, M.A., Mercadier, G., Poprawski, T.J., 2003. The impact of nutrition on spore yields for various fungal entomopathogens in liquid culture. World J. Microbiol. Biotechnol. 19 (4), 363–368.

Vestergaard, S., Butt, T.M., Bresciani, J., Gillespie, A.T., Eilenberg, J., 1999. Light and electron microscopy studies of the infection of the western flower thrips *Frankliniella occidentalis* (Thysanoptera: Thripidae) by the entomopathogenic fungus *Metarhizium anisopliae*. J. Invertebr. Pathol. 73 (1), 25–33.

Villamizar, L., Arriero, C., Bosa, O.C.F., Marina Cotes, A., 2004. Development of preformulated products based on *Nomuraea rileyi* for control of *Spodoptera frugiperda* (Lepidoptera: Noctuidae). Rev. Colombiana Entomol. 30 (1), 99–105.

Vimala Devi, P.S., Chowdary, A., Prasad, Y.G., 2000. Cost-effective multiplication of the entomopathogenic fungus *Nomuraea rileyi* (F.) Samson. Mycopathologia 151 (1), 35–39.

Visvesvara, G.S., Moura, H., Leitch, G.J., Schwartz, D.A., 1999. Culture and propagation of Microsporidia. In: Willmer, M., Weiss, L.M. (Eds.), The Microsporida and Microsporidiosis, American Association for Microbiology, Washington DC, pp. 363–392.

Wang, C.S., St. Leger, R.J., 2007. A scorpion neurotoxin increases the potency of a fungal insecticide. Nat. Biotechnol. 25, 1455–1456.

Whisler, H., 1985. Life history of species of *Coelomomyces*. In: Couch, J.N., Bland, C.E. (Eds.), The Genus Coelomomyces, Academic Press, Orlando, FL, pp. 9–22.

Wikipedia contributors, 2012. Diatomaceous Earth. Wikipedia, the Free Encyclopedia. http://en.wikipedia.org/w/index.php?title=Diatomaceous_earth&oldid=528917162 (accessed 04.12.12.).

Wolf FT. The cultivation of two species of *Entomophthora* on synthetic media. Bulletin Torrey Botanical Club 1951;78:211–20.

Wraight, S.P., Ramos, M.E., 2002. Application parameters affecting field efficacy of *Beauveria bassiana* foliar treatments against Colorado potato beetle *Leptinotarsa decemlineata*. Biol. Control 23, 164–178.

Wraight, S.P., Galaini-Wraight, S., Carruthers, R.I., Roberts, D.W., 2003. *Zoophthora radicans* (Zygomycetes: Entomophthorales) conidia production from naturally infected *Empoasca kraemeri* and dry-formulated mycelium under laboratory and field conditions. Biol. Control 28 (1), 60–77.

Xavier-Santos, S., Lopes, R.B., Faria, M., 2011. Emulsifiable oils protect *Metarhizium robertsii* and *Metarhizium pingshaense* conidia from imbibitional damage. Biol. Control 59 (2), 261–267.

Ye, S., Ying, S., Chen, C., Feng, M., 2006. New solid-state fermentation chamber for bulk production of aerial conidia of fungal biocontrol agents on rice. Biotechnol. Lett. 28, 799–804.

Ying, S.H., Feng, M.G., 2004. Relationship between thermotolerance and hydrophobin-like proteins in aerial conidia of *Beauveria bassiana* and *Paecilomyces fumosoroseus* as fungal biocontrol agents. J. Appl. Microbiol. 97 (2), 323–331.

Ying, S.H., Feng, M.G., 2006. Medium components and culture conditions affect the thermotolerance of aerial conidia of fungal biocontrol agent *Beauveria bassiana*. Lett. Appl. Microbiol. 43 (3), 331–335.

Zattau, W.C., McInnis Jr., T., 1987. Life cycle and mode of infection of *Leptolegnia chapmanii* (Oomycetes) parasitizing *Aedes aegypti*. J. Invertebr. Pathol. 50 (2), 134–145.

Zhang, S., 2001. Studies on sporulation of *Metarhizium* in submerged culture. Sci. Sil. Sinicae 37 (5), 134–139.

Zhang, Y.J., Li, Z.H., Luo, Z.B., Zhang, J.Q., Fan, Y.H., Pei, Y., 2009. Light stimulates conidiation of the entomopathogenic fungus *Beauveria bassiana*. Biocontrol Sci. Technol. 19 (1), 91–101.

Zhang, S., Peng, G., Xia, Y., 2010. Microcycle conidiation and the conidial properties in the entomopathogenic fungus *Metarhizium acridum* on agar medium. Biocontrol Sci. Technol. 20 (8), 809–819.

Zhou, X., Feng, M.G., 2009. Sporulation storage and infectivity of obligate aphid pathogen *Pandora nouryi* grown on novel granules of broomcorn millet and polymer gel. J. Appl. Microbiol. 107, 1847–1856.

Zhu, Y., Pan, J., Qiu, J., Guan, X., 2008. Optimization of nutritional requirements for mycelial growth and sporulation of entomogenous *fungus Aschersonia aleyrodis* Webber. Braz. J. Microbiol. 39 (4), 770–775.

Zimmermann, G., 2007a. Review on safety of the entomopathogenic fungi *Beauveria bassiana* and *Beauveria brongniartii*. Biocontrol Sci. Technol. 17 (5/6), 553–596.

Zimmermann, G., 2007b. Review on safety of the entomopathogenic fungus *Metarhizium anisopliae*. Biocontrol Sci. Technol. 17 (9/10), 879–920.

Zimmermann, G., 2008. The entomopathogenic fungi *Isaria farinosa* (formerly *Paecilomyces farinosus*) and the Isaria fumosorosea species complex (formerly *Paecilomyces fumosoroseus*), biology, ecology and use in biological control. Biocontrol Sci. Technol. 18 (9), 865–901.

This page intentionally left blank

Chapter 12

Commercial Production of Entomopathogenic Bacteria

Terry L. Couch[1] **and Juan Luis Jurat-Fuentes**[2]

[1]*Becker Microbial Products Inc., Parkland, FL, USA,* [2]*Department of Entomology and Plant Pathology, University of Tennessee, Knoxville, TN, USA*

12.1. BIOLOGY OF COMMERCIAL BACTERIAL ENTOMOPATHOGENS

Bacteria may be generally defined as unicellular and ubiquitous microbes possessing a single chromosome that is not surrounded by a nuclear cell membrane and having ribosomes of the 70S type. These organisms proliferate through binary fission, a process resulting in daughter cells that are essentially identical copies of the mother cell. Consequently, genetic variation in bacteria is the result of mutations and/or acquisition of external genetic material and is driven by selection pressure. Relationships between bacteria and insects are the result of hundreds of millions of years of interactions and co-evolution, ranging from symbiosis to pathogenesis. The focus of this chapter is on large-scale production methods for bacterial species lethal to insects as pesticidal agents amenable to commercialization.

The first step in the decision to produce an entomopathogenic bacterium is a determination that there is actually a commercial market for the bacterium. This decision is important because there are considerable financial resources required to scale up, produce, and formulate an entomopathogenic bacterium on a commercial scale. For the purpose of this chapter, we define commercial scale as deep tank fermentation in a fermentor with a minimum working volume of 30,000 l.

Although a number of bacteria have been described as entomopathogenic (Jurat-Fuentes and Jackson, 2012), only a reduced group of species have been considered for commercialization. Three major groups of entomopathogenic bacteria are recognized based on their cell wall: Gram-positive type cell wall, Gram-negative cell wall, and bacteria lacking cell wall. Among the Gram-positive bacteria, many species are capable of producing an endospore—a dispersal form of the organism that is viable for extended time periods under adverse conditions. Production of this endospore greatly facilitates commercial development and application of these bacteria to pest control.

Mass Production of Beneficial Organisms. http://dx.doi.org/10.1016/B978-0-12-391453-8.00012-1

There are no examples of commercial development of entomopathogenic bacteria lacking a cell wall. This observation reflects their difficult propagation and, in some cases, the potential for pathogenicity in vertebrates. Among the Gram-negative entomopathogens, only *Serratia entomophila* has been developed into a biocontrol product. This bacterium was isolated and commercially developed for control of *Costelytra zealandica* (White), the New Zealand grass grub (Jackson et al., 1992). Because this bacterium does not produce an endospore, improvements in production and formulation were necessary to improve stability in storage and delivery (Johnson et al., 2001). In contrast to the other bacterial groups, a number of Gram-positive bacterial species have been developed commercially, specifically in the spore-forming genus *Bacillus*. In fact, the first commercial biopesticide, Sporeine, commercialized in 1938 in France, was based on the Gram-positive *Bacillus thuringiensis* (Berliner). Moreover, the first bacterial entomopathogen used in a major insect control program was *Paenibacillus popilliae* (Dutky, previously *Bacillus popilliae*; Klein and Jackson, 1992), the causative agent for milky disease in Japanese beetle (*Popillia japonica* Newman). However, and despite successful use, problems related to the mass production of viable *P. popilliae* spores (Stahly and Klein, 1992) reduced commercial interest in this bacterium. Although commercial products based on *P. popilliae* are currently available, their use is limited to control of grubs, especially in organic agriculture (Jackson et al., 1992; Johnson et al., 2001). Currently, the most commercially relevant entomopathogenic bacteria are *B. thuringiensis* and *Lysinibacillus (Bacillus) sphaericus* (Meyer and Neide), which will be the focus of this chapter.

Like other microbes, entomopathogenic bacteria require a source of carbon and nitrogen complemented with mineral salts for growth. Importantly, *L. sphaericus* lacks relevant biochemical pathways and cannot use sugars as fermentation metabolites. Three main growth phases are observed in cultures of entomopathogenic bacteria: exponential (vegetative growth), transition, and sporulation. Commercial production involves culture growth to sporulation, when each cell produces an endospore and a complement of soluble and/or crystalline insecticidal proteins. Two main methods have been reported for growth of *B. thuringiensis* and *L. sphaericus* cultures during production of biopesticides: submerged (liquid) and solid-state fermentation.

Because raw materials may comprise 30–40% of the overall production costs (Lisansky et al., 1993), relevant efforts have focused in identifying cost-effective raw materials that support adequate growth yields while maintaining high insecticidal activity (Morris et al., 1997), especially in developing countries. A number of reports have presented the production of *B. thuringiensis* and *L. sphaericus* biopesticides using wastewater or alternative industrial or agricultural waste derivatives (Tirado Montiel et al., 1998; El-Bendary, 2006). For instance, wastewater sludge was shown to support growth of diverse *B. thuringiensis* serovars, yielding lower cell counts but higher entomotoxicity per spore compared to synthetic media (Yan et al., 2007). Both cell counts and entomotoxicity in wastewater sludge systems have been shown to be affected by sludge pretreatment (Tirado Montiel et al., 2001), the amounts of sludge solids (Brar et al., 2009), addition of adjuvants (Brar et al., 2006b), and culture recovery process (Brar et al., 2006a). Production of *L. sphaericus* biopesticides using diverse local raw materials from Ghana was reported to support bacterial growth and mosquito larvicidal activity to similar levels detected in pesticides produced using a synthetic medium (Ampofo, 1995). However, these raw materials have a number of disadvantages that prevent their common use in high-scale commercial production of bacterial biopesticides, including pretreatment costs, variability in composition, seasonality, and local availability.

Lower waste output and capital investments have increased interest in solid-state fermentation as alternative production method for bacterial biopesticides. Although these methods generally require more intensive labor, they are expected to facilitate the use of bacterial biopesticides in developing countries with low labor costs (Devi et al., 2005). Despite the potential applications of these pilot-plant settings and alternative raw materials that may be of interest to local and regional markets in developing countries, our intent in this chapter is to concentrate on those techniques that are used in actual commercial production of bacterial bioinsecticides. For more information on production of entomopathogenic bacteria appropriate to less developed countries, the reader is referred to Chapter 15.

Because the majority of data dealing with actual scale up of commercial products based on entomopathogenic bacteria reside in the files of the companies selling these products, little literature is available. The actual commercial fermentation methods are proprietary or covered by patents; thus, information in this chapter related to this aspect will mostly rely on several excellent topic reviews and the professional experience of the authors. Both Beegle et al. (1990) and Lisansky et al. (1993) produced the most accurate reviews in their description of the entire commercialization process of these agents. The most recent review was written by Ravensberg (2011); although he focused on microbial pest control agents in general, his observations on selection, scale up and application also apply to entomopathogenic bacteria.

12.2. BIOLOGY OF COMMERCIAL BACTERIAL ENTOMOPATHOGENS

Both *B. thuringiensis* and *L. sphaericus* are ubiquitous soil microbes, and isolates have been obtained from multiple environments worldwide (Guerineau et al., 1991; Bernhard et al., 1997). Two main phases—vegetative (exponential) growth and sporulation (stationary)—can be easily differentiated in the life cycle of both species, which are also observed during commercial production. Vegetative cells (Fig. 12.1(a)) are bacilliform (rod shaped); in some cases, they highly motile due to the existence of flagella. These cells divide by fission and grow exponentially until nutrients are depleted or adverse environmental

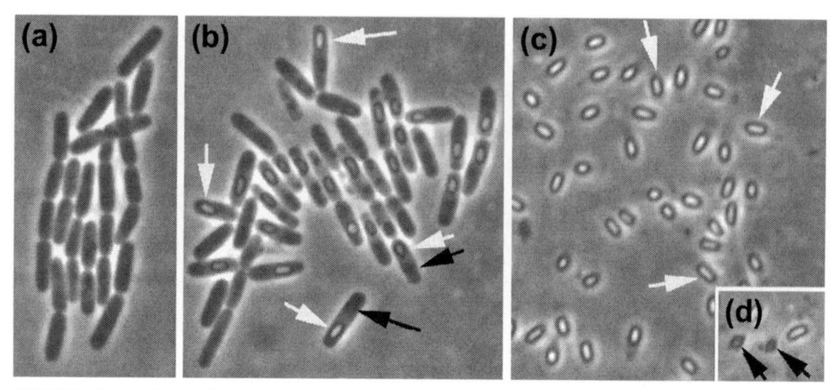

FIGURE 12.1 Vegetative (a) and sporulation (b) stages in a *Bacillus thuringiensis* subsp. *kurstaki* strain HD-73 culture documented using phase contrast microscopy (photos courtesy of Dr Ernest Bernard, Department of Entomology and Plant Pathology, University of Tennessee). White bars at the bottom of the figures represent 5 μm for reference. Vegetative cells observed during early stages (a) appear dark and bacilliform, mostly forming long chains. When nutrients are depleted, cells undergo sporulation (b). During this stage, endospores (white arrows) and bipyramidal parasporal crystals containing the Cry1Ac toxin (black arrows) may be observed in the sporangia. Later in the sporulation stage, sporangia disrupt and both endospores (c) and crystals (d) are released and can be observed in the medium.

conditions occur. At this time, specific genes are activated that drive sporulation and synthesis and formation of crystalline protein parasporal inclusions (Fig. 12.1(b)). Once sporulation is completed, sporangia lyse and the endospore and crystal are released into the medium (Fig. 12.1(c) and 12.1(d)).

Despite their bacilliform shape during the vegetative phase, *B. thuringiensis* and *L. sphaericus* can be easily distinguished during sporulation by the position of their endospore and the sporangium shape. Thus, although *B. thuringiensis* produces an ovoidal spore that does not deform the shape of the cell (sporangium), *L. sphaericus* produces a spherical endospore that locates in a terminal position, resulting in a swollen sporangium. The crystals may represent up to 20–30% of the dry cell weight; they are composed of diverse insecticidal proteins that are necessary for entomotoxicity. Consequently, improvement of these insecticidal proteins has been of interest for developing biopesticides with increased efficacy. Different crystal morphologies, depending on protein composition, have been described, with bipyramidal and spherical being the most common shapes (Bernhard et al., 1997).

Diverse methods have been used to classify *B. thuringiensis* and *L. sphaericus* isolates, with flagellar H-serotyping being the most widely used in the literature. For *B. thuringiensis,* 85 serotypes have been described (Jurat-Fuentes and Jackson, 2012), although most commercial biopesticides are based on serovars (or subsp.) *kurstaki* (Btk), *aizawai*, *israelensis* (Bti), and *tenebrionis* (Btt). Each of these serovars includes isolates expressing crystal (Cry) parasporal proteins with entomotoxicity against specific insect taxonomic orders, including

Lepidoptera, Coleoptera, and Diptera. Additional noncrystal toxins and other virulence factors that enhance entomotoxicity may also be produced by diverse isolates. Important for entomotoxicity of Bti are the cytolytic (Cyt) toxins they produce, which synergize activity of Cry proteins against mosquito larvae (Wu and Chang, 1985). On the other hand, potential secretion during the vegetative stage of some strains of thermostable insecticidal toxins called beta exotoxins is of concern during commercial production of *B. thuringiensis* biopesticides. These exotoxins inhibit protein production and were deemed unsafe for public use, resulting in a ban by the World Health Organization (WHO, 1999) for their use in biopesticides. Chromatographic purification and/or bioassays are commonly used to detect the production of these exotoxins by *B. thuringiensis* strains.

In the case of *L. sphaericus*, insecticidal activity is limited to dipteran larvae, specifically mosquitoes. Susceptibility to *L. sphaericus* amongst mosquito genera varies, with *Culex* being the most susceptible, followed by *Anopheles*, *Mansonia*, and *Aedes* (Yap, 1990). Within the 49 flagellar *L. sphaericus* serotypes initially described (de Barjac et al., 1980), only nine of them contained mosquitocidal strains. Nowadays, most commercial *L. sphaericus* biopesticides against mosquito larvae are based on strains 2362 and C3-41 (Park et al., 2010), although continuous prospecting may help identify new isolates with higher entomotoxicity (Park et al., 2007). Mosquitocidal activity in *L. sphaericus* strains is mostly determined by the production of crystal binary (Bin) and soluble mosquitocidal toxin (Mtx) insecticidal proteins.

The ecology of *B. thuringiensis* and *L. sphaericus* and the evolutionary advantage of producing insecticidal crystalline protein inclusions are still matters of debate. Supporting the identity of these bacteria as bona fide insect pathogens is their commitment of relevant energy resources to production of high amounts of insecticidal parasporal proteins. However, although spores of *B. thuringiensis* are very persistent in the environment, the bacterium displays low rates of horizontal transmission between hosts, with rare epizootics. Horizontal transmission of *B. thuringiensis* was reported to be tightly related to the production of urease (Martin et al., 2009), but this is not observed in commercially used strains. Thus, secondary infections are uncommon after spray with biopesticides based on Btk and subsp. *aizawai* (Smith and Barry, 1998), which are urease positive (Martin et al., 2010). In contrast, persistent activity was reported for applications of biopesticides containing the urease-negative Bti (Tilquin et al., 2008). Persistence of *L. sphaericus* biopesticides is related to the low settling of *L. sphaericus* spores in the water column (Nicolas et al., 1987) and recycling in the host (Charles and Nicolas, 1986).

12.3. PATHOGENESIS AND PEST CONTROL IMPACT

Bacterial biopesticides target the larval stage of susceptible pests. Generally, susceptibility to these biopesticides decreases with larval growth (Wraight et al., 1987;

Huang et al., 1999), although larvae surviving exposure usually present fitness costs in later stages, such as smaller pupae that fail to develop or lower fecundity as adults (Nyouki et al., 1996; Flores et al., 2004). Larvae that recover from treatment are more susceptible to a sequent exposure (Moreau and Bauce, 2003).

Although the mode of action of *B. thuringiensis* insecticidal proteins is still a matter of debate (Vachon et al., 2012), some common features exist among diverse strains and insect hosts. A generalized pathogenic process for *B. thuringiensis* biopesticides is initiated with ingestion of the product, which contains spores and crystal insecticidal proteins as active ingredients. Typically, larval paralysis and feeding interruption are quickly observed after ingestion. The protein crystals are solubilized and processed in the midgut fluids of the larvae to activated insecticidal toxins, which interact with specific receptors on the gut cells (Pigott and Ellar, 2007). This interaction results in formation of a pore on the gut cell membrane, leading to osmotic cell death. As gut cells die, the integrity of the gut epithelium is compromised, allowing mixing between the main body cavity (hemocoel) and the gut fluids, resulting in a lower pH that promotes spore germination to vegetative cells. The possibility that interactions between toxins and receptors also promote spore germination has also been suggested (Du and Nickerson, 1996). If damage to the gut epithelium is extensive enough to overcome any healing mechanism, vegetative cells are able to invade the hemocoel and proliferate to cause septicemia (Raymond et al., 2008). Once nutrient resources in the host cadaver are used, vegetative cells undergo sporulation.

The mode of action of their insecticidal proteins differ (Berry, 2012), but some of the main features observed for *B. thuringiensis* pathogenesis are common to mosquitocidal biopesticides based on *L. sphaericus*. Inhibition of larval feeding is quickly observed after ingestion of *L. sphaericus*, and death occurs within 2 days (Singer, 1981). The *L. sphaericus* crystals are solubilized and the resulting proteins are processed to become activated toxins. These toxins interact with specific receptors on the cells located in the gastric cecum and posterior midgut (Charles et al., 1996). In case of the binary (Bin) toxins, one of the toxins (BinB) binds to the cell surface and acts as binding domain for the other toxin (BinA), which is then internalized to kill the gut cell in a process involving autophagy (Opota et al., 2011). Upon gut epithelium disruption, toxins can affect cells in the neural and skeletal tissues, explaining gut paralysis observed between 24 and 48 h (Davidson et al., 1975). After host death, *L. sphaericus* spores germinate and vegetative cells proliferate until nutrient depletion, when sporulation occurs. This amplification and release of spore numbers, together with the lower spore settling (Yousten et al., 1992), contribute to the longer environmental persistence of *L. sphaericus* compared to Bti. As observed for *B. thuringiensis*, sublethal exposure results in long-term effects on development of the population (Lacey et al., 1987).

Although biopesticides based on *L. sphaericus* target mosquito larvae, products based on *B. thuringiensis* target pests of Lepidoptera (butterflies and moths),

Coleoptera (beetles), or Diptera (mosquitoes and blackflies), depending on the specific isolate used. Thus, Btk produces Cry1A and Cry2 proteins and *aizawai* produces Cry1A, Cry1C, and Cry1D proteins, but both serovars are active against lepidopteran insects. In contrast, Bti strains producing Cry4 and Cry11 are active against dipteran pests (mosquitoes and blackflies), and Btt producing Cry3 toxin is active against coleopteran insects. Commercial *L. sphaericus* pesticides are commonly based on strains 2362 or C3-41 (Park et al., 2010).

The majority of commercialized bacterial insecticidal products are used in specialty crops and for vector control, representing a small niche (1–2%) of total insecticide sales (Nester et al., 2002). Diverse types of formulations have been developed to optimize delivery against specific insect pests. Factors that may limit the efficacy of bacterial bioinsecticides include short persistence and residual activity or poor control of tunneling or root-feeding pests. Most experimental advancements in formulation have concentrated on protection from environmental degradation by the use of melanin as ultraviolet (UV) protectant (Saxena et al., 2002; Zhang et al., 2008) and protective delivery systems (Elcin, 1995a, b). The efficacy of *B. thuringiensis* formulations has been demonstrated in a variety of agricultural and environmental habitats against multiple target species. In the case of mosquito control, biopesticides based on Bti have demonstrated their efficacy in controlling diverse target species, with a concomitant lack of field-evolved resistance. However, these pesticides display lower persistence and activity in rich organic habitats. In contrast, *L. sphaericus* formulations are very effective in these and other habitats, and they present longer persistence due to recycling in the environment. Limitations for *L. sphaericus* products include their narrow host range and lack of activity in some cases (e.g. *Stegomyia* spp.), and the development of field resistance. The relative efficacy of mosquitocidal products based on either bacteria depends on a number of biotic and abiotic factors, including larval density and feeding behavior, rate of ingestion, and environmental conditions (Lacey, 2007).

12.4. CULTURE SELECTION AND MAINTENANCE

Commercial production implies that the best isolate has already been selected for the applicable commercial market. However, before commercial production begins, several important steps involving the culture occur. A spore stock is prepared and, after a series of tests for purity, the spores are stored in freeze-dried ampoules or under liquid nitrogen (both work well). Cultures are grown at 30 °C in a shake flask with Lemco broth or Trypticase soy broth for 48–72 h (Lisansky et al., 1993). A culture may also be grown in an inoculation medium similar to the actual production medium. Observations are made by culturing the spore stock on agar plates and incubating these at 30 °C for 5 days to ensure the absence of contaminating bacteria and other pollutants or bacteria phage. Aliquots of the broth made from the parent are also tested for the presence or absence of beta exotoxin if the bacterium is a *B. thuringiensis* isolate. This determination may be done using chromatographic detection (Hernández et al., 2001) or bioassays (Mac Innes and Bouwer, 2009). Generally, once it is determined to have met all the purity parameters, the culture is stored in lyophilized vials or in liquid nitrogen refrigeration vials. Usually, batches of 100 vials are made from the parent culture. Whenever new batches of the stock culture are made, the starting culture must undergo testing to ensure purity and insecticidal activity at levels equivalent to the parent. This repeated testing is extremely important to ensure product integrity and potency.

Maintenance of the parent culture and subsequent inoculation vials to prevent contamination from phage and the prevention of the loss and exchange of plasmids during subculturing is of paramount importance, especially for *B. thuringiensis* strains (González et al., 1982). The latter is the primary reason that manufactures limit the number of transfers from the parent culture. Finally, the culture selected for industrialization must meet several important commercial criteria: insecticidal spectrum, potency per unit volume of fermentation broth, fermentation media requirements, ease of production, genetic stability, and storage stability (Couch, 2000).

12.5. INOCULUM PREPARATION FOR ENTOMOPATHOGENIC BACTERIA

The preparation of the inoculation that is to be used in the main fermentor is the most important step in the commercial process. If the inoculum is not prepared properly, it could affect the yield of the main fermentor. As previously described (Beegle et al., 1991; Couch, 2000), one or several lyophils or nunc vials are aseptically transferred to several 2-l flasks containing trypticase phosphate broth for *B. thuringiensis* or nutrient broth supplemented with 0.05% yeast extract for *L. sphaericus*. The flasks are then placed on a standard rotary shaker and kept at 28 °C for 24 h. Flasks are checked by microscopic examination for purity. These are then used to inoculate the seed tank. The size of the inoculum will vary between 1% and 5%, depending on the size of the commercial fermentor (Couch, 2000).

The fermentation medium in the seed fermentor is generally similar to that used in the main fermentation tank (Table 12.1). When thoroughly mixed, the media is sterilized. Following sterilization, the medium should have a milky or creamy color. If the medium appears to be brown, this usually indicates that the glucose has caramelized because of overheating. The bacteria will not grow in caramelized glucose and the medium will have to be remade. The actual procedure for handling the seed fermentor varies by manufacturer. In a typical process for a *B. thuringiensis* seed tank, the inoculum is fermented for no more than 12 h and is normally used within 8–10 h under the following conditions: dissolved oxygen is kept above 20%, fermentation temperature is 30–32 °C, and pH is 5.5–8.0 (pH adjustments are not normally required). If pH > 8 is reached, the inoculum is not used. Transfer criteria for the culture are the same for the

TABLE 12.1 An Example of Inoculation (Seed) Medium Components

Components	Concentration (g/l)
Soy flour	25.0–35
Corn steep solids	12.5
Glucose	15.0–35
Yeast extract	2.0
N-Z Amine B	2.0
KH_2PO_4	0.05

Adjust pH to 6.8–7.2.

seed flask. Vegetative cells should be motile, nongranulated, and actively dividing. The parameters for *B. thuringiensis* also apply to *L. sphaericus,* except for the comments regarding glucose changes because *L. sphaericus* cannot use glucose as a carbon source.

12.6. FERMENTATION MEDIUM SELECTION

The choice of the proper medium ingredients is critical to ensure that the maximum insecticidal activity is derived from the fermentation process. Table 12.2 lists the most commonly used materials in the *B. thuringiensis* and *L. sphaericus* commercial fermentation process. This list of ingredients was obtained from available articles describing fermentation media to maximize the attributes of the entomopathogenic bacterium selected for development. With the exception of the trace minerals, the ingredients are undefined, commercially available, and relatively inexpensive.

The selection of the commercial medium is accomplished by setting up a series of experiments using small fermentors to select the optimum concentration of carbon and nitrogen sources and the selection of the proper concentration of trace minerals. The balance of the carbon and nitrogen concentrations is extremely important. Usually the results obtained from 150- to 500-l working volume pilot tanks tend to directly scale up to commercial production. As mentioned in a previous review on this subject (Couch, 2000), a commercial fermentation means that the size of fermentation batch is never less than 30,000 l. Anything smaller is considered a pilot plant and the process described below would not be cost-effective. The runs in the small tanks are thoroughly analyzed for insecticidal activity, cycle length, and the yield of the active ingredient per unit volume of the fermentation medium. Effects of temperature, cycle length, agitation, pH, dissolved oxygen, and concentration of glucose (*B. thuringiensis*) are constantly monitored. After a medium is selected (a typical fermentation

TABLE 12.2 A Listing of Some Typical Commercial Fermentation Ingredients and Their Concentrations

Ingredients	Concentration (g/l)
Carbon Sources	
Glucose	10–30
Corn syrup (dextrose equivalent 95)	20–45
Molasses	1.0–18.6
Glycerol	2.0–10
Corn starch	10–15
Nitrogen Sources	
Soy flour	20–40
Cotton seed flour	14–30
Potato protein	15–40
Fish meal	15–20
Peptone	2.0–5.0
Corn steep liquor or solids	15–30
Yeast	2.0
Trace Minerals	
KH_2PO_4	1.0
K_2HPO_4	1.0
$FeSO_4$	0.02
$FeSO_4 \cdot 7H_2O$	0.0005–0.02
$MgSO_4 \cdot 7H_2O$	0.3
$MnSO_4 \cdot H_2O$	0.02
$ZnSO_4 \cdot 7H_2O$	0.02
$(NH_4)2SO_4$	2.0
$CuSO_4 \cdot 5H_2O$	0.005
$CaCO_3$	1.0–1.5
Polypropylene glycol 2000	2.0–50
Silicone antifoam	0.1–0.2

(Note: Ensure through testing that the silicone antifoam does not interfere with the normal growth of the bacterium and its insecticidal activity.)
Source: Adapted from Couch (2000), Beegle et al. (1990), Lisansky et al. (1993), and Rowe et al. (1987).

TABLE 12.3 An Example of a Commercial Fermentor Media Composition

Components	Concentration (g/l)
Soy flour	35.0–45.0
Corn steep solids	15.0–20.0
Glucose (dextrose equivalent 95)	10.0–20.0
Yeast extract	0.5
N-Z Amine B	0.1
KH_2PO_4	0.8
$MgSO_4$	0.01
$ZnSO_4$	0.01
$FeSO_4$	0.01
PPG 2000 antifoam	2.5

medium is presented in Table 12.3), at least five consecutive fermentation cycles using this medium are run in the pilot-plant tanks. If results are repeatable, the medium will be used in commercial production.

A typical commercial process description is listed here. However, the exact fermentation processes used by commercial companies are closely guarded proprietary secrets. These processes may use either batch or fed batch systems for managing applicable nutrients. Previous authors (Lisansky et al., 1993) effectively described the processes involved in commercial fermentation, including media and the process conditions.

First, the fermentor is charged and the medium sterilized. When cool, the pH is adjusted to 7.2 and the medium inoculated. The sterilized medium can be held until the contents of the inoculation tank meet all transfer criteria. The total fermentor volume will reach 80% of its rated capacity. Running conditions include maximum aeration to ensure thorough oxygenation of the medium (air feed 0.1–1.0 volumes of air per minute per volume of batch) as confirmed by dissolved oxygen measurements. The pH should be maintained between 6.8 and 7.2, and the glucose feed should begin 1 h after the initiation of fermentation to maintain the glucose level above 2.0 gm/l. The sterilized glucose feed consists of 2000–3000 lbs of glucose in a total volume of 3000 gallons. The feed rate usually stabilizes around 2.0–5.0 gallons per minute, and it is stopped when the culture has reached its maximum oxygen utilization rate (OUR) and the log growth phase is over (8–10 h).

Some fermentors are not set up for automatic pH control; therefore, the media used in these fermentors typically contain a buffer. There is some evidence (Smith, 1982; Yousten and Wallis, 1987) that lack of pH control of *B. thuringiensis* and *L. sphaericus* cultures resulted in a higher potency of the endotoxins when compared to a process in which the pH was tightly controlled. However, currently most production facilities control pH.

The process described above is a fed batch system in which the carbon source is fed throughout the fermentation cycle until the culture meets its growth peak as confirmed by pH and OUR. The optimum temperature for the commercial process would have been identified in the pilot-plant experiments. The optimum temperature for Bti and *L. sphaericus* is between 28 °C and 32 °C. Cooler temperatures extend the fermentation cycle and add to the factory costs, and neither culture grows well at temperatures above 32 °C. The published literature is extensive on the effects of dissolved oxygen and the glucose concentration of the fermentation medium on the growth of *B. thuringiensis* (Scherrer et al., 1973; Foda et al., 1985), but this literature reflect data from research done in laboratory equipment, which do not typically scale up to commercial conditions.

The fermentation process takes 28–32 h and is terminated when >80–90% of the culture has lysed as determined by microscopic examination; at this point, the recovery process is initiated (see below). As reported elsewhere (Couch, 2000), in *B. thuringiensis* cultures endospore and crystal formation can be observed after 18 h. The final whole culture (FWC) is cooled to 4 °C and the pH is then adjusted to pH 4.5. Complete lysis of the culture is accomplished during cooling and only free spores and crystals will be present in the FWC. The entire cycle from seed tank to the point just before recovery of the active ingredient is typically between 62 and 92 h (Couch, 2000).

12.7. RECOVERY AND CONCENTRATION STEPS

12.7.1. Recovery

The recovery process must be efficient and at least 80% effective in the removing the active ingredient from the FWC. The FWC for *B. thuringiensis* and *L. sphaericus* generally contains 6–8% solids, of which 1–3% will be spores and delta endotoxin and the remainder of the solids will be soluble and insoluble carbohydrates and proteins (Couch, 2000). The FWC can be recovered using either or a combination of three techniques depending on the strain of *B. thuringiensis* or *L. sphaericus* that is being recovered. These methods are centrifugation, microfiltration, and evaporation.

Continuous centrifugation at >8000 × *g* is the most prevalent method for active ingredient recovery from the *L. sphaericus* FWC, and it is also the least expensive. Lisansky et al. (1993) defined the basic parameters for centrifugation recovery conditions. The starting solid content of the FWC should be 4–6% percent. Prior to centrifugation, a flocculant may be used. Also, lowering the pH to 4.5 seems to aid recovery of the solids. The final concentration following centrifugation will range between 15% and 30% solids. Lisansky et al. (1993) stated that if a starting FWC has a lower solid concentration, centrifugation may be inefficient because the recovery time of this method is longer and more costly. It is at this point when combining centrifugation with evaporation may be a viable option. Centrifugation can also be made more efficient by diluting the FWC before centrifugation to remove more of the dissolved solids. If the FWC is viscous, enzyme treatment may also improve recovery efficiency. Average loss of the active ingredient through the centrifugation step is typically 10–15%.

Microfiltration is a method that removes the active ingredients from the FWC in a very efficient manner. Almost 95–100% recovery of the active solids is possible. This is particularly true for *B. thuringiensis* because of the small size of the delta endotoxin. The filters can be membrane or ceramic and a pore size of 0.1–0.2 microns is the preferred size. During recovery, the FWC is continuously washed and cooled to optimize efficiency and potency. This method has the additional benefit of removing any undesirable metabolites. The downside of microfiltration is that it is very expensive, although its efficiency often negates the expense of the start up and operation of the system.

Every commercial company has its own recovery procedures, which generally are not shared. It is safe to assume commercial companies exercise extreme care during the recovery step to preserve the insecticidal activity of the entomopathogenic bacteria. High temperature and shear will destroy activity quickly and must always be avoided. The recovery process must also be structured to prevent exogenous microbial contamination of the recovered concentrate.

12.8. FORMULATION SELECTION

The formulation of the recovered FWC powder or liquid concentrate is the final important step in the commercialization of the entomopathogenic bacterium. The formulation must be made from readily available inert ingredients, which have the important function of facilitating the application of the bacterium through all standard application systems. Couch and Ignoffo (1981) stated that a basic commercial formulation constitutes the form and contents of the insecticide as supplied by the manufacturer to distributor and ultimately to the end user; a tank mix formulation is the commercial formulation plus the spray vehicle (e.g. water, oil) added by the end user and applicator.

The technical formulation of the entomopathogenic bacterium must be converted into a useable formulation that is compatible in all applications systems, both ground and air. The insecticidal activity of the formulation must be stable under normal storage conditions found in typical warehouses. The shelf life expectation for these formulations is 18–36 months when stored in a cool dry place below 30 °C. Prolonged storage at higher temperatures is discouraged and will have an adverse effect on the activity of the formulation and its performance in the field.

The formulation must be optimized for the distribution of the entomopathogenic bacterium on the target crop or other site depending on the strain of bacterium and target insect. In the case of aquatic sites, a dispersant is often used in the liquid Bti to facilitate the dispersion of the delta endotoxin through the feeding horizon of the mosquitoes or blackflies. In agricultural applications, it is not only important to develop a formulation that is easily mixed and applied, but it should have appropriate ingredients to spread the formulation on the target surface to provide sticking properties to the formulation and some UV protection.

Formulations of bacteria are exactly like those needed for chemical pesticides. The latter have been in the market longer than entomopathogenic bacteria and most equipment has been developed for these products. Therefore, the formulations of entomopathogenic bacteria have to be adaptable to this equipment. These are liquid and solid formulations, which include granules, pellets, briquettes, and donut formulations and emulsifiable suspensions using mineral oil as a carrier.

The aqueous liquid formulations are the least expensive to produce and formulate. The concentrated fermentation slurry (FC) is the base for these formulations. The FC is stabilized using appropriate bacteriostatic and fungistatic agents to prevent secondary fermentation. Also included are usually a dispersant, a suspending agent, and sometimes a thickening agent when the FC has a very low viscosity. The exact formula for the FC is a closely guarded company secret and is listed only in the confidential statement of formula for each aqueous suspension. Several articles present extensive lists of typical liquid and solid formulation components (Couch and Ignoffo, 1981; Beegle et al., 1991; Lisansky et al., 1993; Burges and Jones, 1998; Couch, 2000). For liquids, these lists include the most commonly used dispersants, suspending agents, fungistats, buffers, stabilizers, preservatives, thickeners, and antievaporation agents. Aqueous liquid formulations have very strict storage requirements because they are more susceptible to temperature effects and exogenous contamination when the containers are opened. Prolonged storage temperatures must not exceed 30 °C and storage at 10–15 °C is preferred. Generally, the aqueous liquid formulations when stored properly will work at label rates for two seasons. The potency of the product may be less than when released, but the recommended rates on the label have been adjusted to take into account a 10–15% decrease in label potency.

Solid formulations are produced from the dry technical grade active ingredient (TGAI), which is made from the FC by spray-drying. The drying is a critical step because the type of drier and temperatures used can destroy the insecticidal activity of the TGAI. Experimentation on a laboratory scale drier is used to select the inlet and outlet temperature for the commercial drier. The handling of the FC during drying and the powder after drying is very important. The inlet and outlet temperatures on the spray drier must be continually monitored so it is below the critical temperature that is destructive to the insecticidal activity. Once dried, the powder is recovered by collecting it with cool dehumidified air. Because the powder is an insulator, it will retain heat for a long time in the drums into which it is packed for storage and sale, so it must be less than 30 °C when packed. Each company has its own methodology, and drying and recovery conditions are proprietary. If done properly, an optimized spray drier can recover 85–95% of the active ingredient from the FC. The spray dried powder is easily stored and very stable.

To make a solid formulation from the TGAI, it is mixed with an appropriate diluent. The types and kinds of diluents are extensive. Wettable powders (WP), granules, briquettes, pellets, and donut formulations produced by companies selling entomopathogenic formulations are proprietary and confidential. Each of the dry formulation types has specific uses. Initially, the most common of the dry formulations was the WP, which contains a diluent (usually neutral clay) and appropriate surfactants and dispersants to improve handling in the field. These inerts should be compounded to optimize product deposition to ensure performance against the target insect. The powders also include dust formulations, which are rarely used today.

The most common and preferred dry formulation of entomopathogenic bacteria is the water-dispersible granule; in contrast to the WP, they are not dusty and mix almost instantly. Water-dispersible granules are designated as WG (water granules), WDG (water dispersible granules), or DF (dry flowable). The abbreviation WDG will be used here to describe this type of formulation. Like the powders, the WDG has excellent shelf life with no noticeable decrease in activity when stored for up to 3 years in a cool dry environment. The WDG formulations contain many of the same ingredients as the WP, but they are granulated in a pan granulator or fluid bed drier. Couch (2000) prepared a comprehensive list of the specific types of dry diluents, carrier, and surfactants used in this type of formulation.

The liquid, WP, and WDG formulation parameters apply to Btk, Bti, and *L. sphaericus*. However, BTI and *L. sphaericus* have several unique formulations because they are applied to water. Specifically, these are unique granules, briquettes, and a donut formulation. Both BTI and *L. sphaericus* are formulated on corn cob granules of various mesh sizes. The most common mesh sizes are 10/14 and 5/8. The BTI and *L. sphaericus* TGAI are formulated on the corn cob carrier with an edible mineral oil or soy bean oil binder. The granules, depending on the concentration of TGAI formulated onto the corn cob, are applied using aerial and ground equipment at concentrations of 1–9 kg per hectare. The granules are able to penetrate heavy cover (e.g. trees, bushes, grass, crops), which prevent the penetration of an aqueous liquid suspension. They are applied dry with no mixing required with water. BTI can also be formulated on site with sand and an oil binder to provide a dense granule with even better penetration of a heavy over story. For special applications of the *L. sphaericus* granule, it is packed into water soluble pouches for application to containers and storm water catch basins, for example.

The briquettes are formulated to provide prolonged residual control. Some formulations last up to 180 days in standing water, including abandoned swimming pools, cisterns, open ditches, sewage ponds, water tanks, and other standing water sites. There is a combination briquette using both BTI and *L. sphaericus* as its active ingredients. The donut (which looks like a donut) is also used in sites such as those described for the briquettes. These formulations are patented by their manufacturer.

Transgenic plants are not included in this discussion because they are not considered a formulation by regulatory agencies.

12.9. FORMULATION STANDARDIZATION

Once a formulation is selected, it must be standardized to ensure there is no variation in the concentration of the active ingredient or the potency of the formulation. Standard operating procedures include determination of color, odor, particle size, specific gravity/density, viscosity (liquids only), particle size, wetting times (powders and dispersible granules), pH, and suspensibility. Each lot of the finished formulation produced is checked for conformance to the above parameters. If the batch has any of the parameters that are not the adopted standard for the formulation, it is rejected and sent for reformulation and recovery.

12.10. QUALITY ASSURANCE METHODS

As mentioned previously, the formulations are standardized around specific physical and chemical parameters. The methods to determine these parameters are standard techniques and will not be discussed here. Entomopathogenic bacteria products are unique because there is no chemical or reliable biochemical method to determine the insecticidal potency of the TGAI and formulated products. Therefore, insect bioassays must be used to standardize these products. Insect bioassay procedures have been used since the first *B. thuringiensis* based product was commercialized. An initially proposed method (Dulmage et al., 1971) was later standardized for Btk isolates (Beegle et al., 1986). For Btt, a method of standardization was described using Colorado potato beetle (*Leptinotarsa decemlineata* (Say)) larvae (Riethmüller and Langenbruch, 1989).

Although no transconjugate entomopathogenic bacterium that contains toxins from more than one *B. thuringiensis* serotype is currently registered with the U.S. Environmental Protection Agency, standardization is difficult using insects. In this case, enzyme-linked immunosorbent assays and high-performance liquid chromatography methods are frequently employed to measure potency. However, these methods will often give a positive result for the presence of the insecticidal protein, even if it has been denatured and no longer possesses insecticidal activity. This is the primary reason why current commercial products using entomopathogenic bacteria all use insect bioassay techniques for standardization.

In 1999, WHO convened a panel to develop a set of guidelines for the mosquito biolarvicides BTI and *L. sphaericus* (WHO, 1999). Although these guidelines (WHOPES) were to be applied only to entomopathogenic bacteria used in public health, they can also be applied to agricultural isolates of *B. thuringiensis*. In fact, these guidelines have been and are used as the basis for many of the quality control parameters for the final formulations of commercial entomopathogenic bacteria. A common presence through all the bioassay methodologies and guidelines is the need to conduct all bioassays of the commercial formulations against an internal standard of known potency. The bioassay methods cited describe these standards and how they are maintained.

After the potency determinations have been completed, a mouse safety test must be run on the TGAI used in the commercial formulations. Five mice are injected with $10e^6$ spores subcutaneously and observed for 7 days. The test was designed to screen for infectious or exogenous contaminants.

Although not required by the regulatory authorities, every batch of TGAI and the formulated product are also screened as per the WHOPES guidelines. The exogenous microbial contaminant report must show that the following contaminants do not exceed WHOPES limits: coliforms <10/gm; *Staphylococcus aureus* Rosenbach absent in 1 g; *Salmonella* spp. absent in 10 g; *Enterococcus (Streptococcus) faecalis* Andrews and Horder (Schleifer and Kilpper-Balz) 1×10^4 per gram; viable yeasts and molds <100/gram and no human pathogens. Once the products meet the physical chemistry standard, potency standard, and exogenous contaminant standard, the products are packaged and cleared for sale.

12.11. CONCLUSION

The commercial fermentation methodologies for the bacteria strains currently on the market are highly evolved and optimized to produce the maximum insecticidal proteins per unit volume of the fermentation media. However, manufacturers continually strive to improve the yields and lower manufacturing costs to ensure that the products can compete in the marketplace. It is also obvious from the literature that the most active isolates continue to be naturally occurring bacteria and not transconjugates or other engineered strains, which fuels interest in prospecting for new, more active isolates.

The formulation of these commercial products to ensure stability and ease of application is continually being evaluated to increase shelf life. Many of the new formulations are considered to be "green formulations" and are used in organic farming. Finally, advances have been made to increase residual activity of the formulations; in the case of Bti and *L. sphaericus*, formulations have been developed that have extended release properties. These are effective in controlling mosquito larvae from 30 to 180 days.

REFERENCES

Ampofo, J.A., 1995. Use of local raw materials for the production of *Bacillus sphaericus* insecticide in Ghana. Biocontrol Sci. Technol. 5, 417–423.

Beegle, C.C., Couch, T.L., Alls, R.T., Versoi, P.L., Bulla, L.A., 1986. Standardization of HD-1-S-1980: U.S. standard for assay of lepidopterous-active *Bacillus thuringiensis*. Bull. Entomol. Soc. Amer. 32, 44–45.

Beegle, C.C., Rose, R.I., Ziniu, Y., 1990. Mass production of *Bacillus thuringiensis* and *B. sphaericus* for microbial control of insect pests. In: Maramorosch, K.B. (Ed.), Biotechnology for Biological of Pests and Vectors.. CRC Press, Boca Raton, Florida, 32, pp.195–216.

Bernhard, K., Jarrett, P., Meadows, M., Butt, J., Ellis, D.J., Roberts, G.M., Pauli, S., Rodgers, P., Burges, H.D., 1997. Natural isolates of *Bacillus thuringiensis*: worldwide distribution, characterization, and activity against insect pests. J. Invertebr. Pathol. 70, 59–68.

Berry, C., 2012. The bacterium, *Lysinibacillus sphaericus*, as an insect pathogen. J. Invertebr. Pathol. 109, 1–10.

Brar, S.K., Verma, M., Tyagi, R.D., Valero, J.R., Surampalli, R.Y., 2006a. Efficient centrifugal recovery of *Bacillus thuringiensis* biopesticides from fermented wastewater and wastewater sludge. Water Res. 40, 1310–1320.

Brar, S.K., Verma, M., Tyagi, R.D., Valero, J.R., Surampalli, R.Y., 2006b. Screening of different adjuvants for wastewater/wastewater sludge-based *Bacillus thuringiensis* formulations. J. Econ. Entomol. 99, 1065–1079.

Brar, S.K., Verma, M., Tyagi, R.D., Valéro, J.R., Surampalli, R.Y., 2009. Entomotoxicity, protease and chitinase activity of *Bacillus thuringiensis* fermented wastewater sludge with a high solids content. Bioresour. Technol. 100, 4317–4325.

Burges, H.D., Jones, K.A., 1998. Formulation of bacteria, viruses and protozoa to control insects. In: Burges, H.D. (Ed.), Formulation of Microbial Biopesticides; Beneficial Microorganisms, Nematodes and Seed Treatments, Kluwer Academic Publishers, Dordrecht, pp. 33–127.

Charles, J.F., Nicolas, L., 1986. Recycling of *Bacillus sphaericus* 2362 in mosquito larvae: a laboratory study. Ann. Inst. Pasteur Microbiol. 137, 101–111.

Charles, J.F., Nielsen-LeRoux, C., Delécluse, A., 1996. *Bacillus sphaericus* toxins: molecular biology and mode of action. Annu. Rev. Entomol. 41, 451–472.

Couch, T.L., 2000. Industrial fermentation and formulation of entomopathogenic bacteria. In: Charles, J.F., Delecluse, A., Nielsen-Leroux, C. (Eds.), Entomopathogenic Bacteria: From Laboratory to Field Application, Kluwer Academic Publishers, Dordrecht, pp. 297–314.

Couch, T.L., Ignoffo, C.M., 1981. Formulation of insect pathogens. In: Burges, H.D. (Ed.), Microbial Control of Pests and Plant Diseases 1970–1980, Academic Press, New York and London, pp. 621–634.

Davidson, E.W., Singer, S., Briggs, J.D., 1975. Pathogenesis of *Bacillus sphaericus* strain SSII-1 infections in *Culex pipiens quinquefasciatus* (= *C. pipiens fatigans*) larvae. J. Invertebr. Pathol. 25, 179–184.

de Barjac, H., Véron, M., Cosmao Dumanoir, V., 1980. Biochemical and serological characterization of *Bacillus sphaericus* strains, pathogenic or non-pathogenic for mosquitos. Ann. Microbiol. B131, 191–201.

Devi, P.S.V., Ravinder, T., Jaidev, C., 2005. Cost-effective production of *Bacillus thuringiensis* by solid-state fermentation. J. Invertebr. Pathol. 88, 163–168.

Du, C., Nickerson, K.W., 1996. *Bacillus thuringiensis* HD-73 spores have surface-localized Cry1Ac toxin: physiological and pathogenic consequences. Appl. Environ. Microbiol. 62, 3722–3726.

Dulmage, H.T., Boening, O.P., Rehnborg, C.S., Hansen, G.D., 1971. A proposed standardized bioassay for formulations of *Bacillus thuringiensis* based on the international unit. J. Invertebr. Pathol. 18, 240–245.

El-Bendary, M.A., 2006. *Bacillus thuringiensis* and *Bacillus sphaericus* biopesticides production. J. Basic Microbiol. 46, 158–170.

Elcin, Y.M., 1995a. *Bacillus sphaericus* 2362-calcium alginate microcapsules for mosquito control. Enzyme. Microb. Technol. 17, 587–591.

Elcin, Y.M., 1995b. Control of mosquito larvae by encapsulated pathogen *Bacillus thuringiensis* var. *israelensis*. J. Microencapsul. 12, 515–523.

Flores, A.E., Garcia, G.P., Badii, M.H., Rodriguez Tovar, M.A., Fernnadez Salas, I., 2004. Effects of sublethal concentrations of Vectobac® on biological parameters of *Aedes aegypti*. J. Am. Mosq. Control Assoc. 20, 412–417.

Foda, M.S., Salama, H.S., Selim, M., 1985. Factors affecting growth physiology of *Bacillus thuringiensis*. Appl. Microbiol. Biotechnol. 22, 50–52.

González, J.M.J., Brown, B.J., Carlton, B.C., 1982. Transfer of *Bacillus thuringiensis* plasmids coding for delta-endotoxin among strains of *Bacillus thuringiensis* and *Bacillus cereus*. Proc. Natl. Acad. Sci. U. S. A. 79, 6951–6955.

Guerineau, M., Alexander, B., Priest, F.G., 1991. Isolation and identification of *Bacillus sphaericus* strains pathogenic for mosquito larvae. J. Invertebr. Pathol. 57, 325–333.

Hernández, C.S., Ferré, J., Larget-Thiéry, I., 2001. Update on the detection of β-exotoxin in *Bacillus thuringiensis* strains by HPLC analysis. J. Appl. Microbiol. 90, 643–647.

Huang, F.N., Buschman, L.L., Higgins, R.A., 1999. Susceptibility of different instars of European corn borer (Lepidoptera: Crambidae) to diet containing *Bacillus thuringiensis*. J. Econ. Entomol. 92, 547–550.

Jackson, T.A., Pearson, J.F., O'Callaghan, M., Mahanty, H.K., Willcocks, M.J., 1992. Pathogen to product—development of *Serratia entomophila* (Enterobacteriaceae) as a commercial biological agent for the New Zealand grass grub (*Costelytra zealandica*). In: Jackson, T.A., Glare, T.R. (Eds.), Use of Pathogens in Scarab Pest Management, Intercept Press, Andover, pp. 191–198.

Johnson, V.W., Pearson, J.F., Jackson, T.A., 2001. Formulation of *Serratia entomophila* for biological control of grass grub. N. Z. Plant Prot. 54, 125–127.

Jurat-Fuentes, J.L., Jackson, T.A., 2012. Bacterial entomopathogens. In: Vega, F.E., Kaya, H.K. (Eds.), Insect Pathology, second ed. Academic Press, San Diego, pp. 265–349.

Klein, M.G., Jackson, T.A., 1992. Bacterial diseases of scarabs. In: Jackson, T.A., Glare, T.R. (Eds.), Use of Pathogens in Scarab Pest Management, Intercept Ltd, Andover, UK, pp. 43–61.

Lacey, L.A., 2007. *Bacillus thuringiensis* serovariety *israelensis* and *Bacillus sphaericus* for mosquito control. J. Am. Mosq. Control Assoc. 23, 133–163.

Lacey, L.A., Day, J., Heitzman, C.M., 1987. Long-term effects of *Bacillus sphaericus* on *Culex quinquefasciatus*. J. Invertebr. Pathol. 49, 116–123.

Lisansky, S.G., Quinlan, R., Tassoni, G., 1993. The *Bacillus thuringiensis* Production Handbook. CPL Press, Newbury, Great Britain.

Mac Innes, T.C., Bouwer, G., 2009. An improved bioassay for the detection of *Bacillus thuringiensis* β-exotoxin. J. Invertebr. Pathol. 101, 137–139.

Martin, P.A., Farrar Jr., R.R., Blackburn, M.B., 2009. Survival of diverse *Bacillus thuringiensis* strains in gypsy moth (Lepidoptera: Lymantriidae) is correlated with urease production. Biol. Control 51, 147–151.

Martin, P.A., Gundersen-Rindal, D.E., Blackburn, M.B., 2010. Distribution of phenotypes among *Bacillus thuringiensis* strains. Syst. Appl. Microbiol. 33, 204–208.

Moreau, G., Bauce, E., 2003. Lethal and sublethal effects of single and double applications of *Bacillus thuringiensis* variety *kurstaki* on spruce budworm (Lepidoptera: Tortricidae) larvae. J. Econ. Entomol. 96, 280–286.

Morris, O.N., Kanagaratnam, P., Converse, V., 1997. Suitability of 30 agricultural products and by-products as nutrient sources for laboratory production of *Bacillus thuringiensis* subsp. *aizawai* (HD133). J. Invertebr. Pathol. 70, 113–120.

Nester, E.W., Thomashow, L.S., Metz, M., Gordon, M., 2002. 100 years of *Bacillus thuringiensis*: A Critical Scientific Assessment. American Academy of Microbiology, Washington.

Nicolas, L., Dossou-Yovo, J., Hougard, J.M., 1987. Persistence and recycling of *Bacillus sphaericus* 2362 spores in *Culex quinquefasciatus* breeding sites in West Africa. Appl. Microbiol. Biotechnol. 25, 341–345.

Nyouki, F.F.R., Fuxa, J.R., Richter, A.R., 1996. Spore–toxin interactions and sublethal effects of *Bacillus thuringiensis* in *Spodoptera frugiperda* and *Pseudoplusia includens* (Lepidoptera: Noctuidae). J. Entomol. Sci. 31, 52–62.

Opota, O., Gauthier, N.C., Doye, A., Berry, C., Gounon, P., Lemichez, E., Pauron, D., 2011. *Bacillus sphaericus* binary toxin elicits host cell autophagy as a response to intoxication. PLoS One 6, e14682.

Organization, W.H., 1999. Guidelines specification for bacterial larvicides for public health use. Report of the W.H.O. Informal Consultation, World Health Organization, Geneva, Switzerland-Publication W.H.O./CDS/CPC/WHOPES/99.92.

Park, H.W., Bideshi, D.K., Federici, B.A., 2010. Properties and applied use of the mosquitocidal bacterium, *Bacillus sphaericus*. J. Asia Pac. Entomol. 13, 159–168.

Park, H.W., Mangum, C.M., Zhong, H.E., Hayes, S.R., 2007. Isolation of *Bacillus sphaericus* with improved efficacy against *Culex quinquefasciatus*. J. Am. Mosq. Control Assoc. 23, 478–480.

Pigott, C.R., Ellar, D.J., 2007. Role of receptors in *Bacillus thuringiensis* crystal toxin activity. Microbiol. Mol. Biol. Rev. 71, 255–281.

Ravensberg, W.J., 2011. A Roadmap to the Successful Development and Commercialization of Microbial Pest Control Products for Control of Arthropods. Springer, Dordrecht, The Netherlands.

Raymond, B., Elliot, S.L., Ellis, R.J., 2008. Quantifying the reproduction of *Bacillus thuringiensis* HD1 in cadavers and live larvae of *Plutella xylostella*. J. Invertebr. Pathol. 98, 307–313.

Riethmüller, U., Langenbruch, G.A., 1989. Two bioassay methods to test the efficacy of *Bacillus thuringiensis* subspec. *tenebrionis* against the larvae of the Colorado potato beetle (*Leptinotarsa decemlineata*). Entomophaga 34, 237–245.

Rowe, G.E., Margaritis, A., Dulmage, H.T., 1987. Bioprocess developments in the production of bioinsecticides by *Bacillus thuringiensis*. Crit. Rev. Biotechnol. 6, 87–127.

Saxena, D., Ben-Dov, E., Manasherob, R., Barak, Z., Boussiba, S., Zaritsky, A., 2002. A UV tolerant mutant of *Bacillus thuringiensis* subsp. *kurstaki* producing melanin. Curr. Microbiol. 44, 25–30.

Scherrer, P., Luthy, P., Trumpi, B., 1973. Production of δ-endotoxin by *Bacillus thuringiensis* as a function of glucose concentrations. Appl. Microbiol. 25, 644–646.

Singer, S., 1981. Potential of *Bacillus sphaericus* and related spore-forming bacteria for pest control. In: Burges, H.D. (Ed.), Microbial Control of Pests and Plant Diseases 1970–1980, Academic Press Inc., London, pp. 283–298.

Smith, R.A., 1982. Effect of strain and medium variation on mosquito toxin production by *Bacillus thuringiensis* var. *israeliensis*. Can. J. Microbiol. 28, 1089–1092.

Smith, R.A., Barry, J.W., 1998. Environmental persistence of *Bacillus thuringiensis* spores following aerial application. J. Invertebr. Pathol. 71, 263–267.

Stahly, D.P., Klein, M.G., 1992. Problems with in vitro production of spores of *Bacillus popilliae* for use in biological control of the Japanese beetle. J. Invertebr. Pathol. 60, 283–291.

Tilquin, M., Paris, M., Reynaud, S., Despres, L., Ravanel, P., Geremia, R.A., Gury, J., 2008. Long lasting persistence of *Bacillus thuringiensis* subsp. *israelensis* (Bti) in mosquito natural habitats. PLoS One 3, e3432.

Tirado Montiel, M.L., Tyagi, R.D., Valero, J.R., 2001. Wastewater treatment sludge as a raw material for the production of *Bacillus thuringiensis* based biopesticides. Water Res. 35, 3807–3816.

Tirado Montiel, M.L., Tyagi, R.D., Valéro, J.R., 1998. Production of *Bacillus thuringiensis* Biopesticides Using Waste Materials. International Thompson Publishing, Great Britain.

Vachon, V., Laprade, R., Schwartz, J.L., 2012. Current models of the mode of action of *Bacillus thuringiensis* insecticidal crystal proteins: a critical review. J. Invertebr. Pathol. 111, 1–12.

Wraight, S.P., Molloy, D.P., Singer, S., 1987. Studies on the culicine mosquito host range of *Bacillus sphaericus* and *Bacillus thuringiensis* var. *israelensis* with notes on the effects of temperature and instar on bacterial efficacy. J. Invertebr. Pathol. 49, 291–302.

Wu, D., Chang, F.N., 1985. Synergism in mosquitocidal activity of 26 and 65 kDa proteins from *Bacillus thuringiensis* subsp. *israelensis* crystal. FEBS Lett. 190, 232–236.

Yan, S., Mohammedi, S., Tyagi, R.D., Surampalli, R.Y., Valéro, J.R., 2007. Growth of four serovar of *Bacillus thuringiensis* (var. *kurstaki*, *israelensis*, *tenebrionis*, and *aizawai*) in wastewater sludge. Pract. Period. Hazard., Toxic, Radioact. Waste Manage. 11, 123–129.

Yap, H.H., 1990. Field trials of *Bacillus sphaericus* for mosquito control. In: Barjac, H. d., Sutherland, D.J. (Eds.), Bacterial Control of Mosquitoes and Black Flies, Rutgers University Press, New Brunswick, pp. 307–320.

Yousten, A.A., Genthner, F.J., Benfield, E.F., 1992. Fate of *Bacillus sphaericus* and *Bacillus thuringiensis* serovar *israelensis* in the aquatic environment. J. Am. Mosq. Control Assoc. 8, 143–148.

Yousten, A.A., Wallis, D.A., 1987. Batch and continuous culture of the mosquito larval toxin of *Bacillus sphaericus* 2363. J. Ind. Microbiol. 2, 277–283.

Zhang, J.T., Yan, J.P., Zheng, D.S., Sun, Y.J., Yuan, Z.M., 2008. Expression of mel gene improves the UV resistance of *Bacillus thuringiensis*. J. Appl. Microbiol. 105, 151–157.

FURTHER READING

Ahmed, I., Yokota, A., Yamazoe, A., Fujiwara, T., 2007. Proposal of *Lysinibacillus boronitolerans* gen. nov. sp. nov., and transfer of *Bacillus fusiformis* to *Lysinibacillus fusiformis* comb. nov. and *Bacillus sphaericus* to *Lysinibacillus sphaericus* comb. nov. Int. J. Syst. Evol. Microbiol. 57, 1117–1125.

This page intentionally left blank

Chapter 13

Production of Entomopathogenic Viruses

Steve Reid,[1] Leslie Chan[2] and Monique M. van Oers[3]
[1]School of Chemistry and Molecular Biosciences, The University of Queensland, Brisbane, QLD, Australia, [2]Australian Institute of Bioengineering and Nanotechnology, The University of Queensland, St Lucia, QLD, Australia, [3]Laboratory of Virology, Wageningen University, Wageningen, The Netherlands

13.1. INTRODUCTION

Any document addressing the topic of the production of entomopathogenic viruses will inevitably concentrate on baculoviruses; this chapter is no different. Indeed, this chapter concentrates on the production of baculoviruses in vitro using cell culture technology. The point of view taken in this chapter is that wild-type baculoviruses have an important role to play in the control of insect pests, but a major limitation to their wider use is the lack of a cost-effective in vitro production technology. This point has been well made by a number of previous authors over the past decades (Black et al., 1997; Murhammer, 1996; Ravensberg, 2011b; Weiss and Vaughn, 1986). However, although the topic of in vitro production of baculoviruses has been covered in general terms, the specific yield targets and issues for the realization of in vitro production have been poorly addressed. This chapter attempts to rectify that deficiency in the literature.

Following a brief introduction of entomopathogenic viruses and an explanation of why baculoviruses have received the most attention, the status of in vivo production of baculoviruses will be described. A number of companies have successfully developed products using this mode of manufacturing for baculoviruses. Low-cost in vivo production of baculoviruses also receives further attention in Chapter 15 of this book. The current status of in vitro production of baculoviruses and the limitations to its successful application in the marketplace, along with suggestions for further research, are then addressed in detail. Finally, some comments are made about what needs to happen for in vitro production of baculoviruses to be realized.

This chapter does not address the many issues of product formulation to help protect the virus from ultraviolet (UV) damage and to allow better sticking and

Mass Production of Beneficial Organisms. http://dx.doi.org/10.1016/B978-0-12-391453-8.00013-3

spreading of the virus on various plant surfaces, nor farmer education requirements to encourage the use of integrated pest management strategies—all of which are required to make baculovirus biopesticides a success in the marketplace. These topics have been covered in other publications, and companies currently selling baculovirus products produced in vivo have successfully addressed many of these issues (Black et al., 1997; Harrison and Hoover, 2012; Payne, 1982; Ravensberg, 2011b). See also Chapter 14 for a discussion of the formulation of bioinsecticides by Behle and Birthisel. Although in vitro production of baculoviruses faces many challenges, the sterile harvest of the final product is not one of them. The final concentrated harvest from such processes can readily take advantage of the formulation and application technologies already developed for the in vivo produced virus products.

13.1.1. Entomopathogenic Viruses

Many viruses have been shown to infect insects, including DNA and RNA viruses. It is beyond the scope of this chapter to review them and the reader is referred to other references for detailed discussions of the full range of entomopathogenic viruses (Asgari and Johnson, 2010; Harrison and Hoover, 2012; Miller and Ball, 1988; Payne, 1982). What is clear from all reviews of insect viruses is that baculoviruses have received more attention as pest control agents than any of the other groups because they have never been found to cause disease in any organism outside the phylum Arthropoda; they also are responsible for most of the natural epizootics observed in insects (Miller, 1997; Payne, 1982).

Baculoviruses are occluded, which allows them to persist for long periods outside the host. These occlusions protect the virions from damage when applied to crops using traditional spray equipment. Harrison and Hoover (2012) provided a timely and current review of baculoviruses and other occluded insect viruses, pointing out that occlusion of virions has been observed predominately in viruses of insects, which may be related to the transient nature of insect populations. The other occluded insect viruses are the entomopoxviruses (EPVs) and the cypoviruses. Nudiviruses are nonoccluded viruses (hence the "nudi" name) that are closely related to baculoviruses. Although EPVs and cypoviruses have been assessed as potential biopesticides, the vast majority of viruses registered for use as biopesticides are baculoviruses that target lepidopteran insects (Kabaluk et al., 2010).

13.1.2. Baculoviruses

13.1.2.1. Taxonomy

The family *Baculoviridae* contains rod-shaped, double-stranded DNA viruses with a large circular genome varying in size between 80 and 180 kbp; for a review, see van Oers and Vlak (2007). The family is divided into four genera: *Alphabaculovirus* (containing nucleopolyhedroviruses (NPVs) that infect lepidopteran insects), *Betabaculovirus* (containing the granuloviruses (GVs) found in lepidopteran insects), *Gammabaculovirus* (NPVs infecting hymenoperan insects), and *Deltabaculovirus* (NPVs infecting insects in the order Diptera) (Herniou et al., 2012; Jehle et al., 2006). The alphabaculoviruses are taxonomically separated into group I and group II NPVs (Zanotto et al., 1993). The *Autographa californica* multicapsid NPV (AcMNPV) is the type species of the genus *Alphabaculovirus* and belongs to group I. It is extensively used in biotechnology to produce recombinant proteins, including vaccines. The group I NPV A*nticarsia gemmetalis* NPV, used in soybean crops, is the most widely used baculovirus for biocontrol. *Helicoverpa armigera* (Hear) NPV is used to control the cotton boll worm and belongs to group II. *Cydia pomonella* GV is the type species of the genus *Betabaculovirus*; it is successfully used in Europe in apple orchards to control the coddling moth. In this chapter, we focus on the NPVs and GVs that have commercial potential against lepidopteran insects.

13.1.2.2. Baculovirus Phenotypes and Their Function

Most baculoviruses adopt two genetically identical but phenotypically different structures (Fig. 13.1) that perform different functions in the infection cycle (see reviews by Rohrmann, 2008; Slack and Arif, 2007; van Oers and Vlak, 2007). The form that is required in the final formulation of biocontrol products needs to infect larvae orally. Hence, the virus has to be produced in the form of viral occlusion bodies (OBs), proteinaceous capsules that each contain many so-called occlusion derived viruses (ODVs). These OBs are ingested by the larval stages of the (pest) insect and fall apart in the insect's alkaline midgut. The released ODVs then infect columnar cells in the midgut epithelium. Once inside the cells, the viral nucleocapsids move along actin cables to the nucleus, where DNA replication is initiated (Ohkawa et al., 2010). Meanwhile, viral capsid proteins are being produced that encapsulate the new DNA to form progeny nucleocapsids. At the cell surface, these nucleocapsids acquire an envelope by budding through the plasma membrane to form budded viruses (BVs). Most baculoviruses contain the viral fusion protein F in their BV envelope (van Strien et al., 2000; Westenberg et al., 2004), except for group I NPVs of the genus *Alphabaculovirus*, in which this function is replaced by the GP64 protein (Blissard and Wenz, 1992; Monsma et al., 1996; Oomens and Blissard, 1999). These proteins mediate the pH-dependent fusion of the viral envelope with the membrane of a target cell, the first step in BV infection.

As gut epithelial cells are frequently renewed in a process called sloughing, BV production in the gut needs to be quick to prevent clearing of the infection by removal of infected cells (Washburn et al., 2003b). The current hypothesis is that some incoming nucleocapsids may move directly from the apical to the basal membrane, pick up GP64 expressed from an early promoter, and form a BV (Washburn et al., 2003a).

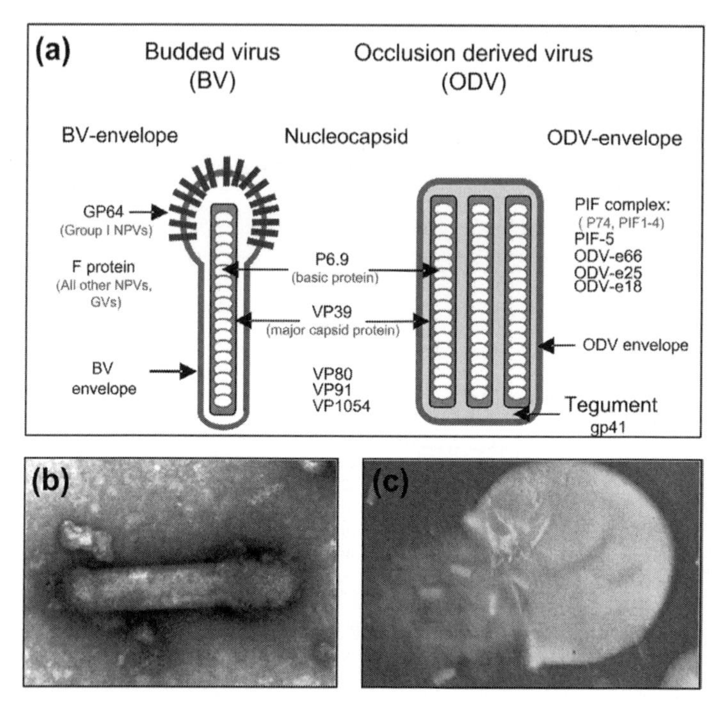

FIGURE 13.1 Baculovirus phenotypes. Schematic (a) of budded viruses (BVs) that spread the infection through the insect body and occlusion-derived viruses (ODVs) that initiate infection in midgut epithelium cells after oral uptake of occlusion bodies (OBs). Important structural proteins present in the BV envelope, the nucleocapsid (orange structure), and the ODV envelope and tegument (light blue), are indicated. GP64 is the BV envelope fusion protein of alphabaculovirus group I NPVs. In all other NPVs and GVs, the F protein serves this function (except for gammabaculoviruses, which do not make BVs). The ODV envelope contains the per os infectivity factors (PIFs), five of which are known to form a complex. PIF are crucial for oral infection. Electromicroscopy shows BV (b) and OB releasing ODVs under alkaline conditions (c). *Modified from van Oers and Vlak (2007); Copyright Bentham Science Publishers.* (For interpretation of the references to color in this figure legend, the reader is referred to the online version of this book.)

In secondary infected cells, the process of infection and virus budding is repeated, but it is followed at a later stage by retention of nucleocapids inside the nucleus. These nucleocapsids obtain an envelope derived from the inner nuclear membrane (Braunagel et al., 2009) and become the new ODVs. These ODVs become occluded in the major OB protein, polyhedrin (NPVs), or granulin (GVs) (Rohrmann, 1986). Internal body cells, especially fat body cells, are the place where massive amounts of ODVs are made and become occluded in OBs. The OBs are released into the environment when the insect dies as a consequence of the infection and serve to horizontally transmit the virus to new individuals. ODVs contain either one (GVs, single-nucleocapsid NPVs [SNPVs]) or multiple nucleocapsids (MNPVs) surrounded by a single envelope (Fig. 13.2).

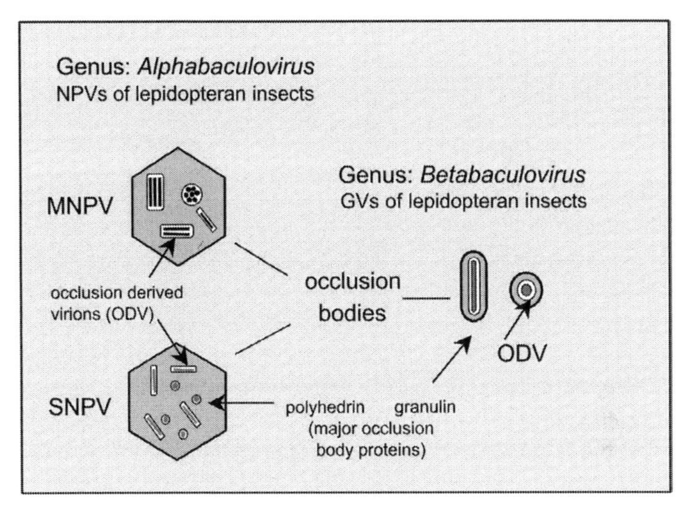

FIGURE 13.2 Baculovirus occlusion body (OB) morphology of lepidopteran-infecting baculoviruses. Comparison of OB morphology in multiple (M) and single (S) nucleocapsid nucleopolyhderoviruses (NPVs; genus *Alphabculovirus*) and granuloviruses (GVs; genus *Betabaculovirus*). *Modified from van Oers and Vlak (2007); Copyright Bentham Science Publishers.* (For color version of this figure, the reader is referred to the online version of this book.)

13.1.2.3. ODV–Midgut Interactions

The first step in a baculovirus infection of an insect is mediated by ODV envelope proteins. The ODV envelope contains so-called per os infectivity factors (PIFs) (see Fig. 13.1). Eight PIF proteins have been identified so far (Fang et al., 2009, 2006; Huang et al., 2012; Kikhno et al., 2002; Nie et al., 2012; Pijlman et al., 2003a; Simon et al., 2012; Song et al., 2008; Yao et al., 2004). PIF proteins are encoded by conserved baculovirus genes and are absolutely required for oral infection in the gut. Hence, they are a prerequisite for the use of these viruses as biocontrol agents. On the other hand, PIFs are not required for BV formation. The BVs of PIF deletion viruses are infectious upon injection into the hemocoel. Currently, little information is available on what these proteins do. We know that P74, PIF-1, and PIF-2 are involved in binding ODV to midgut epithelium (Haas-Stapleton et al., 2004; Ohkawa et al., 2005), but so far we do not know which protein is responsible for fusion between the viral and cellular membrane.

Recently, a PIF protein complex was identified on the surface of AcMNPV ODV particles containing at least five PIF proteins: P74 (PIF-0), PIF-1, PIF-2, PIF-3, and PIF-4, meaning that these probably work in concert to promote cell entry (Peng et al., 2010, 2012). Other proteins, such as P95 and probably AC5, PIF-6 (AC68), and AC108 are associated with this complex. PIF-5 (ODV-E56) was not found in this complex and may perform a separate function (Peng et al., 2012). The *Spodoptera frugiperda* homologue of AC108, SF58, was recently identified as PIF-7 (Simon et al., 2012). PIF proteins appear to be highly species specific and cannot simply be interchanged (Pijlman et al., unpublished data), which may be partly due to the interactions with other proteins in the PIF complex but may also be due to interactions with cellular factors.

Small peptides with homology to the ODV envelope protein ODV-E66 are able to block AcMNPV infection in *Helicoverpa (Heliothis) virescens* (Fabricius) larvae (Sparks et al., 2011), which may indicate that ODV-E66 also binds to a cellular factor and plays an important role in oral infectivity. Deletion of ODV-E66 indeed decreases the virulence of the virus as seen by an increase in LD_{50} (Xiang et al., 2011), but strictly speaking this is not a PIF protein because it is not absolutely required for oral infection.

13.1.2.4. Baculoviruses Encode Apoptosis Inhibitors

A major defense mechanism for insects to infection with a (baculo-)virus is apoptosis or programmed cell death, which leads to removal of infected cells from the body. Apoptosis may also occur in cell culture and may limit or completely block the replication of the virus, thereby reducing BV and OB production levels. Baculoviruses, however, encode proteins that are able to counteract the induction of apoptosis such as P35 and its relative P49 from AcMNPV and SpliNPV, respectively, and inhibitors of apoptosis (IAPs) found in many other baculoviruses. These apoptosis inhibitors often show a certain degree of host specificity and baculoviruses may not be able to prevent apoptosis in nonhost cells. See Clem (2007) for a review of this topic.

13.2. IN VIVO PRODUCTION OF BACULOVIRUS-BASED BIOPESTICIDES

A number of examples of successful programs to produce baculovirus biopesticide products using infected larvae have been documented over the past 40 years (Buerger et al., 2007; Harrison and Hoover, 2012; Ignoffo, 1973, 1999; Moscardi et al., 2011; Shapiro, 1986). A recent review of the use and regulation of microbial pesticides worldwide lists a number of baculovirus products listed for use on various crops in China, India, Europe, Argentina, Brazil, Canada, USA and Australia (Kabaluk et al., 2010). The current in vivo-produced baculoviruses of major economic importance target heliothine larvae (*H. virescens* and *Helicoverpa zea* Boddie in North America; *Helicoverpa Armigera* Hübner in China, India, and Australia; *Anitcarsia gemmatalis* Hübner in Brazil and Argentina; *Cydia pomonella* Linnaeus in Europe, Argentina, Canada, the United States, and New Zealand) and various Spodoptera pests (*Spodoptera litura* Fabricius in China, India, and Europe; *Spodoptera exigua* Hübner in China, Europe, and the United States).

Helicoverpa zea NPV (HzNPV) was developed as the first commercial viral pesticide in the United States, with label approval in 1975 (Ignoffo, 1973, 1999). This product eventually traded as Elcar, but sales fell when low-cost pyrethroids became available. The Elcar isolate currently trades as Gemstar (Harrison and

Hoover, 2012). A number of *Helicoverpa armigera* NPV (HearNPV) products are produced in China and India, and an HearNPV product has been successfully commercialized in Australia (Harrison and Hoover, 2012; Hauxwell et al., 2010; Rabindra and Grzywacz, 2010; Wang and Li, 2010). HearNPV and HzNPV are effectively the same species despite isolates being obtained from *H. armigera* and *H. zea* caterpillars on different continents (Harrison and Hoover, 2012; Jehle et al., 2006).

It has been reported that there are 35 registered viral biopesticides in China, 14 of which are developed from HearNPV (Wang and Li, 2010). The annual output of viral pesticides in China has been estimated at 2000 tons, with HearNPV accounting for perhaps as much as 80% or 1600 tons (Sun and Peng, 2007; Wang and Li, 2010). Although India produces a number of HearNPV products in vivo, the products suffer from quality control issues (Kambrekar et al., 2007). A very successful HearNPV product is produced in Australia using a highly automated insect handling system (Buerger et al., 2007), which was estimated to have been used to treat 500,000 ha of crops (mostly sorghum) in 2008 (Hauxwell et al., 2010).

A very successful in vivo-based baculovirus production process has been developed in Brazil for the *Anticarsia gemmatalis* NPV (AgMNPV), which is used to control the velvetbean caterpillar on soybean. This process was recently reviewed by Moscardi et al. (2011). Although annual treatment of crops was as high as 2 million hectares a decade ago, current treatments are currently around 300,000 ha/year. This case draws attention to the market limitations of baculoviruses, in which their narrow specificity can lead to a loss of market share when secondary pests become more prevalent. Integrated pest management programs preserving natural enemies of secondary pests are critical for the successful use of baculoviruses in many cases (Moscardi et al., 2011).

Similarly, a successful program to produce and treat the codling moth *C. pomonella* (a global pest of a number of fruit crops) with a *Cydia pomonella* GV (CpGV) product produced in vivo has been in place for many years (Vincent et al., 2007). Peak use of this baculovirus has been reported to be as high as 250,000 ha/year, but sustained use of a single CpGV isolate has led to the development of resistance to this product by some populations of codling moth (Harrison and Hoover, 2012).

Other smaller volume baculovirus products are also produced against other pests globally, including SeMNPV against *Spodoptera exigua*, SpltMNPV against *Spodoptera litura,* and PlxyGV against *Plutella xylostella* (Linnaeus) (Kabaluk et al., 2010; Sun and Peng, 2007). It is particularly difficult to produce in vivo baculovirus products against Spodoptera pests due to the extreme liquefaction of the dead larvae, making collection of the product difficult (Moscardi et al., 2011), as well as those against Plutella viruses because of the small size of the larvae (Harrison and Hoover, 2012).

In very general terms, 10^{11}–10^{12} OB/ha are required to control a caterpillar pest on a crop and 10^{9}–10^{10} OB can be produced per infected caterpillar. This means 10–1000 infected caterpillars are required to make sufficient virus per treated hectare. Many production problems need to be overcome in order to produce baculoviruses effectively in vivo, including optimization of the diet used to ensure good growth rate of the larvae (Elvira et al., 2010) and effective harvesting of the virus from a complex mixture of insect bodies and diet, while keeping the final product reasonably free of contaminants (Buerger et al., 2007). Although many challenges remain for in vivo production of baculoviruses, including ongoing quality assurance issues and high labor costs (in developed countries at least), significant global expertise exists and groups interested in accessing low-cost in vivo-produced virus can find a number of successful commercial operations in this regard in Europe, the United States, Australia, Brazil, China, India, and possibly elsewhere.

Assuming a global market for wild-type baculoviruses exists of 100,000–500,000 ha for each of the major pests currently targeted by in vivo produced baculoviruses (HearNPV, AgMNPV, CpGV, PlxyGV, and various *Spodoptera* NPV) at a farm gate price of US $20/ha, then a case can be made for in vitro production of baculoviruses. Given the fluctuating market potential for any one virus, it is important that a production facility produces multiple viruses. In vitro production would certainly be preferable to in vivo production if a company wanted to produce multiple viruses at scale because the final production process would be generic and could be conducted relatively easily in a single facility with a single large bioreactor. Frozen stocks of multiple production cell lines and relatively small stocks of virus would be easier to manage than the maintenance of multiple large-volume insect colonies. In vitro production also avoids problems such as the development of diseases in a production colony and contamination of the final product with other microorganisms and potentially allergenic insect parts (Harrison and Hoover, 2012). In addition, should recombinant baculoviruses expressing nerve toxins resulting in faster rates of kill ever be allowed onto the market, caterpillar-based production would struggle to deliver high OB yields and in vitro production would likely have a significant advantage (Ravensberg, 2011b).

13.3. IN VITRO PRODUCTION: CURRENT STATUS

13.3.1. Cell Lines Available

Many insect cell lines have been developed for studies of insect cell physiology, developmental biology, and microbial pathology, and many have been assessed for their ability to produce various baculoviruses, particularly recombinant baculoviruses producing various foreign proteins (Goodman et al., 2001; Mena and Kamen, 2011; van Oers and Lynn, 2010). For a cell line to be seriously considered for in vitro production of a baculovirus, it needs to be shown to be capable of a cell doubling time of around 24 h and to have the capacity to produce a useful virus at a yield of at least 300 OB/cell. Few cell lines have been demonstrated to have these capabilities.

Because of their demonstrated value for producing recombinant proteins for research purposes and for veterinary and human medical applications, the Sf9 (*Spodoptera frugiperda* Smith) and High-Five (*Trichoplusia ni* Hübner) cell lines have been studied extensively in relation to their ability to be grown in suspension culture using serum-free media (Mena and Kamen, 2011). In addition, the *H. zea* (HzAM1) cell line produced by Arthur McIntosh (McIntosh and Ignoffo, 1981) has received a lot of attention for its ability to produce HearNPV in vitro (Lua et al., 2002; Nguyen et al., 2011; Pedrini et al., 2011). The Sf9 cell line shows potential for the production of an SfMNPV virus (Almeida et al., 2010) and an *Anticarsia gemmatalis* cell line has been established that shows potential to produce an AgMNPV product (Micheloud et al., 2011, 2009). No cell lines with strong potential to produce either CpGV or PlxyGV (*Plutella xylostella* GV) have been reported to date to our knowledge, although the High-Five/*T. ni* cell line may have potential to produce a *Plutella xylostella* MNPV (PxMNPV; Kariuki et al., 2000).

13.3.2. Virus Isolates Available

For in vitro production of baculoviruses, as for in vivo production, the best source of viruses are natural field isolates that show good kill potential for the pest of interest (field isolates from natural epizootics). Because viruses passaged in cell culture are subject to genome instability (see Section 13.4.4), in practice the best approach to preserve good field performance for an in vitro-produced virus is likely to be to minimize the period the virus spends in culture (Reid and Lua, 2005). The approach taken by Reid and Lua (2005) was to generate master and working stocks of virus using caterpillar colonies infected by a natural field isolate and then develop a production process that involves infecting cells using ODVs from the OB working stocks produced and allowing the virus to be in culture for only three passages (see Fig. 13.3 and Table 13.2).

For the HearNPV product developed by the Reid group, various virus isolates supplied by Commonwealth Scientific and Industrial Research Organisation's Division of Entomology (Canberra, Australia) were assessed for production in 50 ml shaker cultures using a HZea cell line grown in serum-free media over five passages. Eleven isolates were tested; three isolates showed more than 200 OB/cell at the fifth passage with good bioactivity. The best isolate from these three was chosen for further work, but all three were likely to be equally suitable for production. Virus isolates that had been purified by many passages within caterpillar colonies or plaques purified via cell culture showed very poor yields at the fifth passage in culture.

Approaches to stabilizing viruses in culture are discussed in Section 13.5.4.4. However, if recombinant viruses are to be avoided, our recommendation for obtaining virus isolates suitable for in vitro production would be to acquire as many natural isolates as possible against a given pest, screen them using shaker

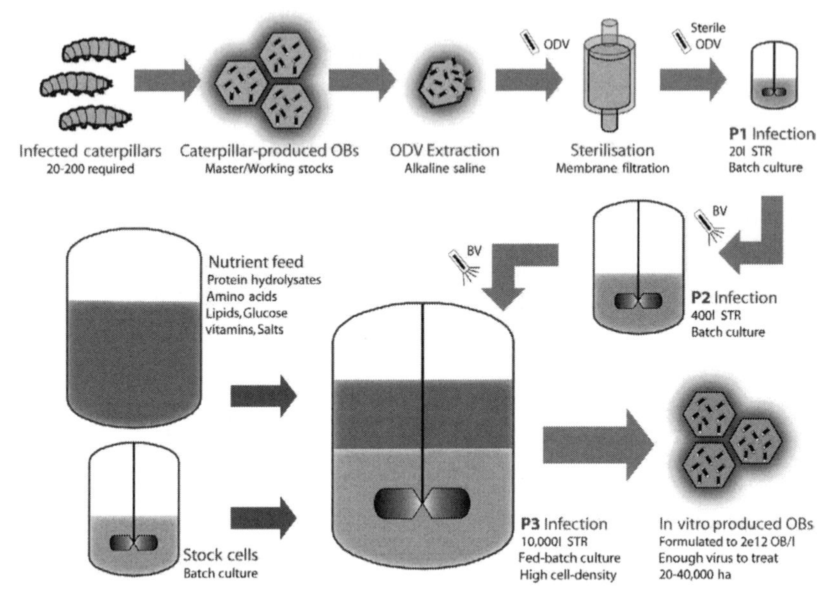

FIGURE 13.3 Current proposed scale-up process for HearNPV (see Table 13.2 for process details). Schematic diagram demonstrates the use of caterpillar-produced HearNPV occlusion bodies (OB) as a source of occlusion-derived virus (ODV) to generate the budded virus (BV) stocks at passage 1 (P1) and passage 2 (P2), for infection of HzAM1 cells in the production-scale stirred tank reactor (STR) at passage 3 (P3). A high-density fed-batch process is used to deliver high yields of the in vitro produced HearNPV OBs for use as a biopesticide. (For color version of this figure, the reader is referred to the online version of this book.)

TABLE 13.1 Baculovirus Biopesticides Production in Insect Cell Suspension Batch Cultures Using Shaker or Spinner Flasks, As Reported in the Literature. For Each Case, the Peak Cell Density (PCD) Postinfection and the Volumetric Yield (VY) of Virus Occlusion Bodies (OBs) Are Indicated

Virus	Cell-Line	Flask	PCD Cells/l	VY OB/l	References
AcMNPV	Tn5B14	Spinner	1×10^9	2×10^{11}	McKenna et al. (1997)
AgMNPV	UFLAG286	Spinner	1×10^9	6×10^{10}	Gioria et al. (2006)
HaSNPV	HzAM1	Shaker	2.6×10^9	7×10^{10}	Chakraborty et al. (1996)
SfMNPV	Sf9	Shaker	1×10^9	5×10^{11}	Almeida et al. (2010)

cultures with an appropriate cell line for five passages in culture, and store OB from those isolates that continue to produce more than 200 OB/cell at the fifth passage.

13.3.3. Low-Cost Media

The development of insect cell media was originally informed by the chemical composition of insect hemolymph, which is used to design a chemically defined basal medium containing inorganic salts, vitamins, amino acids, organic acids, and sugars (Wyatt, 1956). However, the basal medium can only support insect cell growth if it is supplemented with heat-treated hemolymph or vertebrate serum (e.g. fetal bovine serum). Such complex undefined supplements are thought to supply essential growth-promoting components, including fatty acids, sterols, and hormones (Schlaeger, 1996). Serum supplementation also allowed the cultivation of insect cells in suspension cultures, including scaling up the cultures in stirred-tank and gas-sparged bioreactors, because serum components can protect cells from shear damage due to mixing and bubble rupture (Goodwin, 1991).

Serum is certainly a more suitable medium supplement than hemolymph, owing to its wide availability. However, serum is undesirable for large-scale production processes because of its high cost, lot-to-lot variability, potential harboring of adventitious organisms, and protein content (which interferes with downstream processing). Hence, the next breakthrough in insect cell media was the development of serum-free and protein-free media (SFM), whereby the growth factors, lipids, and shear protectants in serum are supplied alternatively by protein hydrolysates and a manufactured lipid emulsion (containing the shear protectant Pluronic F-86; Inlow et al., 1989; Schlaeger, 1996). This configuration forms the basis of the well-established IPL41 SFM for lepidopteran insect cell lines (Inlow et al., 1989), and its improved commercial derivatives such as EX-CELL® 401 (Sigma–Aldrich) and Sf-900™ II (Life Technologies) SFM (Schlaeger, 1996). Further optimization has led to next-generation formulations with reduced hydrolysate content, such as Sf-900™ III SFM (Life Technologies).

Conventional lepidopteran SFM are expensive due to the high number of purified amino acids at substantial concentrations in their basal formulations. For example, IPL41 medium contains 21 added amino acids (Schlaeger, 1996), which account for more than 80% of the total ingredients cost of the complete medium (in-house estimation, based on a uniform assumption of the cheapest nonbulk retail prices for preferably cell culture-grade ingredients from Sigma–Aldrich, September 2012). Low-cost serum-free media (LC-SFM) will be required for the development of a low-cost insect cell technology (ICT) manufacturing platform, suitable for the production of biopesticides, animal vaccines, or biosimilars.

LC-SFM can be formulated by replacing most of the added amino acid content in the basal medium with protein hydrolysates, which are generally inexpensive rich sources of free amino acids and oligopeptides, as well as lipids, carbohydrates, vitamins, trace elements, and undefined growth factor analogs (Siemensma et al., 2010). Protein hydrolysates (or peptones) are generally enzymatic digests of crude animal or plant proteins. For convenience, yeast extract is considered a protein hydrolysate in this article, although it is technically an autolysate of yeast cells. Protein hydrolysates from different animal/plant/microbial origins are likely to have divergent amino acid profiles; hence, it is prudent to combine various hydrolysates in the low-cost formulation to ensure that there are no amino acid deficiencies.

At least five LC-SFM have been described in the literature for lepidopteran insect cell cultures. The ISYL SFM (Donaldson and Shuler, 1998) contains 6 g/l yeast extract and 4 g/l soy hydrolysate, and supports 6×10^9 cells/l (High-Five cells). The YPR SFM (Ikonomou et al., 2001) contains 6 g/l yeast extract and 5 g/l Primatone RL (meat peptone), and supports $5–6 \times 10^9$ cells/l (Sf9 and High-Five cells). Although these two SFM are efficacious for insect cell growth, they are not truly low cost because their basal medium is IPL41, which contains 21 purified amino acids. On the other hand, the SF-1 SFM (Schlaeger, 1996) formulation is low cost because it contains only one added amino acid (glutamine) and three hydrolysates (5.6 g/l each of yeast extract, Primatone RL, and lactalbumin hydrolysate). However, a fortified version of SF-1 was devised for supporting higher cell densities and better infections ($8–11 \times 10^9$ cells/l, Sf9 and High-Five cells), which contains 10% IP301 (a variant of IPL41), methionine, cystine, asparagine, choline chloride, and inosine. In addition, the Reid group (at The University of Queensland) separately developed the low-cost VPM3 SFM in collaboration with Stefan Weiss (Weiss et al., 1992) for HearNPV baculovirus biopesticides production (Huynh et al., 2012; Reid and Lua, 2005). VPM3 contains five added amino acids (asparagine, cystine, glutamic acid, lysine, methionine) and five hydrolysates (3 g/l yeast extract, 2.5 g/l Primatone RL, and 0.5 g/l each of soy hydrolysate, casein hydrolysate, and lactalbumin hydrolysate; Reid and Lua, 2005). VPM3 supported cell densities of around 5×10^9 cells/l (HzAM1 cells, which are approximately twice the volume of Sf9 cells (Marteijn et al., 2003)). In addition, UNL10 was developed for AgMNPV baculovirus biopesticides production (Micheloud et al., 2009); it contains 20 amino acids (mostly at much reduced levels than IPL41) and three hydrolysates (3 g/l yeast extract, 3 g/l casein hydrolysate, and 1 g/l tryptose phosphate broth). UNL10 can support peak cell densities (PCDs) of up to 3×10^9 cells/l (saUFL-AG-286 cell line).

When compared to the existing LC-SFM, VPM3 has the disadvantage of containing a higher number of hydrolysates. Each hydrolysate introduces lot-to-lot variability to the production process; hence, the quality control monitoring tasks are amplified with each additional hydrolysate. One-hydrolysate VPM3 variants were shown to be significantly worse than their two-hydrolysate counterparts, suggesting that the latter is the lower limit for low-cost formulations with a reduced number of added amino acids (Huynh et al., 2012). When a full complement of added amino acids are present (e.g. in IPL41 basal medium), a single added hydrolysate such as yeast extract is sufficient to derive an efficacious SFM (Inlow et al., 1989; Maiorella et al., 1988).

13.3.4. Current Status of Bioreactor-Based Production: HearNPV as a Case Study

13.3.4.1. Overview of Bioreactor-Based Processes

Insect cells can be propagated using bioreactor technologies that are optimized for animal cell cultures. Most of these technologies were developed first for mammalian cell cultures, which were in turn adapted from those used for microbial cultures (Agathos, 1991). Suitable bioreactor types for suspension-adapted animal cells include the stirred-tank reactor (STR), airlift reactor (ALR), and wave/rocking-motion reactor (Warnock and Al-Rubeai, 2006). Anchorage-dependent (adherent) animal cells can also be grown using suspension-based reactors, including fluidised-bed systems, if they are first immobilized on a growth surface (e.g. microcarriers; Warnock and Al-Rubeai, 2006). The stainless steel STR is by far the most prominent bioreactor type for animal cell-based biologics manufacturing (Warnock and Al-Rubeai, 2006), with the largest reactors having a working volume (V_W) of 20,000 l (De Jesus and Wurm, 2011). Plastic bag-based single-use bioreactors (SUBs) are becoming increasingly popular, but they are currently limited in scale, with the largest systems at 300–500 l V_W for rocking-motion reactors (BIOSTAT® RM, Sartorius-Stedim; WAVE Bioreactor™, GE Healthcare) and 2000 l V_W for STRs (XDR®, Xcellerex). In addition, a novel orbitally-shaken SUB is available at 200 l V_W (BIOSTAT® ORB, Sartorius-Stedim), which could potentially be increased to the 2000 l scale (De Jesus and Wurm, 2011). Insect cell cultures have been reported at scales of at least 600 l in STRs (Cox, 2004) and 20 l in rocking-motion SUBs (Cecchini et al., 2011).

Apart from bioreactor selection, another important consideration for ICT is the culture mode, which includes batch, fed-batch, and perfusion. Batch culture is the simplest mode to implement; it involves adding fixed volumes of medium and cell inoculum together in the bioreactor to achieve the desired seeding cell density. No further nutrient supplementation is carried out; hence, the batch culture volume would be constant if not for periodical sampling. Fed-batch culture is initiated as a batch culture; however, when it reaches an appropriate cell density (feeding cell density), one or more nutrient supplements (feeds) are added either pulse-wise or continuously, leading to an increase in the culture volume (Elias et al., 2000; Yang et al., 2007). The feeds are usually in the form of nutrient concentrates, which allow higher PCDs to be achieved when compared to batch cultures (10^{10} vs 10^9 cells/l orders of magnitude; Elias et al., 2000).

Perfusion culture is also initiated as a batch culture, but the subsequent nutrient supplementation is in the continuous mode, with inlet (medium) and outlet (harvest) streams at equal flow rates, leading to a constant culture volume (Crowley et al., 2012). Crucially, perfusion culture involves a cell retention device at the outlet stream, which allows the cell density to be increased dramatically (Crowley et al., 2012). In general, fed-batch culture is preferred for yield maximization on account of its relative simplicity, but challenges may be encountered in terms of formulating feed concentrates and the prospect of toxic metabolite buildup. On the other hand, perfusion culture provides an optimal physico-chemical environment. Through continuous removal of spent medium and toxic metabolites, perfusion culture can achieve much higher PCDs at the 10^{11} cells/l range (Adams et al., 2011; Crowley et al., 2012), but the tradeoff is higher process complexity.

Bioreactor-based ICT processes for recombinant proteins have been developed in batch (5×10^9 cells/l, uninfected; Maiorella et al., 1988), fed-batch (up to 5.2×10^{10} cells/l, uninfected; Elias et al., 2000; Bedard et al., 1997) and perfusion (up to 5.5×10^{10} cells/l, uninfected; Caron et al., 1994; Deutschmann and Jager, 1994) modes since the late 1980s. ICT processes for baculovirus biopesticides have been developed mainly in shaker/spinner flask batch cultures at relatively low cell densities and modest OB yields (Table 13.1). However, some of these batch processes have been scaled up in bench-scale bioreactors, such as wt-AgMNPV (2×10^{11} OB/l; Micheloud et al., 2011) and wt-CfMNPV (6×10^9 OB/l; Meghrous et al., 2000). In one particular case, wt-AcMNPV has been scaled up in a bench-scale perfusion bioreactor (1.6×10^{10} OB/l; Kloppinger et al., 1990).

13.3.4.2. Maximum Fed-Batch Yields Reported

Batch cultures are productive only up to an optimum PCD (postinfection), which is medium specific. For example, Sf9 cells grown in Sf-900 II SFM and infected with recombinant AcMNPV (expressing nonsecreted β-Galactosidase) can only support optimal specific yields (SY, units/cell) up to a PCD of around 4×10^9 cells/l; at greater PCD, a dramatic decline in productivity is observed, even though the uninfected PCD for this medium is two- to threefold higher (Wong et al., 1996). In other words, the volumetric yield (VY, units/ml) initially increases linearly with PCD, then reaches a maximum, and finally declines rapidly at higher PCDs (Radford et al., 1997; Wong et al., 1996). This phenomenon is largely correlated to the PCD and appears to be independent of the multiplicity of infection (MOI) and the infection cell density (ICD), which informs the cell yield concept (Wong et al., 1996).

The decline in productivity of batch cultures can be reversed by adding more nutrients. Hence, the fed-batch mode is a means of maximizing the VY of a production system by elevating the PCD at which the optimal SY is maintained. Over the past two decades, much progress has been made to improve the yield of ICT-expressed recombinant proteins using fed-batch cultures. One of the nt Proposed Scale-Up Process (HearNPV)*

ocess	Virus Form Added to Initiate Infection	No. Virus Particles/Cell Added	ICD 10^9 Cells/l	PCD 10^9 Cells/l	Product of Interest	Yield/Cell	Yield/l
)	ODV**	~200**	1	1.0–2.0	BV	50–100	10^{11} PFU
h)	BV	5.0	1	1.0–1.5	BV	70–100	10^{11} PFU
edbatch)	BV	0.5	4	7.0–9.0	OB	250–300	$\sim 2 \times 10^{1}$

irus; ICD-infection cell density; OB-occlusion body; ODV-occlusion derived virus; PCD-peak cell density; PFU-plaque forming unit.
)0-l run would produce $\sim 2 \times 10^{16}$ OB, which is sufficient virus to treat 20,000–40,000 ha (assuming an application rate of 0.5–1.0 × 10
would produce an equivalent number of OBs that would be produced by 2–20 million infected caterpillars. (3) The scale-up factor in
required to initiate the process is 100,000X. (4) Each run would take 2–3 weeks to complete.
schematic of the proposed process.
g 10 OB/cell (~200 ODV/cell, assuming ~20 ODV/OB). Initial studies in the Reid lab using polymerase chain reaction to quantify ODV
ODV/OB, so the envisaged passage 1 infection would require 2×10^{11} caterpillar derived OB (assuming 10^9–10^{10} OB/caterpillar, indi
lars would be required).

earliest fed-batch studies involved a single pulse-addition of a multicomponent serum-free feed to Sf9 cells grown in Sf-900 II SFM (Bedard et al., 1994), which could extend the optimal infected PCD to around 7×10^9 cells/l, for β-Gal production in shaker flasks. Similar feeds were subsequently tested, which extended the optimal infected PCD to around $7–14\times10^9$ cells/l (Bedard et al., 1997; Chan et al., 1998, 2002). The fed-batch bioreactor processes developed resulted in specific yields that were approximately 20% and 50% lower than optimal yields when tested at higher PCDs than optimal—14×10^9 cells/l and 17×10^9 cells/l, respectively (Elias et al., 2000). Other fed-batch studies also tested the efficacy of the reported feeds, showing that optimal yields can be maintained at PCDs of $9–10\times10^9$ cells/l in bioreactors for ICT-expressed influenza vaccine (Meghrous et al., 2009) and rAAV production (Mena et al., 2010).

In the Reid group, more than a decade has been devoted to the development of a scalable fed-batch bioreactor process for HearNPV biopesticides produced in HzAM1 cells, in collaboration with Agrichem (an Australian liquid fertilizer company). This work followed extensive studies in elucidating and modeling ICT processes for the Sf9/rAcMNPV system using batch/fed-batch modes (Chan et al., 1998; Haas and Nielsen, 2005; Jang et al., 2000; Power et al., 1994; Radford et al., 1997; Rosinski et al., 2002; Wong et al., 1996) and in characterizing the HzAM1/HearNPV system in batch mode (Chakraborty et al., 1999; Chakraborty and Reid, 1999; Lua et al., 2002; Lua and Reid, 2000; Pedrini et al., 2006, 2011). A major difference between the fed-batch processes for HearNPV and those in the past is that both the medium and feed were low-cost formulations, due to the narrow economic margins required for a biopesticide product.

As described in Section 13.3.3, our in-house LC-SFM (VPM3) relies heavily on protein hydrolysates to minimize the requirement for expensive pure amino acids, and this reliance was carried forward for the low-cost serum-free feed designs. The HearNPV fed-batch process was optimized by employing a heuristics-based optimization strategy informed by the cell yield concept (Wong et al., 1996), design of experiments (DoE) techniques (e.g. fractional factorial and response surface designs), and the Power-Nielsen model of baculovirus infection (Power et al., 1994). A major heuristic states that the maximum volumetric yield occurs at an optimal infected PCD that is approximately half of the uninfected PCD supported by a particular batch or fed-batch system, based on in-house observations (Chan et al., 1998; Radford et al., 1997; Wong et al., 1996). Hence, the key strategy for yield maximization is to develop feeds that support increasingly higher uninfected PCDs, so that the optimal infected PCD is also correspondingly elevated. In particular, the feeds were designed so that yeast extract and lipid levels were increased in the same proportion as the desired increase in PCD. The levels of added amino acids and other hydrolysates are designed so that the estimated free concentration of most amino acids (essential and nonessential) in the complete feed increases in approximately the same proportion as the desired increase in PCD. The feeds developed by the

Reid group supported HearNPV occlusion body yields of 2×10^{12} OB/l, which (to the best of our knowledge) is the highest reported for baculovirus OBs generated via an in vitro production system.

13.3.4.3. Economic Feasibility

The use of baculoviruses for insect pest control has a promising future in consideration of the rapidly growing biopesticides market, which is estimated to be around US $400 to 1600 million per year in 2008–2010, depending on the source of market analysis (Glare et al., 2012; Ravensberg, 2011a). Baculovirus biopesticides are estimated to have a market share of between 5% and 10%, with a compounded annual growth rate of 14% (Glare et al., 2012). Environmental degradation, pest resistance, and high development costs are some of the incentives for switching from chemical to biological pesticides. In addition, an increasing demand for organic food, which have to be free of chemical pesticides and transgenics, will further drive the uptake of biopesticides (Zehnder et al., 2007; Thakore, 2006). For example, approximately 37×10^6 ha of farmland are under organic management worldwide (Willer, 2011), with organic sales of US $$28\times10^9$ in 2004 (Zehnder et al., 2007), increasing to US $$55\times10^9$ in 2009 (Willer, 2011). In China, the organic food market is growing rapidly, with 2–3×10^6 ha of certified organic farmland generating US $$1.1\times10^9$ of domestic sales in 2008 (Portocarrero, 2011).

Significant examples of the deployment of in vivo-produced baculovirus biopesticides were mentioned above (Section 13.2). Hence, there are sizable existing markets for baculovirus products, which may expand with the advent of more scalable and reliable in vitro production (Ravensberg, 2011b). The economic feasibility of an ICT-based in vitro production process for baculovirus biopesticides was analyzed previously, using an established fermentation production cost model (Rhodes, 1996). In this case, an economic process is obtained at a final production bioreactor scale of 225,000 l, if four bioreactors are employed and 162 batch cultures are run annually. Assuming a volumetric OB yield of 4×10^{11} OB/l per run, this process was estimated to have a manufacturing cost of US $2/l culture, an operating margin (return on sales) of 29%, and enough product to treat 11.7×10^6 ha/year. However, such a process is unlikely to be implemented because animal cell culture STRs are currently limited to a maximum scale of 20,000 l working volume, as discussed previously (De Jesus and Wurm, 2011; Warnock and Al-Rubeai, 2006). Another baculovirus biopesticide production scheme (for AfMNPV) was postulated to be economically viable at a production scale of 20,000–50,000 l, with a manufacturing cost of US $2–3/l product, but the production targets are not specified (Jem et al., 1997).

The Reid group also developed a comprehensive business plan (unpublished) for the production of HearNPV in HzAM1 cell cultures, using a high cell density fed-batch process (as described in Section 13.3.4.2). At currently achieved fed-batch yields (2×10^{12} OB/l), a production facility based on a final bioreactor scale of 10,000 l working volume and 23 production runs per year appears to be economically feasible. In this case, around 200,000 l/year of formulated product (2×10^{12} OB/l) will be manufactured, which is estimated to have a manufacturing cost of AU $19/l and a sale price of AU $50/l, resulting in an operating margin of 39%. Formulated HearNPV OBs (from 100 l fed-batch bioreactor runs) were shown to be as efficacious as in vivo-produced controls in field trials; they were registered as Heliocide with the Australian Pesticides and Veterinary Medicines Authority in 2008.

Of course, there is much room to improve the economics of the current HearNPV process. For example, the capital and operating costs may be substantially reduced by locating the production facility in lower cost economies. Further cost efficiencies will be obtained by increasing the existing volumetric OB yield. This may be achieved by increasing the PCD of fed-batch infections, which requires the development of improved nutrient feeds. Because current feeds are already at the solubility threshold (in relation to the hydrolysates content), further feed improvements will need to be more targeted by incorporating elements of chemically defined formulations, as discussed later. Another strategy to increase PCDs is to implement perfusion cultures—a technology that has matured considerably and is now capable of supporting ultra-high mammalian cell densities of around 150×10^9 cells/l (Adams et al., 2011; Crowley et al., 2012), about threefold higher than the best reported fed-batch cell densities.

13.4. LIMITATIONS TO BIOREACTOR PRODUCTION OF BACULOVIRUS-BASED PESTICIDES

13.4.1. Lack of a Chemically Defined Media

The reliance of low-cost media on protein hydrolysates to supply growth factors and a majority of amino acids has a major disadvantage—the lot-to-lot variability of such undefined components (Kasprow et al., 1998; Zhang et al., 2003). Because the source material and the manufacturing process of a hydrolysate may be variable, the chemical composition of the end product may also be variable, which leads to variability in the medium and ultimately the cell culture process (Pasupuleti and Braun, 2010; Zhang et al., 2003). Ultrafiltration (e.g. 10,000 MW cutoff) of the hydrolysate may help improve process consistency (by removing endotoxins) (Pasupuleti and Braun, 2010), but the underlining variability in chemical composition is still unresolved.

Apart from this concern, hydrolysates are an untargeted form of nutrient supplementation because many other components—some of which may be undesirable—are added together with the desirable components, such as amino acids, oligopeptides, and small-molecule growth factors. For example, hydrolysates can have high inorganic salt content (Pasupuleti and Braun, 2010), and some undefined components may impart cellular toxicity (Lu et al., 2007). Furthermore, hydrolysates can have widely divergent aqueous solubilities, whether they are from the same or a different source material. For example, the yeast extract HyYest® 444 (Sheffield Biosciences) may exhibit solubility of between 50 and 300 g/l depending on the lot number (as tested by the Reid group). Hence, constraints are likely to be encountered as the concentration of hydrolysate(s) is increased, particularly when formulating concentrated low-cost nutrient feeds for fed-batch cultures in terms of osmolality, toxicity, and solubility limits.

The problem of hydrolysate variability may be addressed by first developing a chemically defined medium (CDM), from which knowledge is gained on what chemical(s) can be used to supplement a "bad" batch of hydrolysate so that it matches the performance of a "good" batch. In this way, hydrolysates can continue to be used in media and feeds to preserve the low-cost nature of the process, which is necessary for baculovirus biopesticide production. Currently, commercial CDM formulations are well established for mammalian cell cultures but are not yet available for insect cell cultures. This situation likely reflects the industrial importance of mammalian cell expressed biologics (Mena and Kamen, 2011). Strategies for developing an insect CDM are described in Section 13.5.1 below.

13.4.2. Low Budded Virus Titers

The baculovirus form required for efficient cell culture infections is the BV. The BVs to be used as infectious seed for bioreactors can be produced in larvae and extracted from hemolymphs, but this is not a feasible process for large-scale production. Therefore, an initial round of OB-derived ODV can be used to generate BVs in cell culture (Almeida et al., 2010; Lynn, 1994; Reid and Lua, 2005), which can be followed by an extra round of amplification in small-size bioreactors (see Fig. 13.3).

A major limitation to reproducible production is the variability in the initial infection resulting from the poor recovery of infectious ODV from the ODV extraction/filtration process used. Optimizing this step has been neglected due to poor options to assay ODV numbers and quality during the extraction/filtration process. However, polymerase chain reaction (PCR) has emerged as a reliable and relatively simple way to monitor ODV numbers (George et al., 2012; Pedrini et al., 2011). Work to date by the Reid group indicates that significant numbers of ODV are lost during the filtration process, and electron microscopy studies suggest severe clumping of the ODV occurs. Optimization of this process needs to be addressed and is likely to improve the consistency of the genetic material supplied into the production process at passage 1, potentially leading to more reproducible BV and OB yields from the subsequent passages.

For large-scale production in bioreactors, the infectious titer of the BV seed stock is important. The amount of infectious virions per volume is normally determined in a plaque assay (plaque-forming units) or an endpoint dilution assay (tissue culture infective dose 50 [$TCID_{50}$]), and it may differ considerably from the number of nucleocapsids measured as genome copy numbers using qualitative PCR. For AcMNPV, high BV titers are obtained in Sf9 (Vaughn et al., 1977) and *T. ni* High Five (Wickham et al., 1992) cell cultures (routinely up to $1–10\times10^{11}$ PFU/l for Sf9 cells). For many viruses, though, the BV titers obtained after even a single passage in cell culture are often lower than in hemolymph. For HearNPV, infectious BV titers are generally 10 times higher in hemolymph than in culture medium, for which titres of $1\times10^{10}–5\times10^{10}$ $TCID_{50}$/l are obtained with HzAM1 cells. In Hz2e cells, titers up to 1×10^{11} $TCID_{50}$/l are seen (Fei Deng, personal communication). Another example is *T. ni* NPV, for which virus titers are much lower than for HearNPV and drop even further after one passage in *T. ni* High Five cells (e.g. ~1.75×10^{9} $TCID_{50}$/l compared to ~5.27×10^{8} $TCID_{50}$/l; Martin Erlandson, personal communication).

In addition, the same batch of virus may give varying levels of infection in various cell lines. For example, the same batch of AcMNPV BV virus gives higher titers when tested on *T. ni* High Five than on Sf9 cells (unpublished data). Variation in BV titers may also exist between virus strains as shown for *Mamestra configurata* (Maco) NPV strains in the *Mamestra brassicae* (Linnaeus) cells IZD-MD-0503 (American Tissue Culture Collection). These MacoNPV plaque isolates have variable efficiency of replication in the *M. brassicae* cells, but the best isolates produce titers of a maximum of 4.5×10^{9} $TCID_{50}$/l; some were as low as 1.5×10^{8} $TCID_{50}$/l (Martin Erlandson, personal communication).

In conclusion, we can say that the infectious BV titer produced in cell culture varies strongly per virus (species/isolate) and may depend strongly on the cell line used for amplification and for testing. This means that to obtain optimal infectious BV titers, careful selection of cell lines and procedures may improve performance. However, as baculoviruses in general have a restricted host range, the choice of permissive cell lines is often limited. For GVs, much fewer cell lines have been shown to be permissive (6 lines for 5 viruses, compared to over 150 cell lines for more than 30 NPVs; Dwight Lynn personal communication). For an overview of different insect cell lines, see van Oers and Lynn (2010).

13.4.3. ODVs Produced in Cell Culture May Have a Lower Speed of Kill

ODVs produced in larvae may show an increased speed of kill compared to ODVs produced in cell culture, as was shown for AcMNPV (Bonning et al., 1995). The lethal viral dose needed for 50% mortality, however, does not differ between ODVs produced in larvae and cell culture. The reason for the difference in lethal time is not clear, but an important contribution may be the status of one of the PIF proteins. AcMNPV ODVs derived from infected larvae contain a proteolytically processed P74 (PIF-0) protein, consisting of two fragments of about equal size (Peng et al., 2011). This proteolytic cleavage of P74 occurs as soon as OBs derived from infected larvae are dissolved under alkaline conditions. On the contrary, ODVs derived from cultured cell lines did not show this cleavage of P74. Identification of the responsible alkaline protease could provide a method to genetically modify cultured cells so that these cells can also produce the protease necessary to cleave P74 upon ODV release form OBs. In this way, cultured cells may be programmed to produce viral OBs with a similar speed of kill as OBs produced in larvae.

The cleavage of P74 into two equally sized parts, as described previously, precedes a second cleavage mediated by midgut trypsin. This second cleavage occurs more towards the N-terminus of the P74 protein in both cell and larvae-derived ODVs. This second cleavage is needed to fully activate the P74 protein (Slack et al., 2008). The fact that P74 is an ODV surface protein that is cleaved in the midgut prior to virus entry suggests that P74 is the viral fusion protein or is a crucial component of a viral fusion complex.

13.4.4. Viral Genome Instability during In Vitro Passaging

13.4.4.1. Natural Virus Populations

Baculovirus species are often found in nature as genotypic mixtures. Such mixed genotypes are common, such as for *Spodoptera exigua* (Se) MNPV (Munoz et al., 1998), *Spodoptera frugiperda* (Sf) MNPV (Simon et al., 2006, 2005), and *Malacosoma californica pluviale* (Mcpl) NPV infecting the western tent caterpillar (Cooper et al., 2003). For McplNPV, a relationship between host density and composition of the viral genetic pool has been reported (Cooper et al., 2003). Genetic mixtures may even occur within a single larvae, as reported for *Panolis flamea* (Pf) NPV, where up to 24 variants were seen within a single insect (Cory et al., 2005).

The individual isolates may vary in speed of kill and/or lethal dose; sometimes, an isolate has completely lost oral infectivity capacity, meaning this isolate was dependent on other genotypes in the population to start a new round of infection (Cory and Myers, 2003; Kemp et al., 2011; Lopez-Ferber et al., 2003). A large screen among virus isolates from heliothine pest insects (*H. armigera, H. virescens,* and *H. zea)* allowed the grouping of isolates based on a selection of sequenced genes and also detected several mixed populations (Rowley et al., 2011). A selection of the isolates was tested against *H. zea* larvae and compared to the commercially available Gemstar® biocontrol product. The commercial product killed the larvae faster than most other Hz/HearSNPV isolates. On the other hand, for Gemstar® and two HearSNPV isolates, more virus was needed to kill 50% of the larvae than for the other Hz/HearSNPV isolates tested (Rowley et al., 2011), demonstrating that speed of kill and lethal dose are two different parameters of virus isolates that are not per se correlated. When developing biocontrol products, one has to determine whether the natural mixture is preferable in terms of dose needed and time to kill or whether one of the isolates performs better. As the composition of mixtures may change during production in bioreactors, methods may be needed to control and determine the composition of mixtures.

13.4.4.2. Selection and Generation of Mutants in Cell Culture

When baculoviruses are passaged in cell culture (of course, in the form of BVs), certain genotypes are easily selected for and previously nonexisting deletion mutants may be generated. One obvious reason is that many functions can be omitted when the virus does not have to be transmitted to a new host insect. For instance, in cell cultures, there is no need to retain ODV-specific proteins such as the PIF proteins, the major occlusion body protein, or viral proteins that regulate OB yield.

Amplification of SeMNPV in *S. exigua* Se301 or SeIZD2109 cells leads to large deletions in the *Xba*I-A fragment, roughly the region from open reading frame (ORF) 16 to 36 (Heldens et al., 1996; IJkel et al., 1999; Pijlman et al., 2003c). Often, such mutants have lost the ability to orally infect larvae; this finding actually led to the discovery of oral infectivity factor PIF-2 (Se35, homologous to Ac22) (Pijlman et al., 2003a). To avoid the problem of obtaining viruses without in vivo infectivity, an alternating cell culture/larval cloning strategy was developed. This procedure resulted in a mutant (Se-XD1) with a deletion affecting ORFs 15–28. This mutant was able to infect larvae orally and was already present in the original virus isolate (Dai et al., 2000). Mutant Se-XD1 lacked, among others, the *egt*, chitinase, and cathepsine genes. The *egt* gene encodes ecdysteroid uridine 5′-diphosphate (UDP)-glucosyltransferase, an enzyme that increases OB yields by suppressing the molting of infected larvae (Cory et al., 2004; O'Reilly and Miller, 1989). Chitinase and cathepsin break down the chitin skeleton of diseased insects and, as such, they assist in the release of OBs from cadavers (Hawtin et al., 1997). These three enzymes are important for efficient virus transmission in nature, but there is apparently not as strong selection pressure on these genes under laboratory settings.

Therefore, in cell culture, genetic variants may be selected from already existing genotypes or new genotypes may be generated. An extreme example of the latter is the formation of defective interfering (DI) particles. DIs appear as soon as baculoviruses are passaged outside insect larvae in cultured insect cells (Pijlman et al., 2001) and easily accumulate in bioreactor runs (Kool et al., 1991). DI particles have a reduced size due to large deletions in the viral genome (Kool et al., 1991; Lee and Krell, 1992). DIs show parasitic behavior toward the wild-type virus because they depend on functions encoded by full-length genomes in the population to replicate. At the same time, DIs provide strong competition to the replication of full-length genomes because the smaller DIs complete their replication quicker. When AcMNPV expression vectors are used in bioreactors, DI accumulation leads to a reduction of recombinant protein production and prevents the application of continuous baculovirus-insect cell production systems (van Lier et al., 1990). Because the heterologous genes are normally inserted in the polyhedrin locus, one can predict that, when using wild-type virus as starting material, the level of polyhedrin production may drop in a similar way, leading to a reduction in OB formation.

For AcMNPV, defective genomes of approximately 50 kbp have been reported, which contained reiterations of baculovirus sequences homologous to 2.2% of the viral genome (Lee and Krell, 1994). As these genomes were replicating, it was concluded they had to contain an origin of replication (ori). Because this origin was not similar to the already known origins in the form of homologous regions (hr), which consist of tandem and inverted repeats and are dispersed over the genome, the new sequence was named a *non-hr* ori (Kool et al., 1994). In AcMNPV, this region is located within the nonessential *p94* gene. For SeMNPV, intracellular DNA circles were reported with iterated *non-hr* sequences in Se301 cells (Pijlman et al., 2002). When SeMNPV and AcMNPV were engineered to remove the *non-hr*, the viruses replicated well and the genomes were stabilized (Pijlman et al., 2002, 2003b). Deleting the *non-hr* element appears to be a good method to stabilize baculovirus genomes in general, but so far has hardly been applied.

From the above, it is clear that the most simple method to keep DI accumulation to a minimum is to limit the number of passages in cell culture and to use low multiplicities of infection when preparing BV seed stocks, so that the wild-type virus cannot provide helper functions for deletion mutants. Deletion mutants enriched in *non-hr* sequences were effectively removed within two passages when larvae were infected with OBs produced in cell culture, which clearly indicates that DIs are something artificial (Zwart et al., 2008a). In vivo passaging may therefore be used to clean up virus stocks propagated in cell culture. Using OBs produced in larvae as a seed stock may help to avoid problems, but this process needs technological development in terms of ODV extraction, filtration to remove bacteria and fungi, and quality control (see Section 13.4.2).

Methods to measure accurately the accumulation of DIs are needed to monitor baculovirus infections in cell culture and to see how quickly DI formation becomes a problem in a particular combination of virus and cell line. A quantitative PCR method was developed for AcMNPV that compares levels of the immediate early gene 1 (*ie-1*) with that of *non-hr* sequences (ideally in the ratio of 1:1; Zwart et al., 2008b); this tool can easily be adopted for other viruses. Such methods will also help to determine whether adaptations in the production process or modifications to the genome have a positive effect by delaying DI accumulation.

13.4.4.3. Instability through Transposable Elements

Important factors that contribute to the plasticity of baculovirus genomes include homologous recombination between related baculoviruses, internal recombination events facilitated by the reiteration of origins of replication *(hrs)* and the presence of baculovirus repeat ORFs, as well as transposon insertions.

Transposons are mobile genetic elements that are mediators of genomic plasticity by promoting recombination between genomes; the hypothesis is that they have played an important role in establishing the composition of baculovirus genomes (van Oers and Vlak, 2007). Transposable elements can result in gene insertions, gene interruptions, and even inversions of genome fragments. Transposons may also alter the expression of genes flanking the insertion site. The transposons in genetic variants of *Cydia pomonella* GV (CpGV) are members of the *Tc1-like/mariner* transposon family. They are no longer mobile because they encode defective transposase enzymes (Jehle et al., 1995, 1998) and comprise insertions of several kbp derived from the hosts *Thaumatotibia (Cryptophlebia) leucotreta* (Meyrick) and *C. pomonella.* The transposon from *C. pomonella* leads to variation in the form of inversions between CpGV strains through homologous recombination between its inverted terminal repeats (Arends and Jehle, 2002).

Insertion of mobile elements in baculovirus genomes also occurs during propagation in lepidopteran cell cultures, contributes to genome instability, and may lead to reduced OB formation. A hotspot for transposon insertions in AcMNPV is the *fp25K* ORF (Beames and Summers, 1988; Fraser et al., 1985).

Insertions in this locus lead to the few polyhedra (FP) phenotype, which is often seen when the virus is passaged in cell culture; it is characterized by fewer OBs with fewer or no ODVs inside (Fraser and Hink, 1982; Hink, 1976; Knudson and Harrap, 1976). As the name implies, FP mutants have a reduced yield of OB production, which is accompanied by reduced levels of the FP25K protein. The FP25K protein plays a role in the intracellular transport of ODV envelope proteins (Braunagel et al., 1999, 2004); as such, it is involved in the biosynthesis of biologically active ODVs. As a consequence, insertions in this locus affect the efficacy of OB preparations.

The most frequent transposon-mediated insertions in baculoviruses are relatively small and are inserted specifically at TTAA target sites, such as the 634-bp insertion found in AcMNPV isolate E amplified in *S. frugiperda* cells (Schetter et al., 1990). The hitchhiker element, derived from *T. ni* cells, is 579 bp long and inserts at TTA sites (Bauser et al., 1996). Other examples with high specificity for TTAA sites are the tagalong (TFP3) (Wang et al., 1989) and piggyBac (IFP2) elements (Cary et al., 1989), which may be inserted when the virus is amplified in *T. ni* cell lines. These elements normally do not have ORFs and are characterized by small inverted repeats. These types of insertions play a major role in the appearance of the FP phenotype mentioned above. After insertion of tagalong and piggyBac, revertants have been recorded that result from precise excision (Fraser et al., 1996).

To stabilize the *fp25k* locus, recent experiments aimed to alter the TTAA motifs in the *fp25k* gene that are prone to transposon insertion, while leaving the encoded amino acid sequence intact. This indeed stabilized the synthesis of the FP25K protein and appeared to delay the accumulation of the FP phenotype, but the envelopment of nucleocapsids and the occlusion of virus particles were aberrant (Giri et al., 2011). The FP phenotype still appeared despite continued FP25K protein synthesis, indicating that other events can also lead to an FP phenotype. This is in line with reports of transposon insertions in the AcMNPV *94k* and *da26* genes, which also gave an FP phenotype (Friesen and Nissen, 1990; O'Reilly et al., 1990).

Larger insertions in the FP25K locus may result from transposition of a *copia-like* retro-transposable element called TED (gypsy family), which is derived from the *T. ni* genome (Friesen and Nissen, 1990; Miller and Miller, 1982). TED insertions are characterized by long terminal repeats and carry *gag, pol,* and *env* genes (Hajek and Friesen, 1998; Ozers and Friesen, 1996). Excision of the TED element from an AcMNPV FP mutant left behind a long terminal repeat that showed promoter activity in both directions (Friesen et al., 1986). This is an example of how a transposon insertion and imperfect excision may alter gene expression of flanking genes. Errantiviruses have probably evolved from a TED retrotransposon that obtained a copy of the baculovirus F gene, thereby converting the retrotransposon into an insect retrovirus (Malik et al., 2000; Pearson and Rohrmann, 2004). Although data on transposons are available only for a limited number of cell types, it is likely that similar elements act in other cell/virus systems.

13.4.5. Complications with High-Density Cell Culture

High-density cell culture is attractive in terms of saving space and medium. On the other hand, complications may arise in providing nutrients and oxygen while avoiding shear stress. Under starvation conditions, which may occur under high cell density, cell division is impaired and apoptosis is easily induced (Laken and Leonard, 2001). Controlling important metabolic processes is therefore extremely important when growing high-density cultures; for a review, see Ikonomou et al. (2003). Baculovirus infection requires dividing cells, and high cell densities give a drop in specific production of recombinant proteins in the baculovirus expression systems (see Section 13.3.4.2). Similar effects are known to occur for OB production (Chakraborty et al., 1996). Metabolic flux analysis, whereby levels of many different metabolic compounds were followed, revealed that the central metabolism of Sf9 cells is progressively inhibited when cell densities increase (Bernal et al., 2009). High cell density affected the incorporation of amino acid carbon backbones into the citric acid cycle and downregulated glycolysis. Modulation of the energetic status of the culture by adding pyruvate or alpha-ketoglutarate at the time of infection resulted in six- to sevenfold higher yields of budded virus at high cell density when compared to control cultures and resulted in higher titres than classical low cell density infections (Carinhas et al., 2010). A recent proteomic study has revealed more cellular components that may be modulated to improve baculovirus infections under high cell density conditions (Carinhas et al., 2011).

13.5. FUTURE RESEARCH DIRECTIONS FOR BIOREACTOR PRODUCTION OF BACULOVIRUS-BASED PESTICIDES

13.5.1. Chemically Defined Media for Insect Cell Culture

As described in Section 13.3.3, conventional insect cell media are composed of a chemically defined basal medium, which has to be supplemented with complex undefined additives, such as vertebrate sera or protein hydrolysates, and an undefined lipid emulsion for serum-free formulations to support cell growth and baculovirus production. In principle, it should be feasible to develop a completely chemically defined medium for insect cell cultures that contains chemically-defined lipids (fatty acids and cholesterol) and that has no requirement for serum or hydrolysate supplementation.

The rationale for this hypothesis is that commercial CDM formulations are widely available for mammalian cell cultures (e.g. CD OptiCHO™, Life Technologies; PowerCHO™, Lonza; IS CHO-CD™, Irvine Scientific). Insects have generally similar nutritional requirements to that of vertebrates, and insect and mammalian basal media share many similarities in chemical composition (Schlaeger, 1996). Furthermore, the development of insect cell media was highly influenced by that of mammalian cell media in terms of the substitution of vertebrate sera with protein hydrolysates and the use of Pluronic F-68 as a cell protectant to enable cell growth in suspension cultures (Schlaeger, 1996). In addition, studies have shown that insect cells can be adapted to grow in mammalian media and even cocultured with mammalian cells, such as *Agallia constricta* Van Duzee (order Hemiptera) and *T. ni* (order: lepidoptera) cells cocultured with mouse fibroblast cells in TC199-MK serum-containing medium (Epstein and Gilula, 1977; McIntosh et al., 1973). Nevertheless, insect cells do require certain conditions for optimal cell growth and baculovirus infection, which are divergent from those for mammalian cells (Drugmand et al., 2012; Schlaeger, 1996). Firstly, sterols are not synthesised by insect cells; hence, they have to be supplemented in the medium (e.g. cholesterol in serum or in a lipid emulsion). Secondly, insect cells prefer lower pH (6.2–6.9 vs 7.0–7.3), lower temperature (22–29 °C vs 33–37 °C), and higher osmolality (340–390 vs 290–330 mOsmol/kg) settings compared to their mammalian counterparts. Furthermore, insect media are usually formulated with higher levels of free amino acids and organic acids to mimic the composition of hemolymph (Schlaeger, 1996).

For Sf9 insect cell cultures, an efficacious serum-free medium can be prepared by supplementing IPL41 basal medium with chemically defined lipids (Life Technologies) and yeastolate ultrafiltrate (Life Technologies), which can support PCDs of around 14×10^9 cells/l in our hands. However, the PCD declines with decreasing yeastolate content, and no cell growth is recorded after the second passage if yeastolate is omitted (unpublished data). Hence, yeastolate is highly critical for insect cell proliferation in IPL41 medium, and the main challenge in developing an insect cell CDM is to replace the complex functionalities of such undefined hydrolysates. One potential solution is to fractionate and determine the active components of yeastolate (Shen et al., 2007) or other hydrolysates, which can then be chemically identified for inclusion in the basal medium. Another approach is to combine aspects of existing fully disclosed mammalian CDM formulations with basal insect media such as IPL41, while taking into account the specific sterol and physicochemical requirements of insect cells.

Examples of mammalian CDM formulations include CDSS (Qi et al., 1996), MET1.5 (Epstein et al., 2009), and CD-CHO (Gorfien et al., 2012). MET1.5 and its affiliated formulations are demonstrated to be efficacious for hybridoma cell culture and monoclonal antibody production (20×10^9 cells/l) but are also claimed to be suitable for the culture of eukaryotic cells in general, including insect cells (Epstein et al., 2009). Suitable chemically defined growth proliferation factors (as hydrolysate substitutes) may include efficient forms of chelated iron (Epstein et al., 2009; Gorfien et al., 2012), synthetic oligopeptides (Franek et al., 2003), trace elements such as sodium selenite (Popham and Shelby, 2007; Qi et al., 1996), and antitoxic factors (Qi et al., 1996).

Additionally, there are a limited number of fully disclosed (uncommercialized) insect cell CDM formulations in the literature (Becker and Landureau, 1981; Mitsuhashi, 1996; Mitsuhashi and Goodwin, 1989; Wilkie et al., 1980). However, these CDM formulations are generally preliminary, complex, not well validated, and tested using poorly defined stationary cultures. In summary, it should be feasible to formulate an insect cell CDM for industrial applications that is preferably animal component-free and protein-free. The recent commercialization of two ICT-expressed human biologics, Cervarix® and Provenge® (Mena and Kamen, 2011), may spur the industrial development of insect cell CDM in the future. An insect cell CDM may then allow the development of more reproducible hydrolysate-based low-cost media and feeds, which would be beneficial in the development of a commercially viable baculovirus biopesticide process.

13.5.2. Genomics/Transcriptomics of Insect Cell Lines

Further improvements in OB yields via in vitro production processes will likely require gene modifications of the host cell lines to reduce the cell lines' ability to attenuate viral pathology (Harrison and Hoover, 2012). It is not an option to genetically modify the virus used if the product is to be accepted as wild type and therefore maintain the ability to avoid severe registration and environmental concerns. Genetically modifying the host cell line to improve yields requires first identifying the best host gene targets to upregulate or downregulate. This in turn requires a detailed assessment of the host cell gene response to a virus infection. Ideally, the genome of the host insect cell lines would be available to allow appropriate studies to be made in this regard. Unfortunately, few insect genomes are available. However, the decreasing cost of next-generation sequencing together with improved bioinformatics methods is making it possible to generate transcriptomes of insect cells (Harrison and Hoover, 2012).

Recently, the transcriptome of an *H. zea* cell line was determined and initial investigations were conducted on in vitro insect cell–baculovirus interactions (Nguyen et al., 2012). Although it is early in the investigations, this approach holds promise to rapidly advance knowledge of such systems, which may lead to two- to fourfold improvements in OB/cell yields. The potential of this approach was illustrated previously with the Sf9 cell line (Fath-Goodin et al., 2009). To keep insect cells viable for a longer time, the vankyrin gene, isolated from the *Campoletis sonorensis* ichnovirus, was used (Fath-Goodin et al., 2006). Vankyrin prolongs the lifetime of infected cells by preventing apoptosis (Fath-Goodin et al., 2009). Whether this can also lead to increased yields of biologically active OBs needs to be tested. Knocking down proapoptotic genes in the virus is not really an option because these often encode crucial proteins for the infection process and as such are a good target for antiviral defense.

Regulated cell death can involve mechanisms other than apoptosis (Degterev and Yuan, 2008) and may involve yet to be discovered pathways. Thus, it is essential to take a transcriptomic approach to investigating how the insect cell in culture responds to a baculovirus infection, particularly at high cell densities where stress may be detected even before the virus is added.

13.5.3. Metabolomics of Insect Cell Lines

Detailed studies on tissue culture requirements and metabolism have been performed with the aim of producing recombinant proteins, virus-like particles, or gene therapy vectors. These studies were performed with polyhedrin-negative viruses, meaning no OBs were made (Bernal et al., 2009; Carinhas et al., 2010; Ikonomou et al., 2003; Vicente et al., 2011). It is unlikely that the results of these studies can simply be translated into optimal conditions for OB production. Also, the cell lines that were used for these studies are normally not the ones that one would need for producing agents to control major insect pests. An exception is a study on *H. zea* HzM1 cells, where culture media adaptations allowed viable cell densities to increase to more than five times over commercial media (Marteijn et al., 2003). However, this study was not combined with infection studies. Metabolic analyses are needed for cell/virus systems relevant for producing biocontrol agents. In these analyses, one needs to look at the effects on the number of OBs, the amount of ODVs occluded, and the infectivity in terms of lethal dose and lethal time. Methods are being developed that will allow the assessment of intracellular metabolite levels of insect cells in culture, both uninfected and following baculovirus infections (Tran et al., 2012).

13.5.4. Genetically Modified Cell Lines/Viruses

Various genetic modification strategies to increase virus production or the efficiency of the virus in the field are discussed below. For more information, we refer the reader to an extensive review about genetically modified baculoviruses for biocontrol of insect pests published several years ago (Inceoglu et al., 2006).

13.5.4.1. Viruses with Increased Speed of Kill

One of the downsides of baculovirus-based biopesticides is their relatively low speed of kill. A fast killing natural virus such as SeMNPV still needs 3–5 days to kill its hosts. A nice overview of various attempts to reduce time to death is presented by (Inceoglu et al., 2001). Time to death can be reduced by deleting the viral *egt*-gene (Cory et al., 2004). EGT reduces the developmental progress of the larvae by inactivation of the molting hormone ecdysone. Deletion of *egt* also leads to reduced OB yields in larvae, but cell cultures infected with an *egt*-deleted mutant will probably not have this limitation. Speed of kill may also be increased by including a copy of the insect juvenile hormone esterase (JHE) gene (especially a stabilized version, with a longer half-life time) in the viral genome under a viral promoter (Bonning et al., 1997; Eldridge et al., 1992; Hammock et al., 1990). JHE inactivates juvenile hormone and this leads to cessation of feeding and the induction of pupation in insect larvae.

Very promising data have been obtained with incorporation of eukaryotic insect toxin genes, such as the venom from the scorpion *Androctonus australis* Linnaeus (aaIT), which paralyzed the insects and killed them faster (Stewart et al., 1991). Incorporation of the gene for a neurotoxin from the straw itch mite *Pyemotes tritici* (LaGrèze-Fossat and Montagné) in AcMNPV (Txp-1) paralyzed *T. ni* larvae (Tomalski and Miller, 1991) and reduced the time to death by 50–60% (Burden et al., 2000). The lethal doses in the examples described above were the same as for the wild-type viruses. Strong reduction in time to death was also observed with a combination of two cooperating neurotoxins from the scorpion *Leiurus quinquestriatus* Hemprich and Ehrenberg (Regev et al., 2003). For HearNPV, hormonal and toxin-based strategies have been combined by replacing the *egt* gene with the *aaIT* toxin gene, resulting in viruses with a much higher speed of kill (Chen et al., 2000). A major advantage of the paralyzing toxins is the reduced feeding damage caused by the pest insect because they stop feeding well before they succumb to the infection.

13.5.4.2. Viruses with Increased Infectivity

Attempts to express *Bacillus thuringiensis* Berliner (Bt) toxins in baculovirus vectors were successful (Martens et al., 1990; Merryweather et al., 1990) but at first did not significantly improve insecticidal ability. The Bt toxins are normally active at the cell surface of midgut epithelium cells. Thus, expression of Bt toxins by recombinant baculovirus within the cell did not lead to increased virulence. By making a trypsin-digestible fusion of BT toxin to the AcMNPV polyhedrin protein, the toxin was targeted to OBs and released in the midgut. This resulted in a 100-fold reduction in LD_{50} as well as a 60% reduction in medial survival time of the *Plutella xylostella* larvae.

Incorporation of enhancin genes also leads to viruses with greater virulence. Enhancins are metaloproteases that were first found in granuloviruses (Kuzio et al., 1999; Lepore et al., 1996; Roelvink et al., 1995) and later also in several NPVs (Bischoff and Slavicek, 1997; Jakubowska et al., 2006; Li et al., 2002). Enhancins make larvae more susceptible by degrading the peritrophic membrane that lines and protects the insect's gut (Toprak et al., 2012). Deletion of the two enhancin genes from *Lymantria dispar* (Ld) MNPV made the virus 12-fold less potent (Popham et al., 2001). An experiment in which the *Mamestra configurata* (Maco) NPV enhancin gene was incorporated into AcMNPV showed that enhancins may work across species because the LD_{50} for *T. ni* larvae was reduced more than 4 times (Li et al., 2003). The lethal time when applied at an LD_{90} was not affected.

13.5.4.3. Changing the Ratio of BV and ODV Production

Mutagenesis to increase the budded virus titre may seem to be a solution for low BV titers. However, this may give adverse effects, such as the reduced infectivity in vivo seen for an HearNPV ORF107 deletion mutant (Pan et al., 2007). Whether this is (partly) the consequence of a reduced number of ODVs in the OBs when the balance is shifted towards BV production was not analyzed precisely in this case. Using such a mutant for preparing seed stock in order to get high BV titers and applying ORF107 in a transgenic cell during OB production has not been tested yet. In a similar way, FP25K may be supplied in trans by a transgenic cell during OB production to prevent the FP phenotype. Removing it from the baculovirus will give more BVs when used in a nontransgenic cell.

13.5.4.4. Stabilized Genomes

As detailed above, major components that destabilize the genome or the *non-hr* origin of replication and the *fp25k* gene that serves as a transposon insertion hotspot have been identified (see also Section 13.4.4.3). Stabilized baculovirus genomes may be achieved by deleting the *non-hr* region; however, whether *non-hr* mutants are as infectious for larvae as the wild-type virus needs to be analyzed. Providing the FP25K protein in trans to prevent transposon insertions in this area, such as with a transgenic cell line, may be an option, especially when this *fp25k* gene is expressed from an inducible promoter.

13.5.5. Future Potential

When the yield of virions per cell in vitro is considered, the potential of genomic/metabolomic studies to improve OB yields is realized. Infected Sf9 cells in culture produce 40,000 virions/cell for AcMNPV (Rosinski et al., 2002) and *H. zea* cells produce a similar number for HearNPV (Pedrini et al., 2011). Most of the virions stay in the nucleus (Pedrini et al., 2011; Rosinski et al., 2002), and OBs contain only around 40 virions/cell (Reid et al., unpublished data), so it should be possible to produce up to 600–800 OB/cell. The best yields reported to date are approximately 400–600 OB/cell for low cell density infections of $0.5–1.0 \times 10^9$ cells/l (Almeida et al., 2010; Micheloud et al., 2009), whereas the best yields reported for high-density fed-batch infections are 250–300 OB/cell at PCDs of $7–9 \times 10^9$ cells/l (see Section 13.3.4.2). An optimistic research objective would be to achieve yields of ≥600 OB/cell at peak fed-batch or perfusion cell densities of 1.5×10^{10} cells/l. If such yields can be achieved for a range of key viruses (HearNPV, SfMNPV, AgMNPV, PxMNPV/PxGV, and CpGV), then in vitro production of baculoviruses would be an attractive option for industry.

13.6. CONCLUSIONS

To be economically viable, a fed-batch in vitro production process for baculoviruses will need to generate yields such that 1 l of broth will produce sufficient virus to treat 2–4 ha of an appropriate crop at a farm gate price of around US $20/ha. At current yields for HearNPV (see Fig. 13.3, Table 13.2, and Section 13.3.4.3), this is achievable if production is conducted at the 10,000 l scale. At this scale, the annual market that would need to be accessed at this price would be of the order of 800,000 ha. To go to the market at this stage with the current HearNPV process requires the following conditions to be met:

- Industry must be confident that the market for wild-type baculoviruses is of the order of 800,000 ha/year at US $20/ha, and the product's manufacturing price is competitive with existing in vivo supplied competitor products.
- The manufacturing process needs to be made more reproducible.
- OB yields per liter of broth in culture need to be improved two- to fourfold.

As indicated in Section 13.3.4.3, the current cost of manufacturing assumed for an in vitro product would be AU $19/l. This cost assumes a virus concentration of 2×10^{12} OB per liter of bioreactor broth (see Section 13.3.4.2), and an application rate of 7.5×10^{11} OB/ha (Buerger et al., 2007). Hence, each liter of final formulated product would be sufficient to treat 2.7 ha at a cost of manufacturing of approximately AU $7/ha. For simplicity, to make the economics more widely applicable, one can assume that the Australian and U.S. dollars are equivalent, which is roughly correct at this time. The cost to produce a HearNPV product in vivo is not publically available, but is known to sell in Australia at around AU $20/ha. A price to produce an AgMNPV product in vivo in Brazil has been reported as US $0.42/ha (Moscardi et al., 2011), so clearly the in vitro manufacturing price would not be competitive in this case assuming similar OB/ha application rates and similar in vivo manufacturing prices apply for the AgMNPV and HearNPV products.

At this stage, it seems reasonable to expect bioreactor yields of OBs to improve twofold further, from 2×10^{12} OB/l to 4×10^{12} OB/l, over the next few years; a fourfold increase to 8×10^{12} OB/l may be possible, but higher yields than this seem unlikely. Hence an in vitro price of US $2–4/ha to produce wild-type baculoviruses may be achieved in the next 5–10 years. This would bring in vitro costs of manufacturing close to in vivo costs. Given the advantage of in vitro production in relation to the production of multiple products within the same facility (see Section 13.2), a role for in vitro production is envisaged. The key to yield improvements is the successful application of systems biology studies involving transcriptomics and metabolomics, as discussed in Sections 13.5.2 and 13.5.3.

Figure 13.3 and Table 13.2 provide an overview of an envisaged 10,000 l process for HearNPV. This process has been demonstrated to the 100 l pilot scale by the Reid group. However, even if yields can be improved a further two- to fourfold, the process would struggle commercially due to concerns with reproducibility.

The dominant concern is the quality control issues associated with using yeast extracts in the media and feeds for the production process. A chemically defined media must be developed for insect manufacturing processes; hence, this matter was discussed in detail in Section 13.5.1 previously. We believe a CDM is essential to aid future research. Applying transcriptomic and metabolomics studies to cells grown in a nondefined media will limit the potential of such studies to lead to yield increases. It is believed that once key components have been discovered that allow cells to be grown without the addition of yeast, these components can be supplied independently to a culture. Then, it should be possible to return to the use of yeast or other low-cost hydrolysates to supply a low cost source of amino acids for use in insect cell processes.

The second most important quality control concern relates to genome stability of the virus in vitro, as discussed extensively in Section 13.4.4. Genetic manipulation of the genome to ensure stable production of OBs does not look feasible at this stage given the range of transposable elements that can act to destabilize the virus genome. The most practical way at this stage to avoid this problem is to passage the virus for as short a period as possible in culture. To ensure this, it is necessary to produce master and working stocks of virus as OBs using caterpillars, then extract and use the ODV form of the virus from these OBs to initiate a bioreactor scale-up process. Hence, optimization of the ODV extraction process and of ODV infections are important research objectives.

Finally, Table 13.3 provides a summary of our suggested research priorities for developing a HearNPV product, which would likely apply to any baculovirus that is to be produced in vitro. Industry needs to effectively engage with the research community in an effort to determine whether the global market justifies the still-considerable research and development that are required to pursue the goal of in vitro production of baculoviruses.

TABLE 13.3 Summary of Suggested Research Priorities (HearNPV)

Objective	Approach	Key Challenge	Passage Number**
Develop chemically defined media (CDM)*	Test CDM formulations for mammalian cells using insect cells (see Section 13.5.1)	Involvement of media companies/proprietary nature of formulation developed	1, 2, 3
Develop chemically defined feed (CDF)*	Test CDF formulations informed by CDM knowledge	Involvement of media companies/proprietary nature of formulation developed	3
Ensure good recovery of sterile, nonclumped ODV from caterpillar-derived OB	Optimize ODV recovery from OB extraction/filtration process using PCR to monitor ODV numbers	Development of an assay to monitor degree of ODV clumping	1
Improve ODV uptake by cells	Test alternative cell lines for improved ODV uptake rates using PCR to monitor ODV binding rates	Access or develop alternate cell lines (van Oers and Lynn, 2010)	1
	Moderate host genes that affect ODV infection process	Identify appropriate gene targets using transcriptomics studies (Nguyen et al., 2012)	
Improve BV yield/cell	Test alternative cell lines for improved BV production	Access or develop alternative cell lines (van Oers and Lynn, 2010)	1, 2
	Moderate host genes that affect BV release	Identify appropriate gene targets using transcriptomics studies (Nguyen et al., 2012)	
Maintain maximum OB/cell at high cell densities (see Section 13.3.4.2)	Identify gene expression and intracellular metabolic differences of cells infected at high and low cell densities	Identify appropriate gene targets using transcriptomics studies (Nguyen et al., 2012) and measure intracellular metabolite levels of virus infected cells (Tran et al., 2012) at high and low cell densities	3

Continued

TABLE 13.3 Summary of Suggested Research Priorities (HearNPV)—cont'd

Objective	Approach	Key Challenge	Passage Number**
Improve OB/cell yield	Identify gene expression and intracellular metabolic differences of high versus low OB-producing cell lines	Identify appropriate gene targets using transcriptomics studies (Nguyen et al., 2012) and measure intracellular metabolite levels of virus infected cells (Tran et al., 2012) for high versus low OB-producing cell lines (Nguyen et al., 2011)	3
Improve OB quality	Express ODV-associated proteins in host cells associated with infectivity of the caterpillar (see Section 13.4.3)	Lack of a simple assay for OB quality to reduce reliance on bioassays	3

Acronyms: BV-budded virus; OB-occlusion body; ODV-occlusion derived virus.
A key purpose of this table is to highlight the ability to use a different cell line (genetically modified or otherwise) for each passage during the scale up process, in order to optimize the three processes of scale-up: ODV to BV conversion, BV expansion, and OB production.
**Developing a CDM is listed as the first objective for improving the outcome of all three passages during the virus production process because the use of an undefined yeast extract in the media limits the ability to apply fundamental studies to understand the process limitations during each stage. Developing a CDF would be reasonably straightforward following the development of a CDM.*
***Passage number of scale-up process shown in Fig. 13.3 that this research objective applies to.*

ACKNOWLEDGMENTS

We acknowledge Agrichem for supporting the HearNPV scale-up project and Prof. Lars Nielsen from the Centre for Systems and Synthetic Biology (University of Queensland) for his leadership in our systems biology studies. Martin Erlandson (Sasketoon Research Centre, Agriculture and Agri-Food Canada) and Dwight Lynn (INSell) are acknowledged for supplying unpublished data for this review. Gorben Pijlman (Laboratory of Virology, Wageningen University) is acknowledged for critical reading of sections of this manuscript prior to submission.

REFERENCES

Adams, T., Noack, U., Frick, T., Greller, G., Fenge, C., 2011. Increasing efficiency in protein supply and cell production by combining single-use bioreactor technology and perfusion. Biopharm Int., S4–S11.

Agathos, S.N., 1991. Production scale insect cell culture. Biotechnol. Adv. 9, 51–68.

Almeida, A.F., Macedo, G.R., Chan, L.C.L., Pedrini, M.R.S., 2010. Kinetic analysis of in vitro production of wild-type *Spodoptera frugiperda* nucleopolyhedrovirus. Braz. Arch. Biol. Technol. 53, 285–291.

Arends, H.M., Jehle, J.A., 2002. Homologous recombination between the inverted terminal repeats of defective transposon TCp3.2 causes an inversion in the genome of *Cydia pomonella* granulovirus. J. Gen. Virol. 83, 1573–1578.

Asgari, S., Johnson, K.N., 2010. Insect Virology. Caister Academic, Norfolk, UK.

Bauser, C.A., Elick, T.A., Fraser, M.J., 1996. Characterization of hitchhiker, a transposon insertion frequently associated with baculovirus FP mutants derived upon passage in the TN-368 cell line. Virology 216, 235–237.

Beames, B., Summers, M.D., 1988. Comparisons of host cell DNA insertions and altered transcription at the site of insertions in few polyhedra baculovirus mutants. Virology 162, 206–220.

Becker, J., Landureau, J.C., 1981. Specific vitamin requirements of insect cell-lines (*P. americana*) according to their tissue origin and in vitro conditions. In Vitro Cell. Dev. Biol. 17, 471–479.

Bedard, C., Kamen, A., Tom, R., Massie, B., 1994. Maximization of recombinant protein yield in the insect-cell baculovirus system by one-time addition of nutrients to high-density batch cultures. Cytotechnology 15, 129–138.

Bedard, C., Perret, S., Kamen, A.A., 1997. Fed-batch culture of Sf-9 cells supports 3 × 10(7) cells per ml and improves baculovirus-expressed recombinant protein yields. Biotechnol. Lett. 19, 629–632.

Bernal, V., Carinhas, N., Yokomizo, A.Y., Carrondo, M.J., Alves, P.M., 2009. Cell density effect in the baculovirus-insect cells system: a quantitative analysis of energetic metabolism. Biotechnol. Bioeng. 104, 162–180.

Bischoff, D.S., Slavicek, J.M., 1997. Molecular analysis of an enhancin gene in the *Lymantria dispar* nuclear polyhedrosis virus. J. Virol. 71, 8133–8140.

Black, B.C., Brennan, L.A., Dierks, P.M., Gard, I.E., 1997. Commercialisation of baculoviral insecticides. In: Miller, L.K. (Ed.), The Baculoviruses, Plenum Press, New York, pp. 341–388.

Blissard, G.W., Wenz, J.R., 1992. Baculovirus gp64 envelope glycoprotein is sufficient to mediate pH-dependent membrane fusion. J. Virol. 66, 6829–6835.

Bonning, B.C., Hoover, K., Duffey, S., Hammock, B.D., 1995. Production of polyhedra of the *Autographa californica* nuclear polyhedrosis virus using the Sf21 and Tn5B1-4 cell lines and comparison with host-derived polyhedra by bioassay. J. Invertebr. Pathol. 66, 224–230.

Bonning, B.C., Ward, V.K., van Meer, M.M., Booth, T.F., Hammock, B.D., 1997. Disruption of lysosomal targeting is associated with insecticidal potency of juvenile hormone esterase. Proc. Natl. Acad. Sci. U. S. A. 94, 6007–6012.

Braunagel, S.C., Burks, J.K., Rosas-Acosta, G., Harrison, R.L., Ma, H., Summers, M.D., 1999. Mutations within the *Autographa californica* nucleopolyhedrovirus FP25K gene decrease the accumulation of ODV-E66 and alter its intranuclear transport. J. Virol. 73, 8559–8570.

Braunagel, S.C., Cox, V., Summers, M.D., 2009. Baculovirus data suggest a common but multifaceted pathway for sorting proteins to the inner nuclear membrane. J. Virol. 83, 1280–1288.

Braunagel, S.C., Williamson, S.T., Saksena, S., Zhong, Z., Russell, W.K., Russell, D.H., Summers, M.D., 2004. Trafficking of ODV-E66 is mediated via a sorting motif and other viral proteins: facilitated trafficking to the inner nuclear membrane. Proc. Natl. Acad. Sci. U. S. A. 101, 8372–8377.

Buerger, P., Hauxwell, C., Murray, D., 2007. Nucleopolyhedrovirus introduction in Australia. Virol. Sin. 22, 173–179.

Burden, J.P., Hails, R.S., Windass, J.D., Suner, M.M., Cory, J.S., 2000. Infectivity, speed of kill, and productivity of a baculovirus expressing the itch mite toxin txp-1 in second and fourth instar larvae of *Trichoplusia ni*. J. Invertebr. Pathol. 75, 226–236.

Carinhas, N., Bernal, V., Monteiro, F., Carrondo, M.J., Oliveira, R., Alves, P.M., 2010. Improving baculovirus production at high cell density through manipulation of energy metabolism. Metab. Eng. 12, 39–52.

Carinhas, N., Robitaille, A.M., Moes, S., Carrondo, M.J., Jenoe, P., Oliveira, R., Alves, P.M., 2011. Quantitative proteomics of *Spodoptera frugiperda* cells during growth and baculovirus infection. PLoS One 6, e26444.

Caron, A.W., Tom, R.L., Kamen, A.A., Massie, B., 1994. Baculovirus expression system scaleup by perfusion of high-density Sf-9 cell cultures. Biotechnol. Bioeng. 43, 881–891.

Cary, L.C., Goebel, M., Corsaro, B.G., Wang, H.G., Rosen, E., Fraser, M.J., 1989. Transposon mutagenesis of baculoviruses: analysis of *Trichoplusia ni* transposon IFP2 insertions within the FP-locus of nuclear polyhedrosis viruses. Virology 172, 156–169.

Cecchini, S., Virag, T., Kotin, R.M., 2011. Reproducible high yields of recombinant adeno-associated virus produced using invertebrate cells in 0.02-to 200-liter cultures. Hum. Gene Ther. 22, 1021–1030.

Chakraborty, S., Greenfield, P., Reid, S., 1996. In vitro production studies with a wild-type *Helicoverpa* baculovirus. Cytotechnology 22, 217–224.

Chakraborty, S., Monsour, C., Teakle, R., Reid, S., 1999. Yield, biological activity, and field performance of a wild-type *Helicoverpa* nucleopolyhedrovirus produced in *H. zea* cell cultures. J. Invertebr. Pathol. 73, 199–205.

Chakraborty, S., Reid, S., 1999. Serial passage of a *Helicoverpa armigera* nucleopolyhedrovirus in *Helicoverpa zea* cell cultures. J. Invertebr. Pathol. 73, 303–308.

Chan, L.C.L., Greenfield, P.F., Reid, S., 1998. Optimising fed-batch production of recombinant proteins using the baculovirus expression vector system. Biotechnol. Bioeng. 59, 178–188.

Chan, L.C.L., Young, P.R., Bletchly, C., Reid, S., 2002. Production of the baculovirus-expressed dengue virus glycoprotein NS1 can be improved dramatically with optimised regimes for fed-batch cultures and the addition of the insect moulting hormone, 20-Hydroxyecdysone. J. Virol. Methods 105, 87–98.

Chen, X., Sun, X., Hu, Z., Li, M., O'Reilly, D.R., Zuidema, D., Vlak, J.M., 2000. Genetic engineering of *Helicoverpa armigera* single-nucleocapsid nucleopolyhedrovirus as an improved pesticide. J. Invertebr. Pathol. 76, 140–146.

Clem, R.J., 2007. Baculoviruses and apoptosis: a diversity of genes and responses. Curr. Drug Targets 8, 1069–1074.

Cooper, D., Cory, J.S., Myers, J.H., 2003. Hierarchical spatial structure of genetically variable nucleopolyhedroviruses infecting cyclic populations of western tent caterpillars. Mol. Ecol. 12, 881–890.

Cory, J.S., Clarke, E.E., Brown, M.L., Hails, R.S., O'Reilly, D.R., 2004. Microparasite manipulation of an insect: the influence of the egt gene on the interaction between a baculovirus and its lepidopteran host. Funct. Ecol. 18, 443–450.

Cory, J.S., Green, B.M., Paul, R.K., Hunter-Fujita, F., 2005. Genotypic and phenotypic diversity of a baculovirus population within an individual insect host. J. Invertebr. Pathol. 89, 101–111.

Cory, J.S., Myers, J.H., 2003. The ecology and evolution of insect baculoviruses. Annu. Rev. Ecol. Evol. Syst. 34, 239–272.

Cox, M.M.J., 2004. Commercial production in insect cells. One company's perspective. BioProcess Int. 2 (Suppl.), 34–38.

Crowley, J., Wubben, M., Coco Martin, J.M., 2012. Process for cell culturing by continuous perfusion and alternating tangential flow.

Dai, X., Hajos, J.P., Joosten, N.N., van Oers, M.M., WF, I.J., Zuidema, D., Pang, Y., Vlak, J.M., 2000. Isolation of a *Spodoptera exigua* baculovirus recombinant with a 10.6 kbp genome deletion that retains biological activity. J. Gen. Virol. 81, 2545–2554.

De Jesus, M., Wurm, F.M., 2011. Manufacturing recombinant proteins in kg-ton quantities using animal cells in bioreactors. Eur. J. Pharm. Biopharm. 78, 184–188.
Degterev, A., Yuan, J., 2008. Expansion and evolution of cell death programmes. Nat. Rev. Mol. Cell Biol. 9, 378–390.
Deutschmann, S.M., Jager, V., 1994. Optimization of the growth conditions of Sf21 insect cells for high-density perfusion culture in stirred-tank bioreactors. Enzyme Microb. Technol. 16, 506–512.
Donaldson, M.S., Shuler, M.L., 1998. Low-cost serum-free medium for the BTI-Tn5B1-4 insect cell line. Biotechnol. Prog. 14, 573–579.
Drugmand, J.C., Schneider, Y.J., Agathos, S.N., 2012. Insect cells as factories for biomanufacturing. Biotechnol. Adv. 30, 1140–1157.
Eldridge, R., O'Reilly, D.R., Hammock, B.D., Miller, L.K., 1992. Insecticidal properties of genetically engineered baculoviruses expressing an insect juvenile hormone esterase gene. Appl. Environ. Microbiol. 58, 1583–1591.
Elias, C.B., Zeiser, A., Bedard, C., Kamen, A.A., 2000. Enhanced growth of Sf-9 cells to a maximum density of 5.2 × 10(7) cells per mL and production of beta-galactosidase at high cell density by fed batch culture. Biotechnol. Bioeng. 68, 381–388.
Elvira, S., Gorria, N., Munoz, D., Williams, T., Caballero, P., 2010. A simplified low-cost diet for rearing *Spodoptera exigua* (Lepidoptera: Noctuidae) and its effect on *S. exigua* nucleopolyhedrovirus production. J. Econ. Entomol. 103, 17–24.
Epstein, D., Monsell, R., Horwitz, J., Lenk, S., Ozturk, S., Marsh, C., 2009. Chemically defined media compositions. US Patent Number 7,598,083 B2. Filing date 27 October, 2005. Publication date 6 October, 2009.
Epstein, M.L., Gilula, N.B., 1977. Study of communication specificity between cells in culture. J. Cell Biol. 75, 769–787.
Fang, M., Nie, Y., Harris, S., Erlandson, M.A., Theilmann, D.A., 2009. *Autographa californica* multiple nucleopolyhedrovirus core gene ac96 encodes a *per os* infectivity factor (PIF-4). J. Virol. 83, 12569–12578.
Fang, M., Nie, Y., Wang, Q., Deng, F., Wang, R., Wang, H., Vlak, J.M., Chen, X., Hu, Z., 2006. Open reading frame 132 of *Helicoverpa armigera* nucleopolyhedrovirus encodes a functional *per os* infectivity factor (PIF-2). J. Gen. Virol. 87, 2563–2569.
Fath-Goodin, A., Kroemer, J., Martin, S., Reeves, K., Webb, B.A., 2006. Polydnavirus genes that enhance the baculovirus expression vector system. Adv. Virus Res. 68, 75–90.
Fath-Goodin, A., Kroemer, J.A., Webb, B.A., 2009. The *Campoletis sonorensis* ichnovirus vankyrin protein P-vank-1 inhibits apoptosis in insect Sf9 cells. Insect Mol. Biol. 18, 497–506.
Franek, F., Eckschlager, T., Katinger, H., 2003. Enhancement of monoclonal antibody production by lysine-containing peptides. Biotechnol. Prog. 19, 169–174.
Fraser, M.J., Brusca, J.S., Smith, G.E., Summers, M.D., 1985. Transposon-mediated mutagenesis of a baculovirus. Virology 145, 356–361.
Fraser, M.J., Ciszczon, T., Elick, T., Bauser, C., 1996. Precise excision of TTAA-specific lepidopteran transposons piggyBac (IFP2) and tagalong (TFP3) from the baculovirus genome in cell lines from two species of Lepidoptera. Insect Mol. Biol. 5, 141–151.
Fraser, M.J., Hink, W.F., 1982. The isolation and characterization of the MP and FP plaque variants of *Galleria mellonella* nuclear polyhedrosis virus. Virology 117, 366–378.
Friesen, P.D., Nissen, M.S., 1990. Gene organization and transcription of TED, a lepidopteran retrotransposon integrated within the baculovirus genome. Mol. Cell Biol. 10, 3067–3077.

Friesen, P.D., Rice, W.C., Miller, D.W., Miller, L.K., 1986. Bidirectional transcription from a solo long terminal repeat of the retrotransposon TED: symmetrical RNA start sites. Mol. Cell Biol. 6, 1599–1607.

George, S., Sokolenko, S., Aucoin, M.G., 2012. Rapid and cost-effective baculovirus sample preparation method as a viable alternative to conventional preparation for quantitative real-time PCR. J. Virol. Methods 182, 27–36.

Gioria, V.V., Jager, V., Claus, J.D., 2006. Growth, metabolism and baculovirus production in suspension cultures of an *Anticarsia gemmatalis* cell line. Cytotechnology 52, 113–124.

Giri, L., Feiss, M.G., Bonning, B.C., Murhammer, D.W., 2011. Production of baculovirus defective interfering particles during serial passage is delayed by removing transposon target sites in fp25k. J. Gen. Virol.

Glare, T., Caradus, J., Gelernter, W., Jackson, T., Keyhani, N., Kohl, J., Marrone, P., Morin, L., Stewart, A., 2012. Have biopesticides come of age? Trends Biotechnol. 30, 250–258.

Goodman, C.L., McIntosh, A.H., El Sayed, G.N., Grasela, J.J., Stiles, B., 2001. Production of selected baculoviruses in newly established lepidopteran cell lines. In Vitro Cell. Dev. Biol. Anim. 37, 374–379.

Goodwin, R.H., 1991. Replacement of vertebrate serum with lipids and other factors in the culture of invertebrate cells, tissues, parasites and pathogens. In Vitro Cell. Dev. Biol. 27A, 470–478.

Gorfien, S.F., Fike, R.M., Godwin, G.P., Dzimian, J.L., Epstein, D.A., Gruber, D., McClure, D., Price, P.J., 2012. Serum-free mammalian cell culture medium, and uses thereof. US Patent Number 8,198,084 B2. Filing date 14 June, 2005. Publication date 12 June, 2012.

Haas-Stapleton, E.J., Washburn, J.O., Volkman, L.E., 2004. P740 mediates specific binding of *Autographa californica* M nucleopolyhedrovirus occlusion-derived virus to primary cellular targets in the midgut epithelia of *Heliothis virescens* larvae. J. Virol. 78, 6786–6791.

Haas, R., Nielsen, L.K., 2005. A physiological product-release model for baculovirus infected insect cells. Biotechnol. Bioeng. 91, 768–772.

Hajek, K.L., Friesen, P.D., 1998. Proteolytic processing and assembly of gag and gag-pol proteins of TED, a baculovirus-associated retrotransposon of the gypsy family. J. Virol. 72, 8718–8724.

Hammock, B.D., Bonning, B.C., Possee, R.D., Hanzlik, T.N., Maeda, S., 1990. Expression and effects of the juvenile hormone esterase in a baculovirus vector 344, 458–461.

Harrison, R., Hoover, K., 2012. Baculoviruses and other occluded insect viruses. In: Vega, F., Kaya, H. (Eds.), Insect Pathology, second ed. Elsevier, Amsterdam, pp. 73–131.

Hauxwell, C., Tichon, M., Buerger, P., Anderson, S., 2010. Australia. In: Kabaluk, J.T., Svircev, A.M., Goettel, M.S., Woo, S.G. (Eds.), The Use and Regulation of Microbial Pesticides in Representative Jurisdictions Worldwide, IOBC Global, pp. 80–88.

Hawtin, R.E., Zarkowska, T., Arnold, K., Thomas, C.J., Gooday, G.W., King, L.A., Kuzio, J.A., Possee, R.D., 1997. Liquefaction of *Autographa californica* nucleopolyhedrovirus-infected insects is dependent on the integrity of virus-encoded chitinase and cathepsin genes. Virology 238, 243–253.

Heldens, J.G., van Strien, E.A., Feldmann, A.M., Kulcsar, P., Munoz, D., Leisy, D.J., Zuidema, D., Goldbach, R.W., Vlak, J.M., 1996. *Spodoptera exigua* multicapsid nucleopolyhedrovirus deletion mutants generated in cell culture lack virulence in vivo. J. Gen. Virol. 77, 3127–3134.

Herniou, E.A., Arif, B.M., Becnel, J.J., Blissard, G.W., Bonning, B., Harrison, R., Jehle, J.A., Theilmann, D.A., Vlak, J.M., 2012. Family *Baculoviridae*. In: King, A.M.Q., Adams, M.J., Carstens, E.B., Lefkowitz, E.J. (Eds.), Virus Taxonomy, Classification and Nomenclature of Viruses, Ninth Report of the International Committee on Taxonomy of Viruses, Elsevier Academic Press, Amsterdam, pp. 163–173.

Hink, W.F.S.E., 1976. Replication and passage of alfalfa looper nuclear polyhedrosis virus plaque variants in cloned cell cultures and larval stages of four host species. J. Invert. Path. 27, 49–55.

Huang, H., Wang, M., Deng, F., Wang, H., Hu, Z., 2012. ORF85 of HearNPV encodes the per os infectivity factor 4 (PIF4) and is essential for the formation of the PIF complex. Virology 427, 217–223.

Huynh, H.T., Chan, L.C.L., Tran, T.T.B., Nielsen, L.K., Reid, S., 2012. Improving the robustness of a low-cost insect cell medium for baculovirus biopesticides production, via hydrolysate streamlining using a tube bioreactor-based statistical optimization routine. Biotechnol. Prog. 28, 788–802.

Ignoffo, C.M., 1973. Development of a viral insecticide—concept to commercialization. Exp. Parasitol. 33, 380–406.

Ignoffo, C.M., 1999. The first viral pesticide: past, present, and future. (Reprinted from Developments in Industrial Microbiology, 20, pg 105–115, 1979) J. Ind. Microbiol. Biotechnol. 22, 407–417.

Ijkel, W., van Strien, E.A., Heldens, J.G., Broer, R., Zuidema, D., Goldbach, R.W., Vlak, J.M., 1999. Sequence and organization of the *Spodoptera exigua* multicapsid nucleopolyhedrovirus genome. J. Gen. Virol. 80, 3289–3304.

Ikonomou, L., Bastin, G., Schneider, Y.J., Agathos, S.N., 2001. Design of an efficient medium for insect cell growth and recombinant protein production. In Vitro Cell. Dev. Biol. Anim. 37, 549–559.

Ikonomou, L., Schneider, Y.J., Agathos, S.N., 2003. Insect cell culture for industrial production of recombinant proteins. Appl. Microbiol. Biotechnol. 62, 1–20.

Inceoglu, A.B., Kamita, S.G., Hammock, B.D., 2006. Genetically modified baculoviruses: a historical overview and future outlook. Adv. Virus Res. 68, 323–360.

Inceoglu, A.B., Kamita, S.G., Hinton, A.C., Huang, Q., Severson, T.F., Kang, K., Hammock, B.D., 2001. Recombinant baculoviruses for insect control. Pest Manage. Sci. 57, 981–987.

Inlow, D., Shauger, A., Maiorella, B., 1989. Insect cell culture and baculovirus propagation in protein-free medium. J. Tissue Cult. Methods 12, 13–16.

Jakubowska, A.K., Peters, S.A., Ziemnicka, J., Vlak, J.M., van Oers, M.M., 2006. Genome sequence of an enhancin gene-rich nucleopolyhedrovirus (NPV) from *Agrotis segetum*: collinearity with *Spodoptera exigua* multiple NPV. J. Gen. Virol. 87, 537–551.

Jang, J.D., Sanderson, C.S., Chan, L.C.L., Barford, J.P., Reid, S., 2000. Structured modeling of recombinant protein production in batch and fed-batch culture of baculovirus-infected insect cells. Cytotechnology 34, 71–82.

Jehle, J.A., Blissard, G.W., Bonning, B.C., Cory, J.S., Herniou, E.A., Rohrmann, G.F., Theilmann, D.A., Thiem, S.M., Vlak, J.M., 2006. On the classification and nomenclature of baculoviruses: a proposal for revision. Arch. Virol. 151, 1257–1266.

Jehle, J.A., Fritsch, E., Nickel, A., Huber, J., Backhaus, H., 1995. TC14.7: a novel lepidopteran transposon found in *Cydia pomonella* granulosis virus. Virology 207, 369–379.

Jehle, J.A., Nickel, A., Vlak, J.M., Backhaus, H., 1998. Horizontal escape of the novel Tc1-like lepidopteran transposon TCp3.2 into Cydia pomonella granulovirus. J. Mol. Evol. 46, 215–224.

Jem, K.J., Gong, T., Mullen, J., Georgis, R., 1997. Development of an industrial insect cell culture process for large scale production of baculovirus biopesticides. In: Maramorosch, K., Mitsuhashi, J. (Eds.), Invertebrate Cell Culture: Novel Directions and Biotechnology Applications, Science Publishers, New Hampshire, pp. 173–180.

Kabaluk, J.T., Svircev, A.M., Goettel, M.S., Woo, S.G. (Eds.), 2010. The Use and Regulation of Microbial Pesticides in Representative Jurisdictions Worldwide, IOBC Global.

Kambrekar, D.N., Kulkarni, K.A., Giraddi, R.S., 2007. Assessment of quality of HaNPV samples produced by private firms. Karnataka J. Agric. Sci. 20, 417–419.

Kariuki, C.W., McIntosh, A.H., Goodman, C.L., 2000. In vitro host range studies with a new baculovirus isolate from the diamondback moth *Plutella xylostella* (L.) (Plutellidae: Lepidoptera). In Vitro Cell. Dev. Biol. Anim. 36, 271–276.

Kasprow, R.P., Lange, A.J., Kirwan, D.J., 1998. Correlation of fermentation yield with yeast extract composition as characterized by near-infrared spectroscopy. Biotechnol. Prog. 14, 318–325.

Kemp, E.M., Woodward, D.T., Cory, J.S., 2011. Detection of single and mixed covert baculovirus infections in eastern spruce budworm, *Choristoneura fumiferana* populations. J. Invertebr. Pathol. 107, 202–205.

Kikhno, I., Gutierrez, S., Croizier, L., Croizier, G., Ferber, M.L., 2002. Characterization of pif, a gene required for the per os infectivity of *Spodoptera littoralis* nucleopolyhedrovirus. J. Gen. Virol. 83, 3013–3022.

Kloppinger, M., Fertig, G., Fraune, E., Miltenburger, H.G., 1990. Multistage production of *Autographa californica* nuclear polyhedrosis-virus in insect cell-cultures. Cytotechnology 4, 271–278.

Knudson, D.L., Harrap, K.A., 1976. Replication of a nuclear polyhedrosis virus in a continuous cell culture of *Spodoptera frugiperda*—Microscopy study of sequence of events of virus infection. J. Virol. 17, 254–268.

Kool, M., Goldbach, R.W., Vlak, J.M., 1994. A putative non-hr origin of DNA replication in the HindIII-K fragment of *Autographa californica* multiple nucleocapsid nuclear polyhedrosis virus. J. Gen. Virol. 75, 3345–3352.

Kool, M., Voncken, J.W., van Lier, F.L., Tramper, J., Vlak, J.M., 1991. Detection and analysis of *Autographa californica* nuclear polyhedrosis virus mutants with defective interfering properties. Virology 183, 739–746.

Kuzio, J., Pearson, M.N., Harwood, S.H., Funk, C.J., Evans, J.T., Slavicek, J.M., Rohrmann, G.F., 1999. Sequence and analysis of the genome of a baculovirus pathogenic for *Lymantria dispar*. Virology 253, 17–34.

Laken, H.A., Leonard, M.W., 2001. Understanding and modulating apoptosis in industrial cell culture. Curr. Opin. Biotechnol. 12, 175–179.

Lee, H., Krell, P.J., 1994. Reiterated DNA fragments in defective genomes of *Autographa californica* nuclear polyhedrosis virus are competent for AcMNPV-dependent DNA replication. Virology 202, 418–429.

Lee, H.Y., Krell, P.J., 1992. Generation and analysis of defective genomes of *Autographa californica* nuclear polyhedrosis virus. J. Virol. 66, 4339–4347.

Lepore, L.S., Roelvink, P.R., Granados, R.R., 1996. Enhancin, the granulosis virus protein that facilitates nucleopolyhedrovirus (NPV) infections, is a metalloprotease. J. Invertebr. Pathol. 68, 131–140.

Li, Q., Donly, C., Li, L., Willis, L.G., Theilmann, D.A., Erlandson, M., 2002. Sequence and organization of the *Mamestra configurata* nucleopolyhedrovirus genome. Virology 294, 106–121.

Li, Q., Li, L., Moore, K., Donly, C., Theilmann, D.A., Erlandson, M., 2003. Characterization of *Mamestra configurata* nucleopolyhedrovirus enhancin and its functional analysis via expression in an *Autographa californica* M nucleopolyhedrovirus recombinant. J. Gen. Virol. 84, 123–132.

Lopez-Ferber, M., Simon, O., Williams, T., Caballero, P., 2003. Defective or effective? Mutualistic interactions between virus genotypes. Proc. Biol. Sci. 270, 2249–2255.

Lu, C., Gonzalez, C., Gleason, J., Gangi, J., Yang, J.D., 2007. A T-flask based screening platform for evaluating and identifying plant hydrolysates for a fed-batch cell culture process. Cytotechnology 55, 15–29.

Lua, L.H.L., Pedrini, M.R.S., Reid, S., Robertson, A., Tribe, D.E., 2002. Phenotypic and genotypic analysis of *Helicoverpa armigera* nucleopolyhedrovirus serially passaged in cell culture. J. Gen. Virol. 83, 945–955.

Lua, L.H.L., Reid, S., 2000. Virus morphogenesis of *Helicoverpa armigera* nucleopolyhedrovirus in *Helicoverpa zea* serum-free suspension culture. J. Gen. Virol. 81, 2531–2543.

Lynn, D.E., 1994. Enhanced infectivity of occluded virions of the gypsy-moth nuclear polyhedrosis-virus for cell-cultures. J. Invertebr. Pathol. 63, 268–274.

Maiorella, B., Inlow, D., Shauger, A., Harano, D., 1988. Large scale insect cell-culture for recombinant protein production. Bio-Technology 6, 1406–1410.

Malik, H.S., Henikoff, S., Eickbush, T.H., 2000. Poised for contagion: evolutionary origins of the infectious abilities of invertebrate retroviruses. Genome Res. 10, 1307–1318.

Marteijn, R.C.L., Jurrius, O., Dhont, J., de Gooijer, C.D., Tramper, J., Martens, D.E., 2003. Optimization of a feed medium for fed-batch culture of insect cells using a genetic algorithm. Biotechnol. Bioeng. 81, 269–278.

Martens, J.W., Honee, G., Zuidema, D., van Lent, J.W., Visser, B., Vlak, J.M., 1990. Insecticidal activity of a bacterial crystal protein expressed by a recombinant baculovirus in insect cells. Appl. Environ. Microbiol. 56, 2764–2770.

McIntosh, A.H., Ignoffo, C.M., 1981. Replication and infectivity of the single-embedded nuclear polyhedrosis-virus, Baculovirus-heliothis, in homologous cell-lines. J. Invertebr. Pathol. 37, 258–264.

McIntosh, A.H., Maramorosch, K., Rechtoris, C., 1973. Adaptation of an insect cell line (*Agallia constricta*) in a mammalian cell culture medium. In Vitro Cell. Dev. B 8, 375–378.

McKenna, K.K., Shuler, M.L., Granados, R.R., 1997. Increased virus production in suspension culture by a *Trichoplusia ni* cell line in serum-free media. Biotechnol. Prog. 13, 805–809.

Meghrous, J., Kamen, A., Palli, S.R., Sohi, S.S., Caputo, G.F., Bedard, C., 2000. Production of *Choristoneura fumiferana* nucleopolyhedrovirus in *C. fumiferana* (CF-2C1 cells) in a 3 litre bioreactor using serum-free medium. Biocontrol Sci. Technol. 10, 301–313.

Meghrous, J., Mahmoud, W., Jacob, D., Chubet, R., Cox, M., Kamen, A.A., 2009. Development of a simple and high-yielding fed-batch process for the production of influenza vaccines. Vaccine 28, 309–316.

Mena, J.A., Aucoin, M.G., Montes, J., Chahal, P.S., Kamen, A.A., 2010. Improving adeno-associated vector yield in high density insect cell cultures. J. Gene Med. 12, 157–167.

Mena, J.A., Kamen, A.A., 2011. Insect cell technology is a versatile and robust vaccine manufacturing platform. Expert Rev. Vaccines 10, 1063–1081.

Merryweather, A.T., Weyer, U., Harris, M.P., Hirst, M., Booth, T., Possee, R.D., 1990. Construction of genetically engineered baculovirus insecticides containing the *Bacillus thuringiensis* subsp. kurstaki HD-73 delta endotoxin. J. Gen. Virol. 71, 1535–1544.

Micheloud, G.A., Gioria, V.V., Eberhardt, I., Visnovsky, G., Claus, J.D., 2011. Production of the *Anticarsia gemmatalis* multiple nucleopolyhedrovirus in serum-free suspension cultures of the saUFL-AG-286 cell line in stirred reactor and airlift reactor. J. Virol. Methods 178, 106–116.

Micheloud, G.A., Gioria, V.V., Perez, G., Claus, J.D., 2009. Production of occlusion bodies of *Anticarsia gemmatalis* multiple nucleopolyhedrovirus in serum-free suspension cultures of the saUFL-AG-286 cell line: influence of infection conditions and statistical optimization. J. Virol. Methods 162, 258–266.

Miller, D.W., Miller, L.K., 1982. A virus mutant with an insertion of a copia-like transposable element. Nature 299, 562–564.

Miller, L.K., 1997. Introduction to the baculoviruses. In: Miller, L.K. (Ed.), The Baculoviruses, Plenum Press, New York, pp. 1–6.

Miller, L.K., Ball, L.A. (Eds.), 1988. The Insect Viruses, Plenum Press, New York.
Mitsuhashi, J., 1996. Preliminary formulation of a chemically defined medium for insect cell cultures. Methods Cell Sci. 18, 293–298.
Mitsuhashi, J., Goodwin, R.H., 1989. The serum-free culture of insect cells in vitro. In: Mitsuhashi, J. (Ed.), Invertebrate Cell System Applications, vol. 1. CRC Press, Boca Raton, pp. 31–43.
Monsma, S.A., Oomens, A.G., Blissard, G.W., 1996. The GP64 envelope fusion protein is an essential baculovirus protein required for cell-to-cell transmission of infection. J. Virol. 70, 4607–4616.
Moscardi, F., de Souza, M.L., de Castro, M.E.B., Moscardi, M.L., Szewczyk, B., 2011. Baculovirus pesticides: present state and future perspectives. In: Ahmad, I., Ahmad, F., Pitchtel, J. (Eds.), Microbes and Microbial Technology: Agricultural and Environmental Applications, first ed. Springer, New York, pp. 415–445.
Munoz, D., Castillejo, J.I., Caballero, P., 1998. Naturally occurring deletion mutants are parasitic genotypes in a wild-type nucleopolyhedrovirus population of *Spodoptera exigua*. Appl. Environ. Microbiol. 64, 4372–4377.
Murhammer, D.W., 1996. Use of viral insecticides for pest control and production in cell culture. Appl. Biochem. Biotechnol. 59, 199–220.
Nguyen, Q., Palfreyman, R.W., Chan, L.C.L., Reid, S., Nielsen, L.K., 2012. Transcriptome sequencing of and microarray development for a *Helicoverpa zea* cell line to investigate in vitro insect cell–baculovirus interactions. PLos One 7, e36324.
Nguyen, Q., Qi, Y.M., Wu, Y., Chan, L.C.L., Nielsen, L.K., Reid, S., 2011. In vitro production of *Helicoverpa* baculovirus biopesticides-automated selection of insect cell clones for manufacturing and systems biology studies. J. Virol. Methods 175, 197–205.
Nie, Y., Fang, M., Erlandson, M.A., Theilmann, D.A., 2012. Analysis of the *Autographa californica* multiple nucleopolyhedrovirus overlapping gene pair lef3 and ac68 reveals that AC68 is a per os infectivity factor and that LEF3 Is critical, but not essential, for virus replication. J. Virol. 86, 3985–3994.
O'Reilly, D.R., Miller, L.K., 1989. A baculovirus blocks insect molting by producing ecdysteroid UDP-glucosyl transferase. Science 245, 1110–1112.
O'Reilly, D.R., Passarelli, A.L., Goldman, I.F., Miller, L.K., 1990. Characterization of the DA26 gene in a hypervariable region of the *Autographa californica* nuclear polyhedrosis virus genome. J. Gen. Virol. 71, 1029–1037.
Ohkawa, T., Volkman, L.E., Welch, M.D., 2010. Actin-based motility drives baculovirus transit to the nucleus and cell surface. J. Cell Biol. 190, 187–195.
Ohkawa, T., Washburn, J.O., Sitapara, R., Sid, E., Volkman, L.E., 2005. Specific binding of *Autographa californica* M nucleopolyhedrovirus occlusion-derived virus to midgut cells of *Heliothis virescens* larvae is mediated by products of pif genes Ac119 and Ac022 but not by Ac115. J. Virol. 79, 15258–15264.
Oomens, A.G., Blissard, G.W., 1999. Requirement for GP64 to drive efficient budding of *Autographa californica* multicapsid nucleopolyhedrovirus. Virology 254, 297–314.
Ozers, M.S., Friesen, P.D., 1996. The Env-like open reading frame of the baculovirus-integrated retrotransposon TED encodes a retrovirus-like envelope protein. Virology 226, 252–259.
Pan, X., Long, G., Wang, R., Hou, S., Wang, H., Zheng, Y., Sun, X., Westenberg, M., Deng, F., Vlak, J.M., Hu, Z., 2007. Deletion of a *Helicoverpa armigera* nucleopolyhedrovirus gene encoding a virion structural protein (ORF107) increases the budded virion titre and reduces in vivo infectivity. J. Gen. Virol. 88, 3307–3316.
Pasupuleti, V.K., Braun, S., 2010. State of the art manufacturing of protein hydrolysates. In: Pasupuleti, V.K., Demain, A.L. (Eds.), Protein Hydrolysates in Biotechnology, .
Payne, C.C., 1982. Insect viruses as control agents. Parasitology 84, 35–77.

Pearson, M.N., Rohrmann, G.F., 2004. Conservation of a proteinase cleavage site between an insect retrovirus (gypsy) Env protein and a baculovirus envelope fusion protein. Virology 322, 61–68.

Pedrini, M.R.S., Christian, P., Nielsen, L.K., Reid, S., Chan, L.C.L., 2006. Importance of virus-medium interactions on the biological activity of wild-type Heliothine nucleopolyhedroviruses propagated via suspension insect cell cultures. J. Virol. Methods 136, 267–272.

Pedrini, M.R.S., Reid, S., Nielsen, L., Chan, L.C.L., 2011. Kinetic characterization of the group II *Helicoverpa armigera* nucleopolyhedrovirus propagated in suspension cell cultures: implications for development of a biopesticides production process. Biotechnol. Prog. 27, 614–624.

Peng, K., van Lent, J.W., Boeren, S., Fang, M., Theilmann, D.A., Erlandson, M.A., Vlak, J.M., van Oers, M.M., 2012. Characterization of novel components of the baculovirus per os infectivity factor complex. J. Virol. 86, 4981–4988.

Peng, K., van Lent, J.W., Vlak, J.M., Hu, Z., van Oers, M.M., 2011. In situ cleavage of baculovirus occlusion-derived virus receptor binding protein P74 in the peroral infectivity complex. J. Virol. 85, 10710–10718.

Peng, K., van Oers, M.M., Hu, Z., van Lent, J.W., Vlak, J.M., 2010. Baculovirus per os infectivity factors form a complex on the surface of occlusion-derived virus. J. Virol. 84, 9497–9504.

Pijlman, G.P., Dortmans, J.C., Vermeesch, A.M., Yang, K., Martens, D.E., Goldbach, R.W., Vlak, J.M., 2002. Pivotal role of the non-hr origin of DNA replication in the genesis of defective interfering baculoviruses. J. Virol. 76, 5605–5611.

Pijlman, G.P., Pruijssers, A.J., Vlak, J.M., 2003a. Identification of pif-2, a third conserved baculovirus gene required for per os infection of insects. J. Gen. Virol. 84, 2041–2049.

Pijlman, G.P., van den Born, E., Martens, D.E., Vlak, J.M., 2001. *Autographa californica* baculoviruses with large genomic deletions are rapidly generated in infected insect cells. Virology 283, 132–138.

Pijlman, G.P., Van Schijndel, J.E., Vlak, J.M., 2003b. Spontaneous excision of BAC vector sequences from bacmid-derived baculovirus expression vectors upon passage in insect cells. J. Gen. Virol. 84, 2669–2678.

Pijlman, G.P., Vermeesch, A.M., Vlak, J.M., 2003c. Cell line-specific accumulation of the baculovirus non-hr origin of DNA replication in infected insect cells. J. Invertebr. Pathol. 84, 214–219.

Popham, H.J., Bischoff, D.S., Slavicek, J.M., 2001. Both *Lymantria dispar* nucleopolyhedrovirus enhancin genes contribute to viral potency. J. Virol. 75, 8639–8648.

Popham, H.J.R., Shelby, K.S., 2007. Effect of inorganic and organic forms of selenium supplementation on development of larval *Heliothis virescens*. Entomol. Exp. Appl. 125, 171–178.

Portocarrero, E., 2011. Organic food products in China: market overview. ITC Publications, International Trade Centre, Geneva, Switzerland.

Power, J.F., Reid, S., Radford, K.M., Greenfield, P.F., Nielsen, L.K., 1994. Modeling and optimization of the baculovirus expression vector system in batch suspension culture. Biotechnol. Bioeng. 44, 710–719.

Qi, Y.M., Greenfield, P.F., Reid, S., 1996. Evaluation of a simple protein free medium that supports high levels of monoclonal antibody production. Cytotechnology 21, 95–109.

Rabindra, R.J., Grzywacz, D., 2010. India. In: Kabaluk, J.T., Svircev, A.M., Goettel, M.S., Woo, S.G. (Eds.), The Use and Regulation of Microbial Pesticides in Representative Jurisdictions Worldwide, IOBC Global, pp. 12–17.

Radford, K.M., Reid, S., Greenfield, P.F., 1997. Substrate limitation in the baculovirus expression vector system. Biotechnol. Bioeng. 56, 32–44.

Ravensberg, W.J., 2011a. Critical factors in the successful commercialization of microbial pest control products. Roadmap to the Successful Development and Commercialization of Microbial Pest Control Products for Control of Arthropods, vol. 10, pp. 295–356.

Ravensberg, W.J., 2011b. Mass production and product development of a microbial pest control agent. Roadmap to the Successful Development and Commercialization of Microbial Pest Control Products for Control of Arthropods, vol. 10, pp. 59–127.

Regev, A., Rivkin, H., Inceoglu, B., Gershburg, E., Hammock, B.D., Gurevitz, M., Chejanovsky, N., 2003. Further enhancement of baculovirus insecticidal efficacy with scorpion toxins that interact cooperatively. FEBS Lett. 537, 106–110.

Reid, S., Lua, L.H.L., 2005. Method of producing baculovirus. International publication number WO2005/045014 A1. Filing date 10 November, 2004. Publication date 19 May, 2005.

Rhodes, D.J., 1996. Economics of baculovirus—insect cell production systems. Cytotechnology 20, 291–297.

Roelvink, P.W., Corsaro, B.G., Granados, R.R., 1995. Characterization of the *Helicoverpa armigera* and *Pseudaletia unipuncta* granulovirus enhancin genes. J. Gen. Virol. 76 (Pt 11), 2693–2705.

Rohrmann, G.F., 1986. Polyhedrin structure. J. Gen. Virol. 67 (Pt 8), 1499–1513.

Rohrmann, G.F., 2008. Baculovirus Molecular Biology. National Library of Medicine (US) National Center for Biotechnology Information, Bethesda (MD). http:\\www.ncbi.nlm.nih.gov/bookshelf/br.fcgi?book=bacvir.

Rosinski, M., Reid, S., Nielsen, L.K., 2002. Kinetics of baculovirus replication and release using real-time quantitative polymerase chain reaction. Biotechnol. Bioeng. 77, 476–480.

Rowley, D.L., Popham, H.J.R., Harrison, R.L., 2011. Genetic variation and virulence of nucleopolyhedroviruses isolated worldwide from the heliothine pests *Helicoverpa armigera*, *Helicoverpa zea*, and *Heliothis virescens*. J. Invertebr. Pathol. 107, 112–126.

Schetter, C., Oellig, C., Doerfler, W., 1990. An insertion of insect cell DNA in the 81-map-unit segment of *Autographa californica* nuclear polyhedrosis virus DNA. J. Virol. 64, 1844–1850.

Schlaeger, E.J., 1996. Medium design for insect cell culture. Cytotechnology 20, 57–70.

Shapiro, M., 1986. In vivo production of baculoviruses. In: Granados, R.R., Federeci, B.A. (Eds.), The Biology of Baculoviruses, vol. II. CRC Press, Boca Raton, pp. 31–61.

Shen, C.F., Kiyota, T., Jardin, B., Konishi, Y., Kamen, A., 2007. Characterization of yeastolate fractions that promote insect cell growth and recombinant protein production. Cytotechnology 54, 25–34.

Siemensma, A., Babcock, J., Wilcox, C., Huttinga, H., 2010. Towards an understanding of how protein hydrolysates stimulate more efficient biosynthesis in cultured cells. In: Pasupuleti, V.K., Demain, A.L. (Eds.), Protein Hydrolysates in Biotechnology, Springer Science + Business Media B.V., Dordrecht, The Netherlands, pp. 33–54.

Simon, O., Palma, L., Williams, T., Lopez-Ferber, M., Caballero, P., 2012. Analysis of a naturally-occurring deletion mutant of *Spodoptera frugiperda* multiple nucleopolyhedrovirus reveals sf58 as a new per os infectivity factor of lepidopteran-infecting baculoviruses. J. Invertebr. Pathol. 109, 117–126.

Simon, O., Williams, T., Caballero, P., Lopez-Ferber, M., 2006. Dynamics of deletion genotypes in an experimental insect virus population. Proc. Biol. Sci. 273, 783–790.

Simon, O., Williams, T., Lopez-Ferber, M., Caballero, P., 2005. Functional importance of deletion mutant genotypes in an insect nucleopolyhedrovirus population. Appl. Environ. Microbiol. 71, 4254–4262.

Slack, J., Arif, B.M., 2007. The baculoviruses occlusion-derived virus: virion structure and function. Adv. Virus Res. 69, 99–165.

Slack, J.M., Lawrence, S.D., Krell, P.J., Arif, B.M., 2008. Trypsin cleavage of the baculovirus occlusion-derived virus attachment protein P74 is prerequisite in per os infection. J. Gen. Virol. 89, 2388–2397.

Song, J., Wang, R., Deng, F., Wang, H., Hu, Z., 2008. Functional studies of per os infectivity factors of *Helicoverpa armigera* single nucleocapsid nucleopolyhedrovirus. J. Gen. Virol. 89, 2331–2338.

Sparks, W.O., Rohlfing, A., Bonning, B.C., 2011. A peptide with similarity to baculovirus ODV-E66 binds the gut epithelium of *Heliothis virescens* and impedes infection with *Autographa californica* multiple nucleopolyhedrovirus. J. Gen. Virol. 92, 1051–1060.

Stewart, L.M., Hirst, M., Lopez Ferber, M., Merryweather, A.T., Cayley, P.J., Possee, R.D., 1991. Construction of an improved baculovirus insecticide containing an insect-specific toxin gene. Nature 352, 85–88.

Sun, X.L., Peng, H.Y., 2007. Recent advances in biological control of pest insects by using viruses in China. Virol. Sin. 22, 158–162.

Thakore, Y., 2006. The biopesticide market for global agricultural use. Ind. Biotechnol. 2, 194–208.

Tomalski, M.D., Miller, L.K., 1991. Insect paralysis by baculovirus-mediated expression of a mite neurotoxin gene. Nature 352, 82–85.

Toprak, U., Harris, S., Baldwin, D., Theilmann, D., Gillott, C., Hegedus, D.D., Erlandson, M.A., 2012. Role of enhancin in *Mamestra configurata* nucleopolyhedrovirus virulence: selective degradation of host peritrophic matrix proteins. J. Gen. Virol. 93, 744–753.

Tran, T.T.B., Dietmair, S., Chan, L.C.L., Huynh, H.T., Nielsen, L.K., Reid, S., 2012. Development of quenching and washing protocols for quantitative intracellular metabolite analysis of uninfected and baculovirus-infected insect cells. Methods 56, 396–407.

van Lier, F.L., van den End, E.J., de Gooijer, C.D., Vlak, J.M., Tramper, J., 1990. Continuous production of baculovirus in a cascade of insect-cell reactors. Appl. Microbiol. Biotechnol. 33, 43–47.

van Oers, M.M., Lynn, D.E., 2010. Insect cell culture. Encyclopedia of Life Science, vol. A22748. Elsevier.

van Oers, M.M., Vlak, J.M., 2007. Baculovirus genomics. Curr. Drug Targets 8, 1051–1068.

van Strien, E.A., IJkel, W.F., Gerrits, H., Vlak, J.M., Zuidema, D., 2000. Characteristics of the transactivator gene ie1 of *Spodoptera exigua* multiple nucleopolyhedrovirus. Arch. Virol. 145, 2115–2133.

Vaughn, J.L., Goodwin, R.H., Tompkins, G.J., McCawley, P., 1977. The establishment of two cell lines from the insect *Spodoptera frugiperda* (Lepidoptera; Noctuidae). In Vitro 13, 213–217.

Vicente, T., Roldao, A., Peixoto, C., Carrondo, M.J., Alves, P.M., 2011. Large-scale production and purification of VLP-based vaccines. J. Invertebr. Pathol. 107 (Suppl.), S42–S48.

Vincent, C., Andermatt, M., Valéro, J., 2007. Madex® and VirosoftCP4®, viral biopesticides for codling moth control. In: Vincent, C., Goettel, M., Lazarovits, G. (Eds.), Biological Control: A Global Perspective, CABI, Wallingford, UK, pp. 336–343.

Wang, B., Li, Z., 2010. China. In: Kabaluk, J.T., Svircev, A.M., Goettel, M.S., Woo, S.G. (Eds.), The Use and Regulation of Microbial Pesticides in Representative Jurisdictions Worldwide, IOBC Global, pp. 7–11.

Wang, H.H., Fraser, M.J., Cary, L.C., 1989. Transposon mutagenesis of baculoviruses: analysis of TFP3 lepidopteran transposon insertions at the FP locus of nuclear polyhedrosis viruses. Gene 81, 97–108.

Warnock, J.N., Al-Rubeai, M., 2006. Bioreactor systems for the production of biopharmaceuticals from animal cells. Biotechnol. Appl. Biochem. 45, 1–12.

Washburn, J.O., Chan, E.Y., Volkman, L.E., Aumiller, J.J., Jarvis, D.L., 2003a. Early synthesis of budded virus envelope fusion protein GP64 enhances *Autographa californica* multicapsid nucleopolyhedrovirus virulence in orally infected *Heliothis virescens*. J. Virol. 77, 280–290.

Washburn, J.O., Trudeau, D., Wong, J.F., Volkman, L.E., 2003b. Early pathogenesis of *Autographa californica* multiple nucleopolyhedrovirus and *Helicoverpa zea* single nucleopolyhedro virus in *Heliothis virescens*: a comparison of the 'M' and 'S' strategies for establishing fatal infection. J. Gen. Virol. 84, 343–351.

Weiss, S.A., Vaughn, J.L., 1986. Cell culture methods for large scale propagation of baculoviruses. In: Granados, R.R., Federeci, B.A. (Eds.), The Biology of Baculoviruses, vol. II. CRC Press Inc., Boca Raton, pp. 63–87.

Weiss, S.A., Whitford, W.G., Godwin, G.P., Reid, S., 1992. Media design: optimizing of recombinant proteins in serum-free culture. In: Vlak, J.M., Schlaeger, E.J., Bernard, A.R. (Eds.), Baculovirus and Recombinant Protein Production Processes, Editiones Roche, Basel, Swizerland, pp. 306–314.

Westenberg, M., Veenman, F., Roode, E.C., Goldbach, R.W., Vlak, J.M., Zuidema, D., 2004. Functional analysis of the putative fusion domain of the baculovirus envelope fusion protein F. J. Virol. 78, 6946–6954.

Wickham, T.J., Davis, T., Granados, R.R., Shuler, M.L., Wood, H.A., 1992. Screening of insect cell lines for the production of recombinant proteins and infectious virus in the baculovirus expression system. Biotechnol. Prog. 8, 391–396.

Wilkie, G.E., Stockdale, H., Pirt, S.V., 1980. Chemically-defined media for production of insect cells and viruses in vitro. Dev. Biol. Stand. 46, 29–37.

Willer, H., 2011. The World of Organic Agriculture 2011: summary. In: Willer, H., Kilcher, L. (Eds.), The World of Organic Agriculture, IFOAM, Bonn and FiBL, FrickStatistics and Emerging Trends 2011. FiBL-IFOAM Report.

Wong, K.T.K., Peter, C.H., Greenfield, P.F., Reid, S., Nielsen, L.K., 1996. Low multiplicity infection of insect cells with a recombinant baculovirus: the cell yield concept. Biotechnol. Bioeng. 49, 659–666.

Wyatt, S.S., 1956. Culture in vitro of tissue from the silkworm, *Bombyx mori* L. J. Gen. Physiol. 39, 841–852.

Xiang, X., Chen, L., Hu, X., Yu, S., Yang, R., Wu, X., 2011. *Autographa californica* multiple nucleopolyhedrovirus odv-e66 is an essential gene required for oral infectivity. Virus Res. 158, 72–78.

Yang, J.D., Lu, C., Stasny, B., Henley, J., Guinto, W., Gonzalez, C., Gleason, J., Fung, M., Collopy, B., Benjamino, M., Gangi, J., Hanson, M., Ille, E., 2007. Fed-batch bioreactor process scale-up from 3-L to 2,500-L scale for monoclonal antibody production from cell culture. Biotechnol. Bioeng. 98, 141–154.

Yao, L., Zhou, W., Xu, H., Zheng, Y., Qi, Y., 2004. The *Heliothis armigera* single nucleocapsid nucleopolyhedrovirus envelope protein P74 is required for infection of the host midgut. Virus Res. 104, 111–121.

Zanotto, P.M., Kessing, B.D., Maruniak, J.E., 1993. Phylogenetic interrelationships among baculoviruses: evolutionary rates and host associations. J. Invertebr. Pathol. 62, 147–164.

Zehnder, G., Gurr, G.M., Kuhne, S., Wade, M.R., Wratten, S.D., Wyss, E., 2007. Arthropod pest management in organic crops. Annu. Rev. Entomol, vol. 52, pp. 57–80.

Zhang, J.Y., Reddy, J., Buckland, B., Greasham, R., 2003. Toward consistent and productive complex media for industrial fermentations: studies on yeast extract for a recombinant yeast fermentation process. Biotechnol. Bioeng. 82, 640–652.

Zwart, M.P., Erro, E., van Oers, M.M., de Visser, J.A., Vlak, J.M., 2008a. Low multiplicity of infection in vivo results in purifying selection against baculovirus deletion mutants. J. Gen. Virol. 89, 1220–1224.

Zwart, M.P., van Oers, M.M., Cory, J.S., van Lent, J.W., van der Werf, W., Vlak, J.M., 2008b. Development of a quantitative real-time PCR for determination of genotype frequencies for studies in baculovirus population biology. J. Virol. Methods 148, 146–154.

Chapter 14

Formulations of Entomopathogens as Bioinsecticides

Robert Behle[1] **and Tim Birthisel**[2,*]
[1]USDA-ARS-NCAUR, Crop Bioprotection Research Unit, Peoria, IL, USA, [2]The Andersons Inc., Turf and Specialties Group, Asheville, NC, USA

Mention of trade names or commercial products in this article is solely for the purpose of providing specific information and does not imply recommendation or endorsement by the US Department of Agriculture. USDA is an equal opportunity provider and Employer.

14.1. INTRODUCTION

Formulation technology is seen as an enabling technology for all pesticides, adding value to the product (Knowles, 2009). Microbes have great potential to provide effective management of insect pests when properly developed as biological insecticides. Improving formulations may provide key benefits for microbes to be effective pest control agents in the field (Auld and Morin, 1995). Bioinsecticides have a recognized ecological advantage over conventional broad-spectrum chemical insecticides because specific pest insects can be strategically controlled by biological agents with little or no effect on other plants or animals inhabiting the same environment. Additionally, bioinsecticides generally leave no harmful residues on food crops, alleviating this public concern relative to food safety. The popularity and use of bioinsecticides is expected to increase as applications of chemical pesticides decrease due to development of insect resistance to chemical agents, tighter government regulations of chemical insecticides that restrict their application, retail pressure for ecological pest control, and increasing public awareness of the benefits when using biological agents (Van Lenteren, 2012).

*Retired.

Mass Production of Beneficial Organisms. http://dx.doi.org/10.1016/B978-0-12-391453-8.00014-5

Biological agents for control of insects were broadly categorized by Copping and Menn (2000) as (1) living organisms, such as predatory insects, parasitoids, nematodes, and micro-organisms; (2) naturally occurring substances, such as plant extracts and insect pheromones; and (3) genetically modified plants expressing introduced genes for protection against pests. The US Environmental Protection Agency (EPA) defines a biopesticide as one "derived from such natural materials as animal, plant, bacteria, and certain minerals" (EPA, 2012). This definition includes plant extracts such as neem oil, fermentation products such as spinosyn, and certain genetic modifications to plants.

In this chapter, we focus on formulations for mass-produced organisms, specifically fungi, bacteria, viruses, and nematodes. As bioinsecticides, these organisms are intended to be applied as an augmentative biological control of insects using an inundative strategy of activity. Inundative activity, as opposed to inoculative activity (Hajek, 2004), is expected to achieve rapid pest control by the mass application of the microbial agent only, with no expectation of control by subsequent progeny of the microbe. Plant extracts, fermentation products, and insect pheromones also require formulation to be used as biopesticides, but these agents fall more in line with chemical pesticides than with microbial agents relative to formulation considerations. For beneficial microbes intended for classical biological control or for inoculative applications, biopesticide formulation technologies may be easily transferred to the benefit of these pest control strategies.

A formulation consists of a specific combination of ingredients, processes, and equipment to form a defined commercial product. The diversity of potential microbial agents and target environments combined with the vast range of product forms, processes, and ingredients results in a multitude of potential combinations for the formulation scientist to consider. Potential formulation considerations are further complicated by requirements of industrial manufacturing for product handling and worker safety. The task becomes manageable only with a thorough understanding of the entire system from the biology of the microbe through various manufacturing processes and including the intended application environment (Fig. 14.1) so that the key factors for each phase are adequately addressed.

14.1.1. Goals and Benefits of Formulations

"Making useful products usable" was a slogan for the Association of Formulation Chemists and is the primary goal of formulation research for bioinsecticides. Microbial agents can be formulated in different ways to improve their delivery and efficacy (Ash, 2010), and the benefits provided by formulation improvements will contribute toward the successful adoption of applied biological control by consumers. More specifically, developing formulations for biological insecticides requires the formulation scientist to select ingredients and processes to address biological considerations relative to the microbe and

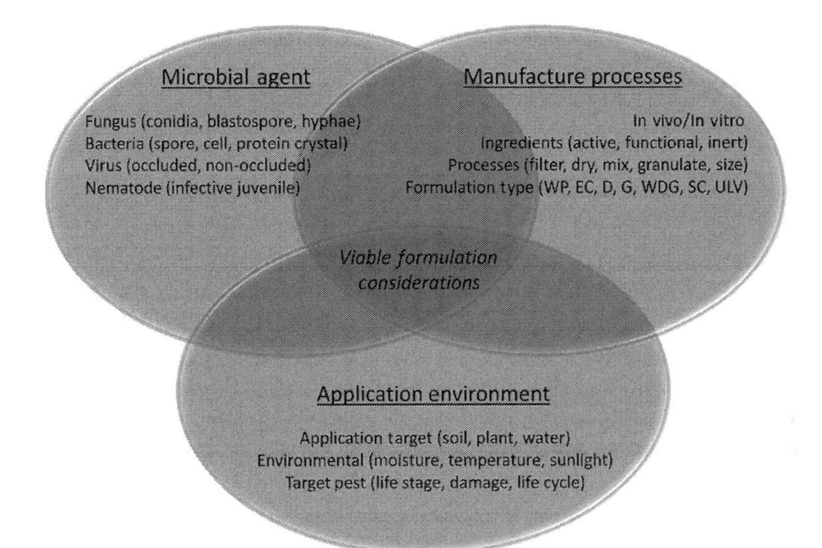

FIGURE 14.1 Spheres of influence depicting formulation considerations while developing microbial-based biological insecticides, which are restricted by specific overlapping knowledge of the microbial agent, manufacturing process, and application environment. (For color version of this figure, the reader is referred to the online version of this book.)

worker safety, as well as address a number of physical considerations related to handling and application. The proper formulation has the potential to provide numerous benefits for bioinsecticides such as longer storage, easier handling, and greater field efficacy, while at the same time maintaining the positive ecological attributes inherent with biological pest control.

Often, successful commercial formulations for bioinsecticides are simple formulations that fit between the microbial production system and the requirements of the end-use application. Successful formulations are often protected by trade secrecy provisions in the legal systems of the United States, and the breadth of technology involved with formulations is so wide that many consider it "more art than science." A significant amount of information about formulations is available in the scientific literature and through such organizations as the American Society for Testing and Materials International (ASTM; West Conshohocken, PA) Committee E35 on Pesticides, Antimicrobials, and Alternative Control Agents, which publishes annual reports of their symposia and published ASTM standards; the American Chemical Society (main offices in Washington, DC and Columbus, OH), which publishes works based on symposia, such as the AGRO Division (formerly Division of Pesticide Chemistry); and the patent literature, which identifies compositions, techniques, and ingredients.

Expectations for biological insecticides are generally aligned with expectations for chemical insecticides relative to ease of use and pest control characteristics. Ease of use relates directly to the physical characteristic of the product, including product form (liquid or dry), storage stability, mixing ability (with water for spray application), compatibility with other agents as tank mixes, and application method (granule, dust, or spray). Efficacy characteristics refer to biological characteristics, including pathogen viability, speed of kill, and residual activity after application. Both ease of use and control characteristics are impacted by product formulation, making selection of the formulation a critical component for product success in the marketplace. Often, the primary goal of the formulation is to allow the application of the biopesticide to fit seamlessly with current chemical application techniques in terms of mixing, application, and pest control expectations.

14.1.2. Challenges of Microbial Pesticides

Years of economical and effective pest control strategies based on chemical insecticides have established a high bar of expectation for biological insecticides. These expectations create several notable challenges for bioinsecticide formulations. It is often expected that the formulation will help to fill gaps between the innate ability of the microbe and the expectation of the bioinsecticide. Microbes have a particulate structure, which often presents a problem with mixing in the spray tank. Disease cycles for microbial agents require significant amounts of time to progress, resulting in slower speed of kill of the target pest, and can contribute to a perception of low efficacy. Maintaining stability of a biopesticide (microbe viability) during storage is a primary requirement for subsequent activity after application, but it is more difficult than maintaining product stability for a chemical insecticide. Environmental degradation after application shortens residual insecticidal activity, requiring more frequent applications to maintain pest control. The high cost of production for many bioinsecticides relative to costs for chemical insecticides can limit the monetary margin available for formulation costs, adding constraints on biopesticide development. Additional challenges include maintaining a product that qualifies for certification as organic, which restricts ingredients that can be used in the formulation. Addressing these and other deficiencies are the challenges of the formulation development process.

Within the realm of pesticide formulations, there exists some confusing terminology about adjuvants. The term *adjuvant* may refer to a component of a formulation or to a stand-alone spray-tank additive. As a component of a registered biopesticide formulation, adjuvants are chemicals added to improve mixing of the formulation components in the product or to improve mixing of the formulation with water in the spray tank. Selection of the adjuvants added to a formulation is not trivial because the formulation of a biopesticide registered by the US EPA is specifically defined; components cannot be changed without notification and submission of a revised Confidential Statement of Formula, which may result in a requirement for re-registration of the product. If the formulation change involves ingredients that are already recognized as "lower concern approved inert ingredients," then the EPA may allow the change until such time that the periodic re-registration process requires full data support and formal review.

By contrast with formulation components, tank-mixed adjuvants are stand-alone products that are sold separate from the biopesticide. These products are intended to be mixed with one or more pesticides to provide added benefit to the treatment and do not have US EPA registration requirements. Benefits may be in the form of drift reduction, spreading over the sprayed surface, resistance to wash-off by rain, slower degradation when exposed to sunlight, or feeding stimulation. Disadvantages of using a tank-mix adjuvant include the added cost to the application and an added step to the mixing process. The benefits provided by tank-mixed adjuvants may be difficult to measure or perceive at the user level of application. Because this chapter covers biopesticide formulations, some of the text may be relevant to "tank-mixed adjuvant formulations" intended for tank mixing specifically with microbial pesticides. Other text may refer to "adjuvant components" as a specific part of the biopesticide formulation.

A major challenge of microbial pesticides lies in the ability of the biopesticide industry to overcome previous biopesticide failures, which will require education of consumers and growers to realistic expectations for bioinsecticides while positioning products into viable markets. Biopesticide development has been most successful when addressing specific target pests in niche markets and in countries with regulatory, economic, and social conditions that favor biopesticides over chemicals. In reality, many attempts to develop microbes as bioinsecticides for large-scale applications fail to achieve all the desired expectations for ease of use and broad pest control. Difficulties with fermentation and formulation of the microorganism or inadequate market research has resulted in "orphaned" biopesticides or products that have been withdrawn from the market (Hynes and Boyetchko, 2006). Production costs for many agents are relatively high, which in turn limits costs that can be allocated to formulation ingredients and processes and may result in problems with application and efficacy. Unsuccessful ventures in the marketplace contribute to a lack of consumer confidence when new products become available. Allocating sufficient resources to developing proper formulations should help to minimize the risk of market failure, improve product performance, and improve consumer confidence in biopesticides as a viable pest control tactic.

Each microbe tends to impose a unique set of challenges related to the formulation required for application. The uniqueness of each microbial agent adds complication when considering competing ingredients, processes, and formulation types. For example, baculoviruses are most commonly produced using non-sterile in vivo systems, and bioinsecticides made using the homogenized insect cadavers have a great need for control of contaminating microbes in the final product. Many fungal bioinsecticides use conidia produced by solid substrate culture, which are hydrophobic and need careful formulation with appropriate surfactants to mix easily with water for spray application. The bacterium

Bacillus thuringiensis Berliner is known to deter feeding by target insects and benefit greatly by the addition of ingredients known to be feeding stimulants (Broza et al., 1986; Farrar et al., 2005). For nematodes, the infective juvenile or dauer has the ability to actively seek a susceptible host, but it generally requires high moisture content from the time of production through application to the point of host contact. These differences among the microbe requirements demonstrate the need for unique formulation solutions among different microbial bioinsecticides.

Formulation research for bioinsecticides faces many challenges, but it has the potential to provide significant benefits. Both physical and biological attributes need to be addressed. Some attributes are dictated by the end-use application, some by the production system, and some by the formulation process itself.

14.2. BIOLOGICAL CONSIDERATIONS

Formulation ingredients and formulation processes address many biological considerations for the microbe as well as the people who handle product from production to end-use application. Relative to the microbe, the choices concerning formulation will impact activity, viability, efficacy, and storage stability characteristics. Relative to persons handling the product, formulations impact issues of worker safety/exposure, allergenicity, and product contamination.

14.2.1. Biological Attributes for the Microbe

Many factors directly and indirectly affect efficacy of biopesticides. It is necessary for the formulation scientist to recognize the need to enhance (or at least, not to inhibit) the desired disease cycle intended with the application of the biopesticide. The formulation should support favorable interactions among the pathogen, host, and environment to support the desired development of disease.

14.2.1.1. Activity

Activity of a biopesticide is measured as the amount of the agent required to initiate infection and the length of time to kill the target pest; these characteristics vary widely among individual microbial pathogens. Pathogens differ in their path of infection, which can be per os, through the host cuticle, pleural connections of sclerites, or through natural insect openings (anus, spiracles, or oral), and may also impact formulation choices to optimize the activity of each microbe. Formulation choices are determined partly by the physical structure of the active agent and partly by the mode of action of the microbial agent. The structure of baculoviruses consists of one or more DNA virions encapsulated in crystalline protein occlusion bodies and initiates infection through the digestive tract after being eaten by a susceptible insect. Bacterial spores also infect through the digestive tract or kill target insects with toxins released by the digestive action on protein crystals found in plasmids formed during sporulation (Stahly, 1984; Beegle and Yamamoto, 1992; Xu et al., 2006). By contrast, fungal agents can be produced in the form of several structures including conidia, blastospores, mycelia, or microsclerotia. Fungal activity results not only from direct contact by spray applications, but also by infection resulting from exposure to treated substrates (Behle, 2006; Douro Kpindou et al., 2011). Conidia and blastospores infect susceptible insects when the spores contact the insect, actively penetrating the integument, whereas mycelia and microsclerotia must form the infective conidia or blastospores after application. The infective juvenile nematode, which can either "ambush" or actively search for hosts, enters the susceptible insect usually through the mouth, anus, or spiracles, encompassing a completely unique set of characteristics to consider for formulation development. These juvenile nematodes require a moist, but not saturated, soil for a week or more following application to allow for their mobility (Gothro and Berry, 2012).

Formulations can optimize activity by improving the chances for association between the microbe and target pest, such as adding feeding stimulants to the formulation to entice pest insects to feed directly on spray residue containing baculoviruses or bacteria, adding oils to improve adhesion of fungal conidia to cuticles, or adding gels to maintain moisture and provide a medium for nematode movement. Each of these examples improves the chance for infection of susceptible insects, benefiting the activity of the bioinsecticide application.

A variety of ingredients can directly improve the activity for specific microbes by decreasing the amount of inoculum required to infect pests or by increasing the speed of kill. Speed of kill of the cattle tick, *Rhipicephalus microplus* (Canestrini) (formerly *Boophilus microplus*), by *Metarhizium anisopliae* (Metschnikoff) Sorokin was improved by the addition of 10% oil emulsion (Leemon and Jonsson, 2008). Additives to stimulate feeding (gustatory stimulation) and chitinase enzymes have been shown to enhance activity of bacterial agents when included in formulations (Brar et al., 2006). Chitinase enzymes increase entomotoxicity of *B. thuringiensis* by perforating the peritorphic membrane in the larval midgut, increasing accessibility of the toxin molecule to receptors on the epithelial cells (Kramer and Muthukrishnan, 1997). Sugars, amino acids, and starches can generally stimulate insect feeding, although a few specific compounds can be used to target specific pests. For example, cucurbitacins provide strong feeding stimulation for corn rootworm beetles (*Diabrotica* spp.) (Arruda et al., 2005). Commercial products based on cucurbitacins have been marketed as tank-mixed adjuvants under names such as Coax, Entice, Gusto, Konsume, Mo-Bait, Cidetrak, and Invite to stimulate insects to feed on spray residue (Farrar and Ridgway, 1994; Lopez and Lingren, 1994). Notably for baculoviruses, adding optical brighteners (stilbene derivatives) enhances their activity, reducing the amount of virus necessary for infection (Shapiro and Hamm, 1999). The brighteners inhibit apoptosis (cell death) of infected midgut cells, allowing the virus to initiate infection more efficiently (Dougherty et al., 2006).

Although formulations can increase activity as described, more often the goal is simply to maintain conditions to prevent microbe degradation before, during, and after application. For fungal conidia, oil formulations are often desirable for viability, storage stability, and residual activity, providing greater efficacy when compared with aqueous formulations (Kaaya and Hassan, 2000). Formulations address critical factors, such as water adsorption and retention necessary for activity of some fungal agents (Lyn et al., 2010). The activity of aqueous formulations may suffer if the fungal conidia imbibe water too rapidly or initiate metabolic activity prematurely before contacting the target pest. Imbibitional damage of conidia increased with decreasing water activity (a_w = measurement of the energy status of water in a system) of the conidia and decreasing temperature of the immersion water (Faria et al., 2009). In other words, conidia germination decreased when the conidia are dryer and when mixed with colder water. Therefore, it is conceivable that adding ingredients to promote rapid wetting (a desirable physical characteristic) of a wettable powder product could adversely affect conidia when adding to a spray tank. When mixed with water for spray applications, conidia may become metabolically active and are susceptible to damage by subsequent drying after application (Leef and Mazur, 1978). Thus, a portion of the spores may be essentially wasted, resulting in reduced activity.

Formulations have the obligation to maintain maximum activity of the microbe for infecting the target pest. Occasionally, proper selection of formulation type and beneficial ingredients can improve activity of the biopesticide, providing improved pest control with less active agent.

14.2.1.2. Viability and Storage Stability

Maintaining microbe viability is often considered to be synonymous with storage stability of biopestides. Obviously, loss of microbe viability results in reduced efficacy when applied. Storage stability may also refer to the physical condition of the product irrespective of microbe viability. Maintaining viability of the pathogen is required for activity of the formulation and often is more difficult than maintaining the physical characteristics of the formulation. The formulation scientist must recognize that bacteria, fungi, and nematodes need to maintain a minimal level of metabolism to remain viable. (Exceptions are *B. thuringiensis* products that rely on the protein crystal as the active agent rather than viable bacterial spores.)

Assuming production techniques provide technical products with high microbe viability, then the formulation processes and ingredients are expected to maintain microbe viability as long as possible during storage. Acceptable storage for microbial agents can be defined as maintaining >90% microbe viability when compared with initial viability, maintaining a predetermined minimal measure of viable microbes per unit of product, or maintaining a minimal level of potency or activity based on a standardized bioassay. Currently, only *B. thuringiensis* products have a standardized bioassay that is used to evaluate product potency (Arora et al., 2006). Fungal products generally claim a minimal concentration of colony-forming units (CFUs) or viable spores. Chemical insecticides are expected to have storage stability for a minimum of 2 years, but microbial insecticides often measure storage stability in terms of months when stored at room temperatures (near 25 °C).

Longer storage of microbial agents as biopesticides may be accomplished by stabilizing cellular membranes, inducing dormancy or slowing metabolic processes of the microbial propagule (Hynes and Boyetchko, 2006). Dormancy is often induced by drying and metabolic processes are slowed by refrigeration. Certain ingredients impact the rate of moisture loss from granules and thus impact survival of microbes (Lyn et al., 2010). Baculovirus formulations stored frozen or refrigerated maintained activity, but storage at 25 °C resulted in significant loss of insecticidal activity and degradation of viral DNA (Lasa et al., 2008).

Formulation types impact storage stability of microbes. Hydrophobic fungal conidia often remain viable for months when stored in vegetable oils at room temperatures (Kim et al., 2011), which was longer than for dried preparations. Variability among oils has also been noted. *Isaria fumosorosea* Wize SFP-198 conidia stored in corn oil maintained germination and viability longer than conidia stored in other vegetable oils (Kim et al., 2011). Dried *B. thuringiensis* products often maintain insecticidal activity for years, even when stored at room temperatures. Baculoviruses tend to retain insecticidal activity when stored in aqueous formulations better than in dry formulations. Contaminating microbes can alter conditions of stored products by consuming resources required for survival of the beneficial microbe. Oxidation may cause a loss of biological activity, but that can be minimized by the physical form of the formulation (solid vs liquid) and by chemical additives, such as antioxidants or free radical scavengers (Li et al., 1995). In summary, the biological aspect of storage stability can be adversely affected by high or low moistures, high temperatures, oxidation, and contamination, which directly degrade the microbial agent.

Changes in the physical characteristics of biopesticide formulations, such as caking of powders, separation of liquid components, flocculation of ingredients, or settling of suspended particles, often result in poor application and reduced biological activity. These problematic conditions can result from excessive moisture, incompatible ingredients, improper particle size distribution, density differences among ingredients, microbial contamination of the product, or improper processes or handling. Addressing problems of the product's physical status tend to follow solutions developed by food, cosmetic, and pharmaceutical industries. Additional information relative to physical considerations of formulations is presented later in this chapter. Relative to biological considerations, adverse alterations in the physical structure of the formulation can impede proper application of the bioinsecticide, resulting in poor pest control.

14.2.1.3. Residual Activity

Numerous environmental exposures contribute to the rapid loss of bioinsecticide application in field environments. Ingredients and formulations are available to address many of these adverse conditions. Adverse exposures include ultraviolet (UV) light energy, excessive or insufficient moisture, wash off by rain, competition by other microbes, and antibiotic leaf chemistries.

Sunlight exposure continues to be considered a major factor at reducing the efficacy of microbial pesticide applications to field-grown plants. Numerous studies have demonstrated the adverse impact of sunlight exposure, resulting in rapid loss of microbe viability (Ignoffo et al., 1977; Daoust and Pereira, 1986; Inglis et al., 1993) and/or loss of insecticidal activity of baculovirus (Ignoffo, 1992; McGuire et al., 2000). UV-B light energy reduces spore viability of *Bacillus sphaericus* Neide and subsequent larvicidal activity against mosquitoes (Hadapad et al., 2008). For baculoviruses, UV degradation has been linked to damage of the viral DNA (Salamouny et al., 2009). Increasing doses of UV-B exposure to fungal conidia reduced conidia viability and subsequent mortality of exposed aphids (Yao et al., 2010). For entomopathogenic nematodes, exposure to UV-B energy killed *Steinernema kushidai* Mamiya juveniles, decreased the density of their viable symbiotic bacteria (*Xenorhabdus japonicas* Nishimura) and reduced nematode infectivity of cupreous chafer (*Anomala cuprea* Hope) larvae (Fujiie and Yokoyama, 1998).

Microbes have been protected from harmful short wavelength light energy by adding ingredients to formulations or as tank-mixed adjuvants. Ingredient protect the microbes in the following ways: (1) as a chemical sunscreen by absorbing short wavelength energy, such as optical brighteners, chemical dyes, and absorbers (sunscreens) or (2) as a physical sunscreen by reflecting light energy, such as clays and titanium dioxide (TiO_2). Ingredients include chromophores such as congo red (Hadapad et al., 2011) for aqueous-based formulations. Stilbene-derived optical brighteners that absorb UV energy and convert it to safe visible light wavelengths can be included as components of the biopesticide formulation or added to the spray tank as adjuvants (Shapiro, 1992; Shapiro and Argauer, 1997, 2001; Vail et al., 1999; Reddy et al., 2008). Optical brighteners not only provide UV protection for microbes (Chaparro et al., 2010), but they also interact synergistically with many baculoviruses, resulting in greater than expected efficacy (Caballero et al., 2009). For oil formulations, organic compounds commonly used in sunscreens (e.g. oxyl methoxycinnimate) (Inglis et al., 1995) and experimental sunscreens (modified soybean oil known as soyscreen) (Behle et al., 2009) protect viability of fungal conidia when exposed to natural and simulated sunlight. Chen et al. (2010) demonstrated UV protection provided by zinc oxide (ZnO) nanoparticles for *Beauveria bassiana* (Balsamo) Vuillemin conidia; they recommended adding 1–2% to oil formulations made with fungal conidia.

Moisture affects the biological activity of bioinsecticides in several ways. For nematodes, moisture is required to prevent desiccation and provide a medium for host searching. Likewise, a certain amount of moisture is necessary for germination of fungal conidia to initiate infection. Relative to residual activity, excessive moisture in the form of heavy dew or rain can effectively remove the bioinsecticide treatment from the application target. Several approaches are able to reduce the impact of rain. One approach has been to add sticking agents as tank-mix adjuvants to bioinsecticide applications, whereas another approach has been to include a sticking agent in the formulation. These agents generally consist of natural (carbohydrate, protein, or other polymers; Morales Ramos et al., 2000) or synthetic chemistries (ethoxylated phenoxy alcohols, latex polymer) intended to stick the microbe to the treated surface. The mechanism of protection is provided by molecules that are hydrophobic or that polymerize or precipitate to form water-insoluble residue to entrap the microbe and effectively reduce wash-off.

Leaf exudates are known to adversely affect residual activity of baculoviruses (Young and Yearian, 1974; Ellerman and Entwistle, 1985; Stevenson et al., 2012). Aldehydes, ketones, carboxylic acids, and their derivatives produced by plants have antibiotic effect on the *B. thuringiensis* (Maksymiuk, 1970). Insecticidal activity of entomopathogens can be further compromised by competing microbes present in the environment. It is probable that encapsulating formulations could protect beneficial microbes by separating them from harmful leaf chemistries and competing microbes. Unfortunately, countering the impact of these adverse conditions has received little attention in the scientific literature.

14.2.1.4. Efficacy

Efficacy is the culmination of the bioinsecticide treatment, and it is often measured in terms of insect control as pest mortality and/or in terms of crop protection. The pathogen, formulation, pest, and environment all impact efficacy. Before application, many contributions of the formulation have already been realized by maintaining activity and viability of the microbe through processing and storage as well as maintaining the physical properties of the product. Improved efficacy provided by experimental formulations are best evaluated by comparing the efficacy of formulated treatments with the efficacy of unformulated treatments applied under field conditions. Successful formulations are subsequently evaluated commercially by comparing efficacy of the bioinsecticide treatment with alternative pest control applications. Additional contributions provided by the formulation now focus on the practical aspects of application, including mixing and coverage by spray applications, flowability of granules, or ease of application for dusts to target pest habitats. The formulation also provides the built-in benefits of targeting the pest with feeding stimulants or extending residual activity.

14.2.1.5. Integration with Production

Developmental research on biopesticides often focuses on production techniques to maximize the yield of infective propagules. It is recognized that proper formulation of these propagules is necessary for storage stability and easy application, and it is an essential component for product development (Jackson et al., 2010). Formulation development is most efficient when conducted as an integral part of production. However, the eventual use of the product also imparts a set of considerations necessary for easy application and efficient pest control. Thus, the formulation is the technology that bridges microbe production with pest control and must address requirements of both the production side and control side of the application. Formulation technology plays a key role in the commercial success of many potential microbial insecticides.

Production and formulation are not uniquely separate processes when developing a bioinsecticide. Many of the processes, such as drying, can be considered as both a step of production and/or as a step of formulation. Thus, greater integration between production and formulation will likely increase efficiency of the overall process. Whether using either in vitro or in vivo production techniques, the resulting pathogen propagule usually needs to be separated from other production components and concentrated to form a technical product, which is then processed to the final product form. Fermentation broth, solid media substrate, or insect parts may need to be separated from the pathogen to provide a suitable concentrated technical product. These production components can be removed by screening, filtration, centrifuge, or drying, thus establishing the initial conditions for subsequent steps of formulation.

At this point, the formulation process has a multitude of options in terms of ingredients and processes that can be performed. Simply drying the microbe may be all that is necessary for a usable product. Many *B. thuringiensis* products are the result of fermentation production, centrifuging to reduce liquid, and spray drying to form technical material consisting of a fine powder. This technical material may be the final formulation or it may receive additional processing. A carrier such as clay may be added by simply mixing it with the technical material (e.g. using a ribbon blender) to form a dust formulation for homeowner application. Granules may be formed by agglomerating the powder using a pan granulator, fluid bed dryer, or extruder. Granule formulations differ in that some are intended for direct application whereas others are intended to be dissolved in water for spray application. Alternatively, the technical powder may be mixed with aqueous or organic liquids to provide concentrated liquid formulations for application as direct or diluted sprays. Selecting among these possible formulations is often dictated by the conditions of the intended end-use application environment.

14.2.2. Potential Hazards

A large amount of basic research is initially directed at identifying, characterizing, and developing production techniques for microbes to be used as biopesticides. When basic research associated with an inundative biological control comes to fruition and efforts shift to commercial production, then worker's safety and quality control considerations need to be addressed relative to proposed manufacturing processes. Even though entomopathogens are generally considered to pose extremely small health risks (Laird et al., 1990), the active agent, formulation ingredients, and production processes must be evaluated as legally required for any manufacturing process.

14.2.2.1. Contamination Issues

Contamination of biopesticide formulations with unwanted microorganisms during production, storage, and application is a concern for the producer and the general public. Maintaining conditions for viable beneficial microorganisms generally equates to favorable conditions for contaminating microbes. Recognition and quantification of contaminating microbes are important components in the quality control of biopesticide production (Jenkins and Grzywacz, 2000). One adverse effect of contamination is the fouling of the product, resulting in either a product that is ineffective (destruction of the active agent) or a product that is impossible to apply (powder caking, emulsion flocculation). A serious problem is the potential to produce human pathogenic microorganisms in the formulation and disseminate them by application to a food crop. Although not a human example, a bioinsecticide application to honeybee hives for control of Varroa mite resulted in greater death of bees because the biopesticide was contaminated with bacteria (Meilke et al., 2012). Microbial contamination of bioinsecticide products represents a hazard for both the product and the public.

Bioinsecticides can become contaminated from multiple sources. Most in vitro (liquid fermentation) procedures effectively produce beneficial microbes with minimal contamination. By contrast, in vivo production may result in greater amounts of contaminating microbes from the colony insects used for these production systems. Formulation processing steps between production and packaging are often conducted under clean (not sterile) conditions and offer opportunities for additional microbial contamination. Another source of contamination is from nonsterile ingredients added to the formulation. Operating under sterile conditions and adding sterilized ingredients are often cost-prohibitive for economic processing of microbial insecticides. As a result, the goal of formulation processing is to minimize and/or manage product contamination by unwanted microbes.

Managing contamination of bioinsecticide formulations is accomplished by selecting procedures, conditions, and ingredients to mitigate contaminating organisms. Maintaining clean conditions and ingredients may be sufficient to prevent excessive contamination from developing in the final product. Cold storage and/or drying are commonly used to prevent proliferation of contaminating microbes. The measurement of water activity (a_w) is commonly used as an indicator of product perishability with respect to microbial growth, chemical and biochemical reaction rates, and physical properties. Controlling pH and adding selective antimicrobial ingredients also contribute to maintaining acceptable low levels of contamination.

14.2.2.2. Biohazard Worker Exposure Concerns

Hazards to employees are a major concern for all industries, including the biopesticide industry. Relative to the microbial agents used in bioinsecticides, the hazards associated with the production, handling, and application of products are relatively benign. Human safety is well documented for fungi (Zimmermann, 2007), bacteria (Peng et al., 2008), and baculoviruses (Burges et al., 1980) used as bioinsecticides. Nematodes go a step further along the safety path, in that nematode products are not required to be registered as pesticides with the US EPA and many other countries (Ehlers, 2003). These beneficial agents simply do not pose a significant toxicity risk to most workers, although they may induce allergic reactions to susceptible individuals (detailed below).

Given the safety of these agents, additional concerns about the biohazard for worker exposure often reflect the hazards associated with the formulation ingredients. Generally, the ingredients added to formulate bioinsecticides also pose minimal toxicity hazards to humans. Many ingredients are well characterized because of their use in other industries and are generally regarded as safe. Carbohydrates, proteins, and vegetable oils can be pulled from food industry applications; organic and inorganic fillers can be adopted from pharmaceutical or cosmetic industries. The wide availability of ingredients allows for selection of those with minimal toxicity hazard. Ingredients listed by the US Food and Drug Administration as generally regarded as safe (GRAS) or those included on List 4A (US EPA) are often selected to be used with biopesticide formulations. For the bioinsecticide to be certified for organic applications, all ingredients need to be listed in the Organic Materials Review Institute (OMRI) database. As for the US EPA registered bioinsecticide products, safety regulations covering works and handlers of pesticides are outlined by the Worker Protection Standard 40 DFR part 170.

14.2.2.3. Allergenicity

One biohazard of worker exposure is the ability of microbes, ingredients, and final formulations to induce allergic responses by exposed/susceptible individuals. Allergic responses have been reported for persons exposed to bioinsecticides and insect pathogenic microbes. Most commonly, reports focus on the most common bacterial and fungal agents, including *B. thuringiensis* (Doekes et al., 2004), *B. bassiana* (Westwood et al., 2006), and *M. anisopliae* (Ward et al., 2009, 2011). Much of the research was conducted using inhalation exposure of laboratory animals (Ward et al., 2009, 2011; Barfod et al., 2010). Several reports documented responses in human subjects as a result of bioinsecticide application (Petrie et al., 2003; Jensen et al., 2002; Doekes et al., 2004). End users can reduce the possibility of exposure by using personal protection equipment as outlined on product labels when handling, mixing, and applying treatments. Formulations with less dust also reduce the risk of inhalation exposure to allergens.

14.2.2.4. *Combustion, Thermal Degradation, and Dust Explosion Hazards*

The organic nature of bioinsecticide formulations warrants serious consideration of fire hazards. In the United States, the National Fire Protection Association (NFPA) is the leading source of advice and technical information in this area, and their labeling and hazard categories are conveyed by Material Safety Data Sheets (MSDS) issued in compliance with the Occupational Safety and Health Administration (OSHA). Some fire hazards may not be obvious. For examples, seemingly innocuous materials, such as powdered milk, can carry a weak dust explosion rating by NFPA; when suspended in air in a confined space and subjected to a static electricity spark or a loose metal part making impact with a bin wall, these materials may release dangerous explosive energy. Dried and milled grains, sawdust, corn cobs, and rice hulls have a higher rating class than coal dust and account for many plant explosions.

Often, an MSDS for a given organic material will indicate that no special precautions are needed, which may be true on an as-is basis. The data in an MSDS includes lower and upper dust explosion limits, flash point temperature, and the autoignition temperature at which a material self-ignites in the presence of air. What is not documented on the MSDS is that seemingly simple operations, such as milling and drying, performed upon relatively harmless materials can result in autoignition of storage piles inbins due to common conditions of insulation, compaction, and oxidation. Stored products and ingredients may be degraded or begin to burn as a result of excessive moisture. Heat pockets can develop deep in a product stored in bulk and can move moisture within the pile. Older storage bins may leak rainwater. As a result, autoignition may be triggered by the heat of microbial degradation in the case of localized wet conditions. Formulators need to be aware of potential hazardous conditions associated with scale-up processes and work with fire prevention experts and testing laboratories to prevent disastrous events.

14.3. PHYSICAL CONSIDERATIONS

A number of physical considerations for formulations of bioinsecticides, such as cost and physical form, are obvious to the observer. However, others are not as obvious, such as the specific ingredients and processes required to create the formulation. Currently, production of the active agents is relatively expensive and often limits potential markets to high-value crops, consumer markets, or situations where less expensive control measures are not available (organic production).

14.3.1. Cost

The cost of formulating a bioinsecticide can be determined by simple modeling of the proposed formulation process, including costs of ingredients, equipment, labor, and facilities. The costs relative to the benefits provided by formulation ingredients and processes are an integral part of industrial formulation research. Cost considerations for bioinsecticide formulations tend to be more stringent than for related industries (food, cosmetic, or pharmaceutical industries), predominantly because retail profit margins for production products tend to be less than for consumer products.

Several basic concepts are worth mentioning relative to formulations of bioinsecticides. The unit price of a product increases rapidly, perhaps doubling through each step of a typical distribution chain from the producer through the wholesaler, distributor, and retailer to the consumer. Each small increase at the production/formulation level results in a proportionally greater cost for the consumer. The costs of ingredients tends to garner greater scrutiny than costs of processing equipment because unit cost of ingredients is directly related to the units of product made, whereas the unit costs related to equipment decreases with increased production. In addition to ingredients and equipment, formulation-related costs can result from added bulk or weight, requirements for refrigerated storage, excessive waste disposal, extended handling time, safety risk, or extra labor requirements. Therefore, the goal of the formulation scientist is to minimize each of these costs through the appropriate selection of cost-effective ingredients and processes.

Obviously, low costs are desirable when selecting among ingredients and processes used to formulate microbial pesticides. However, formulations that improve efficacy, extend storage, or lengthen residual activity may allow lower application rates and thus may offset some costs. For example, Chaparro et al. (2010) demonstrated that adding 0.5% optical brightener to a granulovirus treatment for *Tecia solanivora* (Povolny) larvae increased toxicity 50-fold. Adding the brightener not only reduces the amount of active ingredient required and cost of the treatment, but it could also reduce costs storage and shipment compared with higher treatment rates of unformulated virus.

One benefit for bioinsecticides is that costs of registration are generally less than costs for registration of chemical insecticides. For example, from 1995 to 2000, new synthetic chemicals required over $185 million and a 10-year development period before reaching the market, compared with a biopesticide that required about $6 million and only 3 years (Kennedy et al., 2007).

14.3.2. Formulation Form

Components of a biopesticide formulation are generally categorized into three parts: active ingredient, carrier, and adjuvant (Ash, 2010; Burges, 1998). The active agent relative to this chapter is the infective form of the microbe (e.g. the infective juvenile nematode, fungal conidia, baculovirus occlusion body, bacterial spore). Unlike chemical insecticides, microbial agents are particulate and are characterized as suspensions in the formulation or spray tank, not solutions. The carriers are components of the formulation that generally serve as inert ingredients used to dilute the active agent with little or no added benefit. Adjuvant components comprise a wide variety of agents that improve one or more formulation characteristics, such as storage (anticaking, humectants, oxygen scavengers), mixing (dispersing or wetting agents), application (viscosity control, spreaders, stickers), or efficacy (protection from degradation by UV or desiccation) of the bioinsecticide.

The three components of the biopesticide products are combined to form liquid or dry formulations. Aqueous and oil-based liquids are often concentrated products that are intended to be applied directly or mixed with additional water for application through spray equipment. Traditional liquid formulations include emulsifiable oils (oil miscible flowable concentrate), suspension concentrates, and oil-based liquids intended for direct application as an ultra-low volume (ULV) application. Dry formulations include a variety of product formulations from dry granules (Quiroga et al., 2011) and briquettes for direct application to water-dispersible granules and wettable powders, which are intended to be diluted in water for spray application. Recently, Faria and Wraight (2007) reviewed mycoinseticides and mycoacaracides and defined nine formulation types, adding three to those listed above: bait, contact powder, and oil dispersion. These designations cover most formulations used for other microbial pathogens. Characteristics of these formulations are listed in Table 14.1.

Encapsulation of microbial agents has been evaluated to provide a variety of benefits for microbial biopesticides. Two common benefits addressed for application to field-grown plants are moisture retention for nematodes and UV protection for fungi, bacteria, and baculoviruses. Encapsulation could be used to mask microbes known to deter feeding, link synergistic ingredients with the microbe, or protect the microbe from enzymatic degradation on the leaf surface

TABLE 14.1 Common Formulations of Biopesticides

Formulation	Abbreviation	Application
Emuslifiable concentrate	EC	Spray
Suspension concentrate	SC	Spray
Ultra-low volume	ULV	Spray
Wettable powder	WP	Spray
Capsule suspension	CS	Spray
Emulsion in water	EW	Spray
Dust	D, P	Dust
Granule	G	Dry
Water-dispersable granule	WDG, WG, SG	Spray

(Stevenson et al., 2012). Many encapsulation technologies have been developed for pharmaceutical, cosmetic, and food industries but may not be sufficiently economical for adaptation to the less lucrative biopesticide industry. Gouin (2004) reviewed encapsulation technologies relative to the food industry, listing spray drying, spray cooling/chilling, spinning disk and centrifugal coextrusion, extrusion, fluidized bed, coacervation, alginate beads, liposomes, rapid expansion of supercritical solutions, the supercritical anti-solvent method, and inclusion encapsulation as encapsulation techniques. These techniques vary in physical structure of the encapsulated product. Both true encapsulation (consisting of a shell around a core) and matrix encapsulation (consisting of aggregates of microbes stuck together by a matrix material) can be effective forms of protecting microbes. Propagules can be encapsulated by chemical or physical processes. Chemical process may include coacervation or forming calcium alginate beads. Physical encapsulation uses processes such as spray drying or fluid bed drying to encapsulate propagules; these processes are described in this chapter.

Specialized formulations often address one or more key characteristics of the microbe or application, such as sponges for storage stability of nematodes (Strauch et al., 2000; Chen and Glazer, 2005), floating briquettes for applications to control mosquito larvae (Brar et al., 2006), or foams intended to expand and fill termite chambers inside structures (Dunlap et al., 2007). These specialized formulations are intended to maximize the potential benefits of the respective microbes as a practical treatment by targeting unique characteristics of the pest environment. A biopesticide application may be followed by posttreatment irrigation as an aid to penetrating a crop canopy, residue, or mulch material (Anderson et al., 2006). After noting these specialized applications, most general bioinsecticide formulations emulate chemical insecticide formulations in an effort to fit with current consumer perceptions for application and to use available application equipment.

14.3.3. Ingredients

The ingredients used in bioinsecticide formulations are often simple and inexpensive, even though considerations for ingredient selection offer vast possibilities. Lists of ingredients with reduced risks or for organic certification can be found at the web sites listed in Table 14.2. Ideally, each ingredient included as part of the formulation will add value with minimal cost. Value can be the result of biological benefits to the microbe during processing, handling, storage, and application or provide physical benefits in terms of easier product handling, mixing, or consumer appeal. Costs associated with ingredients are not always monetary; rather, they can result from indirect costs related to the ingredient. Examples of indirect costs include an explosion hazard handling dusty material or volatile compounds, introducing contaminating microbes to the product, or added legal requirements associated with use of regulated ingredients such as antibiotics. Most benefits can be realized and indirect costs can be avoided by proper ingredient selection.

TABLE 14.2 Sources for Formulation Ingredients with Reduced Risk or Organic Certification

Source	Purpose	Website
Organic Materials Review Institute (OMRI)	Organic certification	http://www.omri.org/omri-lists
U.S. Food and Drug Administration (FDA)	Generally regarded as safe (GRAS)	http://www.fda.gov/Food/FoodIngredientsPackaging/GenerallyRecognizedasSafeGRAS/default.htm
US Environmental Protection Agency (EPA)	Minimal risk inert ingredients (List 4A)	http://www.epa.gov/opprd001/inerts/section25b_inerts.pdf

Scientific literature reports a wide variety of ingredients that have been evaluated in bioinsecticide formulations to provide benefits directly to one or more aspects of the bioinsecticide, many of which having been presented previously in this chapter. ULV application may consist of kerosene–peanut oil combinations (Douro Kpindou et al., 2011), although many organic and vegetable oils may be suitable (Kim et al., 2011). Oil formulations often include surfactants to allow the oils to mix with water for aqueous spray applications (Santi et al., 2011). Various surfactants included as a component of an oil formulation have been studied for their direct impact on conidia dispersal, germination, and colony growth, with results that varied from no impact to significant inhibition of fungal growth (Santos et al., 2012). Larena et al. (2003) demonstrated that fungal conidia survived freeze drying better with skim milk, peptone, and sucrose. Similarly, conidia of *B. bassiana* retained greater viability after spray drying when vegetable oil was added to the feed stock (Fig. 14.2). Proper selection among adjuvants is necessary to maintain the viability of the pathogen and provide for a stable emulsion/suspension of the biopesticide in the spray tank.

However, fewer reports are available concerning ingredients added to address physical properties of the product formulation. Experimental wettable powder formulations have been reported to include inert powders such as talc, sand, diatomaceous earth, and various clays (Arthurs et al., 2008). Jin et al. (2008) demonstrated that conidia from different fungal strains required surfactants with different hydrophilic–lipophylic balance (HLB) values for optimal wetting time and conidia suspension. Surfactants are categorized based on their HLB value, which is a relative measure of a chemical's ability to mix with water or oil. Ingredients with HLB values less than 11 tend to be lipid soluble and form water in oil emulsions, whereas those with HLB values greater than 11 tend to be water soluble and form oil in water emulsions.

Natural ingredients are often used in formulations in order to maintain the environmentally "green" concept associated with bioinsecticides. Polymers are

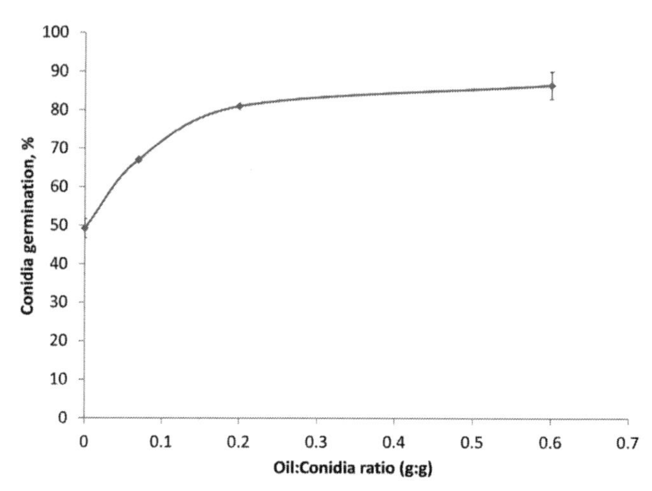

FIGURE 14.2 Example of how an ingredient can benefit from microbial agents during processing. Germination of *Beauveria bassiana* conidia increased after spray drying with increasing concentrations of corn oil added to the dryer feed stock. Bars represent standard deviation. *Robert W. Behle, unpublished data.* (For color version of this figure, the reader is referred to the online version of this book.)

necessary for encapsulation and often include natural carbohydrate and/or protein polymers such as starch, sodium alginate, gelatin, acacia gum (Hadapad et al., 2011), and lignin (Tamez-Guerra et al., 2000). Dust formulations require fine powders (often clays or ground corn cobs) to dilute microbes for applications. Natural plant extracts have been shown to protect baculovirus from degradation by exposure to UV energy and could be certified for use on organic crops (El-Salamouny et al., 2009). Also, feeding stimulants may be included to target specific pests of cropping systems, using natural components such as leaf powders (Rosas-Garcia and De Luna-Santillana, 2006).

14.3.4. Processing

A wide variety of equipment and processes are available to formulate microbes into biopesticide products. Commercial-scale equipment may be specialized for biopesticides or adapted from other industries processing cosmetic, pharmaceutical, or food products. Homogeneous mixing of ingredients is necessary for both liquid and dry formulations. Liquids can be added to large containers and stirred using a variety of available paddles. Dry ingredients can be mixed with a rotary shell, fluidizing paddle, ribbon blender, or a rotating V-mixer. Pasty ingredients can be mixed using an extruder or industrial paddle mixer. These mixing techniques are mostly batch processes, although both liquid and dry metering systems are available to combine ingredients for continuous flow systems. Selecting appropriately sized equipment that is easily cleaned helps to streamline the formulation process and minimizes potential contamination of formulated product.

Drying is another common and necessary process for many beneficial microbes to slow metabolism and extend storage stability. Drying equipment comes in many forms, including spray driers, ovens, circulating air driers, drum driers, vacuum driers, or freeze driers. Each system has unique features, and not all are suitable for processing viable microbes. For example, freeze dryers have the advantage of being gentle drying systems with little negative impact on microbe viability. Vacuum driers require sealed vessels and vacuum pumps, but they allow moisture reduction at milder temperatures than normal sea-level pressure. For disadvantages, freeze driers tend to be slower with greater drying expense while operating as a batch system. By contrast, spray driers are more economical to operate with continual flow operation, but they may reduce microbe viability. Vast amounts of information on driers, drying systems, capacity, and costs of operation are available from equipment suppliers; such technical information can be accessed through technical publishing houses, such as *Powder and Bulk Solids* (see http://www.powderbulksolids.com/manufacturers/drying-thermal-solids-processing).

Centrifuges, filters, and screens are often used to concentrate microbial propagules by separating the active agent from unnecessary production components, such as fermentation broth or insect parts. Liquids and solids can be separated using a centrifuge, inline filtration, rotary-drum vacuum-filter, aspirators, or vibrating screen separators. The major advantage of concentrating the pathogen is to reduce downstream processing costs, such as drying. However, this separation produces waste material that may result in disposal issues, so inline recycling is an important overall system design consideration.

Many production systems yield a product with relatively large particle sizes that are too large for subsequent processing and/or efficient application. A variety of milling techniques can be used to reduce particle size to meet physical specifications, even though milling may be detrimental to the microbe. For example, Kim and Je (2012) reported on milling of *B. thuringiensis* technical powder and found the following: (1) smaller particles had higher control efficacy against diamondback moth *Plutella zylostella* (L.) in the laboratory but not in greenhouse evaluations, (2) air-jet-milled powder had higher control efficacy than hammer-milled powder, and (3) excessive processing (second milling) reduced efficacy. This is an example in which a relatively simple process, such as milling, both positively and negatively affects the biological and physical properties of the final formulation.

Dust is a concern for many formulation processes and products. Various methods are used to reduce dust by forming the product into granules. Reducing dust improves worker safety and consumer appeal. In addition to reducing dust, effective granule formulations improve product flow, dispersability, and application dosing. Granulation can be formed by numerous methods including extrusion, fluid bed processing, or pan granulation. Small and dusty particles can be agglomerated using various natural and synthetic chemical polymers.

The benefits of encapsulating microbial agents were discussed previously in this chapter and can be accomplished using a variety of techniques and equipment (Table 14.3). Spray drying is a process of spraying an aqueous suspension of propagules into a chamber of heated air that rapidly vaporizes the water, leaving the solids as a powder to be collected. Spray drying has long been used for drying and formulating *B. thuringiensis*. A recent example of a spray-dried formulation of *B. thuringiensis* was reported to contain tapioca starch, sucrose, milk powder, silica fume, polyvinyl alcohol, Tween 20, rice bran, oil, and antifoam solution (Teera-Arunsiri et al., 2003). Kassa et al. (2004) found that spray drying maintained viability of submerged conidia better than freeze drying. Also, alternative ingredients, such as wastewater and wastewater sludge, have been studied as formulation ingredients with spray drying (Adjalle et al., 2011). The final product is a fine powder and the drying process can be controlled to provide one or a few propagules per particle.

TABLE 14.3 Summary of Characteristics of Potential Encapsulation Techniques

Method	Simplicity	Cost	Advantage	Disadvantage
Spray drying	Simple	Low	Versatile, continuous, simple	Limited to aqueous feed stock, high drying temperatures
Spray chilling/ cooling	Simple	Low	Hydrophobic process	Untried technology for microbes
Extrusion	Simple	Low	Simple continuous process	Limited scope, low payload
Fluidized bed	Simple	Medium	Variable shell materials	Batch process
Coacervation	Complex	High	High payloads	Challenging batch process
Spinning disk	Simple	Medium	High payload, high throughput, continuous	Difficult scale-up
Liposome entrapment	Simple	High	High encapsulation efficiency	Costs of process scale-up and delivery form
Rapid expansion of supercritical solutions	Complex	?	No water, mild process	Relatively unexplored technology

Fluid bed drying can be used to remove moisture, agglomerate dust into granules, or apply coatings to particles or granules. The process uses heated-air flow to "fluidize" solid particles in a chamber while applying a liquid spray to the solid substrate. The active agent can be either in the liquid or solid phase. This process is considered to be gentler than spray drying. Final products can vary widely in terms of physical attributes such as size, particle density, and particle disintegration.

Encapsulation in calcium alginate (as beads or granules) has been evaluated for a variety of microbial propagules. Gel beads are formed by dripping a mixture containing a dilute alginate solution and the control agent, such as infective juvenile nematodes, into a dilute solution of calcium salt (Chen and Glazer, 2005). The calcium ions quickly cross-link alginate to entrap the control agent in the newly formed drop-sized gel. Gel beads can be dried to form hardened granules. A calcium alginate formulation (compared with sodium alginate, gelatin, and acacia gum) provided controlled release of *B. sphaericus* for control of mosquito larvae (Hadapad et al., 2011). The bacterial symbiont, *Photorhabdus luminescens* Thomas and Poinar, of the *Heterorhabditis* spp. nematode was encapsulated in sodium alginate beads and successfully infected *Spodoptera litura* (F.) larvae exposed in a sterile soil environment (Rajagopal et al., 2006).

Coacervation is well known for the encapsulation of ink into small beads that are then impregnated into paper to form carbonless copy paper. Coacervantion has been used extensively to encapsulate flavors for the food industry (Gouin, 2004) and has shown promise for encapsulating cells to extend viability (Baruch and Machluf, 2006), but it has yet to be reported as a method to encapsulate biopesticides. The process uses dilute solutions of polymers, often carbohydrates and/or proteins, that are deposited around the active ingredient suspended in the mixture. Once capsules are formed, the shells are hardened using chemical or enzymatic cross-linking. Capsules can then be separated from the solution and dried. Coacervation has not been reported for microbial bioinsecticide formulations because the process is relatively expensive, complex, and relatively harsh (often using gluteraldehyde to crosslink the shell)—three undesirable characteristics (Chilvers and Morris, 1987; Burges, 1990).

Cydia pomonella granulovirus was encapsulated in particles from gas-saturated solutions (PGSS) with no loss of biological activity (Pemsel et al., 2010). These particles included ingredients to provide protection from degradation by UV light and phagostimulants to improve uptake. The PGSS process runs at low temperatures (65 °C), using supercritical fluid carbon dioxide. The virus is mixed in a melted fat matrix, then mixed with the supercritical fluid under pressure and sprayed in a spray tower at ambient conditions. The cooling effect of reduced pressure causes the matrix to solidify, and the particles are collected as the carbon dioxide gas is vented. This process has been documented for producing pharmaceuticals, food, and food-related products.

14.3.5. Mixing, Handling, and Packaging

Although mixing, handling, and packaging are often associated with marketing, these three physical characteristics are partly determined by the biopesticide formulation. Mixing refers to the ability of the product to mix with water for application; this characteristic can be measured in terms of wetting time, suspension/settling, or emulsion characteristics. Methods to measure mixing ability of formulations are published as standards by American Society for Testing and Materials (West Conshohocken, PA). For example, water-dispersible granules should be evaluated for dispersibility, suspension characteristics, particle hardness, and dust properties (Al Amin and Bin, 1994). Desirable mixing properties include quick wetting without forming clumps for dry products and forming uniform suspensions (or emulsions) with minimal agitation for dry and liquid products. Effective formulations have little or no settling, creaming, flocculation, foaming, or phase separation in the spray tank.

Products with good handling characteristics often go unrecognized, either in the factory or by the end user. Good handling attributes are those that make the product easy to use, such as pouring, measuring, and mixing. Those with poor handling characteristics are quickly identified by persons working with products. Fine powders often flow poorly and pack tightly in containers from the vibration of processing or shipping. As a result, accurate dispensing and measuring of these powders is more difficult than for free-flowing powders. Flow of powders can be improved by adding ingredients such as silica. For liquid products, the terms “thick” or “sticky” describe poor handling properties that can result in waste of the material remaining in packaging. Propagules that settle or separate in the packing container may not resuspend before being dispensed; thus, they contribute to inaccurate application and waste by remaining in the package. Bulky or dilute products increase handling costs. Finally, products requiring refrigeration to maintain microbe viability incur additional handling consideration, such as refrigerated transport, and the associated added expense. Yet, refrigeration may be necessary to maintain propagule viability (Lasa et al., 2008).

Product packaging often provides simple convenience for shipment of the bioinsecticide, but it must address the needs of the end user. Cardboard cylinders are typical for marketing dust formulations to home owners, and 5-gallon plastic drums may be suitable for emulsifiable oil formulations for farmers. The package must provide sturdy, cost-effective product containment and provide easy dispensing of the formulation. Most common package materials (paper, plastic, glass, cardboard, and metal) are suitable. Selection among these materials is often directly associated with the form of the formulation.

14.3.6. Consumer Aesthetics

Consumer aesthetics will often reflect mixing, handling, and packaging as described above. The formulation contributes favorably to consumer aesthetics when it provides the convenience expected of the product. The amount of convenience varies with the target market. Homeowners operate under the simplest concepts associated with applications of pest controls and generally prefer ready-to-use products. Trained commercial applicators want highly concentrated products that mix easily for application through spray equipment. Just as the formulation scientist must work with the production system, the scientist must also work with the marketing sector to provide the characteristics necessary for the end-user of the proposed product.

14.3.7. Application

As discussed earlier in this chapter, for crops grown under modern agricultural production systems, microbial insecticides have a better chance for adoption when they fit into the current chemical pesticide application systems. If novel equipment or techniques are needed for application of a biopesticide, then the likelihood of adopting the product for widespread use becomes greatly reduced (Chapple and Bateman, 1997; reported in Bateman, 1999). Most pesticide treatments are applied as sprays, often using equipment specialized for the target crop. Examples include aerial application for large acreages or tall crops; air blast sprayers for orchard crops; irrigation systems for row, orchard, and greenhouse crops (Wang et al., 2009); and ground sprayers for vegetable or specialty crops. Some application systems require specific formulations of the pesticide because spray parameters can affect the efficacy of the treatment. For fungus, higher spray volume and shorter time between spray intervals improved inoculation of *B. bassiana* conidia onto thrips infesting greenhouse-grown plants, whereas increasing the concentrations of conidia applied at a constant spray volume had no benefit. The most effective application program for control of flower thrips consisted of multiple applications at intervals less than 7 days at the highest rate of the fungus and at the highest application volume (Ugine et al., 2007).

Research supports use of conventional equipment for application of biopesticides, such as hydraulic pumps, selection among nozzle tips, and pumping systems (Chapple et al., 1996; Guimaraes et al., 2004). Some concerns include loss of nematode viability with repeated passage through centrifugal pumps (Fife et al., 2007; Brusselman et al., 2008), nonuniform discharge of nematodes through drip irrigation (Wang et al., 2009), and potential problems with flow for viscous virus formulations in spray equipment and large drop size resulting in poor distribution and reduced control (Steinke and Akesson, 1993). Spray volume and nozzles affect deposition and infectivity of nematodes applied as foliar sprays on vegetables (Brusselman et al., 2012a, b). Limitations among application techniques for a product are listed on the product label. Application equipment and formulations can directly impact the efficacy of biopesticides, most notably by affecting spray deposition.

Occasionally, unique circumstances may negate potentially effective biopesticide applications. For example, unengorged ticks removed from cattle treated with emulsified conidia died from infection by *M. anisopliae* (Leemon et al., 2008), although there was little effect of this treatment on ticks remaining on the cattle. The conclusion was that an interaction between the environment of the ticks on the cattle with the biopesticide application reduced the efficacy of the treatment. As an alternative, unengorged ticks may be controlled by application of fungal formulations as sprays or granules to vegetative ground cover to infect ticks before they attach to hosts. This research demonstrates the need for direct applications of biopesticides to target the susceptible environment of the pest.

14.4. ADDITIONAL CONSIDERATIONS ON FORMULATION

14.4.1. Sources of Technologies

Many sources of information concerning formulations have been identified previously in this chapter. Formulations typical to the chemical pesticide industry may not be adequate to fulfill the unique requirements for microbial-based insecticides. Gaugler (1997) correctly suggested that biopesticide formulators establish research linkages with scientists in the food processing and pharmaceutical industries. We further include formulation techniques used by the cosmetic industry. These industries use ingredients and techniques that are more adapted to biological systems compared with chemical pesticide industries. As consumer-driven industries, they often develop in-depth formulation research to address aspects of product functionality and consumer appeal.

14.4.2. Legal Requirements

Requirements for registration of biopesticides vary among countries around the world. In the United States, obtaining EPA registration as a biopesticide is a major and necessary requirement for all microbial-based active agents (except nematodes, which do not require registration in the United States or Europe; Ehlers, 2003). Regulations for registration of microbial pest control agents in Canada are defined by Pest Management Regulatory Agency regulatory directive DIR 2001–02 and are “essentially harmonized” with data requirements of the US EPA. In Europe, the EU commission approves active substances, then EU countries authorize products for use in their territories and ensure compliance with EU rules. Among countries in Central and South America (Cotes, 2011), only Colombia has specific regulations for biopesticides, established by Decree 1840 of 1994. Regulations in Brazil, Argentina, and Chile follow chemical pesticide requirements. Other countries in South America do not register biopesticides because the products are not backed by a regulatory agency.

Meeting label requirements for registration are based on the final formulation to be marketed. One requirement is the identification of the concentration of the active agent—a measurement that differs among active agents. Concentrations of *B. thuringiensis* are identified as International Units based on standard bioassay activity or estimation of protein content (Arora et al., 2006). Conidia counts or CFUs may be listed for fungal products. Virus concentrations are reported as occlusion body (OB or PIB) counts, whereas nematodes concentrations are viable infective juveniles. The remaining ingredients are often lumped together under the listing as "inert" ingredients on the registration label. These inert ingredients are scrutinized by government agencies and may be subject to restrictions. For example, some commonly used surfactants (nonylphenol ethoxylates) have been banned from use in European markets, prompting a search for more environmentally friendly ingredients (Lowe and Milbradt, 2011).

14.4.3. Current Effective Formulations

Currently, effective formulations for microbial biopesticides tend to be simple mixtures. Baculoviruses are produced using in vivo techniques resulting in a technical product containing occlusion bodies suspended in aqueous liquid juice of the production host. A common and effective formulation is made by the addition of glycerol to limit growth of contaminating microbes. The resulting product stores well and mixes easily with water for an aqueous application. Similarly, formulations of hydrophobic fungal conidia tend to have good storage stability in oils. Emulsifiers are added to aid with aqueous spray applications. In contrast to oil formulations, the fungal conidia produced using solid substrate techniques are easily formulated as wettable powder formulations by adding dry ingredients to improve wetting of the hydrophobic spores with water for spray applications. Bacterial-based products (predominantly *B. thuringiensis*) are often marketed as sprayable powders. After fermentation, the bacteria are concentrated by centrifuging, mixed with aqueous dispersing agents and spray dried to form powder. Commercial products store for years, maintaining activity of the spore and efficacy by the protein crystal.

14.4.4. Unique Applications

Specialized formulations have been successfully developed for specific control situations. Solid controlled-delivery compositions (briquettes) are able to direct *B. thuringiensis* or *B. sphaericus* at surface and/or subsurface feeding zones for mosquito larvae (Levy et al., 1997). Conidia of *B. bassiana* have been formulated in a wax powder to control varroa mites on honeybee (Meikle et al., 2008). Oil emulsions have been formulated for application to cattle using a motor-driven spray unit for tick control (Leemon et al., 2008). Injections of foaming formulations have been used to coat termite galleries with blastospores of entomopathogenic fungi in living trees (Dunlap et al., 2007). Insect cadavers containing nematodes have been coated to maintain the integrity of the cadaver until being applied to pots for insect control of soil pests (Shapiro-Ilan et al., 2001, 2003, 2010). Wettable granules have been evaluated for simultaneous application with a water spray, to stick granules to the foliage. Baits have been evaluated to induce feeding of housefly adults to initiate infection by the housefly salivary gland hypertrophy virus, which shuts down fly reproduction (Geden, 2012). These examples of highly specific formulations will not likely result in a grand expansion in the application of bioinsecticides, but they may fill niche pest control opportunities.

A promising new application system that is being developed for use is a combination of foliar granules (FGs). FGs are designed to adhere to aerial parts of plants when applied; the analogous description of sticky "glue balls" is apt. They are applied in a dry state and wetted at the time of application with a wet applied granule system (WAGS), which is actually a granular applicator equipped with a water spray system that wets the granules after they have left the applicator. Once contacted by water, the FG instantly becomes very sticky via partial dissolution of biopolymers, such as proteins or various carbohydrates. The FG products are generally smaller than ~0.5 mm; however, this is much larger than the particle size of solids that are sprayed through the typical nozzles used for foliar treatment, so spray drift is virtually eliminated. Because the viscosity of an airborne particle can be much higher than what is sprayed, the adhesion functionality of these wetted granules can allow for positive adhesion without much runoff of the formulation. Tank mix and pumping stability of the FG bioinsecticide is not an issue, nor is disposal of leftover mixtures and dealing with bioactive liquid spills.

14.5. CONCLUSIONS AND FUTURE OF BIOPESTICIDE FORMULATIONS

Microbial-based insecticides continue to be promoted worldwide as a favorable alternative to applications of chemical pesticide for pest control. Developing proper formulations is a key component for transforming biological control agents into a biopesticide. Just as pest control is best achieved by implementing integrated control tactics, so is the development of a formulation best achieved when based on an integrated research approach that spans from production through final application. Cross-disciplinary education among chemistry, microbiology, and biological pest control could prove most beneficial by specifically training future scientists to tackle integrated biopesticide formulation research. Many physical and biological deficiencies associated with microbial agents can be addressed with proper formulation development; formulations that address these deficiencies will assist in the effective adoption of microbial biopesticide products.

Successful commercial biopesticides will support development of new technologies in the future. Future efforts should continue to study basic biology, ecology, and epidemiology aspects of beneficial microbes in order to identify formulations that benefit the viability and efficacy of these microbes as biopesticides. Boyetchko et al. (1999) suggested that the commercial success of bioinsecticides has been limited by inadequate biomass scale-up and formulation technologies. Production and formulation tend to go hand-in-hand, so continued advancement in both areas is necessary for successful development of current or newly identified microbes as bioinsecticides. As for formulations, diligent reviews of new techniques, processes, and ingredients (including those adapted from other industries) requires constant scrutiny to identify those that are suitable for development as microbial-based insecticides.

REFERENCES

Adjalle, K.D., Vu, K.D., Tyagi, R.D., Brar, S.K., Valero, J.R., Surampalli, R.Y., 2011. Optimization of spray drying process for *Bacillus thuringiensis* fermented waste water and wastewater sludge. Bioprocess Biosyst. Eng. 34, 237–246.

Al Amin, A.I., Bin, A.K., 1994. Application of the fluidized bed process for formulation of WG-type pesticide granules. Powder Technol. 79, 135–146.

Anderson, C.W., Birthisel, T.D., Lynch, J.R., 2006. Development of a dry dispersible granular pesticide carrier. J. ASTM Int. 3 (2).

Arora, S., Koundal, K.R., Gopal, M., 2006. Efficacy of commercial formulations of *Bacillus thuringiensis* subsp. *kurstaki* against *Plutella xylostella*. Indian J. Agric. Sci. 76, 335–337.

Arruda, I.C., Ventura, M.U., Scarminio, I.S., 2005. Feeding and arrestment responses of *Diabrotica speciosa* to cucurbitacin-content formulations. Pesqui. Agropecu. Bras. 40, 639–643.

Arthurs, S.P., Lacey, L.A., De La Rosa, F., 2008. Evaluation of a granulovirus (PoGV) and *Bacillus thuringiensis* subsp. *kurstaki* for control of the potato tuberworm (Lepidoptera: Gelechiidae) in stored tubers. J. Econ. Entomol. 101, 1540–1546.

Ash, G.J., 2010. The science, art and business of successful bioherbicides. Biol. Control 52, 230–240.

Auld, B.A., Morin, L., 1995. Constraints in the development of bioherbicides. Weed Technol. 9, 638–652.

Barfod, K.K., Poulsen, S.S., Hammer, M., Larsen, S.T., 2010. Sub-chronic lung inflammation after airway exposure to *Bacillus thuringiensis* biopesticides in mice. BMC Microbiol. 10, article no. 233.

Baruch, L., Machluf, M., 2006. Alginate-chitosan complex coacervation for cell encapsulation: effect on mechanical properties and on long-term viability. Biopolymers 82, 570–579.

Bateman, R., 1999. Delivery systems and protocols for biopesticides. In: Hall, Franklin R., Menn, Julius J. (Eds.), Biopesticides: Use and Delivery, pp. 509–528.

Beegle, C.C., Yamamoto, T., 1992. History of *Bacillus thuringiensis* Berliner research and development. Can. Entomol. 124, 587–616.

Behle, R.W., 2006. Importance of direct spray and spray residue contact for infection of *Trichoplusia ni* larvae by field applications of *Beauveria bassiana*. J. Econ. Entomol. 99, 1120–1128.

Behle, R.W., Compton, D.L., Laszlo, J.A., Shapiro-Ilan, D.I., 2009. Evaluation of Soyscreen in an oil-based formulation for UV protection of *Beauveria bassiana* conidia. J. Econ. Entomol. 102, 1759–1766.

Boyetchko, S., Pedersen, E., Punja, Z., Reddy, M., 1999. Fomulations of biopesticides. In: Hall, F.R., Menn, J.J. (Eds.), Biopesticides: Uses and Delivery, Humana Press, Totowa, NJ, pp. 487–508.

Brar, S.K., Verma, M., Tyagi, R.D., Valero, J.R., 2006. Recent advances in downstream processing and formulations of *Bacillus thuringiensis* based biopesticides. Process Biochem. 41, 323–342.

Broza, M., Sneh, B., Levi, M., 1986. Evaluation of the effectiveness of *Bacillus thuringiensis* var. *entomocidus* as a pest control agent to replace chemical pesticides in alfalfa fields in Israel. Anz. Schäd. Pfl. Umwelt. 59, 152–156.

Brusselman, E., Beck, B., Pollet, S., Temmerman, F., Spanoghe, P., Moens, M., Nuyttens, D., 2012a. Effect of spray volume on the deposition, viability and infectivity of entomopathogenic nematodes in a foliar spray on vegetables. Pest Manag. Sci. 68, 1413–1418.

Brusselman, E., Beck, B., Pollet, S., Temmerman, F., Spanoghe, P., Moens, M., Nuyttens, D., 2012b. Effect of the spray application technique on the deposition of entomopathogenic nematodes in vegetables. Pest Manag. Sci. 68, 444–453.

Brusselman, E., Nuyttens, D., De Sutter, N., Viaene, N., Steurbaut, W., Moens, M., 2008. Effect of several centrifugal pump passages on the viability and activity of *Steinernema carpocapsae*, a biopesticide. Commun. Agric. Appl. Biol. Sci. 73, 705–708.

Burges, D.J., 1990. Practical analysis of complex coacervate systems. J. Colloid Interface Sci. 140, 227–238.

Burges, H.D., 1998. In: Burges, H.D. (Ed.), Formulation of Microbial Biopesticides: Beneficial Microorganisms, Nematodes and Seed Treatments, Kluwer Academic, Dordrecht, pp. 1–6.

Burges, H.D., Croizier, G., Huber, J., 1980. A review of safety tests on baculoviruses. Entomophaga 25, 329–339.

Caballero, P., Murillo, R., Munoz, D., Williams, T., 2009. The nucleopolyhedrovirus of *Spodoptera exigua* (Lepidoptera: Noctuidae) as a biopesticide: analysis of recent advances in Spain. Rev. Colomb. Entomol. 35, 105–115.

Chaparro, M., Carlos Espinel, C., Alba Marina Cotes, P.L., Laura Villamizar, R., 2010. Photostability and insecticidal activity of two formulations of granulovirus against *Tecia solanivora* larvae. Rev. Colomb. Entomol. 36, 25–30.

Chapple, A.C., Bateman, R.P., 1997. Application systems for microbial pesticides: necessity not novelty. Microbial Insecticides: Novelty or Necessity? British Crop Protection Council Proceedings/Monograph Series No. 68, pp. 160–173.

Chapple, A.C., Downer, R.A., Wolf, T.M., Taylor, R.A.J., Hall, F.R., 1996. The application of biological pesticides: limitations and a practical solution. Entomophaga 41, 456–474.

Chen, P.R., Wang, B., Way, H.Z., Lin, H.F., Huang, B., Li, Z.Z., 2010. Protection of ZnO nanoparticles to *Beauveria bassiana* conidia from ultraviolet radiation and their biocompatibility. Chem. J. Chin. Univ. 31, 2322–2328.

Chen, S., Glazer, I., 2005. A novel method for long-term storage of the entomopathogenic nematode *Steinernema feltiae* at room temperature. Biol. Control 32, 104–110.

Chilvers, G.R., Morris, V.J., 1987. Coacervation of gelatin-gellan gum mixtures and use in microencapsulation. Carbohydr. Polym. 7, 111–120.

Copping, L.G., Menn, J.J., 2000. Biopesticides: a review of their action, applications and efficacy. Pest Manag. Sci. 56, 651–676.

Cotes, A.M., 2011. Registry and regulation of biocontrol agents on food commodities in South America. Acta Hortic. 905, 301–306.

Daoust, R.A., Pereira, R.M., 1986. Stability of entomopathogenic fungi *Beauveria bassiana* and *Metarhizium anisopliae* on beetle attracting tubers and cowpea foliage in Brazil. Environ. Entomol. 15, 1237–1243.

Doekes, G., Larsen, P., Sigsgaard, T., Baelum, J., 2004. IgE sensitization to bacterial and fungal biopesticides in a cohort of Danish greenhouse workers: the BIOGART study. Am. J. Ind. Med. 46, 404–407.

Dougherty, E.M., Narang, N., Loeb, M., Lynn, D.E., Shapiro, M., 2006. Fluorescent brightener inhibits apoptosis in baculovirus-infected gypsy moth larval midgut cells in vitro. Biocontrol Sci. Tech. 16, 157–168.

Douro Kpindou, O.K., Djegui, D.A., Glitho, I.A., Tamo, M., 2011. Dose transfer of an oil-based formulation of *Metarhizium anisopliae* (Hypocreales: Clavicipitaceae) sprays to cotton bollworm in an arena trial. Int. J. Trop. Insect Sci. 31, 262–268.

Dunlap, C.A., Jackson, M.A., Wright, M.S., 2007. A foam formulation of *Paecilomyces fumosoroseus*, and entomopathogenic biocontrol agent. Biocontrol Sci. Tech. 17, 513–523.

Ehlers, R.U., 2003. Entomopathogenic nematodes in the European biocontrol market. Commun. Agric. Appl. Biol. Sci. 68 (4 Pt A), 3–16.

Ellerman, C.J., Entwistle, P.F., 1985. Inactivation of a nuclear polyhedrosis-virus on cotton by the substances produced by the cotton leaf surface glands. Ann. Appl. Biol. 106, 83–92.

El-Salamouny, S., Ranwala, D., Shapiro, M., Shepard, B.M., Farrar Jr., R.R., 2009. Tea, coffee, and cocoa as ultraviolet radiation protectants for the beet armyworm nucleopolyhedrovirus. J. Econ. Entomol. 102, 1767–1773.

EPA, 2012. http://www.epa.gov/pesticides/biopesticides/whatarebiopesticides.htm. Updated May 9, 2012.

Faria, M., Hajek, A.E., Wraight, S.P., 2009. Imbibitional damage in conidia of the entomopathogenic fungi *Beauveria bassiana*, *Metarhizium acridum* and *Metarhizium anisopliae*. Biol. Control 51, 346–354.

Faria, M.R. de, Wraight, S.P., 2007. Mycoinsecticides and mycoacaricides: a comprehensive list with worldwide coverage and international classification of formulation types. Biol. Control 43, 237–256.

Farrar Jr., R.R., Ridgway, R.L., 1994. Enhancement of activity of *Bacillus thuringiensis* Berliner against four Lepidopterous insect pests by nutrient based phagostimulants. J. Entomol. Sci. 30, 29–42.

Farrar Jr., R.R., Shapiro, M., Shepard, B.M., 2005. Enhanced activity of the nucleopolyhedrovirus of the fall armyworm (Lepidoptera: Noctuidae) on Bt-transgenic and nontransgenic sweet corn with a fluorescent brightener and a feeding stimulant. Environ. Entomol. 34, 825–832.

Fife, J.P., Ozkan, H.E., Derksen, R.C., Grewal, P.S., 2007. Effects of pumping on entomopathogenic nematodes and temperature increase within a spray system. Appl. Eng. Agric. 23, 405–412.

Fujiie, A., Yokoyama, T., 1998. Effects of ultraviolet light on the entomopathogenic nematode, *Steinernema kushidai* and its symbiotic bacterium, *Xenorhabdus japonicas*. Appl. Entomol. Zool. 33, 263–269.

Gaugler, R., 1997. Alternative paradigms for commercializing biopesticides. Phytoparasitica 25, 179–182.

Geden, C.J., 2012. Status of biopesticides for control of house flies. J. Biopestic. 5, 1–11.

Gothro, P., Berry, R., 2012. Entomopathogenic nematodes—IPM, Symps, biocontrol, Slugs (revised 11/09). In: Hollingsworth, C.S. (Ed.), Pacific Northwest Insect Management Handbook, Extension and Station Communications, Oregon State University, Corvalis, OR.

Gouin, S., 2004. Microencapsulation: industrial appraisal of existing technologies and trends. Trends Food Sci. Tech. 15, 330–347.

Guimaraes, C.D.O., Correia, A.D.C.B., Ferreira, M.D.C., 2004. Application pressure in boom sprayer and efficiency of commercial fungal bioinsecticides. Presq. Agropec. Bras. 39, 117–1182.

Hadapad, A.B., Hire, R.S., Vijayalakshmi, N., Dongre, T.K., 2011. Sustained-release biopolymer based formulations for *Bacillus sphaericus* Neide ISPC-8. J. Pest Sci. 84, 249–255.

Hadapad, A.B., Vijayalakshmi, N., Hire, R.S., Dongre, T.K., 2008. Effect of ultraviolet radiation on spore viability and mosquitocidal activity of an indigenous ISPC-8 *Bacillus sphaericus* Neide strain. Acta Trop. 107, 113–116.

Hajek, A., 2004. Natural Enemies: An Introduction to Biological Control. Cambridge University Press, Cambridge, UK.

Hynes, R.K., Boyetchko, S.M., 2006. Research initiatives in the art and science of biopesticide formulations. Soil Biol. Biochem. 38, 845–849.

Ignoffo, C.M., 1992. Environmental factors affecting persistence of entomopathogens. Fla. Entomol. 75, 516–525.

Ignoffo, C.M., Hostetter, D.L., Sikorowski, P.P., Sutter, G., Brooks, W.M., 1977. Inactivation of representative species of entomopathogenic viruses, a bacterium, fungus, and protozoan by an ultraviolet light source. Environ. Entomol. 6, 411–415.

Inglis, G.D., Goettel, M.S., Johnson, D.L., 1993. Persistence of the entomopathogenic fungus, *Beauveria bassiana*, on phylloplanes of crested wheatgrass and alfalfa. Biol. Control 3, 258–270.

Inglis, G.D., Goettel, M.S., Johnson, D.L., 1995. Influence of ultraviolet light protectants on persistence of the entomopathogenic fungus, *Beauveria bassiana*. Biol. Control 5, 581–590.

Jackson, M.A., Dunlap, C.A., Jaronski, S.T., 2010. Ecological considerations in producing and formulating fungal entomopathogens for use in insect biocontrol. BioControl 55, 129–145.

Jenkins, N.E., Grzywacz, D., 2000. Quality control of fungal and viral biocontrol agents—assurance of product performance. Biocontrol Sci. Tech. 10, 753–777.

Jensen, G.B., Larsen, P., Jacobsen, B.L., Madsen, B., Wilcks, A., Smidt, L., Andrup, L., 2002. Isolation and characterization of *Bacillus cereus*-like bacteria from faecal samples from greenhouse workers who are using *Bacillus thuringiensis*-based insecticides. Int. Arch. Occup. Environ. Health 75, 191–196.

Jin, X., Streett, D.A., Dunlap, C.A., Lyn, M.E., 2008. Application of hydrophilic-lipophilic balance (HLB) number to optimize a compatible non-ionic surfactant for dried aerial conidia of *Beauveria bassiana*. Biol. Control 46, 226–233.

Kaaya, G.P., Hassan, S., 2000. Entomogenous fungi as promising biopesticides for tick control. Exp. Appl. Acarol. 24, 913–926.

Kassa, A., Stephan, D., Vidal, S., Zimmerman, G., 2004. Production and processing of *Metarhizium anisopliae* var. *acridum* submerged conidia for locust and grasshopper control. Mycol. Res. 108, 93–100.

Kennedy, I.R., Solomon, K.R., Gee, S.J., Crossan, A., Wang, S., Sanchez-Bayo, F., 2007. Achieving rational use of agrochemicals: environmental chemistry in action. In: Kennedy, I.R. (Ed.), Rational Environmental Management of Agrochemicals, ACS Symposium Series 966, pp. 2–12.

Kim, J.S., Je, Y.H., 2012. Milling effect on the control efficacy of spray-dried *Bacillus thuringiensis* technical powder against diamondback moths. Pest Manag. Sci. 68, 321–323.

Kim, J.S., Je, Y.H., Woo, E.O., Park, J.S., 2011. Persistence of *Isaria fumosorosea* (Hypecreales: Cordycipitaceae) SFP-198 conidia in corn oil-based suspension. Mycopathologia 171, 67–75.

Knowles, A., 2009. Global trends in pesticide formulation technology: the development of safer formulations in China. Outlooks Pest Manag. 20, 165–170.

Kramer, K.J., Muthukrishnan, S., 1997. Insect chitinases: molecular biology and potential uses as biopesticides. Insect Biochem. Mol. Biol. 27, 887–900.

Laird, M., Lacey, L.A., Davidson, E.W., 1990. Safety of Microbial Insecticides. CRC Press, Inc., Boca Raton, FL.

Larena, I., Melgarejo, P., De Cal, A., 2003. Drying conidia of *Penicillium oxalicum*, a biological control agent against Fusarium wilt of tomato. J. Phytopathol. 151, 600–606.

Lasa, R., Williams, T., Caballero, P., 2008. Insecticidal properties and microbial contaminants in a *Spodoptera exigua* multiple nucleopolyhedrovirus (Baculoviridae) formulation stored at different temperatures. J. Econ. Entomol. 101, 42–49.

Leef, J.L., Mazur, P., 1978. Physiological response of Neurospora conidia to freezing in the dehydrated, hydrated, or germinated state. Appl. Environ. Microbiol. 35, 72–83.

Leemon, D.M., Jonsson, N.N., 2008. Laboratory studies on Australian isolates of *Metarhizium anisopliae* as a biopesticide for the cattle tick *Boophilus microplus*. J. Invertebr. Pathol. 97, 40–49.

Leemon, D.M., Turner, L.B., Jonsson, H.N., 2008. Pen studies on the control of cattle tick (*Rhipicephalus* (*Boophilus*) *microplus*) with *Metarhizium anisopliae* (Sorokin). Vet. Parasitol. 156, 248–260.

Levy, R., Nichols, M.A., Opp, W.R., 1997. Targeted Delivery of Pesticides from Matricap™ 2 Compositions. ASTM Spec. Tech. Pub. 1328, 63–93.

Li, S., Schoneich, C., Borchardt, R.T., 1995. Chemical instability of protein pharmaceuticals: mechanisms of oxidation and strategies for stabilization. Biotechnol. Bioeng. 48, 490–500.

Lopez Jr., J.D., Lingren, P.D., 1994. Feeding response of adult *Helicoverpa zea* (Lepidoptera: Noctuidae) to commercial phagostimulants. J. Econ. Entomol. 87, 1653–1658.

Lowe, E.H., Milbradt, R., 2011. Emulsification performance used for NPE replacement in solvent-based formulations. J. ASTM Int. 8, 7.

Lyn, M.E., Burnett, D., Garcia, A.R., Gray, R., 2010. Interaction of water with three granular biopesticide formulations. J. Agric. Food Chem. 58, 1804–1814.

Maksymiuk, B., 1970. Occurrence and nature of antibacterial substances in plants affecting *Bacillus thuringiensis* and other entomogenous bacteria. J. Invertebr. Pathol. 15, 356–371.

McGuire, M.R., Behle, R.W., Goebel, H.N., Fry, T.C., 2000. Calibration of a sunlight simulator for determining solar stability of *Bacillus thuringiensis* and *Anagrapha falcifera* nuclear polyhedrovirus. Environ. Entomol. 29, 1070–1074.

Meilke, W.G., Mercadier, G., Guermache, F., Bon, M.C., 2012. Pseudomonas contamination of a fungus-based biopesticide: implications for honey bee (Hymenoptera: Apidae) health and Varroa mite (Acari: Varroidae) control. Biol. Control 60, 312–320.

Meikle, W.G., Mercadier, G., Holst, N., Nansen, C., Girod, V., 2008. Impact of a treatment of *Beauveria bassiana* (Deuteromycota: Hyphomycetes) on honeybee (*Aphis mellifera*) colony health and on Varroa destructor mites (Acari: Varroidae). Apidologie 39, 247–259.

Morales Ramos, L.H., McGuire, M.R., Galan Wong, L.J., Castro Franco, R., 2000. Evaluation of pectin, gelatin and starch granular formulations of *Bacillus thuringiensis*. Southwest. Entomol. 25, 59–67.

Pemsel, M., Schwab, S., Scheurer, A., Freitag, D., Schatz, R., Schlucker, E., 2010. Advanced PGSS process for the encapsulation of the biopesticide *Cydia pomonella* granulovirus. J. Supercrit. Fluids 53, 174–178.

Peng, D., Zhou, C., Chen, S., Ruan, L., Yu, Z., Sun, M., 2008. Toxicological safety assessment of genetically modified *Bacillus thuringiensis* with additional N-acyl homoserine lactonase gene. Environ. Toxicol. Chem. 27, 188–195.

Petrie, K.J., Thomas, M., Broadbent, E., 2003. Symptom complaints following aerial spraying with biological insecticide Foray 48B. N. Z. Med. J. 116, 7.

Quiroga, I., Martha Gomex, A., Laura Villamizar, R., 2011. Stability of formulations based on granulovirus for controlling *Tecia solanivora* (Lepitoptera: Gelechiidae) in the field. Rev. Colomb. Entomol. 37, 27–35.

Rajagopal, R., Mohan, S., Bhatnagar, R.K., 2006. Direct infection of *Spodoptera litura* by *Photorhabdus luminescens* encapsulated in alginate beads. J. Invertebr. Pathol. 93, 50–53.

Reddy, N.P., Kahn, P.A.A., Devi, K.U., Victor, J.S., Sharma, H.C., 2008. Assessment of the suitability of Tinopal as an enhancing adjuvant in formulations of the insect pathogenic fungus *Beauveria bassiana* (Bals.) Vuillemin. Pest Manag. Sci. 64, 909–915.

Rosas-Garcia, N.M., De Luna-Santillana, E.D.J., 2006. Microencapsulating matrix design from biodegradable compounds for bioinsecticide preparation. Rev. Int. Contam. Ambiental 22, 135–142.

Salamouny, S.E., Shapiro, M., Ling, K.S., Shepard, B.M., 2009. Black tea and lignin as ultraviolet protectants for the beet armyworm nucleopolyhedrovisus. J. Entomol. Sci. 44, 50–58.

Santi, L., e Silva, L.A.D., da Silva, W.O.B., Correa, A.P.F., Rangel, D.E.N., Carlini, C.R., Schrank, A., Vainstein, M.H., 2011. Virulence of the entomopathogenic fungus *Metarhizium anisopliae* using soybean oil formulation for control of the cotton stainer bug, *Dysdercus peruvianus*. World J. Microbiol. Biotechnol. 27, 2297–2303.

Santos, P. de S., da Silva, M.A.Q., Monteiro, A.C., Gava, C.A.T., 2012. Selection of surfactant compounds to enhance the dispersion of *Beauveria bassiana*. Biocontrol Sci. Tech. 22, 281–292.

Shapiro, M., 1992. Use of optical brighteners as radiation protectants for the gypsy moth (Lepidoptera: Lymantriidae) nuclear polyhedrosis virus. J. Econ. Entomol. 85, 1682–1686.

Shapiro, M., Argauer, R., 1997. Components of the stilbene optical brightener Tinopal LPW as enhancers for the gypsy moth (Lepidoptera: Lymantriidae) baculovirus. J. Econ. Entomol. 90, 899–904.

Shapiro, M., Argauer, R., 2001. Relative effectiveness of selected stilbene optical brighteners as enhancers of the beet armyworm (Lepidoptera: Noctuidae) nuclear polyhedrosis virus. J. Econ. Entomol. 94, 339–343.

Shapiro, M., Hamm, J.J., 1999. Enhancement in activity of homologous and heterologous baculoviruses infection to fall armyworm (Lepidoptera: Noctuidae) by selected optical brighteners. J. Entomol. Sci. 34, 381–390.

Shapiro-Ilan, D.I., Lewis, E.E., Behle, R.W., McGuire, M.R., 2001. Formulation of entomopathogenic nematode-infected cadavers. J. Invertebr. Pathol. 78, 17–23.

Shapiro-Ilan, D.I., Lewis, E.E., Son, Y., Tedders, W.L., 2003. Superior efficacy observed in entomopathogenic nematodes applied in infected-host cadavers compared with application in aqueous suspension. J. Invertebr. Pathol. 83, 270–272.

Shapiro-Ilan, D.I., Morales-Ramos, J.A., Rojas, M.G., Tedders, W.L., 2010. Effects of a novel entomopathogenic nematode-infected host formulation on cadaver integrity, nematode yield, and suppression of *Diaprepes abbreviates* and *Aethina tumida*. J. Invertebr. Pathol. 103, 102–108.

Stahly, D.P., 1984. Biochemical genetics of the bacterial insect-control agent *Bacillus thuringiensis*: basic principles and prospects engineering. Biotechnol. Genet. Eng. Rev. 2, 341–363.

Steinke, W.E., Akesson, N.B., 1993. Atomization and Applications of Biopesticide. ASTM Spec. Tech. Pub. 1146, 257–271.

Stevenson, P.C., D'Cunha, R.F., Grzywacz, D., 2012. Inactivation of baculovirus by isoflavonoids on chickpea (*Cicer arietinum*) leaf surfaces reduces the efficacy of nucleopolyhedrovirus against *Helicoverpa armigera*. J. Chem. Ecol. 36, 227–235.

Strauch, O., Niemann, I., Neumann, A., Schmidt, A.J., Peters, A., Ehlers, R.U., 2000. Storage and formulation of the entomopathogenic nematodes *Heterorhabditis indica* and *H. bacteriophora*. BioControl 45, 483–500.

Tamez-Guerra, P., McGuire, M.R., Behle, R.W., Hamm, J.J., Sumner, H.R., Shasha, B.S., 2000. Sunlight persistence and rainfastness of spray-dried formulations of baculovirus isolated from *Anagrapha falcifera* (Lepidoptera: Noctuidae). J. Econ. Entomol. 93, 210–218.

Teera-Arunsiri, A., Suphantharika, M., Ketunuti, U., 2003. Preparation of spray-dried wettable powder formulations of *Bacillus thuringiensis*-based biopesticides. J. Econ. Entomol. 96, 292–299.

Ugine, T.A., Wraight, S.P., Sanderson, J.P., 2007. Effects of manipulating spray-application parameters on efficacy of the entomopathogenic fungus *Beauveria bassiana* against western flower thrips, *Fankliniella occidentalis*, infesting greenhouse impatients crops. Biocontrol Sci. Tech. 17, 193–219.

Vail, P.V., Hoffmann, D.F., Tebbets, J.S., 1999. Influence of fluorescent brighteners on the field activity of the celery looper nucleopolyhedrovirus. Southwest. Entomol. 24, 87–98.

Van Lenteren, J.C., 2012. The state of commercial augmentative biological control: plenty of natural enemies, but a frustrating lack of uptake. BioControl 57, 1–20.

Wang, X., Zhu, H., Reding, M.E., Locke, J.C., Leland, J.E., Derksen, R.C., Spongberg, A.L., Krause, C.R., 2009. Delivery of chemical and microbial pesticides through drip irrigation systems. Appl. Eng. Agric. 25, 883–893.

Ward, M.D.W., Chung, Y.J., Copeland, L.B., Doerfler, D.L., 2011. Allergic responses induced by a fungal biopesticide *Metarhizium anisopliae* and house dust mite are compared in a mouse model. J. Toxicol., article no. 360805.

Ward, M.D.W., Chung, Y.J., Haykal-Coates, N., Copeland, L.B., 2009. Differential allergy responses to *Metarhizium anisopliae* fungal component extracts in BALB/c mice allergy-like responses to fungal components. J. Immunotoxicol. 6, 62–73.

Westwood, G.S., Juang, S.W., Deyhani, N.O., 2006. Molecular and immunological characterization of allergens from the entomopathogenic fungus *Beauveria bassiana*. Clin. Mol. Allergy 4, article no. 12.

Xu, J., Liu, Q., Yin, X.D., Xhu, S.D., 2006. A review of recent development of *Bacillus thuringiensis* ICP genetically engineered microbes. Entomol. J. East China 15, 53–58.

Yao, S.L., Ying, S.H., Feng, M.G., Hatting, J.L., 2010. In vitro and in vivo responses of fungal biocontrol agents to gradient doses of UV-B and UV-A irradiation. BioControl 55, 413–422.

Young, S.Y., Yearian, W.E., 1974. Persistence of Heliothis NPV on foliage of cotton, soybean, and tomato. Environ. Entomol. 3, 253–255.

Zimmermann, G., 2007. Review on safety of the entomopathogenic fungus *Metarhizium anisopliae*. Biocontrol Sci. Tech. 17, 879–920.

This page intentionally left blank

Chapter 15

Mass Production of Entomopathogens in Less Industrialized Countries

David Grzywacz,[1] David Moore[2] and R.J. Rabindra[3]

[1]*Natural Resources Institute, University of Greenwich, Central Avenue, Chatham, Kent, UK,* [2]*CABI, Bakeham Lane, Egham, Surrey, UK,* [3]*College of Post Graduate Studies, Central Agricultural University, Umiam, Shillong, Meghalaya, India*

15.1. INTRODUCTION

This chapter focuses on the mass production of entomopathogenic organisms as the active ingredients of biopesticides in parts of the world other than the major industrialized nations. Broadly, this may be viewed as anywhere but Australia and New Zealand, Japan and South Korea, the United States and Canada, and many countries in Europe. The countries that we focus on form a heterogeneous group, including many of the poorest countries in the world, as well as Brazil, Russia, India, China, and South Africa (the BRICS nations), which are already (or are rapidly becoming) major world economies. Low-intensity agriculture is one common characteristic in these nations, as reflected in the high percentage of the labor force involved. Even here the range is enormous, with the countries referred to in this chapter varying between 9% of their labor force working in agriculture (South Africa) to 75% of their labor force working in agriculture (Kenya) (Central Intelligence Agency, 2010).

Despite (or possibly because of) the variations between countries, the less industrialized countries (LICs) provide a wide range of production examples. What may be considered implicit in the title of the chapter is that low-cost, small-scale production systems (often with a large work force) are characteristic of mass production efforts in LICs, and many such examples exist in these countries. It is also true that if biopesticides take their potential market share, demand will be so great that industrial-scale production will be at least a part of production in LICs.

Mass Production of Beneficial Organisms. http://dx.doi.org/10.1016/B978-0-12-391453-8.00015-7

15.2. ISSUES AND OPPORTUNITIES FOR ENTOMOPATHOGEN UPTAKE IN LESS INDUSTRIALIZED COUNTRIES

The control of pests in LICs has always posed a special series of challenges that have created opportunities for entomopathogens (EPs). In tropical countries such as Thailand, where pests have continuous generations, resistance to chemicals develops rapidly such that chemical pesticides have a long history of failures (Sukonthabhirom et al., 2011). Many LICs also have weak agro-input systems, underdeveloped agricultural research systems, and poor agricultural extension advice services. Consequently, farmers, who face some of the most difficult pests to control, lack adequate resources for assistance. Biological control agents (BCAs) have long been considered as potentially very useful tools for pest control in LICs and wealthier countries. In particular, entomopathogens such as fungi, viruses, and nematodes offer great potential for a number of reasons:

1. Good biopesticides can offer economic pest control equivalent to that achieved by chemical pesticides (Langewalde et al., 1999). Their specificity and environmentally benign nature pose much less of a health hazard than the use of toxic chemical pesticides to farmers lacking adequate safety equipment. EP residues pose no hazard to consumers and produce treated with BCAs is not excluded by maximum residue limits legislation/regulations from local or export markets.
2. The facility with which many EPs can be produced using low-cost technology has made them attractive options for local production in countries where technical resources are limited.
3. Local production of EPs is also seen as a way of ensuring costs are lower than that of imported pesticide products, suiting them to the needs of many small farmers who are only able to afford cheaper pest control technologies.
4. Some key global pests of LICs that have become resistant to chemical pesticides, such as *Heliothis/Helicoverpa* spp., *Spodoptera* spp., and diamondback moths (*Plutella xylostella* L.), have well-known EPs. Such EPs can provide key tools in the development of insect resistance management (IRM) programs (Marrone, 2007). Although some argue the ideal would be to develop biologically based integrated pest management (IPM) systems using no chemicals, in some cases this is not yet practicable. However, incorporating EPs into IPM systems is a proven way to overcome pesticide resistance and/or delay its appearance.

Thus, the rationale for developing EPs for IPM is strong. Indeed, the development of EPs was enacted as coherent national plan in some countries, such as India (Rabindra and Grzywacz, 2010).

15.3. PRACTICAL CONSTRAINTS FOR ENTOMOPATHOGEN UPTAKE IN DEVELOPING COUNTRIES

Although at a policy level the logic of developing local production of EPs in LICs may seem convincing, implementation faces some serious, but not insurmountable, challenges. The size of markets in many LICs is seen as a significant constraint (apart from Brazil, India, and China, where single markets seem very large). Many LICs have markets with relatively low numbers of customers and/or value. The presently small and fragmented nature of markets for EPs in Africa makes it less attractive to set up production. Smaller countries can still provide significant markets; for example, Malaysia, Indonesia, and the Philippines have enormous areas under oil palm that would be appropriate for EP use. In reality, the market need exists in LICs but presently is largely occupied by chemical pesticides.

In many LICs, there is a dearth of knowledge about key pests and the EPs available to support potential producers. In sub-Saharan Africa, for example, there has been a serious lack of national funding for agricultural research systems, agriculture in tertiary education institutes, and extension services since the early 1990s. Therefore, the technical knowledge base needed to support production and promote the wider use of EPs in farming is lacking. Even where EPs have been registered and are being promoted, poor product regulation in some countries has undermined the market by the proliferation of poor-quality products. Sometimes, a weak understanding of EPs and IPM by the farmer, rather than failure of the products themselves, can lead to inadequate control and a loss of consumer confidence in EPs. Many farmers rely upon chemical suppliers for advice; this can be a problem where these suppliers lack knowledge of EPs, but it can be an opportunity if the suppliers take up an interest in supplying EPs.

The lack of appropriate infrastructure can also be an important issue for EP supply. Some EPs, as with predators and parasitoids (whose use is also expanding), require cooling along the supply chain to ensure they are in good condition. Marked improvements in the shelf-life of fungal products has ameliorated this constraint to a degree, but improved infrastructure is important. For example, in Africa, outside of horticultural hot spots (e.g. Nairobi, Kenya or South Africa) where growing produce for export is a mainstay of agriculture, cool chain facilities may be absent. The cost of registering an EP pesticide in some countries can be substantial and can deter potential producers. This issue is compounded in LICs, which may have small markets, each with a different and separate registration system for pesticides.

Regulation can have also a positive impact. For instance, the requirements for low-residue produce in major markets, such as the Organisation for Economic Co-operation and Development (OECD), has recently become a powerful driver promoting use of BCAs and EPs and encouraging local production. The refusal of major aid donors to fund the use of chemical pesticides for migratory pest control was a stimulus to the development and use of EPs, such as Green Muscle in Africa, and continues to feed the search for new EP products.

15.4. PRODUCTION OF ENTOMOPATHOGENS IN LESS INDUSTRIALIZED COUNTRIES

Most LIC production is based on entomopathogenic fungi and insect viruses. Although there is some production of bacterial insecticides, such as *Bacillus thuringiensis* (Bt) Berliner, these are produced using the same high-technology fermenter systems used in industrialized countries. Production of entomopathogenic nematodes (EPNs) and protozoa has not yet appeared in any scale in LICs outside China and the Republic of Korea (see Chapter 10); in the former case, this is because the technological and cost barriers that need to be surmounted have been too great and the lack of a specific market pull means that production has not appeared commercially. There is a renewed interest in the use of EPNs in India, partly because EPNs do not need to be registered. Following successful field trials on the control of the cardamom white grubs, *Basilepta fulvicorne* (Jacoby) with *Heterorhabditis indica* (Poinar), and *Steinernema carpocapsae* (Weiser), the Project Directorate of Biological Control developed an in vivo mass production method using *Galleria mellonella* L. larvae and sold the technology to three commercial producers who started selling the EPN formulations. The uptake, however, will be only for niche areas with high-value crops, such as the cardamom.

Mycoinsecticides (biopesticides using entomopathogenic fungi) are commonly produced in LICs; in this chapter, the emphasis is on low to intermediate technology levels of their production. This is partly to reduce overlap with Jaronski (Chapter 11) and partly because industrial-scale production occurring in China, for example, is bedeviled with the same secrecy as in other parts of the world. Although some public producers are quite open with their methodologies, dissemination is often through gray literature, which is difficult to access.

The production of insect viruses has almost exclusively focused on baculoviruses; again, there are few studies of the various commercial production systems, although there are some better documented cases, such as the *Anticarsia gemmatalis* (Hübner) nucleopolyhedrovirus (NPV) (Moscardi, 1999; Moscardi, 2007; Moscardi et al., 2011). Many producers closely guard their production methodologies, which they see as commercially important intellectual property; this severely limits the accessibility of information.

15.5. PRODUCTION OF ENTOMOPATHOGENIC FUNGI

In the late 1950s and the 1960s, China and Brazil (then developing countries) pioneered extensive use of fungi (Prior, 1989), with *Metarhizium anisopliae* (Metchinikoff) Sorokin being used over many thousands of hectares. The sugarcane froghopper *Mahanarva posticata* (Stål) was used in Brazilian sugarcane and *Beauveria bassiana* (Balsamo-Crivelli) Vuillemin was used in China against *Ostrinia nubilalis* (Hübner) on massive scales. The increasing wealth and technological sophistication of these nations could lead to transformations in the production and use of mycoinsecticides. Li et al. (2010) described the history and present status of the use of mycoinsecticides in both these countries, whereas Feng et al. (1994) described in some detail the production of *B. bassiana* in China. Useful information can also be found in Roberts and St Leger (2004).

In India, *M. anisopliae, B. bassiana*, and *Lecanicillium lecanii* (Zare) and Gams are produced by several companies and have been registered. (Taxonomic advances may make some species names tenuous; registered names will be maintained here, although *Metarhizium* spp. may prove to be more accurate in many situations.) Poor shelf-life and low persistence in the field have been major constraints in enhancing the use of these fungi. There is an increased interest in development of oil formulations with enhanced shelf-lives. The Department of Biotechnology India has funded several research and development projects to facilitate this because there is considerable interest in using these EPs in IPM of vegetable pests.

Despite a number of recent publications, there is little real detail about mass production methodologies at the industrial scale necessary to meet major markets. There is a vast amount of gray literature, but these publications mostly describe minor variations on well-established themes, types of solid substrate, methods of sterilizing, or types of containers used and are often difficult to source. The more modern LIC producers tend to keep at least some of their production processes secret. However, although 20 years ago focus was on quantity of production, more recently it seems to have been acknowledged that product quality is also vital.

Brazil, India, and China have worked with entomopathogens extensively. Recent publications demonstrate the extent of interest (Koul, 2011; Leng et al., 2011; Li et al., 2010) and anticipated developments. Faria and Wraight (2007) assembled a comprehensive list of 171 mycoinsecticides and mycoacaricides developed since the 1960s; 15% were no longer available and a further 10% were of an uncertain status. South American institutions and companies were responsible for the development of around 43% of the products.

Roberts and St Leger (2004) highlighted the fact that there is relatively little detailed information on specific mass production facilities in general—and almost none from commercial companies. However, noncommercial organizations fill some of this gap (Alves and Pereira, 1989; Feng et al., 1994; Jaronski (Chapter 11); Jenkins and Goettel, 1997; Jenkins et al., 1998; Cherry et al., 1999). An excellent description of production processes has been given by Grace and Jaronski (2005), with a clear step-by-step account of mass production techniques; most importantly, it emphasizes the need to view each isolate individually. Jenkins et al. (1998) and Cherry et al. (1999) gave great detail about the mass production of *Metarhizium acridum* (Humber) at the International Institute for Tropical Agriculture (IITA) in Benin, West Africa, during the course of the LUBILOSA (*Lutte Biologique contre les Locustes et les Sauteriaux*) program, which was aimed at developing an alternative to chemical insecticides for use against locust plagues. Over a 13-year period, the program developed the commercial mycoinsecticide Green Muscle® (Lomer et al., 2001; Moore, 2008). To supply conidia for major field trials, a mass production facility was built at IITA, becoming operational towards the end of 1996 and producing into 1998, allowing economic assessments to be made.

15.5.1. The LUBILOSA System

The production capacity adapted systems that were used widely in Brazil and China, but the basic principles were used much earlier. For example, Rorer (1910, 1913) gave details of production using a cabinet method, before the age of plastic bags. The LUBILOSA method was based on low capital investment and high labor inputs; it was robust, simple, and appropriate to developing country situations (Cherry et al., 1999) while still capable of producing high-quality fungal material. The locust control project required the complete separation of conidia from substrate to allow formulation for controlled droplet application, making the LUBILOSA system somewhat more complicated than some other mass production systems may need to be.

In summary, the LUBILOSA process required the maintenance of the isolate stock (achieved by good storage of material passaged through the desert locust at 6-month intervals), which was used to inoculate sterile liquid medium prepared from cheap and locally available materials, such as waste brewer's yeast and sugar (Fig. 15.1). The liquid inoculum was incubated for a few days and then added to bags of sterile rice which, in turn, was incubated for approximately 10–12 days.

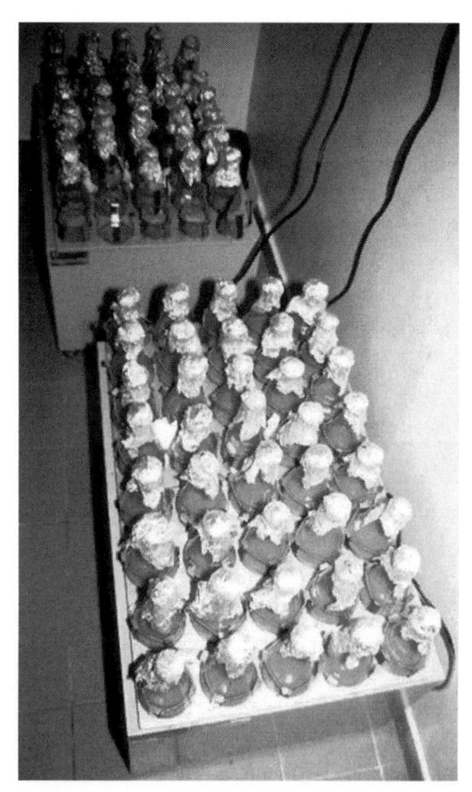

FIGURE 15.1 Flasks of *Metarhizium acridum* in liquid media on shakers at the IITA facility in Benin.

These bags were cut open, the conidiated rice was dried to around 20% moisture, and then the conidia were harvested and dried to 5% or less before packing in plastic lined foil sachets. Production capacity was around 300–350 kg of conidia per year, allowing the treatment of about 7000 ha. The system is highly flexible and many process variations can be used. These may include the local availability of substrates, nutrients and equipment, energy and labor costs and, critically, the fungal isolate(s) used and pest systems in which the mycoinsecticide is to be used. One common characteristic of relatively low intensity mass production units is that, usually, there is loose control over certain production parameters, such as temperature for substrate incubation and relative humidity and temperature for the drying processes. In addition, the exact composition of the substrate used may vary over time, from harvest onwards and perhaps from cultivar used.

The LUBILOSA process resulted in major variations in production, especially yield, apparently occurring in a cyclical trend. Temperature and incubation period were key factors, but they explained less than 40% of the yield variation (Cherry et al., 1999); the nutritional quality of the rice substrate was also considered likely to be important, but this was not examined in depth. Even minor variations can significantly alter spore characteristics, as represented by shelf-life or thermal tolerance (Hong et al., 2001; McClatchie et al., 1994), and can be of considerable importance to quality. Consequently, with low-intensity production, quite major variations in yield should be expected over time; variations in the intrinsic quality of the spores are also inevitable. This variation is not unique to loosely controlled systems; even systems with accurate process control over aspects such as aeration, humidity, and temperature can show considerable variation (Bradley et al., 1992).

15.5.2. The Caroni System

Metarhizium anisopliae was developed for the management of the sugarcane froghopper, *Aeneolamia varia saccharina* (Distant), a serious wet season pest of sugarcane in Trinidad. This project was conducted at the Caroni Research and Development Division of the national sugarcane research institute in Trinidad and Tobago and described in Jenkins et al. (2007). The production facility was located in the central region of Trinidad and Tobago. The main plant equipment included two locally built steam-reliant type autoclaves, a diesel-fuelled boiler with accompanying water treatment system, incubators, clean benches, microscopes, an in-house designed and built spore-dryer, and a cyclone harvester.

The production process was based on the LUBILOSA system and included the discrete activities of sterilization, inoculation, incubation under strain-specific environmental conditions, drying, cyclone extraction of conidia, packaging, storage, and distribution. However, the scale of the operation was quite different, with a handling capacity of 250 kg per batch and a weekly throughput of 3000 kg. The production unit operated in two 12-h shifts, with each shift consisting of 20 persons including supervisory, technical, and skilled labor.

A combination of unpolished rice and tap water was packaged in autoclavable plastic bags manufactured within the facility. The bags were partially sealed, stacked in the autoclave, and sterilized for 1 h. Three- to five-day-old liquid inoculum, produced in yeast and glucose, was used to inoculate the sterilized rice within the plastic bags. The bags of inoculated rice were shaken to thoroughly mix the inoculum amongst the rice and then resealed with micropore tape to allow airflow. For conidia production, the inoculated rice was incubated for 14 days in stacked plastic containers (Fig. 15.2), during which time the bags were intermittently shaken to break up clumps and encourage aeration. (Some isolates do not grow well if the bags are disturbed, in which case this stage would be omitted.) For harvesting, one face of the bag was cut away and the bags were placed in a cool dehumidified room for 48 h. Conidia were extracted with an industrial cyclone extractor (www.mycoharvester.info/) and placed in the dryer for further drying. The conidia were then sealed in plastic bags and stored at 4 °C. Although not optimal for storage because plastic bags are not totally waterproof or airtight, this was appropriate for a system in which the conidia were used promptly.

At the conclusion of the 10-year period of research and development (1991–2001), the annual production increased to 1500–2000 kg of spore powder, which showed a high level of viability, virulence, and purity and was sufficient for treating around 40,000 ha. The rice substrate yielded an average of 45 g of dry conidia per kilogram of rice, with an end product concentration of 5×10^{10} conidia per gram. Very high quality conidia were produced. Research then showed that *A. varia saccharina* could be controlled using preconidiated rice taken from the plastic bags and applied to the base of sugarcane plants; conidiation occurred in the field. This effectively halved the time period required for

FIGURE 15.2 Inoculated rice incubating in plastic containers at the IITA facility in Benin.

production. This project ended with the privatization of sugarcane production in Trinidad and Tobago.

15.6. ADDITIONAL EXAMPLES FROM OTHER COUNTRIES

15.6.1. China

The development of fungal biopesticides was strongly supported in China in terms of research, production, and use programs by local and central government (Wang and Li, 2010). Details of production methods are given in Li et al. (2010) and Feng et al. (1994). Modern and sophisticated production facilities have begun operating, reflecting the true industrial nature required for mass use over the scale required.

15.6.2. Brazil

Useful information is given in Li et al. (2010) for both Brazil and China (Rangel and Faria, 2010). Biological pesticides remain economically advantageous in specific pest control situations. A significant number (~40) of commercial mycoinsecticides are produced by 19 commercial companies (Rangel and Faria, 2010). Additionally, 20 laboratories operated by sugar/ethanol mills produce *M. anisopliae* for local use, using substrates from sugarcane. Much of the success of the industry reflects significant government and university research support.

15.6.3. Cuba

In 1982, IPM was adopted by the Cuban state as official policy for pest control, and a small number of industrial scale production centers were established (Henderson and Sinfontes, 2010). By the late 1990s, these centers had fermenters ranging from 5 to 500 l capacity in which they produced Bt and entomopathogenic fungi, although solid state production was preferred for fungi (D. Moore, unpublished data).

Biological control production is centered on *Centros de Reproducción de Entomófagos y Entomopatógenos* (CREEs), which are centers for mass production of both macrobial and microbial agents. Consuegra (2004) stated that in the 10-year period of 1990–1999, approximately 5650 tons of entomopathogens were produced, with *B. thuringiensis* making up 60% of the production and *B. bassiana* accounting for 30%, with the remainder being *Lecanicillium, Metarhizium,* and *Isaria.* Even the lower figure suggests a production of 1.5 tons of *B. bassiana* per year from each of the 130 entomopathogen-producing CREEs. It is unlikely that these levels were achieved, even if these figures include the solid substrate (D. Moore, unpublished data). Certified microbial strains are provided to the CREEs by a network of Provincial Plant Health Laboratories (Henderson and Sinfontes, 2010), who also inspect each CREE three or four times a year (D. Moore, unpublished data).

Production methods vary in CREEs, especially in terms of the solid substrate used, which reflects local availability, such as rice, coffee, sugarcane, or maize (Henderson and Sinfontes, 2010; D. Moore unpublished). A range of isolates are produced and used throughout the country as appropriate for local pests. Mycoinsecticides are used as stand-alone controls or in conjunction with other biorational methods, such as pheromones (Henderson and Sinfontes, 2010).

15.6.4. Honduras

The *Fundación Hondureña de Investigación Agrícola* (FHIA) produces a product consisting of conidia formed on rice. It is dried and sold to farmers who wash the conidia off and apply them as a water spray (J.T. Kabaluk, personal communication). With basic personal protective equipment, this method is quite safe, although buying a liquid formulation would remove any respiratory hazard presented by powder. Production was based around standard texts (e.g. Jenkins and Goettel, 1997; Grace and Jaronski, 2005), but the need for customizing production according to isolate was emphasized. Consequently, the need for the country's own production manual was stressed. Tap water was found to be better than purified water for production; this could well be a very localized phenomenon because tap water is highly variable, but it demonstrates that customizing production is very important. Suggestions for improving production from J. T. Kabaluk (personal communication) included giving the laboratory coordinator experience in an established mycoinsecticide laboratory, obtaining an autoclave for sterilization, and developing a dry product with long shelf-life.

15.6.5. Kenya and South Africa

Microbial pesticides in Africa are mainly available in areas of good horticultural production, including for export, with Kenya and South Africa being the leading champions. Although the present market for fungi is small (less than $2.0 million of biopesticide sales in Kenya, with most sales for Bt), the environment is positive (Gwynn and Maniania, 2010). Products may be of indigenous or exotic isolates as the regulatory environment is quite pragmatic, but fungi production is generally achieved by conventional methods. However, some liquid formulations involve innovative techniques and some solid substrate techniques have employed clay beads.

Although there are many descriptions of the mass production of entomopathogenic fungi in LICs, the major references cited in this chapter (and references therein) cover most of the issues. Jenkins et al. (1998) and Cherry et al. (1999) gave possibly the most detailed accounts of mass production in commercial-scale situations. Jenkins et al. (2007) illustrated scaling up, whereas Grace and Jaronski (2005) emphasized the central truth that systems must adapt to the isolate, not the other way around.

15.7. OTHER SYSTEMS

Some innovative schemes use less common techniques. Mass production inside coconuts exploits the sterility and complex nutritional nature of coconut water. Difficulties involve maintaining sterility. Although good production is possible, the logistics make it impractical for most circumstances. Other substrates have been used as mentioned previously, including waste from aquaculture and poultry units and the far more conventional waste from breweries. For example, Torres et al. (1993), working in Peru, used wheat, pearl barley, moss (also in combination with prawn residues), coffee residue (postpulping), sugarcane residue, maize, and brewery residue. Torres et al. (1993) referred to coworkers focusing on residues of local crops of another area in Peru, such as inflorescences and grains of *Chenopodium quinoa* (Willdenow) and *Amaranthus caudatus* L., as well as husks of barley oats and wheat.

A rare attempt at low-intensity production in liquid culture was described by Srikanth and Singaravelu (2011). *Beauveria brongniartii* (Saccardo) Petch is produced on dilute molasses broth medium in autoclavable polyvinyl chloride bottles or glass jars. The fungal mats that grow on the surface are harvested after about 3 weeks. Another system using *B. brongniartii* involved collecting the larvae of the Andean potato weevil, *Premnotrypes* spp., and placing them in fungus-infected soil (Cisneros and Vera, 2001). A short while later, infected insects and soil were applied to the soil of local potato stores. This is a very niche-oriented method and could well be one of the few examples of truly local production that could be sustainable. One interesting feature was that the fungus survived for long periods in dried cadavers.

Some success has been achieved at the experimental level with producing conidia on nonnutrient substrates (this assumes that most substrates supply a valuable nutrient source). Yields of up to 3×10^8 conidia per square centimeter of cloth were harvested from absorbent cellulose cloth inoculated with a culture containing brewer's yeast and sucrose (Jenkins and Goettel, 1997).

15.8. MASS PRODUCTION OF BACULOVIRUSES

The production of baculoviruses (BVs) has been reported to be established in a number of LICs (Table 15.1). The commercial production of baculoviruses has been entirely through in vivo production in insect hosts. Although in vitro production has been intensively researched and is seen by many as a key technology to expand BV use, large-scale production systems for BV have yet to be developed enough to be commercially viable (Granados et al., 2007; Szewczyk et al., 2006; Moscardi et al., 2011; Reid et al., Chapter 13). A significant difference between BV production in OECD and LICs is that mechanization in production is used widely in Europe and the United States by producers, reducing the need for expensive labor. In LICs, such labor costs are much lower and capital is often harder to access, so labor-intensive production

TABLE 15.1 Production of Baculoviruses (BVs) in Less Industrialized Countries Either on Commercial Scale or Pilot Production*

BV Species	Countries with Commercial Production	Countries with Pilot or NGO Production
Adoxophyes orana GV	Thailand	
Amsacta albistriga NPV		India
Anticarsia gemmatalis MNPV	Brazil, Mexico	Paraguay, Bolivia, Argentina
Autographa californica MNPV	El Salvador, Guatemala	
Biston suppressaria NPV	China	
Autographa californica mNPV + *Spodoptera albula* NPV	Guatemala	
Cryptophlebia leucotreta GV	South Africa	
Cydia pomonella GV	Argentina	Chile
Dendrolimus punctatus CPV	China	
Ectropis oblique NPV	China	
Erynnyis ello GV	Brazil	Columbia
Gynaephora sp. NPV	China	
Helicoverpa armigera NPV	China, India, Kenya, Thailand,	Vietnam
Helicoverpa zea NPV	Mexico	
Homona magnamina GV	Thailand	
Hyblea peura NPV		India
Leucania separate NPV	China	
Phthorimaea operculella GV	Bolivia, Peru	Venezuela
Pieris rapae GV	China	
Plutella xylostella GV	China, Kenya	
Pseudaletis separate GV	China	
Spodoptera albula mNPV	El Salvador, Guatemala	
Spodoptera exempta NPV	Tanzania	
Spodoptera exigua NPV	China, Mexico, Thailand	Nicaragua
Spodoptera frugiperda mNPV	Brazil	Nicaragua

TABLE 15.1 Production of Baculoviruses (BVs) in Less Industrialized Countries either on Commercial Scale or Pilot Production*—cont'd

BV Species	Countries with Commercial Production	Countries with Pilot or NGO Production
Spodoptera litura NPV	China, India, Thailand	
Spodoptera litura NPV + cypermethrin	China	

Abbreviations: GV, granuloviruses; NGO, nongovernmental organizations; NPV, nucleopolyhedrovirus.
**Key sources,* *Sosa-Gómez (2008), CPL consultants (2010), Yang et al. (2012).*

systems are more common and more financially viable, especially in small start biopesticide companies that have appeared in countries such as India.

In vivo production relies on the controlled infection of susceptible hosts and their subsequent rearing to allow optimal multiplication of the BV. The insects are then harvested and the infective occlusion bodies extracted. This in vivo system is cost-effective because BV systemically infects a high proportion of the host cells, so the production of viral occlusion bodies can reach 15% of the dried weight of the cadavers. This level of multiplication is not seen with other families of insect viruses, which have more restricted tissue multiplication and thus lower productivity per insect. In North America and the EU, such systems were developed during the 1970s and 1980s, notably for *Helicoverpa (Heliothis) zea* (Boddie) NPV (Ignoffo, 1973), and *Lymantria dispar* L. NPV (Shapiro et al., 1981), and the generic issues of industrial production were identified (Shapiro, 1986; Shieh, 1989). These basically simple in vivo systems for BV have been adopted in South America, Asia, and Southeast Asia for producing a range of BV—mostly NPV but also some granuloviruses (GV). However, production in Africa has only begun within the last few years. In India, baculoviruses, particularly the NPVs of *Helicoverpa armigera* (Hübner) and *Spodoptera litura* F. (SlNPV), have been reported to be effective on different crops, and commercial-scale production of NPVs began in the late 1990s (Rabindra et al., 2003).

Detailed descriptions of commercial production systems for BV are scarce. A few have been produced by public sector researchers (Shapiro et al., 1981; Grzywacz et al., 2004; Moscardi et al., 2011) as generic training material for small-scale commercial producers, but only one has been produced commercially (Van Beek and Davies, 2009). A comprehensive source of technical advice on the production and handling of baculoviruses was gathered together in Hunter-Fujita et al. (1997). The BV production system contains a number of separate components, including insect rearing, insect infection, virus multiplication, insect harvest, processing, and formulation. A key element to successful systems is the quality control needed at all stages to ensure that production is continuous and efficient (Jenkins and Grzywacz, 2000). Although the system is simple in concept, a high level of competence and close attention to protocols is essential if a quality product is to be produced cost effectively.

The quality and quantity of larvae used for the production of BV are a major factor in in vivo production. A number of studies have established that the optimum age/weight of larvae is crucial in production efficacy for individual species (Cherry et al., 1997; Grzywacz et al., 1998; Senthil Kumar et al., 2005; Biji et al., 2006). If larvae are inoculated when either too small or too large, production of occlusion bodies is reduced. Mass production is thus optimized if there is a controlled supply of larvae available at a specific standard age/weight. To achieve this consistently, most commercial operations use a supply from custom-built larval production plants that ensure an adequate supply of healthy larvae. Attempts to use insects collected from the wild have been tried (Ranga-Rao and Meher, 2004); although this is a potentially cheap solution, it can compromise production by inadvertently including other insect species in which the BV will not replicate. It may also introduce contamination by other insect pathogens whose presence may interfere with successful BV replication. The product from such wild-collected insects is not recommended unless the insect supply is adequately monitored and selected.

The rearing of large numbers of healthy standardized larvae over a long period is a challenge. Such cultures are vulnerable to infection by other lethal parasites or pathogens; consequently the culture can collapse and halt virus production. Culture lines can also be contaminated by parasites or pathogens, which have a more subtle impact, causing gradual loss of vitality and a decline in target BV production. The systemic spread in cultures of parasites such as *Nosema* and *Variamorpha* spp. are often the cause of gradual colony decline in some small-scale BV production laboratories (Grzywacz et al., 2004). This also can have a serious impact on BV purity and productivity, as harvested insects come to contain increasing levels of the contaminant and reduced levels of the target BV. To ensure a healthy culture for production, it is essential to initiate development from clean insect lines and to rigorously apply adequate sanitation measures to exclude contamination once a clean culture is established. The selection and selective rearing of clean culture lines is an important task that can be very time-consuming. Insects from the wild are commonly infected with a variety of pathogens and parasites, and it may take many months of careful rearing to eliminate these by selective breeding (Grzywacz et al., 2004). When starting new production systems, accessing already established clean cultures is recommended and can save a great deal of time and effort. To guard against catastrophic culture contamination, rearing the insects at a physically separate facility to the main virus production facility is often adopted. Establishing separate backup insect cultures with research partners or collaborators is also strongly advised as insurance against problems with the main colony.

The provision of a hygienic, cost-effective insect diet is important for many small-scale or low-technology producers. The use of diet from grown plant material may be needed by some species that are not adapted to artificial diets, but such material needs careful monitoring because it could bring in pesticide residues or pathogens that are dangerous to the health of the insect production culture. In India, some use is made of soaked legumes, such as chickpeas, as a food source for insects such as *H. armigera* (Ranga-Rao and Meher, 2004). The preferred option for insect rearing is an artificial diet made from standardized ingredients and preferably heat sterilized; details of diets for various species are available (Singh, 1977; Singh and Moore, 1985; Anderson and Leppla, 1993). Diet cooking protocols that involve autoclaving have the advantage of effectively sterilizing as well as cooking insect diet. Diet production costs are a major item in of in vivo BV production, and a very significant part of this can be the gelling agents used in standard diets. Agar is the agent of choice in research cultures, but this is expensive and many producers have developed cheaper solutions using gelling agents from local food industries.

A very important issue is the quality of the virus inoculum. The use of highly purified inoculum of the selected strain is one of the most important measures contributing to trouble-free virus production. Inoculum should ideally be purified by established centrifugation protocols (Hunter-Fujita et al., 1998) and the genetic identity checked using restriction endonuclease or polymerase chain reaction analysis. This is not only to ensure that the selected strain alone is produced but that other potentially contaminating viruses are excluded. Some BV species, such as the GV of *Spodoptera littoralis,* can, when in a mixed inoculum, infiltrate cultures and outcompete the faster killing NPV, thus contaminating the product and lowering its performance (Hunter-Fujita et al., 1997). The recycling of product as inoculum without purification and strain verification should be avoided because this can result in the buildup of contaminating parasites and pathogens; it is a flawed approach that has probably done the most to degrade BV product quality of small producers in the past (Kennedy et al., 1999). Once again, selected strains should be lodged in separate culture collection facilities both as a product reference and as a backup in case of accidents at the main facility.

Effective multiplication of viruses in vivo is usually dependent upon the provision and maintenance of appropriate rearing conditions for the host. Temperature and humidity need to be kept within defined limits to avoid stressing infected insects such that they die before virus replication is completed. Although some small-scale producers rely on ambient conditions in rearing facilities, this can have very deleterious consequences for product quality, especially during periods of high humidity. One interesting development to increase productivity is the use of juvenile hormones in the insect diet to improve the insects' growth and BV productivity (Lasa et al., 2007). Inoculated insects need to be reared in containers during the virus multiplication period. In production in northern countries, disposable container handling systems are common to facilitate good hygiene. In LICs, low-cost producers often opt for reusable containers, which require effective cleansing procedures if contamination is to be avoided.

Some insect species, such as *Cydia pomonella* (L.), can be reared communally during virus multiplication. However, some of the most important hosts, such as *Heliothis/Helicoverpa* species and *Spodoptera* spp., are cannibalistic in later instars, so they must be reared individually to avoid excessive loss of virus. Compartmentalized trays are generally used to rear such insects, and a number of different designs have been developed and adopted (Fig. 15.3), which are then incubated in stacks (Fig. 15.4), although detailed descriptions are sparse (Grzywacz et al., 2004; Van Beek and Davies, 2009). As often is the case in BV production, details of these systems and their relative merits are usually not publically available because producers rightly perceive these to be crucial commercial intellectual property that underpins market advantage.

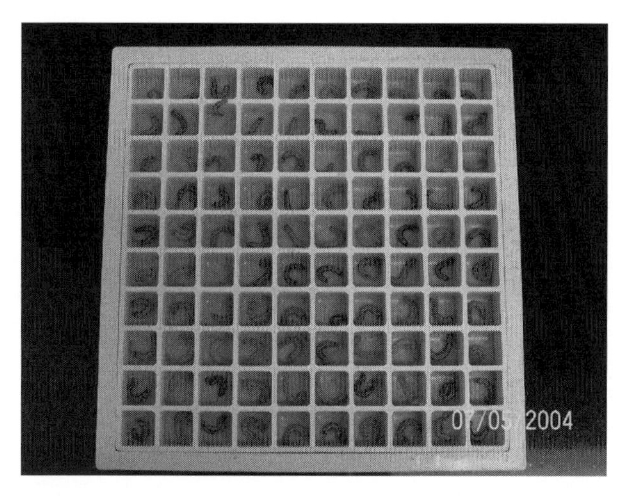

FIGURE 15.3 Compartmentalized tray used for rearing baculovirus infected insects.

FIGURE 15.4 Compartmentalized rearing trays being incubated in stacks to mass propagate baculovirus (U. Kentunuti).

Infected larvae are harvested once BV multiplication is complete; in most tropical pest species, 5–7 days is sufficient to reach optimal BV production (Cherry et al., 1997; Gupta et al., 2007; Van Beek and Davies, 2009). The harvesting of the larvae is commonly accomplished by whole containers on a preselected day. However, in some small production units, optimally infected insects are harvested individually over 2–3 days. Harvesting insects at or near death can maximize virus occlusion body production while limiting the microbial contamination load (Grzywacz et al., 1997). Although there is little evidence that such contaminants pose a significant risk (Podgwaite et al., 1983), registered virus products usually have maximum allowable levels of bacterial contamination that must be adhered to. In virus-infected insects after death, saprophytic bacteria multiply quickly, so harvesting delays increase the load of contamination (Lasa et al., 2008). The harvesting of infected insects can be done with suction devices or automated harvesting systems, but individual hand collecting of infected insects is also common. In some systems, dead insects, diet, and feces are all harvested and processed together, which may exacerbate product contamination issues.

Once harvested, the cadavers and virus are frozen and stored prior to processing and formulation. Exact details of specific processing and formulation systems are sparse from commercial sources, although information from public sector producers is more available (Grzywacz et al., 2004: Van Beek and Davies, 2009). Larger insect parts, such as skin and jaws, are generally removed during processing by filtering to avoid blocking spray apparatus, but other insect debris is almost always left in formulations because it is known to improve the UV stability of BV products (Burges and Jones, 1998).

Some BV systems include a primary centrifugation step to further remove gross insect debris prior to preparation of BV for formulation. The subsequent formulation of BV varies between producers; a generic discussion of formulation issues is presented in Burges and Jones (1998) and also can be found in Behle and Birthisel (Chapter 14), but details of formulations used by producers are generally kept secret for commercial reasons. Early formulations were simple aqueous suspensions of homogenized filtered insects. These aqueous suspensions needed to be stored frozen and distributed in cool chains, which was a significant barrier to commercial uptake in LICs. To improve storage and handling characteristics, some production is as wettable powders formulated with clays or talc or to more stable liquid suspensions that often incorporate glycerol as an antibacterial carrier (Lasa et al., 2008).

Crude suspensions of NPV need drying to prevent loss of activity or adverse effect on storage characteristics; air drying, freeze drying, and spray drying are all possible (Tamez-Guerra et al., 2002; Grzywacz et al., 2004). These formulations can extend active shelf-life at ambient conditions and generally benefit from freezing/refrigeration for long-term storage; they are also more amenable to short-term ambient storage prior to use by farmers. However, it remains a challenge to produce BV formulations that could match the 2-year stability at ambient temperature that remains the benchmark for chemical pesticide stability.

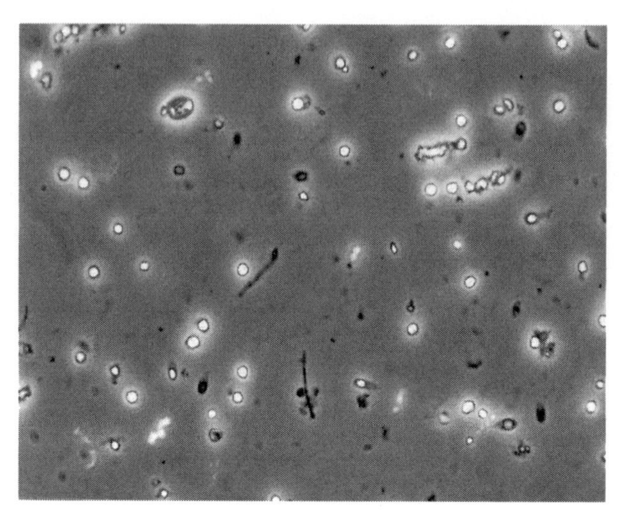

FIGURE 15.5 Infective particles or occlusion bodies (OB) of nucleopolyhedrovirus under ×400 phase contrast microscopy. OB are seen as bright refractive crystals.

Successful BV production needs an effective quality control system, but there is no single method for quickly and accurately determining BV activity. The physical numbers of virus occlusion bodies (OBs; Fig. 15.5) can be counted by standardized microscopical methods (Evans and Shapiro, 1997a, b), but activity can only be estimated through insect bioassays. A number of standardized bioassay methods for insects exist (Evans and Shapiro, 1997a, b; Jones, 2000), but establishing and maintaining a bioassay program is demanding. However, it is only by incorporating real-time counting of OB and bioassays that production quality can be continuously monitored and failings or problems in production can be detected and resolved. One solution for smaller producers is to link up with research institutes to carry out routine quality control operations, such as a system used in Brazil where *Empresa Brasileira de Pesquisa Agropecuária* (EMBRAPA) operates a quality control laboratory for a number of production companies (Moscardi, 2007).

15.8.1. Country Case Studies

15.8.1.1. China

Although China might have been classified as a developing country at the end of the last century, today it is clearly not so currently; however, it seems possible that China is the greatest user and producer of BV worldwide (Yang et al., 2012) and is a very interesting case study in the scale-up of BV use. The situation of BV production in China was reviewed by Entwistle (1998) when BV production was recently established. BV production has since expanded significantly, with at least 12EP virus species registered (Sun and Peng, 2007; Yang et al., 2012). Production has been estimated to be 500 metric tons of product, although most of this is *H. armigera* nucleopolyhedrovirus (HearNPV) products (Xiulian Sun, personal communication). Other authors have estimated the total of product material may have reached 2000 metric tons (Wang and Li, 2010; Yang et al., 2012). However, details of production systems, quality, and regulation (which would certainly prove to be extremely interesting) are poorly reported in non-Chinese literature. One unusual feature is the production of combination Bt and BV products in China; these include combinations of GV plus Bt against *Pieris rapae* (L.), NPV plus Bt against *S. litura,* and NPV plus Bt against *Biston suppressaria* (Guenée) (CPL, 2010). The effectiveness of these combinations products is unclear, and their sustainability and genuine utility in IPM or IRM packages remain to be demonstrated.

15.8.1.2. India

There has been a considerable body of research into microbial pesticides in India, and a wide range of pathogens have been investigated in a variety of cropping systems (Koul et al., 2003) dating back to the 1960s. There are at least 32 commercial companies active in biopesticide production, with additional IPM centers under the Ministry of Agriculture producing selected biocontrol agents (Rabindra and Grzywacz, 2010). In India, a number of BV were researched, but the usefulness of only a few has been demonstrated on a field scale. These include the NPVs of *H. armigera, S. litura, Amsacta albistriga* (Walker), the teak defoliator *Hyblea puera* (Cramer), the GV virus of the sugarcane borer (*Chilo infuscatellus* Snellen), and diamondback moth *P. xylostella*.

The NPV of *H. armigera* is the most widely studied, followed by that of *S. litura,* due to their nationwide pest status on economically important crops such as cotton, pulses, and oilseeds. During the early period of development, the virus was produced in field-collected larvae, but subsequently EPs were brought under the purview of the Insecticide Act, necessitating registration with the Central Insecticides Board; thus, the emphasis shifted towards production in larvae mass produced with a semisynthetic diet to address the quality issues (Rabindra et al., 2003).

The registration system for EPs has been favorable and flexible. Provisional registration of approved agents is allowed before full dossiers are submitted, which has facilitated production and registration by small and medium enterprise producers (Kulshrestha, 2004; Rabindra and Grzywacz, 2010). The marketing was initially largely through the government subsidy programs; to date, there has been only limited direct sale to farmers. Due to erratic pest occurrence, the demand has not been predictable; therefore, some of the companies that registered their NPVs have since ceased to produce the virus and at present uptake is limited.

Quality standards for products have been established, but the regulation and enforcement is rather poor and has affected consumer confidence (Kennedy et al., 1999; Alam, 2000). Surveys to test the quality of biopesticides are conducted; although some manufacturers clearly meet accepted standards (Kambrekar et al., 2007; Srinivasa et al., 2008), other reports indicate quality concerns, especially from new and inexperienced producers. A system of referral laboratories accredited by the Indian Department of Biotechnology for quality testing has been established, but enforcement of standards remains uncertain (Shetty et al., 2008). The Kerala Forest Research Institute established pilot plant production of the NPV of *Hyblaea peura* (C.) (Biji et al., 2006), but as yet there is no significant market among the growers.

15.8.1.3. Thailand

Biopesticide production in Thailand began in the late 1980s when researchers produced and studied BV and Bt for control of insecticide-resistant pests, such as *H. armigera* and *Spodoptera exigua* (Hübner) (Jones et al., 1998). Field research demonstrated the efficacy of BV for controlling key pests (Kolodny-Hirsh et al., 1997). Subsequently, production facilities for HearNPV, *S. exigua* NPV, and SlNPV were set up by the Department of Agriculture in 1995 at Kasetsart University (Jones et al., 1998; Skovmand, 2007), with capacity of around 1000 l BV product per annum (Ratanasatien et al., 2005). After 2004, the Department of Agriculture collaborated with the National Science and Technology Development Agency to set up a BV pilot plant. The BV pilot plant was established in 2007 in the area of Thailand Science Park, Bangkok. The new pilot plant can expand the production capacity of BV product up to 3000 l per annum. Since 2008, BV pilot plant has also produced BV for the control of tea leaf rollers (*Homona magnamina* [Diakonoff] and *Adoxophyes orana* [Fisher von Rosteramm]) under a collaboration with Arista Life Science Corporation, Japan.

15.8.1.4. South Africa

Research into biopesticides and insect viruses is well established in South Africa, dating back to pioneering work by L.L.J. Ossowaski on the control of bagworm (*Kotochalia junodi* [Heylaerts]) in the 1950s and V.H. Whitlock on *H. armigera* NPV in the 1970s (Kunjeku et al., 1998). A strong entomological research base and the high economic value of the horticultural export sector were undoubtedly important factors that have contributed to the establishment of biopesticide production in South Africa. A major focus was the control of false codling moth (*Thaumatotibia (Cryptophlebia) leucotreta* [Meyrick]), a pest refractory to control by chemicals. Research on the potential of the *T. leucotreta* granulovirus (CrleGV) as a control agent was initiated (Moore, 2002b), which was followed by the establishment of local in vivo production of CrleGV and its subsequent registration by River Bioscience (Moore et al., 2004). Production of this virus is now well established and further research into its efficacy continues (Moore et al., 2011).

15.8.1.5. Africa (Kenya and Tanzania)

In East Africa, both Kenya and Tanzania have been centers of research into BV for pest control. Efforts in Kenya focused on the diamondback moth (*P. xylostella*) and other lepidopteran pests of the horticultural industry (Wabule et al., 2004). In Tanzania, the focus was on the use of endemic BV for control of African armyworm *Spodoptera exempta* (Walker) (Grzywacz et al., 2008). Production and registration of *P. xylostella* GV and HearNPV is underway in Kenya; some details of the production systems are published (Van Beek and Davies, 2009). A facility created for *S. exempta* NPV in Tanzania is due to start production in 2013. The Tanzanian production system will use field production of the virus using a modification of the EMBRAPA *A. gemmatalis* NPV (AgNPV) system; this is deemed technically feasible because of the large synchronous outbreaks of larvae at high densities, which are a feature of this pest (Mushobozi et al., 2005). However, it remains to be determined if this is economically viable or can meet the requirement for the large amounts of virus that will be needed when periodic major outbreaks occur, as these can cover hundreds of thousands of hectares.

15.8.1.6. South America and Potato Tuber Moth

Production of the GV of *Phthorimaea operculella* (Zeller; the potato tuber moth) for control of this pest in potato stores is well established in South America through a series of production centers in Peru, Bolivia, Ecuador, and Colombia (Winters and Fano, 1997; Sporleder and Kroschal, 2008). The larvae used for virus production are reared on potatoes, then placed on fresh potato surfaces inoculated with GV, where they are allowed to grow for 20–30 days for infection to develop and the GV to multiply. The dead larvae are harvested, then macerated and mixed with a clay or talc carrier, which is then air dried and powdered to produce a formulation. Production is carried out at the International Potato Center in Peru, PROINPA in Bolivia, and various commercial producers; it is sold under a variety of trade names, including Matapol. This product is mainly targeted at protecting seed potatoes in the nonrefrigerated stores of small-scale farmers. Thus, total production is presently limited, but expansion to larger scale producers is underway (Lacey et al., 2010). Production of other BV is established in a number of South American countries (Table 15.1); however, published details of production are lacking, so detailed discussion of the production is not possible (Oliveira, 1998; Moscardi, 1999; Sosa Gomez et al., 2008). One very interesting development is the production by the Agricola el Sol company of Guatemala of a BV product containing both the NPV of *Autographa californica* (Speyer) and the NPV of *S. albula* (W.).

15.9. OTHER PRODUCTION SYSTEMS

Most BV biopesticide products are produced by infecting and rearing insects from laboratory cultures. However, with some pests, the concept of infecting and producing the BV from insects in the wild is feasible for mass production. Many host species are seemingly not suitable for this approach, but it has been commercially viable for a few species. Insects that have larvae appearing at a high density, and for which generations are highly synchronized and visible, are probably the most suitable for this system. The system may also be adopted for species in which a synthetic rearing diet either has not been developed or is too expensive for commercial mass production.

The largest and best documented field production system was developed in Brazil by EMBRAPA for *A. gemmatalis* NPV, in a program that produced enough BV to treat 1 million ha of crops (Moscardi, 1999). The system involves scouting for natural pest infestations in farmers' fields where high concentrations of the target pest are identified at a suitable stage, and then these insects are used for production. At an appropriate larval stage, these chosen pest outbreaks are sprayed with AgNPV and the plots inspected daily. When the peak appearance of virus-killed larvae occurs, pickers are sent to harvest infected larvae. Peak harvest is 8–10 days postinoculation and pickers are paid per volume of larvae collected. Production areas may require 200–300 pickers and may yield up to 600 kg of larvae per location (Moscardi, 2007). The franchising of this system by EMBRAPA to a number of private producers underpinned an expansion of AgNPV production, which involved the collection of up to 20 tons of caterpillars. Infected larvae were formulated with a kaolin carrier and air dried to produce enough formulated product to apply to 2 million ha per annum. EMBRAPA remained central to production, acting as a quality control center for the production companies.

The use of insects collected from the field is also part of a very interesting initiative developed in India. To meet the challenge of producing low-cost biopesticides for use by the poorest farmers, a system of "IPM villages" was developed. HearNPV had been shown to be effective in controlling *H. armigera* on a number of crops in India (Rabindra et al., 1992; Visalakshmi et al., 2005) but the high cost of commercially produced NPV discouraged its use by poor farmers. To overcome this, village-level production of HearNPV was established as part of a program of IPM promotion in 2005–2007. This involved collecting *H. armigera* by shaking infested pigeon pea plants and collecting the dislodged larvae. The larvae were then taken to the village where a local NPV multiplication unit had been established (Ranga-Rao and Meher, 2004). The cost of the simple equipment needed was about US $500 and consisted of larval-rearing containers, a homogenizer, a simple centrifuge, and the facilities themselves, which were shared by participating farmers (Fig. 15.6). During the program, 96 village NPV production units were established in India and Nepal (Ranga Rao et al., 2007). However, despite this progress, it has been suggested that such farmer production has issues ensuring the financial sustainability of these initiatives in the absence of outside support (Tripp and Arif, 2001). A simple field production technique was also adopted for propagating the NPV of the red hairy caterpillar, *A. albistriga*. The virus was applied on the crop that was heavily infested by the larvae, after which the diseased larvae were collected and processed (Veenakumari et al., 2007). This use of the NPV by the local community and self-help groups avoids the necessity for registration, but so far has not spread further due to lack of follow-up programs.

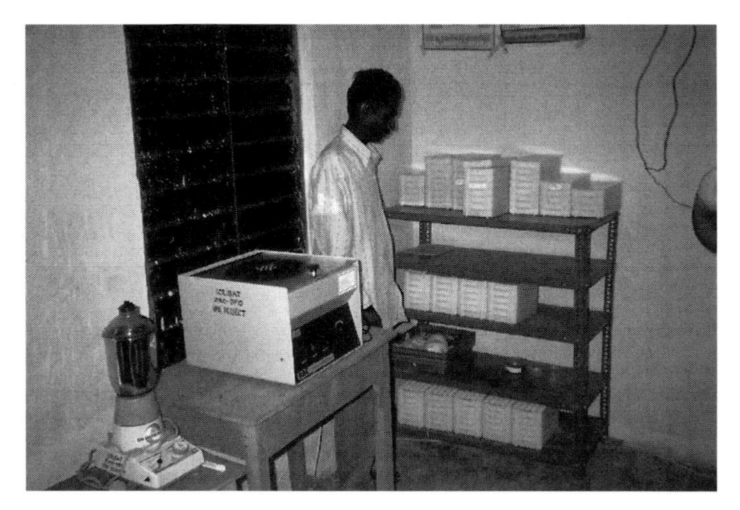

FIGURE 15.6 Village biopesticide production unit in India for production of NPV.

15.10. GENERIC PRODUCTION ISSUES

15.10.1. Product Quality

The quality of biopesticides products in LICs can be extremely variable, which has been a significant issue impairing the uptake of biopesticides in some countries (Alam, 2000). In a survey of three *Metarhizium*, two *Beauveria*, and one *Trichoderma* products (two from China, two from Indonesia, one from India, and one from Colombia), only one product, a *Beauveria* sp., showed low levels of contamination and a high germination level (Jenkins and Grzywacz, 2000). Surveys of BV products have also shown up for products lacking in active ingredient and highly contaminated with bacteria and other microbial contaminants (Kennedy et al., 1999). Objective published data on this issue is sparse, and it is probable that this is a seriously underreported issue in many LICs. The situation can be exacerbated where biopesticides are being sold unregistered and therefore unregulated. Additionally, dealers may market cheap illegal imports because legitimate local supplies are insufficient or because local product registration systems are too slow or uncertain to facilitate registration, causing users to obtain essential inputs illegally.

Large-scale use of biopesticides began in Brazil and China in the 1970s (Li et al., 2010), but many small companies disappeared in the 1980s because of poor-quality products in Brazil and reduced government support in China. Interestingly, neither Brazil nor China published much on mass production systems, which may be a partial explanation for poor quality.

In some countries, the problem has been about production systems, which have been established or rapidly expanded without sufficient attention to quality control issues. Small producers may set up without in-house capacity to monitor active ingredient or microbial contamination. Improved product safety greatly reduces losses resulting from contaminated product being rejected (as well as leading to a failure to supply a product), increasing both financial return and user confidence (Jenkins and Grzywacz, 2003). Such producers could overcome this by linking with local research institutes, such as is the case in Brazil with the EMBRAPA system for AgNPV production. A number of papers detailing acceptable quality standards have been published (Jenkins and Grzywacz, 2000, 2003), providing essential guides for quality control of fungal and viral biocontrol agents.

15.10.2. Product Quantity

It is often underestimated how much fungus may be required for pest control, and variable conidial yields can greatly alter substrate volumes required (Table 15.2). The 25.9 million ha treated in the African locust plague of 1986–1989 would have required 1300 tons of conidia (produced from perhaps 30,000 tons of rice). Posada-Flórez (2008), working with *B. bassiana* against the coffee berry borer *Hypothenemus hampei* (Ferrari) in Colombia, similarly calculated that application to 1000 ha would require 92.3 tons of rice to produce the spores. Further examples are given by Bartlett and Jaronski (1988), who calculated conidial numbers required at standard field rates of 1×10^{13}/ha; extrapolating, the treatment of 4–5 million ha of, in their example, Illinois corn would require 500–1000 tons of *B. bassiana* conidia.

Even where mycoinsecticides have been used on what is considered a large area, the figures reveal a massive scale-up is required. China used *B. bassiana* on up to 1.3 million ha until the 1980s and uses probably half that figure now (Li et al., 2010); this represents 13,000 km^2 out of the 1.4 million km^2 of arable land (with a further 1.8 million km^2 of forestry) in China. More intensive production is required. Li et al. (2010) reported that new private companies equipped with modern large liquid and solid fermenters or other fermentation facilities have begun producing mycoinsecticides in China in recent years. Where scale-up requirements are substantial, the use of low-technology processes is likely to fail. According to Roberts and St Leger (2004) "Difficulties in producing *M. anisopliae* described in Krassilstschik (1888) are woefully similar to those expounded by DWR (Don Roberts) in Brazil in 2002.'' Yields, as shown by conidia per gram, can vary greatly (Table 15.2), but 5×10^{10} conidia per gram of substrate may not be unreasonable by incremental improvements. Some truly innovative approaches may add substantially to this, at which point the economics begin to favor biopesticides overwhelmingly.

For baculoviruses, field application rates may be in the range of 5×10^{11} OB/ha for certain viruses, such as AgNPV, but for most NPVs the field rates lie in the 1–5×10^{12} OB/ha range (Copping, 2009). Even with the relatively low rate used for AgNPV, 40 tons of infected insects were needed to treat 1 million ha of crop in a season (Moscardi, 2007). Thus, the mass production of very large les of Conidial Yields in a Range of Low-Intensity Production Systems

Substrate	Yield/g*	Reference	Notes
Rice bran/husk	2.3×10^{10}	Dorta et al., 1996	Considered 80% of maximum theoretical yield
Rice	2.5×10^{9}	Barajas et al., 2010	
Rice	5.3×10^{10}	Prakash et al., 2008	Yields of 4.6 and 4.2×10^{10} for sc and barley respectively
Rice	2.7×10^{9}	Chen et al., 2009	
Rice	2.2×10^{9}	Chen et al., 2009	Characterized as *M. anisopliae* va *acridum* by Fernandes et al. (200 (*M. acridum*)
Rice	1.5×10^{9}	Jenkins et al., 1998	Small-scale mass production of 3 year
Rice	$1.8–2.7 \times 10^{9}$	Ye et al., 2006	
Sorghum	4.2×10^{9}	Rajanikanth et al., 2010	Six isolates tested, ranging from 2 4.2×10^{9}
Rice + wheat bran	2.9×10^{10}	Dhar and Kaur, 2011	At 35% moisture and 1.5% yeast extract in liquid nutrient
Wheat	1.2×10^{9}	Sahayaraj and Namasivayam, 2008*	Range of grains tested, yielding a 1×10^{9}

:xamples of Conidial Yields in a Range of Low-Intensity Production Systems—cont'd

	Substrate	Yield/g*	Reference	Notes
	Rice	$4.6–5.4 \times 10^9$	Chen et al., 2009	Two isolates examined
	Rice	2.0×10^{11}	Alves and Pereira, 1989	Harvesting method may ha exaggerated yield
	Not specified	2.6×10^{10}	Bradley et al., 1992	Mycotech pilot productior
∕i	Crushed sorghum	2.8×10^9	Devi et al., 2000	Air exchange required
ɔsea	Sorghum	1.0×10^{10}	Sahayaraj and Namasivayam, 2008*	Range of grains tested, yie $0.8–1.0 \times 10^9$
ɔsea	Rice	$1.3–1.8 \times 10^9$	Chen et al., 2009	Two isolates tested
ecanii	Sorghum	1.1×10^{10}	Sahayaraj and Namasivayam, 2008*	

in yield were recorded between text and tables in this reference, but these do not alter general patterns.
j and Namasivayam (2008).

numbers of insects will be needed if BVs are to meet the pest control needs of major field crops in the future. Currently, it is uncertain if existing in vivo systems can be scaled up reliably to meet this need; thus, there is interest in large-scale tissue culture production (Granados et al., 2007). On the other hand, the expansion of BV production in China in recent years may have reached 500 tons per year (Yang et al., 2012), which shows that larger scale production is attainable with existing systems.

15.10.3. Safety

EPs selected for commercial production by virtue of passing a registration system should be inherently safe for use as specified, both as formulations and active ingredients. Therefore, the main safety issues concern hazards in the production of the BCA rather than its use. Generally, the genera of entomopathogenic fungi used in mycoinsecticides are known to be very safe (Zimmermann, 2007a; 2007b, 2008). It is possible to obtain adverse reactions from isolates, but under circumstances that would not occur in field use. Goettel and Jaronski (1997) described serious responses to pulmonary (intranasal) tests of a *M. anisopliae* isolate from Madagascar.

In reality, safety is an issue at production, during formulation and use, and during registration. Conidial powder becomes airborne very easily, presenting a respiratory risk—a longstanding issue in fungal production (Rorer, 1913). Today, this can be controlled through appropriate safety measures. Chinese production units developed vacuum extracting equipment to harvest conidia, reducing allergy problems resulting from inhalation in the 1980s (Li et al., 2010). Subsequent to production, farmers could be exposed to respiratory hazards from product formulated as a powder. Different formulation types can create or reduce hazard. In Brazil in the 1970s and 1980s, *M. anisopliae* was washed from the rice substrate with water for use against the sugarcane froghopper *M. posticata* (Mendonça, 1992). The liquid contained spores, mycelium, rice fragments, starch, and bacteria, which rapidly became a significant contamination problem if the product was not applied rapidly after harvest. The problems of contamination developing and of respiratory hazard of powders can be removed by formulating in oils. The LUBILOSA project used an oil suspensible flowable concentrate—essentially a very concentrated spore sludge—which could be diluted easily with oils to achieve the correct concentration for application.

BV has been characterized as inherently safe for use in pest control (OECD, 2002), so safety issues here concern production and incidental microbial contamination. Because BVs are produced in insects, there are issues around the microbial flora that occur in live insects and proliferate in dead insects. Several studies have characterized these conditions (Podgwaite et al., 1983; Grzywacz et al., 1997; Lasa et al., 2008). To date, none of the studies have identified human or veterinary pathogens, but such microbiological studies are an essential safety measure that should not be neglected. The various systems for low-technology production that do not include microbiological screening as a routine part of quality control must remain a real cause for concern.

15.10.4. Economics

It is often assumed that a labor-intensive model for mass production of mycoinsecticides is the more appropriate model for developing countries and the capital-intensive model is for wealthy countries. Swanson (1997) examined the economics, based on net present value, of two production systems; the labor-intensive LUBILOSA model and a capital-intensive model based on the Mycotech Corporation. The LUBILOSA production system was better for markets of 20,000 ha per annum, whereas the capital-intensive model was more profitable at markets of 80,000 ha per annum. The two main points are that labor-intensive models can be financially viable, but to satisfy potential demand, large production capacity will be required. Most LICs have industries that require a similar level of technology and sophistication (e.g. brewing, pharmaceuticals), and it is likely that any country seriously wishing to use fungal EP will eventually turn to factory systems for the quantities required for broad-acre agriculture. Nonetheless, this may still leave niches for specialized smaller producers. An economic study of Green Muscle use in West Africa suggested that the product was expensive, being produced at US$17 per hectare at 100 g/ha, compared with the chemical competitor sold in the market at US$12 per hectare (Groote, 1997). Industrial-scale production would lead to reductions in cost but, more importantly, the effective dose rate was later reduced to 50 g/ha with greater experience of using the product.

Published studies of the economics of BV production are unfortunately very rare; an exception is the data on the AgNPV field production in Brazil estimating a BV production of $1.28 per hectare, which made it highly competitive with chemical insecticides (Moscardi, 1999; Moscardi, 2007). However, the unusually low application rate for this NPV and the field-based system make it unlikely that this system would be appropriate for many other BV (which need to be produced through mass insect rearing and thus will be much more costly). Generally, to date, BV products have not been produced at a cost less than alternative chemical insecticides. BV production is, for the most part, labor intensive; thus, production in low-cost countries such as China, Brazil, India, Kenya, and South Africa is seen as a feasible way to reduce product costs.

15.11. REQUIREMENTS FOR ESTABLISHING BIOPESTICIDE INDUSTRIES IN LICS

15.11.1. Research and Information

One common feature with mass production in many LICs is a poor understanding of principles and a lack of awareness of fundamental work. With the internet, there should be less of a problem in obtaining sources of information. Although it may be true that some very important information is not freely available, there is sufficient accessible material that will allow mass production to be carried to a reasonable level while the isolates are being researched for optimization.

With fungi, a major issue is the need to have solid substrate for the production of aerial conidia, which is the basis of most production processes (Roberts and St Leger, 2004). Although liquid systems have been developed, the logistics of many supply chains require better shelf-life than many liquid systems provide at present. Although blastospores can show improved shelf-life while retained in their original production liquid, there is a belief that these propagule types are not of practical value (Li et al., 2010), despite some evidence for a number of species that good field results can be obtained. The more exotic propagules, such as microsclerotia and submerged conidia (Jaronski, Chapter 11), may well have a valuable role in the future, but major advances could be obtained from properly understanding the solid substrate.

There is a belief that the substrate has a vital role in supplying nutrition to a fungal culture, beyond that provided by a liquid nutrient culture, but the substrate's physical characteristics may be more important. Very little is known about the physical requirements, the relationships between surface area, and appropriate spaces to optimize conidial production. Relatively poor production from coarse material, such as ears of cereals or other husk material, may well be due to physical space more than an inadequate carbon:nitrogen ratio. Particles the size of cracked rice or millet usually seem to provide good spaces, but increasing the surface area of a substrate may result in disrupting the benefits of a correct particle size. If a solid substrate could be optimized for surface area and air spaces when stacked and made of an easily sterilizable material so it can be recycled, a number of problems could be solved.

Certain themes are apparent with promoting improved biopesticide production in LICs (Table 15.3). In countries where markets are relatively large (e.g. India, China, Brazil) or where the export horticulture sector provides an active market for biopesticides and biocontrol agents (e.g. Thailand, Kenya, South Africa), progress in developing an EP industry has been most significant. Government support was pivotal in establishing most biocontrol programs in countries such as India, Brazil, and China. Funding for research and development of products has been effective where it brought together researchers and industry into productive partnerships. Government/public support is necessary to allow biopesticide businesses to establish. This is clear at research levels to optimize products, support biopesticide businesses, and create a commercial environment where biopesticides are viewed positively for their many advantages (Marrone, 2007).

In Brazil, India, Thailand, and China, national research systems had significant specific biopesticide research programs aimed to generate new knowledge, build local expertise and research capacity, and disseminate research findings to industry (Wahab, 2004). The Indian Council of Agricultural Research, Department of Biotechnology, of the Ministry of Science and Technology supports xamples of Features Promoting Improved Mycoinsecticide Production, Their Registration, and Adop

Feature	Reference
Government support essential for most biocontrol programs at initial stages	Li et al., 2010
Avoid contamination by lowering pH with lactic or acetic acids to suppress bacteria	Prior, 1989
Avoid contamination by lowering pH with 3% lime water to suppress bacteria	Li et al., 2010
Avoid contamination by incorporating large amounts of inoculum into the substrate	Prior, 1989
Avoid contamination by incorporating selective fungicides into substrate	Prior, 1989
Lower costs by use of locally made polypropylene carrier bags, which are autoclavable instead of laboratory standard bags	Jenkins et al., 1998
Reduce production costs by applying inoculated rice in sugarcane, allowing conidiation to occur in the field	Bibi Ali, personal commu
Use tap water instead of purified water for increased blastospore production at liquid phase, in Honduras	Todd Kabaluk, personal communication
Substrate inoculum in bags done above a pot of steaming water if no laminar flow is available	Torres et al., 1993

mmon fungi such as *Beauveria bassiana* and *Trichoderma* do not need ecific safety data in Ecuador	S. Jaronski, personal communic
China and Brazil, many products are sold and used without registration	Li et al., 2010
e Pest Management Regulatory Agency does not charge a fee for biopesticide gistration and the right to sell in Canada. Also, there is no requirement for idue studies in crops because generally there are no maximum residue limits biopesticides	Kabaluk et al., 2010b
alified presumption of safety may be granted on grounds of taxonomic ouping, intended use, and pest knowledge or organism	Kabaluk et al., 2010b
rean government is aiming for a 40% reduction (from 2004 levels) in the use chemical insecticides by 2013	Jeong Jun Kim et al., 2010
nistry of forestry pest management in China required to use biological control easures	Bin Wang and Zengzhi Li, 2010
conidiated rice can be distributed at the early growth stage, with conidiation curring naturally in the field (in a humid environment/crop) (see above)	Jenkins et al., 2007; Bibi Ali, pe communication

research and development projects for EP products. The focus has been to enhance quality. The Department of Biotechnology is providing funds to the public sector for generation of toxicological data and has made a provision for sharing the data generated among several potential entrepreneurs, so that the cost of registration can be reduced substantially. International donor support has also been important; the LUBILOSA program was funded by a number of donor countries, and the scientific results were published and made available so that both Brazil and China benefited from this program. The UK Department for International Development, U.S. Agency for International Development IPM programs and other European donors have funded various biopesticide projects in Asia and Africa (Sweetmore et al., 2006; Skovmand, 2007).

Having locally available researchers with biopesticide expertise is a major factor in promoting the development of biopesticide companies. In India, Thailand, and South Africa, the presence of a core of active biopesticide researchers in academia, national research centers, or international research centers played a key role by promoting biopesticides, influencing policy makers, providing competent contract research services, and training students able to join the industry. In addition, development funding schemes, which enabled northern research expertise to be made available to support local researchers and industry, were also of benefit. Research institutes in LICs are often weakest at scale up and manufacturing, which is essential to turn research of local biocontrol agents into practical biopesticide products (Harris and Dent, 2000). Although there are often nuclei of researchers doing early-stage research on identifying new biocontrol agents and preliminary laboratory and field evaluations, the specific expertise to develop these agents into products can be absent (Grzywacz et al., 2009).

Existing research funding and peer review reward systems can focus overly on academic research in these early-stage activities. Many in the biopesticides industry believe that researchers may have limited experience at working with the private sector. This leads to both sides having inappropriate expectations and reduces the value of collaborations between industry and academics—a situation not confined to LICs alone (Lisansky, 1997). There is a clear need for research funding agencies to build researchers' capacities, not only in technical aspects of research but also in enterprise skills; in this way, research can more effectively be translated into practice and products that are beneficial to the wider community. Some agencies have specific programs to facilitate academia–industry collaboration; these have been effective in some LICs (Wahab, 2004).

However, existing research is creating improvements. Ye et al. (2006) described the development of a solid-state apparatus consisting of an upright multitray conidiation chamber (and explained how a 200-m^2 factory could produce material for 80,000 ha). Using this apparatus in series, Chen et al. (2009) produced six different isolates of EPs simultaneously and achieved yields of some isolates that were three times that of Ye et al. (2006). Optimization of moisture content and yeast extract concentration enabled Dhar and Kaur (2011) to give precise figures for these requirements with different substrates; optimization of pH and moisture content for other substrates can also be accomplished (Prakash et al., 2008). The figures can vary greatly among substrates (Table 15.2); for example, moisture contents of 73% and 22% were optimal for production on sorghum and rice, respectively (Prakash et al., 2008).

15.11.2. Registration and Regulation in LIC

Commercial companies are clear that registration issues in many countries remain a significant barrier to expanding the role of biopesticides (Ehlers, 2011; Ravensberg, 2011). Products are not produced or registered in many smaller markets if registration requirements are too costly or the processes are uncertain or lack transparency. An issue in many countries has been that registration systems are focused on the registration of synthetic chemical pesticides and authorities lack the expertise or experience in EPs, biocontrol agents, and biopesticides. One pragmatic solution—apparently quite common in some countries—is to sell unregistered products. This can all too easily lead to the appearance of spurious, fake, and poorly performing products, which can quickly result in a reputational loss and kill customer confidence in biopesticides (Alam, 2000). Although simple, this approach is far from ideal. However, this issue noted by Alam (2000) is not just one of registration; rather, it is regulation and enforcement. Registration alone does not prevent pirated products from being sold illegally.

The adoption of registration systems that fast-track biopesticides and take a pragmatic and flexible approach, including harmonizing registration requirements with other countries and acceptance of waivers and publicly available data, is a necessary step to improving the supply of biopesticides in many LICs (Grzywacz et al., 2009). Biopesticides do not readily fit into a standard pesticides registration model, so registration systems in LICs require adaptation to fit the specific characteristics of biopesticides. There is evidence that the adoption of registration protocols that are favorable to biopesticides has greatly increased the supply of new products and stimulated the local biopesticide industry in some LICs (Cherry and Gwynn, 2007; Gwynn and Maniania, 2010). Fundamentally, for some species, registration should be easy, based only on a detailed characterization of the isolate to be used; if these are known species of *Beauveria, Metarhizium*, or *Trichoderma*, this should be enough to allow sale and field use. For all other fungal species, a simplified acute oral and dermal test may be required. Various regulatory models are presented and discussed in Kabaluk et al. (2010a), who note that different regions have different philosophies; South and Central America work with a belief that indigenous organisms are safer than the chemicals they replace; Cuba and parts of Asia prioritize the protection of consumers and farmers while encouraging local industries; and Europe is employing Qualified Presumption of Risk for organisms from specific groups.

However, regardless of the registration system, existing small market sizes are serious constraints. In Africa and many parts of Asia and South America, national markets for all pest management products are small, which inhibits the registration and commercialization of all biopesticides. There is a need to develop larger markets through harmonization of regulations and trading blocs (Cherry and Gwynn, 2007). It seems likely that biopesticides were able to become established commercially in Brazil, India, and China partly because these countries represented large trading opportunities with many product niches, all accessible through a single regulatory system.

15.11.3. Responsibility

Although registration should be quick, cheap, and easy, subsequently there needs to be greater emphasis on the quality and efficacy of the products. Regulation should be very stringent to ensure that products are of the specified quality and meet label specifications, and all products in use should be tested regularly. This must be effective enough that users become confident to apply the products. Finally, responsibility by manufacturers for their products should be comprehensive and stringent. Producers must be responsible for significant problems, including health issues and deviations from label claims; a simple and cheap registration process should not allow producers to avoid responsibilities for their products.

Government/public support is necessary to allow biopesticide businesses to establish. This is clear at research levels to optimize products, provide support for biopesticide businesses, and create a commercial environment where biopesticides are viewed positively for their many advantages (Marrone, 2007). In this context, it is notable that in India, the Department of Biotechnology, some state governments, and the Indian Council Agricultural Research have worked to implement a range of pro-biopesticide policies, including low-cost provisional registration and supported biopesticide research and IPM programs that promote biopesticides to farmers; these efforts have facilitated EP industry start-up and expansion (Rabindra and Grzywacz, 2010).

15.11.4. Future

Production of EPs in LICs, as illustrated by the cases of fungal EP and (to a lesser extent) BV, is now established in parts of Asia, Africa, and South America. However, production is not yet on the scale that is required to meet the needs of agriculture over the next decade as the use chemical pesticides become increasingly restricted. Although many of the producers have started with a low-technology production deemed to be appropriate for LICs, this approach may be a key constraint to expanding production to the scale needed. Although it is possible to produce high-quality material using simple systems, almost invariably sophistication of knowledge and equipment is required in maintaining a pure stock culture, ensuring sterility, achieving a safe harvest, and ensuring a good shelf-life. Despite the various models of production and local production from microproducers that can be profitable (and hence sustainable), the potential need to replace chemical pesticides on major field crops if legislation further restricts the use of chemical pesticides will require mass production of EP on the industrial scale. For some types of products, such as several fungal EP, good mass production systems with a comprehensive knowledge base capable of supporting large-scale expansion already exist, although knowledge of these has not yet spread widely to producers in LICs. However, for other promising agents, such as BV, reliable large-scale production systems remain to be developed further. For all EPs, it is not just the adoption of good production norms but the rigorous application of effective quality control that are the vital components for success.

Any increase in the adoption of EPs requires a shift in thinking away from seeing EP products only as straight substitutes for chemical pesticides. They should be seen as components in holistic, ecologically sound IPM systems (Moore, 2002a, b). This approach has already appeared in the export horticulture sector of some LICs, where maximum residue level requirements and consumer pressure have driven the adoption of IPM.

Despite these constraints, good biological pesticides can be more effective, economic, and more environmentally sustainable than chemical insecticides, and their use in LICs is increasing. The changing face of global agriculture is opening new niches as well as broad-acre markets for EPs, and this offers new opportunities for farmers in LICs to produce and use EPs. To achieve success, however, governments need to enact and implement new pro-biopesticide policies that facilitate EP production, registration, and use.

ACKNOWLEDGMENTS

Thanks are due to Todd Kabaluk for the provision of a number of interesting reports and Lesley Ragab for detective work with some early papers.

REFERENCES

Alam, G., 2000. A Study of Biopesticides and Biofertilisers in Haryana, India Gatekeeper series no. SA93. International Institute for Environment and Development, London.

Alves, S.B., Pereira, R.M., 1989. Produção do *Metarhizium anisopliae* (Metsch.) Sorok e *Beauveria bassian* (Bals.) Vuill em bandejas. [Production of *Metarhzium anisopliae* (Metsch.) Sorok and *Beauveria bassiana* (Bals.) Vuill in plastic trays]. Ecossistema 14, 188–192.

Anderson, T.E., Leppla, N.C. (Eds.), 1993. Advances in Insect Rearing for Research and Pest Management, Westview Press Inc., Boulder.

Barajas, C.G., Well, E.M., Garcia, I., Mendez, A., 2010. Conidia ma-002 isolation of *Metarhizium anisopliae* (Metsch.) Sorokin growing through an alternative biphasic [in Spanish]. Rev. Prot. Veg. 25, 174–180.

Bartlett, M.C., Jaronski, S.T., 1988. Mass production of entomogenous fungi for biological control of insects. In: Burge, M.N. (Ed.), Fungi in Biological Control Systems, Manchester University Press, Manchester, pp. 61–85.

Biji, C.P., Sudheendrakumar, V.V., Sajeev, T.V., 2006. Quantitative estimation of *Hyblaea puera* NPV production in three larval stages of the teak defoliator, *Hyblaea puera* (Cramer). J. Virol. Methods 136, 78–82.

Bradley, C.A., Black, W.E., Kearns, R., Wood, P., 1992. Role of production technology in mycoinsecticide development. In: Leatham, G.F. (Ed.), Frontiers in Industrial Mycology, Chapman and Hall, London, pp. 160–173.

Burges, H.D., Jones, K.A., 1998. Formulation of bacteria, viruses and protozoa to control insects. In: Burges, H.D. (Ed.), Formulation of Microbial Biopesticides, Kluwer Academic Publishers, Dordrecht, pp. 33–127.

Central Intelligence Agency, 2010. The World Factbook. www.cia.gov/library/publications/the-world-factbook.

Chen, C., Wang, Z., Ye, S., Feng, M., 2009. Synchronous production of conidial powder of several fungal biocontrol agents in series fermentation chamber system. Afr. J. Biotechnol. 8, 3649–3653.

Cherry, A.J., Gwynn, R.L., 2007. Perspectives on the development of biological control agents in Africa. Biocontrol Sci. Technol. 17, 665–676.

Cherry, A.J., Parnell, M., Grzywacz, D., Brown, M., Jones, K.A., 1997. The optimization of in vivo nuclear polyhedrosis virus production of *Spodoptera exempta* (Walker) and *Spodoptera exigua* (Hübner). J. Invertebr. Pathol. 70, 50–58.

Cherry, A.J., Jenkins, N.E., Heviefo, G., Bateman, R., Lomer, C.J., 1999. Operational and economic analysis of a West African pilot-scale production plant for aerial conidia of *Metarhizium* spp. for use as a mycoinsecticide against locusts and grasshoppers. Biocontrol Sci. Technol. 9, 35–51.

Cisneros, F., Vera, A., 2001. Mass-producing *Beauveria brongniartii* inoculum, an economical, farm-level method. Scientist and Farmer: Partners in Research for the 21st Century, Centro Internacional de la Patata, pp. 155–160.

Consuegra, N.P., 2004. Manejo Ecológico de Plagas. Centro de Estudios de Desarrollo Agrario y Rural, La Habana, Cuba.

Copping, L.G., 2009. Manual of Biocontrol Agents, fourth ed. British Crop Protection Council, Alton.

CPL Business Consultants, 2010. North America: Biopesticides Market. CPL Business Consultants, Oxfordshire, UK.

Devi, P.S.V., Chowdary, A., Prasad, Y.G., 2000. Cost-effective multiplication of the entomopathogenic fungus *Nomuraea rileyi* (F) Samson. Mycopathologia 151, 35–39.

Dhar, P., Kaur, G., 2011. Response surface methodology for optimizing process parameters for the mass production of *Beauveria bassiana* conidiospores. Afr. J. Microbiol. Res. 4, 2399–2406.

Dorta, B., Ertola, R.J., Arcas, J., 1996. Characterization of growth and sporulation of Metarhizium anisopliae in solid-substrate fermentation. Enzyme Microbial Tech. 19, 434–439.

Ehlers, R., 2011. Regulation of Biological Control Agents. Springer, Dordrecht.

Entwistle, P.F., 1998. Africa. In: Hunter-Fujita, F.R., Entwistle, P.F., Evans, H.F., Crook, N.E. (Eds.), Insect Viruses and Pest Management, Wiley and Sons, New York, pp. 189–200, 258–268.

Evans, H.F., Shapiro, M., 1997a. Viruses. In: Lacey, L.A. (Ed.), Manual of Techniques in Insect Pathology (Biological Techniques), Academic Press Inc., San Diego, pp. 17–53.

Evans, H.F., Shapiro, M., 1997b. Viruses. In: Lacey, L.A. (Ed.), Manual of Techniques in Insect Pathology, Academic Press, London, pp. 17–53.

Faria, de M.R., Wraight, S.P., 2007. Mycoinsecticides and mycoacaricides: a comprehensive list with worldwide coverage and international classification of formulation types. Biol. Control 43, 237–256.

Feng, M.G., Poprawski, T.J., Khachatourians, G.G., 1994. Production, formulation and application of the entomopathogenic fungus *Beauveria bassiana* for insect control: current status. Biocontrol Sci. Technol. 4, 3–34.

Fernandes, É.K.K., Keyser, C.A., Chong, J.P., Rangel, D.E.N., Miller, M.P., Roberts, D.W., 2009. Characterization of *Metarhizium* species and varieties based on molecular analysis, heat tolerance and cold activity. J. Appl. Microbiol. 108, 115–128.

Goettel, M.S., Jaronski, S.T., 1997. Safety and registration of microbial agents for control of grasshoppers and locusts. In: Goettel, M.S., Johnson, D.L. (Eds.), Microbial Control of Grasshoppers and Locusts, Mem. Entomol. Soc. Can. vol. 171, pp. 83–99.

Grace, J., Jaronski, S., 2005. The Joy of Zen and the Art of Fermentation or the Tao Fungi. Solid Substrate Fermentation Workshop Manual. February 16–18, 2005 USDA/ARS/NPARL, Sidney, Mt., pp. 23. http://www.ars.usda.gov/SP2UserFiles/Place/54360510/The%20Art%20of%20Fermentation%204-06.pdf.

Granados, R.R., Li, G., Blissard, G.W., 2007. Insect cell culture and biotechnology. Virol. Sin. 22, 83–93.

Groote, H. de, 1997. Potential for Mycopesticide Use in Africa: Socioeconomic Analysis. LUBILOSA Socioeconomic Working Paper Series No. 98/5, pp. 27.

Grzywacz, D., Jones, K.A., Moawad, G., Cherry, A., 1998. The *in vivo* production of *Spodoptera littoralis* nuclear polyhedrosis virus. J. Virol. Methods 71, 115–122.

Grzywacz, D., McKinley, D., Jones, K.A., Moawad, G., 1997. Microbial contamination in *Spodoptera littoralis* nuclear polyhedrosis virus produced in insects in Egypt. J. Invertebr. Pathol. 69, 151–156.

Grzywacz, D., Rabindra, R.J., Brown, M., Jones, K.A.J., Parnell, M., 2004. *Helicoverpa armigera* Nucleopolyhedrovirus Production Manual. Natural Resources Institute, p. 107. http://www.fao.org/docs/eims/upload/agrotech/2011/HaNPVmanual-pt1.pdf.

Grzywacz, D., Mushobozi, W.L., Parnell, M., Jolliffe, F., Wilson, K., 2008. The evaluation of *Spodoptera exempta* nucleopolyhedrovirus (SpexNPV) for the field control of African armyworm (*Spodoptera exempta*) in Tanzania. Crop Prot. 27, 17–24.

Grzywacz, D., Cherry, A.C., Gwynn, R., 2009. Biological pesticides for Africa: why has so little of the research undertaken to date led to new products to help Africa's poor? Pestic. Outlook 20, 77–81.

Gupta, R.K., Raina, J.C., Monobrullah, M.D., 2007. Optimisation of in vivo production of nucleoplohedrovirus in homologous host larvae of *Helicoverpa armigera*. J. Entomol. 4, 279–288.

Gwynn, R.L., Maniania, J.N.K., 2010. Africa with special reference to Kenya. In: Kabaluk, J.T., Svircev, A.M., Goettel, M.S., Woo, S.G. (Eds.), The Use and Regulation of Microbial Pesticides in Representative Jurisdictions Worldwide, IOBC Global, pp. 1–6.

Harris, J., Dent, D., 2000. Priorities in Biopesticide Research in Developing Countries. CABI Publishing, Wallingford.

Henderson, D.E., Sinfontes, J.L.A., 2010. Cuba. In: Kabaluk, J.T., Svircev, A.M., Goettel, M.S., Woo, S.G. (Eds.), The Use and Regulation of Microbial Pesticides in Representative Jurisdictions Worldwide, IOBC Global, pp. 51–58.

Hong, T.D., Gunn, J., Ellis, R.H., Jenkins, N.E., Moore, D., 2001. The effect of storage environment on the longevity of conidia of *Beauveria bassiana*. Mycol. Res. 105, 597–602.

Hunter-Fujita, F.R., Vasiljevic, S., Jones, K.A., Cherry, A.C., 1997. Effects of mixed infection with GV and NPV on the biology of Egyptian cotton leafworm *Spodoptera littoralis*. Microbial Insecticides: Novelty or Necessity?, pp. 271–278 British Crop Protection Council Proceedings/Monograph Series No. 68.

Hunter-Fujita, F.R., Entwistle, P.F., Evans, H.F., Crook, N.E. (Eds.), 1998. "Insect Viruses" and Pest Management. Wiley and Sons, New York.

Ignoffo, C.M., 1973. The development of viral insecticide; concept to commercialization. Exp. Parasitol. 33, 380–406.

Jenkins, N.E., Goettel, M.S., 1997. Methods for mass-production of microbial control agents of grasshoppers and locusts. In: Goettel, M.S., Johnson, D.L. (Eds.), Microbial Control of Grasshoppers and Locusts, vol. 171. Mem. Entomol. Soc. Can. pp. 37–48.

Jenkins, N.E., Gryzwacz, D., 2000. Quality control of fungal and viral biocontrol agents—assurance of product performance. Biocontrol Sci. Technol. 10, 753–777.

Jenkins, N.E., Gryzwacz, D., 2003. Towards the standardization of quality control of fungal and viral biocontrol agents. In: van Lenteren, J.C. (Ed.), Quality Control and Production of Biological Control Agents: Theory and Testing Procedures, CAB International, Wallingford, pp. 247–263.

Jenkins, N.E., Heviefo, G., Langewalde, J., Cherry, A.C., Lomer, C.J., 1998. Development of mass production technology for aerial conidia for use as mycopesticides. Biocon. News Inform. 19, 21N–31N.

Jenkins, N.E., Ali, B.S., Moore, D., 2007. Mass production of entomopathogenic fungi for biological control of insect pests. In: Parker, B., Skinner, M., El-Bouhssini, M., Kumari, S. (Eds.), Sunn Pest Management: A Decade of Progress 1994–2004, Arab Society of Plant Protection, Beirut, pp. 287–294.

Jones, K.A., 2000. Bioassays of entomopathogenic viruses. In: Navon, A., Ascher, K.R.S. (Eds.), Bioassays of Entomopathogenic Microbes and Nematodes, CABI Publishing, Wallingford, pp. 95–140.

Jones, K.A., Zelazny, B., Ketunuti, U., Cherry, A.C., Grzywacz, D., 1998. South east Asia and Western Pacific. In: Hunter-Fujita, F.R., Entwistle, P.F., Evans, H.F., Crook, N.E. (Eds.), Insect Viruses and Pest Management, Wiley and Sons, New York, pp. 244–258.

Kabaluk, J.T., Svircev, A.M., Goettel, M.S., Woo, S.G. (Eds.), 2010a. The Use and Regulation of Microbial Pesticides in Representative Jurisdictions Worldwide, IOBC Global, pp. 99, Online, www.IOBC-Global.org.

Kabaluk, J.T., Brookes, V.R., Svircev, A.M., 2010b. Canada. In: Kabaluk, J.T., Goettel, M.S., Svircev, A.M., Woo, S.G. (Eds.), The use and regulation of microbial pesticides in representative jurisdictions worldwide. IOBC Global, pp. 59–73.

Kambrekar, D.N., Kulkarni, K.A., Girradi, R.S., 2007. Assessment of quality of HaNPV samples produced by private firms. Karnataka J. Agric. Sci. 20, 417–419.

Kennedy, J.S., Rabindra, R.J., Sathiah, N., Grzywacz, D., 1999. The role of standardisation and quality control in the successful promotion of NPV insecticides. In: Sen, A. (Ed.), Biopesticides in Insect Pest Management, Phoenix Publishing House, New Delhi, pp. 170–174.

Kim, J.J., Lee, S.G., Lee, S., Jee, H.E., 2010. South Korea. In: Kabaluk, J.T., Svircev, A.M., Goettel, M.S., Woo, S.G. (Eds.), The Use and Regulation of Microbial Pesticides in Representative Jurisdictions Worldwide, IOBC Global, pp. 18–23.

Kolodny-Hirsch, D.M., Sitchawat, T., Jansiri, T., Chenrchaivachirakul, A., Ketunuti, U., 1997. Field evaluation of a commercial formulation of the *Spodoptera exigua* (Lepidoptera: Noctuidae) nuclear polyhedrosis virus for control of beet armyworm on vegetable crops in Thailand. Biocontrol Sci. Technol. 7, 475–488.

Koul, O., 2011. Microbial biopesticides: opportunities and challenges. Cab reviews: Perspect. Agric. Vet. Sci., Nutr. Resour.. 6 (056), pp. 26. www.cababstractsplus.org/cabreviews.

Koul, O., Dhaliwal, G.S., Marwaha, S.S., Arora, J.K., 2003. Biopesticides and Pest Management. vol. 2. Campus Books, New Delhi.

Krassilstchik, J., 1888. La production industrielle des parasites végétaux pour la destruction des insectes nuisibles. Bull. Biologique de la France de la Belgique 19, 461–472, Downloaded from www.biodiversitylibrary.org.

Kujeku, E., Jones, K.A., Moawad, G.M., 1998. Africa near and Middle East. In: Hunter-Fujita, F.R., Entwistle, P.F., Evans, H.F., Crook, N.E. (Eds.), Insect Viruses and Pest Management, Wiley and Sons, New York, pp. 281–302.

Kulshrestha, S., 2004. The status of regulatory norms for biopesticides in India. In: Kaushik, N. (Ed.), Biopesticides for Sustainable Agriculture: Prospects and Constraints, The Energy Research Institute, New Delhi, pp. 67–72.

Lacey, L.A., Headrick, H.L., Horton, D.R., Schriber, A., 2010. Effect of granulovirus on the mortality and dispersal of potato tuber worm (*Lepidoptera: Gelechiidae*) in refrigerated storage warehouse conditions. Biocontrol Sci. Technol. 20, 437–447.

Langewalde, J., Ouambama, Z., Mamadou, A., Peveling, R., Stolz, I., Bateman, R., Attignon, S., Blanford, S., Arthurs, S., Lomer, C., 1999. Comparison of an organophosphate insecticide with a mycoinsecticide for the control of *Oedaleus senegalensis* (Orthoptera: Acrididae) and other Sahelian grasshoppers at an operational scale. Biocontrol Sci. Technol. 9, 199–214.

Lasa, R., Caballero, P., Williams, T., 2007. Juvenile hormone analogs greatly increase the production of a nucleopolyhedrovirus. Biol. Control 41, 389–396.

Lasa, R., Williams, T., Caballero, P., 2008. Insecticidal properties and microbial contaminants in a *Spodoptera exigua* multiple nucleopolyhedrovirus (Baculoviroidae) formulation stored at different temperatures. J. Econ. Entomol. 101, 42–49.

Leng, P., Zhang, Z., Pan, G., Zhao, M., 2011. Applications and development trends in biopesticides. Afr. J. Biotechnol. 10, 19864–19873.

Li, Z., Alves, S.B., Roberts, D.W., Fan, M., Delalibera Jr., I., Tang, J., Lopes, R.B., Faria, M., Rangel, D.E.N., 2010. Biological control of insects in Brazil and China: history, current programs and reasons for their successes using entomopathogenic fungi. Biocontrol Sci. Technol. 20, 117–136.

Lisansky, S., 1997. Microbial biopesticides. Microbial Insecticides: Novelty or Necessity?, pp. 3–10. British Crop Protection Council Proceedings/Monograph Series No. 68.

Lomer, C.J., Bateman, R.P., Johnson, D.L., Langewald, J., Thomas, M., 2001. Biological control of locusts and grasshoppers. Ann. Rev. Entomol. 46, 667–702.

Marrone, P.G., 2007. Barriers to adoption of biological control agents and biological pesticides. CAB Rev. Perspect. Agric. Vet. Sci. Nutr. Nat. Resour. 2 (51). www.cababstractsplus.org/cabreviews.

McClatchie, G.V., Moore, D., Bateman, R.P., Prior, C., 1994. Effects of temperature on the viability of the conidia of *Metarhizium flavoviride* in oil formulations. Mycol. Res. 98, 749–756.

Mĕndonça, A.F., 1992. Mass production, application and formulation of *Metarhzium anisopliae* for control of sugarcane froghopper, *Mahanarva posticata*, in Brazil. In: Lomer, C.J., Prior, C. (Eds.), Biological Control of Locusts and Grasshoppers, CAB International, Wallingford, UK, pp. 239–244.

Moore, D., 2002a. Fungal control of pests. In: Pimentel, D. (Ed.), Encyclopedia of Pest Management, Marcel Dekker, Inc., New York, pp. 320–324.

Moore, D., 2008. A plague on locusts—the LUBILOSA story. Outlook Pest Manag. 19, 14–17.

Moore, S.D., 2002b. The Development and Evaluation of *Cryptophlebia leucotreta* Granulovirus (CrleGV) as a Biological Control Agent for the Management of False Codling Moth, *Cryptophlebia leucotreta*, On Citrus. PhD thesis, Rhodes University. http://eprints.ru.ac.za/2329/.

Moore, S.D., Kirkman, W., Stephen, P., 2004. CRYPTOGRAN: a virus for the biological control of false codling moth. S. Afr. Fruit J. 3, 35–39.

Moore, S.D., Hendry, D.A., Richards, G.I., 2011. Virulence of a South African isolate of the *Cryptophlebia leucotreta* granulovirus to *Thaumatotibia leucotreta* neonate larvae. BioControl 56, 341–352.

Moscardi, F., 1999. Assessment of the application of baculoviruses for the control of Lepidoptera. Ann. Rev. Entomol. 44, 257–289.

Moscardi, F., 2007. Development and use of the nucleopolyhedrovirus of the velvetbean caterpillar in soybeans. In: Vincent, C., Goettel, M.S., Lazarovits, G. (Eds.), Biological Control: A Global Perspective, CAB International, Wallingford, Oxfordshire, pp. 344–353.

Moscardi, F., de Souza, M.L., de Castro, M.E.B., Moscardi, M.L., Szewczyk, B., 2011. Baculovirus pesticides: present state and future perspectives. In: Ahmad, I., Ahmad, F., Pichtel, J. (Eds.), Microbes and Microbial Technology, Springer, Dordrecht, pp. 415–445.

Mushobozi, W.L., Grzywacz, D., Musebe, R., Kimani, M., Wilson, K., 2005. New approaches to improve the livelihoods of poor farmers and pastoralists in Tanzania through monitoring and control of African armyworm, *Spodoptera exempta*. Asp. Appl. Biol. 75, 37–45.

O.E.C.D., 2002. Consensus document on information used in assessment of environmental applications involving baculoviruses. Series on harmonisation of regulatory oversight in biotechnology No. 20. ENV/JM/MONO(2002)1.

Olivera, M. R. V. de, 1998. South America. In: Hunter-Fujita, F.R., Entwistle, P.F., Evans, H.F., Crook, N.E. (Eds.), Insect Viruses and Pest Management, Wiley and Sons, New York, pp. 339–355.

Podgwaite, J.D., Bruen, R.B., Shapiro, M., 1983. Microorganisms associated with production lots of the nucleopolyhedrosis virus of the gypsy moth, *Lymantria dispar* (Lepidoptera: Lymantriidae). Entomophaga 28, 9–15.

Posada-Flórez, F.J., 2008. Production of *Beauveria bassiana* fungal spores on rice to control the coffee berry borer, *Hypothenemus hampei*, in Colombia. J. Insect Sci. 8, pp. 13. Available online: insectscience.org/8.41.

Prakash, G.V.S.B., Padmaja, V., Kiran, R.R.S., 2008. Statistical optimization of process variables for the large-scale production of *Metarhizium anisopliae* conidiospores in solid-state fermentation. Biores. Technol. 99, 1530–1537.

Prior, C., 1989. Biological pesticides for low external-input agriculture. Biocontrol News Inform. 10, 17–22.

Rabindra, R.J., Grzywacz, D., 2010. India. In: Kabuluk, T., Svircev, A., Goettel, M., Woo, S.G. (Eds.), Use and Regulation of Microbial Pesticides in Representative Jurisdictions Worldwide, IOBC Global, pp. 12–17.

Rabindra, R.J., Sathiah, N., Jayaraj, S., 1992. Efficacy of nuclear polyhedrosis virus against *Heliothis armigera* (Hubner) on *Helicoverpa* resistant and susceptible varieties of chickpea. Crop Prot. 11, 320–322.

Rabindra, R.J., Sathiah, N., Jayanth, K.P., Gupta, R.K., 2003. Commercial scale production and formulation of nuclear polyhedrosis viruses of *Helicoverpa armigera* and *Spodptera litura*, recent advances in integrated pest management. Proceedings of the NATP Interactive Workshop on Integrated Pest Management, February 26–28, pp. 32–40.

Rajanikanth, P., Subbaratnam, G.V., Rahaman, S.J., 2010. Evaluation of economically viable substrates for mass production of *Beauveria bassiana* (Balsamo) Vuillemin. J. Biol. Cont. 24, 322–326.

Ranga Rao, G.V., Meher, K.S., 2004. Optimization of in vivo production of *Helicoverpa armigera* NPV and regulation of Malador associated with the process. Indian J. Plant Prot. 32, 15–18.

Ranga Rao, G.V., Rupela, O.P., Wanil, S.P., Jyothsna, J.S., Rameshwar Rao, V., Humayun, P., 2007. Bio-intensive management reduces pesticide use in India. Pestic. News 76, 16–17.

Rangel, D.E.N., Faria, M., 2010. Brazil. In: Kabaluk, J.T., Svircev, A.M., Goettel, M.S., Woo, S.G. (Eds.), The Use and Regulation of Microbial Pesticides in Representative Jurisdictions World-wide, IOBC Global, pp. 46–50.

Ratanasatien, P., Ketunuti, U., Tantichodok, A., 2005. Positioning of biopesticides in Thailand. In: Côté, J.C., Otvos, I.S., Schwartz, J.L., Vincent, C. (Eds.), 6th Pacific Rim Conference on the Biotechnology of *Bacillus thuringiensis* and Its Environmental Impact, Biocontrol Network of Canada, Victoria, pp. 100–107. http://www.erudit.org/livre/pacrim/2005/000224co.pdf.

Ravensberg, W.J., 2011. A Roadmap to the Successful Development and Commercialization of Microbial Pest Control Products for Control of Arthropods. Progress in Biological Control, vol. 10. Springer, Dordrecht.

Roberts, D.W., St Leger, R.J., 2004. *Metarhizium* spp., cosmopolitan insect-pathogenic fungi: mycological aspects. Adv. Appl. Microbiol. 54, 1–70.

Rorer, J.B., 1910. The green muscardine of froghoppers. Proc. Agric. Soc. Trinidad Tobago X, 467–482.

Rorer, J.B., 1913. The Green Muscardine Fungus and Its Use in Cane Fields. Board of Agriculture, Trinidad and Tobago Circular No. 8, pp. 11.

Sahayaraj, K., Namasivayam, S.K.R., 2008. Mass production of entomopathogenic fungi using agricultural products and by products. Afr. J. Biotechnol. 7, 1907–1910.

Senthil Kumar, C.M., Sathiah, N., Rabindra, R.J., 2005. Optimizing the time of harvest of nucleopolyhedrovirus infected *Spodoptera litura* (Fabricius) larvae under in vivo production systems. Curr. Sci. 88, 1682–1684.

Shapiro, M., 1986. In vivo production of baculoviruses. In: Granados, R.R., Federici, B.A. (Eds.), The Biology of Baculoviruses, vol. 2. CRC Press, Boca Raton, pp. 31–62.

Shapiro, M., Bell, R.A., Owens, C.D., 1981. In vivo mass production of gypsy moth nucleopolyhedrovirus. In: Doane, C.C., MacManus, M.L. (Eds.), The Gypsy Moth: Research towards Pest Management, Forest Service Technical Bulletin 1584, United State Department of Agriculture, Washington, pp. 633–655.

Shetty, P.K., Murugan, M., Sreeja, K.G., 2008. Crop protection stewardship in India: wanted or unwanted. Curr. Sci. 95, 457–464.

Shieh, T.R., 1989. Industrial production of viral pesticides. Adv. Virus Res. 36, 315–343.

Singh, P., 1977. Artificial Diets for Insects, Mites, and Spiders. Plenum Publishing Corporation, New York.

Singh, P., Moore, R.F., 1985. Handbook of Insect Rearing. Elsevier, Amsterdam.

Skovmand, O., 2007. Microbial control in South East Asia. J. Invertebr. Pathol. 95, 168–174.

Sosa-Gómez, D.R., Moscardi, F., Santos, B., Alves, L.F.A., Alves, S.B., 2008. Produção e uso de vírus para o controle de pragas na América Latina. In: Alves, S.B., Lopes, R.B. (Eds.), Controle Microbiano de Pragas na América Latina: avanços e desafios, FEALQ, Piracicaba, pp. 49–68.

Sporleder, M., Kroschel, J., 2008. The potato tuber moth granulovirus (PoGV): use, limitations and possibilities for field applications. In: Kroschel, J., Lacey, L. (Eds.), Integrated Pest Management for the Potato Tuber Moth, *Phthorimaea operculella* Zeller a Potato Pest of Global Importance, Margraf Publishers, Weikersheim, pp. 49–71. Tropical Agriculture 20, Advances in Crop Research 10.

Srikanth, J., Singaravelu, B., 2011. White Grub (*Holotrichia serrata*) as a Pest of Sugarcane and Its Management. Technical Bulletin No 197. Sugarcane Breeding Institute, Coimbatore.

Srinivasa, M., Jagadeesh Babu, C.S., Anitha, C.N., Girish, G., 2008. Laboratory evaluation of available commercial formulations of HaNPV against *Helicoverpa armigera* (Hub.). J. Biopestic. 1, 138–139.

Sukonthabhirom, S., Dumrongsak, D., Jumroon, S., Saroch, T., Chaweng, A., Tanaka, T., 2011. Update on DBM diamide resistance from Thailand: causal factors and learnings. In: Srinivasan, R., Shelton, A.M., Collins, H.I. (Eds.), The Sixth International Workshop on Management of the Diamondback Moth and Other Crucifer Insect Pests, AVRDC—The World Vegetable Center, Taiwan. http://203.64.245.61/fulltext_pdf/EB/2011-2015/eb0170.pdf.

Sun, X., Peng, H., 2007. Recent advances in control of insect pests by using viruses in China. Virol. Sin. 22, 158–162.

Swanson, D., 1997. Economic feasibility of two technologies for production of a mycopesticide in Madagascar. In: Goettel, M.S., Johnson, D.L. (Eds.), Microbial Control of Grasshoppers and Locusts, Mem. Entomol. Soc. Can., vol. 171, pp. 101–113.

Sweetmore, A., Kimmins, F., Silverside, P., 2006. Perspectives on Pests II: Achievements of Research under UK Department for International Development Crop Protection Programme 2000–05. Natural Resources International Ltd, Ayesford.

Szewcyk, B., Hoyos-Carvajal, L., Paluszek, M., Skrzecz, I., Lobo de Souza, M., 2006. Baculoviruses re-emerging biopesticides. Biotechnol. Adv. 24, 143–160.

Tamez-Guerra, P., McGuire, M.R., Behle, R.H., Shashad, B.S., Pingel, R.L., 2002. Storage stability of *Anagrapha falcifera* nucleopolyhedrovirus in spray dried formulations. J. Invertebr. Pathol. 79, 7–16.

Torres, H., Ortega, A.M., Alcázar, J., Ames, T., Palomino, L., 1993. Control biológico del gorgojo de los Andes (*Premnotrypes* spp.) con *Beauveria brongniartii*. Guía de Investigación CIP8. Centro Internacional de la Papa, Lima.

Tripp, R., Arif, A., 2001. Farmers Access to Natural Pest Products: Experience from an IPM Project in India. ODI Agriculture and extension network paper 113. ODI, London.

Van Beek, N., Davies, D.C., 2009. Baculovirus production in insect larvae. In: Murhammer, D.W. (Ed.), Methods in Molecular Biology, Baculovirus and Insect Cell Expression Protocols, vol. 338. Humana Press, Towata.

Veenakumari, K., Rabindra, R.J., Srinivasa Naik, C.D., Shubha, M.R., 2007. In situ field mass production of *Amsacta albistriga* (Lepidoptera: Arctiidae) nucleopolyhedrovirus in a groundnut ecosystem in South India. Int. J. Trop. Insect Sci. 27, 48–52.

Visalakshmi, V., Ranga Rao, G.V., Arjuna Rao, P., 2005. Integrated pest management strategy against *Helicoverpa armigera* Hüber in chickpea. Indian J. Plant Prot. 33, 17–22.

Wabule, M.N., Ngaruiya, P.N., Kimmins, F.K., Silverside, P.J., 2004. Registration for biocontrol agents in Kenya. Proceedings of National Workshop on Legislation of Biopesticides in Kenya. Nukuru Kenya May 14–16, 2003, Natural Resources International, Alyesford.

Wahab, S., 2004. The Department of Biotechnology initiates towards the development and use of biopesticides in India. In: Kaushik, N. (Ed.), Biopesticides for Sustainable Agriculture: Prospects and Constraints, The Energy and Resources Institute, New Delhi, pp. 73–90.

Wang, B., Li, Z., 2010. China. In: Kabaluk, J.T., Svircev, A.M., Goettel, M.S., Woo, S.G. (Eds.), The Use and Regulation of Microbial Pesticides in Representative Jurisdictions Worldwide, IOBC Global, pp. 7–11.

Winters, P., Fano, H., 1997. The Economics of Biological Control in Peruvian Potato Production. International Potato Center Working Paper No. 1997. CIP, Lima, Peru.

Yang, M.M., Meng, L.L., Zang, Y.A., Wang, Y.Z., Qu, L.J., Wang, Q.H., Ding, J.Y., 2012. Baculoviruses and insect pest control in China. Afr. J. Microbiol. Res. 6, 214–218.

Ye, S.D., Ying, S.H., Chen, C., Feng, M.G., 2006. New solid state fermentation chamber for bulk production of aerial conidia of fungal biocontrol agents on rice. Biotechnol. Lett. 28, 799–804.

Zimmermann, G., 2007a. Review on safety of the entomopathogenic fungi *Beauveria bassiana* and *Beauveria brongniartii*. Biocontrol Sci. Technol. 17, 553–596.

Zimmermann, G., 2007b. Review on safety of the entomopathogenic fungus *Metarhzium anisopliae*. Biocontrol Sci. Technol. 17, 879–920.

Zimmermann, G., 2008. The entomopathogenic fungi *Isaria farinosa* (formerly *Paecilomyces farinosus*) and the *Isaria fumosorosea* species complex (formerly *Paecilomyces fumosoroseus*); biology, ecology and use in biological control. Biocontrol Sci. Technol. 18, 865–901.

This page intentionally left blank

Section III

This page intentionally left blank

Chapter 16

Insect Protein as a Partial Replacement for Fishmeal in the Diets of Juvenile Fish and Crustaceans

Eric W. Riddick
National Biological Control Laboratory, USDA-ARS, Stoneville, MS, USA

16.1. INTRODUCTION

16.1.1. The Need for Alternatives to Fishmeal

The availability of land to produce livestock is rapidly dwindling as the human population continues to increase throughout many regions of the world. Similarly, the availability of wild aquatic organisms (fish and crustaceans) is declining due to overharvesting and increased demand by humans (Sargent and Tacon, 1999). Fish farming (cultured fish) could potentially satisfy the increasing demands for animal protein (Naylor et al., 2001). However, rising prices for feed for aquatic organisms destined for human consumption is a reason for concern. One of the main sources of animal protein in commercial feed for juvenile fish and crustaceans is fishmeal, which is dried, ground tissues of undecomposed marine baitfish (herring, anchovy, or menhaden). Fishmeal is becoming less available (Tacon and Metian, 2008; Ng, 2000; Abowei and Ekubo, 2011; van Huis, 2013). The safety of fishmeal has also been under scrutiny. Alternative sources of protein for fish are necessary to decrease dependence on fishmeal. However, the aquaculture industry may need economic incentives to encourage the transition from fishmeal-based feedstuffs to alternative sources of food for fish and crustaceans (Naylor et al., 2009). Researchers are currently investigating the potential for protein from plants as partial or even complete replacements for fishmeal (Le Boucher et al., 2011, 2012); similar research with protein from yeast is ongoing (Peterson et al., 2012). In recent years, the value of insect protein as partial or complete replacements for fishmeal has been studied (van Huis, 2013).

Mass Production of Beneficial Organisms. http://dx.doi.org/10.1016/B978-0-12-391453-8.00016-9

16.1.2. Aims of this Chapter

Although there is some published information on the usefulness of insects as sources of protein in feed for terrestrial livestock (poultry and swine), the purpose of this chapter is to highlight research on insects as sources of protein for juvenile fish and crustaceans (prawn). This chapter attempts to review and synthesize the evidence that protein from insects can partially replace animal protein in fishmeal in feed for juvenile stages of cultured fish. This research is in support of the aquaculture industry and its efforts to provide plentiful fish and prawns for human consumption and remain competitive in a global economy.

16.1.3. Overview of the Content

This chapter presents a review of the scientific literature on insects as important sources of protein to incorporate into fish and prawn diets. In this chapter, the coverage is restricted to insects within three orders—Lepidoptera, Diptera, and Coleoptera. In the Lepidoptera, the focus is on the Oriental silkworm moth *Bombyx mori* (L.) (Family Bombycidae) and other silkworm moth species (Family Saturniidae). In the Diptera, the focus is on the common house fly *Musca domestica* (L.) (family Muscidae) and the black soldier fly *Hermetia illucens* (L.) (family Stratiomyidae). In the Coleoptera, the focus is on the yellow mealworm *Tenebrio molitor* (L.) and the superworm *Zophobas morio* (F.) (family Tenebrionidae). This chapter also mentions some challenges and opportunities that could hinder the expansion of the market for insect feed for commercial production of livestock. Topics will include the challenges of creating artificial diets to facilitate rearing insects at a commercial scale and opportunities (or lack thereof) to scale up production of insects to meet the needs of the aquaculture industry. The production of insects of high quality is an important topic to those that rear them professionally (Schneider, 2009).

16.2. MODEL INSECTS AND POTENTIAL AS FEED FOR FISH

16.2.1. Oriental Silkworm Moth and Relatives

The Chinese have reared silkworm moths for centuries for their silk (Zhang et al., 2008). Even today, a sericulture industry for production of silk for human clothing is ongoing in China, India, and elsewhere. The Oriental (mulberry) silkworm moth *B. mori* is used for this trade. Other silkworm moths including the silkworm *Anaphe infracta* Walsingham, muga silkworm *Antheraea assamensis* Helfer, eri silkworm *Samia ricini* (Donovan), and Chinese oak silkworm *Antheraea pernyi* Guérin-Méneville are also reared for their silk. Traditionally, Oriental silkworm larvae are reared on leaves of mulberry *Morus alba* (L.) and close relatives, which have origins in Asia. Due to human transport, the mulberry tree inhabits many countries around the globe in association with silkworm and the sericulture industry. Interestingly, humans consume mulberry leaves as a vegetable and tea. We eat the fruit (berries) fresh, in preserves, or drink the juice. Only the cocoon of the silkworm moth pupal stage is necessary in silk production. Most growers extract the pupae from cocoons and discard them as waste. Utilization of "waste" silkworm pupae to feed livestock is therefore a good way of capitalizing on the unused product of silk production. This practice of using waste silkworm pupae to support the production of fish is an effective use of resources.

The idea of using silkworm pupae as food for fish is not novel. Akiyama et al. (1984) supplemented a fishmeal-based diet for chum salmon *Oncorhynchus keta* (Walbaum) with 5% powdered silkworm pupae, dried beef liver, krill meal, or powdered earthworm. Over a 6-week feeding experiment, weight gain, feeding efficiency, and fat content of salmon were greatest when the authors incorporated powdered earthworm rather than any other treatments or control (untreated fishmeal) into the fishmeal-based diet. The protein content of salmon did not differ between treatments in this study (Akiyama et al., 1984). Whether the addition of larger percentages of protein from silkworm pupae would increase its value in the diet of chum salmon is not clear.

Nandeesha et al. (1990) measured the growth of common carp *Cyprinus carpio* (L.) over 140 days on diets containing variable percentages of nondefatted silkworm pupae (41% crude protein) and fishmeal (68% crude protein) in dry pellets, using the methods of Jayaram and Shetty (1981). They used the following treatments: 10% silkworm plus 20% fishmeal, 20% silkworm plus 10% fishmeal, 30% silkworm plus 0% fishmeal, and 0% silkworm plus 25% fishmeal. All diets contained approximately 20%, 40%, and 10% groundnut cake, rice bran, and tapioca flour, respectively. The diet with 30% silkworm plus 0% fishmeal stimulated the most growth of juvenile carp over the test period. The authors concluded that nondefatted silkworm pupae could replace fishmeal in diet formulations as feed for common carp, especially because the texture, flavor, odor, and color of this fish does not change with a change from the standard fishmeal diet (Nandeesha et al., 1990).

In a companion study, Nandeesha et al. (2000) increased the percentage of silkworm pupae in experimental diets. In dry pellets, they formulated diets containing 30%, 40%, or 50% silkworm pupae without any fishmeal; the control diet consisted of fishmeal without any silkworm. All diets contained variable amounts of groundnut cake, rice bran, and tapioca flour to balance the protein content between treatment diets. After a test period of 90 days, the authors discovered no changes in weight gain, food conversion ratio, or protein conversion ratio when they fed carp juveniles any percentage of silkworm-based diet or the control (fishmeal). However, carp retained increasing amounts of protein in tissues with an increase (30–50%) in the percentage of silkworm in the diets. Carp fed the 50% silkworm diet contained more protein (but less fat) than those fed lower percentages of silkworm. Apparently, carp diets can contain up to 50% silkworm pupae without changing the growth and quality of the product (Nandeesha et al., 2000). Adding oil from silkworm pupae or sardines improved the growth, feed efficiency, and biomass production of common carp (Nandeesha et al., 1999). In outdoor tanks, 9% addition of either oil increased biomass production up to almost 50% over the control (fishmeal without added oil). They did not detect differences between fish oil versus silkworm oil at 3%, 6%, or 9% concentrations in dry feed pellets. The authors concluded their study by stating that adding more fat into diets can incur economic benefits to carp production.

Silkworm (*B. mori*) pupae were a supplement rather than a complete replacement of fishmeal in fish diets (Jintasataporn et al., 2011). Silkworm pupae could replace up to 50% of the protein from fishmeal (equivalent to 14.6% silkworm pupae by weight) in formulated diets for the snakeskin gourami *Trichogaster pectoralis* (Regan), an omnivorous fish. Greater amounts of silkworm pupae (22–30% by weight) caused a decrease in growth and egg production of females, seen after 5 months of feeding. Farmers in Southeast Asia culture this species (and others, such as common carp) in temporary bodies of water in rice fields for human consumption and for export to the aquarium industry (Little et al., 1996).

Commercial production of shrimp (i.e. prawns) relies on fishmeal as feed. A search for alternatives is ongoing. Langer et al. (2011) compared the effects of diets containing silkworm pupae (41%), fishmeal (49%), soybean meal (42%), or earthworm meal (46%) on freshwater prawn *Macrobrachium dayanum* (Henderson) for 90 days. All diets contained variable amounts of rice bran and mustard oil cake to balance the proximate composition of protein in formulated feeds. The researchers found that the diet containing earthworm meal (rather than any other diet) proved most beneficial for prawn growth, survival, and food conversion ratio. The diet containing silkworm pupae was second best, followed closely by fishmeal. The diet with soybean meal was the least nutritious for freshwater prawn in this study. Biochemical analysis showed that protein content did not differ between diets based on silkworm pupae, fishmeal, or earthworm meal; protein content was greater in all three of these diets when compared to the diet based on soybean. Lipid content was approximately the same for diets containing silkworm pupae versus earthworm meal; lipid content was lowest in the fishmeal diet (Langer et al., 2011).

16.2.2. House Fly and Black Soldier Fly

The common house fly *M. domestica* has a long history of interactions with humans. It is a nuisance pest and has the capacity to vector diseases, several of which are highly virulent to humans (Nazni et al., 2005; Fasanella et al., 2010). Nevertheless, researchers have proven that house fly maggots are suitable sources of protein for poultry (Ocio et al., 1979; Hwango et al., 2009). Maggots are also good sources of protein for catfish. Fasakin et al. (2003) evaluated methods of processing maggots—hydrolyzed (defatted), not removing fat (full-fatted), and dried in an oven or dried in the sun—on growth and utilization of African catfish *Clarias gariepinus* (Burchell). The authors compared the diets containing processed maggots with diets containing fishmeal (control). Over a 56 day timeframe, catfish fingerlings were fed the treatments diets. The drying method (oven versus sun-dried) did not affect the suitability of defatted maggots for fish; defatted maggots were not less nutritious than fishmeal. The sun-dried, full-fatted maggots were less nutritious; fish exhibited less growth and weight gain when fed this treatment rather than fishmeal. Fasakin et al. (2003) suggested that defatted maggots provide animal protein comparable in quality to that of fishmeal.

Aniebo et al. (2009) also assessed the potential of house fly maggots, in the form of a meal later compacted into sun-dried pellets, on growth and nutrient utilization of the African catfish *C. gariepinus*. They monitored catfish juveniles fed feed composed of 0%, 50%, or 100% animal protein from maggot meal rather than fishmeal (control) over a 10 week period. Maggot meal contained 48% crude protein. Soybean meal was a major constituent (34–43%) in the three treatment feeds. The authors discovered that feed composed of 50% or 100% animal protein from maggots had no negative effects on fish growth or nutrient utilization when compared to feed composed of fishmeal. Maggot meal (in compacted pellets) is an alternative protein source that should reduce costs and lead to sustainable production of catfish (Aniebo et al., 2009).

In another study with catfish, Sogbesan et al. (2006) compared two techniques to harvest house fly maggots and evaluated the potential of these maggots, formulated into a maggot meal, as dietary protein for catfish. They compared a screen method in which they thinly spread maggot-infested dung on a screen net in the sunlight. In attempts to avoid sunlight, the maggots crawled out of the dung and down into a basin positioned beneath the screen net. The other harvesting technique, the flotation method, involved placing maggot-infested dung in a basin of water; the maggots floated to the surface of the water. The authors used a 3mm mesh size net to sieve the maggots from the water. The catfish was a hybrid between *Heterobranchus longifilis* and *C. gariepinus*. They evaluated the effects of five maggot meal-based diets designed to replace 0%, 25%, 50%, 75%, and 100% of the fishmeal typically present in fish feed. Catfish fingerlings in their experiment were in concrete tanks for 70days. They found that harvest method (screen vs flotation) had no effect on the quantity of maggots collected within 4h. Catfish fed the diet that replaced 25% (rather than 0%, 50%, 75%, or 100%) of fishmeal with a maggot meal displayed the greatest total final body weight, mean weight gain, feed conversion ratio, and protein efficiency rate. Sogbesan et al. (2006) recommend replacing 25% of fishmeal for maggot meal for better growth performance of hybrid catfish.

Nsofor et al. (2008) compared the growth and food utilization of the catfish *C. gariepinus* when fed diets (pellet form) containing 25% fishmeal, 25% maggot meal, or 12.5% fishmeal plus 12.5% maggot meal for 10weeks. They discovered that all diets were effective. Fish did not differ significantly in final body weight, specific growth rate, food conversion ratio, or protein efficiency ratio. The authors state that 25% maggot meal could replace the 25% fishmeal in diets for catfish. Omoyinmi and Olaoye (2012) measured the growth performance of Nile tilapia *Oreochromis niloticus* (L.) fed feed containing protein from house fly *M. domestica*, earthworm *Eudrilus eugeniae* (Kinberg), palm grub *Oryctes rhinoceros* (L.), garden snail *Limocolaria aurora* (Jay), or fishmeal. They subjected tilapia fingerlings to the diets (as crushed pellets) for 70 days. The results revealed that diets with protein from earthworm, garden snail, or house fly were as effective as the fishmeal diet on tilapia growth and survival. Diet with protein from palm grub was less effective than fishmeal (control) for tilapia growth and survival (Omoyinmi and Olaoye, 2012). Earthworm and maggot protein appear to be very suitable alternatives to fishmeal for tilapia growth and development.

St-Hilaire et al. (2007a) evaluated the importance of house fly pupae and black soldier fly *H. illucens* prepupae in feed for rainbow trout *Oncorhynchus mykiss* (Walbaum). Treatment diets (feed) involved substituting 0%, 25%, or 50% fly protein for fishmeal (anchovy) and feeding this to juvenile trout for up to 9 weeks. All diets contained soybean meal (16%) and corn gluten meal (8%). Total weight gain of fish did not differ for those fed 25% soldier fly diet and the control (fishmeal). Total weight gain was less for fish fed 50% soldier fly or 25% house fly diet rather than the control. The feed conversion ratio (grams of diet fed to fish/grams of weight gained by fish) was the highest for fish fed the 50% soldier fly diet compared to the other treatments. The protein content (whole body proximate composition) of fish after the end of the experiment did not differ significantly between treatments. The lipid content of fish fed diets containing soldier fly (25% or 50% pupae) did not significantly differ from the control (fishmeal); lipid content in fish fed 25% house fly diet was significantly greater than in fish fed 50% soldier fly diet. Protein from house fly or black soldier fly can replace up to 25% of the fishmeal in fish feed without negatively affecting growth performance and quality of rainbow trout.

Sealey et al. (2011) were concerned that supplementing fish diets with insect protein from the black soldier fly could alter the aroma, taste, and quality of fillets of rainbow trout. In a previous study, adding protein from soldier fly pupae into fish feed altered the fatty acid profile of rainbow trout (St-Hilaire et al., 2007a). Changes to the fatty acid concentration can affect fish flavor and aroma (Turchini et al., 2003). Using black soldier fly prepupae reared on manure from dairy cows, Sealey et al. (2011) replaced either 25% or 50% of fishmeal (anchovy) protein with soldier fly protein or with soldier fly protein enriched with fish offal (discarded visceral organs and fat from rainbow trout). Note that all four diets contained considerable amounts of soybean meal (16%), wheat gluten meal (7.8%), and corn gluten meal (7.0%). They fed rainbow trout the treatment diets (pellet form) for 8 weeks and measured fish growth, fillet taste, and aroma. Replacing fishmeal with either 25% or 50% insect meal from soldier fly protein reduced the growth rate of fish. However, the combination of soldier fly protein with fish offal did not affect growth of fish. The fish that were fed protein from black soldier fly—with or without the fish offal—did not differ in taste or aroma from fish fed protein from fishmeal (Sealey et al., 2011).

Bondari and Sheppard (1987) compared diets formulated with either 10% fishmeal, 10% dried prepupae of black soldier fly, or a commercial feed (with an unknown percentage of fishmeal as the animal protein) as dry pellets for the channel catfish *Ictalurus punctatus* (Rafinesque). When they reared catfish juveniles in cages floating on the surface of the water in a 2 ha reservoir, the growth rate (weight gain, total body weight) of males and females was reduced over a 15-week study period when using the 10% soldier fly diet rather than the 10% fishmeal diet or commercial diet. The crude animal protein content was 24%, 30%, and 37.5% in the soldier fly diet, fishmeal diet, and commercial diet, respectively. The authors concluded that black soldier fly could not replace fishmeal as the sole source of animal protein in fish feed.

Kroeckel et al. (2012) examined insect meal derived from prepupae of the black soldier fly as a substitute for fishmeal (herring) for juvenile turbot *Psetta maxima* (L.), a carnivorous flatfish in marine waters. The prepupae were defatted before combining with other components in the feed. Wheat starch and fish oil were key components. They formulated six diets with 0%, 17%, 33%, 49%, 64%, or 76% insect meal. The diets were in pellet form. The authors discovered that growth performance of turbot decreased, generally, as the percentage of insect meal in test diets increased. Fish fed the control diet (no insect meal) showed higher specific growth rate and final body weight than those fed any of the diets containing insect meal. The feed conversion ratio (food intake/weight gain) was approximately the same for diets with 0%, 17%, and 33% insect meal. Greater percentages of insect meal (49%, 64%, and 76%) gave a significantly higher food conversion ratio. The authors concluded that juvenile turbot could have some difficulty digesting chitin (from the soldier fly); they did not detect chitinase activity in fish intestines. They concluded that insect meal derived from black soldier fly can replace only a limited percentage of fishmeal in diets for turbot. Despite these results, the ease with which this fly can be reared on agricultural wastes—effectively converting these wastes into body protein and fat—warrants further study on using it as a protein and fat replacement in food for fish or other livestock (Sheppard et al., 1994; St-Hilaire et al., 2007b; Diener et al., 2011; Kroeckel et al., 2012).

16.2.3. Yellow Mealworm and Superworm

The yellow mealworm is common in pet stores and other places that sell and market exotic organisms (birds, reptiles, amphibians, and fish) and their nutritional supplies. Yellow mealworms can provide essential amino acids and fats in the diet of humans that consume them (Ghally and Alkoaik, 2009; Oonincx and de Boer, 2012). Unfortunately, published research on the nutritional value and suitability of the mealworm as feed to replace or supplement fishmeal in the diets of fish cultured for human consumption is scant. One study compared the potential of mealworm in the form of a dry powder formulation (mealworm meal) as a partial replacement for fishmeal (Ng, 2000). The authors replaced

0%, 20%, 40%, 60%, 80%, or 100% of the fishmeal with mealworm meal in diets for the African catfish *C. gariepinus*. They were able to replace up to 80% of the fishmeal in the catfish diet without negatively affecting growth rate or feed utilization efficiency (Ng, 2000).

The superworm is also a common insect in pet stores. Similarly, published research on the usefulness and nutritional value of this insect as feed for fish that are destined for human consumption is difficult to find. One study formulated diets (dry pellets) using protein from superworm to replace from 0% to 100% of the protein from fishmeal as feed for the Nile tilapia *O. niloticus* (Jabir et al., 2012a). At the conclusion of an 8-week feeding trial, they found that fish fed the diets in which protein from superworm replaced 25% or 50% of the fishmeal experienced the greatest weight gain, specific growth rates, feed conversion ratio, and protein efficiency ratio. Higher percentages of superworm in the fish diets resulted in a decrease in the parameters relating to growth. Note that soybean meal and rice bran were major components in all of the diet formulations (ranging from 22% to 31%); the total crude protein in each formulation was nearly the same (ranging from 31% to 34%) in this study (Jabir et al., 2012a). In a companion study, Jabir et al. (2012b) compared the nutrient profiles of diets (dry pellets) containing protein from fishmeal or superworm meal as feed for the red tilapia *Oreochromis* spp. Superworm meal had lower protein but higher fat content than fishmeal. Seventeen amino acids (including the eight essential amino acids) were found in the superworm and fishmeal diets. Only two amino acids, arginine and tyrosine, were found in higher levels in the superworm meal than in the fishmeal. Based on apparent digestibility coefficients of protein and lipids, the authors discovered that fish had more difficulty digesting the superworm meal than the fishmeal. Slight improvements to the superworm diets are necessary before they can totally replace fishmeal as feed for red tilapia (Jabir et al., 2012b).

16.3. CHALLENGES AND OPPORTUNITIES TO EXPANSION OF MARKET FOR INSECTS AS FEED

16.3.1. Artificial Diets

The desire and necessity to rear insects on artificial diets rather than their natural food to reduce costs has been one of the most important topics in the insect-rearing community (Vanderzant, 1974; Singh, 1977; Cohen, 2004). Researchers are working to develop an artificial diet for silkworms (*B. mori* and relatives) to expedite its rearing. *Bombyx mori* depends on mulberry leaves for proper growth, development, and reproduction. Researchers in China and India have developed crude artificial diets of mulberry leaf powder as food for silkworm moth larvae in attempts to reduce costs of maintaining live plants (Cappellozza et al., 2005). Efforts to increase the suitability of the artificial diet (as compared to fresh mulberry leaves) for silkworm are ongoing. The lack of ascorbic acid (vitamin C) in the powdered diet could limit its suitability for optimum growth and development of silkworm moth larvae in lieu of fresh mulberry (Cappellozza et al., 2005). Supplementing natural diets (fresh mulberry) with ascorbic acid improves larval growth, development, and cocoon weight in *B. mori* (Singh and Bandey, 2012). The eri silkworm *S. ricini*, which is indigenous to northeast India, performs well on foliage from castor or tapioca (Longvah et al., 2011). There is no crude artificial diet for the eri silkworm.

There is limited need to develop an artificial diet for the common house fly *M. domestica* because researchers have reared it for decades on natural substances such as garbage, waste products, and animal feces under field conditions. Ocio et al. (1979) grew *M. domestica* larvae in residues of municipal organic waste. Aniebo et al. (2008) reared larvae on a mixture of cattle blood and wheat bran. For the sake of laboratory rearing, investigators have formulated several crude artificial diets (Spiller, 1963). For example, Hogsette (1992) showed that *M. domestica* reared well on a mixture of alfalfa meal, wheat bran, and corn meal.

Tomberlin et al. (2002) tested the suitability of three diets as food for black soldier fly larvae in the laboratory. The first diet was a mixture of alfalfa meal, wheat bran, and corn meal, whereas the second contained alfalfa meal, wheat bran, and brewer's dried grain. The third diet contained undisclosed proprietary components. The protein contents in these diets were 15%, 19%, and 15%, respectively. There were no significant differences between diets in terms of their effects on larval development or survivorship. All three diets were suitable for rearing the soldier fly; more than 96% of larvae survived to become prepupae, and up to 27% emerged as adults of slightly female-biased sex ratio (55–60% females). Unfortunately, adult soldier flies were smaller in size when reared on the three diets rather than under natural field conditions in swine and poultry manure (Tomberlin et al., 2002). Black soldier flies mated well inside screen cages (2×2×4 m) in a small greenhouse (7×9×5 m), accessible to natural sunlight (Sheppard et al., 2002). Mating even occurred during the winter months in Georgia (the southeastern United States) when cloud cover did not block penetration of sunlight. Adults required water but not food.

Davis (1975) reared yellow mealworm on a mixture of amino acids in proportions occurring in larval tissues as the only source of protein. He commented that his diet was inferior to one based simply on wheat plus brewer's yeast. Zinc and potassium might be essential nutrients to incorporate into an artificial diet (based on casein protein) for normal development of *T. molitor* over multiple generations (Fraenkel, 1958). Culturists have reared yellow mealworms and the superworm on brewer's yeast, rolled oats, white potato, wheat germ, and other stored products for decades. Lipke and Fraenkel (1955) found that corn germ inhibits growth of *T. molitor*. Because of the low cost of many stored products, despite the fact that humans also consume these items, there has been limited progress on developing and refining artificial diets for mealworms and super worms.

Effective and economical artificial diets might facilitate the scale up of rearing operations to provide enough insect protein to support commercial aquaculture. Assessing the quality or health of insects reared on artificial diets against those reared on natural food is essential to any successful mass rearing operation (Cohen, 2004).

16.3.2. Scale-Up of Production

The major concern at this point is increasing the cost-effectiveness of rearing systems for the insect species described previously, considering all components (artificial diets, enclosures, environmental conditions). Designing cages that maximize the production per unit of space, automation of rearing, and use of machinery to replace human labor could reduce some costs associated with rearing on a large scale (Nordlund, 1998; Smith and Nordlund, 1999). Only then can one begin to consider increasing their production on a scale that will meet the needs of the commercial industry. Researchers and industry representatives have begun to engineer rearing factories with specific environmental conditions (light, temperature, humidity) to support large-scale rearing of insects indoors. Published information on the size, design, and engineering of factories that produce the insects discussed in this chapter is limited.

A commercial producer of mealworms (van de Ven Insectenkwekerij) operates a facility in the Netherlands (see Oonincx and de Boer, 2012). At this facility, the producer grows the yellow mealworm and superworm in large quantities and markets them live or freeze-dried to wholesalers, zoos, and import and export companies (http://www.insectenkwekerij.nl/). A commercial producer of the common house fly in Shan Dong Province, China produces 10 and 30 metric tons of dried and live maggots, respectively, per month in a 1000 m^2 plant (http://shengyang2006.en.busytrade.com). They market their products to the medical and pharmaceutical industries and pet food manufacturers around the globe.

A number of companies in Asia produce and market silkworms, particularly the Oriental silkworm moth *B. mori*. Bratac (http://www.bratac.com.br/bratac/pt/index.php), a company in Parana, Brazil, has rapidly expanded its production of silk. The requisite of feeding on mulberry foliage could hamper the scale-up of production of *B. mori* in some countries where mulberry is less abundant. A potential market for the eri silkworm *S. ricini* is growing in northeast India. This species feeds well on castor and tapioca plants rather than mulberry. Longvah et al. (2011) stated that silk production in India reached approximately 18,475 metric tons from 2006 to 2007. Pupae, which are a waste byproduct of sericulture, represent approximately 60% of the weight of cocoons. These waste pupae could generate 4000 and 2000 metric tons of protein and oil, respectively, each year (Longvah et al., 2011). Therefore, collection and utilization of silkworm pupae could provide added economic benefits to farmers and the sericulture industry if silkworm protein can replace a significant percentage of fishmeal in feed for fish and other livestock.

Researchers have produced the black soldier fly outdoors in sufficient quantities to suggest that scale-up of production to meet the needs of the aquaculture industry is a possibility. A side benefit of this system of rearing is the management of manure (Diener et al., 2011). The soldier fly ingests some wastes and thereby converts it into proteins and fats that are suitable for consumption by fish and other livestock (Sheppard et al., 1994; St-Hilaire et al., 2007b). The soldier fly can also convert organic waste into biodiesel fuel (Li et al., 2011). No special equipment is necessary to contain the soldier fly; larvae essentially develop in waste in the near vicinity of the livestock. This system of rearing the soldier fly is crude, but it could be economically and environmentally sustainable to the farmer with limited resources.

The capacity to manage fish wastes using the house fly and soldier fly should help facilitate the scale-up potential of rearing systems of both species (see Sheppard et al., 1994; Diener et al., 2011). On the other hand, we rarely consider techniques for managing insect wastes (feces) in connection with scale up of insect production. Elaborate techniques may or may not be necessary in this regard, depending on the habitat of the insect. Clearly, more research on this topic is necessary. Feces from terrestrial species (mealworms) could be sieved away from developing larvae and their food medium. For insects living within aquatic systems or very humid terrestrial systems, special filtration systems may be necessary to separate insect waste from fish waste. Feces can be air-dried, disinfected to eliminate pathogens (e.g. bacteria), then used as a biodegradable fertilizer for plants in terrestrial systems. One company in southern California sells insect feces to gardeners for this purpose (see http://www.pchydro.com/news_article/article_id/50).

16.4. QUALITY CONTROL AND PRODUCTION

16.4.1. Maintenance of Long-Term Insect Colonies

Techniques that could prevent the decline or eventual deterioration of the colony are necessary in rearing all species highlighted in this review. Methods used to prevent decline in colonies may include manipulative breeding between hybrids to restore and maintain normal growth and reproduction of populations in culture. Mating between siblings should be discouraged (Mackauer, 1976; Roush and Hopper, 1995). Introducing wild (field) adults into well-established colonies to mate with domesticated adults could prevent inbreeding depression and any subsequent reduction in fitness. (Careful screening of wild adults is necessary to prevent inadvertent spread of pathogens and parasitoids into the colonies.) Periodic monitoring of colonies for apparent undesirable changes in behavior of adults (compared to wild adults) is necessary (Boller, 1972; Huettel, 1976).

16.4.2. Colony Hygiene and Preventing Disease Transmission

As mentioned briefly in the introduction of this chapter, the production of high-quality insects is an important topic to those that rear them professionally (Schneider, 2009). Ensuring that insects are produced using the highest standards of care and nutrition (food sources) is profitable to everyone involved. Rearing professionals must maintain healthy, disease-free colonies in laboratories or rearing facilities. Most insects held in confinement at the moderate to high densities necessary for mass production can potentially harbor pathogens. Consequently, culturists must routinely check the health of their colonies and maintain hygienic rearing conditions. In this chapter, the common house fly—more than the other species—is probably most often revered as a vector of pathogens that can harm humans (Banjo et al., 2005; Fasanella et al., 2010). Culturists must use constant vigilance in ensuring that adult house flies do not escape from rearing facilities. The inadvertent transmission of disease is not a concern for the black soldier fly; they do not vector human pathogens. In fact, the black soldier fly can limit populations of the house fly through competitive exclusion under natural conditions in the field (Sheppard, 1983; Bradley and Sheppard, 1984). The black soldier fly can also help reduce infectious bacterial populations in chicken and cow manure (Erickson et al., 2004; Liu et al., 2008).

The other species highlighted in this chapter—yellow mealworm, superworm, and several species of silkworm—are not known to vector human pathogens. However, they all have their own complement of parasites and pathogens (Barnes and Siva-Jothy, 2000; Valtonen et al., 2010; Guo-Ping and Xi-Jie, 2011; Isaiarasu et al., 2011), and several could reduce the health of a colony if left unchecked. The effects of feeding diseased or contaminated insects—whether in a natural form or formulated into an insect meal or pellet—to fish is unknown. We know very little about the interactions of insect pathogens and fish health, so this subject requires research. The likelihood that fish would become contaminated after feeding on diseased/unhealthy insects and therefore become unfit for human consumption is very remote, but it cannot be ruled out entirely (see Noonin et al., 2010).

16.5. CONCLUSIONS AND RECOMMENDATIONS

16.5.1. Synthesis

The reliance on fishmeal arose, in part, from a need to supply mass quantities of animal protein in the diet to meet the nutritional requirements of fish in culture. Today, rising costs, unreliable availability, and questionable quality (safety) of fishmeal has encouraged many feed production companies to search for alternative sources of animal protein to incorporate into fish feeds. In this review, I have attempted to demonstrate that all model insects have merits as potential sources of protein to incorporate into feed for fish destined for human consumption. The model species examined in this review have positive and negative aspects that could affect their adoption as sources of insect protein in feed for fish and crustaceans on a large scale. The differences in protein and fat content between these species are important considerations. It varies between species and between stages of development within a species. Most research shows that the yellow mealworm and the superworm are at the high end of the scale for these nutrients; the Oriental silkworm is at the low end (see Finke, 2002). However, other factors such as ease of rearing, availability of artificial diets to scale-up rearing, potential of vectoring infectious diseases, and societal approval are just as important. The black soldier fly is a prime candidate for large-scale production. It is easily reared in the same outdoor facilities alongside fish and has an added benefit of recycling animal wastes. Although it is not known to vector human pathogens, societal acceptance of this fly—or any fly species, for that matter—is low, particularly in the more affluent societies around the globe. Thus, we will need to re-educate the public on the value of insects and their potential role in aquaculture in the near future.

Most studies show that insect protein can be formulated into a dry powdered meal or into pellets. Researchers mix plant protein with insect protein, along with various lipids, into the meal or pellets. The combination of the insect and plant-based nutrients into the same feed does not appear to affect its acceptance by fish, at least not for omnivorous species such as catfish and carp. These fish accept the insect meal or pellets as readily as live or whole insects. This suggests that the procedures used to process the insect-based feeds do not destroy essential nutrients that fish require for growth and reproduction.

The innate feeding preferences of fish and prawn may determine the suitability of insects as protein in feed. Research shows that omnivorous species (e.g. carp and catfish) that typically feed on insects found on the bottom of lakes, ponds, and other temporary or permanent bodies of water survive on feed containing a considerable percentage of insect protein. Therefore, omnivorous fish (especially freshwater species) are less dependent on fishmeal. For example, fishmeal represents a small percentage (less than 5%) of the total composition of commercial feed for channel catfish farmed in the United States (Li et al., 2008; Robinson and Li, 2012). In 2007, fishmeal only represented 5% of the total composition of feed for Chinese carp (Tacon and Metian, 2008; Naylor et al., 2009). Highly carnivorous species (e.g. trout and salmon) also eat insects, particularly in the juvenile stages, but research shows their growth and feeding efficiency declines when their feed contains a considerable amount of insect protein. Highly carnivorous fish and marine fish generally have a high requirement (400–500 g/kg dietary dry matter) for protein rich in essential amino acids (Sargent and Tacon, 1999). These fish can easily fulfill this requirement with the consumption of fishmeal.

Several crude artificial diets are available to rear some insects that show promise as sources of protein to support the aquaculture industry. However, more research is necessary to refine these diets. A systems approach to rearing of target insects is necessary; this would encompass artificial diets, efficient design of enclosures, and automation (mechanization) of essential operations to decrease the costs associated with human labor. Establishment of insect factories that can produce the massive quantities of insects needed to support the increasing demands of the aquaculture industry is also necessary.

16.5.2. Future Research

1. More research in the area of feeding preferences of fish that show some affinity for insect protein (in comparison to species that do not) is necessary. We need to determine if some fish can adapt after several generations of exposure to diets that contain more protein derived from insect meal rather than fishmeal. There is reason to believe that a carnivorous species could adapt to an insect-based diet. Note that carnivorous fish can potentially adapt to a plant-based diet after a few generations of artificial selection (see Le Boucher et al., 2011, 2012).
2. The potential of producing insects to meet the increasing needs of the fish industry is unclear. Knowledge on this subject has rarely progressed beyond experimental stages in small plots or enclosures using juvenile fish. More research is necessary to prove that insects can be produced effectively and efficiently to support the nutrition of juvenile and adult fish to meet the demands of this burgeoning market.
3. Expansion of the efforts to promote insect protein to niche markets, such as producers/sellers of exotic fish and other pets, is encouraged. The pet industry may be a near-term (rather than long-term) option to using insect protein rather than fishmeal in artificial diets designed for tropical and temperate fish sold as pets. Pet stores sell insects (e.g. mealworms) as live food for birds, reptiles, amphibians, and fish (Barker et al., 1998; Finke, 2002). Formulation of insect protein into a meal or dry pellet to reduce costs and increase shelf life (compared to live food) would be advantageous. This idea would be problematic, however, for species that only accept live prey (e.g. many amphibians in the adult stage).
4. Development of cost-effective artificial diets to rear insects on a large scale, accompanied by design of machinery to automate rearing, will be a prerequisite to expanding the market for insect protein. Automation could potentially reduce labor costs.
5. Disclosure of information on the size, design, and engineering of rearing facilities (factories, warehouses) is necessary. Presently, very few companies that produce and market insects as feed for vertebrates furnish details on the design of their rearing facilities in the published literature. Financial incentives might be necessary to encourage companies to disclose and share information.
6. More collaboration between the feed industry, government and academic researchers, and local farmers is necessary to identify novel sources of insect protein and design rearing systems that are efficient, cost-effective, and sustainable. For example, Neptune Industries (http://www.neptuneindustries.net) was collaborating with Mississippi State University to patent Ento-Protein, a product formulated with protein from laboratory-reared insects (http://www.dafvm.msstate.edu/landmarks/07/fall/8-9.pdf). The company intended to market Ento-Protein as a replacement for fishmeal.

ACKNOWLEDGMENTS

John C. Schneider and Jeffrey Tomberlin reviewed an earlier version of this article. The U.S. Government has the right to retain a nonexclusive, royalty-free license in and to any copyright of this article. This article reports the results of research only. Mention of a commercial or proprietary product does not constitute an endorsement of the product by the U.S. Department of Agriculture.

REFERENCES

Abowei, J.F.N., Ekubo, A.T., 2011. A review of conventional and unconventional feeds in fish nutrition. Br. J. Pharmacol. Technol. 2, 179–191.

Akiyama, T., Murai, T., Hirasawa, Y., Nose, T., 1984. Supplementation of various meals to fish diet for chum salmon fry. Aquaculture 37, 217–222.

Aniebo, A.O., Erondu, E.S., Owen, O.J., 2008. Proximate composition of housefly larvae (*Musca domestica*) meal generated from mixture of cattle blood and wheat bran. Livestock Res. Rural Develop. 20, 5, Available online: www.lrrd.org/lrrd20/12/anie20205.htm.

Aniebo, A.O., Erondu, E.S., Owen, O.J., 2009. Replacement of fishmeal with maggot meal in African catfish (*Clarias gariepinus*) diets. Rev. UDO Agric. 9, 666–671.

Banjo, A.D., Lawal, O.A., Adeduji, O.O., 2005. Bacteria and fungi isolated from housefly (*Musca domestica* L.) larvae. Afr. J. Biotechnol. 4, 780–784.

Barker, D., Fitzpatrick, M.P., Dierenfeld, E.S., 1998. Nutrient composition of selected whole invertebrates. Zoo Biol. 17, 123–134.

Barnes, A.I., Siva-Jothy, T., 2000. Density-dependent prophylaxis in the mealworm beetle *Tenebrio molitor* L. (Coleoptera: Tenebrionidae): cuticular melanization is an indicator of investment in immunity. Proc. R. Soc. Lond. B. 267, 177–182.

Boller, E., 1972. Behavioral aspects of mass-rearing of insects. Entomophaga 17, 9–25.

Bondari, K., Sheppard, D.C., 1987. Soldier fly, *Hermetia illucens* L., larvae as feed for channel catfish, *Ictalurus punctatus* (Rafinesque), and blue tilapia, *Oreochromis aureus* (Steindachner). Aquacult. Fish Manage. 18, 209–220.

Bradley, S.W., Sheppard, D.C., 1984. House fly oviposition inhibition by larvae of *Hermetia illucens*, the black soldier fly. J. Chem. Ecol. 10, 853–859.

Cappellozza, L., Cappellozza, S., Saviane, A., Sbrenna, G., 2005. Artificial diet rearing system for the silkworm *Bombyx mori* (Lepidoptera: Bombycidae): effect of vitamin C deprivation on larval growth and cocoon production. Appl. Entomol. Zool. 40, 405–412.

Cohen, A.C., 2004. Insect Diets: Science and Technology. CRC Press LLC, Boca Raton, FL.

Davis, G.R.F., 1975. Essential dietary amino acids for growth of larvae of the yellow mealworm, *Tenebrio molitor* L. J. Nutr. 105, 1071–1075.

Diener, S., Solano, N.M.S., Gutiérrez, F.R., Zurbrügg, C., Tockner, K., 2011. Biological treatment of municipal organic waste using black soldier fly larvae. Waste Biomass Valorization 2, 357–363.

Erickson, M.C., Islam, M., Sheppard, C., Liao, J., Doyle, M.P., 2004. Reduction of *Escherichia coli* 0157:H7 and *Salmonella enterica* serovar enteritidis in chicken manure by larvae of the black soldier fly. J. Food Prod. 67, 685–690.

Fasakin, E.A., Balogun, A.M., Ajayi, O.O., 2003. Evaluation of full-fat and defatted maggot meals in the feeding of clariid catfish *Clarias gariepinus* fingerlings. Aquacult. Res. 34, 733–738.

Fasanella, A., Scasciamacchia, S., Garofolo, G., Giangaspero, A., Tarsitano, E., Adone, R., 2010. Evaluation of the house fly *Musca domestica* as a mechanical vector for an anthrax. PLoS One 5 (8), e12219, doi:10.137/journal.pone.0012219.

Finke, M.D., 2002. Complete nutrient composition of commercially raised invertebrates used as food for insectivores. Zoo Biol. 21, 269–285.

Fraenkel, G.S., 1958. The effect of zinc and potassium in the nutrition of *Tenebrio molitor*, with observations on the expression of a carnitine deficiency. J. Nutr. 65, 361–395.

Ghally, A.E., Alkoaik, F.N., 2009. The yellow mealworm as a novel source of protein. Am. J. Agric. Biol. Sci. 4, 319–331.

Guo-Ping, K., Xi-Jie, G., 2011. Overview of silkworm pathology in China. Afr. J. Biotechnol. 10, 18046–18056.

Hogsette, J.A., 1992. New diets for the production of house flies and stable flies (Diptera: Muscidae) in the laboratory. J. Econ. Entomol. 85, 2291–2294.

Huettel, M.D., 1976. Monitoring the quality of laboratory-reared insects: a biological and behavioral perspective. Environ. Entomol. 5, 807–814.

Hwango, J., Hong, E.C., Jang, A., Kang, H.K., Oh, J.S., Kim, B.W., Park, B.S., 2009. Utilization of house fly-maggots, a feed supplement in the production of broiler chickens. J. Environ. Biol. 30, 609–614.

Isaiarasu, L., Sakthivel, N., Ravikumar, J., Samuthiravelu, P., 2011. Effect of herbal extracts on the microbial pathogens causing flacherie and muscardine diseases in the mulberry silkworm, *Bombyx mori* L. J. Biopestic. 4, 150–155.

Jabir, M.D.A.R., Razak, S.A., Vikineswary, S., 2012a. Nutritive potential and utilization of super worm (*Zophobas morio*) meal in the diet of Nile tilapia (*Oreochromis niloticus*) juvenile. Afr. J. Biotechnol. 11, 6592–6598.

Jabir, M.D.A.R., Razak, S.A., Vikineswary, S., 2012b. Chemical composition and nutrient digestibility of super worm meal in red tilapia juvenile. Pak. Vet. J. 32, 489–493.

Jayaram, M.G., Shetty, H.P.C., 1981. Formulation, processing and water stability of two new pellet fish feeds. Aquaculture 23, 355–359.

Jintasataporn, O., Chumkam, S., Jintasataporn, O., 2011. Substitution of silkworm pupae (*Bombyx mori*) for fishmeal in broodstock diets for snakeskin gourami (*Trichogaster pectoralis*). J. Agric. Sci. Technol. A 1, 1341–1344.

Kroeckel, S., Harjes, A.G.E., Roth, I., Katz, H., Wuertz, S., Susenbeth, A., Schulz, C., 2012. When a turbot catches a fly: evaluation of a pre-pupae meal of the black soldier fly (*Hermetia illucens*) as fish meal substitute—growth performance and chitin degradation in juvenile turbot (*Psetta maxima*). Aquaculture 2012, 345–352.

Langer, S., Bakhtiyar, Y., Lakhnotra, R., 2011. Replacement of fishmeal with locally available ingredients in diet composition of *Macrobrachium dayanum*. Afr. J. Agric. Res. 6, 1080–1084.

Le Boucher, R., Quillet, E., Vandeputte, M., Lecalvez, J.M., Goardon, L., Chatain, B., Médale, F., Dupont-Nivet, M., 2011. Plant-based diet in rainbow trout (*Oncorhynchus mykiss* Walbaum): are there genotype-diet interactions for main production traits when fish are fed marine vs. plant-based diets from the first meal? Aquaculture 321, 41–48.

Le Boucher, R., Dupont-Nivet, M., Vandeputte, M., Kerneïs, T., Goardon, L., Labbé, L., Chatain, B., Bothaire, M.J., Larroquet, L., Médale, F., Quillet, E., 2012. Selection for adaptation to dietary shifts: towards sustainable breeding of carnivorous fish. PLoS One 7 (9), e44898. http://dx.doi.org/10.1371/journal.pone.0044898.

Li, M.H., Robinson, E.H., Peterson, B.C., Bates, T.D., 2008. Growth and feed efficiency of juvenile channel catfish reared at different water temperatures and fed diets containing various levels of fish meal. North Am. J. Aquacult. 70, 347–352.

Li, Q., Zheng, L., Cai, H., Garza, E., Yu, Z., Zhou, S., 2011. From organic waste to biodiesel: black soldier fly, *Hermetia illucens*, makes it feasible. Fuel 90, 1545–1548.

Lipke, H., Fraenkel, G., 1955. The toxicity of corn germ to the mealworm, *Tenebrio molitor*. J. Nutr. 55, 165–178.

Little, D.C., Surintaraseree, P., Innes-Taylor, N., 1996. Fish culture in rainfed rice fields of northeast Thailand. Aquaculture 140, 295–321.

Liu, Q., Tomberlin, J.K., Brady, J.A., Sanford, M.R., Yu, Z., 2008. Black soldier fly (Diptera: Stratiomyidae) larvae reduce *Escherichia coli* in dairy manure. Environ. Entomol. 37, 1525–1530.

Longvah, T., Mangthya, K., Ramulu, P., 2011. Nutrient composition and protein quality evaluation of eri silkworm (*Samia ricini*) prepupae and pupae. Food Chem. 128, 400–403.

Mackauer, M., 1976. Genetic problems in the production of biological control agents. Annu. Rev. Entomol. 21, 369–385.

Nandeesha, M.C., Gangadhara, B., Manissery, J.K., 1999. Silkworm pupa oil and sardine oil as an additional energy source in the diet of common carp, *Cyprinus carpio*. Asian Fish Sci. 12, 207–215.

Nandeesha, M.C., Gangadhara, B., Varghese, T.J., Keshavanath, P., 2000. Growth response and flesh quality of common carp, *Cyprinus carpio* fed with high levels of nondefatted silkworm pupae. Asian Fish Sci. 13, 235–242.

Nandeesha, M.C., Srikanth, G.K., Keshavanath, P., Varghese, T.J., Basavaraja, N., Das, S.K., 1990. Effects of non-defatted silkworm pupae in diets on the growth of common carp, *Cyprinus carpio*. Biol. Wastes 33, 17–23.

Naylor, R.L., Goldburg, R.J., Primavera, J., Kautsky, N., Beveridge, M.C.M., Clay, J., Folke, C., Lubchenco, J., Mooney, H., Troell, M., 2001. Effects of aquaculture on world fish supplies. Issues Ecol. Winter 2001, 1–12.

Naylor, R.L., Hardy, R.W., Bureau, D.P., Chiu, A., Elliott, M., Farrell, A.P., Forster, I., Gatlin, D.M., Goldburg, R.J., Hua, K., Nichols, P.D., 2009. Feeding aquaculture in an era of finite resources. Proc. Natl. Acad. Sci. U. S. A. 106, 15103–15110.

Nazni, W.A., Seleena, B., Lee, H.L., Jeffery, J.T., Rogayah, T.A.R., Sofian, M.A., 2005. Bacteria fauna from the house fly, *Musca domestica* (L.). Trop. Biomed. 22, 225–231.

Ng, W.K., 2000. Worms: a potential feed source for cultured aquatic animals. The Advocate, 82–83, June 2000.

Noonin, C., Jiravanichpaisal, P., Söderhöll, I., Merino, S., Tomás, J.M., Söderhöll, K., 2010. Melanization and pathogenicity in the insect, *Tenebrio molitor*, and the crustacean, *Pacifastacus leniusculus*, by *Aeromonas hydrophila* AH-3. PLoS One 5 (12), e15728, doi:10.137/journal.pone.0015728.

Nordlund, D.A., 1998. Capacity and quality: keys to success in the mass rearing of biological control agents. Nat. Enemies Ins. 20, 169–179.

Nsofor, C.I., Osayamwen, E.M., Ewuim, S.C., Etaga, H.O., 2008. Effects of varying levels of maggot and fish meal on food utilization and growth of *Clarias gariepinus* (Burchell, 1822) fingerlings reared in net hapas in concrete ponds. Nat. Appl. Sci. J. 9 (1), 7, Available online at www.nasjournal-nq.org.

Ocio, E., Viñaras, R., Rey, J.M., 1979. House fly larvae meal grown on municipal organic waste as a source of protein in poultry diets. Anim. Feed Sci. Technol. 4, 227–231.

Omoyinmi, G.A.K., Olaoye, O.J., 2012. Growth performance of Nile tilapia-*Oreochromis niloticus* fed diets containing different sources of animal protein. Libyan Agric. Res. Cent. J. Intl. 3, 18–23.

Oonincx, D.G.A.B., de Boer, I.J.M., 2012. Environmental impact of the production of mealworms as a protein source for humans—a life cycle assessment. PLoS One 7 (12), e51145. http://dx.doi.org/10.1371/journal.pone.0051145.

Peterson, B.C., Booth, N.J., Manning, B.B., 2012. Replacement of fish meal in juvenile channel catfish, *Ictalurus punctatus*, diets using a yeast-derived protein source: the effects on weight gain, food conversion ratio, body composition and survival of catfish challenged with *Edwardsiella ictaluri*. Aquacult. Nutr. 18, 132–137.

Robinson, E.H., Li, M.H., 2012. Composition and formulation of channel catfish feeds. Miss. Agric. For. Exp. Stn. Bull. 1200, 10.

Roush, R.T., Hopper, K.R., 1995. Use of single family lines to preserve genetic variation in laboratory colonies. Ann. Entomol. Soc. Am. 88, 713–717.

Sargent, J.R., Tacon, A.G.J., 1999. Development of farmed fish: a nutritionally necessary alternative to meat. Proc. Nutr. Soc. 58, 377–383.

Schneider, J.C., 2009. Principles and Procedures for Rearing High Quality Insects. Mississippi State University, Mississippi State, MS.

Sealey, W.M., Gaylord, T.G., Barrows, F.T., Tomberlin, J.K., McGuire, M.A., Ross, C., St-Hilaire, S., 2011. Sensory analysis of rainbow trout, *Oncorhynchus mykiss*, fed enriched black soldier fly prepupae, *Hermetia illucens*. J. World Aquacult. Soc. 42, 34–45.

Sheppard, C., 1983. House fly and lesser fly control utilizing the black soldier fly in manure management systems for caged laying hens. Environ. Entomol. 12, 1439–1442.

Sheppard, D.C., Newton, G.L., Thompson, S.A., Savage, S., 1994. A value added manure management system using the black soldier fly. Bioresour. Technol. 50, 275–279.

Sheppard, D.C., Tomberlin, J.K., Joyce, J.A., Kiser, B.C., Sumner, S.M., 2002. Rearing methods for the black soldier fly (Diptera: Stratiomyidae). J. Med. Entomol. 39, 695–698.

Singh, P., 1977. Artificial Diets for Insects, Mites, and Spiders. Plenum Press, New York, NY.

Singh, A., Bandey, S.A., 2012. Supplementation of synthetic vitamin C in the fifth instars bivoltine hybrid larvae of NB4 $D_2 \times SH_6$ of silkworm, *Bombyx mori* L. Intl. J. Food Agric. Vet. Sci. 2, 54–57.

Smith, R.A., Nordlund, D.A., 1999. Automation of insect rearing—a key to the development of competitive augmentative biological control. Nat. Enemies Ins. 21, 70–81.

Sogbesan, A.O., Ajuonu, N., Musa, B.O., Adewole, A.M., 2006. Harvesting techniques and evaluation of maggot meal as animal dietary protein source for 'Heteoclarias' in outdoor concrete tanks. World J. Agric. Sci. 2, 394–402.

Spiller, D., 1963. Procedure for rearing houseflies. Nature 199, 405.

St-Hilaire, S., Sheppard, C., Tomberlin, J.K., Irving, S., Newton, L., McGuire, M.A., Mosley, E.E., Hardy, R.W., Sealey, W., 2007a. Fly prepupae as a feedstuff for rainbow trout, *Oncorhynchus mykiss*. J. World Aquacult. Soc. 38, 59–67.

St-Hilaire, S., Cranfill, K., McGuire, M.A., Mosley, E.E., Tomberlin, J.K., Newton, L., Sealey, W., Sheppard, C., Irving, S., 2007b. Fish offal recycling by the black soldier fly produces a foodstuff high in omega-3 fatty acids. J. World Aquacult. Soc. 38, 309–313.

Tacon, A.G.J., Metian, M., 2008. Global overview on the use of fish meal and fish oil in industrially compounded aquafeeds: trends and future prospects. Aquaculture 285, 146–158.

Tomberlin, J.K., Sheppard, D.C., Joyce, J.A., 2002. Selected life-history traits of black soldier flies (Diptera: Stratiomyidae) reared on three artificial diets. Ann. Entomol. Soc. Am. 95, 379–386.

Turchini, G.M., Mentati, T., Caprino, F., Panseri, S., Moretti, V.M., Valré, F., 2003. Effects of dietary lipid sources on flavor volatile compounds of brown trout (*Salmo trutta* L.) fillet. J. Appl. Ichthyol. 20, 71–75.

Valtonen, T.M., Viitaniemi, H., Rantala, M., 2010. Copulation enhances resistance against an entomopathogenic fungus in the mealworm beetle *Tenebrio molitor*. Parasitology 137, 985–989.

Vanderzant, E.S., 1974. Development, significance, and application of artificial diets for insects. Annu. Rev. Entomol. 19, 139–160.

van Huis, A., 2013. Potential of insects as food and feed in assuring food security. Annu. Rev. Entomol. 58, 563–583.

Zhang, C.X., Tang, X.D., Cheng, J.A., 2008. The utilization and industrialization of insect resources in China. Entomol. Res. 38, S38–S47.

Chapter 17

Insects as Food for Insectivores

Mark D. Finke[1] and Dennis Oonincx[2]
[1]*Mark Finke LLC, Rio Verde, AZ, USA,* [2]*Laboratory of Entomology, Department of Plant Sciences, Wageningen University, Wageningen, The Netherlands*

17.1. INTRODUCTION

Insects are generally considered to be a good source of most nutrients (DeFoliart, 1992), and many insect species have been analyzed for their nutrient composition. These analyses include insects taken from the field (Bukkens, 1997; Gullan and Cranston, 2005; Punzo, 2003) and those commercially reared (Dierenfeld and King, 2008; Finke, 2002, 2013; Oonincx et al., 2010; Oonincx and Dierenfeld, 2012; Ramos-Elorduy et al., 2002; Simpson and Raubenheimer, 2001). In nature, most insectivores will feed on a variety of arthropod prey species. As a general rule, it is better for insectivores to be offered a varied diet rather than a single insect species (Bernard and Allen, 1997). When the insectivore is able to select between different insects offered, this allows for self-selection of nutrients (Oonincx et al., 2010; Senar et al., 2010).

A number of variables can influence the chemical composition of insects, such as gender (Ali and Ewiess, 1977; Hoffmann, 1973; Sönmez and Gülel, 2008), stage of development (McClements et al., 2003), diet (Calvez, 1975; Oonincx and van der Poel, 2011; Ramos-Elorduy et al., 2002; Simpson and Raubenheimer, 2001) and environmental factors such as temperature (Hoffmann, 1973; Sönmez and Gülel, 2008), day length (Ali and Ewiess, 1977; Koc and Gulel, 2008; Shearer and Jones, 1996), and humidity (Ali et al., 2011; Han et al., 2008; Nedvěd and Kalushkov, 2012). Additionally, other factors, such as light intensity and spectral composition, might also affect the chemical composition of insects.

Nutrient densities can be expressed on a fresh matter basis. However, although insectivores consume live prey, in most publications data are expressed on a dry matter basis. Variation in moisture content can strongly influence the reported amount of nutrients when these data are expressed as fresh material. Although water is a critical nutrient, in most cases providing it is not the primary function of offering insects as food. Therefore, in this chapter nutrient densities are expressed on a dry matter basis to be consistent with the existing literature.

Mass Production of Beneficial Organisms. http://dx.doi.org/10.1016/B978-0-12-391453-8.00017-0

Large variations in dry matter content exist between species and developmental stages. As a rule of thumb, the dry matter content typically ranges from 15% to 40% of the fresh weight of a live insect (Barker et al., 1998; Bernard and Allen, 1997; Finke, 2002, 2013; Oonincx and van der Poel, 2011; Oonincx and Dierenfeld, 2012; Punzo, 2003).

17.2. NUTRIENT CONTENT OF INSECTS

17.2.1. Protein and Amino Acids

The protein content of insects is highly variable and ranges between 7.5% and 91%, with many species containing approximately 60% protein on a dry matter basis (Barker et al., 1998; Bernard and Allen, 1997; Bukkens, 1997; Finke, 2002, 2013; Oonincx et al., 2010; Oonincx and van der Poel, 2011; Oonincx and Dierenfeld, 2012; Punzo, 2003; Ramos-Elorduy et al., 2002). Protein content is commonly determined by multiplying the amount of nitrogen by 6.25, known as the crude protein content. Although convenient, this method assumes an even distribution of amino acids and may lead to a slight overestimation of true protein content due to the presence of other nitrogen-containing compounds (such as chitin). In most papers in which all amino acids are properly quantified, protein recovery (sum of the amino acids divided by crude protein content) is generally quite good, suggesting that crude protein is a reasonable estimate of true protein in most species of insects (Finke, 2002, 2007, 2013).

Amino acids are the building blocks of proteins. Certain amino acids are known as essential amino acids because they cannot be synthesized from simpler molecules by most animal species (Bender, 2002). It seems plausible that amino acid composition is fairly constant in a given species in a given developmental stage relative to the total amount of protein (Sealey et al., 2011), although plasticity has been suggested (Ramos-Elorduy et al., 2002).

Prediction of an insect's amino acid profile based on species or diet provided is difficult at best. In addition to these factors, various analytical techniques are used, which increases variation (Bukkens, 1997; Finke, 2002, 2013). However, certain trends regarding protein composition can be distinguished. Table 17.1 (Finke, 2002, 2005, 2007, 2013) provides expected ranges for specific amino acids. The amino acids in Table 17.1 together typically account for more than 90% of the crude protein in most insect species. The small amount of nitrogen that is not accounted for by these amino acids is likely due to other nitrogen-containing compounds, such as other amino acids, chitin, and uric acid, plus the nitrogen that is lost in the conversion of glutamine to glutamic acid or asparagine to aspartic acid, during protein hydrolysis.

The nutritional quality of insect protein has generally been described as good, but the quality depends on the digestibility of the amino acids and the match of the amino acid profile to the requirements of the insectivore in question (Finke et al., 1987, 1989; Ramos-Elorduy, 1997). Unfortunately, the amino

TABLE 17.1 Typical Amino Acids Content of Some Common Feeder Insects (All Values Expressed as mg Amino Acid/g Crude Protein)

Amino Acid	Mean	Range
Alanine	71.4	40.9–101.1
Arginine	56.6	41.4–75.5
Aspartic acid	82.0	65.6–95.0
Cystine	9.1	5.6–21.3
Glutamic acid	114.5	96.2–138.3
Glycine	54.7	42.1–84.4
Histidine	27.8	22.1–37.9
Isoleucine	42.6	32.3–50.3
Leucine	76.2	48.9–106.4
Lysine	56.9	44.3–68.0
Methionine	17.0	10.7–29.6
Phenylalanine	34.9	26.2–43.7
Proline	58.1	33.9–77.0
Serine	47.8	35.1–77.0
Threonine	36.6	28.4–41.8
Tryptophan	9.0	5.2–17.3
Tyrosine	56.3	31.2–79.9
Valine	56.1	39.8–73.7
Taurine	1.5	0.0–8.1

acid requirements of most insectivores are unknown, so most comparisons have been made using more common laboratory animals, such as rats or chickens (Finke, 2002, 2013).

The crude protein digestibility of housefly (*Musca domestica* (L.)) larvae and pupae is between 70% and 80% in chickens, and the nine amino acids tested had a digestibility between 86% and 100% (Pretorius, 2011). Crude protein digestibility of the larvae of the black soldier fly (*Hermetia illucens* (L.)) fed to mountain chicken frogs (*Leptodactylus fallax* Müller) was quite low (44%) when the larvae were offered alive; however, when the larvae were mashed, the digestibility increased to 77% (Dierenfeld and King, 2008). The crude protein digestibility of intact house crickets (*Acheta domesticus* (L.)) in these frogs was

95% (Dierenfeld and King, 2008). The lower protein digestibility of intact larvae versus mashed larvae could be due to the protein in the exoskeleton, which might need to be mechanically or enzymatically treated before all of the digestible nutrients can be accessed. The apparent protein digestibility of dried yellow mealworm larvae (*Tenebrio molitor* (L.)) was 75% when fed to rats, which was slightly lower than that for both soy protein concentrate (84%) and casein (88%) (Goulet et al., 1978). It is likely that the digestibility of protein from insects is highly variable. Insects that have a larger proportion of their amino acids in cuticular protein complexed with chitin might be expected to have a lower protein digestibility than those that do not (Finke, 2007).

The first limiting amino acid depends on both the insect species being consumed and the species consuming the prey. Because different insectivore species might require different proportions of amino acids, the first limiting amino acids could differ between them. In chickens fed purified diets in which Mormon crickets (*Anabrus simplex* Haldeman) or house crickets were the sole source of protein, the first limiting amino acids were methionine and arginine (Finke et al., 1985; Nakagaki et al., 1987). When fed to growing rats, the first limiting amino acids in insect protein from yellow mealworm larvae, common housefly larvae, and adult Mormon crickets was methionine (Finke et al., 1987; Goulet et al., 1978; Onifade et al., 2001). Similarly, methionine and cystine were calculated to be the first limiting amino acid for rats in superworm larvae (*Zophobas morio* F.), larvae of the greater wax moth (*Galleria mellonella* (L.)), domesticated silkworm larvae (*Bombyx mori* (L.)), honey bee larvae and pupae (*Apis mellifera* (L.)), adult common house flies, black soldier fly larvae, rusty red roaches (*Blatta lateralis* Walker), and butterworm larvae (*Chilecomadia moorei* Silva) (Finke, 2002, 2005, 2013). As such, it seems likely that for most mammalian insectivores, the sulphur amino acids are first-limiting. For fish, birds, and other species, especially those without a functioning urea cycle, other amino acids such as arginine may be important. It has been proposed that adult passerines have a preference for spiders, driven by a need for enhanced levels of methionine, cystine, or perhaps taurine by their growing chicks (Ramsay and Houston, 2003).

17.2.2. Fats and Fatty Acids

The most common way to estimate insect fat content is by determining the total weight of all fat-soluble molecules (mostly lipids, but also waxes and some other compounds). Fat tissue is used for energy storage in the body. It is either obtained and stored directly from a dietary source or synthesized from carbohydrates (Bender, 2002; Fast, 1970; Hanson et al., 1983). The main storage site for insect lipids is the fat body (Beenakkers et al., 1985). As a dietary component, fat is not only an important energy source, but it may also play a role in the palatability of the insect when fed to insectivores. Large variations in the lipid content (4.6–64% dry matter) of insects have been reported (Barker et al., 1998;

Bukkens, 1997; Finke, 2002, 2013; Punzo, 2003; Yang et al., 2006). As a rule of thumb, most adult wild insects contain less than 10% fat on a fresh weight basis (Fast, 1970; Lease and Wolf, 2011).

There are indications that a certain amount of protein is needed to reach adulthood in migratory locusts (*Locusta migratoria* (L.)). The adult weight in this species varies, but this is mainly due to differences in fat content (Oonincx and van der Poel, 2011; Simpson and Raubenheimer, 2001). This could indicate that differences in weight upon reaching adulthood are a crude indication of differences in fat content.

The behavioral ecology of a species likely influences the amount of fat stored (Thompson, 1973). In certain species, the males have a higher fat content than the females (Fast, 1970; Nakasone and Ito, 1967). For species in which male combat is customary (e.g. Odonata), this can be explained by a need for greater energy reserves. In several families within the order Lepidoptera, males contain greater fat reserves than females (Beenakkers et al., 1985). In most other species, females have greater fat reserves than males (Lease and Wolf, 2011; Nestel et al., 2005; Zhou et al., 1995). In species that use accumulated energy reserves obtained during their larval stages for reproduction, greater reserves would have a positive effect on fecundity because eggs have a high lipid content and therefore provide fitness advantages (Beenakkers et al., 1985; Downer and Matthews, 1976; Lease and Wolf, 2011). Before oviposition, these females would have a higher fat content than after oviposition (Nestel et al., 2005). Insects collected from the wild seem to have a lower fat content than insects that are commercially produced (Finke, 2002, 2013; Oonincx and Dierenfeld, 2012; Yang et al., 2006). This might be a result of decreased movement in captivity, access to high-energy diets, or a combination of the two.

Fatty acids are the building blocks of fat. Two or three fatty acids are coupled to glycerol to form diglycerides and triglycerides, respectively. These fatty acids are stored in the insect fat body, making up more than 90% of the total fat body lipid (Beenakkers et al., 1985; Bender, 2002; Downer and Matthews, 1976). Based on the degree of saturation, fatty acids can be subdivided into saturated fatty acids (those with no double bonds) and unsaturated fatty acids (those with one or more double bonds). Unsaturated fatty acids can be divided in monounsaturated fatty acids and polyunsaturated fatty acids. Polyunsaturated fatty acids can be further subdivided into omega-3, -6, or -9 unsaturated fatty acids based on the relative position of the first double bond. Both the absolute occurrence of unsaturated fatty acids (Haglund et al., 1998) and the relative occurrence of specific unsaturated fatty acids (Schmitz and Ecker, 2008) is associated with health in humans, and these proportions might also play a role in the health of some species of insectivores.

For most insect species, more than half of the fatty acids are unsaturated (a notable exception being the Hemiptera; Thompson, 1973). The main saturated fatty acids found in insects are C16:0 and C18:0, and, as observed with most other land animals, C16:0 is normally present in larger quantities than

C18:0 (Fast, 1970; Majumder and Sengupta, 1979; Thompson, 1973; Yang et al., 2006). The most prevalent unsaturated fatty acids found in insects are C16:1, C18:1, C18:2, and C18:3 (Beenakkers et al., 1985; Bukkens, 1997; Cookman et al., 1984; Ekpo et al., 2009; Fast, 1970; Majumder and Sengupta, 1979; Yang et al., 2006). Polyunsaturated fatty acid content has been reported to vary between 0.4% and 52.4%, but it is usually a large proportion of the unsaturated fatty acids (Bukkens, 1997; Yang et al., 2006). For instance, the fat of the Chinese grasshopper (*Acrida cinerea* Thunberg) contains 41% C18:3 and 12% C18:2 (Yang et al., 2006). In yellow mealworm larvae, C18:1 seems to be the most prevalent fatty acid, but C18:2 and C18:3 constituted 25.5% and 0.3% of the total fat, respectively (Aguilar-Miranda et al., 2002). A similar trend was reported for commercially raised feeder insects (Finke, 2002, 2013). The fatty acid composition of insects is affected by four main variables: (1) species, (2) developmental phase, (3) diet, and (4) environmental factors such as temperature, light, and humidity.

Phylogeny is not the main determining factor for fatty acid composition, although some general distinctions can be made (Fast, 1970; Fontaneto et al., 2011). For instance, aphids and other Hemiptera tend to contain large amounts of the short-chained fatty acids C14:0, reaching up to 80% of total fatty acids (Thompson, 1973). Coccids contain large amounts of C10:0 and C12:0 (Fast, 1970). Most fatty acids found in Diptera are shorter than 18 carbons, with C16:1 being especially abundant (Fast, 1970; Thompson, 1973). In adult house flies, less than half of the fatty acids were shorter than 18 carbons, whereas C18:1 and C18:2 made up about half of the fatty acids (Finke, 2013). Black soldier fly larvae seem to contain high levels of C12:0, although the exact amounts are affected by the insect's diet (St-Hilaire et al., 2007). Lepidopterans tend to have a higher C18:3 content than other insect species (Fast, 1970; Fontaneto et al., 2011; Majumder and Sengupta, 1979), whereas Dictyoptera, such as cockroaches, contain little or no C18:3 (Finke, 2013; Thompson, 1973).

There also seems to be a difference between the fats found in aquatic and terrestrial insects, presumably as a result of their diet. Aquatic insects have a higher monounsaturated fatty acid content than terrestrial insects, whereas terrestrial insects tend to have a higher polyunsaturated fatty acid content, especially the omega-6 fatty acids C20:3 and C20:4 (Fontaneto et al., 2011).

Insects contain higher amounts of C18:3 than most mammals. Similar to vertebrates, de novo synthesis of saturated fatty acids and polyunsaturated fatty acids by elongation and desaturation occurs in insects (Beenakkers et al., 1985; Tietz and Stern, 1969). Certain polyunsaturated fatty acids, such as C18:2 and C18:3, are considered to be essential because most animal species are unable to synthesize them (Bender, 2002). This also appears to be true for most insect species, which makes these essential fatty acids a dietary requirement (Cookman et al., 1984; Fast, 1970; Thompson, 1973). A few noticeable exceptions have been identified, such as the American cockroach (*Periplaneta americana* (L.)) and the house cricket (Beenakkers et al., 1985; Blomquist et al., 1991). It was originally assumed that bacterial fermentation was the source for C18:2 in these insect species, but more recent studies have shown that it can be synthesized de novo in these and other insect species distributed over four different orders (Blomquist et al., 1991; Borgeson et al., 1991; Borgeson and Blomquist, 1993; de Renobales et al., 1987).

17.2.3. Carbohydrates

Few publications have focused on the carbohydrate content of insects. In general, carbohydrates are present in relatively small amounts in insects (Finke, 2002, 2013). In the field cricket (*Gryllus bimaculatus* De Geer), polysaccharide and carbohydrate content are approximately 0.3% of the fresh weight (Hoffmann, 1973) and free carbohydrate content of the fat body in females of this species is less than 0.5% dry matter (Lorenz and Anand, 2004). In yellow mealworm larvae, the carbohydrate content can vary between 1% and 7% dry matter, depending on the diet provided (Ramos-Elorduy et al., 2002), although some of these differences may be the result of the food remaining in the gastrointestinal tract.

17.2.4. Fiber and Chitin

Insects contain significant amounts of fiber as measured by crude fiber (CF), acid detergent fiber (ADF), and neutral detergent fiber (NDF) (Barker et al., 1998; Finke, 1984, 2002, 2007, 2013; Lease and Wolf, 2010; Oonincx and Dierenfeld, 2012; Pennino et al., 1991; Punzo, 2003). Although insects contain significant amounts of fiber, the components that make up these fibers are unknown. It has been suggested that the fiber in insects represents chitin because chitin (linear polymer of β-(1 → 4) N-acetyl-D-glucosamine units) is structurally similar to cellulose (linear polymer of β-(1 → 4)-D-glucopyranose units) found in plant material. In reality, the fiber in whole insects likely represents a variety of different compounds, including chitin, sclerotized proteins, and other substances that are bound to chitin (Finke, 2007). Chitinase activity has been reported in certain species of frogs, lizards, tortoises, fish, and bats, suggesting that certain insectivores might be able to digest chitin (Donoghue, 2006; Fujimoto et al., 2002; German et al., 2010; Whitaker et al., 2004).

In insects, chitin exists in a matrix with proteins, lipids, and other compounds, which together comprise the cuticle (Kramer et al., 1995). Because chitin is present only in the insect's exocuticle, the chitin content of insects is likely not high. Little quantitative data exists concerning the chitin content of whole insects, but using an enzymatic assay, aquatic insect larvae were determined to contain between 2.9% and 10.1% chitin on a dry weight basis (Cauchie, 2002). In contrast, in most insects studied, protein—not chitin—is the predominant compound in the cuticle (Kramer et al., 1995). The amino acid patterns of whole insects and the proteins in the ADF fractions were different, and the amino acid patterns of the ADF fraction also differed between insect species (Finke, 2007).

These differences likely reflect specificity in cuticular proteins that are present in insects, which contribute to their unique properties.

Although harder-bodied insects such as adult beetles contained higher levels of ADF than softer-bodied insects such as yellow mealworm larvae, silkworm larvae, or cricket nymphs, those fiber levels were a result of higher levels of amino acids in the ADF fraction (Finke, 2007). This suggests that insects with harder cuticles do not necessarily contain more chitin than softer-bodied insects, but rather they contain higher levels of cross-linking proteins that are essential for sclerotization.

17.2.5. Minerals

Minerals can be broadly classified as macrominerals and microminerals (or trace minerals) based on the amounts needed to meet requirements. The essential macrominerals include calcium, phosphorus, magnesium, sodium, potassium, and chloride. The essential microminerals include iron, zinc, copper, manganese, iodine, and selenium. The macrominerals calcium, phosphorus, and magnesium play a primary role in helping maintain the skeletal structure in vertebrates, whereas sodium, potassium, and chloride function as electrolytes and help maintain acid–base balance. The trace minerals play wide-ranging roles, from oxygen transport to functioning as cofactors in a variety of enzymes systems.

Most species of insects contain little calcium because insects as invertebrates do not have a mineralized skeleton. Calcium levels are typically less than 0.3% dry matter (Barker et al., 1998; Finke, 2002, 2013; Oonincx and van der Poel, 2011; Oonincx and Dierenfeld, 2012; Punzo, 2003). The higher levels of calcium occasionally reported for feeder crickets likely reflect calcium in the gut contents (Barker et al., 1998; Finke, 2003; Hatt et al., 2003; Punzo, 2003). The exoskeleton of most insects is primarily composed of protein and chitin, although some insects have a mineralized exoskeleton in which calcium and other minerals are incorporated into the cuticle (Dashefsky et al., 1976). Examples include larvae of the face fly, *Musca autumnalis* De Geer, and the black soldier fly (Dierenfeld and King, 2008; Finke, 2013). Black soldier fly larvae contain a high level of calcium and are now commonly sold commercially as food for captive insectivores. Concerns have been raised concerning the availability of the calcium in black soldier fly larvae, especially for insectivores that swallow their prey whole. When fed to mountain chicken frogs, the calcium digestibility of whole black soldier fly larvae was only 44% compared to 88% for larvae that had been mashed (Dierenfeld and King, 2008). Even though calcium availability was relatively low, given the high level of calcium in this species, they should provide sufficient calcium for most insectivores. Other invertebrates, such as millipedes and isopods, also have a mineralized exoskeleton and likely serve as a source of calcium for wild insectivores (Gist and Crossley, 1975; Graveland and van Gijzen, 1994; Oonincx and Dierenfeld, 2012; Reichle et al., 1969). However, one way of supplying this mineral might be offering calcium grit, powder, or other calcium-rich materials separately to insectivores (Rich and Talent, 2008; Tordoff, 2001). Wild birds seek out calcium-rich invertebrates when calcium requirements are high, such as during egg laying and nestling growth (Graveland and van Gijzen, 1994; Bureš and Weidinger, 2003).

The phosphorus content of feeder insects is much higher than calcium levels in most insect species, except black soldier fly larvae (Barker et al., 1998; Finke, 2002, 2013; Hatt et al., 2003; Jones et al., 1972; Martin et al., 1976; Oonincx and van der Poel, 2011; Oonincx and Dierenfeld, 2012; Punzo, 2003). Most insects would likely contain adequate levels of phosphorus to meet the requirements of insectivores, especially because the phosphorus in insects is likely to be readily available, as was shown for face fly pupa (Dashefsky et al., 1976).

Most species of feeder insects contain levels of magnesium ranging from 0.08% to 0.30% dry matter. These levels would likely be sufficient to meet the dietary requirements of most species of insectivores. Like calcium, the magnesium content of black soldier fly larvae was 3–10 times higher than that of other feeder insects. It seems likely that both calcium and magnesium form a complex with chitin in the larval cuticle in this species.

There are few reports on the sodium and potassium content of captive bred insects (Finke, 2002, 2013; Oonincx and van der Poel, 2011; Oonincx and Dierenfeld, 2012), but these data are similar and comparable to the values obtained for wild-caught insects (Levy and Cromroy, 1973; Oyarzun et al., 1996; Reichle et al., 1969; Studier et al., 1991; Studier and Sevick, 1992). Levels of potassium generally range from 0.6% to 2.0% dry matter, whereas sodium levels are somewhat lower, ranging from 0.1% to 0.6% dry matter. There are limited data concerning the chloride content of feeder insects, with values ranging from 0.16% to 0.97% dry matter (Finke, 2002, 2013). These data suggest that most insects likely contain adequate amounts of these three minerals to meet the needs of most species of insectivores (Finke, 2002, 2013; Oonincx and Dierenfeld, 2012).

Most insects appear to contain relatively high levels of the trace minerals iron, zinc, copper, and manganese. Although the high-fat larval stage of some species of feeder insects (e.g. greater wax moth larvae, yellow mealworm larvae, butterworm larvae) might be marginally low in iron relative to energy content, most insects likely supply adequate amounts for the typical insectivore (Barker et al., 1998; Finke, 2002, 2013; Hatt et al., 2003; Oonincx and van der Poel, 2011; Oonincx and Dierenfeld, 2012; Punzo, 2003). For unknown reasons, both adult house flies and fruit flies contain relatively high levels of iron (Barker et al., 1998; Finke, 2013; Oonincx and Dierenfeld, 2012). House crickets can contain up to 200 mg/kg dry matter, although other reports documented iron content to be around 60 mg/kg dry matter (Barker et al., 1998; Bernard and Allen, 1997; Finke, 2002). These variations again are likely to be due to food remaining in the gastrointestinal tract when the insects were analyzed. Wild-caught insects also appear to contain significant amounts of iron (Punzo, 2003; Studier and Sevick, 1992).

Insects are generally a good source of zinc, with values for commercially raised insects ranging from 61.6 to 340.5 mg/kg dry matter (Barker et al., 1998; Finke, 2002, 2013; Oonincx and Dierenfeld, 2012; Punzo, 2003). These values are similar to those obtained for wild-caught species (Punzo, 2003; Levy and Cromroy, 1973).

Copper in commercially raised feeder insects ranged from 3.1 to 51.2 mg/kg dry matter (Barker et al., 1998; Finke, 2002, 2013; Punzo, 2003). The lowest value seen in captive-raised insects was 3.1 mg/kg dry matter for greater wax moth larvae, as reported by Barker (Barker et al., 1998), whereas Finke (2002) found a much higher level for this species (9.2 mg/kg dry matter). All other species analyzed had values greater than 7 mg/kg dry matter, suggesting that insects are typically good sources of copper. Wild-caught insects also appear to contain significant amounts of copper (Punzo, 2003; Levy and Cromroy, 1973).

Levels of manganese in feeder insects range from 1.5 to 364 mg/kg dry matter (Barker et al., 1998; Bernard and Allen, 1997; Dierenfeld and King, 2008; Finke, 2002, 2013; Oonincx and Dierenfeld, 2012; Punzo, 2003). Like calcium and magnesium, the highest levels of manganese observed were seen in black soldier fly larvae (Dierenfeld and King, 2008; Finke, 2013), although the reasons for elevated levels of manganese in this species are unclear. Some species of stored product insects contain elevated levels of zinc and manganese in their mandibles, presumably to harden them in order to better penetrate whole seeds (Morgan et al., 2003). Wild-caught insects also appear to contain significant amounts of manganese (Punzo, 2003).

There are very little data regarding the iodine content of insects. Of the 12 species of feeder insects analyzed, only six had any detectable iodine, with levels ranging from 0.45 to 1.22 mg/kg dry matter (Finke, 2002, 2013). Bee brood (pupae and larvae) did not contain detectable levels of iodine (Finke, 2005). No other data are available concerning the iodine content of insects.

As is the case for iodine, there are only limited data on the selenium content of feeder insects. Although butterworm larvae did not contain any selenium, the other 11 species of feeder insects contain selenium at levels ranging from 0.27 to 0.97 mg/kg dry matter (Finke, 2002, 2013).

Because the contents of the gastrointestinal tract can represent a significant percentage of the total weight of the insect (Finke, 2003), it can have a significant effect on the mineral content of the insect if it is analyzed when fully fed. Additionally studies of wild insects show both seasonal variation as well as variation between different populations of the same species living in the same general area (Finke, 1984; Studier et al., 1991; Studier and Sevick, 1992).

17.2.6. Vitamins and Carotenoids

17.2.6.1. Vitamin A

Vitamin A plays a role in a wide variety of physiological processes, including vision, cell differentiation, immune response, reproduction, and growth. There are limited data regarding the vitamin A content of wild insects, and most species of captive-bred insects contain relatively low levels of vitamin A/retinol (typically less than 300 μg retinol/kg dry matter; Barker et al., 1998; Finke, 2002, 2013; Hatt et al., 2003; Oonincx and van der Poel, 2011; Pennino et al., 1991; Punzo, 2003). Although migratory locusts fed a grass diet supplemented with wheat bran and fresh carrots contained significantly more retinol than those fed only a grass diet, the retinol levels for all locusts were well below the requirements of the rat (Oonincx and van der Poel, 2011).

In fruit flies, only the eyes contained measurable quantities of retinoids, and the amount detected was a function of the carotenoid content of the larval diet (Goldsmith and Warner, 1964; Seki et al., 1998; Von Lintig, 2012). The low values reported in the literature for the vitamin A content of captive-bred insects may be a result of a several factors. First, one analytical method used for vitamin A analysis of insects is specific for retinol and may not detect the other retinoids (retinal and 3-hydroxyretinal) found in insect eyes (Goldsmith and Warner, 1964; Seki et al., 1998; Smith and Goldsmith, 1990). It is unclear if 3-hydroxyretinal can serve as a source of vitamin A because it is unknown if 3-hydroxyretinal can be converted into retinal by insectivores. Honey buzzards (*Pernis apivorus* (L.)) a primarily insectivorous bird, were shown to contain high levels of 3-4-didehydroretinol in their plasma, suggesting insectivorous birds might be able to use some of the atypical retinoids found in adult insects (Müller et al., 2012). Second, because the retinoid levels in fruit fly eyes was a function of dietary carotenoid content, it may be that diets fed to commercially raised insects do not contain sufficient levels or types of carotenoids to optimize the retinol, retinal, and 3-hydroxyretinal content of insects. Third, many insect species raised for food are fed to insectivores as larva. Because insect larvae do not possess compound eyes, where retinoids are synthesized from carotenoids, the larval stage of these insects would not be expected to contain large quantities of retinoids (Von Lintig, 2012).

In addition to retinol, certain carotenoids can be converted into vitamin A in many species of animals, although it is unclear if all insectivores have the ability to convert beta-carotene to retinol (Bender, 2002; Levi et al., 2012; McComb, 2010; Olson, 1989). Carotenoids are found at high levels in many species of wild insects, whereas captive-bred insects contain significantly lower quantities (Eeva et al., 2010; Finke, 2013; Isaksson and Andersson, 2007; Oonincx and van der Poel, 2011). The reason for this discrepancy is unclear but is likely a function of dietary carotenoid content. It may be that wild insectivores use a combination of retinoids (retinol, retinal, and 3-hydroxyretinal) as well as carotenoids to meet their vitamin A requirements. A better understanding of the retinol, retinal, and 3-hydroxyretinal content of insects and the utilization of the retinoids and carotenoids found in insects as a source of vitamin A in insectivores is important because vitamin A deficiency has been reported in several species of captive insectivores (Ferguson et al., 1996; Hoby et al., 2010; Miller et al., 2001; Pessier et al., 2005).

17.2.6.2. Vitamin D

Vitamin D can be considered to be conditionally essential for many species of insectivores because it can be synthesized by most animal species that are provided the proper environmental conditions. An analysis of the vitamin D content of a variety of species of commercially raised insects detected no vitamin D, although the threshold for detection was 250IU/kg (as is) or roughly 595–1445IU/kg dry matter (Finke, 2002). Using a more sensitive technique, black soldier fly larvae, butterworm larvae, and rusty red roaches were shown to contain 388–633IU vitamin D_3/kg dry matter, while house flies contained no detectable vitamin D_3 (Finke, 2013). These values are similar to those obtained for yellow mealworms (150IU vitamin D/kg dry matter) and house crickets (934IU vitamin D/kg dry matter) (Oonincx et al., 2010). There is no analysis of the vitamin D content of wild insects available for comparison. Further studies on the vitamin D content of both captive-raised and wild-caught insects should be conducted to provide a better understanding of insect vitamin D content. However, it seems likely that exposure to ultraviolet light is an appropriate way to ensure an adequate vitamin D status in most insectivores (Ferguson et al., 1996; Oonincx et al., 2010).

17.2.6.3. Vitamin E

Vitamin E serves as an antioxidant and therefore helps maintain the functionality of a variety of lipid-soluble compounds in the body. Insects contain varying amounts of vitamin E. Both house crickets and yellow mealworms have been shown to contain widely varying levels of vitamin E. Values for house crickets ranged from 5 to 79mg/kg dry matter (Barker et al., 1998; Finke, 2002; Hatt et al., 2003; Pennino et al., 1991), whereas yellow mealworm larvae ranged from less than 15mg/kg to 50mg/kg dry matter (Barker et al., 1998; Finke, 2002; Pennino et al., 1991). The large variations found could well be due to the diet provided to the insect, both the vitamin E incorporated into the body tissue as well as the vitamin from any diet remaining in the insect's gastrointestinal tract.

The vitamin E content of other commercially raised insects was relatively low (typically less than 15mg/kg dry matter; Barker et al., 1998; Finke, 2002, 2013; Oonincx and Dierenfeld, 2012; Punzo, 2003). Several species, including butterworm larvae and silkworm larvae, had somewhat higher levels (33–35mg/kg dry matter), whereas much higher levels (110–120mg/kg dry matter) were reported for fruit flies, house flies, and false katydids (*Microcentrum rhombifolium* Saussure; Finke, 2002, 2013; Oonincx and Dierenfeld, 2012). There are few data on the vitamin E content of wild insects, although they appear to contain vitamin E at levels on the higher end of the range or exceeding that seen in captive-bred insects (Pennino et al., 1991; Punzo, 2003).

17.2.6.4. B Vitamins

There is limited comprehensive information regarding the B vitamin content of most insects. Recent research has focused on the B vitamin content of the most common commercially raised feeder insects (Finke, 2002, 2013). There is much less information regarding the B vitamin content of unprocessed wild-caught insects for comparison. Complicating a comparison between the data for captive-bred insects and that for various dried insects are differences in the analytical methods used (microbiological vs chemical techniques) and the method used for sample preparation. Some B vitamins are relatively unstable when exposed to heat, light, or oxygen. Therefore, values obtained by analyzing dried insects that have been processed for human consumption may not be representative of the values found in live whole insects. This also means that commercially available dried whole insects may contain lower levels of some B vitamins due to processing, drying, and storage.

Thiamine (vitamin B_1) is needed for the function of several important enzymes associated with energy metabolism (Thurnham et al., 2000). A number of species of feeder insects including house crickets, adult yellow mealworms, superworms, butterworms, and Turkistan roaches contained relatively low levels of thiamine (0.8–2.9 mg/kg dry matter), whereas other species of feeder insects (black soldier fly larvae, adult house flies, silkworms, yellow mealworm larvae, and waxworms) contained much higher levels (5.6–44.8 mg/kg) (Finke, 2002, 2013). Similarly, for a selection of collected Nigerian insects comprising five orders, large variations were reported (0.3–32.4 mg/kg dry matter; Banjo et al., 2006). When analyzed using a microbiological method, high levels of thiamine were reported for the African palm weevil larva (*Rhynchophorus phoenicis* F.) and the larvae of the cavorting emperor moth (*Usta terpsichore* Maassen and Weymer; 30.2 and 36.7 mg thiamine/kg dry matter, respectively), whereas the values for termites (*Macrotermes subhyalinus* Rambur) were very low (1.3 mg/kg dry matter; Santos Oliveira et al., 1976). Dried smoked Attacidae caterpillars from Zaire analyzed using a microbiological method also contained low levels of thiamine (1.5–2.7 mg/kg dry product; Kodondi et al., 1987). Because thiamine is relatively unstable and the values reported are for dried insects processed using a variety of traditional methods (drying, smoking, and frying), it is unclear how representative these values are of raw whole insects (Banjo et al., 2006; Kodondi et al., 1987; Santos Oliveira et al., 1976).

Riboflavin (vitamin B_2) functions as a coenzyme required in the metabolism of a variety of other nutrients (Thurnham et al., 2000). Most species of commercial feeder insects contain relatively high levels of riboflavin, ranging from 17.6 to 306.3 mg/kg dry matter (Finke, 2002, 2013; Jones et al., 1972). A similar concentration 32.4 mg/kg dry matter was found for honey bee brood (larvae and pupae) (Banjo et al., 2006; Finke, 2005). Dried smoked Attacidae caterpillars also contained high levels of riboflavin (32–51 mg/kg dry product) (Kodondi et al., 1987). These values are slightly higher than those for termites, palm weevil larvae, and a species of Saturniid larvae processed for human consumption (11.4–22.4 mg riboflavin/kg dry matter) (Santos Oliveira et al., 1976). Banjo reported highly variable levels of riboflavin in 14 species of insects from Nigeria, ranging from 0.9 to 32.4 mg/kg dry matter (Banjo et al., 2006). It should be noted that after collection these insects were kept dry (adults) or stored in 70% alcohol (larvae). Because riboflavin is degraded by light, it is unclear if these represent real differences or resulted from losses during storage prior to analysis.

Niacin (vitamin B_3) plays a role in metabolism and tissue respiration (Thurnham et al., 2000). It appears to be abundant in insects, with commercially reared feeder insects containing levels ranging from 76.7 to 359.1 mg/kg dry matter (Finke, 2002, 2013). A variety of dried insect species typically consumed in Africa, including termites, palm weevil larvae, and several different species of lepidopteran larvae, have been analyzed for niacin. Even though these insects were processed using a variety of different methods (drying, smoking, and frying), they contained high levels of niacin, ranging from 52 to 110 mg/kg dry matter (Kodondi et al., 1987), whereas larvae of the cavorting emperor moth contained only 3 mg niacin/kg dry matter (Santos Oliveira et al., 1976). Niacin is a relatively stable B vitamin; therefore, the values in these studies are likely more representative for live insects when compared to other B vitamins.

Pantothenic acid (vitamin B_5) functions as a component of coenzyme A. As such, it plays a role in the citric acid cycle, fatty acid synthesis, and oxidation reactions. It is widely distributed in most foodstuffs and commercially raised insects appear to be no exception, with levels ranging from 46.1 to 179.8 mg/kg dry matter (Finke, 2002, 2013). Bee brood also appears to be a good source of pantothenic acid, containing 51.3 mg/kg dry matter (Finke, 2005). There are little additional data available on the pantothenic acid content of insects, although using a microbiological assay, dried smoked Attacidae caterpillars were found to contain very low levels of pantothenic acid (0.073–0.102 mg/kg dry product) (Kodondi et al., 1987).

Pyridoxine (vitamin B_6) plays an important role in a variety of metabolic reactions, most notably those involved in amino acid metabolism. Commercially raised insects appear to be good sources of pyridoxine, with values ranging from 3.1 mg/kg dry matter for waxworm larvae to 22.3 mg/kg dry matter for yellow mealworm beetles (Finke, 2002, 2013; Jones et al., 1972). Most insect species, however, fall in a fairly narrow range between 6 and 10 mg pyridoxine/kg dry matter. Using a microbiological technique, very low levels of pyridoxine (0.37–0.63 mg/kg) were detected in three species of dried smoked Attacidae caterpillars (Kodondi et al., 1987). Because pyridoxine is unstable, it is likely that the low values obtained are a result of the smoking and drying process used to preserve the dried insect product.

Biotin (vitamin B_8) is a carrier of carboxyl groups in reactions involving adenosine triphosphate. The biotin content of commercially raised insects is highly variable, ranging from 0.23 to 2.69 mg/kg dry matter; however, the values for most species ranged from 0.7 to 1.4 mg/kg dry matter (Finke, 2002, 2013). Dried smoked meal from Attacidae caterpillars have been reported to contain 0.23–0.45 mg/kg (Kodondi et al., 1987). It is unclear if these lower values reflect biological differences, biotin loss during processing (drying and smoking), or analytical techniques used.

Folic acid (vitamin B_9) plays an essential role in DNA synthesis and one carbon metabolism (Thurnham et al., 2000). Using a chemical method, commercially raised insects were shown to contain high levels of folic acid, ranging from a low of 1.57 mg/kg for silkworm larvae to 7.22 mg/kg for house flies (Finke, 2002, 2013). In contrast, using a microbiological method, very low folic acid values (0.20–0.63 mg/kg) for smoked and dried product from three species of Attacidae caterpillars were found (Kodondi et al., 1987). Because folic acid is susceptible to degradation both by light and oxidizing agents, the low values are likely a result of the smoking and drying process used to preserve the dried insect product.

Cobalamin (vitamin B_{12}) is found exclusively in products of animal origin and plays a key role in reactions involving methyl donors. Although silkworm and waxworm larvae contain no detectable vitamin B_{12} (less than 3–4 µg/kg dry matter), most other species are good sources of vitamin B_{12} (Finke, 2002, 2013). In particular, crickets, soldier fly larvae, and roaches contain extremely high levels, ranging from 143 to 767 µg/kg dry matter. Vitamin B_{12} levels of 140–250 µg/kg dry matter were reported for dried and smoked products from three species of Attacidae caterpillars (Kodondi et al., 1987), levels similar to those obtained for commercially raised species.

17.2.6.5. Vitamin C

Vitamin C is needed to form connective tissue, and it functions as an antioxidant. Insects contain some vitamin C. It has been suggested that certain insects, such as the grasshopper *Melanoplus sanguinipes* (F.), use enzymes to reuse antioxidants such as vitamin C and E (Barbehenn, 2003). Honey bees contain relatively high amounts of vitamin C (102.5–163.8 mg/kg dry matter; Banjo et al., 2006). Most other species in which vitamin C was determined contained low levels, from 0 to about 50 mg/kg dry matter, although both adult house crickets and adult mealworms contained levels similar to those reported for honey bees (Finke, 2002, 2013, 2007; Banjo et al., 2006).

17.3. EFFECTS OF INSECT SIZE/LIFE STAGE ON NUTRIENT COMPOSITION

In general, the lipid content of wild insects is approximately 30% for larvae and 20% for adults (Fast, 1970). Fat stores are usually largest in the final larval stage prior to metamorphosis (Fast, 1970). This depends to some degree on whether species undergo a complete metamorphosis (holometabolous species; e.g. yellow mealworms) or an incomplete metamorphosis (hemimetabolous species; e.g. house crickets).

Larvae of holometabolous species have a higher fat content than adults (Finke, 2002; Lease and Wolf, 2011; Punzo, 2003). Yellow mealworms, for example, increase their fat reserves during larval growth (McClements et al., 2003). These fat reserves are used as an energy source during metamorphosis, resulting in a lower fat content of adults and a subsequent increase in the relative content of protein and ash (Barker et al., 1998; Bernard and Allen, 1997; Downer and Matthews, 1976; Finke, 2002; Lease and Wolf, 2011; Oonincx and Dierenfeld, 2012). A similar trend is seen in fruit flies. Furthermore, during metamorphosis, the protein composition of yellow mealworms change: glycine, tryptophan and taurine content increase, while tyrosine content decreases (Finke, 2002). The taurine content of both fruit flies as well as a species of noctuid moth (*Mamestra configurata* Walker) also increased after metamorphosis (Bodnaryk, 1981; Massie et al., 1989). During pupation of the fly *Agria affinis* (Fallen), two-thirds of their fat reserve is used. This coincides with a relative decrease in C16:0, C16:1, and C18:1, and a relative increase in C18:2 and C22:1; the relative concentration of C18:0 is not affected (Barlow, 1965).

Different changes in fat composition occur in silkworms. During larval development, both fat content and composition change; the total fat content increases and relative increases of C16:0, C16:1, and C18:1 and decreases of C18:0 and C18:2 are observed. During pupation, C18:1 continues to increase, while C16:0 continues to decrease. The levels of C18:3 in silkworm larvae differed between males and females (Nakasone and Ito, 1967). In the velvet bean caterpillar, total lipid content decreases during metamorphosis, as would be expected. However, C16:0 and C18:1 levels increase, C18:3 levels decrease, and C18:2 levels remain stable (Cookman et al., 1984). To what extent these changes are species-specific or constitute differences between Diptera and Lepidoptera needs further investigation. Similar to species preparing for pupation, preparation for diapause or hibernation results in increased energy (fat) reserves (Ali and Ewiess, 1977; Downer and Matthews, 1976). In three species of aquatic insects, the concentration of C18:3 increased while C20:4 and C20:5 decreased after the larval phase, when the insect left the aquatic environment (Hanson et al., 1985).

Unlike holometabolous insects, hemimetabolous adults tend to have a higher fat content than nymphs (Lease and Wolf, 2011). This holds true for certain studies conducted on migratory locusts, but not all (Oonincx et al., 2010; Oonincx and van der Poel, 2011). Probably the time of sampling, whether directly after adult emergence or later during adulthood, explains these differences because fat continues to accumulate after adult emergence (Beenakkers et al., 1985). Generally, flying and migratory insects have a tendency to accumulate more fat, which is used as an energy source for their flight muscles (Downer and Matthews, 1976). In this locust species, dry matter content increases and ash content decreases between the penultimate instar and adulthood, although copper and iron levels increase (Oonincx and van der Poel, 2011). Similar to these locusts, the dry matter content in house crickets significantly increases (from 23% to 30%) during the penultimate stage of development (Roe et al., 1980). However, adults have a higher protein and a lower fat content than small and medium house crickets (Finke, 2002; McClements et al., 2003). The amino acid composition seems similar between nymphs and adults of this species (Finke, 2002).

In a study on three cockroach species—rusty red roaches, six-spotted roaches (*Eublaberus distanti* Kirby), and hissing roaches (*Gromphadorhina portentosa* Schaum)—an increase in dry matter content and crude protein content, and a concomitant decrease in fat content, was noted between small and medium specimens. The opposite change was noted when six-spotted roaches reached adulthood (Oonincx and Dierenfeld (2012)).

Besides size differences, certain gender differences seem apparent in insects. Males often have a lower body weight than females, which might be caused by a lower fat content, as explained previously (Ali and Ewiess, 1977; Hoffmann, 1973; Sönmez and Gülel, 2008).

17.4. EFFECTS OF INSECT DIET ON INSECT NUTRIENT COMPOSITION

Diet can have a significant effect on the nutrient composition of an insect. Because the entire insect is normally consumed, dietary effects described in the literature can partially be due to diet remaining in the gut, as was mentioned previously. This is discussed in more detail later in this chapter. The body composition of the insect itself can also be altered to a certain extent through the diet. For instance, a higher water content of bran (8.7% vs 6.7%) provided to yellow mealworm larvae increases their water content (64% vs 59%) (Machin, 1975).

For the macronutrients, it seems that insect fat content is highly variable and that the content of protein and ash are subsequently affected. If an insect with a certain amount of protein increases its fat reserve, the percentage of protein thereby decreases (Oonincx and van der Poel, 2011; Simpson and Raubenheimer, 2001; St-Hilaire et al., 2007). This might be the reason why the protein content stays constant during adulthood if a protein source is available and protein content decreases if only a carbohydrate source (which can be converted to fat) is available, as was shown in Mediterranean fruit flies (*Ceratitis capitata* Wiedemann; Nestel et al., 2005). Although the percentage of protein can be altered through the diet, it seems unlikely that the amino acid composition is affected. Within a specific life stage, the relative abundance of amino acids is expected to be species-specific because specific parts of the insect body are made from certain amino acids. A study of the black soldier fly found no effects of diet on amino acid composition (Sealey et al., 2011). However, a study of yellow mealworms, reared on different diets, did not show large differences in crude protein content but found differences in amino acid profiles (Ramos-Elorduy et al., 2002). It seems likely that these were not necessarily due to changes in the amino acid composition of the insect, but rather were due to differences in the amino acid composition of the diet present in the gut of the larvae.

As stated previously, the crude fat content of insects can be altered by the diet provided. Two independent studies on fruit flies using the same artificial diet showed a similar nutrient composition, especially regarding fat and iron. A third study on fruit flies found a dissimilar composition, which could indicate a strong dietary influence (Barker et al., 1998; Bernard and Allen, 1997; Oonincx and Dierenfeld, 2012). For house crickets, several studies are available on their chemical composition. Large differences in fat content have been reported, ranging from 17% to 37% dry matter (Barker et al., 1998; Finke, 2002; Hatt et al., 2003; Oonincx et al., 2010). It seems likely that variation in fat content is a result of both the diet provided and the age of the cricket (Hatt et al., 2003). When subadult house crickets were provided with either water, lettuce, or a commercial cricket diet, after 3 weeks the crickets that were fed water or lettuce had higher moisture and protein contents and lower fat content than the crickets that were fed the commercial cricket diet. During this period, the fat content decreased for all three groups, which led to a concomitant rise in the other macronutrients (Hatt et al., 2003). This may have been caused by a lower energy content in the three experimental diets than in the diet provided before the trial, especially because the crickets that were fed water only or lettuce showed the greatest decline. Additionally the subadult house crickets would be expected to become adults during those 3 weeks, and as such their fat content would be expected to decrease (McClements et al., 2003).

Besides the plasticity of total fat content, large differences can be expected in the fatty acid composition when different diets are provided. In nonruminant production animals, short-term provision of polyunsaturated fatty acids have been shown to increase the polyunsaturated fatty acid content of the meat, indicating that these fatty acids were incorporated in the body (Kouba and Mourot, 2011). Studies on insects indicate that fatty acid composition of both larvae and adults tends to reflect the fatty acid composition of the diet provided (Cookman et al., 1984; St-Hilaire et al., 2007). A study with black soldier flies indicated that polyunsaturated fatty acid content can be elevated by supplementing the diet with fish offal during the last month of development (St-Hilaire et al., 2007).

The fatty acid composition of the insect does not always directly match that of the diet, suggesting selective accumulation or synthesis (Cookman et al., 1984). For instance, in migratory locusts, the C18:1 and C18:3 content of the diet strongly relates to the fatty acid content of the locust fat body. However, for C18:2, this was not the case, possibly due to poor absorption or due to saturation of this fatty acid (Beenakkers and Scheres, 1971). Although diet composition influences the fatty acid composition directly, indirect effects have also been reported. For instance, a sufficient supply of C18:2 can be used to synthesize C20:4 and C20:5 (Hanson et al., 1983).

For many feeder insects, carrots are a well-accepted part of the diet or a means of providing moisture. In migratory locusts, the addition of carrots increased carotene content as well as retinol (vitamin A) (Oonincx and van der Poel, 2011). Although the elevated carotene levels could be due to gut loading, it seems likely that the increased retinol content is due to conversion of carotene to retinol. Fruit flies are able to convert β-carotene to retinoids, which might indicate that more insect species have this ability (Von Lintig and Vogt, 2000).

Great tits (*Parus major* (L.)) have been shown to prefer carotenoid-enriched yellow mealworms over nonenriched yellow mealworms (Senar et al., 2010).

Besides differences in required nutrients, undesirable effects can arise through long-term provision of a certain diet. For instance, accumulation of heavy metals, such as cadmium and lead, can occur in certain insect species (Zhang et al., 2012).

17.5. EFFECTS OF ENVIRONMENT ON INSECT COMPOSITION

A number of environmental factors, such as temperature, light, and humidity, can affect the growth, development, and chemical composition of insects.

17.5.1. Temperature

Insects are poikilothermic (i.e. cold-blooded); their body temperature depends to a large degree on the environmental temperature. Therefore, within a range of temperatures suitable for the specific species, their metabolic rate and growth rate should increase with higher temperatures (Akman Gündüz and Gülel, 2002; Ali and Ewiess, 1977; Ali et al., 2011; Angilletta et al., 2004; Krengel et al., 2012). A standard way of quantifying this is the relative difference when the temperature is increased by 10 °C (Q10). For instance, in house crickets, the development time is halved with a 10 °C increase in temperature (Roe et al., 1980). Although growth rates are increased by higher temperatures, adult size generally seems to decrease; a colder environment results in larger animals (Akman Gündüz and Gülel, 2002; Angilletta et al., 2004; Krengel et al., 2012). Exceptions include both desert locusts (*Schistocerca gregaria* Forsskål) and migratory locusts, for which adult body weights are higher when reared at 30 °C and 31 °C respectively, than at 25 °C (Akman Gündüz and Gülel, 2002; Beenakkers et al., 1971). Temperatures between 32° and 43 °C appear most suitable when appropriate humidity is provided (Hamilton, 1936). Last instar female house crickets gain weight quicker and more efficiently at 35 °C than at 25 °C or 30 °C (Roe et al., 1980, 1985). At higher temperatures, lipid content decreases during the latter half of this instar, but at 25 °C carbohydrates are still converted to lipids. It seems that optimal growth occurs at 35 °C, which is similar to the 34 °C for the field cricket. However, for the latter species, mortality is also highest at 34 °C. Both growth rate and mortality are decreased at 27 °C. The lowest adult fresh weight is achieved at ambient temperatures around 10 °C, but the dry matter content is about 20% higher compared to crickets raised at 27 °C. For field crickets, temperature changes during rearing (alterations between 20 °C and −1.5 °C) versus constant temperatures (13 °C) increase water content and decrease protein content (Hoffmann, 1973). Alternating temperatures around the optimal growth temperatures appears to result in a higher protein and fat content in field crickets (Hoffmann, 1973). At low temperatures (13 °C and 20 °C vs 27 °C), fat content is higher, as are the proportions of saturated fatty acids. This seems a likely adaptation to the thermal regimen. In the fly *Pseudosarcophaga affinis* (Fallén), higher proportions of unsaturated fatty acids are known to increase heat tolerance (House et al., 1958). A comparative study on the seven-spotted (*Coccinella septempuctata* (L.)) and the Asian lady beetle (*Harmonia axyridis* Pallas) indicated that the first species accumulates more fat in general and increases fat storage at elevated temperatures compared to normal temperatures (18 °C vs 21 °C). Fat accumulation remains low under both conditions in the latter species. Females of the seven-spotted lady beetle have a higher fat content than males at normal temperatures, but a similar fat content at elevated temperatures. The authors suggested that temperatures before eclosion determine the fat content of adults in both species (Krenge et al., 2012). The carbohydrate and protein content of bean beetles (*Acanthoscelides obtectus* Say) is lower at 20 °C than at 30 °C, but lipid content is the same (Sönmez and Gülel, 2008).

17.5.2. Humidity

Humidity is normally expressed as relative humidity (RH), which is the relative amount of water that can be stored in air of a certain temperature. Higher levels of humidity, within an appropriate range, seem to decrease development time (Ali et al., 2011; Han et al., 2008; Nedvěd and Kalushkov, 2012). However, unlike in the case of increasing temperatures, this increase in growth rate does not seem to lead to a lower body mass in adults. For instance, the body mass of the pine caterpillar, *Dendrolimus tabulaeformis* Tsai and Liu increases with higher humidity during their larval stages (20–100%) (Han et al., 2008).

A comparative study on the effects of humidity and temperature on locust species (*L. migratoria* and *S. gregaria*) indicated development is possible between 35% and 80% RH, and increasing temperatures require a higher RH (Hamilton, 1936). Optimal RH is around 60–65%, whereas alternating the humidity to provide a day–night rhythm showed inconsistent effects on development. Similarly, pine caterpillars require an RH greater than 40% for development, whereas the optimal RH, resulting in increased body mass, seems to be approximately 80%. During diapause, this species can absorb water from the substrate, thereby increasing its fresh weight (Han et al., 2008). Fasting mealworms are capable of absorbing water vapor if relative humidity is above 88% RH (Fraenkel, 1950; Machin, 1975).

Of course, relative humidity has an indirect effect through the feed provided as well. If the feed provided to yellow mealworms is in equilibrium with 70% RH, it allows for rapid growth (Fraenkel, 1950). Mealworms first exposed to high humidity have a higher dry matter weight gain if feed is provided after exposure (Machin, 1975). However, long-term exposure to high humidity (>85%) results in higher larval mortality, probably due to excessive hydration (Machin, 1975). The observed dry matter weight gain could be due to the larvae compensating for the excessive hydration by an increased feed intake. Other problems are likely to occur at high RH, such as the development of fungi and/or mites (Machin, 1975). The optimal growth rate for mealworms at 25 °C is attained at 70% RH (Fraenkel, 1950; Machin, 1975).

As indicated before, the optimal RH is related to temperature, although other variables can play a role as well. The Asian lady beetle tends to grow larger at higher humidity levels (between 30% and 90%) if fed with the aphid *Acyrthosiphon pisum* (Harris), but not when fed on frozen eggs of *Ephestia kuehniella* (Zeller) (Nedvěd and Kalushkov, 2012).

It seems that body weight and development rate of insects is higher at the top of their RH range. Also, the moisture content of insects seems to increase at a higher RH. However, little is known on how variables such as fat content and body composition are affected.

17.5.3. Photoperiod

The effect of photoperiod (daily exposure time to light) on insect composition has not been studied in detail. Most studies have focused on behavioral effects, fecundity, and body weight. For certain locust species, long photoperiods (up to 24 h) increase their growth rate. Possibly, the food intake in this diurnal species is increased if a longer photoperiod is provided. This effect could, however, also be due to concomitantly higher temperatures. At constant temperature, Asian lady beetles tended to develop quicker, with 16 h of light compared to 12 h of light, but adult weight was similar (Berkvens et al., 2008). This seems likely for these diurnal species, but the green stink bug (*Nezara viridula* (L.)), which mainly feeds at night, has an increase in the rate of development with an increased photoperiod (10 vs 14 h) (Ali and Ewiess, 1977; Shearer and Jones, 1996). This increase in developmental rate coincided with an increase in body weight when reaching (and during) adulthood, especially for females (Ali and Ewiess, 1977). Short photoperiods can induce diapause, whereas longer photoperiods are more likely to induce reproduction. It seems that for this species, more energy is accumulated when preparing for reproduction than for diapause, which would be likely to lead to a higher fat content.

However, little is known on the influence of photoperiod on the chemical composition. For adults of the giant wax moth, it has been suggested that the protein content of adults increases more rapidly if kept in constant light compared to constant darkness (Koc and Gulel, 2008). The effects of photoperiod on the nutrient profile of insects mostly seems to be indirect, acting through other processes, such as preparation for reproduction or diapause or concomitant changes in temperature.

17.6. NUTRIENT REQUIREMENTS OF INSECTIVORES, INCLUDING DIET AVAILABILITY

It is difficult to compare the nutritive value of insects as a group to the dietary requirements of insectivores because of the large number of insect species and the large differences in nutrient content between those species (Barker et al., 1998; Bukken, 1997; Finke, 2002, 2013). A proper evaluation of a food/insect requires it to be evaluated in the broader context of a complete diet, made up of a number of different foods/insects.

17.6.1. Availability and Digestibility

It has been suggested that insects with sclerotized exoskeletons might have poor digestibility, which would be lowest in the last larval instar (Dufour, 1987). This lower digestibility would be caused by the higher chitin content. For aquatic insects, this has been reported to be between 3% and 10% (Cauchie, 2002). For the commonly reared house crickets, the larvae of the giant wax moth, and the giant mealworm (hormonally treated *T. molitor*), chitin contents were estimated to be between 1.6% and 2.0% (Finke, 2007). Yellow mealworm adults contain about 7.4% chitin (Finke, 2007), which concurs with the findings of Dufour (1987). A study of the protein digestibility of mopane caterpillars (*Imbrasia belina* Westwood) in rats showed a slightly lower digestibility when compared to other products of animal origin (Dreyer and Wehmeyer, 1982). When fed to rats, the protein digestibility of freeze-dried mealworms was slightly lower than that of casein (Goulet et al., 1978), whereas the protein digestibility of dried caterpillars (*Clanis bilineata*) was shown to be similar to that for casein (Xia et al., 2012).

The Chinese grasshopper *A. cinerea* showed a slightly higher true protein digestibility than fish meal when fed to poultry (Wang et al., 2007). For grasshoppers, a protein digestibility of 62% has been reported in poultry (Ravindran and Blair, 1993). For broilers, rendered beef meal can be replaced by a meal of the termite *Kalotermes flavicollis* (F.) or the cockroach *Blatta orientalis* (L.), with similar or better growth rates of broilers (Munyuli Bin Mushambanyi and Balezi, 2002). In general, the protein digestibility of insects seems to be relatively high and the variability reported in the literature is likely a result of differences in how the insect was prepared prior to being used and the proportion of amino acids that are used for sclerotization.

17.7. ENHANCING THE NUTRIENT COMPOSITION OF INSECTS AS FOOD FOR INSECTIVORES

Although the exact nutrient requirements for most insectivores are unknown, certain nutrient deficiencies are known to occur regularly in captive insectivores. Probably the three most common in captive insectivores are calcium, vitamin A, and vitamin D deficiencies. Nutrients that are expected to be present at low levels in insects can be enhanced in order to increase the concentration of these nutrients in the insectivore diet. There are two primary methods of providing these nutrients: gut loading and dusting. The goal of both gut loading and dusting is to increase the nutrient intake of selected nutrients by the insectivore.

Although this is valid within a certain range, one must be careful that this range is not surpassed, resulting in adverse effects. For instance, oversupplementation with calcium can decrease the absorption of other minerals, leading to secondary trace mineral deficiencies. Likewise, oversupplementation of the fat-soluble vitamins, in particular vitamins A and D, can cause toxicity (Bender, 2002). A difference in size (developmental stage) of the dusted or gut-loaded insect leads to differences in surface-to-volume and gut size-to-volume ratios. Smaller insects have a relatively large surface area to which the dust can adhere. The same powder used for pinhead or adult house crickets could therefore have different effects on their chemical composition (Sullivan et al., 2009). Similarly, for gut-loaded insects, size differences could lead to differences in nutrient delivery of smaller versus larger insects (Finke, 2003, 2005).

17.7.1. Gut Loading

Gut loading is the term used for the provision of a special diet to insects, shortly before the insects will be consumed. When this diet, which contains high levels of the desired nutrient(s), is consumed by the insect it will be present in the gut, thereby increasing the insectivore's nutrient intake when the insect is consumed. Due to the nature of gut loading, it is suitable for almost all nutrients as long as the diet is palatable to the insect and the diet can contain sufficient quantities of the desired nutrient(s) (Hunt-Coslik et al., 2009).

Most research on the effects of gut loading has focused on increasing the calcium content of insects. High-calcium diets containing 4–9% calcium, typically from calcium carbonate, have proven to be effective in increasing the calcium content of wax moth larvae, house crickets, yellow mealworm larvae, and silkworm larvae (Allen and Oftedal, 1989; Anderson, 2000; Finke, 2003, 2005; Klasing et al., 2000; Strzeleqicz et al., 1985). Chemical analysis of the diet provided might be necessary to verify the true calcium content of commercially available gut-loading diets (Finke et al., 2004, 2005). The calcium from gut-loaded yellow mealworms fed high-calcium diets was shown to be readily available to growing chicks, showing its usefulness in providing available calcium (Klasing et al., 2000).

Different studies have found different results in optimal gut-loading times, and the optimal amount of time for properly gut-loading insects seems to vary slightly. This is likely a result of the insect species being studied, the palatability of the gut-loading diet, and the environmental conditions (temperature, light, and humidity). In general, however, gut loading for 24–72h appears to result in similar levels of nutrients in the intact insect. When gut-loading diets are fed for longer periods of time, adverse effects on the viability of the insects have been observed (Klasing et al., 2000). In yellow mealworms, a gut-loading period of 24h raised Ca:P ratios; extending this period to 48 or 72h resulted in a slightly higher Ca:P ratio (Klasing et al., 2000; Anderson, 2000). For house crickets, a period of 48h seems sufficient to attain a significant increase in Ca:P ratios, while extending this to 72h does not affect the ratio (Anderson, 2000). However, other studies report the highest calcium content after 1 day compared to 2, 3, or 7 days (Dikeman et al., 2007). It has been reported that offering certain gut-loading diets longer than 2 days can reduce the initially increased calcium levels (Hunt-Coslik et al., 2009), which could be an effect of the palatability of the gut-loading diet (McComb, 2010).

In addition, the physical form of the diet and the presence of other nutrients such as amino acids and fatty acids, which affect diet palatability, is something to take into account when designing a gut-loading diet (Anderson, 2000). The addition of polyunsaturated fatty acids to the gut-loading diet has been suggested for insectivores stemming from temperate climates, which would be likely to encounter insects with relatively high polyunsaturated fatty acid concentrations (Li et al., 2009).

A simple way of providing extra carotenoids to insectivores is the provision of carrots or certain other fruits or vegetables during the last 24h before feeding the insects to the insectivores. The amount of carotenoids that accumulates through gut loading differs per insect species. The field cricket, for instance, accumulates more carotenoids on a high-carotenoid diet than the house cricket or the banded cricket (*Gryllodes sigillatus* Walker) (Ogilvy et al., 2012).

A study in which yellow mealworms were gut loaded with chicken starter feed increased their vitamin D content by 132IU/kg dry matter, while undetectable levels were present in mealworms provided a wheat bran diet (Klasing et al., 2000).

17.7.2. Dusting

Dusting is the term used for coating an insect with a fine powder containing the desired nutrients, such that the powder adheres to the outside of the insect. When the insect is eaten, the powder on the outside is also ingested. Little data exist on the nutritional effects of this method. In a study in which house crickets were dusted with a fine calcium powder, the Ca:P ratio was raised from 1:5.7 to 5.3:1 and the digestibility of these minerals was high (84% and 94%, respectively; Dierenfeld and King, 2008).

An important factor in the effectiveness of this method is the time between dusting and consumption (Trusk and Crissey, 1987). House crickets can groom off up to half of the amount of adhering powder in 90 seconds (Li et al., 2009). For animals which immediately consume their prey, dusting can be an acceptable method. A downside of dusting is that it is difficult to estimate the amount of dust adhering to the insect. Using two different calcium carbonate dusts, weight increases of 0.8–6.3% for giant wax moth larvae, yellow mealworm larvae, and house crickets when dusted with two different types of calcium carbonate dusts have been reported (Winn et al., 2003).

The physical characteristics of the dust, the relative surface area of the insect, and the physical characteristics of the surface of the insect exoskeleton can all affect the amount of dust adhering to the feeder insect. For aquatic insectivores, this way of enhancing the nutrient content is obviously unsuitable. A study with the Wyoming toad (*Bufo baxteri* Porter) indicated that both the composition and the method of providing the powder or gut-load diet can lead to differences in the weight gain of the insectivore (Li et al., 2009). This might be due to a decreased consumption rate, which in turn might be caused by decreased palatability due to the vitamin powder (Li et al., 2009). A study of Puerto Rican crested toads (*Peltophryne lemur* Cope) compared the effects of direct oral application of vitamin A with gut loading and dusting in house crickets. Retinol blood values were significantly higher in toads offered dusted crickets (McComb, 2010). This indicates the effectiveness of this method. Why the other methods were not effective is not clear, although it may be that retinol absorption was enhanced when ingested with food. Dietary fat enhances the absorption of fat-soluble nutrients such as retinol, and so the fat in the crickets may have increased retinol absorption compared to direct oral supplementation.

17.8. OTHER CONSIDERATIONS

17.8.1. Pathogens/Parasites

There is little information regarding commercial feeder insects as a source of pathogens. House crickets from five commercial suppliers in the United States were shown to be free of oxyurids/pinworms, but little other information is available regarding insects as a source of parasites for insectivores (Klarsfeld and Mitchell, 2005). The lesser mealworm *Alphitobius diaperinus* (Panzer) has been shown to be a potential vector for *Salmonella* spp. with both larvae and adults being able to carry *Salmonella* spp. both externally and internally (Crippen et al., 2009). Additionally, *Salmonella* could be detected in newly emerged adult beetles from infected larvae, suggesting some bacteria are carried through metamorphosis (Crippen et al., 2012). Because several species of grain beetles (*Tenebrio* sp and *Z. morio*) are commonly used as feeder insects, it seems likely that these species could serve as a vector of *Salmonella* spp. and other pathogenic microorganisms. To minimize this risk, feeder insects should be obtained from qualified suppliers, fed an appropriate diet, and maintained under hygienic conditions to minimize the risk of transmitting pathogenic bacteria.

17.8.2. Toxins

It is well known that many species of insects sequester toxic compounds from their diet, making them unpalatable or even toxic to certain insectivores. In the wild, these species are generally brightly colored (aposomatic) to warn potential predators of the consequences of feeding on these species. Monarch butterflies (*Danaus plexippus* L.) and milkweed bugs (*Oncopeltus fasciatus* Dallas) are but a few of the many species that sequester toxins from their food (Berebaum and Miliczky, 1984; Brower, 1969). Because most captive-raised feeder insects are fed controlled diets containing commercial feed ingredients that are used to feed domestic animals, it seems likely that they would accumulate few if any toxins from their diet, as long as the diet was properly made and stored.

Ingredients commonly used in commercial insect diets can become contaminated with various species of molds that can produce mycotoxins, including alflatoxin, fumonisin, and zearalenone. Little is known regarding the effects of these mycotoxins on both insects and insectivores. However, at least some insect species seem to be relatively resistant to moderate levels of aflatoxin (McMillian et al., 1981). None of the insect species commonly used to feed captive insectivores have been studied with respect to mycotoxins, but feeder insects should be sourced from qualified suppliers. Additionally, the diet fed to the insects prior to use should be stored properly and periodically checked to ensure they are free of molds that might produce mycotoxins.

Almost nothing is known regarding the potential antinutritional properties of insects, with the exception of thiaminase. Thiaminase is an enzyme that when ingested splits thiamin, effectively destroying its vitamin properties. Although thiaminase is typically associated with certain species of fish, it has been reported in both domesticated silkworm larvae (*B. mori*) and African silkworm pupae (*Anaphe* spp), although the levels found in domesticated silkworm larvae were only one-third of those found in African silkworm pupae. These authors noted that in addition to thiamine, pyridoxine, taurine, and other nutrients could also serve as a substrate for this enzyme. The consumption of *Anaphe* pupae has been associated with a seasonal ataxia in local populations in Nigeria, which is presumably due to thiamine deficiency (Adamolekun, 1993; Adamolekun et al., 1997; Nishimune et al., 2000). The extent to which thiaminase is found in other species of insects and the potential effects on insectivores are currently unknown.

17.9. CONCLUSIONS

Insects are a good source of many nutrients, although for most nutrients the values vary widely depending on the insect species, the life stage, and the conditions in which they are raised. In general, most species appear to be good sources of amino acids, fatty acids, most minerals, and most B vitamins. Based on analysis of feeder insects and reports of nutrient deficiencies in captive insectivores, the nutrients of concern in a captive insectivore feeding program include calcium and the fat-soluble vitamins A, D, and E. As such, captive insectivores should be fed a mix of invertebrates that have been dusted or gut loaded to provide a wide range of nutrient intakes in an effort to reduce the risk of nutrient deficiencies.

REFERENCES

Adamolekun, B., 1993. *Anaphe venata* entomophagy and seasonal ataxic syndrome in southwest Nigeria. Lancet 341, 629.

Adamolekun, B., McCandless, D.W., Butterworth, R.F., 1997. Epidemic of seasonal ataxia in Nigeria following ingestion of the African silkworm *Anaphe venata*: role of thiamine deficiency? Metab. Brain Dis. 12, 251–258.

Aguilar-Miranda, E.D., Lopez, M.G., Escamilla-Santana, C., Barba de la Rosa, A.P., 2002. Characteristics of maize flour tortilla supplemented with ground *Tenebrio molitor* larvae. J. Agric. Food Chem. 50, 192–195.

Akman Gündüz, N.E., Gülel, A., 2002. Effect of temperature on development, sexual maturation time, food consumption and body weight of *Schistocerca gregaria* Forsk. (Orthoptera: Acrididae). Türk. Zool. Derg. 26, 223–227.

Ali, M., Ewiess, M.A., 1977. Photoperiodic and temperature effects on rate of development and diapause in the green stink bug, *Nezara viridula* L. (Heteroptera: Pentatomidae). Z. Angew. Entomol. 84, 256–264.

Ali, M.F., Mashaly, A.M.A., Mohammed, A.A., El-Magd Mahmoud, M.A., 2011. Effect of temperature and humidity on the biology of *Attagenus fasciatus* (Thunberg) (Coleoptera: Dermestidae). J. Stored Prod. Res. 47, 25–31.

Allen, M.E., Oftedal, O.T., 1989. Dietary manipulation of the calcium content of feed crickets. J. Zoo Wildl. Med. 20, 26–33.

Anderson, S.J., 2000. Increasing calcium levels in cultured insects. Zoo Biol. 19, 1–9.

Angilletta Jr., M.J., Steury, T.D., Sears, M.W., 2004. Temperature, growth rate, and body size in ectotherms: fitting pieces of a life-history puzzle. Integr. Comp. Biol. 44, 498–509.

Banjo, A.D., Lawal, O.A., Songonuga, E.A., 2006. The nutritional value of fourteen species of edible insects in southwestern Nigeria. Afr. J. Biotechnol. 5, 298–301.

Barbehenn, R.V., 2003. Antioxidants in grasshoppers: higher levels defend the midgut tissues of a polyphagous species than a graminivorous species. J. Chem. Ecol. 29, 683–702.

Barker, D., Fitzpatrick, M.P., Dierenfeld, E.S., 1998. Nutrient composition of selected whole invertebrates. Zoo Biol. 17, 123–134.

Barlow, J.S., 1965. Composition of the fats in pupae of *Agria affinis* (Fallen) (Diptera: Sarcophagidae). Can. J. Zool. 43, 291–295.

Beenakkers, A.M.T., Scheres, J.M.J.C., 1971. Dietary lipids and lipid composition of the fat-body of *Locusta migratoria*. Insect Biochem. 1, 125–129.

Beenakkers, A.M.T., Meisen, M.A.H.Q., Scheres, J.M.J.C., 1971. Influence of temperature and food on growth and digestion in fifth instar larvae and adults of Locusta. J. Insect Physiol. 17, 871–880.

Beenakkers, M.T.A., Van der Horst, D.J., Van Marrewijk, W.J.A., 1985. Insect lipids and lipoproteins, and their role in physiological processes. Prog. Lipid Res. 24, 19–67.

Bender, D.A., 2002. Introduction to Nutrition and Metabolism. CRC Press, London, p. 450.

Berebaum, M.R., Miliczky, E., 1984. Mantids and milkweed bugs: efficacy of aposematic coloration against invertebrate predators. Am. Midl. Nat. 111, 64–68.

Berkvens, N., Bonte, J., Berkvens, D., Tirry, L., De Clercq, P., 2008. Influence of diet and photoperiod on development and reproduction of European populations of *Harmonia axyridis* (Pallas) (Coleoptera: Coccinellidae). BioControl 53, 211–221.

Bernard, J.B., Allen, M.E., 1997. Feeding Captive Insectivorous Animals: Nutritional Aspects of Insects as Food. Nutrition Advisory Group Handbook.

Blomquist, G.J., Borgeson, C.E., Vundla, M., 1991. Polyunsaturated fatty acids and eicosanoids in insects. Insect Biochem. 21, 99–106.

Bodnaryk, R.P., 1981. The biosynthesis function, and fate of taurine during the metamorphosis of the *Noctuid moth Mamestra configurata* Wlk. Insect Biochem. 11, 199–205.

Borgeson, C.E., Kurtti, T.J., Munderloh, U.G., Blomquist, G.J., 1991. Insect tissues, not microorganisms, produce linoleic-acid in the house cricket and the American cockroach. Experientia 47, 238–241.

Borgeson, C.E., Blomquist, G.J., 1993. Subcellular location of the [Delta]12 desaturase rules out bacteriocyte contribution to linoleate biosynthesis in the house cricket and the American cockroach. Insect Biochem. Molec. Biol. 23, 297–302.

Brower, L.P., 1969. Ecological chemistry. Sci. Am. 220, 22–29.

Bukkens, S.G.F., 1997. The nutritional value of edible insects. Ecol. Food Nutr. 36, 287–319.

Bureš, S., Weidinger, K., 2003. Sources and timing of calcium intake during reproduction in flycatchers. Oecologia 137, 634–647.

Calvez, B., 1975. Effect of nutritional level of diet on chemical composition of organs and their interrelations in silkworm *Bombyx-mori*-L. Ann. Nutr. Aliment. 29, 259–269.

Cauchie, H.M., 2002. Chitin production by arthropods in the hydrosphere. Hydrobiologia 470, 63–96.

Cookman, J.E., Angelo, M.J., Slansky Jr., F., Nation, J.L., 1984. Lipid content and fatty acid composition of larvae and adults of the velvetbean caterpillar, *Anticarsia gemmatalis*, as affected by larval diet. J. Insect Physiol. 30, 523–527.

Crippen, T.L., Sheffield, C.L., Esquivel, S.V., Droleskey, R.E., Esquivel, J.F., 2009. The acquisition and internalization of *Salmonella* by the lesser mealworm, *Alphitobius diaperinus* (Coleoptera: Tenebrionidae). Vector Borne Zoonotic Dis. 9, 65–71.

Crippen, T.L., Zheng, L., Sheffield, C.L., Tomberlin, J.K., Beier, R.C., Yu, Z., 2012. Transient gut retention and persistence of *Salmonella* through metamorphosis in the lesser mealworm, *Alphitobius diaperinus* (Coleoptera: Tenebrionidae). J. Appl. Microbiol. 112, 920–926.

Dashefsky, H.S., Anderson, D.L., Tobin, E.N., Peters, T.M., 1976. Face fly pupae: a potential feed supplement for poultry. Environ. Entomol. 5, 680–682.

de Renobales, M., Cripps, C., Stanley-Samuelson, D.W., Jurenka, R.A., Blomquist, G.J., 1987. Biosynthesis of linoleic acid in insects. Trends Biochem. Sci. 12, 364–366.

DeFoliart, G.R., 1992. Insect as human food: gene DeFoliart discusses some nutritional and economic aspects. Crop Prot. 11, 395–399.

Dierenfeld, E.S., King, J., 2008. Digestibility and mineral availability of phoenix worms (*Hermetia illucens*) ingested by mountain chicken frogs (*Leptodactylus fallax*). J. Herp. Med. Surg. 18, 100–105.

Dikeman, C.L., Plesuk, S.D., Klimek, D.L., Simmons, L.G., 2007. Effect of gut-loading time on nutrient content of adult feeder crickets. J. Anim. Sci. 85, S195.

Donoghue, S., 2006. Nutrition. In: Mader, D.R. (Ed.), Reptile Medicine and Surgery, second ed. Saunders Elsevier, St. Louis, Missouri, pp. 251–298.

Downer, R.G.H., Matthews, J.R., 1976. Patterns of lipid distribution and utilisation in insects. Integr. Comp. Biol. 16, 733–745.

Dreyer, J.J., Wehmeyer, A.S., 1982. On the nutritive value of mopanie worms. S. Afr. J. Sci. 78, 33–35.

Dufour, D.L., 1987. Insects as food: a case study from the northwest Amazon. Am. Anthropol. 89, 383–397.

Eeva, T., Helle, S., Salminen, J.P., 2010. Carotenoid composition of invertebrates consumed by two insectivorous bird species. J. Chem. Ecol. 36, 608–613.

Ekpo, K.E., Onigbinde, A.O., Asia, I.O., 2009. Pharmaceutical potentials of the oils of some popular insects consumed in southern Nigeria. Afr. J. Pharm. Pharmacol. 3, 051–057.

Fast, P.G., 1970. Insect lipids. Prog. Chem. Fats Lipids 11, 181–242.

Ferguson, G.W., Jones, J.R., Gehrmann, W.H., Hammack, S.H., Talent, L.G., Hudson, R.D., Dierenfeld, E.S., Fitzpatrick, M.P., Frye, F.L., Holick, M.F., Chen, T.C., Lu, Z., Gross, T.S., Vogel, J.J., 1996. Indoor husbandry of the panther chameleon *Chamaeleo* [Furcifer] *pardalis*: effects of dietary vitamins A and D and ultraviolet irradiation on pathology and life-history traits. Zoo Biol. 15, 279–299.

Finke, M.D., 1984. The Use of Nonlinear Models to Evaluate the Nutritional Quality of Insect Protein. University of Wisconsin, Madison, WI.

Finke, M.D., 2002. Complete nutrient composition of commercially raised invertebrates used as food for insectivores. Zoo Biol. 21, 269–285.

Finke, M.D., 2003. Gut loading to enhance the nutrient content of insects as food for reptiles: a mathematical approach. Zoo Biol. 22, 147–162.

Finke, M.D., 2005. Nutrient composition of bee brood and its potential as human food. Ecol. Food Nutr. 44, 257–270.

Finke, M.D., 2007. Estimate of chitin in raw whole insects. Zoo Biol. 26, 105–115.

Finke, M.D., 2013. Complete nutrient content of four species of feeder Insects. Zoo Biol. 32, 27–36.

Finke, M.D., Sunde, M.L., DeFoliart, G.R., 1985. An evaluation of the protein quality of Mormon crickets (*Anabrus simplex* Haldeman) when used as a high protein feedstuff for poultry. Poult. Sci. 64, 708–712.

Finke, M.D., DeFoliart, G.R., Benevenga, N.J., 1987. Use of a four-parameter logistic model to evaluate the protein quality of mixtures of Mormon cricket meal and corn gluten meal in rats. J. Nutr. 117, 1740–1750.

Finke, M.D., DeFoliart, G.R., Benevenga, N.J., 1989. Use of a four-parameter logistic model to evaluate the quality of the protein from three insect species when fed to rats. J. Nutr. 119, 864–871.

Finke, M.D., Dunham, S.D., Cole, J.S., 2004. Evaluation of various calcium-fortified high moisture commercial products for improving the calcium content of crickets, *Acheta domesticus*. J. Herp. Med. Surg. 14, 6–9.

Finke, M.D., Dunham, S.D., Kwabi, C.A., 2005. Evaluation of four dry commercial gut loading products for improving the calcium content of crickets, *Acheta domesticus*. J. Herp. Med. Surg. 15, 7–12.

Fontaneto, D., Tommaseo-Ponzetta, M., Galli, C., Risé, P., Glew, R.H., Paoletti, M.G., 2011. Differences in fatty acid composition between aquatic and terrestrial insects used as food in human nutrition. Ecol. Food Nutr. 50, 351–367.

Fraenkel, G., 1950. The nutrition of the mealworm, *Tenebrio molitor* L. (Tenebrionidae, Coleoptera). Physiol. Zool. 23, 92–108.

Fujimoto, W., Suzuki, M., Kimura, K., Iwanaga, T., 2002. Cellular expression of the gut chitinase in the stomach of frogs *Xenopus laevis* and *Rana catesbeiana*. Biomed. Res. Tokyo 23, 91–99.

German, D.P., Nagle, B.C., Villeda, J.M., Ruiz, A.M., Thomson, A.W., Contreras Balderas, S., Evans, D.H., 2010. Evolution of herbivory in a carnivorous clade of minnows (Teleostei: Cyprinidae): effects on gut size and digestive physiology. Physiol. Biochem. Zool. 83, 1–18.

Gist, C.S., Crossley Jr., D.A., 1975. The litter arthropod community in a Southern Appalachian hardwood forest: numbers, biomass and mineral element content. Am. Midl. Nat. 93, 107–122.

Goldsmith, T.H., Warner, L.T., 1964. Vitamin A in the vision of insects. J. Gen. Physiol. 47, 433–441.

Goulet, G., Mullier, P., Sinave, P., Brisson, G.J., 1978. Nutritional evaluation of dried *Tenebrio molitor* larvae in the rat. Nutr. Rep. Int. 18, 11–15.

Graveland, J., van Gijzen, T., 1994. Arthropods and seeds are not sufficient as calcium sources for shell formation and skeletal growth in passerines. Ardea 82, 299–314.

Gullan, P.J., Cranston, P.S., 2005. The Insects: An Outline of Entomology. Blackwell Publishing, pp. 10–20.

Haglund, O., Wallin, R., Wretling, S., Hultberg, B., Saldeen, T., 1998. Effects of fish oil alone and combined with long chain (n-6) fatty acids on some coronary risk factors in male subjects. J. Nutr. Biochem. 9, 629–635.

Hamilton, A.G., 1936. The relation of humidity and temperature to the development of three species of African locusts—*Locusta migratoria migratoioides* (R. & F.), *Schistocerca gregaria* (Forsk.), *Nomadacris septemfasciata* (Serv.). T. Roy. Ent. Soc. London 85, 1–60.

Han, R.D., Parajulee, M., Zhong, H., Feng, G., 2008. Effects of environmental humidity on the survival and development of pine caterpillars, *Dendrolimus tabulaeformis* (Lepidoptera: Lasiocampidae). Insect Sci. 15, 147–152.

Hanson, B.J., Cummins, K.W., Ii, A.S.C., Lowry, R.R., 1983. Dietary effects on lipid and fatty acid composition of *Clistoronia magnifica* (Trichoptera: Limnephilidae). Freshwater Invert. Biol. 2, 2–15.

Hanson, B.J., Cummins, K.W., Cargill, A.S., Lowry, R.R., 1985. Lipid content, fatty acid composition, and the effect of diet on fats of aquatic insects. Comp. Biochem. Physiol. B 80, 257–276.

Hatt, J.M., Hung, E., Wanner, M., 2003. The influence of diet on the body composition of the house cricket (*Acheta domesticus*) and consequences for their use in zoo animal nutrition. Zool. Garten 73, 238–244.

Hoby, S., Wenker, C., Robert, N., Jermann, T., Hartnack, S., Segner, H., Aebischer, C., Liesegang, A., 2010. Nutritional metabolic bone disease in juvenile veiled chameleons (*Chamaeleo calyptratus*) and its prevention. J. Nutr. 140, 1923–1931.

Hoffmann, K.H., 1973. Effects of temperature on chemical composition of crickets (Gryllus, orthopt.) (In German). Oecologia 13, 147–175.

House, H.L., Riordan, D.F., Barlow, J.S., 1958. Effects of thermal conditioning and of degree of saturation of dietary lipids on resistance of an insect to a high temperature. Can. J. Zool. 36, 629–632.

Hunt-Coslik, A., Ward, A.M., McClements, R.D., 2009. In: Ward, A., Treiber, K., Schmidt, D., Coslik, A., Maslanka, M. (Eds.), Gut-Loading as a Method to Effectively Supplements Crickets with Calcium and Vitamin A, AZA Nutrition Advisory Group, Tulsa OK, pp. 163–171.

Isaksson, C., Andersson, S., 2007. Carotenoid diet and nestling provisioning in urban and rural great tits *Parus major*. J. Avian Biol. 38, 564–572.

Jones, L.D., Cooper, R.W., Harding, R.S., 1972. Composition of mealworm *Tenebrio molitor* larvae. J. Zoo. Anim. Med. 3, 34–41.

Klarsfeld, J.D., Mitchell, M.A., 2005. An evaluation of the gray cricket, *Acheta domestica*, as a source of oxyurids for reptiles. J. Herp. Med. Surg. 15, 18–20.

Klasing, K.C., Thacker, P., Lopez, M.A., Calvert, C.C., 2000. Increasing the calcium content of mealworms (*Tenebrio molitor*) to improve their nutritional value for bone mineralization of growing chicks. J. Zoo Wildl. Med. 31, 512–517.

Koc, Y., Gulel, A., 2008. Age and sex related variations in protein and carbohydrate levels of *Galleria mellonella* (Linnaeus, 1758) (Lepidoptera: Pyralidae) in constant lightness and darkness. Pak. J. Biol. Sci. 11, 733–739.

Kodondi, K.K., Leclercq, M., Gaudin-Harding, F., 1987. Vitamin estimations of three edible species of Attacidae caterpillars from Zaire. Int. J. Vitam. Nutr. Res. 57, 333–334.

Kouba, M., Mourot, J., 2011. A review of nutritional effects on fat composition of animal products with special emphasis on *n*-3 polyunsaturated fatty acids. Biochimie 93, 13–17.

Kramer, K.J., Hopkins, T.L., Schaefer, J., 1995. Applications of solids NMR to the analysis of insect sclerotized structures. Insect Biochem. Mol. Biol. 25, 1067–1080.

Krengel, S., Stangl, G.I., Brandsch, C., Freier, B., Klose, T., Mill, E., Kiowsi, A., 2012. A comparative study on effects of normal versus elevated temperatures during preimaginal and young adult period on body weight and fat body content of mature *Coccinella septempunctata* and *Harmonia axyridis* (Coleoptera: Coccinellidae). Environ. Entomol. 41, 676–687.

Lease, H.M., Wolf, B.O., 2010. Exoskeletal chitin scales isometrically with body size in terrestrial insects. J. Morphol. 271, 759–768.

Lease, H.M., Wolf, B.O., 2011. Lipid content of terrestrial arthropods in relation to body size, phylogeny, ontogeny and sex. Physiol. Entomol. 36, 29–38.

Levi, L., Ziv, T., Admon, A., Levavi-Sivan, B., Lubzens, E., 2012. Insight into molecular pathways of retinal metabolism, associated with vitellogenesis in zebrafish. Am. J. Physiol-Endocrinol. Metab. 302, 626–644.

Levy, R., Cromroy, H.L., 1973. Concentration of some major and trace elements in forty-one species of adult and immature insects determined by atomic absorption spectroscopy. Ann. Entomol. Soc. Am. 66, 523–526.

Li, H., Vaughan, M.J., Browne, R.K., 2009. A complex enrichment diet improves growth and health in the endangered Wyoming toad (*Bufo baxteri*). Zoo Biol. 28, 197–213.

Lorenz, M.W., Anand, A.N., 2004. Changes in the biochemical composition of fat body stores during adult development of female crickets, *Gryllus bimaculatus*. Arch. Insect Biochem. Physiol. 56, 110–119.

Machin, J., 1975. Water balance in *Tenebrio molitor*, L. larvae; the effect of atmospheric water absorption. J. Comp. Physiol. B 101, 121–132.

Majumder, U.K., Sengupta, A., 1979. Triglyceride composition of chrysalis oil, an insect lipid. J. Am. Oil. Chem. Soc. 56, 620–623.

Martin, R.D., Rivers, J.P.W., Cowgill, U.M., 1976. Culturing mealworms as food for animals in captivity. Int. Zoo Yearb. 16, 63–70.

Massie, H.R., Williams, T.R., DeWolfe, L.K., 1989. Changes in taurine in aging fruit flies and mice. Exp. Gerontol. 24, 57–65.

McClements, R.D., Lintzenich, B.A., Boardman, J., 2003. A zoo-wide evaluation into the current feeder insect supplementation program at the Brookfield Zoo. Proceedings of the Nutrition Advisory Group Fifth Conference on Zoo and Wildlife Nutrition, pp. 54–59.

McComb, A., 2010. Evaluation of Vitamin A Supplementations for Captive Amphibian Species. North Carolina State University, Raleigh, North Carolina, p. 129.

McMillian, W.W., Widstrom, N.W., Wilson, D.M., 1981. Rearing the maize weevil on maize genotypes when aflatoxin-producing *Aspergillus flavus* and *A. parasiticus* isolates were present. Environ. Entomol. 10, 760–762.

Miller, E.A., Green, S.L., Otto, G.M., Bouley, D.M., 2001. Suspected hypovitaminosis A in a colony of captive green anoles (*Anolis carolinensis*). J. Am. Assoc. Lab. Anim. Sci. 40, 18–20.

Morgan, T.D., Baker, P., Kramer, K.J., Basibuyuk, H.H., Quick, D.L.J., 2003. Metals in mandibles of stored product insects: do zinc and manganese enhance the ability of larvae to infest seeds? J. Stored Prod. Res. 39, 65–75.

Müller, K., Raila, J., Altenkamp, R., Schmidt, D., Dietrich, R., Hurtienne, A., Wink, M., Krone, O., Brunnberg, L., Schweigert, F.J., 2012. Concentrations of retinol, 3,4-didehydroretinol, and retinyl esters in plasma of free-ranging birds of prey. J. Anim. Physiol. Anim. Nutr. 96, 1044–1053.

Munyuli Bin Mushambanyi, T., Balezi, N., 2002. Utilisation des blattes et des termites comme substituts potentiels de la farine de viande dans l'alimentation des poulets de chair au Sud Kivu, République. Démocratique, du Congo. Tropicultura 20, 10–16.

Nakagaki, B.J., Sunde, M.L., Defoliart, G.R., 1987. Protein quality of the house cricket, *Acheta domesticus*, when fed to broiler chicks. Poult. Sci. 66, 1367–1371.

Nakasone, S., Ito, T., 1967. Fatty acid composition of the silkworm, *Bombyx mori* L. J. Insect Physiol. 13, 1237–1246.

Nedvěd, O., Kalushkov, P., 2012. Effect of air humidity on sex ratio and development of ladybird *Harmonia axyridis* (Coleoptera: Coccinellidae). Psyche.

Nestel, D., Papadopoulos, N.T., Liedo, P., Gonzales-Ceron, L., Carey, J.R., 2005. Trends in lipid and protein contents during medfly aging: an harmonic path to death. Arch. Insect Biochem. Physiol. 60, 130–139.

Nishimune, T., Watanabe, Y., Okazaki, H., Akai, H., 2000. Thiamin is decomposed due to *Anaphe* spp. entomophagy in seasonal ataxia patients in Nigeria. J. Nutr. 130, 1625–1628.

Ogilvy, V., Fidgett, A.L., Preziosi, R.F., 2012. Differences in carotenoid accumulation among three feeder-cricket species: implications for carotenoid delivery to captive insectivores. Zoo Biol. 31, 470–478.

Olson, J.A., 1989. Provitamin A function of carotenoids: the conversion of beta-carotene into vitamin A. J. Nutr. 119, 105–108.

Onifade, A.A., Oduguwa, O.O., Fanimo, A.O., Abu, A.O., Olutunde, T.O., Arije, A., Babatunde, G.M., 2001. Effects of supplemental methionine and lysine on the nutritional value of housefly larvae meal (*Musca domestica*) fed to rats. Bioresour. Technol. 78, 191–194.

Oonincx, D.G.A.B., Stevens, Y., van den Borne, J.J., van Leeuwen, J.P., Hendriks, W.H., 2010. Effects of vitamin D3 supplementation and UVb exposure on the growth and plasma concentration of vitamin D3 metabolites in juvenile bearded dragons (*Pogona vitticeps*). Comp. Biochem. Physiol. B. 156, 122–128.

Oonincx, D.G.A.B., van der Poel, A.F., 2011. Effects of diet on the chemical composition of migratory locusts (*Locusta migratoria*). Zoo Biol. 30, 9–16.

Oonincx, D.G.A.B., Dierenfeld, E.S., 2012. An investigation into the chemical composition of alternative invertebrate prey. Zoo Biol. 31, 40–54.

Oyarzun, S.E., Crawshaw, G.J., Valdes, E.V., 1996. Nutrition of the tamandua: I. Nutrient composition of termites (*Nasutitermes* spp.) and stomach contents from wild tamanduas (*Tamandua tetradactyla*). Zoo Biol. 15, 509–524.

Pennino, M., Dierenfeld, E.S., Behler, J.L., 1991. Retinol, alpha-tocopherol, and proximate nutrient composition of invertebrates used as food. Int. Zoo Yearb. 30, 143–149.

Pessier, A.P., Linn, M., Garner, M.M., Raymond, J.T., Dierenfeld, E.S., Graffam, W., 2005. Suspected hypovitaminosis A in captive toads (*Bufo* spp.). Proceedings of the AAZV AAWV AZAA/NAG Joint Conference Held in Omaha, NE, October 14–21, 2005, p. 57.

Pretorius, Q., 2011. The Evaluation of Larvae of *Musca Domestica* (Common Housefly) as Protein Source for Broiler Production [MSc]. Stellenbosch University, Stellenbosch.

Punzo, F., 2003. Nutrient composition of some insects and arachnids. Fla. Scientist 66, 84–98.

Ramos-Elorduy, J., 1997. Insects: a sustainable source of food? Ecol. Food Nutr. 36, 246–276.

Ramos-Elorduy, J., Gonzalez, E.A., Hernandez, A.R., Pino, J.M., 2002. Use of *Tenebrio molitor* (Coleoptera: Tenebrionidae) to recycle organic wastes and as feed for broiler chickens. J. Econ. Entomol. 95, 214–220.

Ramsay, S.L., Houston, D.C., 2003. Amino acid composition of some woodland arthropods and its implications for breeding tits and other passerines. Ibis 145, 227–232.

Ravindran, V., Blair, R., 1993. Feed resources for poultry production in Asia and the Pacific. 3. Animal protein-sources. World Poult. Sci. J. 49, 219–235.

Reichle, D.E., Shanks, M.H., Crossley, D.A., 1969. Calcium, potassium, and sodium content of forest floor arthropods. Ann. Entomol. Soc. Am. 62, 57–62.

Rich, C.N., Talent, L.G., 2008. The effects of prey species on food conversion efficiency and growth of an insectivorous lizard. Zoo Biol. 27, 181–187.

Roe, R.M., Clifford, C.W., Woodring, J.P., 1980. The effect of temperature on feeding, growth, and metabolism during the last larval stadium of the female house cricket, *Acheta domesticus*. J. Insect Physiol. 26, 639–644.

Roe, R.M., Clifford, C.W., Woodring, J.P., 1985. The effect of temperature on energy distribution during the last-larval stadium of the female house cricket, *Acheta domesticus*. J. Insect Physiol. 31, 371–378.

Santos Oliveira, J.F., Passos De Carvalho, J., Bruno De Sousa, R.F.X., Simao, M.M., 1976. The nutritional value of four species of insects consumed in Angola. Ecol. Food Nutr. 5, 91–97.

Schmitz, G., Ecker, J., 2008. The opposing effects of n-3 and n-6 fatty acids. Prog. Lipid Res. 47, 147–155.

Sealey, W.M., Gaylord, T.G., Barrows, F.T., Tomberlin, J.K., McGuire, M.A., Ross, C., St-Hilaire, S., 2011. Sensory analysis of rainbow trout, *Oncorhynchus mykiss*, fed enriched black soldier fly prepupae, *Hermetia illucens*. J. World Aquacult. Soc. 42, 34–45.

Seki, T., Isono, K., Ozaki, K., Tsukahara, Y., Shibata-Katsuta, Y., Ito, M., Irie, T., Katagiri, M., 1998. The metabolic pathway of visual pigment chromophore formation in *Drosophila melanogaster*: all-trans (3S)-3-hydroxyretinal is formed from all-trans retinal via (3R)-3-hydroxyretinal in the dark. Eur. J. Biochem. 257, 522–527.

Senar, J.C., Møller, A.P., Ruiz, I., Negro, J.J., Broggi, J., Hohtola, E., 2010. Specific appetite for carotenoids in a colorful bird. PLoS One 5, e10716.

Shearer, P.W., Jones, V.P., 1996. Diel feeding pattern of adult female southern green stink bug (Hemiptera: Pentatomidae). Environ. Entomol. 25, 599–602.

Simpson, S.J., Raubenheimer, D., 2001. The geometric analysis of nutrient—allelochemical interactions: a case study using locusts. Ecology 82, 422–439.

Smith, W.C., Goldsmith, T.H., 1990. Phyletic aspects of the distribution of 3-hydroxyretinal in the class Insecta. J. Mol. Evol. 30, 72–84.

Sönmez, E., Gülel, A., 2008. Effects of different temperatures on the total carbohydrate, lipid and protein amounts of the bean beetle, *Acanthoscelides obtectus* Say (Coleoptera: Bruchidae). Pak. J. Biol. Sci. 11, 1803–1808.

St-Hilaire, S., Sheppard, C., Tomberlin, J.K., Irving, S., Newton, L., McGuire, M.A., Mosley, E.E., Hardy, R.W., Sealey, W., 2007. Fly prepupae as a feedstuff for rainbow trout, *Oncorhynchus mykiss*. J. World Aquacult. Soc. 38, 59–67.

Studier, E.H., Keeler, J.O., Sevick, S.H., 1991. Nutrient composition of caterpillars, pupae, cocoons and adults of the eastern tent moth, *Malacosoma americanum* (Lepidoptera: Lasiocampidae). Comp. Biochem. Phys. A. 100, 1041–1043.

Studier, E.H., Sevick, S.H., 1992. Live mass, water content, nitrogen and mineral levels in some insects from south-central lower Michigan. Comp. Biochem. Phys. A. 103, 579–595.

Strzeleqicz, M.A., Ullrey, D.E., Schafer, S.F., Bacon, J.P., 1985. Feeding insectivores: increasing the calcium content of wax moth (*Galleria mellonella*) larvae. J. Zoo Med. 16, 25–27.

Sullivan, K.E., Livingston, S., Valdes, E.V., 2009. In: Ward, A., Treiber, K., Schmidt, D., Coslik, A., Maslanka, M. (Eds.), Vitamin A Supplementation via Cricket Dusting :The Effects of Dusting Fed and Fasted Crickets of Three Sizes Using Two Different Supplements on Nutrient Content, AZA Nutrition Advisory Group, Tulsa OK, pp. 160–162.

Thompson, S.N., 1973. Review and comparative characterization of fatty-acid compositions of 7 insect orders. Comp. Biochem. Physiol. 45, 467–482.

Thurnham, D.I., Bender, D.A., Scott, J., Halsted, C.H., 2000. Water-soluble vitamins. In: Garrow, J.S., James, W.P.T., Ralph, A. (Eds.), Human Nutrition and Dietetics, tenth ed. Churchill Livingstone, London.

Tietz, A., Stern, N., 1969. Stearate desaturation by microsomes on the locust fat-body. FEBS Lett. 2, 286–288.

Tordoff, M.G., 2001. Calcium: taste, intake, and appetite. Physiol. Rev. 81, 1567–1597.

Trusk, A.M., Crissey, S., 1987. Comparison of calcium and phosphorus levels in crickets fed a high calcium diet versus those dusted with supplement. Proceedings of the 6th and 7th Dr. Scholl Conferences on the Nutrition of Captive Wild Animals, pp. 93–99.

Von Lintig, J., Vogt, K., 2000. Filling the gap in vitamin A research. J. Biol. Chem. 275, 11915–11920.

Von Lintig, J., 2012. Metabolism of carotenoids and retinoids related to vision. J. Biol. Chem. 287, 1627–1634.

Wang, D., Zhai, S.W., Zhang, C.X., Zhang, Q., Chen, H., 2007. Nutrition value of the Chinese grasshopper *Acrida cinerea* (Thunberg) for broilers. Anim. Feed Sci. Tech. 135, 66.

Whitaker Jr., J.O., Dannelly, H.K., Prentice, D.A., 2004. Chitinase in insectivorous bats. J. Mammal. 85, 15–18.

Winn, D., Dunham, S., Mikulski, S., 2003. Food for insects and insects as food: viable strategies for achieving adequate calcium. J. Wildl. Rehabil. 26, 4–13.

Xia, Z., Wu, S., Pan, S., Kim, J.M., 2012. Nutritional evaluation of protein from *Clanis bilineata* (Lepidoptera), an edible insect. J. Sci. Food Agric. 92, 1479–1482.

Yang, L.F., Siriamornpun, S., Li, D., 2006. Polyunsaturated fatty acid content of edible insects in Thailand. J. Food Lipids 13, 277–285.

Zhang, Z., Song, X., Wang, Q., Lu, X., 2012. Cd and Pb contents in soil, plants, and grasshoppers along a pollution gradient in Huludao City, Northeast China. Biol. Trace Elem. Res. 145, 403–410.

Zhou, X., Honek, A., Powell, W., Carter, N., 1995. Variations in body length, weight, fat content and survival in *Coccinella septempunctata* at different hibernation sites. Entomol. Exp. Appl. 75, 99–107.

Chapter 18

Insects for Human Consumption

Marianne Shockley[1] and Aaron T. Dossey[2]
[1]*Department of Entomology, University of Georgia, Athens, GA, USA,* [2]*All Things Bugs, Gainesville, FL, USA*

18.1. INTRODUCTION

The utilization of insects as a sustainable and secure source of animal-based food for the human diet has continued to increase in popularity in recent years (Ash et al., 2010; Crabbe, 2012; Dossey, 2013; Dzamba, 2010; FAO, 2008; Gahukar, 2011; Katayama et al., 2008; Nonaka, 2009; Premalatha et al., 2011; Ramos-Elorduy, 2009; Smith, 2012; Srivastava et al., 2009; van Huis, 2013; van Huis et al., 2013; Vantomme et al., 2012; Vogel, 2010; Yen, 2009a, b). Throughout the world, a large portion of the human population consumes insects as a regular part of their diet (Fig. 18.1). Thousands of edible species have been identified (Bukkens, 1997; Bukkens and Paoletti, 2005; DeFoliart, 1999; Ramos-Elorduy, 2009). However, in regions of the world where Western cultures dominate, such as North America and Europe, and in developing countries heavily influenced by Western culture, mass media have negatively influenced the public's perception of insects by creating or reinforcing fears and phobias (Kellert, 1993; Looy and Wood, 2006). Nonetheless, the potentially substantial benefits of farming and utilizing insects as a primary dietary component, particularly to supplement or replace foods and food ingredients made from vertebrate livestock, are gaining increased attention even in Europe and the United States. Thus, we present this chapter to describe this emerging field at all levels, including historical, cultural, agricultural, industrial, and food science perspectives. To accomplish this, the chapter covers the topic of human entomophagy via the following subtopic categories: (1) historic and cultural precedents for insects as food; (2) the nutritional and human health value of insects; (3) insects as a sustainable protein and a secure source of human food; (4) current examples of mass-produced insects with potential as human food; and (5) potential products and byproducts from mass-produced food or feed insects. The definition of entomophagy is the dietary consumption of insects by any organism, but it is commonly used to refer specifically to the human consumption of insects. Thus, throughout this chapter we will use the term "entomophagy" interchangeably with and in referring only to the dietary consumption of insects by humans.

Mass Production of Beneficial Organisms. http://dx.doi.org/10.1016/B978-0-12-391453-8.00018-2

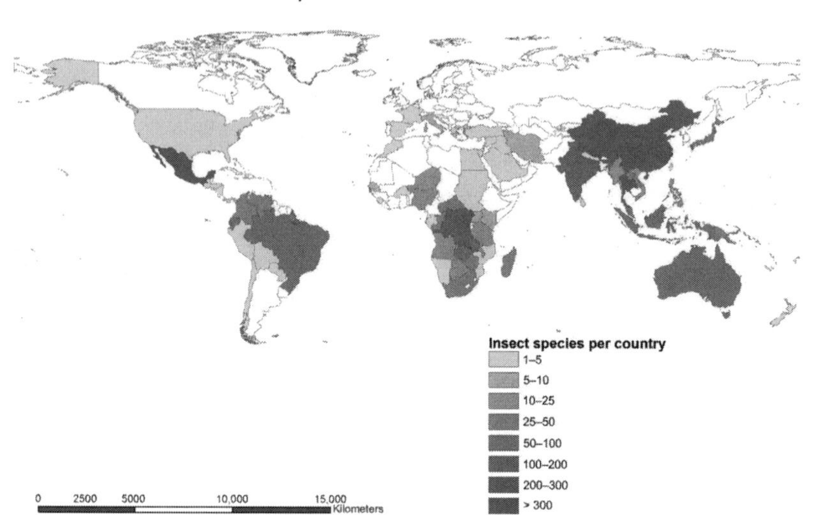

FIGURE 18.1 Edible insects of the world by biogeographical realm. *Wageningen University Laboratory of Entomology (2012).* (For the color version of this figure, the reader is referred to the online version of this book.)

18.2. HISTORIC AND CULTURAL PRECEDENTS FOR INSECTS AS FOOD

Many species of insects are important natural resources. Insects can be consumed as food for humans and animals alike and used for either self-sufficiency or commercial food products in many parts of the world. The use of edible insects varies by local preference, sociocultural significance, and region (Fig. 18.1). Edible insects are often regarded as cultural resources reflecting a rich biodiversity. As well, people who eat insects have established a broad variety of methods for their collection and preparation (Nonaka, 2009).

Wageningen University's Laboratory of Entomology's 2012 inventory of edible insects worldwide listed 2163 species that were reported in the literature as being consumed by humans (Fig. 18.2) (Jongema, 2012). Using the biogeographical realms proposed by Udvardy, high insect consumption of ~350 to ~700 edible insect species can be observed in the African, Neotropical, Oriental, and Palearctic Realms (Figs 18.3 and 18.4). There are several major orders of insects utilized in entomophagy worldwide, including Lepidoptera, Hemiptera, Coleoptera, Diptera, Orthoptera, and Hymenoptera. Worldwide, the three most common orders of insects eaten by humans are Lepidoptera (36 families and 396 species), Hemiptera (27 families and 222 species), and Coleoptera (26 families and 661 species). It is interesting to note that at the species level, the insect families that are most commonly consumed by humans worldwide are the Coleoptera, with the Scarabaeidae

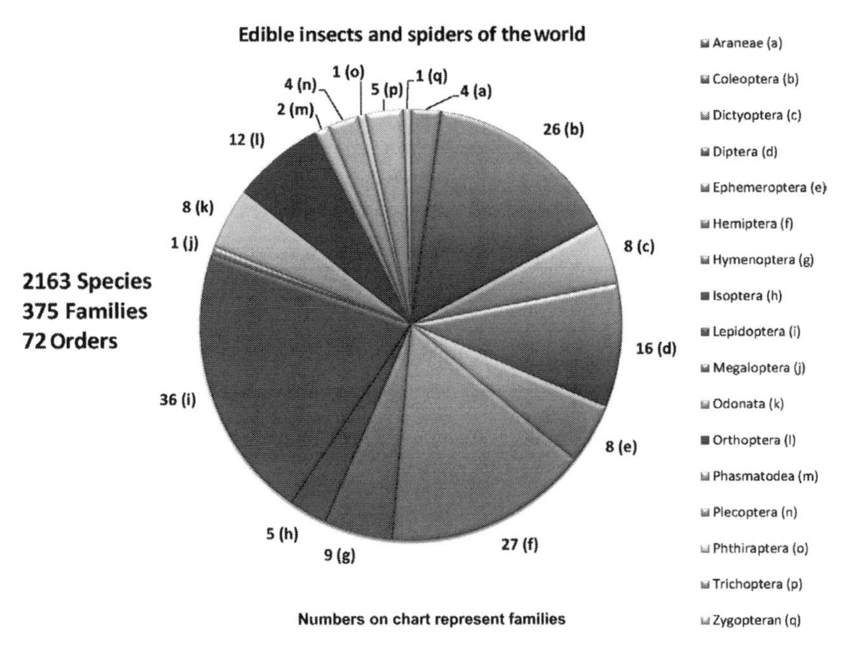

FIGURE 18.2 Edible insects and spiders of the world. *This chart is based on data from Wageningen University Laboratory of Entomology (2012).* (For the color version of this figure, the reader is referred to the online version of this book.)

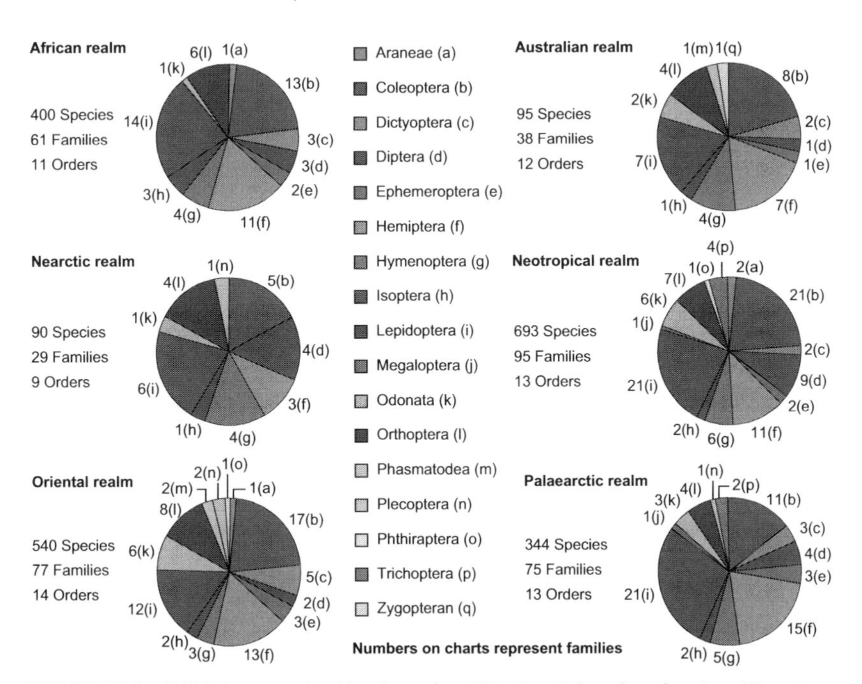

FIGURE 18.3 Edible insects and spiders by region. *This chart is based on data from Wageningen University Laboratory of Entomology (2012).* (For the color version of this figure, the reader is referred to the online version of this book.)

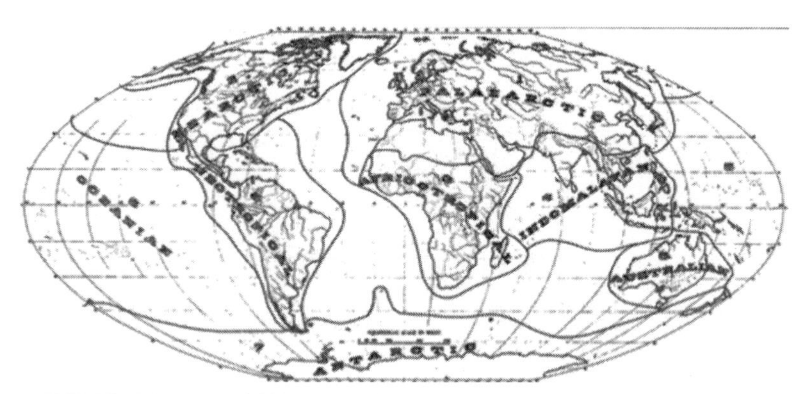

FIGURE 18.4 Terrestrial biogeographic realms of the world M.D.F. *Udvardy (1975)*.

(247 species), Dytiscidae (55 species), and Cerambycidae (129 species) being the most commonly eaten families of insects. As for Lepidoptera, the most commonly consumed families are Saturniidae (109 species), Hepialidae (47 species), and Sphingidae (36 species). Among Hemiptera, the most commonly edible families are Cicadidae (70 species), Pentatomidae (31 species), and Belostomatidae (17 species) (Jongema, 2012).

The following list outlines edible insect species consumed by humans by biogeographical realm, from those with the highest to the lowest frequency: Neotropical (639 species), Oriental (540 species), African (400 species), Palearctic (344 species), Australian (95 species), and Nearctic (90 species) (Fig. 18.4). More specifically, in the Neotropical realm, 13 orders, 95 families, and 693 species of insects are edible. As for the Oriental realm, 14 orders, 77 families, and 540 species of insects are incorporated into human diets. In the African realm, 11 orders, 61 families, and 400 species of insects are eaten. In the Palearctic realm, 13 orders, 75 families, and 344 species of insects are edible. In the Australian realm, 12 orders, 38 families, and 95 species of insects are consumed by humans. Last, in the Nearctic realm, nine orders, 29 families, and 90 species of edible insects have been documented in the literature (Fig. 18.4) (Jongema, 2012). This information is consistent with Western bias and Western influence against entomophagy, with the Nearctic and Australian realms consuming the lowest number of species of insects. It is estimated that as much as 80% of the world's population eats insects intentionally, and 100% do so unintentionally (Srivastava et al., 2009).

18.2.1. History of Human Insect Consumption

An 1877 book by Alpheus Spring Packard dedicated a chapter to edible insects, describing insects such as grasshoppers of the East, eaten by Arabs and in other parts of Africa, and collected by scorching their wings with fire during flight. Packard then described bees and ants eaten in Mexico and used in Sweden to flavor brandy, and discussed the Chinese eating silkworms and the larvae of hawkmoths. Last, he discussed the palm weevils eaten in the West Indies (Packard, 1877). Another book, titled *Why Not Eat Insects?*, is one of the earliest entomophagy-related publications in the United States. In that book, author Vincent M. Holt shares his angst about "a long-existing and deep-rooted public prejudice" to insects as food (Holt, 1885). In a 1916 publication, Leland Ossian Howard and members of the Bureau of Entomology and the Bureau of Biological Survey discussed practical suggestions concerning any new cheap foods such as insects (Howard, 1916). During World War I, food prices worldwide were increasing, and many nations were facing very serious shortages. He and his colleagues prepared and personally ate various edible insect dishes. Howard concluded that colleges of agriculture, with their departments of home economics and entomology, were in an excellent position to conduct entomophagy research (Howard, 1916). In a review of the early literature, Friedrich Simon Bodenheimer's book *Insects as Human Food* cites more than 500 entomophagy references, from prehistory to the 1950s (Bodenheimer, 1951; Dufour, 1990). These studies revealed that many of the primitive peoples of Africa, Asia, and America were underfed or lived on unbalanced, unsatisfactory diets, with a serious shortage of animal fats, animal proteins, and carbohydrates. To date, Bodenheimer's book remains the most comprehensive review of historical entomophagy literature.

18.2.2. Overview of Insect-Eating Cultures in Modern Times

Entomophagy is accepted and practiced by many cultures around the world and constitutes a major source of nutritious food for many people (DeFoliart, 1995; Nonaka, 2009; Ramos-Elorduy, 2009) (Fig. 18.1). It exists in both protocultures and formal cultures (Ramos-Elorduy, 2009). The traditional use of insects as food continues to be widespread in tropical and subtropical countries, and it provides significant nutritional, economic, and ecological benefits for rural communities (Figs 18.3 and 18.4). As many as 3071 ethnic groups in 130 countries (Ramos-Elorduy, 2009) utilize insects as essential elements of their diet (FAO, 2008; Srivastava et al., 2009; Yen, 2009a, b). Many of the poorest populations in the world routinely eat insects as part of their diet, particularly in Africa, Asia, the Neotropics, and the Palearctic (Figs 18.3 and 18.4) (Gahukar, 2011; Manary and Sandige, 2008; Nonaka, 2009; Ramos-Elorduy, 2009). A study in Kenya, where malnutrition is prevalent, found that wheat buns enriched with insects were actually preferred by the locals over ordinary breads (Gahukar, 2011).

Recently, there has been increasing interest in entomophagy in the United States and Europe (FAO, 2008; Gahukar, 2011; Polis, 2011). However, European populations and European-derived populations in North America historically have placed taboos on entomophagous eating practices and continue to do so. Multiple attempts by entomologists have been made to make insects more broadly appealing. A popular example is Ronald Taylor's (1975) book *Butterflies in My Stomach*, and the accompanying recipe guide, *Entertaining with Insects* (1976) (Taylor, 1975; Taylor and Carter, 1976). Several subsequent entomophagy cookbooks have been published, including *Insectes à croquer (Insects to Munch)*, produced by the Montreal Insectarium; and *Cuisine des insectes: À la découverte de l'entomophagie*, by Gabriel Martinez, a French culinary guide offering professional cooking advice. Others, such as the humorous *The Eat-a-Bug Cookbook*, offer readers familiar American recipes such as pancakes, pizza, and alphabet soup altered with the addition of edible invertebrates (Gordon, 1998). Due to recent popularity and interest in entomophagy in the United States, a second edition of *The Eat-a-Bug Cookbook* was released in 2013 (Gordon, 2013). Edible insects are also featured in Peter Menzel and Faith D'Aluisio's photo-essay volume, *Man Eating Bugs: The Art and Science of Eating Insects*, including diverse entomophagy scenes and dishes from around the globe (Menzel and D'Aluisio, 1998). Julieta Ramos-Elorduy's *Creepy Crawly Cuisine* is an introduction to the world of edible insects, complete with recipes and photographs (Ramos-Elorduy and Menzel, 1998). It includes a historical look at the use of edible insects in indigenous cultures and provides information on where to obtain insects and how to store and prepare them. *Het insectenkookboek* or *The Insects Cookbook* was recently published in The Netherlands and features various insect recipes (van Huis et al., 2012). *Edible Bugs: Insects on Our Plate* by Chad Peterson was published in 2012 and is available as an electronic book (Peterson, 2012).

18.2.3. Cultural Acceptance of Edible Insects

In Western-dominated cultures, people are accustomed to eating and willing to eat food items that they perceive as safe and are inherently unafraid of. Eating insects for nutrition is not imperative for most people in the United States and Western Europe, who may have many food options. However, there have been recent advancements in edible insect research in Europe. In The Netherlands a project entitled Sustainable Production of Insect Proteins for Human Consumption investigates the industrial extraction of insect protein to be incorporated into a range of food products. Likewise, cultural acceptance of edible insects in the United States is slowly becoming more prevalent. Although entomophagy may be informally observed in the United States, there is still an attitudinal barrier to the use of insects as human food. During 2011, entomophagy appeared almost weekly in newspapers, magazines, and blogs (Cruz and Johar, 2012). The *Huffington Post* reported insects as the number-one food trend of 2011 (Polis, 2011). Edible insects can also be found in a few restaurants in the United States in Arizona, California, and New York (Oaxaca, Toloache, Typhoon, and Bugs), as well as in food carts (Don Bugito, Jungle George's Exotic Meats, and Bugs) (Table 18.1). Worldwide, there are also entomophagy-based businesses, including Hotlix, Edible Unique, Thailand Unique Edible Insects/Bugs, All Things Bugs, Edible Insectivor, Deli Bugs, Insectes Comestibles, Lazybone, World Entomophagy, BugMuscle, and Chapul. In addition, there are numerous

TABLE 18.1 Additional Entomophagy Resources

Books and Cookbooks	URL
Creepy Crawly Cuisine	NA
Eat-a-Bug Cookbook	NA
Butterflies in My Stomach	NA
Entertaining with Insects	NA
Insectes à croquer (Insects to Munch)	NA
Cuisine des insects: À la decouverte de l'entomophagie	NA
Edible Bugs: Insects on Our Plate	NA
Het Insectenkookboek or *The Insects Cookbook*	http://www.uitgeverijatlas.nl/result titel.asp?id=3368
Cicada-licious	http://www.newdesk.umd.edu/pdf/cicada%20recipes.PDF
Man Eating Bugs	http://www.menzelphoto.com/books/meb.php
Websites, Activists, and Informal Educators	**URL**
All Things Bugs	http://www.allthingsbugs.com/
Food-Insects	http://www.food-insects.com/
Insects are Food	http://insectsarefood.com/
World Entomophagy	https://worldento.com/
David George Gordon	http://www.davidgeorgegordon.com/
Small Stock Foods	http://www.smallstockfoods.com
MiniLivestock	http://minilivestock.org/
Girl Meets Bug	http://girlmeetsbug.com/
Entom Foods	http://entomfoods.com/
Eating Bugs (from the Manataka American Indian Council)	http://www.manataka.org/page160.html
Food Insects Newsletter (through 2009)	http://www.hollowtop.com/finl html/finl.html
Food Insects Newsletter (2010 through present)	http://www.foodinsectsnewsletter.org/

Continued

TABLE 18.1 Additional Entomophagy Resources—cont'd

Websites, Activists, and Informal Educators	URL
Edible Forest Insects: Humans Bite Back (February 2010)	http://www.fao.org/docrep/012/i1380e/i1380e00.pdf
National Geographic, "Bugs as Food: Humans Bite Back"	http://news.nationalgeographic.com/news/2004/04/0416_040416_eatingcicadas.html
International	**URL**
Ento Cargo Collective	http://cargocollective.com/ento
Food Factory Foundation	http://www.foodfactoryfoundation.org
UN Food and Agriculture Organization, Edible Forest Insects	http;//www.fao.org/forestry/65422/en/
Heifer International: Extra Crunch with Lunch	http://www.heifer.org/media/world-ark/archives/2011/fall/extra-cruch-with-lunch
Entomophagy in Japan (2009)	http://tonymcnicol.com/2009/03/15/entomophagy-in-japan/
Food, Supplements, and Vendors	**URL**
All Things Bugs	http://www.allthingsbugs.com/
Chapul	http://chapul.com
Don Bugito	http://www.donbugito.com
World Entomophagy	https://worldentocom/
Bugmuscle	http://www.bugmuscle.com
Insectes Comestibles	http://www.insectescomestibles.fr/
Edible Unique Bugs and Gourmet Exotic	http://www.edibleunique.com/
Hotlix	http://www.hotlix.com/
Thailand Unique Edible Insects/Bugs	http://www.thailandunique.com/store/edible-insects-bugs-c-1.html
Edible Insectivore	http://www.edible-shop.com/shop/insectivore/
Deli Bugs	http://www.delibugs.nl/
Lazybone	http://lazyboneuk.com/categories/Edible-Insects-/

websites and blogs devoted to entomophagy, including Food-Insects, Insects Are Food, All Things Bugs, World Entomophagy, David George Gordon, Small Stock Foods, MiniLivestock, Girl Meets Bug, Entom Foods, Eating Bugs (from the Manataka American Indian Council), *Food Insects Newsletter*, Entomophagy in Japan, National Geographic's "Bugs as Food: Humans Bite Back," the Bay Area Bug Eating Society, the Ento Cargo Collective, the Food Factory Foundation, the UN Food and Agriculture Organization, Edible Forest Insects, Heifer International, Extra Crunch with Lunch, Insect Europe, and Harvesting of Insects in South Africa and Japan: Indigenous Knowledge in the Classroom. For a complete list of websites associated with entomophagy books, cookbooks, websites, activists and informal educators, and blogs; international entomophagy; entomophagy food and supplement products; and insect vendors, refer to Table 18.1.

18.2.4. Social Change, Food Choice, and Perceptions

In areas where entomophagy is not prevalent, food choice and food habits are driving forces of acceptance. Lewin (1952) first researched food habits in the 1940s during World War II, when there were food shortages and increased food costs. He described methods to integrate cultural anthropology, psychology, and sociology into the concepts of group decision making and ultimately social change. Group decision making lies at the interface of many basic problems of group life and individual psychology, and it addresses the relationship of motivation to action with the effect of a group setting on the individual's readiness to change or to keep certain standards. Lewin (1943) observed that the subjects in his study had a desire to be educated about the new food sources and food item alternatives (such as changing sirloin and ribeye to liver and hearts). They also wanted to discuss difficulties in preparation, the concept of group decision making in a household, ideas for presenting the new food to the family, and determining success after the food had been introduced in the home. Lewin (1943) discussed the techniques utilized to change food habits, such as lecture or group discussion or decision making, which were perceived as essential components by the study participants. Similar methods could be used for action research into changing group opinions of food habits and choices related to edible insects. In order to accomplish this, we need a better understanding of the fundamental problems of action research, namely, how to change group conduct in a way that persists (Lewin, 1952).

In order for entomophagy to be popularly accepted, the concepts of group decision making and social change theory must be utilized and further researched. The perceived issue of edible insects being associated only with poor countries, as a low-status problem, must be eliminated (Lewin, 1943). Lewin's food habits research can be the basis for future entomophagy research in the United States related to cultural acceptance of edible insects, relying on the fields of adult education, psychology, and entomology from a multidisciplinary approach. However, for social change to occur so that entomophagy can be accepted in the United States, we must simultaneously decrease the fears and phobias associated with insects while increasing the public's positive perception of insects. Food neophobia (fear of new foods) prevents people from trying new foods, such as insects (Birch, 1999). Phobias such as these need to be addressed for insects to be accepted and incorporated into Western-dominated diets.

Another entomophagy barrier to overcome is the public's perception of invertebrates. Kellert (1993), in his invertebrate perception study, proposed the use of psychological attitude scales in relation to people's perceptions toward invertebrates. Basic attitudes toward invertebrates were summarized by nine scales: aesthetic, dominionistic, ecologistic, humanistic, moralistic, naturalistic, negativistic, scientistic, and utilitarian (Kellert, 1993). This study found that a more positive view of invertebrates was observed when participants were presented with taxa that possessed aesthetic value (e.g. butterflies) or practical value (e.g. bees). It is interesting to note that the most common order of insects eaten by humans worldwide is Lepidoptera (butterflies and moths) (Fig. 18.2). Variation in the results may be attributed to an innate learning disposition, the association of many invertebrates with disease and damage, the ecological scale between us and them, the relative abundance of insects, the lack of a sense of identity and consciousness among invertebrates, the presumption of mindlessness among invertebrates, and the radical autonomy of invertebrates from human control.

Kellert's psychological attitude scale has been widely utilized in animal perception research by multiple disciplines, including entomology and psychology. Looy and Wood (2006) studied the psychology of entomophagy, which examined whether educational bug banquets influenced attitudes toward invertebrates. There is a general informal consensus among entomophagy educators and activists that incorporating edible insects into entomology outreach programs as a form of informal education to introduce the general public to entomophagy may be vital to the acceptance of entomophagy (Looy and Wood, 2006).

18.3. NUTRITIONAL AND HUMAN HEALTH VALUE OF INSECTS

The United Nations (UN) has placed heavy emphasis on alleviating hunger and malnutrition in children. Two of their eight Millennium Development Goals (MDGs), set to be achieved by 2015, are directly related to this area: The first MDG is to "eradicate extreme poverty and hunger," and the fourth is to "reduce child mortality rates'' (Dossey, 2013). Here, we present a number of assets that insects provide to improve sustainable access to highly nutritious food, as well as how insects can help alleviate world hunger. Number 7 on the MDG list is "ensuring environmental sustainability'' (Dossey, 2013). The tremendous benefits of insects in ensuring environmental sustainability are discussed in this chapter.

18.3.1. Animal- versus Plant-Based Food and Protein

Animals, which include insects, are important or even the sole sources of numerous necessary nutrients for humans. Some important examples of these are the eight essential amino acids ("complete" protein), vitamin B_{12}, riboflavin, the biologically active form of vitamin A (retinol, retinoic acid, and retinaldehyde), and several minerals (Bukkens, 1997; Bukkens and Paoletti, 2005; Hoppe et al., 2008; Michaelsen et al., 2009; Singh and Singh, 1991; Yen, 2009a, b). In particular, it is broadly accepted that animal-sourced dietary protein is superior to that derived from plants (Babji et al., 2010; Hoppe et al., 2008; Michaelsen et al., 2009; Singh and Singh, 1991). Animal-based food is important for the nutritional status, growth, and recovery rate as well as cognitive performance of undernourished children (Michaelsen et al., 2009; Neumann et al., 2003). Even small amounts of animal-sourced food ingredients, such as insects, can substantially improve nutrient adequacy (Michaelsen et al., 2009). A review by Michaelsen et al. (2009) cites numerous benefits of animal-sourced foods in general and particularly their importance for children suffering from malnutrition. Specifically, they recommend that these foods contain at least 25–33% animal-sourced food ingredients to significantly improve growth rate. Most ready-to-use therapeutic food (RUTF) used to treat malnutrition in children is "lipid based" in order to provide high energy content (Manary and Sandige, 2008). Michaelsen et al. (2009) recommend a diet, including animal sources of eicosapentaenoic and docosahexaenoic acids, for children born undernourished and at low birth weight.

As described in this chapter, insects can be grown highly efficiently in many areas not amenable to dairy cattle and, thus, help to provide a robust alternative to milk as well as a potential alternative income source for farmers. Consequently, insects are likely the best source of animal-sourced food for people in much of the world.

18.3.2. Nutrient Content of Insects

Insects present a substantial, yet extremely underexplored, alternative opportunity to provide much-needed animal-sourced nutrients, particularly to the developing world. Several authors have examined the nutritional value of insects and their role in human nutrition (Banjo et al., 2006; Bukkens, 1997; Bukkens and Paoletti, 2005; DeFoliart, 1992; Finke, 2002, 2004, 2005, 2012; Jokthan et al., 2007; Michaelsen et al., 2009; Ramos-Elorduy, 1997, 2008, 2009; Ramos-Elorduy and Menzel, 1998). Hundreds of insect species have been formally studied for nutrient composition. Most nutrition studies analyze insects for moisture, protein, fat, ash, and fiber. Insects are generally high in protein and fat at levels comparable to those of milk and meat such as beef (Table 18.2). A review by Bukkens concludes that the amino acid composition of insects compares favorably with the reference standard recommended by the UN Food and Agriculture Organization (FAO), World Health Organization (WHO), and

TABLE 18.2 Nutritional Content of Insects Compared with Other High-Protein Foods

Insect or Food Item	Protein (g/kg)	Fat (g/kg)	Calories (kcal/kg)	Thiamin (mg/kg)	Riboflavin (mg/kg)
Black soldier fly*	175	140	1994	7.7	16.2
House fly*	197	19	918	11.3	77.2
House cricket**	205	68	1402	0.4	34.1
Superworm**	197	177	2423	0.6	7.5
Mealworm**	187	134	2056	2.4	8.1
Giant mealworm**	184	168	2252	1.2	16.1
Waxworm**	141	249	2747	2.3	7.3
Silkworm**	93	14	674	3.3	9.4
Beef†	256	187	2776	0.5	1.8
Milk powder‡	265	268	4982	2.6	14.8

*Finke (2012).
**Finke (2002).
†Current USDA National Nutrient Database for Standard Reference.
‡From the US Dairy Export Council website.
Beef: ground, 75% lean meat and 25% fat, patty, cooked, broiled, milk, whole dry powder (black soldier fly = Hermetia illucens larvae; house fly = Musca domestica adults; house cricket = Acheta domestica adults; Superworm = Zophobas morio larvae; mealworm or giant mealworm = Tenebrio molitor larvae; Waxworm = Galleria mellonella larvae; silkworm = Bombyx mori larvae).

United Nations University (UNU) (Bukkens and Paoletti, 2005). Insects are particularly high in protein, at levels comparable to those of beef and milk (Table 18.2). House crickets, for example, contain approximately 205 g of protein per kilogram of cricket, while ground beef contains about 256 g per kilogram and whole powdered milk contains about 265 g per kilogram. Some estimate that the digestibility of flour made from insects is as high as 91% (Bukkens, 1997). Protein from various sources, in dried form, is most frequently found in village markets of the developing world. Insects are very high in crude protein, with many species ranging above 60% by dry weight (Bukkens, 1997; Bukkens and Paoletti, 2005). Some insect proteins are equivalent to soy protein, whereas others are superior as a source of amino acids at all levels of intake. Whole insects as a source of protein are of somewhat lower quality than vertebrate animal products because of the indigestibility of chitin. Removal of chitin increases the quality of insect protein to a level comparable to that of products from vertebrate animals (see Section 18.4.3 for more information on insect processing). In general, insect protein tends to be low in amino acids such as methionine and cysteine, but it is high in lysine and threonine, one or both of which may be deficient in the wheat-, rice-, cassava-, and maize-based diets that are prevalent in the developing world (DeFoliart, 1992).

Insects are also particularly rich in fat (Table 18.2) (Bukkens, 1997; Bukkens and Paoletti, 2005; DeFoliart, 1992; Finke, 2012; Gahukar, 2011) and can supply a high caloric contribution for energy-dense foods (Table 18.2). In the reviews by Bukkens, all insect species were found to be a "significant source of the essential fatty acids linoleic and linolenic acid" (Bukkens, 1997; Bukkens and Paoletti, 2005). Some insects can also provide a higher caloric contribution to the diet than soy, maize, or beef (Gahukar, 2011). Cholesterol levels in insects vary from low to approximately the levels found in other animals, depending on the species and diet. Insect fatty acids are similar to those of poultry and fish in their degree of unsaturation, with some groups being rather higher in linoleic and/or linolenic acids, which are the essential fatty acids (DeFoliart, 1992).

In addition to protein, fat, and caloric content, many insects are particularly high in a number of vitamins, minerals, and other valuable nutrients (Table 18.2). For example, many species are significantly higher in thiamin and riboflavin than whole-meal bread and hen's eggs (Bukkens, 1997; Bukkens and Paoletti, 2005). The retinol (a biologically active form of vitamin A) and beta-carotene content of many insect species is also high, with levels in some species as high as 356 mg/kg and 1800 mg/kg, respectively (Bukkens and Paoletti, 2005). However, currently very limited data are available for vitamin analyses of insects, and more studies are needed.

18.3.3. Global Malnutrition and How Insects Can Help

Malnutrition affects 178–195 million children worldwide (Black et al., 2008). The WHO estimates that malnutrition accounts for 54% of child mortality worldwide (Duggan et al., 2008). Malnutrition results in wasting, a reduced ability to fight infection, impaired cognitive disorders, and other developmental disorders that can be permanent (Victora et al., 2008). About 60 million children suffer from moderate acute malnutrition, and 13 million from severe acute malnutrition (Collins et al., 2006). Malnutrition is a contributing factor in over 50% of the 10–11 million children under the age of 5 years who die from preventable causes each year (Black et al., 2003; Caulfield et al., 2004; Pelletier and Frongillo, 2003; Rice et al., 2000). Fatality rates from severe acute malnutrition have averaged 20–30% since the 1950s (Schofield and Ashworth, 1996). It is estimated that 9% of Sub-Saharan African children and 15% of Asian children suffer from moderate acute malnutrition (Collins et al., 2006).

As mentioned in Section 18.3, the United Nations has established eight MDGs; the first MDG is to "eradicate extreme poverty and hunger," and the fourth is to "reduce child mortality rates'' (Dossey, 2013). Specifically, the goals involve reducing childhood mortality by two-thirds and worldwide hunger by one-half. Indeed, the topic of the potential for insects to contribute to sustainable human food security has taken the notice of multiple organizations. For example, as mentioned in this chapter, the FAO has taken the initiative and proposed a program of feeding people with alternative food sources, including insects (Gahukar, 2011). Two major meetings on the feasibility and benefits of insects as a food source have resulted from this initiative: (1) a February 2008 workshop in Thailand, which produced an important book in the field called *Forest Insects as Food: Humans Bite Back* (FAO, 2008); and (2) a technical consultation held January 2012 at FAO headquarters in Rome, Italy (Vantomme et al., 2012). At the 2012 meeting, a global summit on insects and food security for 2014 was also proposed. Those in attendance at the 2012 meeting appeared to come away quite optimistic that insect-based food products are indeed an important part of our future. It was also generally recognized that government and industry backing is necessary to support the widespread implementation of insect-based diets.

Malnutrition in developing countries is primarily a problem of protein and calorie deficiency. While the content of particular nutrients and nutrient categories varies widely among the hundreds of species analyzed, insects are in general a very good source of protein, fat, and other nutrients that are valuable to the human diet (see also Section 18.3.2). Additionally, the problem of protein deficiency is one that affects low-income people worldwide. The consumption of some insects can provide both high protein and caloric intake. For example, the nutrient content of the African palm weevil larvae (*Rhynchophorus phoenicis*) was described in a Nigerian study, and they were found to be very high in both carbohydrate and protein (Okaraonye and Ikewuchi, 2008). Because insects are a robust source of animal protein, it is important to improve production and preservation techniques, to market them, and to make them available to all populations (Melo et al., 2011).

Given the fact that insects compare quite favorably in their nutrient content with vertebrate livestock, as described in this chapter, increasing their consumption in places where vertebrate livestock is unavailable or their growth is infeasible can greatly improve the health and nutritional status of a great number of people worldwide. Access to edible insects can be achieved by different methods, including wild harvesting, efficient and/or sustainable farming, or even household rearing or gardening. Insects are much less resource intensive and much more resistant to drought and disease (typically) than cattle and most other vertebrate livestock commonly utilized by humans (see Section 18.4). Research and education programs on insect farming and preparation and on safe selection and preparation of wild-harvested insects are two activities that can greatly improve nutrition and food security worldwide. However, the primary methods of preparing, cooking, or processing insects in many areas of the world, particularly in places where poverty and malnutrition are highest, involve roasting, searing, or frying whole insects. Unfortunately, these methods of processing can remove fat (an important source of calories) and other nutrients, while destroying others or making them indigestible. Education programs on methods of cooking, processing, or preparing insects (either wild harvested or farmed) for human consumption in places where malnutrition rates are high might dramatically improve the nutritional status and overall health of the people who live there.

Relief foods such as RUTF products present another opportunity to begin utilizing insects as a more sustainable source of animal-sourced nutrition. One reason for this is that it is often highly desirable for organizations or companies to grow these products in the same country where they will be consumed and from ingredients sourced as nearby as possible. Whereas milk powder is almost always imported and is rarely locally available in places where malnutrition is prevalent, many types of insects can be locally farmed or harvested from the wild. There is already at least one US company working on just this sort of product (Crabbe, 2012) (Table 18.1). That firm was awarded a grant from the Bill and Melinda Gates Foundation in 2012 to develop an insect-based RUTF (Crabbe, 2012). If this sort of endeavor is successful, it could substantially improve the prospects of the insect-based food industry by drastically increasing demand for insects reared as human food ingredients for the insect-based RUTF.

Another potentially impactful way in which insects can improve food security is strategic utilization of pest species and/or species of high seasonal or regular abundance. This can not only expand the use of insects (currently an extremely underutilized resource) but also be a safer, cleaner, and more sustainable method of pest control—thus turning the proverbial "lemons into lemonade." For example, insects such as Lepidoptera larvae (caterpillars), locusts and other grasshoppers (Orthoptera), and others tend to be of extremely high abundance in places where effective pesticides are not used. Cerritos and Cano-Santana (2008) found that manual harvest of grasshopper pests reduces the density of grasshoppers and suggests that implementation of this mechanical method of control may be an effective substitute for chemical control (Cerritos and Cano-Santana, 2008). Mechanical control provides general advantages that include (1) a second profitable product, (2) savings realized from the reduced cost of insecticides, and (3) a reduced risk of soil and water contamination by insecticides. This sort of dual-function pest control might be highly amenable to farming in developing nations as well as organic farms in the United States. Even in more modern agricultural settings, if insect pests can be allowed to consume the inedible parts of crop plants, or crops that are otherwise deemed no longer useful as a harvest, they can potentially be an additional crop with value added to the already ongoing farming practice. Other species, while not actual pests, can also be highly abundant at certain times of the year or in periodic outbreaks. Some examples of this include "lake flies" in Africa (Ayieko et al., 2010a; Ayieko and Oriaro, 2008; Ayieko et al., 2010b) and even some walking stick insects (Phasmatodea), which defoliate sections of forest from time to time (Campbell, 1960, 1961; Graham, 1937; Hennemann and Conle, 2008). Thus, more efficient and increased use of abundant insect resources results in the reduction of pesticides and creates new economic opportunities for indigenous people (DeFoliart, 1992).

Some specific studies have already identified cases whereby pest insects can have a substantial positive health impact in areas where malnutrition is high, as well as economic benefits. For example, a grasshopper in Mexico known as *chapulines* (*Sphenarium purpurascens* Charpentier) (see Section 18.5) is controlled by harvesting them as a food item in addition to the crops they attack. This activity provides an annual profit of about $3000 per family and a total of 100 metric tons of edible insect mass (Premalatha et al., 2011). Studies on the nutritional value of edible insects from Mexico and Nigeria revealed that some of the insects, which are pests, also have high nutritional qualities (Banjo et al., 2006; Ramos-Elorduy, 1997). Christensen et al. (2006) identified edible insects as a mineral source in Kenya (Christensen et al., 2006). Deficiencies of important minerals such as iron, zinc, and calcium are often widespread among people in developing countries. These deficiencies are caused by the low availability of these nutrients in staple foods such as cereals and legumes due to the considerable amount of phytic acid and other antinutrients present in these foods and due to the lack of animal foods with higher content and bioavailability of these nutrients (Christensen et al., 2006). Iron and zinc deficiency is widespread in developing countries, especially in children and women of reproductive age. Iron deficiency leads to anemia, reduced physical activity, and increased maternal morbidity and mortality. As a result, entomophagy could prove to be a valuable measure to combat iron and zinc deficiency in developing countries. Some caterpillars have been found to be a rich source of iron, copper, zinc, thiamin (vitamin B_1), and riboflavin (vitamin B_2); 100 g of cooked insect provided >100% of the daily requirement of each of these minerals and vitamins. Iron deficiency is a major problem in women's diets in the developing world, particularly among pregnant women, and especially in Africa (DeFoliart, 1992).

18.3.4. Insects in Medicine and Drug Discovery

Arthropods appear to be a largely unexplored and underutilized source of drugs and drug lead compounds for modern medicine and biomedical research (Alves and Alves, 2011; Dossey, 2010; Ratcliffe et al., 2011; Srivastava et al., 2009; Yi et al., 2010). Insects and other arthropods are used as drugs by many cultures around the world, particularly Korea (Pemberton, 1999), China, Brazil, and India (Dossey, 2010). For example, they appear in pharmacopoeias of Korean traditional medicine, but little was known about their use in modern South Korea. There, most of the arthropod drugs were traditionally collected or reared, but now they are imported, mainly from China. Folk logic appears to be the basis for some arthropod drug uses. However, many of the arthropods have venom and other defensive chemicals, which are biologically active (Dossey, 2010). The South Korean use of arthropods as drugs is due, in part, to more positive attitudes toward these animals compared to many other cultures (Yi et al., 2010). Medicinal use of insects is also prevalent in China. For example, in a monograph by Read and Li (1931), they describe numerous notes of the ancient Chinese records about insects for medicinal use, as well as their descriptions, habitat, and folklore (Read and Li, 1931).

Like all other organisms, insects and related arthropods mainly utilize chemistry to adapt to their environments in a wide variety of ways, such as for defense against predation or infection, communication and socialization, life cycle development, and surviving environmental conditions. Arthropods harbor a large variety of chemical substances used for these ecological functions, and this is the overarching theme of the field of chemical ecology. However, the fields of pharmacognosy (primarily involved with drug discovery from natural sources) and natural products chemistry have largely ignored insects as a potential source of pharmaceuticals, favoring plants and, more recently, marine organisms instead. Some potential rationale for this bias, such as the relatively larger amounts of material available from plants, their role as primary biosynthetic producers, and the limitations of the minimum amounts of material needed for analysis by standard analytical chemistry techniques, has been discussed in a review article by Dossey (2010) for the journal *Natural Product Reports*. However, Dossey (2010) also points out that recent advances in analytical chemistry technology and instrumental sensitivity, along with the aforementioned biodiversity (see the "Biodiversity" paragraph of Section 18.4) of insects and other arthropods, merits reconsideration of insects as a source of material for natural products. While the potential role of insects in drug discovery, development, and production is fascinating and represents a quite important set of applications for insects (whether farmed, collected, or cultured in the laboratory), it is beyond the scope of this chapter, and we recommend the following review articles for more information on this topic: Alves and Alves (2011), Dossey (2010), Pemberton (1999), Srivastava et al. (2009), Trowell (2003), and Yi et al. (2010).

18.4. INSECTS AS A SUSTAINABLE SOURCE OF HUMAN FOOD

As the human population grows, it is ever more important to sustain rather than increase our levels of consuming and harvesting materials from the earth and its ecosphere. The world adds about 70 million people each year. The United Nations expects the population to grow to more than 9 billion people by 2050, adding approximately twice the current population of China (Dzamba, 2010; Safina, 2011; Vogel, 2010). Humans consume roughly 40% of the biomass that the land and the coastal seas produce (Safina, 2011). Approximately 70% of agricultural land, and 30% of the total land on earth, is used to raise livestock (Steinfeld et al., 2006). Food reserves are at a 50-year low, yet demand for food is expected to increase by 50% by 2030. The historic US drought of 2012 also throws into sharp relief our need for more sustainable agricultural practices (Smith, 2012). We cannot rely on food production strategies utilizing livestock such as cattle to feed our growing population. Insects hold tremendous promise in addressing this gloomy outlook (Dossey, 2013). Expanding the amount of land used for livestock production is not a feasible or sustainable solution to cover the food and protein needs of the projected increases in population. Therefore, it is important to use sources of high-quality animal protein that reduce the amount of pollution, habitat destruction, and abuse of natural resources. However, new technologies for improving food security, such as producing and processing insects as human food, take time for development and application on a large scale, so it is important to make investments in these innovations sooner rather than later (Gahukar, 2011).

The use of insects as a major human food source presents two important technological challenges: (1) how to turn insects into safe, healthy, and tasty food products; and (2) how to cheaply, efficiently, and sustainably produce enough insects to meet market demand. Many also advocate the wild harvesting of insects as human food as a solution to hunger and global food security. While this may be a reasonable short-term solution in a number of isolated situations in certain localities, we feel strongly that for full realization of the potential benefits derived from utilization of insects as a safe, reliable, and sustainable alternative to vertebrate livestock, the ultimate goal must be the efficient farming and/or mass production of insects. Dependence on wild-harvested insects for feeding large populations involves serious risks, such as overharvesting, ecological damage, and consuming insects contaminated by pesticides, environmental contaminants, and/or exposure to pathogens, parasites, and other disease-causing agents that may exist in the environment but can be eliminated or controlled in farmed or captive-reared stocks. Additionally, data on nutrient content vary widely by species, so using homogeneously farmed stock of individual species provides much greater product quality control (Bukkens, 1997, 2005). As a consequence, this chapter focuses primarily on the latter.

Insects have numerous attributes that make them highly attractive, yet underexplored, sources of highly nutritious and sustainable food (Crabbe, 2012; Dossey, 2013; Dzamba, 2010; FAO, 2008; Gahukar, 2011; Katayama et al., 2008; Nonaka, 2009; Premalatha et al., 2011; Ramos-Elorduy, 2009; Smith, 2012; Srivastava et al., 2009; van Huis, 2013; Vogel, 2010; Yen, 2009a, b). The general categories where insects provide the most substantial benefits for sustainable and secure food supplies are (1) efficiency and (2) biodiversity.

Efficiency. Insects can be produced more sustainably and with a much smaller ecological footprint than most vertebrate livestock such as cattle and swine. They are very efficient at biotransformation of a wide variety of organic matter into edible insect biomass (a high feed conversion ratio) (Nakagaki and Defoliart, 1991; Oonincx et al., 2010). For example, cows consume 8 g of food mass per gram of weight gained, whereas insects can require less than 2 g (Vogel, 2010). This is partly due to insects being poikilothermic ("cold blooded"), thus using less energy for body warmth since they utilize their environment for body temperature regulation (Premalatha et al., 2011). House crickets (*Acheta domesticus* L.) have an "efficiency of conversion of ingested food (ECI) that is twice that of pigs and chickens, 4 times that of sheep and 6 times that of steer" (Gahukar, 2011). This efficiency can also lead to less usage of pesticides on animal feed, thus providing additional environmental, health, and economic incentives.

In addition to their highly efficient feed conversion ratios and diet variability, insect fecundity, rapid growth rates, and short life cycles add to the greater efficiency with which they can be produced as a human food ingredient compared with vertebrate livestock. Insects tend to reproduce quickly, are highly adaptable, have large numbers of progeny per individual (high fecundity), and have a large biomass. For example, house crickets can lay 1200–1500 eggs in a 3–4 week period (Gahukar, 2011).

Biodiversity. The UN FAO estimates that there are well over 1000 edible insects currently used (Vogel, 2010), and others estimate that number to be over 2000 (Figs 18.1–18.3) (Jongema, 2012; Ramos-Elorduy, 1997, 2009). There are over 1 million species described and 4–30 million species estimated to exist on earth, living in every niche inhabited by humans and beyond (Dossey, 2010). With this diversity and their collective adaptability, they are a much safer source for future food security than are vertebrate animals such as cattle, fowl, or even fish. Development of more diversity in animal livestock and protein sources is critical to human food security going forward. Insects, being the largest and most diverse group of organisms on earth, certainly have a substantial role to play. For example, since there are insects of some sort on nearly every patch of land on earth, chances are that some local species in every area can be farmed as human food without the need to import nonnative species for the same purpose. Since insects are prolific in the wild, they can be readily tapped to replenish gene pools of farmed stock, unlike most terrestrial vertebrate livestock. That capability, along with a reliance on multiple farmed insect species, can greatly reduce the impact of livestock shortages on food security.

18.4.1. Environmental Footprint of Insects versus Vertebrate Livestock

Since, as described in this chapter, about 70% of agricultural land, and 30% of the total land on earth, is used to raise livestock (Steinfeld et al., 2006), increasing the efficiency with which we utilize this land to generate food for the growing population is one of the most effective ways we can reduce our impact on the world's ecosystems. Toward the goal of reducing human contribution to climate change, insects, in addition to possessing a high feed conversion ratio (as described in Section 18.4), also produce lower levels of greenhouse gases (GHGs) (Oonincx et al., 2010). Global average temperature has increased by 1.4 °F over the last century, and it is predicted to continue rising, a trend that is expected to lead to significant negative consequences. Global warming is primarily caused by GHG emissions related to human activity (Smith et al., 2001). The livestock sector contributes about 8.2% of anthropogenic GHG emissions (Takle and Hofstrand, 2008). A recent study by Oonincx et al. (2010) has demonstrated through rigorous experimentation and measurement that insects give off lower levels of greenhouse gases, such as methane, carbon dioxide, and nitrous oxide, than do cows. Oonincx et al. (2010) quantified the production of carbon dioxide and average daily gain (ADG) as a measure of feed conversion efficiency, and quantified the production of the GHGs methane and nitrous oxide, as well as ammonia, by edible insect species. Results varied considerably by species, but they generally had a higher relative growth rate and emitted comparable or lower amounts of GHGs and ammonia than pigs and much lower amounts of GHGs than cattle (Oonincx et al., 2010). The same was true for carbon dioxide production per kilogram of metabolic weight and per kilogram of mass gain. Additionally, livestock produce large amounts of ammonia, leading to soil nitrification and acidification. The results of this study can be used as basic information to compare the production of insects versus conventional livestock by means of a life cycle analysis (Oonincx et al., 2010). Thus, the data presented in that report add to the already growing amount of research suggesting that insects could serve as a more environmentally friendly alternative to the production of animal protein with respect to GHG and ammonia emissions.

Insect production also uses much less water than production of vertebrate livestock (van Huis, 2013) because insects obtain their water directly from food. Additionally, the higher feed conversion ratios for insects, described in Section 18.4, also contribute to this water use efficiency since a lower feed requirement means less water is used to grow that feed. Lower water usage also reduces the energy needed to pump or recycle more clean water for crops and vertebrate livestock, adding to the benefits of farming insects rather than larger animals. In fact, many insects can be produced with almost no additional feed crop production. For example, many insects, such as the black soldier fly (*Hermetia illucens* L.) (Bondari and Sheppard, 1981; Popa and Green, 2012), can eat organic biomass such as agricultural and food byproducts (see Section 18.5 for more details on the black soldier fly). Such organic biomass could include corn stalks, pulp from fruit-juicing or wine-making operations, expired produce from grocery stores, yeast from wine or beer production, portions of crops that cannot be converted into human food, and other types of clean, safe, and low- or no-value biomass. For simplicity, this type of biomass will be referred to as "agricultural or food byproducts," "nonfood crops," or simply "byproducts." Insects are able to convert these byproducts, and other biomass otherwise not useful for human food production such as switchgrass and algae, into edible insect mass. Hence, the production of some insects as human food in many cases may neither require feed to be grown especially for that purpose nor compete with the existing human food supply, unlike vertebrate livestock such as cows and chickens, which are often fed a diet primarily consisting of grain such as corn. In addition to production efficiency, insects may provide nutritional value directly to the consumer more efficiently, per kilogram of foodstuff consumed, than other food resources (Ramos-Elorduy, 2008; Ramos-Elorduy et al., 2008). This adds another level of environmental impact reduction from the use of insects as human food over other, currently more popular alternatives such as vertebrate livestock. For example, edible insects provide 217–777 kcal/100 g (insects raised on organic byproducts provide 288–575 kcal/100 g), whereas energetic values for livestock are 165–705 kcal/100 g and for vegetables are 308–352 kcal/100 g. Fats provide the majority of the energy necessary for sustaining life. Immature stages of holometabolous insects have high quantities of polyunsaturated fat, which is stored in preparation for the pupal stage when they do not eat and are developing into adults. While the energy contents of edible insects vary according to the species and region found, coleopteran and lepidopteran species tend to provide more energy. Additionally, the energetic cost of collecting edible insects can be lower than that for vertebrates. Hence, insects may efficiently provide the necessary energy for the vital functions of our organism (Ramos-Elorduy, 2008; Ramos-Elorduy et al., 2008).

18.4.2. Insects as Animal Feed

In addition to the tremendous potential that insects hold as primary sources of human food and other directly consumed products, they also present a promising opportunity in their use as intermediate products such as animal feed. Considering the substantial efficiency, sustainability, and nutritional aspects of insects as described in this chapter, the logic easily follows as to how they might provide similar value when utilized in the vertebrate livestock industry, or even in vertebrate animal production at the subsistence level. For example, early stages of poultry, fish, ostrich, and pig, which were raised on insects, had conversion efficiency values of 1.24:1–2.83:1 (Ramos-Elorduy, 2008). Insects fed with biomass such as manure and certain forms of agricultural or other organic waste may not be safe, suitable, or acceptable if used directly as a human food ingredient. However, such insects can be perfectly safe when used as feed for vertebrate animals, such as fish and chicken, which are more commonly eaten by most human populations. Using insects in this way can also provide some of the aforementioned food production sustainability benefits while avoiding social stigmas or food safety concerns. For example, in some studies, insects such as the black solder fly have been explored and developed for chicken and fish feed in a sustainable and efficient nutrient-recycling paradigm due to their ability to be reared entirely on waste such as animal manure and other agricultural or food industry waste streams (Bondari and Sheppard, 1981; Sheppard et al., 1994; St-Hilaire et al., 2007). Indeed, several companies and organizations around the world, even in the United States, are beginning to develop various insects as animal feed. Black soldier fly (discussed in this chapter) appears to be one of the species favored for this type of application. The topic of use of insects as animal feed is fascinating and critical for the larger picture of how insects can greatly improve human food security. This topic is beyond the scope of the current chapter, but the use of insects as animal feed is covered extensively in Chapter 16 of this book.

18.4.3. Considerations for Insect-Based Food Production, Processing, and Safety

The full realization of aforementioned benefits of edible insects requires the development and implementation of food-processing and preparation methods for incorporation of insects into more standard foodstuffs at the large-scale industrial, cottage industrial, as well as household level. While very little if any literature exists on studies demonstrating the capabilities of such food-processing methodologies, we can begin to discuss what the important features to consider are. First, since insects are poikilothermic, and individuals of many species die naturally with the coming of winter (at least in temperate and subtropical zones), a consensus among insect cooks seems to be that the most humane and efficient way to kill them for use in food is by freezing. Subsequently, most insect preparations will likely involve using whole insects or grinding them into a paste or powder.

There are numerous procedures and methods already in existence in the food industry for dealing with pastes, powders, and liquids, so it is likely that little additional innovation, if any, will be developed to feasibly use many of these methods with insects. Also, the chitin in insects is likely undesirable for many food products due to its indigestibility; its possible function as fiber, which limits nutrient absorption; or simply because it adds a disagreeable texture to foods that are ideally low in fiber. Thus, methods that efficiently extract the valuable nutritional content from insects while leaving behind chitin are highly desirable. Additionally, shelf life is an important consideration for any food product. Much research is needed for this aspect of insect-based foods as very little information on insect shelf life is currently available. Nonetheless, some known features of insects may suggest that insect-based food ingredients may have an advantage of longer shelf life than similar noninsect alternatives, particularly for dry products or pastes with high levels of insect content. For example, a number of insects have been shown to contain antimicrobial substances such as peptides, fatty acids and other secondary metabolites (Dossey, 2010; Finke, 2012; Huang et al., 2011), which may lend themselves to extending the shelf life of insect-containing foods. Additionally, the chitin from insects is known to have antimicrobial properties (Tharanathan and Kittur, 2003). Thus, even though in many cases the texture of insect-based foods is improved by removing chitin (as mentioned here), in other food products there may be advantages in leaving the chitin. In those cases, the texture issues with chitin might be alleviated by grinding the insects into a fine powder, while maintaining the fiber and antimicrobial benefits of chitin.

Food safety is one of the most important considerations for developing methods for incorporating insects into any diet. While a good number of people around the world consume insects in large numbers without suffering any adverse health effects, and proper processing and preparation of insects as a food ingredient can alleviate many hazards that insects may present, it is important to point out some potential hazards. Some of those hazards include entomophobia, nutritional losses, dyspepsia, toxic effects, tumorigenic metabolites, trauma to intestinal mucosa, and allergic reactions (Gorham, 1979). Nearly all food preparations using insects as an ingredient will involve use of the whole insect in some form or another, including its digestive tract and its contents. This is different from most vertebrate livestock-derived food products, in which the intestines and other bacteria-rich portions are removed in early stages of processing the carcass. It has even been suggested that removing the digestive tract ("gutting") of insects is important to prepare them for human consumption, but not if fed to livestock because the livestock benefit from the fiber (Shackleton and Shackleton, 2004). However, robust processing methods and microbial kill-steps, such as pasteurization, will likely mitigate the need for this sort of tedious and inefficient process. Such kill-steps, which eliminate bacteria and other microbes (and their spores) from processed insect material, are quite necessary, and it is highly desirable to incorporate these early in any insect-derived food product operations or protocols. Additionally, as insects become a more popular and widely used food ingredient, regulations on how they are handled will need to evolve, particularly in the United States. The US Federal Food, Drug, and Cosmetic Act of 1976 (FD&C Act) prohibits the "adulteration of any food in interstate commerce" "if it consists of whole or in part of any filthy, putrid, or decomposed substance, or if it is otherwise unfit for food" (Gorham, 1979). Therefore, Title 21, Part 110.110, of the Code of Federal Regulations allows the US Food and Drug Administration to establish maximum levels of natural or unavoidable defects (e.g. insect parts) in foods for human use that present no health hazard; these are known as "Food Defect Action Levels" in the FD&C Act. Finally, educational programs and materials containing protocols, recipes, and methods for safe and effective processing, cooking, and so on can assure the safe and efficient use of insects as food ingredients in general, particularly in the developing world, where more effective food-processing equipment and methods are not readily available (Amadi et al., 2005).

18.5. CURRENT EXAMPLES OF MASS-PRODUCED INSECTS WITH POTENTIAL AS HUMAN FOOD

There are a few large and many small farms in the United States rearing pet feeder insects such as crickets (usually *A. domesticus*), mealworms (larvae of *Tenebrio molitor*), and waxworms (larvae of *Galleria mellonella*). For example, the 10 largest producers of crickets in the United States already collectively produce approximately 2 billion crickets annually. This amounts to about 1.36 million kg (3 million pounds), or 1680 tonnes (1500 tons) of total cricket mass produced each year, assuming a weight of 680 g (1.5 lb) per 1000 crickets. It is likely that similar amounts of feeder mealworms are also being produced at a similar rate. Additionally, there are many more small companies producing from only a few thousand of various feeder species to hundreds of millions of insects per year. Many insect farms in the United States are located in rural communities and contribute to local economies. Additionally, there is substantial interest in, and new companies being formed for the purpose of, mass producing other feeder insects, such as black soldier flies. The current market for insect farms in the United States currently primarily consists of pet feed (for reptiles and amphibians), fish food, zoos, pest control companies (particularly for biocontrol), research labs, and a handful of aquaculture companies. Here we provide a few prominent examples of insects that are mass reared in the United States, which we feel merit further examination as potential targets for use as insect-based food ingredients.

18.5.1. Orthoptera

Around the world, 12 families and 278 species of crickets, grasshoppers, and katydids are recorded as being consumed by humans. Orthoptera is the fifth most consumed insect order worldwide (Fig. 18.2). Acrididae represent the highest frequency of human consumption (171 species), followed by Gryllidae (34 species) and Tettigoniidae (30 species) (Jongema, 2012).

The house cricket (*A. domesticus*) probably has the longest industry history of any insect mass produced in the United States, with the first large cricket farms having started in the 1940s–1950s. As mentioned in Section 18.5, they are also one of the most commonly mass-reared insects in the United States, with at least 1680 tonnes (1500 tons) being produced annually. House crickets are likely one of the least expensive insects to farm since their mass-rearing methods have been refined for several decades. Their nutrient content is also well established (Table 18.2) (Finke, 2002). Crickets and other insects are proving more efficient and sustainable to produce than several types of vertebrate livestock, including chicken, pig, lamb, and steer (Nakagaki and Defoliart, 1991). Currently, a typical wholesale cost per 1000 feeder house crickets in the United States, which weighs about 453–680 g (1–1.5 lb), is approximately $9–15 (data from personal communications with cricket farm owners). This is the price point even without the considerably larger potential demand for these insects that will arise as insect-based foods become more popular and widespread. Minimal shifts in industry practices would need to occur for prices of crickets marketed for human consumption to become more competitive. As an additional Orthopteran example, in Mexico, the grasshopper called *chapulines* has been documented as being the most frequently ingested insect in regions where it is popular, such as Oaxaca (Ramos-Elorduy et al., 2008). Huge baskets full of these *chapulines* can be found at markets throughout southern Mexico.

18.5.2. Diptera

Flies (order Diptera) are the fourth most consumed insect order by humans, with 16 families and 39 total edible examples documented in the literature (Fig. 18.2) (Jongema, 2012). They are probably the insects with the largest reproductive capacity, shortest life cycles, and rapid growth rates and that are able to eat the widest variety of organic material as feed input for mass production of insect biomass. These and other features make them some of the most attractive insects for applications to increase world food security from a production perspective. A number of applied laboratories and companies are already beginning to focus on scaling up production of black soldier flies for use as animal feed, composting, waste mitigation, and other nonfood applications (see Chapter 16 for more information on insects used as animal feed). Black soldier fly nutrient content is well established. For example, their larvae are very high in fat and calories (Table 18.2) (Finke, 2012; Popa and Green, 2012), since, like other holometatolous insects, they store it for their immobile noneating pupal development stage. Black soldier fly larvae have very little hard chitin and are easy to process. This means that we can efficiently remove the chitin without losing protein. The black soldier fly is also high in lauric acid (Finke, 2012), a fatty acid with significant antimicrobial activity (Huang et al., 2011) that is typically found in milk (Beare-Rogers et al., 2001). Fly larvae are known to contain other antimicrobial compounds, some of which might improve the shelf life of food ingredients made from them (Dossey, 2010). Additionally, several aspects of this fly give it the potential for highly efficient and sustainable production. Black soldier fly larvae can develop on almost any kind of nontoxic organic matter, including a wide array of agricultural byproducts (Bondari and Sheppard, 1981; Popa and Green, 2012; Sheppard et al., 2002). Thus, their production costs are likely to decrease over time as methodologies for mass rearing and low-cost feed are identified. Additionally, black soldier fly rearing involves several processes that are highly amenable to automation.

Currently, some of the most substantial examples of mass-reared Diptera in the United States and around the world are flies, such as various tephritid fruit flies (family Tephritidae) that are mass reared for sterile male releases. Mass production methods and facility and equipment designs for these insects have been heavily researched and are already very efficient and highly refined. Thus, these production methods could likely be very easily adapted for human food or animal feed applications. These flies are typically produced as pupae, which are irradiated, and then the resulting reproductively nonviable adults are released in orchards and farms in order to reduce wild populations of these pests; this is known as the "sterile insect technique" (SIT) (Dowell et al., 2005). Some of the most prominent examples include the Caribbean fruit fly (or "carib fly," *Anastrpha suspensa* (Loew)), the Mediterranean fruit fly (or "med fly," *Ceratitis capitata* (Widemann)), the Mexican fruit fly (or "mex fly," *Anastrepha ludens* (Loew)), and the oriental fruit fly (*Bactrocera dorsalis* (Hendel)). However, in the United States and many other places, only government facilities produce these insects and only for SIT. As a consequence, the potential of these insects is tremendously underrealized. While regulatory constraints may make actual pest species impractical for commercial production in areas where they are not already established, the methods developed to grow pest species are easily adaptable to many nonpest species of flies. Thus, commercial enterprises producing these insects as human food, animal feed, or other biomass applications (drugs, biomaterials, neutraceuticals, etc.), even if only in places where these "pests" are already established, can be a tremendous benefit to the local economies of those areas as well as contribute to global sustainability, food security, human health, and technological advances.

18.5.3. Coleoptera

At the species level, the insects most commonly consumed by humans worldwide are the Coleoptera, with 661 documented species being consumed among 26 families (Fig. 18.4). This makes sense, since Coleoptera is by far the largest group of any organism on earth. Scarabaeidae (247 species) demonstrate the greatest diversity, followed by Dytiscidae (55 species) and Cerambycidae (129 species) as being the most commonly eaten by humans (Jongema, 2012). In the United States, mealworms (*T. molitor* L.) and superworms (*Zophobas morio* F.) are currently mass produced for the pet industry as live feeder animals. Their nutrient content is also well established (Table 18.2) (Finke, 2002). Interest in using mealworms and superworms as human food has increased recently in the United States. They are featured as edible insects in many entomology outreach programs and are being purchased by various businesses and incorporated into hard candies as well as baked goods such as cakes, cookies, and cupcakes (see, e.g. http://www.hotlix.com). Mealworms and superworms can be easily reared on multiple types of vegetables and grains by the general public in a small space and are often used as an educational teaching tool in K-12 classrooms to demonstrate complete metamorphosis. Mealworms and superworms may be considered more acceptable for human consumption due to their minimal appendages.

18.5.4. Lepidoptera

Worldwide, more families of Lepidoptera (36) are consumed by humans than any other insect (Fig. 18.2). The most commonly consumed families are Saturniidae (109 species), Hepialidae (47 species), and Sphingidae (36 species) (Fig. 18.4). Their frequency of human consumption is highest in the Neotropical and Palearctic biogeographical realms, where they are the dominant order of insect consumed (Figs 18.3 and 18.4) (Jongema, 2012). In the United States, a commonly consumed Lepidopteran insect by humans is the waxworm (*G. mellonella* L.). They are mass reared for the animal feed industry as well as for fish bait. Their nutrient content is well established (Table 18.2) (Finke, 2002). Due to their holometabolous life cycle and lack of noticeable appendages during the larval stage, they are also a great candidate for roasting, grinding into flour, and incorporating into various food products. They are very high in fat, which makes them an appealing food supplement in resource-limited areas where people are malnourished and underfed.

18.6. POTENTIAL PRODUCTS AND BYPRODUCTS FROM MASS-PRODUCED FOOD OR FEED INSECTS

It is already established that insects can be used as pollinators, biological control agents, vehicles for education and outreach programs, objects of art, pets (particularly in Europe and Asia), and feeder insects for animals (Yi et al., 2010). Indeed, there is also an increasing market for edible insects and insect-based food products worldwide, particularly outside the United States. Some US restaurants, particularly Latin American and Asian restaurants, are now offering insects on their menus (FAO, 2008; Gahukar, 2011). A number of companies, as mentioned in this chapter, are preparing to capitalize on this emerging market in its early stages by being the first food product developers using insects as a primary ingredient in the United States. However, promising technologies for addressing food security and sustainability and meeting new and novel market demands take time to develop, perfect, and bring up to industrial scale. Thus, it is important for these research and industry communities to begin contemplating and developing the most feasible products on which this emerging industry can launch. This chapter has identified prominent examples so far. Once protocols are developed to produce various insect-based food ingredients, they can then be incorporated into numerous consumer items such as "meat" substitutes, protein-fortified bars and nutritional powders, as well as numerous types of snack foods (Dossey, 2013). Section 18.6.1 gives a few examples that we feel merit examination at this time.

18.6.1. Alternative "Meats"

Replacing vertebrate animal meat (muscle and other tissue) in our diet with protein-rich meat-like products derived from insects, the coup de grâce of insects replacing vertebrate livestock, means that our diets do not need to change drastically. We do not need to do without the delicious meaty products that most of us enjoy, including tacos, hotdogs, and breaded meat nuggets (currently made of chicken and fish). With a small amount of innovation, most if not all of these can be made from insects! This can be achieved in part by utilizing processes very similar to those used to make vegetarian or vegan meat substitutes from plant protein (tofu, tempeh, etc.). The internal protein of insects often appears to behave very similar to other proteins and meats when cooked, with similar textures, flavors, and odors. Even many popular vertebrate livestock-derived products such as hotdogs, sausages, ground beef (for tacos, etc.), chicken nuggets, and others often contain substantial amounts of fillers and other nonmeat ingredients as well as seasoning to improve their palatability. Thus, with a little research and development for applying existing meat and meat substitute production methods to mass-produced or mass-farmed insects as starting raw material, many products currently made from vertebrate animals could be made quite easily from insects. Some companies in the United States are already developing strategies to make these types of insect-derived alternative meat products feasible.

18.6.2. Protein and Nutritional Supplements

In the short term, protein supplements (bars, shakes, powders, cereals, etc.) for athletes and others wishing to increase their protein consumption also present a simple and ripe opportunity for incorporation of insect-derived protein. Much of the protein in these products comes from peanuts or soy. As mentioned in this chapter, animal-derived protein is superior to protein from plants, so the best protein supplements also must include some animal protein. Many of these products contain whey protein derived from milk, which has a much larger environmental footprint to produce than insects. These types of products have a relatively low barrier to entry, since they are very simple to produce and are typically sold to nutrition or environmentally conscience consumers, and the protein they contain is not visibly or gustatorially distinguishable from just about any other protein source (e.g. replacing soy powder with insect powder does not change a product's look, flavor, or texture). As a result, simple strategies to produce powders and pastes from insects can constitute high-quality protein ingredients for high-end protein supplement foods and beverages. Some of the aforementioned companies are already exploring some of these types of products in the United States.

18.6.3. Chitin: Opportunities for New Products from Insects

For many insect-based food products, the chitinous exoskeleton must be removed. This is particularly true for products such as those discussed in this chapter to alleviate malnutrition (RUTF, etc.) and others from which high nutrient absorption efficiency is desired, as chitin can reduce the absorption of some nutrients. However, the leftover chitin from industrial-scale insect-based food-producing operations can itself be a desirable high-value product with a number of additional applications, such as a neutraceutical for reduction of fat or cholesterol, a drug carrier, in agricultural pest control, in water purification, in biodegradable materials and plastic alternatives, as an antimicrobial ingredient in food and other perishable materials, to aid in wound healing, in cosmetics, and in a host of other applications. Indeed, this topic has gained the attention of others who have reviewed the potential applied and industrial value of chitin in greater detail (Je and Kim, 2012; Tharanathan and Kittur, 2003). Marketing chitin as a high-value commodity can help subsidize the budding insect-based food industry, particularly in the United States and Europe. However, the vast majority of chitin currently used for these applications comes from unsustainable harvesting of shrimp that, like much of the ocean's other resources, are constantly being overharvested. As a consequence, chitin from farmed and mass-produced insects, particularly those fed with agricultural or food byproducts, can present a much more sustainable alternative source of this chitin.

18.6.4. Novelty Products

Of course, the modern paradigm of insect-containing foods and insects as food ingredients, particularly in the United States and Europe, is still various exotic or novelty items that contain only a small amount of whole insects. These types of products can be sold at relatively high prices relative to ingredient costs. Thus, they are a way for the budding insect-derived food product industry to generate revenue while developing cheaper and more mainstream products and allowing cultural acceptance for insect-based foods to increase. On the other hand, as long as novelty or exotic food items containing whole visible insects remain the primary insect-containing foods available on the market, both cultural acceptability and overall market feasibility on a larger scale of insects as a primary food ingredient will continue to be severely limited. Thus, we propose that an increased focus on developing more standard processed insect-containing foods, similar to those that people are already familiar with and enjoy (described in Sections 18.6.1 and 18.6.2), is essential to truly realizing all of the benefits of humans eating insects that are discussed in this chapter.

18.7. CONCLUSIONS AND A CALL TO ACTION

In cultures that eat insects regularly, they are considered beneficial and are viewed as nutritious, medicinal, environmental, and a sustainable and secure food item. Once insects become more widely accepted as a respectable food item in industrialized countries, the economic implications will have a profound positive impact on businesses, industry, governments, and research. Replacing vertebrate livestock-derived foods and food ingredients with those derived from insects will also substantially improve the health of the earth's natural environment. Incorporating insects into Western food habits will increase and diversify food production, increase food supply and availability, as well as meet the nutritional needs of resource-limited households (Yen, 2009a, b). Insects will form a whole new class of foods for low-input small-business and small-farm production, with tremendous potential to be mass produced for human consumption. The future of worldwide acceptance of entomophagy relies on the commercialization of new food products but must be coupled with patents and revised regulations. International trade in edible insects would almost certainly increase as well (DeFoliart, 1992). Currently, insects in places like the United States are still not widely available and come at a high cost as compared to other animal-based food ingredients. However, as the production of insects as human food and animal feed increases worldwide, particularly in the United States, many aspects related to the feasibility of insect-based foods, such as cost, safety, the efficiency of insect mass production, and availability, will improve. The resulting increase in demand will synergistically drive the status of insects as a standard mainstream food ingredient.

Lewin's *Forces behind Our Food* (1943) gave insight into the food-buying habits of individuals primarily responsible for choosing family foods during wartime, and it can be applied to household and individual food choices today (Lewin, 1943). If the public's positive perception about insects as a mainstream edible food source is our intended outcome, knowledge about why people choose the foods they do is essential while simultaneously addressing the specific processes involved. A multidisciplinary approach is needed to realize the substantial potential benefits of human consumption of insects on a global scale. Engineers are needed to develop appropriate rearing systems for different environments and insects. Food scientists are needed to study the nature of insect foods and nutritional content, the causes of deterioration of insect food products, the principles underlying insect food processing, and the improvement of insect foods. Family and consumer scientists are needed to address the relationship between individuals, families, and communities in relation to food habits and food choices. Nutritionists are needed to advise consumers on insect foods and the nutritional impacts of insect foods on human health. Marketing, promotion, and advertising specialists are needed to promote the nutritional benefits of edible insects while marketing and selling insects for human consumption.

In order to realize their potential as a major source of human food, there are several constraints that must be overcome. These include production cost and efficiency, commercialization, technology, regulation, and social change regarding food habits and food choices. The need for development of multiple-product food–insect systems is pressing (Gahukar, 2011). This is already beginning in the United States and, in its current form, consists of human consumption of edible insects or feeder insects purchased from producers who grow them primarily for livestock and pet food. Although there are several private and public mass insect-rearing facilities in the United States, they do not yet produce insects for human consumption. Limited research and government funding are focused on mass-rearing or mass-processing methodologies of insects for human consumption. However, as mentioned in this chapter, several small companies are beginning to produce food products containing insects as well as develop methods of processing insects for use as safe, efficient food ingredients. There is limited research on insect nutrition, the psychology of insect phobias and fears, as well as insects in food science and engineering. Much of the current entomophagy research is being generated in Europe, Asia, Africa, and Central and South America. Indeed, the call for using insects to improve human food security has become more prominent in the past couple of years as well (Crabbe, 2012; Dossey, 2013; Dzamba, 2010; FAO, 2008; Gahukar, 2011; Katayama et al., 2008; Nonaka, 2009; Premalatha et al., 2011; Ramos-Elorduy, 2009; Smith, 2012; Srivastava et al., 2009; van Huis, 2013; van Huis et al., 2013; Vogel, 2010; Yen, 2009a, b).

Westerners will likely need to be exposed to entomophagy in informal environments such as festivals, fairs, museums, nature centers, parks, and restaurants as well as formal teaching and research environments in order to be more accustomed to the concept of accepting insects as food. DeFoliart (1999) emphasized that Westerners should become more aware of their negative impact on the global natural environment and their bias against consuming insects, and should increase their acceptance of insects as an alternative food source. At the time of this writing in 2013, we argue that this is even truer. In order to achieve greater recognition of insects as a viable alternative food source, a more positive social attitude about insects as human food must occur. With an increasing human population and environmental degradation, many people face a major problem in obtaining adequate protein levels in their diet. Westernized societies are reluctant to use insects, despite being major consumers of other animal proteins. We now need to consider insects as a source of food for humans in a manner that acknowledges both the role of entomophagy in indigenous societies and the need for Westernized societies to reduce the size of their environmental footprint with regard to food production, in part by replacing vertebrate livestock with insects wherever and whenever possible.

There is a need to eliminate or greatly reduce the Western-driven stigma over the use of insects as food. This will help to provide increased opportunities for research on large- and small-scale mass production as well as optimization of ecological benefits and the nutritional benefits of insects. In our global society, entomophagy must play a role in decision making and policies related to agriculture, nutrition, and food security. In order to realize the potential benefits that insects can provide to our food security as a human food and as animal feed, we call for the following: (1) greater support and attention from government funding, agricultural, and regulatory agencies for research on insect production and use as a human food ingredient; (2) support from industry to provide the means to move insect-based foods and other products from the laboratory to the market; and (3) the establishment of a formal international society, an industry association, and a journal for researchers, academics, industry partners, and other practitioners in the field of food and feed insect production. These will help tremendously in moving this emerging field forward. A formal international society would promote and present how entomophagy research could be beneficial to science, society, and industry and, thus, set the stage for what might be one of the most substantial revolutions in modern agriculture and food production: the human utilization of insects for food.

REFERENCES

Alves, R.R.N., Alves, H.N., 2011. The faunal drugstore: animal-based remedies used in traditional medicines in Latin America. J. Ethnobiol. Ethnomed. 7.

Amadi, E.N., Ogbalu, O.K., Barimalaa, I.S., Pius, M., 2005. Microbiology and nutritional composition of an edible larva (*Bunaea alcinoe* Stoll) of the Niger Delta. J. Food Saf. 25, 193–197.

Ash, C., Jasny, B.R., Malakoff, D.A., Sugden, A.M., 2010. Feeding the future. Science 327, 797.

Ayieko, M.A., Ndong'a, M.F.O., Tamale, A., 2010a. Climate change and the abundance of edible insects in the Lake Victoria region. J. Cell Anim. Biol. 4, 112–118.

Ayieko, M.A., Oriaro, V., 2008. Consumption, indigenous knowledge and cultural values of the lakefly species within the Lake Victoria region. Afr. J. Environ. Sci. Technol. 2, 282–286.

Ayieko, M.A., Oriaro, V., Nyambuga, I.A., 2010b. Processed products of termites and lake flies: improving entomophagy for food security within the Lake Victoria region. Afr. J. Food Agric. Nutr. Dev. 10, 2085–2098.

Babji, A.S., Fatimah, S., Ghassem, M., Abolhassani, Y., 2010. Protein quality of selected edible animal and plant protein sources using rat bio-assay. Int. Food Res. 17, 303–308.

Banjo, A.D., Lawal, O.A., Songonuga, E.A., 2006. The nutritional value of fourteen species of edible insects in southwestern Nigeria. Afr. J. Biotechnol. 5, 298–301.

Beare-Rogers, J., Dieffenbacher, A., Holm, J.V., 2001. Lexicon of lipid nutrition. Pure Appl. Chem. 73, 685–744.

Birch, L., 1999. Development of food preferences. Annu. Rev. Nutr. 19, 41–62.

Black, R.E., Allen, L.H., Bhutta, Z.A., Caulfield, L.E., de Onis, M., Ezzati, M., Mathers, C., Rivera, J., 2008. Maternal and child undernutrition: global and regional exposures and health consequences. Lancet 371, 243–260.

Black, R.E., Morris, S.S., Bryce, J., 2003. Where and why are 10 million children dying every year? Lancet 361, 2226–2234.

Bodenheimer, F.S., 1951. Insects as Human Food: A Chapter of the Ecology of Man. The Hague, The Netherlands.

Bondari, K., Sheppard, D.C., 1981. Soldier fly larvae as feed in commercial fish production. Aquaculture 24, 103–109.

Bukkens, S.G.F., 1997. The nutritional value of edible insects. Ecol. Food Nutr. 36, 287–319.

Bukkens, S.G.F., 2005. Insects in the human diet: nutritional aspects. In: Paoletti, M.G. (Ed.), Ecological Implications of Minilivestock: Potential of Insects, Rodents, Frogs, and Snails, Science Publishers, Enfield, NH, pp. 545–577.

Bukkens, S.G.F., Paoletti, M.G., 2005. Insects in the human diet: nutritional aspects. Ecological Implications of Minilivestock: Potential of Insects, Rodents, Frogs, and Snails, Science Publishers, Enfield, NH, pp. 545–57728.

Campbell, K.G., 1960. Preliminary studies in population estimation of two species of stick insects (Phasmatidae: Phasmatodes) occurring in plague numbers in highland forest areas of south-eastern Australia. Proc. Linn. Soc. N. S. Wales 85, 121–141.

Campbell, K.G., 1961. The effects of forest fires on three species of stick insects (Phasmatidae: Phasmatodea) occurring in plagues in forest areas of south-eastern Australia. Proc. Linn. Soc. N. S. Wales 86, 112–121.

Caulfield, L.E., de Onis, M., Blossner, M., Black, R.E., 2004. Undernutrition as an underlying cause of child deaths associated with diarrhea, pneumonia, malaria, and measles. Am. J. Clin. Nutr. 80, 193–198.

Cerritos, R., Cano-Santana, Z., 2008. Harvesting grasshoppers *Sphenarium purpurascens* in Mexico for human consumption: a comparison with insecticidal control for managing pest outbreaks [electronic resource]. Crop Prot. 27, 473–480.

Christensen, D.L., Orech, F.O., Mungai, M.N., Larsen, T., Friis, H., Aagaard-Hansen, J., 2006. Entomophagy among the Luo of Kenya: a potential mineral source? Int. J. Food Sci. Nutr. 57, 198–203.

Collins, S., Dent, N., Binns, P., Bahwere, P., Sadler, K., Hallam, A., 2006. Management of severe acute malnutrition in children. Lancet 368, 1992–2000.

Crabbe, N., May 10, 2012. Local expert gets funding to develop insect-based food for starving children. Gainesville Sun, 1B–6A.

Cruz, M.S., Johar, H.S., 2012. The Emergence of the Entomophagy Food Industry in the U.S. ESASEB/SWB Branch Meeting. Little Rock, Arkansas, USA.

DeFoliart, G., 1992. Insects as human food. Crop Prot. 11, 395–399.

DeFoliart, G.R., 1995. Edible insects as minilivestock. Biodiversity Conserv. 4, 306–321.

DeFoliart, G.R., 1999. Insects as food: why the western attitude is important. Annu. Rev. Entomol. 44, 21–50.

Dossey, A.T., 2010. Insects and their chemical weaponry: new potential for drug discovery. Nat. Prod. Rep. 27, 1737–1757.

Dossey, A.T., 2013. Why insects should be in your diet. Scientist. 27, 22–23.

Dowell, R., Worley, J., Gomes, P.V.D., Hendrichs, J., Robinson, A.S., 2005. Sterile insect supply, emergence, and release. Sterile Insect Technique: Principles and Practice in Area-Wide Integrated Pest Management, Springer, Dordrecht, The Netherlands, pp. 297–324.

Dufour, D., 1990. Insects as food—aboriginal entomophagy in the Great-Basin. Sutton, M. Q. Am. Anthropol. 92, 214–215.

Duggan, C., Watkins, J.B., Walker, W.A., 2008. Nutrition in Pediatrics: Basic Science, Clinical Application. xvii. BC Decker, Hamilton p. 923.

Dzamba, J., 2010. Third Millennium Farming. Is it Time for Another Farming Revolution? Architecture, Landscape and Design, Toronto, CA. http://www.thirdmillenniumfarming.com/.

FAO U, 2008. In: Durst, Patrick B., Johnson, Dennis V., Leslie, Robin N., Shono, Kenichi (Eds.), Forest Insects as Food: Humans Bite Back. Regional Office for Asia and the Pacific, Chiang Mai, Thailand.

Finke, M.D., 2002. Complete nutrient composition of commercially raised invertebrates used as food for insectivores. Zoo Biol. 21, 269–285.

Finke, M.D., 2004. Nutrient content of insects. In: Capinera, J.L. (Ed.), Encyclopedia of Entomology, Kluwer Academic, Dordrecht; London, pp. 1562–1575.

Finke, M.D., 2005. Nutrient composition of bee brood and its potential as human food. Ecol. Food Nutr. 44, 257–270.

Finke, M.D., 2012. Complete nutrient content of four species of feeder insects. Zoo Biol. Available Online as of June 11, 2012.

Gahukar, R.T., 2011. Entomophagy and human food security. Int. J. Trop. Insect Sci. 31, 129–144.

Gordon, D.G., 1998. The Eat-a-Bug Cookbook. Ten Speed Press, Berkeley, Calif. p. 101.

Gordon, D.G., 2013. The Eat-a-Bug Cookbook, second ed. Ten Speed Press.

Gorham, J.R., 1979. Significance for human health of insects in food. Annu. Rev. Entomol. 24, 209–224.

Graham, S.A., 1937. The Walking Stick as a Forest Defoliator. Circular. University of Michigan School of Forestry and Conservation, p. 28.

Hennemann, F.H., Conle, O.V., 2008. Revision of oriental phasmatodea: the tribe Pharnaciini Gunther, 1953, including the description of the world's longest insect, and a survey of the family Phasmatidae Gray, 1835 with keys to the subfamilies and tribes (Phasmatodea: "Anareolatae": Phasmatidae). Zootaxa, 1–311.

Holt, V.M., 1885. Why Not Eat Insects? Field & Tuer, London.

Hoppe, C., Andersen, G.S., Jacobsen, S., Molgaard, C., Friis, H., Sangild, P.T., Michaelsen, K.F., 2008. The use of whey or skimmed milk powder in fortified blended foods for vulnerable groups. J. Nutr. 138, 145S–161S.

Howard, L.O., 1916. Lachnosterna larvae as a possible food supply. J. Econ. Entomol. 9, 390–392.

Huang, C.B., Alimova, Y., Myers, T.M., Ebersole, J.L., 2011. Short- and medium-chain fatty acids exhibit antimicrobial activity for oral microorganisms. Arch. Oral Biol. 56, 650–654.

Je, J.Y., Kim, S.K., 2012. Chitosan as potential marine nutraceutical. Adv. Food Nutr. Res. 65, 121–135.

Jokthan, G.E., Olugbemi, T.S., Jolomi, A., 2007. The nutritive value of some microlivestock and their role in human nutrition. Savannah J. Agric. 2, 52–58.

Jongema, Y., 2012. List of Edible Insects of the World. Wageningen University, Wageningen, The Netherlands. As of 2012 http://www.ent.wur.nl/UK/Edible+insects/Worldwide+species+list/.

Katayama, N., Ishikawa, Y., Takaoki, M., Yamashita, M., Nakayama, S., Kiguchi, K., Kok, R., Wada, H., Mitsuhashi, J., Force, S.A.T., 2008. Entomophagy: a key to space agriculture. Adv. Space Res. 41, 701–705.

Kellert, S.R., 1993. Values and perceptions of invertebrates. Conserv. Biol. 7, 845–855.

Lewin, K., 1943. Forces behind Food Habits and Methods of Change: The Problem of Changing Food Habits, Report of the Committee on Food Habits. National Research Council, National Academy of Sciences, Washington, DC. Bulletin No. 108.

Lewin, K., 1952. Group decision and social change. In: Swanson, G., Newcomb, T.,E.H. (Eds.), Readings in Social Psychology, rev. ed. Henry Holt, New York, NY, USA, pp. 459–473.

Looy, H., Wood J.R., 2006. Attitudes toward invertebrates: are educational "bug banquets" effective? J. Environ. Educ.: Taylor & Francis Ltd., pp. 37–48.

Manary, M.J., Sandige, H.L., 2008. Management of acute moderate and severe childhood malnutrition. BMJ 337, a2180.

Melo, V., Garcia, M., Sandoval, H., Jimenez, H.D., Calvo, C., 2011. Quality proteins from edible indigenous insect food of Latin America and Asia. Emirates J. Food Agric. 23, 283–289.

Menzel, P., D'Aluisio, F., 1998. Man Eating Bugs: The Art and Science of Eating Insects. Ten Speed Press, Berkeley, Calif. p. 191.

Michaelsen, K.F., Hoppe, C., Roos, N., Kaestel, P., Stougaard, M., Lauritzen, L., Molgaard, C., Girma, T., Friis, H., 2009. Choice of foods and ingredients for moderately malnourished children 6 months to 5 years of age. Food Nutr. Bull. 30, S343–S404.

Nakagaki, B.J., Defoliart, G.R., 1991. Comparison of diets for mass-rearing *Acheta domesticus* (Orthoptera: Gryllidae) as a novelty food, and comparison of food conversion efficiency with values reported for livestock. J. Econ. Entomol. 84, 891–896.

Neumann, C.G., Bwibo, N.O., Murphy, S.P., Sigman, M., Whaley, S., Allen, L.H., Guthrie, D., Weiss, R.E., Demment, M.W., 2003. Animal source foods improve dietary quality, micronutrient status, growth and cognitive function in Kenyan school children: background, study design and baseline findings. J. Nutr. 133, 3941S–3949S.

Nonaka, K., 2009. Feasting on insects. Entomol. Res. 39, 304–312.

Okaraonye, C.C., Ikewuchi, J.C., 2008. *Rhynchophorus phoenicis* (F) larva meal: nutritional value and health implications. J. Biol. Sci. 8, 1221–1225.

Oonincx, D.G., van Itterbeeck, J., Heetkamp, M.J., van den Brand, H., van Loon, J.J., van Huis, A., 2010. An exploration on greenhouse gas and ammonia production by insect species suitable for animal or human consumption. PLoS One 5, e14445.

Packard, A.S., 1877. Half Hours with Insects. Estes and Lauriat, Boston, p. 285.

Pelletier, D.L., Frongillo, E.A., 2003. Changes in child survival are strongly associated with changes in malnutrition in developing countries. J. Nutr. 133, 107–119.

Pemberton, R.W., 1999. Insects and other arthropods used as drugs in Korean traditional medicine. J. Ethnopharmacol. 65, 207–216.

Peterson, C., 2012. Edible Bugs—Insects on Our Plates. Amazon Digital Services, Inc., p. 23.

Polis, C., 2011. December 5. The 11 biggest food trends of 2011. The Huffington Post.

Popa, R., Green, T.R., 2012. Using black soldier fly larvae for processing organic leachates. J. Econ. Entomol. 105, 374–378.

Premalatha, M., Abbasi, T., Abbasi, T., Abbasi, S.A., 2011. Energy-efficient food production to reduce global warming and ecodegradation: the use of edible insects. Renewable Sust. Energy Rev. 15, 4357–4360.

Ramos-Elorduy, J., 1997. Insects: a sustainable source of food? Ecol. Food Nutr. 36, 247–276.

Ramos-Elorduy, J., 2008. Energy Supplied by Edible Insects from Mexico and Their Nutritional and Ecological Importance. Ecology of Food and Nutrition. Routledge, pp. 280–297.

Ramos-Elorduy, J., 2009. Anthropo-entomophagy: cultures, evolution and sustainability. Entomol. Res. 39, 271–288.

Ramos-Elorduy, J., Landero-Torres, I., Murguía-González, J., Pino, M.J.M., 2008. Anthropoentomophagic biodiversity of the Zongolica region, Veracruz, Mexico. Rev. Biol. Trop. 56, 303–316.

Ramos-Elorduy, J., Menzel, P., 1998. Creepy Crawly Cuisine: The Gourmet Guide to Edible Insects. Park Street Press, Rochester, VT. p. 150.

Ratcliffe, N.A., Mello, C.B., Garcia, E.S., Butt, T.M., Azambuja, P., 2011. Insect natural products and processes: new treatments for human disease. Insect Biochem. Mol. Biol. 41, 747–769.

Read, B.E., Li, S., 1931. Chinese Materia Medica. Peking Natural History Bulletin, Peiping, China.

Rice, A.L., Sacco, L., Hyder, A., Black, R.E., 2000. Malnutrition as an underlying cause of childhood deaths associated with infectious diseases in developing countries. Bull. World Health Organ. 78, 1207–1221.

Safina, C., 2011. Why Are We Using Up the Earth? CNN Opinion: Carbon Dioxide. CNN, New York.

Schofield, C., Ashworth, A., 1996. Why have mortality rates for severe malnutrition remained so high? Bull. World Health Organ 74, 223–229.

Shackleton, C., Shackleton, S., 2004. The importance of non-timber forest products in rural livelihood security and as safety nets: a review of evidence from South Africa. S. Afr. J. Sci., 658–654.

Sheppard, D.C., Newton, G.L., Thompson, S.A., Savage, S., 1994. A value-added manure management-system using the black soldier fly. Bioresour. Tech. 50, 275–279.

Sheppard, D.C., Tomberlin, J.K., Joyce, J.A., Kiser, B.C., Sumner, S.M., 2002. Rearing methods for the black soldier fly (Diptera: Stratiomyidae). J. Med. Entomol. 39, 695–698.

Singh, B., Singh, U., 1991. Peanut as a source of protein for human foods. Plant Foods Hum. Nutr. 41, 165–177.

Smith, A., 2012. Get Ready to Pay More for Your Steak. CNN Money, New York City, NY. USA.

Smith, A., Schellnhuber, H.J., Mirza, M.M.Q., 2001. Vulnerability to climate change and reasons for concern: a synthesis. In: McCarthy, J.J., White, K.S., Canziani, O., Leary, N., Dokken, D.J. (Eds.), Climate Change 2001: Impacts, Adaptation, and Vulnerability: Contribution of Working Group II to the Third Assessment Report of the Intergovernmental Panel on Climate Change. Cambridge University Press, Cambridge, UK; New York, pp. 913–970.

Srivastava, S.K., Babu, N., Pandey, H., 2009. Traditional insect bioprospecting—as human food and medicine. Indian J. Traditional Knowledge 8, 485–494.

St-Hilaire, S., Cranfill, K., McGuire, M.A., Mosley, E.E., Tomberlin, J.K., Newton, L., Sealey, W., Sheppard, C., Irving, S., 2007. Fish offal recycling by the black soldier fly produces a foodstuff high in omega-3 fatty acids. J. World Aquacult. Soc. 38, 309–313.

Steinfeld, H., Gerber, P., Wassenaar, T.D., Castel, V., Rosales, M.M., Haan, C. d, 2006. Food and Agriculture Organization of the United Nations, and Livestock Environment and Development (Firm). Livestock's Long Shadow: Environmental Issues and Options. xxiv. Food and Agriculture Organization of the United Nations, Rome, p. 390.

Takle, E., Hofstrand, D., 2008. Global warming—agriculture's impact on greenhouse gas emissions. Ag. Decision Maker A Business Newsl. Agric., Iowa State University 12, 1–4.

Taylor, R.L., 1975. Butterflies in My Stomach. Woodbridge Press, California.

Taylor, R.L., Carter, B.J., 1976. Entertaining with Insects. Woodbridge Press, Santa Barbara, California.

Tharanathan, R.N., Kittur, F.S., 2003. Chitin—the undisputed biomolecule of great potential. Crit. Rev. Food Sci. Nutr. 43, 61–87.

Trowell, S., 2003. Drugs from bugs: the promise of pharmaceutical entomology. The Futurist 37, 17–19.

Udvardy, M.D.F., 1975. A Classification of the Biogeographical Provinces of the World. IUCN, Morges, Switzerland.

van Huis, A., 2013. Potential of insects as food and feed in assuring food security. Annu. Rev. Entomol. 58, 563–583.

van Huis, A., Van Gurp, H., Dicke, M., 2012. Het Insectenkookboek. Uitgeverij Atlas, Amsterdam.

van Huis, A.,Itterbeeck, J.v.,Klunder, H., Mertens, E.,Halloran, A.,Muir, G.,Vantomme, P., 2013. Food and Agriculture Organization of the United Nations. Edible Insects: Future Prospects for Food and Feed Security, Food and Agriculture Organization of the United Nations, Rome, 187 pp.

Vantomme, P., Mertens, E., van Huis, A., Klunder, H., 2012. Assessing the Potential of Insects as Food and Feed in Assuring Food Security. United Nations Food and Agricultural Organization, Rome, Italy.

Victora, C.G., Adair, L., Fall, C., Hallal, P.C., Martorell, R., Richter, L., Sachdev, H.S., 2008. Maternal and child undernutrition: consequences for adult health and human capital. Lancet 371, 340–357.

Vogel, G., 2010. For more protein, filet of cricket. Science 327, 811–811.

Yen, A.L., 2009a. Edible insects: traditional knowledge or western phobia? Entomol. Res. 39, 289–298.

Yen, A.L., 2009b. Entomophagy and insect conservation: some thoughts for digestion. Insect Conserv. 13, 667–670.

Yi, C., He, Q., Wang, L., Kuang, R., 2010. The utilization of insect-resources in Chinese rural area. J. Agric. Sci. 2, 146–154.

Chapter 19

Production of Solitary Bees for Pollination in the United States

Stephen S. Peterson[1] and Derek R. Artz[2]
[1]*AgPollen LLC, Visalia, CA, USA,* [2]*US Department of Agriculture, Agricultural Research Service (USDA-ARS), Pollinating Insects Research Unit, Logan, UT, USA*

19.1. INTRODUCTION

Many of our most valuable crops require an insect, usually a bee, for pollination. Bees pollinate some 400 crops worldwide and 130 in the United States (O'Toole, 2008). In addition, agriculture is becoming more dependent on the services of bees because the proportion of crops that require insect pollination has increased in recent years (Aizen et al., 2008). When a managed pollinator is provided, most often it is the familiar honey bee, *Apis mellifera* L., which is employed to do the job. For centuries, the honey bee was cultured for its production of honey and wax, and, more recently, intensive monoculture crop cultivation has made providing bees for a pollination fee an important aspect of beekeeping. Europeans brought the honey bee everywhere they settled, and it is now employed as a crop pollinator on every continent except Antarctica. But the ubiquitous honey bee is just one of about 16,000 described species of bees in 1200 genera worldwide (Michener, 2000). North America alone has 3800 species, of which 21 are introduced (Cane, 2003). Bees are most diverse in the warm-temperate, drier parts of the world; for example, California has 1985 species (Michener, 2000). The bees range from specialists that visit only one species of plant to generalists like the honey bee, and from solitary to eusocial, again like the honey bee. They are grouped into seven major families (Michener, 2000). The most familiar are the Halictidae (sweat bees), Megachilidae (leafcutting bees and mason bees), and Apidae (carpenter bees, bumble bees, and honey bees). Many of the solitary species visit crop plants, and some are now managed for their pollination services. The value of insect-pollinated crops in the United States was estimated to be $15.1 billion in 2009, $3.4 billion of which was attributed to non-*Apis* bees (Calderone, 2012).

Each bee species has its requirements for floral and nesting resources. Many bees species nest in the ground, which does not preclude them from being managed as a pollinator (e.g. *Nomia melanderi* Cockerell), but does present difficulties

Mass Production of Beneficial Organisms. http://dx.doi.org/10.1016/B978-0-12-391453-8.00019-4

when it comes to moving them to a crop. Those species that nest above ground, or will accept structures above ground, are much easier to manipulate. Many of the Megachilidae nest in beetle borings in dead trees in the wild, and these can be mimicked to provide a place for them to nest. Others Megachilids nest in crevices in embankments or use cavities in plant stems, and many of these will also accept artificial nest tunnels.

In this chapter, we will focus on the production of three species of solitary bees that have been employed as managed pollinators in the United States: the alfalfa leafcutting bee, *Megachile rotundata* F.; the alkali bee, *N. melanderi*; and the blue orchard bee, *Osmia lignaria* Say.

19.2. THE ALFALFA LEAFCUTTING BEE

Megachile rotundata (Fig. 19.1) is native to Europe and arrived in North America in 1930 (Free, 1993). Nests of wintering prepupae (the inactive larval stage prior to the pupal stage) were probably inadvertently transported in wooden shipping materials. It gradually spread across the continent and reached the western United States in the 1950s. Stephen and Torchio (1961) observed this bee readily visiting and collecting alfalfa pollen. Honey bees were used for alfalfa seed production at that time, but Stephen (1955) observed that most honey bees would not collect pollen from alfalfa, instead approaching the flower from the side to gain access to the nectar. The reproductive structures of the alfalfa flower pop out explosively when a bee probes them, and honey bees apparently learn to avoid this triggering. The alfalfa leafcutting bee is undeterred by the alfalfa flower's mechanics, tripping 81% of flowers when allowed a single visit (Cane, 2002), and readily accepts artificial nesting sites (Stephen, 1961).

FIGURE 19.1 The alfalfa leafcutting bee, *Megachile rotundata*, on an alfalfa flower.

Stephen (2003) noted that the North American and European *M. rotundata* are different in several ways that make the North American bee a better candidate for domestication. The accidental introduction must have created a genetic bottleneck resulting in a more gregarious population that flies at lower temperatures, completes nests faster, collects pollen from fewer hosts, collects more pollen, and is multivoltine rather than univoltine (Stephen, 2003; see citations within).

Alfalfa seed yield improved greatly where alfalfa leafcutting bees were employed. For example, in Idaho, yields were 110–220 kg/ha prior to the arrival of the alfalfa leafcutting bee, but yields climbed to 280–560 kg/ha after 1956, when the leafcutting bees became commonplace (Olmstead and Wooten, 1987). Mass rearing this bee involves wintering the cocoons in cold storage, followed by incubation for about 4 weeks and release in the field. Cavities in wood, polystyrene, or paper tubes must be provided for nesting, with some protection from rain and direct sun. The females require an abundant source of pollen and nectar for 3–4 weeks for adequate reproduction. The cocoons may be left in their nesting substrate or stripped out, depending on the management system.

The alfalfa leafcutting bee usually has a single generation per year in northern latitudes. It is a relatively small bee of 7–9 mm in length, and it overwinters as a prepupa in its cocoon. In the spring, as temperatures warm, it continues its development through the pupal and adult stages and then chews out of its cocoon and nest. Females mate once (although males can mate more than once) and then search for a suitable place to nest. The female holds the sperm in a sac-like organ called the spermatheca, and as in all Hymenoptera, unfertilized eggs become males and fertilized eggs become females (arrhenotoky). Once a female has selected a cavity for nesting, she lines it with 14–16 cut leaf pieces (Richards, 1984a). Soft leaves or flower petals are preferred, and in alfalfa, leaves are used to line nests (Horne, 1995). A female can carry a leaf piece that is 17% of her body weight, which averages 35 mg (Klostermeyer et al., 1973). Leaves are carefully arranged to overlap, and the edges are chewed to make them adhere together. Next, the bee goes to work gathering pollen, which it packs onto the thick layer of hairs (the scopa) on its abdomen. The maximum pollen load size is about 23% of the bee's body weight. After arriving back at her nest, the female will turn around and back into the hole, removing the pollen with her hind legs. Nectar is regurgitated and added to the pollen provision to create a doughy mass. Females will make 15–27 provisioning trips and take about 5 h, under ideal conditions, to provision a single cell (Richards, 1984a).

When the pollen provision reaches a suitable size, the female deposits an egg on it. The cell is then sealed off with leaf discs, and the process starts over. Depending on the length, a female will construct 8 to 12 cells in a cavity (Richards, 1984a). Female eggs are deposited at the back of the hole, and males toward the entrance. The ratio of males to females has been reported as 2:1 (Richards, 1984a), but recent studies show that the ratio of males to females in commercial populations in the United States is close to 1:1 (Pitts-Singer and James, 2005). The larvae have five instars, although the first instar remains inside the egg (Trostle and Torchio, 1994). After eggs hatch, 14.5 days are required to completed larval development at 23 °C (Whitfield and Richards, 1992). The lower threshold for larval development is 15 °C, and 116 degree-days are required for complete development (Whitfield and Richards, 1992). Kemp and Bosch (2000) found that in diapausing individuals, a fluctuating thermal regime (14:27 °C, 8:16 h, mean 22 °C) reduced the length of the prepupal and pupal stages after diapause compared with a constant temperature (22 °C).

Females will frequently nest in cavities as narrow as 3–4 mm in diameter when offered a choice (O'Neill et al., 2010). However, offspring produced in larger diameter tunnels are larger (Stephen and Osgood, 1965a; O'Neill et al., 2010) and collect more pollen per foraging trip as adults than those reared in smaller diameter tunnels (Stephen and Osgood, 1965a). Studies have shown that cavities that are 6.4–7.2 mm in diameter and 100–150 mm deep are optimum for female and overall cell production (Richards, 1984a). Industry-standard nesting blocks do not always adhere to these specifications. For instance, drilled solid pine boards (120 × 15 × 7 cm) have tunnels that are 5 mm in diameter and 65 mm deep (Peterson et al., 1992). The most commonly used nesting medium is solid polystyrene blocks (M. Wendell, personal communication), and they come in two depths, 7.6 and 9.5 cm (Fig. 19.2, bottom). The second most commonly used nesting medium is grooved wooden laminates (Fig. 19.2, top). When offered a choice, the bees will typically prefer to nest in wood rather than polystyrene nesting materials and previously used nesting material rather than new (Pankiw and Siemens, 1974). Machines that push out the leaf-encased cocoons have been developed and are widely used (e.g. Pinmatic Inc., Pinawa, Manitoba; Fig. 19.3). Richards (1978) found a greater percentage of viable cocoons when the bees nested in grooved wood, Styrofoam, and polystyrene laminates compared to drilled boards, soda straws, particle board, aluminum, and corrugated paper. Tumblers are often used to remove debris-feeding insects, predators, and excess leaf pieces from stripped cocoons (Richards, 1984b).

Once the larval bees have reached the prepupal stage, there are physiological benefits to providing a pre-wintering period of a month or more at a cool temperature (e.g. 16 °C) prior to wintering (Pitts-Singer and James, 2009). Leafcutting bees are best held in cold storage (5–8 °C) over the winter to minimize mortality (Stephen and Osgood, 1965b). Kemp et al. (2004) found that survival of alfalfa leafcutting bees increased sharply with greater than 3 months of wintering and that weight was static during this period. Stripped cocoons or loose cells can be held in boxes, barrels, or trays. It is important to monitor the temperature in the center of the mass of cocoons, especially late in wintering, because they can begin to warm prematurely. The heat given off by respiring immature bees can trigger those nearby to break diapause, and the container can warm considerably despite being in cold storage. This is of particular concern during shipping in the spring. If this situation occurs, the cocoons can be spread out in trays to allow them to cool back down. These cocoons would need to be sampled for developmental stage and mortality. Bees held in their natal nests are

FIGURE 19.2 Alfalfa leafcutting bees nesting in grooved wooden laminates (top); and polystyrene nests strapped back to back and stacked in a portable trailer in central California (bottom). (For color version of this figure, the reader is referred to the online version of this book.)

FIGURE 19.3 An alfalfa leafcutting bee cocoon extractor. This Dual Pinmatic Harvester is capable of harvesting 420 nests per minute (PINMATIC Inc., Pinawa, Manitoba). (For color version of this figure, the reader is referred to the online version of this book.)

less prone to heating during cold storage; however, it is important to allow good air circulation around stacked nesting materials.

As pollination season approaches, an estimate must be made at least 3 weeks in advance as to when the bees will be needed in the field. Since the bees overwinter in the prepupal stage, the immature bees must molt into the pupal stage and then again into the adult stage. This process takes about 18–23 days for males and 21–25 days for females at 30 °C. It is usually best to begin incubation about 30 days prior to the expected release date because adult bees can be cooled and held for up to 14 days at 15 °C. However, they cannot be forced to develop any faster than they do at 30 °C. The lower threshold for postdiapause development is 15.7 °C (Richards and Whitfield, 1988), so at 30 °C, 14.3 Celsius-degree-days accumulate each day. It is useful to monitor temperatures and calculate degree days, which can be compared to Table 19.1 to keep track of developmental events. A recent study showed that providing a 1 h pulse of high temperature during the adult holding period can markedly increase the storage time period (Rinehart et al., 2011).

TABLE 19.1 Alfalfa Leafcutting Incubation Events by Day of Incubation (at 30 °C) and Degree-Days

Day	Degree-Days	Comments
1	14.3	Cocoons to 30 °C. Black-light traps in place.
7	100	Dichlorvos strips in; one strip per 38 m^3.
8	114	*Pteromalus* begin to emerge; bees begin to pupate.
10	143	Pink eye stage.
13	186	Remove dichlorvos strips and turn air out.
14	200	*Megachile relativa* (a native leafcutting bee) begin to emerge. Ignore these.
15	215	Incubator can be closed again. Temperature may be dropped to 10–15 °C from now until day 22 to delay emergence.
18	257	Males begin to emerge.
21	300	Females begin to emerge.
22	315	Peak male emergence.
24	343	Peak female emergence.
28	400	Emergence is complete.

Source: Adapted from a "Calendar of Incubation" for alfalfa leafcutting bees by D. Murrell, Saskatchewan Agriculture and Food.

Usually, nests or loose cocoons that are beginning to emerge are taken to the field to complete the process. It becomes obvious when the bees are ready to be taken to the field because the sound of their chewing is audible. Cool weather in the field can slow emergence, so to avoid this problem, some bee producers release emerged bees in the field in a method called a "bleed-off system" (Stephen, 1981). Incubating bees are allowed to emerge in the incubator, and they are then drawn to a light source, where they fall into a cold room and into containers with excelsior (wood shavings). The excelsior gives the bees something to cling to and prevents them from chewing on each other. This system has the advantage of generating a full population of flying bees immediately, but bees released this way have a greater tendency to abscond from the area.

Production of alfalfa leafcutting bees increased rapidly in the 1970s and 1980s. Alfalfa seed acreage also increased markedly, and from 1978 to 1980, leafcutting bees were being sold for over $185/10,000 bees (10,000 bees = 1 gallon in the alfalfa leafcutting bee industry). As acreage leveled off in the late 1980s and 1990s, prices fluctuated between around $50 to $100 per gallon (Fig. 19.4). In 1999, several seed companies grew more alfalfa seed for markets in Argentina. Then, a financial crisis in that country left the seed companies with a surplus of alfalfa seed. This surplus lasted several years and resulted in a large portion of alfalfa seed coming out of production in the United States. The price of alfalfa leafcutting bees plummeted in 2000 to $25 per gallon and reached a low in 2004 of $15 per gallon. Finally, by the late 2000s, alfalfa seed acreage began to increase, and the price of leafcutting bees likewise has recovered. Since 2009, prices have been over $90 per gallon. Alfalfa leafcutting bees are now also being used in some other crops, especially canola. The

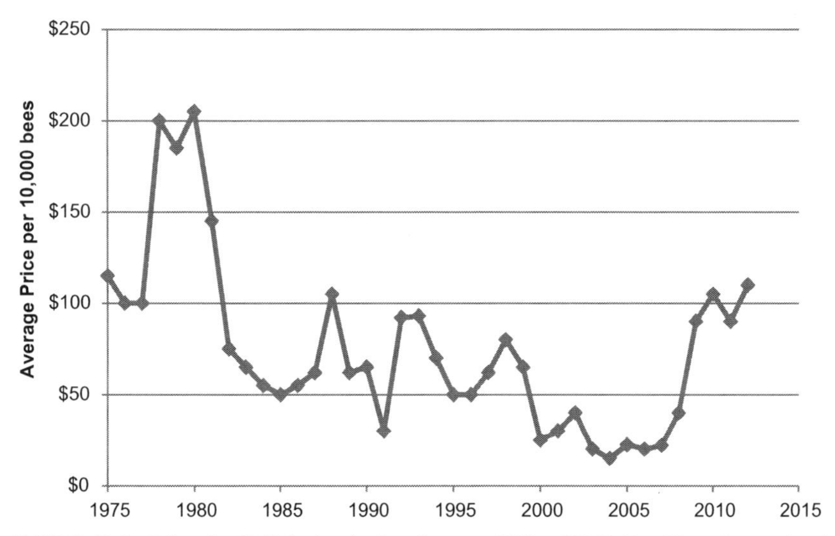

FIGURE 19.4 Prices for alfalfa leafcutting bees between 1975 and 2012 (Ron Bitner, International Pollination Systems). (For color version of this figure, the reader is referred to the online version of this book.)

combination of commodity prices for alfalfa seed and canola can have a large influence on alfalfa leafcutting bee prices. When both commodities reach relatively high prices, as they have recently, more acres are planted; this in turn creates more demand for bees, and the price of bees rises.

In the United States, it is common to have fewer bees produced in alfalfa seed than were released there. Typical bee losses range from 50% to 80% (Pitts-Singer and James, 2005, 2008). In Canada, however, it is common to increase bees in alfalfa seed. Fourfold increases are known to have occurred, usually in small-scale, isolated locations. Usual increases are 1.5-fold to twofold in large-scale Canadian operations (D. Nahuliak, personal communication). The practice of releasing lower densities of bees in Canada, to encourage bee reproduction, seems to be the main explanation. Baird and Bitner (1991) recommend 50,000 to 100,000 bees per hectare (2–4 gallons per acre). Strickler and Freitas (1999) found growers stocking 78,000 to 130,000 bees per hectare in Idaho. At these levels, pollination occurs rapidly, but floral resources decline rapidly as well (Strickler and Freitas, 1999). Bosch and Kemp (2005) found evidence that U.S. alfalfa seed growers often stock too many bees in their fields and that the timing of release is late. When bee density is high compared to floral resources, nest establishment, pollination efficiency, and reproductive success are reduced (Pitts-Singer and Bosch, 2010). In Canada, 25,000 to 50,000 bees per hectare are stocked, although fewer than that may be sufficient for good pollination (Parker et al., 1987). Other factors that help Canadians produce more bees include reduced pest and disease pressure, and a single generation of bees timed for a very intense bloom coinciding with very long days. Also, a lower percentage of cocoons produced in the United States contain live larvae, and survival of females is lower compared to Canadian bees (Pitts-Singer and James, 2005). Thus, a system has developed in which Canadian producers export their excess bees to the United States, where they are stocked at high density and replaced annually (Parker et al., 1987). In 2011, Canada produced 355,126 gallons of leafcutting bees, down from 399,968 in 2006 (Census of Agriculture, 2011). At $90 per gallon, the leafcutting bees produced in 2011 were worth over $31 million. Alberta produced 53%, Saskatchewan 35%, and Manitoba 11% of Canadian leafcutting bees.

Like any organism, the alfalfa leafcutting bee has a guild of natural enemies. Several of these can severely limit bee production if not managed. One of these is *Pteromalus venustus* Walker, a parasitoid in the family Pteromalidae. These small wasps (2.5 mm in length for females) parasitize the prepupal stage of the bee. The wasp has an ovipositor that can pierce through 1 mm to reach the larvae, and an average of 16 adult parasitoids (range = 7–26) will develop in a single bee larva (Richards, 1984a). From 1976 to 1983, parasitism levels averaged about 1% in western Canada, but they can reach as high as 54% (Whitfield and Richards, 1985). Cocoons are most vulnerable during incubation as the adults emerge around the 12th day of incubation. Females then can parasitize other leafcutting bees in the incubator. Fortunately, the wasps are attracted to ultraviolet light, where they can be drowned in a pan of water. In addition, dichlorvos strips introduced to the incubator are very effective in killing wasps (Hill et al., 1984). Strips are put into incubation chambers on day 7 (30 °C incubation temperature) and removed on day 13, then fans are used to air out the chamber for 24–48 h with no ill effects on the bees.

The fungal pathogen, chalkbrood, is caused by *Ascosphaera aggregata* Skow and is the most serious disease for leafcutting bee producers (James, 2011). Infected larvae grow to maturity but die before pupation. The dead larva turns black and is filled with billions of spores. As sibling leafcutting bees chew their way out of the nest, they must chew through these cadavers and become contaminated with the spores (Vandenberg et al., 1980). When these females begin to nest, they inadvertently leave spores in the nest provisions. In commercial pollination, cocoons are often removed from nesting substrates to reduce transmission, but this does not eliminate the disease (James, 2011). In Manitoba and Alberta, chalkbrood is prevalent, while Saskatchewan is relatively free of the disease (M. Wendell, personal communication). In order to keep Saskatchewan free of chalkbrood, it is common practice not to move leafcutting bees from Alberta or Manitoba into the province. In areas where chalkbrood is prevalent, there are methods to limit the disease buildup. Wooden boards can be redrilled and heated to 120 °C to kill the spores (Kish and Stephen, 1991). For polystyrene nests, the fumigant paraformaldehyde is effective, but it is registered for use only in Canada (Goerzen, 1992) and Montana (EPA Reg. No. 4972-43, EPA SLN No. MT-12-0001). Dipping in chlorine bleach (sodium hypochlorite) is also effective on wooden boards and polystyrene nests (Mayer et al., 1988).

It is also common to find failed provision-filled cells or "pollen balls" in nests of the alfalfa leafcutting bee. These losses can vary from 4% to 42% and are more prevalent in the United States than in Canada (Pitts-Singer and James, 2008). These pollen balls appear to have multiple causes, such as hot weather or cool, wet weather (with the presence of fungus). In addition, high bee density can also increase the incidence of pollen balls (Pitts-Singer and Bosch, 2010).

Bee producers in Canada can have bees examined for pests, parasites, number alive per pound, machine damage, and sex ratio by the Canadian Leafcutter Bee Cocoon Testing Centre (Brooks, Alberta). Batches of cocoons are imaged using X-rays, which readily show live and dead larvae, parasites, and diseased individuals (Stephen and Undurraga, 1976). A health certificate for each batch examined is generated that helps the industry monitor diseases and pests and allows buyers and sellers to negotiate a fair price for the bees.

The alfalfa leafcutting bee has a single generation in northern latitudes, but it can have a partial second or even third generation in more southern locations. For producers in Canada, the single generation may benefit bee health because chalkbrood and other parasites have less opportunity to reproduce in the field. In more southern latitudes, as in Idaho and California, a second generation is common; and in the San Joaquin Valley of California, it is not uncommon to have a third generation. With the long bloom of alfalfa seed (from late May to late July), two generations of bees can be beneficial for pollination. In fact, sometimes growers can sell second-generation bees when they have a crop that needs no further pollination to growers who have a late-blooming field. The trade-off in having multiple generations is that reproduction usually suffers with subsequent generations. The partial second generation is more likely to disperse due to waning bloom at that time of year, and the emerging bees may destroy diapausing bees as they exit the nest (Tepedino and Parker, 1986). Also, the second generation has a female-biased sex ratio, while the diapausing generation has a male-biased sex ratio (Tepedino and Parker, 1988). Hobbs and Richards (1976) selected for a single-generation bee with success, but there has yet to be a consistently single-generation bee strain available to producers.

The alfalfa leafcutting bee was once thought of as strictly a pollinator for alfalfa and other legumes, such as sweet clover and birdsfoot trefoil seed production. However, over the years, the bee has been tried on many other crops, with some excellent results. For example, carrot seed in enclosures (Tepedino, 1997), blueberry (Stubbs et al., 1994), and canola (Soroka et al., 2001) have been successfully pollinated with alfalfa leafcutting bees. Recently, hybrid lettuce seed was successfully pollinated by alfalfa leafcutting bees (Gibson et al., 2007).

The alfalfa leafcutting bee is an excellent example of how a solitary bee can be employed to pollinate crops. Research and industry innovation have produced equipment and methods to make the bee a commercially sustainable pollinator. The industry has proven to be highly resilient to perturbations in the market, and agriculturalists continue to find new crops for it to pollinate. It is clear that the alfalfa leafcutting bee will remain an important managed pollinator in the future.

19.3. THE ALKALI BEE

Nomia melanderi (Fig. 19.5), the alkali bee, is a ground-nesting bee that is native to western North America and is in the family Halictidae. It can nest in dense aggregations in favorable habitats. It requires soils with 7–8% clay content, a high water table, and a compacted surface (Mayer and Johansen, 2003) (Fig. 19.6). Mass producing this species can be as simple as planting alfalfa near an existing, natural bed and ensuring that the bees are protected from pesticide applications. Artificial beds can be constructed and “seeded” with cocoons or adult bees.

The alkali bee is just as efficient as the alfalfa leafcutting bee, tripping 78% of alfalfa flowers on single visits (Cane, 2002). Also, alkali bees are more likely to fly below the alfalfa canopy than other bees. Females dig holes up to 21 inches deep and construct individual cells branching from a main downward tunnel. Pollen is carried back to the nest on hairs on the hind legs. A female can construct 18 to 20 cells under ideal conditions, with half of her offspring being female (Stephen, 1960). During an optimal day, a female starts by laying an

FIGURE 19.5 The alkali bee, *Nomia melanderi*, on alfalfa blossoms. *Photograph by J. Cane.* (For color version of this figure, the reader is referred to the online version of this book.)

FIGURE 19.6 A managed alkali bee bed showing tumuli (piles of excavated soil). *Photograph by J. Cane.* (For color version of this figure, the reader is referred to the online version of this book.)

egg on a pollen provision collected the previous day, gathers and makes a new pollen provision, and then digs a new cavity for the next day (Stephen, 1959). A single generation develops annually in the Pacific Northwest, but multiple generations occur in California (Stephen, 1965). Stephen (1960) constructed artificial beds that produced population densities eight times higher than in the best naturally occurring beds after 2 years. Beds can become quite crowded with nests; for instance, one populous nesting bed carried a median density of 278 nests per m^2 (Cane, 2008a). Alkali bees will fly up to 3 miles away when floral resources are not nearby (Stephen, 2003).

During the 1960s, interest in alkali bees stimulated the installation of over 100,000 square feet of artificial bee beds in one water district in central California (Wichelns et al., 1992). Statistical analysis showed that the alkali bees had a positive impact on alfalfa seed yield during that time (Wichelns et al., 1992); however, the beds declined suddenly in the 1970s, most likely due to changing pesticide use in the area.

Because alkali bees are soil-nesting bees, there are some challenges involved in establishing new beds and in moving the bees. Blocks of soil containing nests can be cut from existing beds and transplanted into new ones in the spring (Johansen et al., 1982). A steel coring tool, 25–30 cm in diameter, is driven into the soil using tractor or human power, or a large number of cores can be cut with a circular saw blade mounted on a back hoe. The cores can then be transported and buried into a new bed. Newly emerged bees can also be captured, anesthetized with CO_2, and released at a new site in the evening (Johansen et al., 1982).

The number of managed alkali bee beds peaked in the late 1950s and early 1960s, then began to decline. Mayer and Johansen (2003) attribute the decline to the rise in interest in, and competition from, the alfalfa leafcutting bee. They showed that competition from other bees in the field reduces alkali bee foraging. In addition, success of the alkali bee beds can be set back severely by untimely rain storms, which occurred several years in a row during the 1960s. Stephen (2003) agreed that untimely spring rains led to the decline of managed alkali bees, and added the economics of bed construction and maintenance, predators and disease, pesticides, and the fouling of continuously reused nesting media.

Alkali bees continue to be used as pollinators in the Touchet Valley of Washington state, where there are at least 56 nesting sites (Cane, 2008a) supporting 17 million females. The most populous of these sites grew to a population of 5.3 million females. The continued success of these nesting beds in this area can be attributed to careful management of nesting beds by growers and because the primary crop in the Touchet Valley is alfalfa seed. The other rotational crops grown in the valley (wheat, field peas, and chickpeas) are not sprayed with insecticides. Growers are conscientious about any insecticide application on the alfalfa seed, applying them before or after bloom (Cane, 2008a). Yields in fields pollinated by alkali bees are comparable to those pollinated by alfalfa leafcutting bees. As the pollination service provided by the alkali bees has been sufficient there, growers do not bring in alfalfa leafcutting bees or honey bees, so competition for floral resources is minimized. Growers who utilize these bees find that after the initial investment in bee beds, the management requirements are minimal.

19.4. THE BLUE ORCHARD BEE

The blue orchard bee or orchard mason bee, *O. lignaria* (Fig. 19.7), is native to North America and occurs throughout much of the United States. Two subspecies

FIGURE 19.7 The blue orchard bee, *Osmia lignaria*, on an almond blossom. (For color version of this figure, the reader is referred to the online version of this book.)

are recognized: *O. lignaria lignaria* Say occurs east of the 100th meridian, while *O. lignaria propinqua* Cresson occurs to the west (Bosch and Kemp, 2001). Females are usually slightly larger than a honey bee and dark metallic blue to blue-green in color. These bees nest in preexisting cavities in dead wood (abandoned nests of wood-boring beetles and hollow plant stems), but sometimes in paper or mud nests of wasps, ground nests of bees, or *Xylocopa* (carpenter bee) nests (Cane et al., 2007). Blue orchard bees overwinter as adults inside the cocoon and emerge early in the spring, usually when daytime high temperatures begin to exceed 20 °C. They visit a wide variety of plant species in 34 plant families (Bosch and Kemp, 2001). The blue orchard bee is strictly univoltine.

In 1970, Philip Torchio made a survey of bees present in apple and pear orchards in Utah and surrounding states. He found that *O. lignaria* was omnipresent in these orchards and would be a good candidate for development as a commercial pollinator (Torchio, 2003). Field trials showed that blue orchard bees would successfully nest and pollinate in almonds (Torchio, 1979, 1981a, b, 1982a), apples (Torchio. 1982b, c, 1984a, b, 1985), prunes (Torchio, 1976), and cherries (Bosch and Kemp, 1999; Bosch et al., 2006). Fivefold increases in female populations have been documented in a pear–apple orchard (Torchio, 1985) and in cherries (Bosch and Kemp, 1999). In Oregon and Washington, cherry and apple growers are using blue orchard bees successfully. Other crops, such as meadowfoam (Jahns and Jolliff, 1991), pear, plum, prune, peach, nectarine, strawberry, currant, gooseberry, blackberry, rape, and canola, are also good candidates for commercial pollination by blue orchard bees (Torchio, 1976; Bosch and Kemp, 2001).

Mass producing the blue orchard bee involves providing cold storage for the wintering cocoons followed by release in the orchard. Since this species overwinters as an adult, the incubation period is much shorter than with the alfalfa leafcutting bee. Cavities must be provided in wood or paper tubes, and a source of moist soil must be nearby. For good reproduction, nectar and pollen must be present for 3–4 weeks. Offspring cocoons need to be monitored closely in order to begin the wintering period not long after they have reached the adult stage.

Blue orchard bees will accept artificial tunnels (Levin, 1957), but they prefer to nest in holes 6–7 mm in diameter (Torchio, 2003). In cavities less than 10 cm in length or narrower than 6 mm, nesting females will produce a greater proportion of male offspring. Drilled wooden blocks are the most attractive nests, and nesting is improved when holes are separated by more than 12 mm (Torchio, 2003). Visual as well as olfactory cues are important when selecting a cavity, and the odors from old female cocoons are attractive (Pitts-Singer, 2007). A typical nest will contain 2–3 females and 4–6 male cells (Fig. 19.8). Female cells are constructed first at the back of the nest and are about 1.7 times as large as a male cell. When foraging for nest provisions, blue orchard bees will collect both pollen and nectar. Pollen is packed onto the abundant bristles of the scopa on the underside of the abdomen just as does the alfalfa leafcutting bee. When the female returns to her nest, she enters head first and disgorges nectar onto the lump of pollen (Torchio, 1989). She then backs out, turns around, and backs into the cavity so that she can unload the pollen using her hind legs. Once the pollen provision is large enough, the female will make a final foraging trip for nectar and deposit a small pool of nectar on top of the pollen mass. She then turns around and oviposits on the pollen provision. Adjacent cells are separated by mud partitions, and mud is also used to form a plug at the end of the completed nests (Torchio, 1989).

The egg stage lasts about 7 days at 22 °C (Bosch and Kemp, 2000). The first instar consumes embryonic fluids inside the egg and molts into the second instar as it encloses from the egg chorion (Torchio, 1989). The second instar emerges

FIGURE 19.8 A blue orchard bee nest in a split reed with pollen–nectar provisions, mud partitions, and eggs. Larger provisions will become females (five in this nest). (For color version of this figure, the reader is referred to the online version of this book.)

from the egg and proceeds to feed on the pollen–nectar provision. The larva progresses through five instars in about 21 days at 22 °C (Bosch and Kemp, 2000). Cocooning takes about 5 days, and then a prolonged prepupal stage of about 29 days occurs. After pupation, the immature bee spends about 32 days in the pupal stage before molting to the adult stage. Development from egg to adult takes approximately 87 days at 22 °C (Bosch and Kemp, 2000). Developmental rates are highly dependent on temperature. For example, at 18 °C, egg-to-adult development takes more than 123 days, whereas at 29 °C it takes only about 63 days (Bosch and Kemp, 2000). A fluctuating temperature regime, averaging 22 °C, gave the highest survival, and development was completed in 70 days. Blue orchard bee prepupae take longer to develop at temperatures above or below 26 °C (Kemp and Bosch, 2005). As temperatures begin to drop in the fall, the adult enters diapause, still inside the cocoon. Overwintering as an adult is a derived trait and makes these bees better able to take advantage of early spring blooms (Bosch et al., 2008).

Bees from warmer climates (e.g. at low elevation in California) typically have a longer prepupal stage where they enter a summer diapause, or aestivation, compared to bees from cooler climates (like Utah) (Bosch et al., 2000). Cocoons can be wintered artificially at 3–5 °C successfully (Bosch and Kemp, 2003), and at least 3 months of wintering are required for best survival rates (Kemp et al., 2004). Although blue orchard bee adults reduce their metabolic rate as soon as they become adults, they lose considerable weight during prewintering (Kemp et al., 2004; Bosch et al., 2010; Sgolastra et al., 2011). Sgolastra et al. (2010) found that diapause can be divided into two phases. The first phase lasts until about 100 days of wintering have elapsed and is an intense diapause (when incubated, emergence time is prolonged). In the second phase, after 100 days, increasing the wintering temperature can rapidly decrease the diapause intensity and promote emergence from the cocoon. Even with continued winter temperatures, diapause intensity gradually diminishes in this second phase.

In the wild, warming spring temperatures will break diapause, and the bees will chew their way out of their cocoons and through the mud partitions, with males preceding females. In artificial management, cocoons or nests are warmed up, usually a week or two prior to bloom, to promote emergence. After 150 days of wintering at 4 °C, average male emergence will occur in under 2 days at 20 °C, and under 8 days for females (Bosch and Kemp, 2000). Males will patrol the area, feeding on nectar and sleeping in crevices until the females emerge. Females apparently emit a sex pheromone (virgin females give off a distinctive odor), which attracts the males. Mating usually occurs close to the natal nest, but sometimes on nearby vegetation.

Females take 2 to 3 days to feed on nectar and pollen and search for suitable nesting sites. Once they have selected a suitable cavity, they usually first collect mud and make a partition at the back of the cavity, and then proceed to collect a pollen–nectar provision. Females can be observed spiraling around the tunnel as they exit their holes, apparently marking it with a secretion from the Dufour's gland (Guédot et al., 2006; Pitts-Singer et al., 2012). This marking ensures that they have the correct hole when they return. Changing or cleaning the cavity entrance while the bee is foraging results in a disruption of the unloading behavior; this indicates that odor cues are important for the bee in identifying its nest (Guédot et al., 2006).

One-half to one cell is usually completed in one day during normal field conditions, but two per day are possible (Torchio, 1989). A foraging trip typically takes 5–10 min, and a female will visit about 75 flowers in an average trip (Bosch and Kemp, 2001). A single provision may require 20 to 30 foraging trips. When the series of cells is complete, the female constructs a thick (8–9 mm) plug at the opening of the cell to seal the nest and prevent predators or scavengers from entering. A female bee can live up to 6 weeks, but 3–4 weeks is typical. Under ideal conditions, a female can produce up to 32 eggs, 10 of which would be daughters, but it is more typical for a female to construct about two complete nests with about 4–6 daughters.

Management of these bees is relatively simple, but two life stage events require close monitoring for success when artificial storage conditions are used. These are (1) the onset of wintering in the autumn, and (2) the timing of emergence in the spring (Bosch et al., 2008). If the bees experience a lengthy prewintering period (>3 weeks), they can become weakened, so knowing when the bees reach adulthood is critical for timing winter storage. A sample of nests can be X-rayed to monitor development, or a sample of males can be sacrificed by opening the cells to determine stage of development. Emergence timing can vary greatly with the length and temperature of wintering (Bosch and Kemp, 2003; Bosch et al., 2008). At least 150 days of wintering are required for prompt emergence of females (Bosch and Kemp, 2003). To estimate the amount of time until bee emergence, a sample of bees can be incubated 2–3 weeks prior to the expected bloom (Bosch and Kemp, 2001).

Commercial pollination with blue orchard bees requires large numbers of bees. Managers of the species have three options for increasing their population: (1) Trap wild bees or purchase trapped bees, (2) increase populations in the orchards where the crop is pollinated, and (3) propagate bees away from orchard sites. Acquiring enough blue orchard bees through wild trapping cannot meet customer demands, is often unpredictable, and can have negative consequences for wild bee populations (Roulston and Goodell, 2011). Propagation away from orchards, often called "bee ranching," offers the advantage of longer bloom times, depending on the flowers available, and separation from agricultural chemicals. Blooming plants known to be visited by blue orchards bees can be planted, and they can even be enclosed in a greenhouse or screenhouse (Fig. 19.9). A screenhouse enclosure ensures that the bees do not disperse and eliminates honey bees as competitors for the crop. The annual wildflower, *Phacelia tanacetifolia* Bentham (Fig. 19.9, inset), makes an excellent forage crop for the bees. *Collinsia heterophylla* Buist *ex* Graham and *Nemophila maculata* Bentham

FIGURE 19.9 A 5-acre screenhouse dedicated to propagating blue orchard bees in central California. Inset: *Phacelia tanacetifolia* blossom and a blue orchard bee female. (For interpretation of the references to color in this figure legend, the reader is referred to the online version of this book.)

ex Lindley are also good pollen and nectar producers and are highly attractive to the bees. Increases of up to 4.5-fold have been achieved in a screenhouse with wildflowers (S. Peterson, unpublished data). With a single generation per year, propagating bees on wildflowers means forgoing a pollination fee so that a larger population is available for pollination in subsequent years.

In California, almonds have steadily increased in acreage. By 2011, there were approximately 338,000 ha of almonds, with a production value of $3.5 billion (USDA-NASS, 2011), and California is one of the few places in the world where the climate is favorable for production. Almonds are especially dependent on bees for successful pollination (McGregor, 1976), but they are among the earliest crops to bloom, typically starting around February 15 during California's rainy season. Typically, 4.9 colonies of honey bees are recommended for almond pollination per hectare (Thorp, 1996). This amounts to 1.5 million colonies of honey bees, or about two-thirds of the entire U.S. supply. The cost of renting honey bees in almonds more than doubled from 2004 to 2006 (Sumner and Boriss, 2006) due to rising demand from increasing almond acreage and reductions in honey bee supply, largely caused by honey bee pests (including parasitic mites, fungal and bacterial diseases, and viruses) and colony collapse disorder (CCD). Since that time, prices have remained at these levels (in a range of $120 to $150 per colony). Rental prices are not expected to drop back to previous levels because the bearing acreage of almonds continues to increase, but the number of honey bee colonies is not increasing. The National Research Council (2007) documented the decline in honey bee colonies in the United States and determined that pests, pathogens, and environmental stressors (e.g. pesticides, transportation, and poor nutrition) may be contributing to the decline.

While it had been known since 1975 that blue orchard bees can pollinate almonds efficiently, the cost to provide blue orchards bees was not competitive with costs for honey bees. However, with honey bee rental fees currently at or near $740 per ha, interest in alternative pollinators has increased greatly. Blue orchard bees have shown to be highly effective at pollinating almonds (Torchio, 1979), requiring far fewer bees (Bosch and Kemp, 2002) and possessing the ability to forage under worse conditions than the honey bee (Bosch and Kemp, 1999). In addition, a recent study showed that a combination of blue orchard bees and honey bees produces a greater proportion of fruit set than either species of bees alone (Brittain et al., 2013). The blue orchard bee is clearly an attractive candidate as an alternative or a supplement to honey bee pollination in almonds.

At this early stage of using blue orchard bees in almonds, progress has been hampered by the limited supply of bees and low reproduction levels in large-scale releases, but this may soon be alleviated with the emergence of a growing number of suppliers committed to improved processes and products, as well as reduced costs (see http://www.orchardbee.org). For blue orchard bee pollination to expand further in almonds, additional enabling technology is needed to enhance in-orchard female offspring production, as has been demonstrated in apples, cherries, and pears. A reasonable assumption is that each established female can produce two female offspring in an almond orchard (Bosch and Kemp, 2002). This means that at least one-half of the females released must establish nesting for replacement of the parent population. AgPollen LLC, in collaboration with the US Department of Agriculture's Agricultural Research Service (USDA-ARS), is working on attractants for nesting structures to improve establishment (S. Peterson et al., unpublished data). Methods for releasing bees and providing nesting substrates are improving, but they have not been standardized yet for large-scale applications.

Almonds can be a challenging crop to pollinate even for the honey bee. Almonds bloom for 3–4 weeks at best, with the middle 2 weeks at full bloom, so there is little time for the bees to nest and produce offspring with this resource alone. In addition, weeds and vegetation are typically treated with herbicides or mowed, so there is usually little else in bloom near the orchards. While honey bees can be fed syrup and protein patties before almond bloom, blue orchard bees must have flowers available upon release. Blue orchard bees may abscond if released too early without enough bloom. When blooming is finished, the bees are usually still faced with a lack of other food sources. Efforts are underway to find wildflowers that can be planted in or near almond orchards that can provide supplementary bloom before and after almond bloom (http://www.projectapism.org; www.xerces.org; S. Peterson, unpublished data) to improve bee health and reproduction.

Rain can stimulate brown rot blossom blight and other fungal diseases, so preventative fungicides are commonly applied while bees are in the orchard. Often, foliar fertilizers are added with these applications. More rain falls in the northern almond-growing regions, so this problem is more common in the northern and central regions of California. While most fungicides are not acutely toxic to bees, there may be sublethal effects on nesting females and developing larvae. Commercial fungicide applications have been observed to cause disruptions in foraging and nesting behavior in blue orchard bees (Ladurner et al., 2008) and, when tested in cages, caused significant disorientation after the application (D.R. Artz et al., unpublished data). In addition, trace amounts of insecticides may find their way into bee nests. A sample of California almond pollen collected by blue orchard bees in 2011 was analyzed for chemical residues, and two fungicides, a pyrethroid and three neonicotinoid insecticides, were detected (S. Peterson, unpublished data). None of these pesticides had been sprayed in that orchard. Pesticide drift or other type of movement (e.g. windborne dust particles) is clearly a possibility that can impact bee pollination and reproduction in an orchard. Because mud is used as partition between cells, pesticides in the soil are also of concern. More research is needed on the sublethal effects of pesticides (singly and in combinations) and foliar fertilizers on bees. Pollination providers need to communicate with growers to ensure that fungicides are applied at night and to make sure that the spray tank has been properly cleaned so that no insecticides residues are present.

Blue orchard bees normally fly a month later than almonds bloom. However, populations can be manipulated with warmer or fluctuating temperatures during the summer to complete development as soon as possible, which allows an earlier wintering starting date (Bosch et al., 2000). A battery-powered incubation box was developed that can advance emergence despite cool weather conditions in the orchard (Pitts-Singer et al., 2008). Since the bees require some incubation to emerge, time can be lost by emerging them in the orchard, so releasing emerged bees (like the "bleed-off" in alfalfa leafcutting bees) has been tried successfully in almonds (M. Allan and S. Peterson, unpublished data). Additionally, offering many smaller nest sites improves establishment compared to fewer sites per unit of area (Torchio, 1984b; Artz et al., 2013). Recent work showed that nest box density and the number of cavities within nest boxes influenced blue orchard bee retention in almonds (Artz et al., 2013). Females completed more nests in high-density, low-cavity nest boxes compared to low-density, high-cavity nest boxes. Nut yield was also significantly higher in orchard areas with high-density nest boxes. A strategy of providing both honey bees and blue orchard bees, each at roughly one-half the usual stocking density, allows the artificially fed honey bees to pollinate the very first blossoms. The blue orchard bees can be introduced once enough bloom is available to support the population (e.g. 5–10% bloom). If growers plant wildflowers that bloom ahead of almonds, honey bee colonies would likely be stronger and blue orchard bees could be released earlier.

The recommended stocking density for blue orchard bees in almonds is 740 nesting females per hectare (Bosch and Kemp, 2001). To achieve this rate requires estimating losses from mortality and lack of establishment, which at present can be 50% or more in large-scale applications. Therefore, it is usually best to stock at least twice the recommended rate. If honey bees will also be present, then a reduced rate can be justified.

As is the case with the other pollinator species discussed in this chapter, the blue orchard bee is subject to a guild of parasites, predators, and diseases. Lessons learned with the alfalfa leafcutting bee are relevant in keeping managed blue orchard bees healthy. The fungal disease chalkbrood (*Ascosphaera torchioi* Youssef and McManus) has a similar life cycle in the blue orchard bee compared to chalkbrood of alfalfa leafcutting bees (Bosch and Kemp, 2001). Chalkbrood rates in blue orchard bees over a 15-year period in Utah ranged from 0.4% to 2% (Torchio, 1992). Allowing bees to emerge from natal nests could allow chalkbrood to increase in a population just as in alfalfa leafcutting bees. Loose cell management allows for the visual inspection and removal of infected cadavers. So far, this disease has remained at low levels in managed populations.

Several hymenopteran parasitoids and cleptoparasitoids attack the blue orchard bee. *Monodontomerus obscurus* Westwood and *M. montivagus* Ashmead (Hymenoptera: Torymidae) are commonly associated with blue orchard bees (Bosch and Kemp, 2001). Their ovipositor is about 1 mm in length, so when using paper tubes, walls of at least 1 mm thickness will prevent oviposition (Bosch and Kemp, 2001). Adults emerge from blue orchard bee nests during the spring and summer, and adults can be attracted to a black-light water trap and drowned. A study on the European species *M. aeneus* (Fonscolombe) showed that the odor cues of *Osmia* frass and cocoon are utilized by these parasitoids to find their hosts (Filella et al., 2011). Perhaps such cues can be used to trap and kill these pests.

The sapygid wasp, *Sapyga* spp. (Hymenoptera: Sapygidae), is a cleptoparasite and is commonly associated with blue orchard bees. Cleptoparasites lay their eggs in open bee nests, and the larvae consume the pollen provision and kill the developing blue orchard bee larvae. These wasps have a single generation per year and emerge shortly after or with the female blue orchard bees. To control this pest, nests must be visually inspected and the *Sapyga* cocoons manually destroyed. While *Sapyga* cocoons look somewhat different from blue orchard bee cocoons, they still can be overlooked when one is working with large numbers of cocoons. Careful inspections are necessary to eliminate this species from a population.

Stelis montana Cresson (Hymenoptera: Megachilidae) is a cuckoo bee and, like the sapygid wasp, is a cleptoparasite. Cuckoo bees lack the pollen-collecting scopae found on other Megachilid females and can parasitize several *Osmia* species. *S. montana* has a single generation per year and emerges slightly later than blue orchard bee females. It looks similar to the blue orchard bee, but is smaller in size and has a proportionately smaller head. *S. montana* cocoons are smaller and lighter colored, and they have long, coiled frass that distinguishes them from blue orchard bee cocoons. Visual inspection of cocoons is the best way to remove this species.

The hairy fingered mite, *Chaetodactylus krombeini* Baker (Acari: Chaetodactylidae), is a cleptoparasite with multiple generations per year. They attach themselves to emerging bees and are transferred to new nests in this manner. Infested cocoons can be removed individually, or cocoons can be washed in a bleach solution to remove them. Hairy fingered mites are most commonly a problem in cool, humid environments such as in Oregon and Washington. A similar species of mite on *Osmia cornifrons* (Radoszkowski) was controlled using warm temperatures (30–40 °C) in Japan (Yamada, 1990). A protocol has been developed to control hairy fingered mites on blue orchard bees (E. Sugden, personal communication). For best results, nests are brought in from the field soon after completion, or a little before, and the temperature is increased to 27–32 °C and humidity is reduced to 30–40% for a few days. Nighttime temperatures are then allowed to cycle to ambient until the bees pupate. Tumbling cocoons with coarse sand can also be effective (J. Watts, personal communication). White et al. (2009) showed that the fumigants formic acid and wintergreen oil have the potential to control *C. krombeini* on *Osmia cornifrons* (Radoszkowski). Although we have highlighted the main blue orchard bee pests, several others are found associated with blue orchard bees; for complete descriptions with photographs, see Bosch and Kemp (2001).

Concerns over declining honey bee populations have stimulated interest in the blue orchard bee, and it is rapidly becoming a commercial pollinator of important fruit and nut crops. We expect that managed populations of this bee will grow greatly in the coming years. For this species to gain more widespread use, further research is needed on methods to minimize prenesting dispersal, the effects of sublethal doses of pesticides, the diversification of floral resources in orchards, and methods to advance large-scale management processes.

19.6. OTHER SOLITARY BEES OF INTEREST FOR POLLINATION

A number of other bees besides those discussed in this chapter are used in pollination or have potential for mass rearing toward commercial pollination. For example, studies have shown that a North American species, *Megachile pugnata* Say, is an excellent pollinator of sunflowers (Parker and Frohlich, 1983). *Osmia sanrafaelae* Parker is a good pollinator of alfalfa seed and will trip 44% of alfalfa flowers when allowed single visits (Cane, 2002). *Osmia bruneri* Cockerell is native to western North America, nests in preexisting holes, and will use artificial domiciles (Frohlich, 1983). This species has the potential to be a commercial legume pollinator (Cane, 2008b). *Osmia ribifloris* Cockerell is found throughout the western United States and has the potential to be a commercial pollinator of blueberries (Torchio, 1990; Stubbs et al., 1994). *Osmia aglaia* Sandhouse is an excellent candidate in the western United States for commercial pollination of raspberry and blackberry crops (Cane, 2005a, b).

Outside of North America, other solitary bees are being used for commercial pollination as well. *Osmia cornuta* (Latreille) is a European species, like *O. lignaria*, that has been shown to be commercially manageable for almond and pear pollination in southern Europe (Bosch, 1994a, b; Krunic et al., 1991;

Maccagnani et al., 2007). A highly effective pollinator of almond flowers, the body of *O. cornuta* contacted the stigma 98.7% of the time compared to 67.3% of the time with the honey bee (Bosch and Blas, 1994). Single-visitation fruit set was also higher with *O. cornuta* compared to honey bees.

In Japan, *O. cornifrons* has been commercially managed for apple pollination for about 40 years (Maeta and Kitamura, 1974), and it is a more effective apple pollinator than the honey bee (Maeta and Kitamura, 1981). This species, known as the mame-ko bee or Japanese hornfaced bee, was introduced to the United States in 1977 by the USDA (Parker et al., 1987). This species can double or triple its population each year (Adams, 2004). Thus far, the Japanese hornfaced bee is managed in the eastern and north central United States, but it has not been made available on a large commercial scale.

Osmia rufa (L.), the red mason bee, is another European species that can be used to pollinate seed crops in greenhouses (van der Steen and de Ruijter, 1991), and it has been shown to gather *Brassica* pollen and increase its population on the crop in Poland (Teper and Biliński, 2009). Also, the red mason bee is an effective pollinator of strawberry (Wilkaniec and Radajewska, 1997) and is currently being managed to pollinate cherries, apples, and pears in Great Britain (C. Whittles, personal communication).

19.7. CONCLUDING REMARKS

The honey bee is unmatched in its abilities to pollinate a wide variety of crops, adapt to a wide variety of climates, and provide services year-round in warm climates. However, recent history has shown that reliance on this single species for crop pollination has its risks. For the past 30 years, the number of honey bee colonies has been shrinking in the United States due to a suite of factors, including parasites, diseases, poor nutrition, pesticide contamination, and colony collapse disorder (National Resource Council, 2007; Mullin et al., 2010; Ratneiks and Carreck, 2010), while the acreage of bee-dependent crops has been growing. Honey bees will always be an important part of agriculture, but it is becoming increasingly apparent that there are benefits in managing some other bee species to share in the job of crop pollination. The relatively recent adoption of the alfalfa leafcutting bee and alkali bee as commercial pollinators for alfalfa seed, and the burgeoning commercial use of the blue orchard bee in almonds, apples, and cherries, has demonstrated that solitary bees can be managed in an agricultural setting and provide sustainable pollination services.

Species that make their nests above ground, in preexisting cavities, make particularly good candidates for use in agricultural settings. However, soil-nesting species are also possible to manage. Solitary bees have shorter activity periods compared to honey bees, so adult bees do not need to be managed year-round. With blue orchard bees and alfalfa leafcutting bees, the population is held in storage, preferably under controlled conditions, for a large part of the year. With the alkali bee, developing bees are underground most of the year.

These bees are highly efficient pollinators during their adult stage, and fewer individuals are needed to pollinate a given area compared to honey bees. With few management steps, short activity periods, and fewer bees needed, these bees have minimal negative impacts on the environment. As our scientific and practical knowledge of these species expands, solitary bees will become increasingly important commercial pollinators.

ACKNOWLEDGMENTS

We are grateful to Theresa Pitts-Singer, Jordi Bosch, and William Kemp for their thoughtful reviews of this manuscript. We thank Miles Wendell, Dwight Nahuliak, Ron Bitner, Matt Allan, Evan Sugden, Jim Watts, Chris Whittles, James Cane, Gordon Wardell, and David Moreland for helpful discussions on managing solitary bees.

REFERENCES

Adams, L., 2004. Basic Biology and Management of the Japanese Hornfaced Bee. Mid-Atlantic Apicultural Research & Extension Consortium (MAAREC) Publication 5.5. https://agdev.anr.udel.edu/maarec/wp-content/uploads/2010/03/japhornface.pdf.

Aizen, M.A., Garibaldi, L.A., Cunningham, S.A., Klein, A.M., 2008. Long-term global trends in crop yield and production reveal no current pollination shortage but increasing pollinator dependency. Curr. Biol. 18 (20), 1572–1575.

Artz, D.R., Allan, M.J., Wardell, G.I., Pitts-Singer, T.L., 2013. Nesting site density and distribution affects *Osmia lignaria* (Hymenoptera: Megachilidae) reproductive success and almond yield in a commercial orchard. Insect Conserv. Diver. doi:10.1111/icad.12026.

Baird, C.R., Bitner, R.M., 1991. Loose Cell Management of Alfalfa Leafcutting Bees in Idaho. University of Idaho Current Information Series No. 588.

Bosch, J., 1994a. Improvement of field management of *Osmia cornuta* (Latreille) (Hymenoptera, Megachilidae) to pollinate almond. Apidologie 25, 71–83.

Bosch, J., 1994b. *Osmia cornuta* Latr. (Hym., Megachilidae) as a potential pollinator in almond orchards: releasing methods and nest hole-length. J. Appl. Entomol. 117, 151–157.

Bosch, J., Blas, M., 1994. Foraging behaviour and pollinating efficiency of *Osmia cornuta* and *Apis mellifera* on almond (Hymenoptera, Megachilidae and Apidae). Appl. Entomol. Zool. 29 (1), 1–9.

Bosch, J., Kemp, W.P., 1999. Exceptional cherry production in an orchard pollinated with blue orchard bees. Bee World 80 (4), 163–173.

Bosch, J., Kemp, W.P., 2000. Development and emergence of the orchard pollinator *Osmia lignaria* (Hymenoptera: Megachilidae). Environ. Entomol. 29 (1), 8–13.

Bosch, J., Kemp, W.P., Peterson, S.S., 2000. Management of *Osmia lignaria* (Hymenoptera: Megachilidae) populations for almond pollination: methods to advance bee emergence. Environ. Entomol. 29 (5), 874–883.

Bosch, J., Kemp, W.P., 2001. How to Manage the Blue Orchard Bee as an Orchard Pollinator. Sustainable Agriculture Network, Beltsville, Maryland.

Bosch, J., Kemp, W.P., 2002. Developing and establishing bee species as crop pollinators: the example of *Osmia* spp. and fruit trees. Bull. Entomol. Res. 92, 3–16.

Bosch, J., Kemp, W.P., 2003. Effect of wintering duration and temperature on survival and emergence time in males of the orchard pollinator *Osmia lignaria* (Hymenoptera: Megachilidae). Environ. Entomol. 32 (4), 711–716.

Bosch, J., Kemp, W.P., 2005. Alfalfa leafcutting bee populaiton dynamics, flower availability, and pollination rates in two oregon alfalfa fields. J. Econ. Entomol. 98 (4), 1077–1086.

Bosch, J., Kemp, W.P., Trostle, G.E., 2006. Bee population returns and cherry yields in an orchard pollinated with *Osmia lignaria* (Hymenoptera: Megachilidae). J. Econ. Entomol. 99 (2), 408–413.

Bosch, J., Sgolastra, F., Kemp, W.P., 2008. Life cycle ecophysiology of *Osmia* mason bees used as crop pollinators. In: James, R.R., Pitts-Singer, T.L. (Eds.), Bee Pollination in Agricultural Ecosystems, Oxford University Press, New York, pp. 83–104.

Bosch, J., Sgolastra, F., Kemp, W.P., 2010. Timing of eclosion affects diapause development, fat body consumption and longevity in *Osmia lignaria*, a univoltine, adult-wintering solitary bee. J. Insect Physiol. 56, 1949–1957.

Brittain, C., Williams, N., Kremen, C., Klein, A.M., 2013. Synergistic effects of non-*Apis* bees and honey bees for pollination services. Proc. R. Soc. B. 280, 20122767.

Calderone, N.W., 2012. Insect pollinated crops, insect pollinators and US agriculture: trend analysis of aggregate data for the period 1992–2009. PLoS One 7 (5), e37235.

Cane, J.H., 2002. Pollinating bees (Hymenoptera: Apiformes) of U.S. alfalfa compared for rates of pod and seed set. J. Econ. Entomol. 95 (1), 22–27.

Cane, J.H., 2003. Exotic nonsocial bees (Hymenoptera: Apiformes) in North America: ecological implications. In: Strickler, K., Cane, J.H. (Eds.), For Nonnative Crops, Whence Pollinators of the Future?, Entomological Society of America, Lanham, Maryland, pp. 113–126.

Cane, J.H., 2005a. Pollination potential of the bee *Osmia aglaia* for cultivated raspberries and blackberries (*Rubus*: Rosaceae). HortScience 40 (6), 1705–1708.

Cane, J.H., 2005b. An effective, manageable bee for pollination of *Rubus* bramble fruits, *Osmia aglaia*. Acta Hortic. 777, 459–464.

Cane, J.H., 2008a. A native ground-nesting bee (*Nomia melanderi*) sustainably managed to pollinate alfalfa across an intensively agricultural landscape. Apidologie 39, 315–323.

Cane, J.H., 2008b. Pollinating bees crucial to farming wildflower seed for U.S. habitat restoration. In: James, R.R., Pitts-Singer, T.L. (Eds.), Bee Pollination in Agricultural Ecosystems, Oxford University Press, New York, pp. 48–64.

Cane, J.H., Griswold, T., Parker, F.D., 2007. Substrates and materials used for nesting by North American *Osmia* bees (Hymenoptera: Apiformes: Megachilidae). Ann. Entomol. Soc. Am. 100 (3), 350–358.

Census of Agriculture, 2011. Statistics Canada. www.statcan.gc.ca, (accessed 05.06.12).

Filella, I., Bosch, J., Llusià, J., Seco, R., Peñuelas, J., 2011. The role of frass and cocoon volatiles in host location by *Monodontomerus aeneus*, a parasitoid of megachilid solitary bees. Environ. Entomol. 40 (1), 126–131.

Free, J.B., 1993. Insect Pollination of Crops, second ed. Academic Press Limited, London, pp. 684.

Frohlich, D.R., 1983. On the nesting biology of *Osmia (Chenosmia) bruneri* (Hymenoptera: Megachilidae). J. Kans. Entomol. Soc. 56 (2), 123–130.

Gibson, G.D., Olivas, N.K., Salm, P., 2007. Lettuce breeding method. U.S. Patent 7,569,743.

Goerzen, D.W., 1992. Paraformaldehyde Fumigation of Alfalfa Leafcutting Bee Nest Material. Saskatchewan Agricultural Development Fund. ISBN:0-88656-556-1.

Guédot, C., Pitts-Singer, T.L., Buckner, J.S., Bosch, J., Kemp, W.P., 2006. Olfactory cues and nest recognition in the solitary bee *Osmia lignaria*. Physio. Entomol. 31, 110–119.

Hill, B.D., Richards, K.W., Schaalje, G.B., 1984. Use of dichlorvos resin strips to reduce parasitism of alfalfa leafcutter bee (Hymenoptera: Megachilidae) cocoons during incubation. J. Econ. Entomol. 77 (5), 1307–1312.

Hobbs, G.A., Richards, K.W., 1976. Selection for a univoltine strain of *Megachile (Eutricharaea) pacifica* (Hymenoptera: Megachilidae). Can. Entomol. 108, 165–167.

Horne, M., 1995. Leaf area and toughness: effects on nesting material preferences of *Megachile rotundata* (Hymenoptera: Megachilidae). Ann. Entomol. Soc. Am. 88 (6), 868–875.

Jahns, T.R., Jolliff, G.D., 1991. Survival rate and reproductive success of *Osmia lignaria propinqua* Cresson (Hymenoptera: Megachilidae) in caged meadowfoam, *Limnanthes alba* Benth. (Limnanthaceae). J. Kans. Entomol. Soc. 64 (1), 95–106.

James, R.R., 2011. Chalkbrood transmission in the alfalfa leafcutting bee: the impact of disinfecting bee cocoons in loose cell management systems. Environ. Entomol. 40 (4), 782–787.

Johansen, C., Mayer, D., Stanford, A., Kious, C., 1982. Alkali Bees: Their Biology and Management of Alfalfa Seed Production in the Pacific Northwest. Pacific Northwest Extension Publication 0155.

Kemp, W.P., Bosch, J., 2000. Development and emergence of the alfalfa pollinator *Megachile rotundata* (Hymenoptera: Megachilidae). Ann. Entomol. Soc. Am. 93 (4), 904–911.

Kemp, W.P., Bosch, J., 2005. Effect of temperature on *Osmia lignaria* (Hymenoptera: Megachilidae) prepupa–adult development, survival, and emergence. J. Econ. Entomol. 98 (6), 1917–1923.

Kemp, W.P., Bosch, J., Dennis, B., 2004. Oxygen consumption during the life cycles of the prepupa-wintering *Megachile rotundata* and the adult-wintering bee *Osmia lignaria* (Hymenoptera: Megachilidae). Ann. Entomol. Soc. Am. 97 (1), 161–170.

Kish, L.P., Stephen, W.P., 1991. Chalkbrood disease. Alfalfa Seed Production and Pest Management, Western Regional Extension Publication 12.

Klostermeyer, E.C., Mech Jr., S.J., Rasmussen, W.B., 1973. Sex and weight of *Megachile rotundata* (Hymenoptera: Megachilidae) progeny associated with provision weight. J. Kans. Entomol. Soc. 46 (4), 536–548.

Krunic, M., Brajkovic, M.M., Mihajlovic, L.S., 1991. Management and utilization of *Osmia cornuta* Latr. for orchard pollination in Yugoslavia. Acta Horic. 288, 190–193.

Ladurner, E., Bosch, J., Kemp, W.P., Maini, S., 2008. Foraging and nesting behavior of *Osmia lignaria* (Hymenoptera: Megachilidae) in the presence of fungicides: cage studies. J. Econ. Entomol. 101 (3), 647–653.

Levin, M.D., 1957. Artificial nesting burrows for *Osmia lignaria* Say. J. Econ. Entomol. 50 (4), 506–507.

Maccagnani, B., Burgio, G., Stanisavljevic, L.Z., Maini, S., 2007. *Osmia cornuta* management in pear orchards. Bull. Insectology 60 (1), 77–82.

Maeta, Y., Kitamura, T., 1974. How to Manage the Mame-ko Bee (*Osmia Cornifrons* Radoszkowski) for Pollination of Fruit Crops. Ask Co Ltd.

Maeta, Y., Kitamura, T., 1981. Pollinating efficiency by *Osmia cornifrons* (Radoszkowski) in relation to required number of nesting bees for economic fruit production. Honeybee Sci. 2, 65–72.

Mayer, D.F., Johansen, C.A., 2003. The rise and decline of *Nomia melanderi* (Hymenoptera: Halictidae) as a commercial pollinator for alfalfa seed. In: Strickler, K., Cane, J.H. (Eds.), For Non-native Crops, Whence Pollinators of the Future?, Entomological Society of America, Lanham, Maryland, pp. 139–150.

Mayer, D.F., Lunden, J.D., Kious, C.W., 1988. Effects of dipping alfalfa leaf-cutting bee nesting materials on chalkbrood disease. Appl. Agric. Res. 3, 167–169.

McGregor, S.E., 1976. Insect Pollination of Cultivated Crop Plants. Agriculture Handbook No. 496 USDA Agricultural Research Service.

Michener, C.D., 2000. The Bees of the World. The Johns Hopkins University Press, Baltimore and London, pp. 913.

Mullin, C.A., Frazier, M., Frazier, J.L., Ashcraft, S., Simonds, R., et al., 2010. High levels of miticides and agrochemicals in North American apiaries: implications for honey bee health. PLoS One 5 (3), e9754. http://dx.doi.org/10.1371/journal.pone.0009754.

National Research Council, 2007. Status of Pollinators in North America. The National Academies Press, Washington, DC, pp. 307.

Olmstead, A.L., Wooten, D.B., 1987. Bee pollination and productivity growth: the case of alfalfa. Am. J. Agr. Econ. 69 (1), 56–63.

O'Neill, K.M., Pearce, A.M., O'Neill, R.P., Miller, R.S., 2010. Offspring size and sex ratio variation in a feral population of alfalfa leafcutting bees (Hymenoptera: Megachilidae). Ann. Entomol. Soc. Am. 103, 775–784.

O'Toole, C., 2008. Forward. In: James, R.R., Pitts-Singer, T.L. (Eds.), Bee Pollination in Agricultural Ecosystems. Oxford University Press, New York, pp. v–vii.

Pankiw, P., Siemens, B., 1974. Management of *Megachile rotundata* in northwestern Canada for population increase. Can. Entomol. 106 (9), 1003–1008.

Peterson, S.S., Baird, C.R., Bitner, R.M., 1992. Current status of the alfalfa leafcutting bee, *Megachile rotundata*, as a pollinator of alfalfa seed. Bee Sci. 2 (3), 135–142.

Parker, F.D., Batra, S.W.T., Tepedino, V.J., 1987. New pollinators for our crops. Ag. Zool. Rev. 2, 279–304.

Parker, F.D., Frohlich, D.R., 1983. Hybrid sunflower pollination by a manageable composite specialist: the sunflower leafcutter bee (Hymenoptera: Megachilidae). Environ. Entomol. 12 (2), 576–581.

Pitts-Singer, T.L., 2007. Olfactory response of megachilid bees, *Osmia lignaria*, *Megachile rotundata*, and *M. pugnata*, to individual cues from old nest cavities. Environ. Entomol. 36 (2), 402–408.

Pitts-Singer, T.L., Bosch, J., 2010. Nest establishment, pollination efficiency, and reproductive success of *Megachile rotundata* (Hymenoptera: Megachilidae) in relation to resource availability in field enclosures. Environ. Entomol. 39 (1), 149–158.

Pitts-Singer, T.L., Bosch, J., Kemp, W.P., Trostle, G.E., 2008. Field use of an incubation box for improved emergence timing of *Osmia lignaria* populations used for orchard pollination. Apidologie 39 (2), 235–246.

Pitts-Singer, T., Buckner, J.S., Freeman, T.P., Guédot, C.N., 2012. Structural examination of the Dufour's gland of the cavity-nesting bees *Osmia lignaria* Say and *Megachile rotundata* (Fabricius) (Hymenoptera: Megachilidae). Ann. Entomol. Soc. Am. 105 (1), 103–110.

Pitts-Singer, T.L., James, R.R., 2005. Emergence success and sex ratio of commercial alfalfa leafcutting bees, *Megachile rotundata* Say, from the United States and Canada. J. Econ. Entomol. 98 (6), 1785–1790.

Pitts-Singer, T.L., James, R.R., 2008. Do weather conditions correlate with findings in failed, provision-filled nest cells of *Megachile rotundata* (Hymenoptera: Megachilidae) in western North America? J. Econ. Entomol. 101 (3), 674–685.

Pitts-Singer, T.L., James, R.R., 2009. Prewinter management affects *Megachile rotundata* (Hymenoptera: Megachilidae) prepupal physiology and adult emergence and survival. J. Econ. Entomol. 102, 1407–1416.

Ratnieks, F.L.W., Carreck, N.L., 2010. Clarity on honey bee collapse? Science 327, 152–153.

Richards, K.W., 1978. Comparisons of nesting materials used for the alfalfa leafcutter bee, *Megachile pacifica* (Hymenoptera: Megachilidae). Can. Entomol. 110 (8), 841–846.

Richards, K.W., 1984a. Alfalfa Leafcutter Bee Management in Western Canada. Agriculture Canada Publication No. 1495/E. Agriculture Canada, Ottawa, Ontario.

Richards, K.W., 1984b. Comparison of tumblers used to remove debris from cells of the alfalfa leafcutter bee, *Megachile rotundata* (Hymenoptera: Megachilidae). Can. Entomol. 116 (5), 719–723.

Richards, K.W., Whitfield, G.H., 1988. Emergence and survival of leafcutter bees, *Megachile rotundata*, held at constant incubation temperatures (Hymenoptera: Megachilidae). J. Apic. Res. 27 (3), 197–204.

Rinehart, J.P., Yocum, G.D., West, M., Kemp, W.P., 2011. A fluctuating thermal regime improves survival of cold-mediated delayed emergence in developing *Megachile rotundata* (Hymenoptera: Megachilidae). J. Econ. Entomol. 104 (4), 1162–1166.

Roulston, T.H., Goodell, K., 2011. The role of resources and risks in regulating wild bee populations. Ann. Rev. Entomol. 56, 293–312.

Sgolastra, F., Bosch, J., Molowny-Horas, R., Maini, S., Kemp, W.P., 2010. Effect of temperature regime on diapause intensity in an adult-wintering Hymenopteran with obligate diapause. J. Insect Physiol. 56 (2), 185–194.

Sgolastra, F., Kemp, W.P., Buckner, J.S., Pitts-Singer, T.L., Maini, S., Bosch, J., 2011. The long summer: pre-wintering temperatures affect metabolic expenditure and winter survival in a solitary bee. J. Insect Physiol. 57 (12), 1651–1659.

Soroka, J.J., Goerzen, D.W., Falk, K.C., Bett, K.E., 2001. Alfalfa leafcutting bee (Hymenoptera: Megachilidae) pollination of oilseed rape (*Brassica napus* L.) under isolation tents for hybrid seed production. Can. J. Plant Sci. 81, 199–204.

Stephen, W.P., 1955. Alfalfa pollination in Manitoba. J. Econ. Entomol. 48 (5), 543–548.

Stephen, W.P., 1959. Maintaining alkali bees for alfalfa seed production. Oregon State College Agric. Exp. Sta. Bull. 568, pp. 23.

Stephen, W.P., 1960. Artificial bee beds for the propagation of the alkali bee, *Nomia melanderi*. J. Econ. Entomol. 53 (6), 1025–1030.

Stephen, W.P., 1961. Artificial nesting sites for the propagation of the leaf-cutter bee, *Megachile (Eutricharaea) rotundata*, for alfalfa pollination. J. Econ. Entomol. 54 (5), 989–993.

Stephen, W.P., 1965. Temperature effects on the development and multiple generations in the alkali bee, *Nomia melanderi* Cockerell. Ent. Exp. Appl. 8 (3), 228–240.

Stephen, W.P., 1981. The Design and Function of Field Domiciles and Incubators for Leafcutting Bee Management (*Megachile Rotundata* [Fabricius]). Oregon State University Agricultural Experiment Station, Bulletin No. 654, Corvallis, Oregon.

Stephen, W.P., 2003. Solitary bees in North American agriculture: a perspective. In: Strickler, K., Cane, J.H. (Eds.), For Nonnative Crops, Whence Pollinators of the Future?, Entomological Society of America, Lanham, Maryland, pp. 41–66.

Stephen, W.P., Osgood, C.E., 1965a. Influence of tunnel size and nesting medium on sex ratios in a leaf-cutter bee, *Megachile rotundata*. J. Econ. Entomol. 58 (5), 965–968.

Stephen, W.P., Osgood, C.E., 1965b. The induction of emergence in the leaf-cutter bee *Megachile rotundata*, an important pollinator of alfalfa. J. Econ. Entomol. 58 (2), 284–286.

Stephen, W.P., Torchio, P.F., 1961. Biological notes on the leafcutter bee, *Megachile (Eutricharaea) rotundata* (Fabricius). Pan-Pac. Entomol. 37, 85–93.

Stephen, W.P., Undurraga, J.M., 1976. X-radiography, an analytical tool in population studies of the leafcutter bee, *Megachile pacifica*. J. Apic. Res. 15 (2), 81–87.

Strickler, K., Freitas, S., 1999. Interactions between floral resources and bees (Hymenoptera: Megachilidae) in commercial alfalfa seed fields. Environ. Entomol. 28 (2), 178–187.

Stubbs, C.S., Drummond, F.A., Osgood, E.A., 1994. *Osmia ribifloris biedermannii* and *Megachile rotundata* (Hymenoptera: Megachilidae) introduced into the lowbush blueberry agroecosystem in Maine. J. Kans. Entomol. Soc. 67 (2), 173–185.

Sumner, D.A., Boriss, H., 2006. Bee-conomics and the Leap in Pollination Fees. Agricultural and Resource Economics Update. vol. 9. University of California, Giannini Foundation of Agricultural Economics No. 3, January/February 2006.

Tepedino, V.J., 1997. A comparison of the alfalfa leafcutting bee (*Megachile rotundata*) and the honey bee (*Apis mellifera*) as pollinators for hybrid carrot seed in field cages: Seventh International Symposium on Pollination. Acta Hortic. 437, 457–461.

Tepedino, V.J., Parker, F.D., 1986. Effect of rearing temperature of mortality, second-generation emergence, and size of adult in *Megachile rotundata* (Hymenoptera: Megachilidae). J. Econ. Entomol. 79 (4), 974–977.

Tepedino, V.J., Parker, F.D., 1988. Alternation of sex ratio in a partially bivoltine bee, *Megachile rotundata* (Hymenoptera: Megachilidae). Ann. Entomol. Soc. Am. 81 (3), 467–476.

Teper, D., Biliński, M., 2009. Red mason bee (*Osmia rufa* L.) as a pollinator of rape plantations. J. Apic. Sci. 53 (2), 115–120.

Thorp, R.W., 1996. Bee Management for Pollination in Almond Production Manual. University of California, Division of Agriculture and Natural Resources, Publication 3364.

Torchio, P.F., 1976. Use of *Osmia lignaria* Say (Hymenoptera: Apoidea, Megachilidae) as a pollinator in an apple and prune orchard. J. Kans. Entomol. Soc. 49 (4), 475–482.

Torchio, P.F., 1979. Use of *Osmia lignaria* Say as a pollinator of caged almond in California. Md. Agric. Exp. Sta. Spec. Misc. Publ. 1, 285–293.

Torchio, P.F., 1981a. Field experiments with *Osmia lignaria propinqua* Cresson as a pollinator in almond orchards: I 1975 studies (Hymenoptera: Megachilidae). J. Kans. Entomol. Soc. 54 (4), 815–823.

Torchio, P.F., 1981b. Field experiments with *Osmia lignaria propinqua* Cresson as a pollinator in almond orchards: II 1976 studies (Hymenoptera: Megachilidae). J. Kans. Entomol. Soc. 54 (4), 824–836.

Torchio, P.F., 1982a. Field experiments with *Osmia lignaria propinqua* Cresson as a pollinator in almond orchards: III 1977 studies (Hymenoptera: Megachilidae). J. Kans. Entomol. Soc. 55 (1), 101–116.

Torchio, P.F., 1982b. Field experiments with the pollinator species, *Osmia lignaria propinqua* Cresson, in apple orchards: I, 1975 studies (Hymenoptera: Megachilidae). J. Kans. Entomol. Soc. 55 (1), 136–144.

Torchio, P.F., 1982c. Field experiments with the pollinator species, *Osmia lignaria propinqua* Cresson, in apple orchards: II, 1976 studies (Hymenoptera: Megachilidae). J. Kans. Entomol. Soc. 55 (4), 759–778.

Torchio, P.F., 1984a. Field experiments with the pollinator species, *Osmia lignaria propinqua* Cresson, in apple orchards: III, 1977 studies (Hymenoptera: Megachilidae). J. Kans. Entomol. Soc. 57 (3), 517–521.

Torchio, P.F., 1984b. Field experiments with the pollinator species, *Osmia lignaria propinqua* Cresson, in apple orchards: IV, 1978 studies (Hymenoptera: Megachilidae). J. Kans. Entomol. Soc. 57 (4), 689–694.

Torchio, P.F., 1985. Field experiments with the pollinator species, *Osmia lignaria propinqua* Cresson in apple orchards: V, (1979–1980), method of introducing bees, nesting success, seed counts, fruit yields (Hymenoptera: Megachilidae). J. Kans. Entomol. Soc. 58 (3), 448–464.

Torchio, P.F., 1989. In-nest biologies and development of immature stages of three *Osmia* species (Hymenoptera: Megachilidae). Ann. Entomol. Soc. Am. 82 (5), 599–615.

Torchio, P.F., 1990. *Osmia ribifloris*, a native bee species developed as a commercially managed pollinator of highbush blueberry (Hymenoptera: Megachilidae). J. Kans. Entomol. Soc. 63 (3), 427–436.

Torchio, P.F., 1992. Effects of spore dosage and temperature on pathogenic expressions of chalkbrood syndrome caused by *Ascosphaera torchioi* within larvae of *Osmia lignaria propinqua* (Hymenoptera: Megachilidae). Environ. Entomol. 21 (5), 1086–1091.

Torchio, P.F., 2003. Development of *Osmia lignaria* (Hymenoptera: Megachilidae) as a managed pollinator of apple and almond crops: a case history. In: Strickler, K., Cane, J.H. (Eds.), For

Nonnative Crops, Whence Pollinators of the Future?, Entomological Society of America, Lanham, Maryland, pp. 67–84.

Trostle, G., Torchio, P.F., 1994. Comparative nesting behavior and immature development of *Megachile rotundata* (Fabricius) and *Megachile apicalis* Spinola (Hymenoptera: Megachilidae). J. Kans. Entomol. Soc. 67 (1), 53–72.

USDA National Agricultural Statistics Service, 2011. 2011 California Almond Acreage Report. California Field Office. U.S. Government Printing Office, Washington, DC.

van der Steen, J., de Ruijter, A., 1991. The management of *Osmia rufa* L. for pollination of seed crops in greenhouses. Proc. Exp. Appl. Entomol. 2, 137–141.

Vandenberg, J.D., Fichter, B.L., Stephen, W.P., 1980. Spore load of *Ascosphaera* species on emerging adults of the alfalfa leafcutting bee, *Megachile rotundata*. Appl. Environ. Microbiol. 39 (3), 650–655.

White, J.B., Park, Y., West, T.P., Tobin, P.C., 2009. Assessment of potential fumigants to control *Chaetodactylus krombeini* (Acari: Chaetodactylidae) associated with *Osmia cornifrons* (Hymenoptera: Megachilidae). J. Econ. Entomol. 102 (6), 2090–2095.

Whitfield, G.H., Richards, K.W., 1985. Influence of temperature on survival and rate of development of *Pteromalus venustus* (Hymenoptera: Pteromalidae), a parasite of the alfalfa leafcutter bee (Hymenoptera: Megachilidae). Can. Entomol. 117 (7), 811–818.

Whitfield, G.H., Richards, K.W., 1992. Temperature-dependent development and survival of immature stages of the alfalfa leafcutter bee, *Megachile rotundata* (Hymenoptera: Megachilidae). Apidologie 23, 11–23.

Wichelns, D., Weaver, T.F., Brooks, P.M., 1992. Estimating the impact of alkali bees on the yield and acreage of alfalfa seed. J. Prod. Agric. 5 (4), 512–518.

Wilkaniec, Z., Radajewska, B., 1997. Solitary bee *Osmia rufa* L. (Apoidea, Megachilidae) as pollinator of strawberry cultivated in an unheated plastic tunnel. Acta Hortic. 439 (1), 489–493.

Yamada, M., 1990. Control of Chaetodactylus mite, *Chaetodactylus nipponicus* Kurosa, an important mortality agent of hornfaced Osmia bee *Osmia cornifrons* Radoszkowski. Bull. Aomori Apple Exp. Stn. 26, 39–77.

This page intentionally left blank

Chapter 20

Current and Potential Benefits of Mass Earthworm Culture

Christopher N. Lowe,[1] Kevin R. Butt[1] and Rhonda L. Sherman[2]
[1]*School of Built and Natural Environment, University of Central Lancashire, Preston, UK,*
[2]*Biological and Agricultural Engineering, North Carolina State University, Raleigh, NC, USA*

20.1. INTRODUCTION

Earthworms form a major component of the fauna of fertile soils. The importance of earthworms in improving crop yields and maintaining soil fertility has been recognized for millennia. In ancient Greece, Aristotle referred to earthworms as "the intestines of the earth," and during the reign of Cleopatra in early Egypt, the earthworm was considered a sacred animal, and removing them from the soil was strictly forbidden (Brown et al., 2003). More recently, Charles Darwin was the first to recognize the importance of earthworms in soil formation, referring to "these lowly organized creatures" as "nature's plough" (Darwin, 1881).

Today, earthworms are considered to be "ecosystem engineers" (Lavelle et al., 1997) that are able to regulate natural processes to an extent that overrides organisms in other functional categories. Earthworms play a major role in improving the aeration, drainage, and water-holding capacity of the soil. Through their actions, organic material, such as leaves and crop residues, are incorporated into the mineral soil, increasing availability of nutrients for plants and other soil fauna. Aside from improving soil fertility, earthworms are also a source of protein for many invertebrates like carabid beetles (Harper et al., 2005) and vertebrate species such as badgers (Kruuk, 1978) and moles (Raw, 1966), and they even form a part of the diet of some tropical indigenous human populations (Paoletti et al., 2003). Earthworms are also utilized by humans as a food source for farm animals, including chickens and pigs (Kostecka and Pączka, 2006); as pet food, such as a dietary supplement for ornamental fish; and as bait for recreational fishing (Tomlin, 1983).

Earthworms also have a role in a number of applied areas, including land restoration (Butt et al., 1997), organic waste recycling (Fernández-Gómez et al., 2010), and ecotoxicology (Byung-Tae et al., 2008). A demand for earthworms therefore exists to satisfy these scientific and commercial applications. This has resulted in the development of commercial, often large-scale earthworm collection, storage,

Mass Production of Beneficial Organisms. http://dx.doi.org/10.1016/B978-0-12-391453-8.00020-0

and transportation industries in North America (Tomlin, 1983), predominantly for the established fishing bait market, and the breeding of earthworms (referred to as vermiculture), which forms the focus of this chapter.

Sabine (1988) suggested that vermiculture could be categorized as either low or high technology. Low technology implies small-scale, individual or community-level activity, often in a developing country, that seeks to maximize biological resource use such as (vermi) composting of organic residues with little financial input or gain. In contrast, high-technology vermiculture implies an industrial-scale, commercial approach with high financial input (often associated with risk) but also a high degree of scientific interest.

Before embarking on the details of earthworm culture, there is a need to establish some basic information relating to earthworm ecology.

20.1.1. Ecological Groupings

Globally, some 6000 earthworm species have been described, existing under a range of climatic regimes from boreal to tropical biomes (Butt and Lowe, 2011). All earthworms have a similar life cycle. After a period of weeks to months, an earthworm cocoon will hatch in the soil or leaf litter to produce a hatchling worm. This animal is free living and grows over a period of weeks or months to maturity, the latter shown by the presence of a clitellum (saddle). Earthworms are hermaphrodites (i.e. they act as both male and female), but most species reproduce sexually by exchange of sperm (amphimictic), whilst others reproduce asexually (parthenogenetic). Cocoons are produced periodically by adults through secretions from the clitellum, and they are deposited in the soil (Fig. 20.1). The exact timescales and numbers produced are species specific.

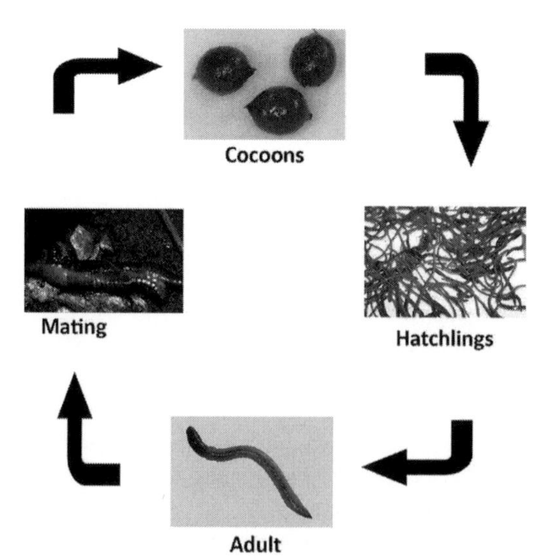

FIGURE 20.1 Stylized life cycle of an amphimictic earthworm, *Lumbricus terrestris*, used for illustration. (For color version of this figure, the reader is referred to the online version of this book.)

Earthworm communities are separated into ecological niches by vertical and horizontal spatial distributions, feeding specialization, and species size (Lavelle, 1983). It is therefore possible to subdivide earthworms into categories based upon these ecological and physical characteristics. Bouché (1972) identified three major ecological categories among earthworms and proposed the widely accepted terminology of anecic, endogeic, and epigeic (Table 20.1).

For the purposes of this chapter, it is more relevant to separate earthworms into two distinct categories—those that inhabit organic matter (litter or compost dwelling) and those that inhabit the mineral soil (soil dwelling).

20.1.2. Selection of Species

A relatively small number of earthworm species have been successfully cultured for scientific or commercial purposes. Vermiculture has often concentrated on culturing a restricted number of litter-dwelling species (including *Eisenia fetida* (Savigny), *Dendrodrilus rubidus* (Savigny), *Dendrobaena veneta* (Rosa), *Lumbricus rubellus* Hoffmeister, *Eudrilus eugeniae* (Kinberg), *Perionyx excavatus* (Perrier), and *Pheretima elongata* (Perrier)) that have commercial applications (like processing organic residues into a potentially saleable product). Litter-dwelling species are ideally suited to large-scale breeding programs due to their high growth, maturation, and reproductive rates. In addition, these species can be relatively easily maintained at high densities in organic media (in the absence of mineral soil). Once processed, these organic media may also have commercial value as soil amendments. As a result, there has been a significant research effort focused on the culture and maintenance of these species (Edwards et al., 2011), which has led to the adoption of litter-dwelling species (often without justification) in emerging fields of earthworm research such as ecotoxicology (Section 20.2.7). Culture of soil-dwelling (endogeic and anecic) species has been almost exclusively restricted to scientific studies (Lowe and Butt, 2005), with commercial availability achievable only through field collection. It is widely accepted that lengthy life cycles, low fecundity, and a requirement for an appropriate soil-based culture medium do not predispose these species to large-scale breeding programs. However, potential commercial applications (such as in land restoration) and an acknowledgment of their ecological relevance (compared with litter dwellers) in environmental monitoring and ecotoxicology have led to an increase in demand for soil-dwelling earthworms of known provenance.

20.1.3. Cultivation Techniques

The primary aim of all earthworm cultivation techniques is to develop a sustainable population while at the same time maximizing output. To achieve this goal requires the maintenance of environmental conditions (abiotic and biotic) that are optimal for earthworm cocoon production, development, and growth.

TABLE 20.1 Generalized Characteristics of the Three Earthworm Ecological Groupings Proposed by Bouché (1972), Including Examples of Temperate and Tropical Species

Characteristic	Epigeic	Endogeic	Anecic
Adult size	Small-medium	Medium	Large
Reproductive rate	High	Intermediate	Low
Longevity	Short-lived	Intermediate	Long-lived
Pigmentation	Predominantly red coloration	Predominantly unpigmented	Predominantly brown to black coloration
Mobility	Highly mobile—selected members of this group are considered pioneer species	Intermediate	Low mobility, generally found in undisturbed habitats, inhabiting semipermanent burrows
Location in the soil profile	Organic horizons in or near the surface litter	Usually found in the soil profile inhabiting horizontal burrows within 15 cm of the surface (in temperate locations)	Construct semipermanent vertical burrows (to a depth of 2 m) in the soil profile. Burrows open onto the soil surface
Food source	Decomposing surface litter	Organic matter ingested with soil (geophagous)	Surface litter drawn into the burrow—usually at night
Temperate examples	*Eisenia fetida, Dendrodrilus rubidus*	*Aporrectodea caliginosa, Aporrectodea rosea*	*Aporrectodea longa, Lumbricus terrestris*
Tropical examples	*Perionyx excavatus, Eudrilius eugeniae*	*Pontoscolex corethrurus, Millsonia anomala*	*Amynthas rodericensis, Eutyphoeus gammiei*

Source: Adapted from Butt and Lowe (2011).

Optimal culture conditions are species specific and predominantly influenced by ecological grouping (Table 20.2). Furthermore, earthworm culture can be labor intensive due to the requirement for the separation of all life stages from the culture medium. This has led to the development of semiautomated systems for the large-scale production of litter-dwelling species, and it is considered the d Life Cycle Characteristics and Culture Requirements for Selected Epigeic, Anecic, and Endogeic Sp

Epigeic		Endogeic		Anecic	
Eisenia fetida	*Eudrilus eugeniae*	*Allolobophora chlorotica*	*Aporrectodea caliginosa*	*Aporrectodea longa*	*Lumbric terrestr*
Parthenogenetic	Amphimictic	Amphimictic	Amphimictic	Amphimictic	Amphim
28–30	40–49	84 (At 15 °C)	No data available	168 (At 15 °C)	112 (At
18–26	12–16	51–59 (At 15 °C)	62–84 (At 15 °C)	54–61(At 15 °C)	90 (At 1
73–80	75–84	62 (At 15 °C)	90 (At 20 °C)	70 (At 15 °C)	83 (At 1
0.35–0.5	0.42–0.51	0.05 (At 15 °C)	0.08 (At 15 °C)	0.05 (At 20 °C)	0.1 (At 1
80–85	80	25	25	25	25
25	25	15	15	15	15

ve and Butt (2005), Butt and Lowe (2011), and Domínguez (2004).

key factor currently restricting the commercial culture of soil-dwelling species and causing the continued reliance on field collection.

Significant advances in the development of culture techniques of temperate soil-dwelling earthworm species have been made in the last 30 years (Lowe and Butt, 2005). This body of research has demonstrated the feasibility of producing sustainable cultures of both anecic and endogeic species for relatively small-scale use in scientific studies (Section 20.2.6). Nevertheless, the potential development of commercially viable, ex situ, large-scale production of soil-dwelling species is a significant challenge. A more feasible alternative may be in situ encouragement of naturally occurring populations, a concept referred to as biostimulation (Brun et al., 1987) (see Section 20.3).

20.2. CURRENT APPLICATIONS

20.2.1. As a Protein Source

Throughout the world, earthworms are used as a protein source for animals and humans. Earthworms are fed to farm stock, poultry, pets (including rodents, amphibians, reptiles, and fish), and zoo animals. Lawrence and Millar (1945) were the first to suggest that earthworms could be used in commercial feed for animals. Earthworm meal is currently sold for animal feed by companies across the globe but in the United States, it is not approved by the American Association of Feed Control Officials as an ingredient of commercial animal feed. Individuals can raise earthworms and feed them to their own animals, but earthworms cannot be sold as an animal feed ingredient.

McInroy (1971) conducted the first compositional analysis of earthworm tissues, concluding that its nutrient contents are suitable for both animal and human diets. Dry matter protein content has been reported in the range of 54.6–71.0% (Sun et al., 1997; Kangmin, 2005; Edwards and Niederer, 1988), and research has indicated that the mean values for essential amino acids fall within the range recommended by the UN Food and Agriculture Organization (FAO) and the World Health Organization (WHO). Compared with commonly consumed human foods and animal feed, earthworm protein and amino acid structures are similar to those recorded in fishmeal and chicken eggs and at higher levels than those in cow milk powder and soybean meal (Sun et al., 1997). Earthworm tissues also contain a wide range of essential vitamins. Kangmin (2005) reported that earthworms are rich in vitamins A and B, with 100 g of earthworms containing 0.25 mg of vitamin B_1 and 2.3 mg of vitamin B_2, and with vitamin D comprising up to 0.07% of earthworm wet weight. In addition, Sun et al. (1997) suggested that earthworm castings have the potential to partially replace cornmeal or wheat bran in animal feed as *E. fetida* casts contained protein levels (7.9% dry matter) comparable to that of cornmeal. These researchers also reported that body fluids of earthworms contained high amounts of vitamins and minerals (especially iron), 9.4% protein, and 78.8% free amino acids per liter.

Sabine (1978) conducted the first animal-feeding trial using earthworms, and since then, numerous scientists have studied the effects of feeding earthworms to a variety of animals. Prayogi (2011) established that substituting 10% of the diet of quails with earthworms caused a significant increase in growth rates. A study by Orozco Almanza et al. (1988) suggested that there was no difference in the feed intake, weight gain, and feed conversion of rabbits fed either 30% earthworm meal or 30% soybean meal as a protein source. Similarly, Taboga (1980) reported that earthworms used as a protein source in the diets of chickens (up to 8 weeks old) produced similar growth rates to those fed maize in a complete grower feed.

People in approximately 113 countries eat earthworms and insects (entomophagy) (Raloff, 2008). Humans in various parts of Asia, Africa, and Latin America and indigenous people in Australia, New Zealand, and North America regularly consume earthworms. Although sometimes collected from the wild, most earthworms consumed by humans are farmed in a controlled environment. Recently, there has been increased interest in consuming earthworms as a sustainable alternative to "regular" meat. Compared with raising cattle and pigs, it takes less food, water, and land to raise earthworms. Kangmin (2005) suggested that earthworms may be an ideal food for humans due to high protein and low lipid levels and the presence of essential amino acids. The indigenous Yekuana people of Venezuela traditionally consume two species of earthworms (*motto* and *kuru*) as a main part of their diet. Earthworms are consumed fresh, after heating in water or smoking over a fire. Paoletti et al. (2003) concluded that the nutrient contents of both earthworm species were appropriate for human sustenance. The earthworms also contain significant amounts of minerals and trace elements required by humans, particularly iron and calcium, in addition to magnesium, potassium, phosphorus, and copper. The iron content of the *motto* and *kuru* earthworms is nearly 10 times greater than that of soybeans.

20.2.2. In Organic Waste Management

Litter-dwelling (epigeic) earthworms can process most organic wastes, with some modifications discussed in this chapter. However, certain organic wastes are more favorable to earthworm growth than others. For example, of seven types of animal manure fed to *E. fetida* by Garg et al. (2005), sheep dung produced the highest earthworm growth rate and weight gain. The manure of donkeys achieved the second highest biomass gain, followed by that of buffalo, goat, cow, horse, and camel. These animal manures also had differing effects on cocoon production; sheep dung yielded the most. In addition to animal manure, an extensive variety of other organic wastes have been shown to be acceptable feed material for earthworms, for example brewery waste, potato-processing residuals, paper, sewage sludge (Benitez et al., 1999; Ndegwa and Thompson, 2001), food waste (Chaudhuri et al., 2000), crop residues (Bansal and Kapoor, 2000), solid textile mill sludge (Kaushik and Garg, 2003), and spent mushroom wastes (Edwards et al., 1998).

Animal manures are produced in a variety of forms ranging from solid to almost liquid slurry. Manures are either used directly from the animal or mixed with a variety of bedding materials, such as straw, sawdust, paper, and wood shavings. These wastes usually require treatment before use, as earthworms are sensitive to ammonia and inorganic salts (Edwards and Niederer, 2011). Precomposting operations can process the material to remove the ammonia and feedstock high in salts; for example, pig manure or kitchen scraps can either be pretreated in this way or washed to reduce the salt content.

Precomposting (using naturally occurring thermophilic microbes) is prevalent in commercial vermiculture operations and usually takes place for 14–21 days. The goal is to achieve sustained, elevated temperatures (to 70 °C) to reduce pathogens that can be harmful to humans, animals, and plants; eliminate weed seeds; and reduce the heating potential of the feedstock. If organic materials are allowed to compost for longer than this period, the inherent nutrition needed by the earthworms is greatly reduced. To further diminish the heating potential of the feedstock, Edwards (2011) suggested that organic materials ought to be applied to worm beds in thin layers (2–3 cm) and that beds should not be deeper than 1 m. Additional feed should not be applied until the previous application has been consumed to avoid overheating and prevent anaerobic conditions. Epigeic earthworms need an aerobic environment, normally live in the top 10–15 cm of a worm bed, and migrate upward to consume feedstock as it is added to the system (Edwards and Niederer, 2011).

Epigeic earthworms require a moist environment; thus, feedstock should be in the 80–90% moisture range, and never lower than 60% (Edwards, 1985). Liquid should not be allowed to pool in a worm bed, and overwatering may be avoided by thoroughly moistening organic materials and squeezing out excess liquid prior to addition, and then periodically misting the bed with water.

The temperature within vermicomposting systems should be maintained at 15–25 °C, and it should not exceed 35 °C (Edwards and Niederer, 2011). During winter, a slightly thicker layer of feed can be applied (5 cm) to provide greater heat insulation, as necessary. Conversely, methods of reducing heat in a worm bed include thinner food layers, precomposting the feedstock, utilizing fans or air conditioning, and light misting of water. In climates with temperature extremes, worm beds will likely need to be established in an enclosed, environmentally controlled structure.

Large volumes of organic residuals are being processed with earthworms using a variety of techniques, ranging from simple methods requiring significant amounts of land and labor to cost-intensive automated systems. These techniques include 1 m high windrows; a "wedge system"; batch schemes in containers, boxes, and trays that may be stacked to utilize vertical space and require less horizontal room; beds with low walls; pits; and automated, continuous-flow elevated reactors.

Outdoor windrows or beds are the most frequently used technique for vermicomposting and can be any length, but they should not exceed 2.4 m in width, so the whole bed may be examined and maintained effectively. Beds are usually covered (with cloth, wood, and bamboo) to reduce the effects of direct sunlight or rain.

An automated, continuous-flow vermicomposting reactor was developed in 1981 by a team of scientists and engineers in the United Kingdom led by Clive Edwards. It consists of a 1 m deep steel vessel on raised legs with a mesh floor. Mobile gantries move horizontally across the top of the reactor, depositing a thin layer of feedstock on top of material in the bed. About 2.5 cm of vermicompost is periodically removed through the basal mesh using a motorized, winch-driven breaker bar. The vermicompost can then be removed from underneath the reactor, either manually or on a conveyor. Organic wastes can be processed in these systems within 30–45 days.

With the exception of automated, continuous-flow reactors, all techniques are labor-intensive as they require the manual separation of earthworms from a culture medium (vermiprocessed organic matter) with high moisture content. One method of harvesting vermicompost is to remove the top 10–15 cm of the worm bed. Most of the earthworms will be in this layer, so it may be placed on top of a new vermicomposting bed. Remaining processed vermicompost can then be removed from the original bed, and any earthworms can be drawn out of this by laying a screen on top covered with food.

Another harvesting technique is sideways separation, which involves creating a new bed beside a working worm bed and applying feedstock to entice the earthworms to move. After feeding for several weeks, most of the earthworms will have moved into the new bed. A rotating trommel screen (0.6–1.3 cm woven mesh) can also be used to separate earthworms from vermicompost.

20.2.3. As Fishing Bait

Eisenia fetida (red wiggler) and *E. eugeniae* (African nightcrawler) have been used as bait worms since the 1940s in North America (Mason et al., 2006), and these species of earthworms, plus *P. excavatus*, are currently used as a nutritional supplement in aquaculture. The primary ingredients in cultured fish feed are wild-caught fish; however, worldwide supplies are shrinking and will soon be unable to meet global demand. The aquaculture industry is expanding rapidly, and annual production has increased from fewer than 1 million tons in the early 1950s to 51.7 million tons in 2006. The FAO estimates that aquaculture production worldwide will need to increase to 80 million tons by 2050 to sustain the current demand (Subasinghe, 2012).

Most cultured aquatic species have high protein requirements, and supplying optimal nutrition for them comprises 40–50% of the costs of operation and production facilities. Thus, fishmeal is the highest cost of an aquaculture operation. The raw materials for cultured fish feed are becoming increasingly scarce and higher in cost, so alternative protein sources are being investigated. The use of earthworms as a protein source for cultured aquatic species (fish and shrimp) is gathering substantial attention worldwide. Producing earthworms locally, using organic wastes, is increasingly seen as an inexpensive method for supplying high-quality protein for farmed fish.

Numerous studies have demonstrated that several species of earthworms may be used as a whole or partial feed in aquaculture operations. Earthworms have been successfully fed to cultured aquatic species as live fare or in desiccated, pelleted form. For example, *E. eugeniae* and *E. fetida* have been effectively fed to carp and sturgeon (Mason et al., 2006), and *P. excavatus* fed to common carp (Pucher et al., 2012).

Earthworm meal has a protein content comparable to that of fishmeal (Hasanuzzaman et al., 2010; Sogbesan et al., 2007; Kostecka and Pączka, 2006), and Wing-Keong (2000) suggests that the combination of palatable and highly nutritious earthworm protein is superior to plant-based proteins, which frequently lack sufficient sulfur-based amino acids and are often unappetizing to cultured aquatic animals. Replacement of fishmeal with earthworm meal may also be beneficial to piscivorous humans. Feeding with frozen earthworms (*E. fetida*) at a rate of 25, 50, and 75% replacement of fishmeal resulted in a decreased lipid content of farmed rainbow trout (Pereira and Games, 1995). Earthworms are also advocated as a replacement feed additive to fishmeal due to concerns regarding contaminants found in conventional sources of protein—as Hasanuzzaman et al. (2010) stated, “The quality of fishmeal is often questioned, with the presence of sand, stone, heavy metals, antibiotics, fine sawdust, poultry by-products and tannery wastes.” The European Food Safety Authority also found polychlorinated biphenyls and dioxin-like contaminants in fishmeal and recommended the replacement of fish products (Environmental Working Group, 2003; EFSA, 2005).

20.2.4. In Soil Restoration

Degradation of soils is a growing global problem and an issue that has arisen from both natural and anthropogenic developments and activities. Where human intervention has been central, for example through soil removal for mineral extraction and a later desire to reinstate the soil, actions have sometimes been taken to try to recreate a living soil. Work undertaken previously often operated only in terms of the physical and chemical constituents of the soil and had not considered the structure and function of the microbiological and macrobiological components, as promoted by Bradshaw (1983). Following major landform corrections, development of a living soil can proceed once the major components have been brought together. Constructing soils usually requires the use of a subsoil and addition of organic matter (OM). This is not new, but the UK

National Building Specifications (2011) now dictate that consideration should be given to earthworm augmentation or inoculation to enhance topsoil function. This long-awaited recognition takes note of research that has been undertaken over recent decades (Butt et al., 1997). Earthworms are now considered for inclusion in soil restoration, due to the ecosystem services that they can provide. In other words, earthworms are able to modify ecosystem functions in ways that are useful to humans. This occurs through the way soil-dwelling species behave in soils. As mentioned in this chapter (Table 20.1), soil-dwelling species have positive effects on (reconstituted) soils. They bring about greater aeration and water infiltration through the provision of macropores (Shipitalo et al., 2004) but also, more importantly, bring an improved crumb structure to the soil through ingestion of mineral and organic components and production of casting material that is highly attractive to microorganisms (Edwards and Bohlen, 1996). Mixing of soil layers and integration of the component parts permit other smaller (micro- and meso-) fauna to access the soil and "breathe life" into initially inert substrates, which no amount of human physical manipulation could achieve.

Provision of appropriate earthworms can be brought about through mass collection (Tomlin, 1983) or even through turf transfer (Stockdill, 1982; see Section 20.2.5), but these may be less sustainable and not as effective as earthworm culture for this purpose. With respect to this chapter, cultivation of earthworms for soil restoration has been shown to be perfectly feasible. Butt et al. (1997), using data from laboratory-based experiments, showed that provision of optimal conditions (Table 20.2) within a unit-based, larger scale system still results in the desired outcome of earthworm production. However, the innovative thinking here was to have production in Earthworm Inoculation Units (EIUs), typically of less than 5l, which served both as a cultivation medium and inoculation product. Starter cultures of less than 10 mature earthworms were kept within polythene bags containing sterilized soil and a suitable OM feedstock. The number of earthworms in the starter culture was a function of population density and EIU size. The EIUs were sealed, provided with air holes, and maintained for a period of approximately 3 months at optimal conditions of temperature in an insulated, darkened polythene greenhouse. The time period was determined by the reproductive rate of the earthworm species (Table 20.2). After this cultivation stage, the intact units were transported to the proposed inoculation site (a semirestored landfill), where they were deposited into holes of corresponding size dug into the soil (Fig. 20.2). Prior to inoculation, the polythene bag was removed and the contents were inserted as a unit, with as little disturbance to the soil matrix as possible. This was undertaken to ensure maintenance of a protective microenvironment for the earthworms, and so that cocoons deposited in the EIU remained at the same level in relation to the soil surface.

In this production system, spacing between EIUs is determined by cost and required rates of spread. At inoculation, all three life stages (adult, cocoon,

FIGURE 20.2 Two-liter (2 l) EIUs transported to an organic matter-enriched landfill cap in the postcultivation phase. The plastic envelopes, tied at the neck, are removed prior to soil inoculation. (For color version of this figure, the reader is referred to the online version of this book.)

and hatchling earthworms) are present, maximizing successful population establishment. Results with monocultures of *Aporrectodea longa* (Ude) and *Allolobophora chlorotica* (Savigny) showed that this technique led to an 8- and 13-fold increase, respectively, in earthworms for inoculation during the cultivation phase and that these figures can be enhanced further by utilizing mixed-species culture (Butt et al., 1997). At the Calvert landfill site, where this work was undertaken, colonization rates from the point of inoculation were approximately 1 m per year and gave rise to typical sustainable earthworm populations after 5–10 years (Butt et al., 2004). Nevertheless, one aspect that needs further consideration here is the economics of such earthworm breeding (Butt, 2011).

20.2.5. In Agro-Ecosystems

In an agricultural context, the value of adding earthworms where they were previously absent became apparent through a number of serendipitous actions. For example, in an upland area of New Zealand, farmers recognized that grassland around introduced fruit trees was more productive than adjacent areas. Close examination revealed the presence of lumbricid earthworms, which were introduced with the soil around tree roots; the earthworms were shown to be responsible for the observed increase in productivity. In their absence, a thick, undisturbed mat of dead plant material had accumulated. The nonindigenous earthworms, mainly *Aporrectodea caliginosa* (Savigny), broke down the mat in the immediate area around the trees, and the release of nutrients initially led to an increased grass production of over 70% (Stockdill and Cossens, 1966). So, here, the actions of an endogeic species assisted plant productivity. To utilize this effect, further deliberate inoculations of similar species were performed using turf (20 cm square and 5–8 cm deep) cut from earthworm-rich pasture and laid at earthworm-deficient sites. This proved successful in New Zealand and Tasmania (Stockdill, 1982; Baker et al., 2002), with up to a threefold economic return.

Such practices, although of value, do not make use of earthworm cultures. More recently, though, and building on the success of the work of Butt et al. (1997), workers in Finland addressed a field-related soil problem by breeding earthworms for inoculation. Nuutinen et al. (2006) bred field-collected *Lumbricus terrestris* and used the EIU technique to produce 71 cultures that were introduced into the margins of an arable heavy clay soil where this species was previously absent. The rationale for this work was to increase macroporosity (through provision of large-diameter, deep burrows) for greater infiltration of water and prevention of oversurface runoff. The inoculation took place in 1997, and the spread of earthworms has been monitored since (Nuutinen et al., 2011). The earthworms have begun to advance into the field, and spread is occurring particularly in no-till plots. Over time, the 1 m deep burrows are likely to link up with subsoil drainage pipes, enhancing water flow through the heavy clay. Thus, the ecosystem engineers (earthworms) are interacting directly with human drainage engineers (Shipitalo et al., 2004). Here, the focus is on improving agricultural systems, similar to that of Section 20.2.4, but more applied in the context of food production.

Another potential way of breeding earthworms for agro-ecosystems (particularly in pasture) is seen more as extensive enhancement. Here, instead of breeding earthworms in small units (larger soil-based units do not work for reasons given by Butt, 2011), the field itself can be utilized for earthworm culture. A suitable patch of pasture can be enriched by application of OM. This is done to increase the breeding rate of earthworms present or species deliberately added. This thinking is not new and was proposed by Curry (1988) prior to turf cutting for enhancing endogeic earthworm numbers for inoculation. This type of earthworm culture was used in County Durham, UK, by the Agricultural Development and Advisory Service (ADAS) to encourage the growth of earthworm populations within soil banks set aside during the process of land restoration by the addition of well-rotted farmyard manure. Populations of up to 4000 earthworms per m^2 were recorded under such conditions prior to inoculation. In Australia, similar measures have been employed to increase populations of *A. longa* under similar field conditions by applying organic matter to a pasture near Canberra, where this earthworm was introduced. The long-term goal is to use this "seed bed" as a source for further inoculations of this species in appropriate areas in southern Australia (Curry and Baker, 1998).

In this agricultural setting, and also in the context of land restoration, a variety of techniques can be used for breeding earthworms; however, as suggested by Butt (2011), intensive production of soil-dwelling species is not really a viable way forward. Soil improvement through earthworm culture has a future based on two potential routes, both of which require the site of inoculation to be made as hospitable as possible for the chosen earthworm species (Section 20.3). Targeted breeding of specific species in small units (using the EIU technique) may be one direction forward, or a more extensive soil enrichment process may be desired for a given site that is then periodically harvested for earthworm-rich turf.

20.2.6. In Laboratory Experimentation

In spite of a recent expansion in earthworm research, predominantly in applied areas of vermicomposting and ecotoxicology, perceived difficulties associated with laboratory culture of earthworms (in particular soil-dwelling species) have led to a reliance on field-collected and commercially purchased earthworms for laboratory-based research. This practice restricts experiments to the study of juveniles or adults, often of unknown age and/or history, which may compromise experimental design and the validity of results (Fründ et al., 2010). The use of laboratory-reared earthworms can provide reliable and replicable experimental data and allows for the study of all stages of the earthworm life cycle under controlled conditions. While it is recognized that extrapolation of laboratory-based results to field conditions is questionable, macrocosm–microcosm studies provide an opportunity to obtain fundamental ecological and biological data that is not possible in field-based studies.

Optimal culture conditions for a range of litter- and soil-dwelling species were determined (see Table 20.2) and were employed in the development of earthworm cultures and in experimental systems. This section concentrates on the control and maintenance of key abiotic and biotic factors used in the culture of soil-dwelling earthworms for laboratory experimentation. Laboratory culture of litter-dwelling species is not considered as this follows well-established practices (Section 20.2.2).

Moisture: Under laboratory conditions, evaporation of moisture from culture substrate can quickly become a limiting factor. To combat water loss, cultures are usually maintained in sealed containers with small air holes for ventilation, which allows inspection every few weeks without the need for rewatering. In several studies (Wever et al., 2001; Baker and Whitby, 2003), water loss was determined by weighing culture vessels on a regular basis and replenishing water as necessary. Water is usually applied to the substrate surface; however, Nuutinen and Butt (1997) cultured *L. terrestris* in PVC cylinders (6.5 cm diameter and 95 cm height) standing in a vessel filled with water to a depth of 1 cm, replacing moisture lost at the surface by capillary action. Similarly, Elvira et al. (1996) kept *L. rubellus* in 5 l cylindrical containers with a damp sponge at the base. Soil moisture conditions can also significantly influence cocoon development, and a method of incubating cocoons on or between moist filter paper(s) in Petri dishes (or similar vessels), with filter paper rehydrated as required, has widely been adopted (Butt, 1991; Holmstrup et al., 1991; Garvin et al., 2002). The filter paper can also serve as a food source for hatchlings. To avoid dehydration, excess water can be added to culture vessels, submerging cocoons, which does not negatively affect cocoon development or survival of hatchlings.

Temperature: Temperature control has been used to manipulate earthworm life cycles; Holmstrup et al. (1991) recorded that *A. chlorotica* developed to maturity in 34–38 days at 20 °C compared with 400 days at 5 °C. Therefore, in experiments requiring large cohorts of hatchlings, low temperatures can be used to inhibit growth (enforced quiescence) until sufficient individuals have hatched (Spurgeon and Hopkin, 1999). Similarly, cocoon development can be manipulated by controlling temperature (Boström and Löfs-Holmin, 1986), as the time required for embryo development increases with decreasing temperature and may be inhibited at 3 °C (Holmstrup et al., 1991), with cocoons becoming nonviable if frozen.

Earthworms kept under constant conditions, especially at temperatures approaching the upper limits of a species' tolerance, can suffer from reproductive fatigue and experience high death rates and loss in body mass compared to conspecifics under fluctuating temperatures (Uvarov, 1995). For example, Butt (1997) maintained *A. chlorotica* adults at 10, 15, and 20 °C for a 12-month period during which mean cocoon production rates were 9.9, 17.8, and 27.3 cocoons per individual per year, respectively; but equivalent survival rates of adults were 73, 93, and 15%. A trade-off is therefore exhibited between production rate and survival. Accordingly, researchers (Baker et al., 2002; Dalby et al., 1998; Fraser et al., 2003; Lowe and Butt, 1999) who successfully cultured soil-dwelling temperate species have used a suboptimal temperature of 15 °C for rearing juveniles and incubating cocoons. Earthworm cultures are usually maintained in temperature-controlled incubators or rooms (Lowe and Butt, 2002); however, other methods have also been used. Baker et al. (2002) kept cultures in a water bath, while Butt et al. (1997) used an insulated polythene greenhouse with subsoil heating cables for the large-scale breeding of soil-dwelling species for use in soil restoration.

Substrate: Litter-dwelling species are usually cultured in 100% organic matter substrates; however, soil-dwelling species require a mix of soil and OM (Butt et al., 1992; Doube et al., 1997). The reliance on field-collected earthworms in laboratory-based studies has led, with some justification, to the use of soils collected from the area of earthworm extraction as a culture medium (Daniel et al., 1996). However, field-collected soils have an inherent soil macro- and microfauna that, if not removed, may compromise experimental results and negatively influence earthworm survival and production. As a result, several methods of soil sterilization and macrofauna removal have been employed, including sieving and handsorting (Fraser et al., 2003), steam sterilization (Butt, 1991), microwaves (Langmaack et al., 2002), or simply air drying (Shipitalo et al., 1988). Soil sterilization is time-consuming and has led to the use of presterilized, commercially available soils. Butt et al. (1994) used presterilized and sieved (<6 mm) Kettering loam soil with an organic content of 5% and a pH of 6.4 to culture *L. terrestris*. This soil has since been used by other researchers as a reliable substrate for the culture of temperate earthworm species (Langdon et al., 2003) and proposed as a standard medium for use in ecotoxicology (Spurgeon et al., 2004).

Feed: The preference for animal dung over other organic materials as a suitable feed for soil-dwelling earthworms in laboratory studies has been recognized since the pioneering work of Evans and Guild (1948). As a result, cow

(Kostecka and Butt, 2001), sheep (Baker and Whitby, 2003), and horse (Spurgeon et al., 2004) dung has been used as a feed source. Fresh or semi-decomposed dung has been utilized as a food source (Elvira et al., 1996). However, fresh dung may contain potential predators and pathogens along with a resident earthworm fauna that may compromise the cultures. To achieve a consistent and reliable feed source, animal dung requires pretreatment. Langdon et al. (2003) froze fresh animal dung to sterilize and maintain its nutritional value, while Lőfs-Holmin (1983) recommended keeping semicomposted cattle dung in 60 l containers with airtight lids at 25 °C for more than a month to kill off invertebrates and earthworm cocoons. Food particle size has also been manipulated to increase earthworm growth rates. Lowe and Butt (2003) demonstrated that the influence of food particle size was both species and life stage specific and inversely related to size. *Allolobophora chlorotica* (adult mass 0.3–0.6 g) and *L. terrestris* (adult mass 3–5 g) were maintained in treatments of Kettering loam with either milled (<1 mm) or unmilled separated cattle solids (SCS). After 18 weeks, *A. chlorotica* fed milled SCS had a mean mass 185% greater than that in unmilled cultures.

Density: Laboratory-based experiments have also shown that earthworm growth, adult mass, and fecundity are significantly influenced by earthworm biomass and density. Butt et al. (1994) maintained mature *L. terrestris* at two densities (4 or 8) in 2 l of soil with paper pulp applied to the surface as feed. In addition, hatchling *L. terrestris* were kept in densities of 1, 2, 4, 8, or 12 in 200 g of soil, and they were surface fed 50 g paper pulp with 0.75 g yeast extract. Field-collected mature *L. terrestris* were also kept in 300 g of soil and surface fed separated cattle solids at densities of 1, 2, 3, 4, 6, or 8 worms. Increased density had a negative effect on growth rate and final mean earthworm mass. The development of full reproductive capacity was also reduced at higher densities. In the 2 l system, it is estimated that a mass in the range of 15–25 live g/l (3–5 adults) may be optimal (under reported experimental conditions); while in the smaller pots (0.3 l) and when fed with a superior feed, an optimal density may be 20–40 g/l, suggesting that the influence of density may be modified by environmental factors.

Löfs-Holmin (1983) recommended that "small vessels should be preferred to large ones for ease of handling and sampling." It is also important that vessels are reusable, easily stacked to maximize available space, and have sealable lids to prevent excess loss of soil moisture; and, if cultures are maintained in the light, vessels with opaque sides should be used. These recommendations have been widely adopted (Holmstrup et al., 1991; Lowe and Butt, 2002), while novel culture vessels have also been developed to counter specific problems and requirements. Butt et al. (1997) cultured earthworms in plastic sealable envelopes (EIUs) specifically designed for inoculating earthworms in soil restoration projects (Section 20.2.4).

Species interactions: Laboratory-based research demonstrated that soil-dwelling earthworms are capable of coexistence in mixed-species cultures

(Butt, 1998) but also that species composition can reduce earthworm production (Lowe and Butt, 1999; Garvin et al., 2002). This "negative" form of interaction is species specific and thought to result from competition for resources (feed and space). The intensity of interaction may be determined largely by the degree of niche overlap and is therefore most intense between species from the same ecological grouping (Lowe and Butt, 1999). Further research by Lowe and Butt (2002) also determined that the stage of individual earthworm development can influence both inter- and intraspecific interactions. These authors cultured hatchling *L. terrestris*, *A. longa*, *A. chlorotica*, and *L. rubellus* in monoculture or in the presence of adults. Early growth of *L. terrestris* hatchlings was significantly greater in the presence of conspecific adults (where a high level of niche overlap would be expected), but such an advantage (possibly mediated by the availability of fragmented organic matter in adult middens) decreased with age. It was also noted that the behavior of anecics (*L. terrestris* and *A. longa*) changed from a subsurface (endogeic) to a deep-burrowing (anecic) mode after 12 weeks (approx. 1 g mass). This would suggest that in the early stages of their growth, anecic earthworms may be in direct competition for space and feed with endogeics.

20.2.7. In Ecotoxicology

Earthworms possess a number of qualities that predispose them for use as test organisms in ecotoxicology, in particular ecological assessment of contaminated soils. They are present in the majority of soils and are relatively sedentary, with natural immigration rates of 5 m/year (Marinnissen and van den Bosch, 1992). Earthworms are both resistant and sensitive to pollutants (Cortet et al., 1999); in intimate contact with the substrate in which they live (soil), they may consume the substrate to gain nutrition and as a result accumulate some pollutants in their tissues at higher levels than the surrounding substrate. Survival, growth, and reproduction rates, in addition to behavior of earthworms, may also be affected by pollutant type and concentration, so earthworms are considered biological sentinel species (Stürzenbaum et al., 2009). Earthworms have played a major role in acute and chronic toxicity testing of chemicals, which has led to several standardized test procedures (e.g. ISO 11268-1) (Lowe and Butt, 2007). Furthermore, the potential of earthworms as bioindicators in field-based ecological assessment and in particular bioaccumulation studies is widely recognized (Sanchez-Hernandez, 2006).

Acute toxicity tests have relied predominantly on the use of *E. fetida*, and this species has also been utilized in chronic toxicity studies. The use of this litter-dwelling earthworm is due mainly to its short life cycle, high fecundity, ease of culture, and comparable availability. However, the continued use of this species, especially in chronic toxicity studies, is now questioned (Lowe and Butt, 2007). In addition, *E. fetida* is more tolerant than most earthworm species to contaminants (Lukkari et al., 2005) and therefore of limited ecological relevance. There is recognition of the need to adopt a species-specific approach in many ecotoxicological studies (Morgan and Morgan, 1998; Van Gestel and Weeks, 2004) and select test species that are representative of the site(s) under scrutiny (Svendsen et al., 2005). This has resulted in the use of a wider range of earthworm species from the three recognized ecological groupings, for example, *A. caliginosa* (Khalil et al., 1996), *Dendrobaena octaedra* (Savigny) (Rożen, 2006), and *L. terrestris* (Svendsen et al., 2005).

While there remains a reliance on commercially purchased earthworms for use as test organisms, there is a growing recognition of the benefits of using laboratory-reared earthworms (reviewed by Lowe and Butt, 2007). Earthworm origin (Lowe and Butt, 2007), age (Svendsen et al., 2005), genotype (Kautenburger, 2006), and/or preexposure (Langdon et al., 2009) can significantly influence their response to pollutants and therefore potentially invalidate results. Laboratory culture of earthworms allows production of cohorts of known age and history through the manipulation of environmental conditions. In addition, Spurgeon and Hopkin (1996) suggested that juvenile growth rates are a more sensitive and ecologically relevant indicator of pollution effects than adult weight change. Controlled laboratory culture allows for the production of juvenile cohorts, which is not feasible via field collection due to difficulties associated with species identification of juveniles (Lowe and Butt, 2007). Earthworm culture may also increase genetic homogeneity within experimental populations, which may be further enhanced by the culture of obligatory parthenogenetic species (Lowe and Butt, 2008). The latter provide the opportunity of culturing cohorts of genetically similar populations, especially in species that exhibit low clonal diversity such as *Octolasion cyaneum* (Savigny) (Terhivou and Saura, 2003). This would enhance the robustness of individual studies as it allows for individuals to be maintained in isolation and provides individually monitored end points in reproduction studies. In addition, the use of parthenogenetic species would also allow the results of geographically distinct studies employing the same species to be compared more reliably. This proposal supports recommendations made by Bouché (1992) that advocated the centralized culture of specific earthworm species and strains (an "earthworm bank"; see Section 20.3) that could be distributed to laboratories to form test cultures. While this proposal has distinct advantages, protocols are required to avoid inbreeding and the production of individuals adapted to laboratory conditions.

20.3. THE FUTURE FOR MASS EARTHWORM CULTURE

The United Nations (2004) predicted the global human population will continue to increase and peak at 9.22 billion in 2075. This will only increase the existing demand for finite natural resources that are required in the production of food crops and the rearing of livestock for human consumption. The FAO (2006) estimated that livestock grazing now uses 30% of the earth's land mass, and 33% of arable land is used to produce feed for livestock. The mass culture of earthworms as a protein replacement in livestock feed and as a direct source of protein for human consumption could make a significant contribution in addressing this issue. Furthermore, as the demand for land increases, earthworms may also have a significant role to play in the restoration of degraded land and the maintenance of fertility in cultivated soils. Here, we have selected two emerging areas that may further enhance the anthropogenic benefits of earthworm culture, with a particular focus on soil amelioration.

Biostimulation and ecosystem rehabilitation: This builds on the extensive field-based approach described in this chapter, but it is wider in its thinking. Biostimulation covers a spectrum, extending from a conservation approach (with respect to existing earthworms), which encourages population development, to more active involvement in earthworm enhancement through direct inoculation into the soil or addition of organic matter (vermicomposted material), which itself may have beneficial effects on soil properties (Airia and Dominguez, 2011).

The concept relies on the assertion that earthworms should be considered as part of the entire system in which they are present, rather than simply as a component part. Equally, each situation to which biostimulation is applied may need to be viewed as a unique setting (with common elements with others) that lies somewhere along the given spectrum or gradient (as shown in Fig. 20.3). Where least damage has occurred, the level of rehabilitation brought about through earthworm enhancement will be minimal and perhaps only requires a change of management practices. At the opposite end of the spectrum, direct inoculation of earthworms may be required to start the process, and this will require active land management.

Recent work (Butt, 2011; Eijsackers, 2011) clearly demonstrates that earthworms can play a significant role in the rehabilitation of soils. Sites that are earthworm-free are likely to be those that have been damaged by anthropogenic activity and therefore require soil improvement rather than simple earthworm introductions. It is also quite reasonable to introduce earthworms to fertile soils, for example Dutch polders reclaimed from the sea, and note rapid earthworm

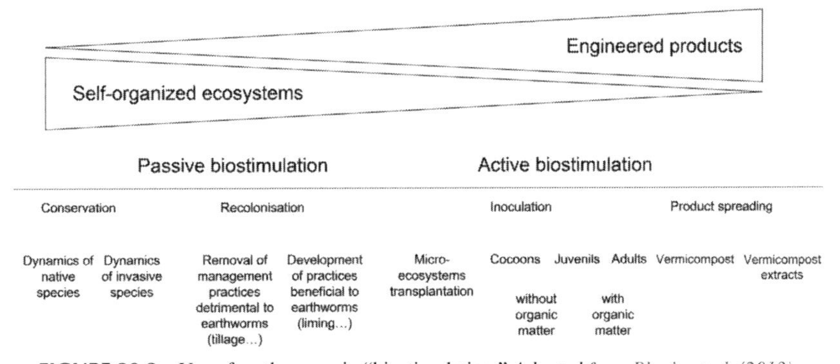

FIGURE 20.3 Use of earthworms in "biostimulation." *Adapted from Blouin et al. (2013).*

population development and expansion. However, in all of these cases, the process will be rendered futile unless the necessary earthworms are available. Biostimulation, through the encouragement of population development at earthworm-rich sites, is likely to be the deciding factor.

Development of an "earthworm bank": The concept of storing, culturing, and/or preserving biological organisms and, more recently, genetic resources (tissues for DNA extraction and somatic cells for culture) is reasonably well established and provides economic, conservation, and long-term environmental-monitoring benefits. In plants, ex situ seed storage follows well-defined internationally agreed standards (Li and Pritchard, 2009). Seed banks have been established to help conserve wild species (Millenium Seed Bank at the Royal Botanical Gardens, Kew, UK) and provide genetic material for reintroduction (Cochrane et al., 2007), with the largest seed banks devoted to economically important crop species (e.g. National Center for Genetic Resources Preservation, Fort Collins, CO, USA) (Li and Pritchard, 2009). A number of countries have also established Environmental Specimen Banks (Germany and Denmark) to collect and store biological and abiotic samples (at low temperature $<-150\,°C$) from the natural environment to allow retrospective studies monitoring pollutants, occurrence of natural substances, and genetic deviations (Poulson and Pritzl, 1993; Rüther and Bandholtz, 2009). As part of this initiative, the German Environmental Specimen Bank developed earthworm sampling and preparation techniques to assist in the monitoring of soil status. Earthworms are collected and processed by cryohomogenization, and heavy metal concentrations are determined. In addition to the storage of genetic resources and samples for environmental monitoring, it has also been suggested that earthworm cultures should be established for use in applied research and commercial applications (like ecotoxicology). The development of techniques for controlled earthworm culture (particularly with respect to soil-dwelling species) and DNA sequencing will, in the future, allow for the culture of specific strains (e.g. those exhibiting resistance to specific pollutants) for a number of commonly utilized earthworm species and potentially gather important life history data on species that have not been studied. These specific strains could then be made available to researchers (as advocated by Bouché, 1992) by adapting the approach already established for the soil nematode *Caenorhabditis elegans* (Maupas). This nematode is the first multicellular organism to have a fully sequenced genome; it is easily cultured and has a short life cycle. While it is recognized that there are significant differences between the culture of *C. elegans* and earthworms, the overarching principle of an earthworm bank could be similar. Studies showed that maintenance of cocoons and juveniles at low temperatures can be used to inhibit hatching and growth, respectively (see Section 20.2.6), and allow for long-term storage. In addition, the efficacy of distributing earthworms through the mail has already been established by commercial suppliers and researchers. The establishment of a centralized earthworm bank would require centralized national and international research council funding and active engagement of research organizations, and this may not be realized or warranted until the complete sequencing of earthworm species genomes is achieved.

Culture of litter-dwelling earthworms is well established, and commercial markets have been developed for the products (earthworms and compost). Fieldson (1988) questioned the economic viability of vermicomposting of organic waste. However, recently, there has been renewed interest, fueled by recognition of the resource value of waste materials (like in horticulture) and diversion of waste away from landfill. In contrast, large-scale culture of soil-dwelling earthworms has not been commercially demonstrated. However, given the key role that these species play in maintenance of soil fertility and land restoration, it is likely that it is only a matter of time before viable commercial operations are established. The importance of earthworms to human society has often been neglected. However, these seemingly primitive organisms may play a significant role in shaping our future existence through direct and indirect services to food provision.

REFERENCES

Aira, M., Dominguez, J., 2011. Earthworm effects without earthworms: inoculation of raw organic matter with worm-worked substrates alters microbial community functioning. PLoS One 6 (1), e16354.

Baker, G.H., Whitby, W.A., 2003. Soil pH preferences and the influences of soil type and temperature on the survival and growth of *Aporrectodea longa* (Lumbricidae). Pedobiologia 47, 745–753.

Baker, G., Carter, P., Barrett, V., Hirth, J., Mele, P., Gourley, C., 2002. Does the deep-burrowing earthworm *Aporrectodea longa*, compete with resident earthworm communities when introduced to pastures in south-eastern Australia. Eur. J. Soil Biol. 38, 39–42.

Bansal, S., Kapoor, K.K., 2000. Vermicomposting of crop residues and cattle dung with *Eisenia foetida*. Bioresour. Technol. 73 (2), 95–98.

Benitez, E., Nogales, R., Elvira, C., Masciandaro, G., Ceccanti, B., 1999. Enzyme activities as indicators of the stabilization of sewage sludges composting with *Eisenia foetida*. Bioresour. Technol. 67 (3), 297–303.

Blouin, M., Hodson, M.E., Aranda Delgado, E., Baker, G., Brussaard, L., Butt, K.R., Dai, J., Dendooven, L., Pérès, G., Tondoh, J., Cluzeau, J., Brun, J.-J., 2013. A review of earthworm impact on soil function and ecosystem services. Eur. J. Soil Sci. 64, 161–182.

Boström, U., Löfs-Holmin, A., 1986. Growth of earthworms (*Allolobophora caliginosa*) fed shoots and roots of barley, meadow fescue and lucerne. Studies in relation to particle size, protein, crude fibre content and toxicity. Pedobiologia 29, 1–12.

Bouché, M.B., 1972. Lombriciens de France. Ecologie et Systématique. Institut National de la Recherche Agronomique, Paris.

Bouché, M.B., 1992. Earthworm species and ecotoxicological studies. In: Greig-Smith, P.W., Becker, H., Edwards, P.J., Heimbach, F. (Eds.), Ecotoxicology of Earthworms. Intercept Press, Andover, pp. 20–35.

Bradshaw, A.D., 1983. The reconstruction of ecosystems. J. Appl. Ecol. 20, 1–17.

Brown, G.G., Feller, C., Blanchart, E., Deleporte, P., Chernyanskii, S.S., 2003. With Darwin, earthworms turn intelligent and become human friends. Pedobiologia 47, 924–933.

Brun, J.J., Cluzeau, D., Trehen, P., Bouché, M.B., 1987. Biostimulation: perspectives et limites de l'amélioration biologique des sols par stimulation ou introduction d'espèces lombriciennes. Rev. Ecol. Biol. Sol. 24, 687–701.

Butt, K.R., 1991. The effects of temperature on the intensive production of *Lumbricus terrestris*. Pedobiologia 35 (4), 257.

Butt, K.R., 1997. Reproduction and growth of the earthworm *Allolobophora chlorotica* (Savigny, 1826) in controlled environments. Pedobiologia 41, 369–374.

Butt, K.R., 1998. Interactions between selected earthworm species: a preliminary, laboratory-based study. Appl. Soil Ecol. 9, 75–79.

Butt, K.R., 2011. The use of vermiculture for land improvement (Chapter 21). In: Edwards, C.A., Aracon, N.Q., Sherman, R. (Eds.), Vermiculture Technology: Earthworms, Organic Matter and Environmental Management, Taylor and Francis, Boca Raton, pp. 335–348.

Butt, K.R., Lowe, C.N., 2011. Controlled cultivation of endogeic and anecic earthworms. In: Karaca, A. (Ed.), Biology of Earthworms, Springer-Verlag, Berlin, pp. 107–122.

Butt, K.R., Fredrickson, J., Morris, R.M., 1992. The intensive production of *Lumbricus terrestris* L. for soil amelioration. Soil Biol. Biochem. 24 (12), 1321–1325.

Butt, K.R., Fredrickson, J., Morris, R.M., 1994. The life cycle of the earthworm *Lumbricus terrestris* in laboratory culture. Eur. J. Soil Biol. 30, 49–54.

Butt, K.R., Frederickson, J., Morris, R.M., 1997. The earthworm inoculation unit (EIU) technique, an integrated system for the cultivation and soil-inoculation of earthworms. Soil Biol. Biochem. 29, 251–257.

Butt, K.R., Lowe, C.N., Frederickson, J., Moffat, A.J., 2004. The development of sustainable earthworm populations at Calvert Landfill Site UK. Land Degrad. Dev. 15, 27–36.

Byung-Tae, L., Kyung-Hee, S., Ju-Yong, K., Kyoung-Woong, K., 2008. Progress in earthworm ecotoxicology. In: Kim, Y.J., Platt, U. (Eds.), Advanced Environmental Monitoring, Springer, The Netherlands, pp. 248–258.

Chaudhuri, P.S., Pal, T.K., Bhattacharjee, G., Dey, S.K., 2000. Chemical changes during vermicomposting (*Perionyx excavatus*) of kitchen wastes. J. Trop. Ecol. 41 (1), 107–110.

Cochrane, J.A., Crawford, A.D., Monks, L.T., 2007. The significance of ex situ seed conservation to reintroduction of threatened plants. Aust. J. Bot. 55 (3), 356–361.

Cortet, J., Gomot-De Vauflery, A., Poisnot-Balaguer, N., Gomot, L., Texier, C., Cluzeau D., 1999. The use of soil fauna in monitoring pollutant effects. Eur. J. Soil Biol. 35, 115–134.

Curry, J.P., 1988. The ecology of earthworms in reclaimed soils and their influence on soil fertility. In: Edwards, C.A., Neuhauser, E.F. (Eds.), Earthworms in Waste and Environmental Management, SPB Academic Publishing, The Hague, pp. 251–261.

Curry, J.P., Baker, G.H., 1998. Cast production and soil turnover by earthworms in soil cores from south Australian pastures. Pedobiologia 42, 283–287.

Dalby, P.R., Baker, G.H., Smith, S.E., 1998. Competition and cocoon consumption by the earthworm *Aporrectodea longa*. Appl. Soil Ecol. 10, 127–136.

Daniel, O., Kohli, L., Bieri, M., 1996. Weight gain and weight loss of the earthworm *Lumbricus terrestris* at different temperatures and body weights. Soil Biol. Biochem. 28, 1235–1240.

Darwin, C.R., 1881. The Formation of Vegetable Mould through the Action of Worms with Observations of Their Habits. Murray, London.

Domínguez, J., 2004. State-of-the-art and new perspectives on vermicomposting research. In: Edwards, C.A. (Ed.), Earthworm Ecology, second ed. CRC Press, Boca Raton, FL, pp. 401–425.

Doube, B.M., Schmidt, O., Killham, K., Correll, R., 1997. Influence of mineral soil on the palatability of organic matter for lumbricid earthworms: a simple food preference study. Soil Biol. Biochem. 29, 569–575.

Edwards, C.A., 1985. Production of feed protein from animal waste by earthworms. Philos. Trans. R. Soc. Lond. Ser. B. B310, 299–307.

Edwards, C.A., 2011. Medium- and high-technology vermicomposting systems. In: Edwards, C.A., Arancon, N.Q., Sherman, R. (Eds.), Vermiculture Technology: Earthworms, Organic Wastes, and Environmental Management, CRC Press, Boca Raton, FL, pp. 91–102 (Chapter 8).

Edwards, C.A., Bohlen, P.J., 1996. Biology and Ecology of Earthworms, third ed. Chapman and Hall, London.

Edwards, C.A., Niederer, A., 1988. The production and processing of earthworm protein. In: Edwards, C.A., Neuhauser, E.F. (Eds.), Earthworms in Waste and Environmental Management, SPB Academic Publishing, The Hague, The Netherlands, pp. 169–179.

Edwards, C.A., Niederer, A., 2011. The production of earthworm protein for animal feed from organic wastes. In: Edwards, C.A., Arancon, N.Q., Sherman, R. (Eds.), Vermiculture Technology: Earthworms, Organic Wastes, and Environmental Management, pp. 323–334 (Chapter 20).

Edwards, C.A., Dominguez, J., Neuhauser, E.F., 1998. Growth and reproduction of *Perionyx excavatus* (Perr.) (Megascolecidae) as factors in organic waste management. Biol. Fertil. Soils 27, 155–161.

Edwards, C.A., Arancon, N.Q., Sherman, R., 2011. Vermiculture Technology: Earthworms, Organic Wastes, and Environmental Management. CRC Press. Boca Raton, FL.

EFSA Journal, 2005. Opinion of the scientific panel on contaminants in the food chain on a request from the European Parliament related to the safety assessment of wild and farmed fish. Question N. EFSA-Q-2004-22 236, 1–118.

Eijsackers, H., 2011. Earthworms as colonizers of natural and cultivated soil environments. Appl. Soil Ecol. 50, 1–13.

Elvira, C., Domínguez, J., Mato, S., 1996. The growth of *Lumbricus rubellus* and *Dendrobaena rubida* in cow manure mixed cultures with *Eisenia andrei*. Appl. Soil Ecol. 5, 97–103.

Environmental Working Group, July 30, 2003. PCBs in Farmed Salmon. http://www.ewg.org.

Evans, A.C., Guild, W.J. Mc. L., 1948. Studies on the relationships between earthworms & soil fertility IV: on the life-cycles of some British Lumbricidae. Ann. Appl. Biol. 35, 471–484.

FAO, 2006. Livestock Impacts on the Environment. The Food and Agriculture Organisation of the United Nations, Agriculture and Consumer Protection Department Spotlight Magazine http://www.fao.org/ag/magazine/0612sp1.htm.

Fernández-Gómez, M.J., Romero, E., Nogales, R., 2010. Feasibility of vermicomposting for vegetable greenhouse waste recycling. Bioresour. Technol. 101 (24), 9654–9660.

Fieldson, R.S., 1988. The economic viability of earthworm culture on animal wastes. In: Edwards, C.A., Neuhauser, E.F. (Eds.), Earthworms in Waste and Environmental Management. SPB Academic Publishing, The Hague, The Netherlands, pp. 145–156.

Fraser, P.M., Beare, M.H., Butler, R.C., Harrison-Kirk, T., Piercy, J.E., 2003. Interactions between earthworms (*Aporrectodea caliginosa*), plants and crop residues for restoring properties of a degraded arable soil. Pedobiologia 47, 870–876.

Fründ, H.C., Butt, K.R., Capowiez, Y., Eisenhauer, N., Emmerling, C., Ernst, G., Potthoff, M., Schädler, M., Schrader, S., 2010. Using earthworms as model organisms in the laboratory: recommendations for experimental implementations. Pedobiologia 53, 119–125.

Garg, V.K., Chand, S., Chhillar, A., Yadav, A., 2005. Growth and reproduction of *Eisenia foetida* in various animal wastes during vermicomposting. Appl. Ecol. Environ. Res. 3 (2), 51–59.

Garvín, M.H., Trigo, D., Hernández, P., Ruiz, M.P., Díaz Cosín, D.J., 2002. Interactions of *Hormogaster elisae* (Oligochaeta, Hormogastridae) with other earthworm species from Reduena (Madrid, Spain). Appl. Soil Ecol. 20 (2), 163–169.

Harper, G.L., King, R.A., Dodd, C.S., Harwood, J.D., Glen, D.M., Bruford, M.W., Symondson, W.O.C., 2005. Rapid screening of invertebrate predators for multiple prey DNA targets. Mol. Ecol. 14, 819–827.

Hasanuzzaman, A.F. Md., Hossian, Sk. Z., Das, M., 2010. Nutritional potentiality of earthworm (*Perionyx excavatus*) for substituting fishmeal used in local feed company in Bangladesh. Mesopot. J. Mar. Sci. 25 (2), 25–30.

Holmstrup, M., Østergaard, I.K., Nielsen, A., Hansen, B.T., 1991. The relationship between temperature and cocoon incubation time for some lumbricid earthworm species. Pedobiologia 35 (3), 179–184.

Kangmin, L., 2005. Vermiculture in Circular Economy. Chinese Academy of Fishery Sciences, Freshwater Fisheries Research Center Asian Pacific Regional Research & Training Center for Integrated Fish Farming.

Kaushik, P., Garg, V.K., 2003. Vermicomposting of mixed solid textile mill sludge and cow dung with the epigeic earthworm *Eisenia foetida*. Bioresour. Technol. 90 (3), 311–316.

Kautenburger, R., 2006. Genetic structure among earthworms (*Lumbricus terrestris* L.) from different sampling sites in western Germany based on random amplified polymorphic DNA. Pedobiologia 50, 257–266.

Khalil, M.A., Abdel-Lateif, H.M., Bayoumi, B.M., van Straalen, N.M., 1996. Analysis of separate and combined effects of heavy metals on the growth of *Aporrectodea caliginosa* (Oligochaeta; Annelida), using the toxic unit approach. Appl. Soil Ecol. 4, 213–219.

Kostecka, J., Butt, K.R., 2001. Ecology of the earthworm *Allolobophora carpathica* from field and laboratory studies. Eur. J. Soil Biol. 37, 255–258.

Kostecka, J., Pączka, G., 2006. Possible use of earthworm *Eisenia fetida* (Sav.) biomass for breeding aquarium fish. Eur. J. Soil Biol. 42, S231–S233.

Kruuk, H., 1978. Foraging and spatial organization of the European badger, *Meles meles* L. Behav. Ecol. Sociobiol. 5, 75–89.

Langdon, C.J., Piearce, T.G., Meharg, A.A., Semple, K.T., 2003. Inherited resistance to arsenate toxicity in two populations of *Lumbricus rubellus*. Environ. Toxicol. Chem. 22 (10), 2344–2348.

Langdon, C.J., Morgan, A.J., Charnock, J.M., Semple, K.T., Lowe, C.N., 2009. As-resistance in laboratory-reared F1, F2 and F3 generation offspring of the earthworm *Lumbricus rubellus* inhabiting an As-contaminated mine soil. Environ. Pollut. 157, 3114–3119.

Langmaack, M., Schrader, S., Rapp-Bernhardt, U., Kotzke, K., 2002. Soil structure rehabilitation of arable soil degraded by compaction. Geoderma 105 (1/2), 141–152.

Lavelle, P., 1983. The structure of earthworm communities. In: Satchell, J.E. (Ed.), Earthworm Ecology: From Darwin to Vermiculture, Chapman & Hall, London, pp. 449–466.

Lavelle, P., Bignell, D., Lepage, M., Wolters, V., Roger, P., Ineson, P., Heal, O.W., Dhillion, S., 1997. Soil function in a changing world: the role of invertebrate ecosystem engineers. Eur. J. Soil Biol. 33, 159–193.

Lawrence, R.D., Millar, R.H., 1945. Protein content of earthworms. Nature 155 (3939), 517.

Li, D.-Z., Pritchard, H.W., 2009. The science and economics of ex situ plant conservation. Trends Plant Sci. 14 (11), 614–621.

Löfs-Holmin, A., 1983. Reproduction and growth of common arable land and pasture species of earthworm (Lumbricidae) in laboratory cultures. Swedish J. Agric. Res. 13, 31–37.

Lowe, C.N., Butt, K.R., 1999. Interspecific interactions between earthworms: potential applications in soil amelioration. Pedobiologia 43, 808–817.

Lowe, C.N., Butt, K.R., 2002. Growth of hatchling earthworms in the presence of adults: interactions in laboratory culture. Biol. Fertil. Soils 35, 204–209.

Lowe, C.N., Butt, K.R., 2003. Influence of food particle size on inter- and intra-specific interactions of *Allolobophora chlorotica* (Savigny) and *Lumbricus terrestris* (L.). Pedobiologia 47, 574–577.

Lowe, C.N., Butt, K.R., 2005. Culture techniques for soil dwelling earthworms: a review. Pedobiologia 49 (5), 401–413.

Lowe, C.N., Butt, K.R., 2007. Earthworm culture, maintenance and species selection in chronic ecotoxicological studies: a critical review. Eur. J. Soil Biol. 43, S281–S288.

Lowe, C.N., Butt, K.R., 2008. Life cycle traits of the parthenogenetic earthworm *Octolasion cyaneum* (Savigny, 1826). Eur. J. Soil Biol. 44, 541–544.

Lukkari, T., Aatsinki, M., Väisänen, A., Haimi, J., 2005. Toxicity of copper and zinc assessed with three different earthworm tests. Appl. Soil Ecol. 30, 133–146.

Marinissen, J.C.M., van den Bosch, F., 1992. Colonisation of new habitats by earthworms. Oecologia 91, 371–376.

Mason, W.T., Rottmann, R.W., Dequine, J.F., September 2006. Culture of Earthworms for Bait or Fish Food. University of Florida IFAS Extension Publication #CIR1053/FAO16.

McInroy, D.M., 1971. Evaluation of the earthworm *Eisenia foetida* as food for man and domestic animals. Feedstuffs 43, 36–47.

Morgan, J.E., Morgan, A.J., 1998. The distribution and intracellular compartmentation of metals in the endogeic *Aporrectodea caliginosa* sampled from an unpolluted and a metal-contaminated site. Envorin. Poll. 99, 167–175.

National Building Specifications, 2011. Topsoil and Soil Ameliorants (Q28). http://www.thenbs.com/.

Ndegwa, P.M., Thompson, S.A., 2001. Integrating composting and vermicomposting in the treatment and bioconversion of biosolids. Bioresour. Technol. 76 (2), 107–112.

Nuutinen, V., Butt, K.R., 1997. The mating behaviour of the earthworm *Lumbricus terrestris* (Oligochaeta Lumbricidae). J. Zool. Lond. 242, 783–798.

Nuutinen, V., Nieminen, M., Butt, K.R., 2006. Introducing deep burrowing earthworms (*Lumbricus terrestris* L.) into arable heavy clay under boreal conditions. Eur. J. Soil Biol. 42, S269–S274.

Nuutinen, V., Butt, K.R., Jauhiainen, L., 2011. Field margins and management affect settlement and spread of an introduced dew-worm (*Lumbricus terrestris*) population. Pedobiologia 54, S167–S172.

Orozco Almanza, M.S., Ortega Cerrilla, M.E., Perez-Gil Romo, F., 1988. Use of earthworms as a protein supplement in diets for rabbits. Arch. latinoam. nutr. 38 (4), 946–955.

Paoletti, M.G., Buscardo, E., VanderJagt, D.J., Pastuszyn, A., Pizzoferrato, L., Huang, Y.-S., Chuang, L.T., Millson, M., Cerda, H., Torres, F., Glew, R.H., 2003. Nutrient content of earthworms consumed by Ye'Kuana Amerindians of the Alto Orinoco of Venezuela. Proc. R. Soc. Lond. 270, 249–257.

Pereira, J.O., Games, E.F., 1995. Growth of rainbow trout fed a diet supplemented with earthworms, after chemical treatment. Aquacult. Int. 3, 36–42.

Poulson, M.E., Pritzl, G., 1993. The Danish Environmental Specimen Bank—status of establishment. Sci. Total Environ. 139/140, 61–68.

Prayogi, H.S., 2011. The effect of earthworm meal supplementation in the diet on quail's growth performance in attempt to replace the usage of fish meal. Int. Poult. Sci. 10, 804–806.

Pucher, J., Tuan, N.N., Yen, T.T.H., Mayrofer, R., El-Matbouli, M., Focken, U., 2012. Earthworm meal as alternative animal protein source for full and supplemental feeds for common carp (*Cyrinus carpio* L.). International Conference "Sustainable Land Use and Rural Development in Mountain Areas." Hohenheim, Stuttgart, Germany, April 16–18, pp. 167–168.

Raloff, J., 2008. Insects: the original white meat. Sci. News 173 (18), 17–21.

Raw, F., 1966. The soil fauna as a food source for moles. J. Zool. 149, 50–54.

Rożen, A., 2006. Effect of cadmium on life-history parameters in *Dendrobaena octaedra* (Lumbricidae: Oligochaeta) in populations originating from forests differently polluted with heavy metals. Soil Biol. Biochem. 38, 489–503.

Rüther, M., Bandholtz, T., 2009. The German Environmental Specimen Bank: Discovering Data and Information on the Web. EnviroInfo 2009 (Berlin) Environmental Informatics and Industrial Environmental Protection: Concepts, Methods and Tools Shaker Verlag, pp. 179–184.

Sabine, J.R., 1978. The nutritive value of earthworm meal. In: Hartenstein, R. (Ed.), Proceedings of the Conference on Utilization of Soil Organisms in Sludge Management, Syracuse, NY, pp. 122–130.

Sabine, J.R., 1988. Vermiculture: bring on the future. In: Edwards, C.A., Neuhauser, E.F. (Eds.), Earthworms in Waste and Environmental Management, SPB Academic Publishing, The Hague, The Netherlands, pp. 3–7.

Sanchez-Hernandez, J.C., 2006. Earthworm biomarkers in ecological risk assessment. Rev. Environ. Contam. Toxicol. 188, 85–126.

Shipitalo, M.J., Protz, R., Tomlin, A.D., 1988. Effect of diet on the feeding and casting activity of *Lumbricus terrestris* and L. *rubellus* in laboratory culture. Soil Biol. Biochem. 20 (2), 233–237.

Shipitalo, M.J., Nuutinen, V., Butt, K.R., 2004. Interaction of earthworm burrows and cracks in a clayey, subsurface-drained soil. Appl. Soil Ecol. 26, 209–217.

Sogbesan, A.O., Ugwumba, A.A.A., Madu, C.T., 2007. Productivity potentials and nutritional values of semi-arid zone earthworm (*Hyperiodrilus euryaulos*; Clausen, 1914) cultured in organic wastes as fish meal supplement. Pak. J. Biol. Sci. 10 (17), 2992–2997.

Spurgeon, D.J., Hopkin, S.P., 1996. The effects of metal contamination on earthworm populations around a smelting works—quantifying species effects. Appl. Soil Ecol. 4, 147–160.

Spurgeon, D.J., Hopkin, S.P., 1999. Life-history patterns in reference and metal-exposed earthworm populations. Ecotoxicology 8, 133–141.

Spurgeon, D.J., Svendsen, C., Kille, P., Morgan, A.J., Weeks, J.M., 2004. Responses of earthworms (*Lumbricus rubellus*) to copper and cadmium as determined by measurement of juvenile traits in a specifically designed test system. Ecotoxicol. Environ. Safe. 57 (1), 54–64.

Stockdill, S.M.J., 1982. Effects of introduced earthworms on the productivity of New Zealand pastures. Pedobiologia 24, 29–35.

Stockdill, S.M.J., Cossens, G.G., 1966. The role of earthworms in pasture production and moisture conservation. Proc. N.Z. Grassl. Assoc. 28, 168–183.

Stürzenbaum, S.R., Andre, J., Kille, P., Morgan, A.J., 2009. Earthworm genomes, genes and proteins: the (re)discovery of Darwin's worms. Proc. Biol. Sci. 276, 789–797.

Subasinghe, R., August 28, 2012. State of World Aquaculture. FAO Fisheries and Aquaculture Department. http://www.fao.org/fishery/topic/13540/en.

Sun, Z.J., Liu, X.C., Sun, L.H., Song, C.Y., 1997. Earthworm as a potential protein resource. Ecol. Food Nutr. 36 (2–4), 221–236.

Svendsen, T.S., Hansen, P.E., Sommer, C., Martinussen, T., Grønvold, J., Holter, P., 2005. Life history characteristics of *Lumbricus terrestris* and effects of the veterinary antiparasitic compounds ivermectin and fenbendazole. Soil Biol. Biochem. 37, 927–936.

Taboga, L., 1980. The nutritional value of earthworms for chickens. Br. Poult. Sci. 21 (5), 405–410.

Terhivuo, J., Saura, A., 2003. Low clonal diversity and morphometrics in the parthenogenetic earthworm *Octolasion cyaneum* (Sav.). Pedobiologia 47, 434–439.

Tomlin, A.D., 1983. The earthworm bait market in North America. In: Satchell, J.E. (Ed.), Earthworm Ecology: From Darwin to Vermiculture, Chapman and Hall, London, pp. 331–338.

United Nations, 2004. World Population to 2300. Department of Economic and Social Affairs Population Division, ST/ESA/SER.A/236 United Nations Publication, New York.

Uvarov, A.V., 1995. Responses of earthworm species to constant and diurnally fluctuating temperature regimes in laboratory microcosms. Eur. J. Soil Biol. 31 (2), 111–118.

Van Gestel, C.A.M., Weeks, J.M., 2004. Recommendations of the 3rd International Workshop on Earthworm Ecotoxicology, Aarhus, Denmark, August 2001. Ecotoxicol. Environ. Safe. 57, 100–105.

Wever, L.A., Lysyk, T.J., Clapperton, M.J., 2001. The influence of soil moisture and temperature on the survival, aestivation, growth and development of juvenile *Aporrectodea tuberculata* (Eisen) (Lumbricidae). Pedobiologia 45, 121–133.

Wing-Keong, N., June 2000. Worms: a potential feed source for cultured aquatic animals. The Advocate, 82–83.

FURTHER READING

Rameshguru, G., Senthilkumar, P., Govindarajan, B., 2011. Vermiwash mixed diet effect on growth of *Oreochromis mossambicus* (Tilapia). J. Res. Biol. 5, 335–340.

This page intentionally left blank

Index

Note: Page numbers with "f" denote figures; "t" tables.

C

D

E

F

G

H

I

J

K

L

M

N

O

P

Q

R

T